THERMODYNAMICS

McGraw-Hill Series in Advanced Chemistry

Espenson: *Chemical Kinetics and Reaction Mechanisms*
McDowell: *Mass Spectrometry*
Pitzer: *Thermodynamics*
Reid: *Chemical Thermodynamics*
Wiggins: *Chemical Information Sources*

THERMODYNAMICS

Third Edition

Kenneth S. Pitzer
University of California
Berkeley

With acknowledgment to
GILBERT NEWTON LEWIS and MERLE RANDALL,
authors of the First Edition,
and to LEO BREWER, coauthor of the Second Edition.

McGraw-Hill, Inc.

New York St. Louis San Francisco Auckland Bogotá Caracas
Lisbon London Madrid Mexico City Milan Montreal New Delhi
San Juan Singapore Sydney Tokyo Toronto

This book was set in Times Roman by Keyword Publishing Services.
The editors were Jennifer B. Speer and John M. Morriss;
the production supervisor was Kathryn Porzio.
Project supervision was done by Keyword Publishing Services.
R. R. Donnelley & Sons Company was printer and binder.

THERMODYNAMICS

Copyright © 1995, 1961, 1923 by McGraw-Hill, Inc. All rights reserved. Copyright renewed 1951 by Mary Lewis and Merle Randall. Printed in the United States of America. Except as permitted under the United States Copyright Act of 1976, no part of this publication may be reproduced or distributed in any form or by any means, or stored in a data base or retrieval system, without prior written permission of the publisher.

 This book is printed on recycled, acid-free paper containing 10% postconsumer waste.

1 2 3 4 5 6 7 8 9 0 DOC DOC 9 0 9 8 7 6 5 4

ISBN 0-07-050221-8

Library of Congress Cataloging-in-Publication Data

Pitzer, Kenneth S. (Kenneth Sanborn), (date).
   Thermodynamics / Kenneth S. Pitzer. — 3rd ed.
      p.   cm.
   Includes index.
   ISBN 0-07-050221-8
     1. Thermodynamics.   2. Chemistry, Physical and theoretical.
   I. Brewer, Leo, (date).   II. Title.
   QC311.P58   1995
   541.3'69—dc20   94–26960

ABOUT THE AUTHOR

Kenneth S. Pitzer was born in Pomona, California, in 1914, received his B.S. degree from the California Institute of Technology in 1935, his Ph.D from the University of California, Berkeley, in 1937, also an LL.D. from the University of California in 1963, and other honorary degrees. He was a member of the faculty at Berkeley from 1937 to 1960, serving also as Dean of the College of Chemistry from 1951 to 1960. On leave from the University of California, he was Technical Director of the Maryland Research Laboratory, a World War II entity, 1943–1944, and was Director of Research for the U.S. Atomic Energy Commission 1949–1951. In 1961, he became President and Professor of Chemistry at Rice University, serving until 1968 when he became President of Stanford University until 1970. In 1971 he returned to the University of California, Berkeley, as Professor and after 1984 as Professor Emeritus but recalled for continued service.

His advisory committee and other part-time service included the General Advisory Committee to the AEC, 1958–1965 (Chairman 1960–1962); President's Science Advisory Committee, 1965–1968; Commission on Chemical Thermodynamics of the International Union of Pure and Applied Chemistry, 1953–1961; NASA Science and Technology Advisory Committee, 1964–1965; Trustee, Harvey Mudd College, 1956–1961, Mills College, 1958–1961, Pitzer College 1966 to the present, Rand Corporation, 1962–1972; Member of the Board of Directors, Owens Illinois, Inc., 1967–1986; Federal Reserve Bank of Dallas, 1965–1968, the American Council on Education, 1967–1971; and Member of the Council of the National Academy of Sciences, 1964–1967 and 1973–1976.

His research has extended through many areas of physical chemistry and into topics of importance also to inorganic and organic chemistry, chemical engineering, and geochemistry. In addition to work on thermodynamics reflected in this volume, his research has included quantum theory and statistical mechanics as applied to chemical problems ranging from the potential restricting rotation about single bonds, to the bonding in polyatomic carbon molecules, and to the effects of relativity on chemical bonds involving very heavy atoms. In addition to many journal articles and several book chapters, his books include *Quantum Chemistry*, 1953; *Selected Values of Physical and Chemical Properties of Hydrocarbons and Related Compounds*, with F. D. Rossini and others, 1947, 2d edition 1953; *Thermodynamics*, 2d edition with L. Brewer, 1961; *Activity Coefficients in Electrolyte Solutions*, 2d edition, editor and chapter author, 1991; and *Molecular Structure and Statistical Thermodynamics*, a selection of his research papers, 1993.

His honors and awards include the National Medal of Science, 1975, the Robert A. Welch Award in Chemistry, 1984, the Alumnus of the Year Award, University of California, Berkeley, 1951, and the Alumni Distinguished Service Award, California Institute of Technology, 1966. From the American Chemical Society he received the Award in Pure Chemistry, 1943, in Petroleum Chemistry, 1949, the Gilbert Newton Lewis Medal, 1965, the Priestley Medal, 1969, and the Willard Gibbs Medal, 1976. Other recognitions include many special lectureships, a Guggenheim Fellowship. 1951, and election to the National Academy of Sciences in 1949 and to the American Philosophical Society in 1954.

To my wife
Jean

CONTENTS

Preface   xiii

Chapters

1	The Basis of Thermodynamics	1
2	Two Laws of Thermodynamics and the Concepts of Energy and Entropy	15
3	Other State Functions and Their Relationships	38
4	Heat Engines, Heat Pumps, Cycles, and Flow Processes	54
5	Absolute Entropies and Statistical Mechanics; Calculations for Simple Molecules	66
6	The Third Law of Thermodynamics; Entropy Calculations	82
7	Chemical Potential, Fugacity, Conditions for Equilibrium	98
8	Properties of Simple Systems; Tables of Thermodynamic Properties	108
9	Real Gases and the Fluid State	122
10	Multicomponent Systems; Basic Relationships	154
11	Ideal Solutions	173
12	Nonideal Solutions, Excess Functions, Activities, and Activity Coefficients	182
13	Mixed Nonideal Gases and Fluids	224
14	Solutions with Composition Given in Molality	244

15	Electrolyte Solutions: Introduction and Simple Examples	254
16	Electrolyte Solutions: Statistical Theory	275
17	Semiempirical Equations for Pure and Mixed Electrolytes	290
18	Electrolytes at Different Temperatures and Pressures: Thermal and Volumetric Properties	322
19	Electrochemical Cells and Electrode Potentials	340
20	Statistical Calculations of Thermodynamic Properties: Advanced Treatment	364
21	Residual Entropies at Zero Kelvin	380
22	Hydrogen, Helium, and Methane at Low Temperatures	386
23	Surface Effects	400
24	Systems Involving Gravitational, Centrifugal, Electrical, or Magnetic Fields	420
25	Irreversible Processes Near Equilibrium; Nonisothermal Systems; Steady States	435
26	Multicomponent Solutions	457
27	Biochemical Systems; Special Aspects	475
28	Multicomponent Solid–Vapor Systems	485

Appendixes

1	Boltzmann Distribution Law and the Translational Entropy of an Ideal Gas	493
2	Debye Functions for the Thermodynamic Properties of Solids	499
3	Estimation of Properties of Normal Fluids	505
4	Equations of State for Normal Fluids	525
5	Fluid Properties Very Near the Critical Point	530
6	Tables of Thermodynamic Properties	534
7	Values of the Debye–Hückel Parameter for Various Temperatures	543
8	Parameters for Aqueous Electrolyte Properties at 25°C	546
9	Theory for Unsymmetrical Mixing of Ions of the Same Sign	554

10	Parameters for Aqueous Electrolytes at High Temperatures	558
11	Equations for Mixed Liquids	564
12	Functions for Restricted Internal Rotation	570
13	Estimation of Entropies	578
14	Tabular Summary of Thermodynamic Formulas	586
15	Symbols	594
16	Equations Involving More Than Two Variables	599
17	Physical Constants and Numerical Factors	602

Answers to Selected Problems   605

Indexes
    Name Index   607
    Subject Index   617

PREFACE

"There are ancient cathedrals which, apart from their consecrated purpose, inspire solemnity and awe. Even the curious visitor speaks of serious things, with hushed voice, and as each whisper reverberates through the vaulted nave, the returning echo seems to bear a message of mystery. The labor of generations of architects and artisans has been forgotten, the scaffolding erected for their toil has long since been removed, their mistakes have been erased, or have become hidden by the dust of centuries. Seeing only the perfection of the completed whole, we are impressed as by some superhuman agency. But sometimes we enter such an edifice that is still partly under construction; then the sound of hammers, the reek of tobacco, the trivial jests bandied from workman to workman, enable us to realize that these great structures are but the result of giving to ordinary human effort a direction and a purpose.

"Science has its cathedrals, built by the efforts of a few architects and of many workers. In these loftier monuments of scientific thought a tradition has arisen whereby the friendly usages of colloquial speech give way to a certain severity and formality. While this may sometimes promote precise thinking, it more often results in the intimidation of the neophyte. Therefore we have attempted, while conducting the reader through the classic edifice of thermodynamics, into the workshops where construction is now in progress, to temper the customary severity of the science in so far as is compatible with clarity of thought. But since it is improbable that we have been successful in this endeavour, to more than a limited extent, we shall take this opportunity of conversing very informally with the reader concerning our book and its purpose.

"There are several kinds of audience to which a book on thermodynamics might be addressed. There is the beginner who, in order that he may decide whether the subject will meet his needs or arouse his interest, asks what thermodynamics is and what sorts of problems in physics, chemistry and engineering can be solved by its aid; there is the reader who looks for the philosophical implications of such concepts as energy and entropy; above all there is the investigator who, attacking problems of pure or applied science, seeks the specific thermodynamic methods which are applicable to his problem and the data requisite for its solution. Perhaps we have been overambitious in attempting, within the confines of a single volume, to meet all these demands—to lead the beginner through the intricacies of thermodynamic theory and to guide the experienced investigator to the extreme limits now set by existing methods and data."

It was with these comments that Gilbert Newton Lewis and Merle Randall introduced to the world in 1923 their book on thermodynamics. The success of this book was so great and its popularity so prolonged that Leo Brewer and I were not easily persuaded to undertake the revision leading to the second edition of 1961. But after full consideration, we realized that those portions that were still of current interest could best be kept available by combining them with a new presentation of recent advances.

After an additional thirty years of continued success of the second edition, the publisher strongly urged another revision. Initially, it was to be a joint project with Professor Brewer, but for various reasons he withdrew from active participation. He has continued to be most helpful with comments and advice on particular topics; also, significant sections that he contributed to the second edition have been carried forward with only minor revision. Chapters 1 and 2 and certain other important sections are revisions of the original presentations of Lewis and Randall. Many chapters, however, are new while others are revisions of material that I contributed to the 1961 edition. And the responsibility has been mine for the decisions about material to be carried forward and for the chapters and sections that are new in this edition.

As for the earlier editions, this book is designed for various types of readers including those beginning their study of thermodynamics as well as those active in research or in the use of thermodynamics in engineering and other applications. Areas of application continue to expand and now include not only physics, chemistry, engineering, metallurgy, and ceramics, but also biochemistry, geology, geochemistry, oceanography, and atmospheric science. Various chapters include examples of interest to one or more of these fields. Indeed, investigators in some of these areas are now active contributors to basic thermodynamic knowledge as well as to applications.

Readers who have already studied the basic principles of thermodynamics will be able to pass quickly over a few of the early chapters. There are many different presentations of the second law that are equally valid. The original presentation of Lewis and Randall is retained; it has both current merit and historical interest. For the thermodynamics of systems of variable composition, however, the presentation has been modified and extended to make it appropriate for systems comprising near-critical or supercritical fluids as well as for systems at ambient temperature and pressure. Thus, the present treatment is closer to that of Gibbs, with emphasis on the chemical potential defined equivalently in terms of either the Helmholtz or the Gibbs energy. It is important to choose the appropriate basis for any particular application, and the factors to be considered in this selection are carefully explained.

Both the statistical viewpoint and precise statistical equations are now an integral part of modern thermodynamics and have been so presented. There are, of course, basic postulates in microscopic terms upon which the science of statistical mechanics may be built, and these postulates yield also the equations of statistical thermodynamics.

Indeed, the best empirical evidence for the truth of statistical mechanics lies in the agreement of calculated and observed thermodynamic properties. But in this book it has seemed preferable to develop these statistical equations simply and directly from thermodynamic laws.

The importance of statistical mechanics to thermodynamics lies both in the understanding of the second and third laws and in the calculation of thermodynamic properties from spectroscopic and other molecular properties. Also, there are complex systems where the available statistical treatments are incomplete but still yield valid theoretical terms. They may give the forms of other terms in an equation. Then the numerical values of parameters or other details can be evaluated empirically to yield a complete treatment. Examples are given of both exact statistical calculations for ideal gas properties and of several types of semi-theoretical, semi-empirical treatments for more complex systems.

With the advances in both statistical theory and computational capacity, modern research has shifted largely from graphical techniques to the use of numerical equations in handling data. Advances in statistical theory are recognized whenever useful, either directly or as a guide to successful semi-empirical equations. But a balance is maintained; graphical methods are still described when useful. In general, methods are presented both for high accuracy and for more convenient calculation when a quick and easy approximation is desired.

Recently, more complex and computationally demanding statistical methods have been implemented that simulate the detailed motion of each molecule in a fluid sample. Initially, these molecular simulations, both of the Monte Carlo and molecular dynamics type, were limited to simplified molecular models. The results were informative, as illustrated in Chapter 16 for electrolytes, but could not be expected to yield accurate results for real systems with complex interparticle potentials. Recently this limitation is being overcome, and molecular simulations are beginning to yield macroscopic properties in accurate agreement with experiment. This extension of statistical methods is expected to play an increasing role in relation to thermodynamic properties in the future.

Symbols and terminology have been revised to be consistent generally with current practice. Dimensions and units for energy and entropy are discussed at the end of Chapter 3. But I do not agree with the effort of some individuals to suppress all units other than those most basic and central in the SI system. Bars are used as units for pressure as well as pascals, and for both historical and practical reasons the reader should know about calories.

Depending upon the interests of the reader, certain chapters may be unimportant, while others will be of primary significance. Thus, if one is not concerned with fluids at high pressure and high reduced temperature, Chapters 9 and 13 can be passed over quickly. But these are very important for much current research and applications to geological, oceanographic, or engineering systems. While most readers will have sufficient interest in electrolyte solutions to study Chapter 15 carefully, portions of Chapters 16 through 19 will be of much less interest to some. All of the Chapters from 21 onward and most of the appendices present specialized topics. In some cases considerable detail is given, while others present only a few typical examples. Thus, the use of thermodynamics for biochemistry and other biological areas is increasing, and a chapter is included discussing appropriate methods with examples. But these are only samples; the full range of biological examples is so wide and diverse that a separate volume would be required for full coverage. My own recent interest in geochemical systems is reflected in the relatively complete treatments in Chapters 17, 18, and 26 where tables and figures show results of geological significance.

I wish to acknowledge with thanks the support for my own research that was received from the American Petroleum Institute in early years and later from the Atomic Energy Commission and its successor agencies, ERDA and the Department of Energy, and was administered through the University of California via the College of Chemistry and the Lawrence Berkeley Laboratory. Although the preparation of this book was independent of the external support, the earlier research contributed to the basis.

It is impractical to list all of my research collaborators whose work is reflected in this book, but their names are in the author index. Of a special character were the contributions of my closest geological advisor and collaborator, James L. Bischoff of the U.S. Geological Survey, and the extended cooperation and collaboration with the group at the Oak Ridge National Laboratory, including H. F. Holmes, R. E. Mesmer, and J. M. Simonson.

I am greatly indebted to colleagues both here at the University of California at Berkeley and elsewhere for their advice, help, and comments. That of Leo Brewer was special as has been noted above. The following individuals reviewed the manuscript and their comments led to many important improvements: Professor S. M. Blinder, University of Michigan, Ann Arbor; Professor Charles F. Curtiss, University of Wisconsin, Madison; and Dr. Robert E. Mesmer, Oak Ridge National Laboratory. John Prausnitz, Norman Phillips, Gabor Somorjai, Ignacio Tinoco, and Robert Goldberg reviewed one or more chapters in areas of their special expertise and made very valuable suggestions and recommendations. Ms. Peggy Southard not only handled all details of manuscript preparation but also assisted with proofreading and index preparation. And my wife, Jean, to whom the book is dedicated, provided not only great encouragement but also very valuable nontechnical editorial improvements.

Kenneth S. Pitzer

THERMODYNAMICS

CHAPTER 1

THE BASIS OF THERMODYNAMICS

Aside from the logical and mathematical sciences, there are several great branches of natural science that stand apart by reason of the variety of far-reaching deductions drawn from a small number of primary postulates. Included are classical and quantum mechanics, electromagnetics, as well as thermodynamics and statistical mechanics.

These sciences are monuments to the power of the human mind; and their intensive study is amply repaid by the aesthetic and intellectual satisfaction derived from a recognition of the order and simplicity that have been discovered among the most complex of natural phenomena. Also, much of the material progress of the past century has sprung from the development of mechanical and electrical engineering and from the application of thermodynamics to the steam engine and other power-generating apparatus.

When it was first discovered that heat and work could be transformed one into the other and the laws governing such transformation were embodied in the science of thermodynamics, it was the primary function of this science to increase efficiency in the design and the use of engines for the production of work. Although this function has grown steadily in importance, it is now overshadowed by numerous applications of thermodynamics to physics and especially to chemistry. Here the methods of thermodynamics have brought quantitative precision in place of the old, vague ideas of chemical affinity, and thus chemistry advanced toward the status of an exact science. In various branches of physical science and even in modern biology, thermodynamics affords an unerring guide to the investigator.

The Power and the Limitations of Thermodynamics

Our book might be introduced by the very words used by Le Chatelier[1] a century ago: "These investigations of a rather theoretical sort are capable of much more immediate practical application than one would be inclined to believe. Indeed the phenomena of chemical equilibrium play a capital role in all operations of industrial chemistry."
Le Chatelier writes further:

> It is known that in the blast furnace the reduction of iron oxide is produced by carbon monoxide, according to the reaction
>
> $$Fe_2O_3 + 3\,CO = 2\,Fe + 3\,CO_2$$
>
> but the gas leaving the chimney contains a considerable proportion of carbon monoxide, which thus carries away an important quantity of unutilized heat. Because this incomplete reaction was thought to be due to an insufficiently prolonged contact between carbon monoxide and the iron ore, the dimensions of the furnaces have been increased. In England they have been made as high as thirty meters. But the proportion of carbon monoxide escaping has not diminished, thus demonstrating, by an experiment costing several hundred thousand francs, that the reduction of iron oxide by carbon monoxide is a limited reaction. Acquaintance with the laws of chemical equilibrium would have permitted the same conclusion to be reached more rapidly and far more economically.

Fortunately, the importance of thermodynamics in designing chemical and metallurgical processes is now generally appreciated in the major industries, although there are still many particular instances of neglect.

Before speaking more in detail of that which can be accomplished through the aid of thermodynamics, we must from the outset recognize its limitations. In mechanics it is possible to foretell by simple laws the minimum expenditure of work by which a certain operation may be effected; but unless we know what frictional resistance may be encountered, we cannot predict how much work will actually be required. So thermodynamics tells us the minimum amount of work necessary for a certain process, but the amount that will actually be used will depend upon many circumstances. Likewise, thermodynamics shows us whether a certain reaction may proceed and what maximum yield may be obtained but gives no information as to the time required. The rate of a reaction is determined by factors that have hitherto eluded any exact analysis. The rate of a reaction is influenced by many factors in addition to those that determine the equilibrium yield. Considerable progress has been made in the development of a general theory of reaction rates; indeed, thermodynamics is one of the foundation stones of the theory. But other basic postulates are required in addition to thermodynamics before a theory of chemical kinetics is possible.

Although subject to these limitations, thermodynamics is an instrument of great power and universality. It shows the engineer the maximum amount of work that a given quantity of fuel can produce in a given type of steam engine; it shows that more work is available when the fuel is burned in an explosion engine and, finally, that still more could be utilized if we were to solve the technical difficulties in obtaining "electricity direct from coal." The maximum efficiency of a turbine or of a refrigerator or of a weapon of ordnance is a subject for thermodynamic calculation.

To the manufacturing chemist, thermodynamics gives information concerning the stability of his substances, the yields that he may hope to attain, the methods of avoiding undesirable substances, the optimum range of temperature and pressure, the proper choice of solvent, the limitations of methods of fractional distillation and crystallization.

To the analytical chemist, it offers the means of predicting the limits of possible error, of avoiding side reactions, of choosing the concentrations best suited to his work.

Examples illustrating these applications will be discussed at various points in the book. Also, the special treatises in these other areas now make extensive use of thermodynamics and thus illustrate the applications more fully than is possible here. However, even those already familiar with the applications will find it rewarding to examine carefully the foundations of thermodynamics and to note the numerous interlocking relationships that may yield additional material of interest.

The Modern Stage of Thermodynamics

The whole foundation of classical thermodynamics was laid before the middle of the nineteenth century. The work of Black, Rumford, Hess, Carnot, Mayer, Joule, Clausius, Kelvin, and Helmholtz established the basic principles of the theory of energy.

Next came the task of building up from these cardinal principles a great body of thermodynamic theorems for systems of variable composition. This was the work of many men, among whom may be mentioned van't Hoff, and especially J. Willard Gibbs, whose great monograph on "The Equilibrium of Heterogeneous Substances"[2] has proved a rich and still unexhausted mine of thermodynamic material.

A third stage of thermodynamic development is characterized by the design of more specific thermodynamic methods and their application to particular chemical processes, together with a systematic accumulation and utilization of the data of thermodynamic chemistry. The first systematic study of all the thermodynamic data necessary for the calculation of the free-energy changes in a group of important reactions was published in Germany by Haber. This book, *Thermodynamik der technischen Gas Reaktionen*,[3] is a model of accuracy and of critical insight. Also in Germany, Nernst and his associates made remarkable contributions to both theory and practice, while in Denmark Bronsted pursued a most valuable series of investigations.

It was, however, particularly the senior author of the first edition of this book, Gilbert N. Lewis, together with his many colleagues and students, who accomplished the major part of this third stage of development. In numerous studies they showed how to apply thermodynamics to various chemical systems, and in an extensive array of experimental investigations they accumulated the necessary data for the common chemical substances. Tables of Gibbs energies and of electrode potentials took their place in basic chemical handbooks alongside the table of atomic masses.

The fourth and modern stage came with the development of statistical mechanical methods for the calculation of thermodynamic properties, together with the understanding and guidance of quantum mechanics. The third law of thermodynamics, which depends in essence on the quantization of molecular energies, was developed, tested, and applied most fruitfully to chemical problems. Also, statistical methods were perfected that make it possible to calculate macroscopic thermodynamic properties from the molecular energy levels observed spectroscopically. These developments are an integral part of modern thermodynamics.

Statistical mechanics constitutes in large part a parallel science to thermodynamics. Frequently the fundamental structure of statistical mechanics is presented in terms of abstract postulates and relatively complex mathematics that seem remote from the laws of thermodynamics. However, Giauque and others developed the important equations of statistical thermodynamics on the basis of thermodynamic postulates, and we shall follow this course as the one appropriate to a book on thermodynamics. The statistical viewpoint gives great insight into the thermodynamic behavior of chemical systems. The concept of entropy, otherwise strange and abstract, is statistically related to familiar ideas. Also, material scientists, metallurgists and ceramists, and earth scientists, geochemists and oceanographers now make full use of thermodynamics and contribute greatly to the array of basic data. Often the same systems are of both geological and industrial interest.

DEFINITIONS; THE CONCEPT OF EQUILIBRIUM

As a science grows more exact, it becomes possible to employ more extensively the accurate and concise methods and notation of mathematics. At the same time it becomes desirable, and indeed necessary, to use words in a more precise sense. For example, if we are to speak, in the course of this work, of a pure substance, or of a homogeneous substance, these words must convey as nearly as possible the same meaning to writer and to reader.

Unfortunately, it is seldom possible to satisfy this need by means of formal definitions, partly because the most fundamental concepts are the least definable, partly because of the inadequacy of language itself, but more particularly because we often wish to distinguish between things that differ rather in degree than in kind. Frequently, therefore, our definitions serve to divide for our convenience a continuous field into more or less arbitrary regions—as a map of Europe shows roughly the main ethnographic and cultural divisions, although the actual boundaries are often determined by chance or by political expediency.

The distinction between a solid and a liquid is a useful one, but no one would attempt to fix the exact temperature at which sealing wax or glass passes from the solid to the liquid state. Any attempt to make the distinction precise makes it the more arbitrary.

Classification of Substances

Whatever part of the objective world is the subject of thermodynamic discourse is customarily called a *system*. Sometimes it is desirable to use this term in a more definite sense, implying a spatial content. If we make an enclosure by means of physical walls, or if we imagine such an enclosure made by a mathematical surface, such an enclosing surface serves as the boundary of the system, which then comprises everything of thermodynamic interest contained within that boundary. The boundary or surface that defines the physical extent of the system can be distorted—as when the system is expanding—or it may have a fixed position. The choice of the boundary in many problems is often arbitrary, and judicious choice of the boundary of a system may often simplify the treatment of changes taking place. Thus, for example, a crystal or any chosen cubic centimeter of that crystal may be chosen as the system. A thermodynamic system may contain no substance at all, in the ordinary sense, but may consist of radiant energy or an electric or magnetic field. Usually, however, a system comprises a substance, which may be homogeneous or heterogeneous.

Systems that can exchange energy with the surroundings but cannot transfer matter across the boundaries are known as *closed systems*. *Open systems* can exchange both energy and matter with surroundings. Our discussion in the first six chapters of this book will be restricted to closed systems, and it is only in Chapter 10 that a general treatment is initiated that includes as variables the concentrations of two or more components of a system.

Homogeneous Systems

At the outset we meet the difficulty of exact definition or classification if we attempt to make this traditional distinction between homogeneous and heterogeneous systems. Nevertheless it will be convenient to define a homogeneous system as one whose properties are the same in all parts, or that at least vary continuously from point to point; a system, in other words, in which there are no apparent surfaces of discontinuity. Cases of variable properties are of thermodynamic interest only if the variation is governed by a recognized force such as a gravity. Such cases are unusual, however, and, ordinarily, by a homogeneous system we shall mean one whose properties are the same throughout.

Heterogeneous Systems

A heterogeneous system consists of two or more distinct homogeneous regions. Thus benzene and water or ice and water form heterogeneous systems. The homogeneous regions, which are called *phases*, appear to be separated from one another by surfaces of discontinuity. (In our usual thermodynamic work it is immaterial whether we have one piece of ice or several pieces of ice in contact with a mass of water. In such cases, it is commonly considered that there are only two phases present, the ice phase and the water phase.)

The boundaries between phases are not surfaces in a strict mathematical sense but are very thin regions in which the properties change with great abruptness from the properties of the one homogeneous phase to those of the other. The effect of surface area on the total thermodynamic properties is so small that it may be neglected unless the subdivision is exceedingly fine. But when we study surface tension, adsorption, and kindred phenomena, these surface regions become of great importance.

Indeed, the classification of systems according to homogeneity or heterogeneity seemed formerly more precise than it does now, when so much attention is being devoted to interesting types of finely divided substances. When, therefore, we speak of a heterogeneous system, without further qualification, it must be understood that the system is one in which each homogeneous region is large.

Components and Solutions

Any homogeneous system, whether solid, liquid, or gaseous, is called a solution if its composition is variable. Thus air, brine, glass, and a mixed crystal of alum and chrome alum are all called solutions. The components are the substances of fixed composition that can be mixed in varying amounts to form the solution. For thermodynamic purposes, the choice of components of a system is often arbitrary and depends upon the range of conditions for the problem under consideration. A system consisting of a single element of fixed isotopic composition is usually considered to be a single-component system, but it is necessary to consider each isotope as an independent component if their amounts are being varied independently and if the properties being studied vary with isotopic content within the accuracy of the measurements.

If the elements are combined in a definite ratio as a compound, so that they are not varied independently, the compound is considered as a single component. A mixture of O_2 and N_2 is a two-component system if the amount of each element can be varied independently, but a system consisting of only NO molecules, or produced from only NO molecules, is considered a one-component system. A system produced by adding NO, O_2, and N_2 molecules independently of one another is a three-component system at low temperatures. However, at high temperatures where these three molecular species are in rapid equilibrium with one another and their relative amounts are fixed by the equilibrium constant for the dissociation of NO, the system is only a two-component system, since it can be produced by the addition of only O_2 and N_2. If the system is restricted to compositions corresponding to equal amounts of nitrogen and oxygen, the high-temperature system can be considered a one-component NO system. Likewise, a mixture of 10 hydrocarbons would be a 10-component system at low temperatures and a two-component system at very high temperatures where the molecular composition is not at one's disposal but is fixed by the equilibrium constants and the amounts of carbon and hydrogen added. A system consisting of HF gas contains many different molecular species. However, they are all in rapid equilibrium with one another under ordinary conditions, and all species can be derived from any one. Thus the system is a one-component system, but the choice of which species, for example, HF, H_2F_2, etc., to use for its designation is arbitrary.

The choice of the molecular species to be considered as components in a solution is not even restricted to those species actually known to be present in appreciable concentration. Thus we can consider the components of a sodium chloride aqueous solution to be H_2O and NaCl even though we know that there is no appreciable concentration of NaCl molecules in dilute solutions at room temperature. We can consider solid solutions of composition in the range $FeO_{1.05-1.19}$ to be formed from the components FeO and Fe_3O_4 even though the composition FeO has no stable existence as a solid phase. We could equally well have chosen FeO and Fe_2O_3, Fe and Fe_3O_4, Fe and O_2, or any pair of arbitrarily chosen compositions between Fe and O_2 as the two components of the solution, depending upon which pair proved to be most convenient for the calculations in mind. Thus Darken and Gurry[4] chose $FeO_{1.0477}$ and O_2 as the components of this system.

Solids, Liquids, and Gases; Crystals and Noncrystals

The ancient categories represented by earth, water, and air have persisted in a simple classification of substances into solids, liquids, and gases. While this useful classification may ordinarily be employed without fear of ambiguity, there are, as we have already pointed out, some substances that are unquestionably solid, like glass, but which, when heated, pass by imperceptible gradations into typical liquids. Also, since the pioneer investigation of the critical state by Andrews,[5] it has been known that a liquid may be changed to a gas by a process in which the substance remains as a pure phase from beginning to end, without the appearance of discontinuity at any stage.

A more fundamental distinction at present seems to be the one between crystalline and noncrystalline states; as yet no substance has succeeded in passing by a continuous process from one of these states to the other. In contrast to the critical point for liquid–gas systems, no similar point corresponding to the solid–liquid transformation has been observed, and there is no clear indication that such a point exists. Crystalline substances, although usually solid, range from hard, rigid substances like diamond, through soft crystals like rubidium, to the extremely fluid crystals.

States and Properties

If it were possible to know all the details of the internal constitution of a system; in other words, if it were possible to find the distribution, the arrangement, and the modes of motion of all the ultimate particles of which it is composed, this great body of information would serve to define what may be called the microscopic state of the system, and this microscopic state would determine in all minutiae the properties of the system.

Statistical mechanics calculates appropriate averages of these properties and, thereby, determines the macroscopic properties such as volume, pressure, and temperature. In thermodynamics we adopt the converse method. The *state* of a system (macroscopic state) is determined by its *properties* just insofar as these properties can be investigated directly or indirectly by experiment. We may therefore regard the state of a substance as adequately described when all its properties that are of interest in a thermodynamic treatment are fixed with a definiteness commensurate with the accuracy of our experimental methods.

The *properties* of a substance describe its present state and do not give a record of its previous history. It is an obvious but highly important corollary of this definition that, when a system is considered in two different states, the difference in volume or in any other property between the two states *depends solely upon those states themselves, and not upon the manner in which the system may pass from one state to the other.*

Extensive and Intensive Properties

Most of the properties we measure quantitatively may be divided into two classes. If we consider two identical systems, let us say two kilogram weights of brass or two exactly similar balloons of hydrogen, the volume, or the internal energy, or the mass of the two is double that of each one. Such properties are called *extensive*.

On the other hand, the temperature of the two identical objects is the same as that of either one, and this is also true of the pressure and the density. Properties of this type are called *intensive*. Some intensive properties are derived from the extensive properties; thus, while mass and volume are both extensive, the density, which is mass per unit volume, and the specific volume, which is volume per unit mass, are intensive properties. These intensive properties are the ones that describe the specific characteristics of a substance in a given state, for they are independent of the amount of substance considered. Indeed, in common usage it is only these intensive properties that are meant when the properties of a substance are being described.

Reproducibility of States

We have tacitly assumed in the preceding paragraphs that a pure substance always exists in one of a few well-defined forms, so that if a few conditions are fixed, all the properties are determined. Indeed, this is true for so large a number of substances that unless otherwise stated it will be taken for granted.

The properties of a given amount of pure water vapor can be completely determined by external conditions. Thus, if the temperature and the pressure are fixed, two equal quantities of water vapor will, by any experimental test, be found identical in all respects. The same is true of liquid water, and it is probably nearly true for ice, as well as for the various other forms of solid water.

On the other hand, certain metals, even when pure, vary greatly according to their previous treatment, and two samples are not identical although all external conditions are the same. In such cases a substance, instead of appearing in only a few well-defined states, may assume any one of an infinite number of states, depending upon its mode of preparation and its mechanical or thermal treatment. Cases of this sort deserve more careful consideration than they usually receive. Thus there is no doubt that many measurements of the electrode potentials of metals have been deprived of value because of the lack of definition of the surface conditions in the electrodes.

In a perfect crystal the atoms are supposed to be arranged in a perfectly definite order. At a given temperature and pressure, we should thus expect the properties to be unambiguously determined. It is, however, doubtful whether there are many actual crystalline substances in which the conditions are so simple.

Let us consider common ice. When pure water is frozen, long crystals first traverse the mass; these are then connected by shorter crystals, until finally a mesh is produced in which the last remaining drops of liquid may not be free to form the same perfect crystals as were produced at the beginning. It is conceivable, therefore, that the material formed at the end of the process has somewhat different properties from those of the more ideal crystals produced at the beginning.

In the case of a typical liquid also we may expect the properties to be definitely determined by external conditions, not because of any ordered arrangement of the particles, but rather because their mobility permits a complete randomness of arrangement, so that with large numbers of molecules the average properties of the mass are constant.

However, in the case of substances of high viscosity, the mobility of the particles is so small that they do not readily assume the positions of symmetry in a space lattice that are characteristic of the perfect crystal or the random arrangement that is characteristic of the mobile fluid. In such a case the particles may remain for long periods of time in strained positions that are determined by their previous treatment or by the fortuitous circumstances of their original assemblage. As examples we may cite, on the one hand, a drawn wire of hard metal or a piece of unevenly cooled glass; on the other, such materials as are obtained when a metal is deposited by electrical spattering. Here each particle may lie as it strikes without later rearrangement.

It is evident that, if we are to treat quantitatively and numerically the properties of substances, the state of a substance must be described with great particularity, unless we can assume that its properties are completely determined by the external conditions. When the properties are so determined, the state of a given amount of a substance can ordinarily be fixed merely by stating the temperature and pressure and, in the case of solutions, the composition. Only in special cases shall we consider the properties of substances as dependent upon the degree of subdivision of the phases, upon gravitational, electric, magnetic, or centrifugal fields, or upon other external influences that have only a minute effect upon the system.

Equilibrium and Reaction Speed

We have developed in some detail the ideas of the preceding section because they lead us directly to the idea of *equilibrium*, and in all thermodynamics there is no concept more fundamental than this.

If a substance such as water is under fixed external conditions that completely determine its state, its properties do not change with time. It is said to be in a state of rest. If the external conditions are momentarily altered, the water returns immediately thereafter to its original state and properties.

If, instead of water, a substance such as soft tar is employed, the same thing happens, but more slowly. If the tar is subjected to some temporary distortion or to some unevenness of pressure, it slowly yields, or flows, until the former state of rest is once more established.

When a system is in such a state that after any slight temporary disturbance of external conditions it returns rapidly or slowly to the initial state, this state is said to be one of equilibrium. A *state of equilibrium is a state of rest* in the sense that no change of macroscopic properties is observed.

Even crystalline substances of the softer sort fail to retain for long any condition differing from the characteristic state of equilibrium. Thus, for example, we may account for the extraordinary reproducibility of the electrode potentials of soft metals, such as sodium or lead, as compared with metals like iron or nickel. Even in the case of substances of great viscosity or rigidity it seems reasonable to suppose that they also behave in a similar manner, although the changes may be imperceptible because of their slowness.

Any change in the properties of a system is called a *process*, and if the process is one that is roughly termed chemical, it is sometimes called a *reaction*, but we shall employ these terms almost interchangeably. The idea we have developed regarding the restoration of equilibrium after a mechanical disturbance may be extended to cases in which chemical reactions are involved.

If we dissolve methyl acetate in water, hydrolysis will set in and the properties of the system will change until a definite state is reached, which is fixed by conditions such as temperature and pressure. In addition to the original substances, methyl alcohol and acetic acid will be found in solution in fixed amounts. We have again a state of equilibrium. If the system is temporarily disturbed, for example by raising the temperature for a short time and then bringing it back to the original value, the solution will once more return to the same state.

Slow establishment of equilibrium, after mechanical disturbance, we have attributed to such factors as viscosity. When the rate of a chemical reaction is involved, the time required to establish an equilibrium depends upon factors that are analogous to viscosity.

We shall therefore consider not only that every state of equilibrium is a state of rest but that *every state of absolute rest is a state of equilibrium*, and therefore that every system that has not reached a state of equilibrium is changing continuously toward such a state with greater or less speed, although possibly at an immeasurably slow speed.

Stable Systems

Frequently one speaks of *stable* systems, and much confusion of thought has resulted from the use of this term for two different ideas that are separable, and indeed must be separated if any clarity is to be obtained in the application of thermodynamics to chemistry. In common usage a system is said to be stable when it undergoes no apparent changes. Now a system that is apparently in a stationary state may be so because it has reached one of the states of equilibrium from which it has no tendency to depart, no matter how great its mobility; or it may be because processes occurring within it are so slow as to be imperceptible, even though the system may be far from a true state of equilibrium. It is only systems of the first kind, which are really in a state of equilibrium, that we shall call stable in any thermodynamic sense. Systems of the second kind may be called *inert* or *unreactive*.

A mixture of oxygen and hydrogen might be kept for a long time without the formation of any measurable quantity of water, but the system is inert, and not thermodynamically stable, as shown by the fact that any one of a number of catalytic substances causes a rapid formation of water. Such a catalyst merely increases the rate of attainment of equilibrium. In the absence of such catalyst, and at room temperature, the rate of the reaction is entirely too slow to be measurable. Nevertheless we can make an approximate calculation of that rate by actually measuring it at a number of high temperatures and employing the method of extrapolation.

Partial Equilibrium

Of the various possible processes that may occur within a system, some may take place with extreme slowness, others with great rapidity. Hence we may speak of equilibrium with respect to the latter processes before the system has reached equilibrium with respect to all the possible processes. Thus, in a system of oxygen, hydrogen, and water, the two gases dissolve rapidly until the water is saturated, and we may say that the system is in equilibrium with respect to the process of solution. It is far from equilibrium with respect to the reaction by which water is formed from oxygen and hydrogen—a process in which the speed is of a far different order of magnitude. As another example we may consider nitrogen tetroxide, which dissociates rapidly until a state of equilibrium is soon reached between N_2O_4 and NO_2. But each of these substances is really unstable with respect to elementary oxygen and nitrogen, although, without catalysts, the process of decomposition into these elements is a very slow one.

Degrees of Stability

A stone lying in a hollow upon a hillside is considered to be in a stable position, although, if pushed over the edge, it will roll to a position of greater stability at the bottom. So in thermodynamics a system may be in a state of rest and, if slightly disturbed, may revert to this same state of rest, but if largely disturbed it may proceed toward some entirely new condition of equilibrium. Thus liquid water, a degree or two below the freezing point, reaches a state of equilibrium to which it will return after a slight disturbance. Any large disturbance, however, may cause it to seek a new condition of equilibrium in the more stable form of ice.

In practice, we often assume the existence of several such equilibrium states toward which a system may tend, all these states being stable but representing higher or lower degrees of stability. From a theoretical standpoint it might be doubted whether there is any condition of real equilibrium, with respect to every conceivable process, except the one that represents the most stable state. This, however, is not a question that need concern us greatly, nor is it one we could discuss adequately at this point, without largely anticipating what we shall later have to say regarding the statistical view of thermodynamics.

Equilibrium as a Macroscopic State

Even here it is desirable to emphasize that by a state of rest, or equilibrium, we mean a state in which the properties of a system, as experimentally measured, would suffer no further observable change even after the lapse of an indefinite period of time. It is not intimated that the individual particles are unchanging. Thus, when sulfuric acid is heated in a closed vessel, a condition is ultimately reached in which definite amounts of the liquid sulfuric acid and of the gaseous sulfuric acid, sulfur trioxide, and water have been produced. These amounts, as determined by any of our quantitative methods, then remain constant. This is what we call the state of equilibrium. If, however, we were in a position to follow the paths of the individual molecules, we should perceive the wildest chaos: molecules of the liquid evaporating, some molecules of vapor entering the liquid phase, others dissociating into molecules of water and sulfur trioxide, and these in turn constantly combining. The absolute number of molecules of each of these species varies from instant to instant, but these variations are so small compared with the total numbers that they would be imperceptible even if the accuracy of our analytical processes were increased a billionfold.

The Mole

The gram and the kilogram are the units of mass that are commonly employed as the units of quantity of material. However, when we are dealing with chemical reactions, it is far more convenient to employ the mole or the equivalent for such a unit. For general purposes the mole is the better unit, since an equivalent of a substance may have different meanings according to the kind of reaction into which the substance enters. Thus, one equivalent of permanganic acid has a variable significance, according as we consider the power to neutralize a base or to act as an oxidizing agent in acid or in alkaline solution.

If M is the molecular mass of the substance, a mole is defined as M grams. This unit is by no means as free from ambiguity as the gram. The atomic weights are subject to revision, but are now very accurately known, and in case of doubt the value assumed may be stated. The more common ambiguity concerns the formula of the molecules actually present in the system. In many cases the composition of these ultimate molecules is not known, and the formula employed represents an estimate or is merely the simplest expression of the stoichiometric proportions.

In general we regard the mole as identical with what has also been called the formula mass. Therefore the mole is not defined unless the chemical formula is established by universal usage or is definitely stated. In this book we shall choose the chemical formula with regard more to convenience than to consistency. For gases the actual molecular mass has frequently been experimentally determined, and in the case of liquids we shall, for the most part, use the same formula as in the gaseous state. In the case of solids we shall sometimes do the same thing. Thus for solid halogens we write I_2, Br_2, etc. On the other hand, we shall use the formula S for solid sulfur, although it is known that in both the rhombic and monoclinic forms the structural unit is the molecule S_8.

Molar Properties

We have defined an extensive property as one whose quantitative measure is proportional to the amount of substance taken. Thus volume is an extensive property, but the volume per mole of any substance is an intensive property. If we denote the volume in general by V, we may denote the volume per mole, or the molar volume, by V_m. Moreover, if Y is any extensive property, Y_m will denote the molar value of Y and is an intensive property. For some sections of the book, all of the equations are on a molar basis and the subscript m is omitted for subsequent equations after an appropriate comment. The word "molal" has also been used instead of "molar" and in this context is synonymous. As a measure of composition of a solution, however, molal means moles per kilogram of solvent, while molar means moles per liter of solution.

Chemical Symbols and Equations

Chemical symbols will frequently be used to indicate not only the substance under consideration but also a definite quantity of that substance. Thus HCl denotes 1 mole of hydrogen chloride, and $\frac{1}{2}O_2$ denotes $\frac{1}{2}$ mole of oxygen. Thus, we may write for water at 4°C, $V_m = 18.02\,\mathrm{cm}^3$. The formula is sometimes placed in a subscript V_{H_2O} or in parentheses $C_P(H_2O)$. Often it is necessary to specify the state of the substance, solid (s), liquid (l), or gas (g). To explicitly designate that a solid is crystalline, (cr) is used. For solutions a measure of composition may be specified; this is discussed further in Chapter 10. The pressure, temperature, or some other condition can be indicated as $V(H_2O, g, 5\,\mathrm{bar})$.

In general the pressure will be assumed to be 1 bar (0.1 MPa) unless otherwise specifically indicated, and the distinguishing marks (l), (cr or s), (g), (aq) may sometimes be omitted when no ambiguity is likely. Thus the plain symbol H_2 will indicate hydrogen gas at unit pressure.

Then, if we write

$$\mathrm{Ag(s)} + \tfrac{1}{2}\mathrm{Cl}_2\,(\mathrm{g},\,10\,\mathrm{bar}) = \mathrm{AgCl(s)}$$

we are considering a process whereby 1 mole of solid silver at unit pressure and $\frac{1}{2}$ mole of chlorine gas at 10 bar disappear and 1 mole of solid silver chloride is produced.

If any quantity, such as the volume, is determined by the state of a system, in other words, if it is a *property* of the system, then when the system changes from one state to another, it will be convenient to designate the increase[†] in that quantity by the symbol Δ. Thus, if we consider the fusion of ice at atmospheric pressure and at 0°C, or approximately 273 K, we write

$$\mathrm{H_2O(s)} = \mathrm{H_2O(l)} \qquad \Delta V_{273} = -2\,\mathrm{cm}^3\,\mathrm{mol}^{-1}$$

[†] This symbol Δ may be used to denote any such increment, whether finite or infinitesimal. It is made to correspond in sign with the symbol d used only for infinitesimal change. Both dx and Δx are to be read as "the increase in x," although this increase may be negative.

The system in the first state consists of 1 mole of ice, the volume of which is about $20\,\text{cm}^3$; the system in the second state consists of 1 mole of water, of which the volume is about $18\,\text{cm}^3$. We could equally well write

$$\text{H}_2\text{O}(l) = \text{H}_2\text{O}(s) \qquad \Delta V_{273} = 2\,\text{cm}^3\,\text{mol}^{-1}$$

In general, when a system passes from state A to state B, $\Delta V = V_B - V_A$.

As we proceed we shall find it necessary to make numerous other conventions that we cannot discuss here without anticipating ideas that will be developed in subsequent chapters.

REFERENCES

1. H. Le Chatelier, *Ann. mines* (8) (**13**), 157 (1888).
2. J. Willard Gibbs, *Trans. Conn. Acad. Sci.*, **3**, 228 (1876).
3. Fritz Haber, *Thermodynamik der technischen Gas Reaktionen*, R. Oldenbourg-Verlag, Munich, 1905 (trans. by A. B. Lamb), Longmans, Green & Co., Inc., New York, 1908.
4. L. S. Darken and R. W. Gurry, *J. Am. Chem. Soc.*, **67**, 1398 (1945).
5. T. Andrews, *Trans. Roy. Soc. (London)*, **159**, 575 (1869).

CHAPTER 2

TWO LAWS OF THERMODYNAMICS AND THE CONCEPTS OF ENERGY AND ENTROPY†

The first law of thermodynamics is just the general principle of physics that energy is conserved. While in thermodynamics we are concerned with macroscopic amounts of matter, the principle also applies at the atomic or molecular level. There are many examples where mechanical energy in one form is transformed to another type and often transformed again back to the original form. Thus, if an object is thrown upward against the earth's gravitational force, its kinetic energy diminishes and at a certain point the object comes momentarily to rest. If it is arrested at this point, the whole of the kinetic energy has disappeared. However, we are accustomed to say that as the object loses in kinetic energy it gains in latent, or potential, energy by an equal amount, and this idea is justified by the fact that, if the object is allowed to drop once again to its original position, it regains the whole of the kinetic energy that it lost in rising.

---

† In the sections introducing the first and second laws of thermodynamics, we have retained substantial portions of the original text of G. N. Lewis and M. Randall in 1923. These paragraphs are beautifully clear and add historical perspective.

If in such a mechanical system there are inelastic collisions, or if frictional processes are at work, there may be a net loss of mechanical energy; in other words, the sum of the kinetic and potential energies may diminish. At the end of the eighteenth century when Count Rumford was observing the boring of a cannon in the Munich arsenal, he noticed that the mechanical energy expended was roughly measured by the amount of heat produced. This idea, developed by Mayer and by Joule, led to the first determinations of the mechanical equivalent of heat. Whenever a system is subjected to a cyclic process whose net result is merely the conversion of work to heat with the system returning to its initial state, it is always found that the heat produced is strictly proportional to the work done. Therefore the units of heat and work can be so chosen that the amount of heat produced is always equal to the amount of mechanical energy lost. This discovery led to the consideration of heat as a manifestation of thermal energy and to the enunciation of the broad principle, that we know as the law of *conservation of energy*, or the first law of thermodynamics.

As far back as 1762 that remarkably accurate thinker and investigator Joseph Black showed, in studying heat alone, that it was necessary to introduce a concept of latent heat (analogous to potential energy). Here again the concept of latent energy is justified by the fact that the amount of heat required to melt 1 g of ice is equal to the heat evolved when 1 g of water freezes.

When the phenomenon of electricity became better understood, it was necessary to define electrical energy, and the brilliant investigations of Maxwell made it possible to follow the course of electrical energy, not only through material bodies, but through space that is empty of all else but an electromagnetic field.

So, as science has progressed, it has been necessary to invent other forms of energy, and indeed an unfriendly critic might claim, with some reason, that the law of conservation of energy is true because we make it true by assuming the existence of forms of energy for which there is no other justification than the desire to retain energy as a conservative quantity. Thus, when the accounting of the energy changes during β-decay of radioactive nuclei did not balance, the existence of the neutrino was postulated to account for the missing energy. Because of our confidence in the law of conservation of energy, we felt sure that this elusive particle would eventually be found, although it required 30 years before the experimental verification of the neutrino and therefore of the conservation of energy in β-decay processes could be obtained. Thus we feel certain that, for every new type of interaction that is discovered, it will be possible to find a characteristic energy function of each system that will allow prediction of the change in one system as a result of a change in the other system. Nevertheless, in spite of our confidence in this law, we must still demand its experimental confirmation when new circumstances arise, and the possibility of its violation under conditions far removed from our present experiments must always be kept open.

THE INTERNAL ENERGY OF A SYSTEM

The observed proportionality of heat and work for any cyclic process requires that the energy contained within a system, or its internal energy, be a *state function*, or a *property* of the system in the technical sense in which this term has been used in Chapter 1. The increase in such energy when a system changes from state A to state B is independent of the way in which the change is brought about. It is simply the difference between the final and the initial energy,

$$\Delta U = U_B - U_A$$

where ΔU for this system must, by the conservation law, be equal and opposite in sign to the total ΔU for all other systems involved as given by summing the heat and work.

Energy and Mass

While the amount of energy given to or taken from the environment in a case like this is easy to measure, the determination of the total energy of any one material system has, until recently, seemed beyond the range of human possibility. For practical purposes it is still necessary to regard as undetermined the total energy that a given system possesses. It is, however, of much theoretical interest to note that the great discovery of Einstein,[1] embodied in the principle of relativity, shows us that every gain or loss of energy by a system is accompanied by a corresponding and proportional gain or loss in mass, and therefore presumably that the total energy of any system is measured merely by its mass.

In other words, mass and energy are different measures of the same thing, expressed in different units, and the law of conservation of energy is but another form of the law of conservation of mass. These units are indeed very different in magnitude, differing by the square of the velocity of light, so that 1 g is equal to 9×10^{13} J. Even the largest amounts of energy evolved in ordinary chemical reactions produce changes in mass that are below the limits of detectability by means of the balance.

PRESSURE AND TEMPERATURE

There are two intensive properties, pressure and temperature, that play an important role in thermodynamics, since they largely affect, and often completely determine, the state of a system. Pressure is too familiar an idea to require definition; it has the dimensions of force per unit area, and therefore pressure times volume has the dimensions of energy.

The official SI unit of pressure is the pascal (Pa), which is one newton per square meter. This is an inconveniently small unit with no obvious relationship to our ordinary experience, which is that with the earth's atmosphere. Until recently, the standard pressure unit was taken as an "atmosphere" defined as 760 mm of mercury in a standard gravitational field; this is 101 325 Pa. Another pressure unit in very common use is called a "bar" and is exactly 10^5 Pa. Thus, the standard atmosphere is 1.013 25 bars, and 1 bar is as typical of real atmospheric pressure as is the "atmosphere." Since the relation 1 bar = 10^5 Pa or 0.1 MPa is simple and exact, it seems preferable to use 1 bar as the standard reference pressure. Indeed, this decision was taken recently by IUPAC. We believe the sense of proportion conveyed by the fact that real atmospheric pressure is near 1 bar is important, and, therefore, we shall normally express pressures in bars rather than Pa or MPa.

The concept of temperature is a little more subtle. When one system loses energy to another by thermal conduction or by the emission of radiant energy, there is said to be a flow of heat, or a *thermal flow*. The consideration of such cases leads immediately to the concept of temperature, which may be qualitatively defined as follows: If there can be no thermal flow from one body to another, the two bodies are at the same temperature, but if one can lose energy to the other by thermal flow, the temperature of the former is the greater. This establishment of a qualitative temperature scale is obviously more than a definition. It involves a fundamental principle, to which we have already given preliminary expression in the discussion of equilibrium, but which we are not yet ready to put in a general and final form. For spontaneous thermal flow, this principle requires that if A loses energy to B, B cannot lose it to A; if A loses to B, and B loses to C, C cannot lose to A.

Expressed briefly, our qualitative laws of temperature are: If $T_A = T_B$ and $T_B = T_C, T_A = T_C$; and if $T_A > T_B$ and $T_B \geq T_C, T_A > T_C$. As in our general discussion of equilibrium, it must be understood that we are dealing with net gains or losses in energy. We do not mean that no thermal energy passes from a cold body to a hot, but only that the amount so transferred is always less than that simultaneously transferred from hot to cold.

When we have established the qualitative laws of temperature, we still have a wide freedom of choice in fixing the quantitative scale. Indeed, temperature, as ordinarily measured, or its square or its logarithm would equally satisfy these qualitative requirements.

Since the volumes of most things change appreciably with the temperature and since volumes are easily measured, it early proved convenient to correlate the temperature with the volume of some chosen thermometric substance. It is obvious that this substance could not be chosen altogether at random, for if water were taken, then, on account of the existence of a maximum density, there would be two temperatures corresponding to one volume. But, even in the case of substances whose volume always increases with the temperature, one choice may be more convenient than another.

If thermometers made of mercury and alcohol, with linear scales, are made to agree at two temperatures, they will not agree at some intermediate temperature; but if two gases such as hydrogen and air are employed, the agreement will be nearly complete over a wide temperature range. Hydrogen and air do not behave exactly alike at atmospheric pressure, but the behavior of any two gases becomes more nearly the same the lower the pressure. We therefore adopt, as our ideal thermometric substance, any gas at very low pressure. Strictly speaking, we measure the pressure–volume (PV) product of a fixed amount of any gas in contact with the system at a series of pressures and then extrapolate the PV product to zero pressure. This extrapolated value of PV is then directly proportional to the absolute temperature.

We must still define in some manner the size of the unit of temperature. The Celsius (also known as centigrade) scale has the values 0°C and 100°C for the freezing and boiling points of water. The corresponding absolute scale was defined to have a 100°C interval between the ice and steam points. Lord Kelvin pointed out in 1848 that this was a clumsy definition and that one should simply define the temperature of a single fixed point. In a sense the absolute zero is the second fixed point, but there is no need to make any measurement at this point. Giauque[2] renewed this proposal in 1939, and it was adopted by the International Unions of Physics and Chemistry. It is now agreed that the triple point of water, where water, ice, and water vapor are in equilibrium, shall be 273.1600 K (where K is for Kelvin and the word "degree" is omitted). (The triple point is more accurately reproducible than the conventional ice point, which is the freezing point in air, because variations from pressure or amount of dissolved air are avoided.) Thus the temperature of a system is given by the expression

$$T = 273.1600 \, \frac{\lim_{P \to 0} (PV)_T}{\lim_{P \to 0} (PV)_{tp}} \tag{2-1}$$

The value 273.16 was selected to make the new degree the same size as that of the Celsius (centigrade) scale. Since the ordinary ice point is 0.010 degree lower than the triple point of water, we say that 0°C = 273.150 K; indeed, this relationship is now exact by definition. The temperature 25°C = 298.15 K has been adopted as a standard in the range of room temperatures at which thermodynamic data will be obtained and tabulated. (Since this is a temperature that we shall very frequently use, we shall designate it commonly, for convenience, 298 K, bearing in mind that in our calculations we are to use the more precise value.)

Later in this chapter, we define the thermodynamic scale of temperature, which fortunately proves to be identical with the ideal-gas scale we have just considered. In that place we shall be able to show that the point we have defined as the absolute zero of temperature is not an arbitrary point, brought into prominence by the properties of any one substance or class of substances. It is in fact the limit of any rational thermodynamic scale, a true zero where all thermal energy† would have been extracted, but as unattainable as the other limit of our scale, the infinite temperature.

Practical Measurement of Temperature

While the perfect-gas basis for temperature is excellent in principle, it is not convenient in practice, even for calibration of working thermometers. Over the years there have been internationally adopted temperature scales using platinum resistance thermometers over a wide intermediate range, together with radiation thermometers at very high temperatures (see Chapter 20), and helium gas or vapor pressure measurements at very low temperatures. The most recent International Temperature Scale is that of 1990 designated ITS-90, which is described in a Technical Note by Mangum and Furukawa.[3]

---

† The zero-point energy of systems in their lowest quantum state remains, of course, at 0 K.

HEAT AND WORK

There are two terms, *heat* and *work*, that have played an important part in the development of thermodynamics, but their use has often brought an element of vagueness into a science that is capable of the greatest precision. We shall discuss particular limitations on their definition a little later. For our present purpose we may say that when a system gains energy by thermal radiation or conduction as a result of a temperature differential, it is absorbing a positive quantity of heat and that when it gains energy by other methods, for example, by the operation of external mechanical forces, a positive quantity of work is being done on the system (or a negative quantity of work is being done by the system).†

The amount of heat transferred can be measured in terms of amount of ice melted or frozen in an ice calorimeter, or in any similar calorimeter, with the volume change due to the melting or freezing of a substance at its melting point used as a measure of the heat transfer.

We have previously defined the amount of work done by a system undergoing a change in state in terms of lifting a mass in a gravitational field. Consider a specific example of a gas in a vertical cylinder that is kept from expanding into the upper evacuated space by a piston of mass m and area a that is locked in place by a stop. If the piston is released, the maximum amount of work that the gas could perform for a given expansion is not generally obtained because of frictional effects. If the expansion process is arranged so that no other work can be done except for the lifting of the piston in the earth's gravitational field and all other energy transfers are heat transfers, the work done by the gas on the piston when the gas raises the piston by a height Δh is $mg\,\Delta h$. The external pressure exerted on the gas by the piston is given by $P_{ex} = mg/a$. Thus the work done on the gas is

$$w = -P_{ex} a\, \Delta h = -P_{ex} \Delta V \qquad (2\text{-}2)$$

where ΔV is the volume increase of the gas. This is a general result, and the compression work done on any substance is given by the external pressure exerted on the substance times the volume decrease. If the external pressure is not constant, we obtain for a differential element of compression or expansion $\delta w = -P_{ex}\, dV$, or

$$w = \int_{V_A}^{V_B} -P_{ex}\, dV \qquad (2\text{-}3)$$

As dV is positive for expansion and negative for compression, the sign of w is opposite to the sign of dV if work done on the system is taken to be positive.

For a given expansion, it can be seen that the work performed depends upon the external pressure resisting the expansion, and in the instance of expansion into a

---

† Some books on thermodynamics define w with the opposite sign as the work done by the system, which is convenient if the primary emphasis is on heat engines. The definition with the same sign for both q and w seems preferable for a more general thermodynamic treatise.

vacuum where P_{ex} is zero, no work at all is done. In order that an expansion may occur, the internal pressure must, of course, be greater than the external, but this difference may be made as small as is desired, and as a limit we may consider the case in which the internal pressure is equal to the external pressure or differs from it by a negligible amount. This will later be termed a reversible process. For such processes, Eq. (2-3) is valid with P_{ex} replaced by P, the pressure of the system.

According to the law of the conservation of energy, any system in a given condition contains a definite quantity of energy, and when this system undergoes change, any gain or loss in its internal energy is equal to the loss or gain in the energy of surrounding systems. If no matter is transferred in any physical or chemical process, the increase in energy of a given system is therefore equal to q, the heat absorbed from the surroundings, plus w, the work absorbed from the surroundings. If U_A represents the initial energy content of the system and U_B the final energy content, then

$$U_B - U_A = \Delta U = q + w \tag{2-4}$$

We shall use δq and δw to represent infinitesimal quantities of heat and work. Thus for an infinitesimal change we write

$$dU = \delta q + \delta w \tag{2-5}$$

The use of δ instead of d will remind us that q and w are not properties but are defined only for a given path of change.

The application of the law of conservation of energy is merely an accounting process. One adds up all the increases in energy of the system due to all the nonthermal interactions, and this sum is the total work done on the system. One then adds up the total energy increase due to thermal interactions, and this represents the total heat absorbed by the system. The need for separate accounting of thermal and nonthermal energies will be clear when the second law of thermodynamics is discussed. It is not feasible to assign absolute values to energies. One merely measures the change in energy that takes place as a result of a change in the state of a system. If one can measure q, the heat absorbed by the system, and w, the work done on the system, one can then, through the first law of thermodynamics, know the change in the energy of the system, U.

The values of q and w depend upon the way in which the process is carried out, and, in general, neither is uniquely determined by the initial and final states of the system. However, their sum is determined, so that if either q or w is fixed by the conditions under which the process occurs, the other is also fixed. Where the work done by the system is the work of expansion against an external pressure, the expansion may be carried out in such a manner that no heat enters or leaves the system. We then say that the process is *adiabatic*, and we may write

$$\Delta U = w = -\int_{V_A}^{V_B} P_{ex} \, dV \tag{2-6}$$

On the other hand, if the process is such that no work is done by the surroundings (as in the case of a chemical reaction taking place in a constant-volume calorimeter), then $\Delta U = q$.

HEAT CAPACITY

When we impart heat to a system and thereby raise its temperature, the average heat capacity between the initial and final temperatures is defined as $q/\Delta T$, and the limit of this ratio, as q is made indefinitely small, or, in other words, as the second temperature is brought indefinitely near to the first, is called the actual heat capacity at that temperature.

If this process takes place at constant volume, no work is done and $w = 0$. Therefore,

$$C_V = \left(\frac{\partial U}{\partial T}\right)_V \tag{2-7}$$

Alternatively, the heating may occur at constant pressure and the heat capacity is reported as C_p. Then there is a work term and the relationship to the internal energy is more complex. This will be discussed in the next chapter.

IDEAL OR PERFECT GAS

At this point it is convenient to state formally the properties of a "perfect" or "ideal" gas. One of these was indicated above in the discussion of temperature. It is the P–V–T relationship

$$PV = nRT \tag{2-8}*$$

where n is the number of moles and R is a constant with the dimension energy divided by temperature. This relationship becomes exact for real gases in the limit as $P \rightarrow 0$, provided we exclude possible dissociation of the molecules. The proportionality with temperature is really a matter of definition of temperature as indicated above.

A less familiar but important second criterion is that

$$\left(\frac{\delta U}{\delta V}\right)_T = 0 \tag{2-9}*$$

In other words the internal energy of the ideal gas is a function only of temperature and is independent of volume.

The work or heat of reversible isothermal expansion or compression of an ideal gas is readily obtained from Eq. (2-6) with $P_{\text{ex}} = P$,

$$q = -w = nRT \ln\left(\frac{V_B}{V_A}\right) = nRT \ln\left(\frac{P_A}{P_B}\right) \tag{2-10}*$$

---

* Equations such as these, which are not true for actual substances but which are useful approximations, will be marked with an asterisk.

The gas constant R is probably the most important numerical constant of thermodynamics. It appears in many equations that are otherwise unrelated to the ideal gas. Its value is now known with high accuracy:

$$R = 8.314\,51 \pm 0.00007 \,\text{J}\,\text{K}^{-1}\,\text{mol}^{-1}$$

In Chapter 3 we will discuss other units of energy in addition to the joule and will give the values of R in those units.

SECOND LAW OF THERMODYNAMICS AND THE CONCEPT OF ENTROPY

We shall see presently that the second law of thermodynamics has a very natural and plausible explanation in terms of the probability of the random motions of enormous numbers of atoms and molecules. But we first examine the development of the second law as it occurred in the mid and late nineteenth century and in terms of macroscopic properties rather than atoms and molecules.

Clausius summed up the findings of thermodynamics in the statement, "Die Energie der Welt ist konstant; die Entropie der Welt strebt einem Maximum zu," and it was this quotation that headed the great memoir of Gibbs on "The Equilibrium of Heterogeneous Substances." What is this entropy, which such masters have placed in a position of coordinate importance with energy, but which has proved a bugbear to so many students of thermodynamics?

The first law of thermodynamics, or the law of conservation of energy, was universally accepted almost as soon as it was stated, not because the experimental evidence in its favor was at that time overwhelming, but rather because it appeared reasonable and in accord with human intuition. The concept of the permanence of things is one which is possessed by all. It has even been extended from the material to the spiritual world. The idea that even if objects are destroyed, their substance is in some way preserved has been handed down to us by the ancients, and in modern science the utility of such a mode of thought has been fully appreciated. The recognition of the conservation of carbon permits us to follow, at least in thought, the course of this element when coal is burned and the resulting carbon dioxide is absorbed by living plants, whence the carbon passes through an unending series of complex transformations.

The second law of thermodynamics, which is known also as the law of the dissipation or degradation of energy, or the law of the increase of entropy, was developed almost simultaneously with the first law through the fundamental work of Carnot, Clausius, and Kelvin. But it met with a different fate, for it seemed in no recognizable way to accord with existing thought and prejudice. The various laws of conservation had been foreshadowed long before their acceptance into the body of scientific thought. The second law came as a new thing, alien to traditional thought, with far-reaching implications in general cosmology.

Because the second law seemed alien to the intuition, and even abhorrent to the philosophy of the time, many attempts were made to find exceptions to this law and thus to disprove its universal validity. But such attempts have served rather to convince the incredulous and to establish the second law of thermodynamics as one of the foundations of modern science. Later, the statistical basis of the second law was recognized, so that now it no longer stands as a thing apart, but rather as a natural consequence of long-familiar ideas.

Preliminary Statement of the Second Law: The Actual, or Irreversible, Process

The second law of thermodynamics may be stated in a great variety of ways. We shall reserve until later our attempt to offer a statement of this law that is free from every limitation and shall confine ourselves for the present to a discussion of the law sufficient to display its character and content.

The second law concerns the direction in which natural systems change and the conditions of equilibrium toward which this change occurs. Many types of processes leading toward equilibrium are familiarly known. The diffusion of material from a concentrated solution into a dilute solution, leading toward a condition of uniform concentration; the passage of heat from a hot body to a cold, leading to uniformity of temperature; the oxidation of organic substances by the atmosphere; the running down of a clock; the easing of strains in a plastic solid; the self-demagnetization of a magnet are all processes that illustrate the kind of change that occurs spontaneously in nature. Sometimes, as in the motion of the planets, this approach to a final state is extremely slow, but their mechanical energy is gradually being converted into heat by unceasing tidal action.

These processes and all other natural processes are alike in one respect, that they are bringing the various systems toward the condition of ultimate equilibrium or rest, and we may think of these systems as thereby losing in some measure their capacity for spontaneous change.

It is not coal but the combustion of coal that causes a steam engine to operate. As a rule, one system affects other systems in consequence of changes that are going on within it. Hence, a system far removed from its condition of equilibrium is the one chosen if we wish to harness its processes for the doing of useful work. A system isolated from all others will always maintain a constant amount of energy, and therefore, if the second law of thermodynamics is spoken of as the law of the dissipation of energy, no loss in energy is meant, but rather a loss in the availability of energy for external purposes. It seems better therefore not to speak of the dissipation or degradation of energy but rather to speak of the degradation of the system as a whole. For in many cases, such as the diffusion of one gas into another, the process does not essentially involve an energy change.

Before proceeding to a more exact characterization of the second law, let us make sure that there is no misunderstanding of its qualitative significance. When we say that heat naturally passes from a hot to a cold body, we mean that, in the absence of other complicating processes, this is the process that inevitably occurs. It is true that by means of a refrigerating machine we may further cool a cold body by transferring heat from it to its warmer surroundings, but here we are in the presence of another dissipative process proceeding in the engine itself. If we include the engine within our system, the whole is moving always toward the condition of equilibrium. A system already in thermal equilibrium may develop large differences of temperature through the occurrence of some chemical reaction, but all such phenomena are but eddies in the general unidirectional flow toward a final state of rest.

The essential content of the second law might be given by the statement that when any actual process occurs it is impossible to invent a means of restoring *every* system concerned to its original condition. Therefore, in a technical sense, any actual process is said to be *irreversible*.

The Ideal, or Reversible, Process

When we speak of an actual process as being always irreversible, we have in mind a distinction between such a process and an ideal process which, although never occurring in nature, is nevertheless imaginable. Such an ideal process, which we shall call *reversible*, is one in which all friction, electrical resistance, or other such sources of dissipation are eliminated. It is to be regarded as a limit of actually realizable processes.

Let us imagine a process so conducted that at every stage an infinitesimal change in the external conditions would cause a reversal in the direction of the process, or, in other words, that every step is characterized by a state of balance. Evidently a system that has undergone such a process can be restored to its initial state without more than infinitesimal changes in external systems. It is in this sense that such an imaginary process is called reversible.

To illustrate, we may consider a system comprising water and water vapor contained in a cylinder with a movable piston. Now in practice the piston cannot be made free from friction, but we have no reason to believe that such friction may not be diminished indefinitely, and therefore we may regard the ideal frictionless piston as a limit that may be approached as nearly as is desired. After a constant temperature is established throughout the system, and when the external pressure upon the piston equals the vapor pressure of water, the system is in equilibrium with respect to internal and external agencies. If, then, the external pressure is increased by any amount, however small, the piston will move inward and the vapor will condense. If the external pressure upon the piston is diminished by any amount, however small, the piston will move outward and the liquid will vaporize. In other words, the work required to condense 1 mole of vapor differs by an infinitesimal amount from the work furnished by the vaporization of 1 mole of liquid.

An excellent example of an actual process that is very nearly reversible is furnished when the emf of a galvanic cell is measured by means of a sensitive potentiometer. Here the driving force of the cell itself is so nicely balanced against an external emf that in favorable cases a current may be made to flow in one direction or the other by external changes of 0.000 001 volt.

Again we may consider a case in which we deal not with the balance of mechanical or electrical forces but with a thermal equilibrium. If two bodies differ in temperature only by an infinitesimal amount, the transfer of heat from one to the other is likewise a reversible process, for evidently it would be possible to restore the system to its original condition without causing more than infinitesimal change in external systems.

A QUANTITATIVE MEASURE OF DEGRADATION

In viewing the reversible process as the limit toward which actual processes may be made to approach indefinitely, it is implied that processes differ from one another in their degree of irreversibility. It is of the utmost importance to establish a quantitative measure of this degree of irreversibility, or this degree of degradation.

When we wish to measure a quantity such as length, we first choose a standard, say a bar of platinum kept at the International Bureau of Weights of Measures, and next we adopt a method of comparing the length of other objects with the length of this standard object. So in defining the degree of irreversibility of a process we shall chose some standard irreversible process and then define the method whereby others may be compared quantitatively with it.

Two familiar types of spontaneous or irreversible processes are (1) the transfer of heat from a higher to a lower temperature and (2) the conversion of work to heat. We have previously seen that the work done on a system can be measured in terms of the lifting of a mass in the earth's gravitational field, while the heat transfer can be measured by an ice calorimeter or some similar fusion calorimeter. We may also use other devices. For example, a spring may be calibrated by allowing it to lift a mass and determining the height to which the mass is lifted for a given degree of compression of the spring. Thus it is clear that we have a number of devices at our disposal that can measure the conversion of work to heat during any spontaneous process by allowing a mechanical device to do work on the system to return it reversibly to its initial state and by using some type of calorimeter to measure the resulting heat. We shall choose a standard system composed of a weight and pulley system and a heat reservoir. In employing this standard system in conjunction with other systems, we are going to use the weight as a source of work and the reservoir as a source or as a sink of heat. It will therefore be desirable to choose them so that the weight will undergo no thermal change and the reservoir will do no work during the processes we are about to consider.

If the weight is allowed to fall and by some frictional process gives up a part of its energy to the reservoir in the form of heat, the process is an irreversible one and, without the intervention of some external agency, the reverse process, whereby the weight would be lifted again at the expense of the heat of the reservoir, is impossible.

As the weight is gradually lowered, its potential energy being transferred into thermal energy in the reservoir, we might measure the extent of this irreversible process by a pointer and scale attached to the weight to indicate its height or by the amount of heat given to the reservoir. We shall in fact take as the measure of the extent of this standard universal process a quantity that is proportional to the energy exchange, but not equal to it, for it is necessary to our purpose to consider also the *temperature* of the reservoir.

To make this clear, we may consider a weight and two separate reservoirs, one at the temperature T_h and one at the lower temperature T_c. If the weight is lowered and a certain amount of heat is given to the reservoir at T_h, and if then this same amount of heat is allowed to flow to the other reservoir at T_c, this latter is also an irreversible process. The net result is the same as if the heat developed by the falling of the weight were given at once to the cold reservoir at T_c. Now the sum of the degradation in two successive irreversible processes must be greater than that in either one alone; otherwise, our definition would not be quantitative. Therefore, if we are to have a genuine scale of irreversibility, the transfer of energy from the weight to the hot reservoir at T_h must be regarded as a less irreversible process than the transfer of the same amount of energy from the weight to the cold reservoir at T_c.

It will therefore be expedient to define the extent of irreversibility of our standard process by making it equal, not to q, but to q/θ, where q is the heat received by the reservoir and θ is some quantity that qualitatively satisfies our definition of temperature. Moreover, when the function θ is determined, it completes the quantitative definition of the degree of degradation. We are going to prove later in this chapter that θ may be completely identified with the absolute temperature, which we have already defined by means of the perfect gas. Lord Kelvin called θ the thermodynamic temperature, and it is interesting to note that it provides a temperature scale that is entirely independent of the properties of any single substance or class of substances.

The Entropy of the Weight–Reservoir System

So far we have not given a name to our measure of the irreversibility of the standard process. The value of q/θ, when this process occurs, we shall call the increase in *entropy* of the weight–reservoir system. If we denote its entropy at the beginning by S_A and at the end by S_B, we write as our definition,

$$S_B - S_A = \frac{q}{\theta} \qquad (2\text{-}11)$$

Our present definition of entropy will be found identical with the definition originally given by Clausius. We have, however, departed radically from some traditional methods of presenting this idea, for we have desired to emphasize the fact that the concept of entropy, as a quantity that is always increasing or is being continuously produced in all natural phenomena, is based upon our recognition of the unidirectional flow of all systems toward the final state of equilibrium. In the ordinary definition of entropy the attention is focused upon the reversible process and not upon the irreversible process, the existence of which necessitates the entropy concept. For this reason we have based our definition immediately upon an irreversible process and shall now employ the reversible process only as a means of comparing the degree of degradation, or the extent of production of entropy, during two irreversible processes.

Comparison of Any Irreversible Process with the Standard Irreversible Process

If, in any system, an irreversible process occurs, it is possible, with sufficient ingenuity, to devise a mechanism by which in actuality, or at least in thought, every part of the system may be restored to its original condition at the expense of a degradation in the standard system.

For example, let the system in question be a mixture of oxygen and hydrogen at the temperature of the standard reservoir, and let the irreversible process consist in the combination of these elements to form water. Then by means of an electric generator, operated by a falling weight, the water can be dissociated by electrolysis, and the various parts of the system can be brought to their original temperature by contact with the large standard reservoir. At the end only the weight–reservoir system has suffered degradation.

Of all methods of restoring to its original condition a system in which some process has occurred, there must be at least one that produces the smallest change in the weight–reservoir. Such a method will be one that consists in a reversible process and therefore causes no *further* degradation. In a reversible process there is no entropy production, and thus there is no change in total entropy. Entropy transfers may take place during a reversible process, but no entropy production will result. A cycle carried out by means of reversible processes can restore all the systems involved to their original states.

When a process has occurred in any system, we may define the increase in entropy of that system as equal to the minimum increase in entropy of the weight–reservoir system necessary to restore the system to its original state. In other words, it is the increase in entropy of the weight–reservoir system when the restoration occurs reversibly, since the entropy transferred from the system is equal to the entropy increase of the weight–reservoir.

The Entropy Change in the Free Expansion of a Perfect Gas

As an illustration of the method of calculating the increase of entropy in a simple irreversible process, let us consider a perfect gas enclosed in a flask of volume V_A. Let this flask be attached by a stopcock to an evacuated flask, such that the volume of the two flasks together is V_B. If these flasks are isolated from other systems and the stopcock is opened, the gas will distribute itself uniformly between them. Since the flasks are isolated, the energy does not change during expansion, and since, by our previous definition of a perfect gas, the temperature and the internal energy uniquely determine one another, the temperature is the same after the expansion as before.

In order to measure the production of entropy during this process, we shall now restore the system to its initial state by means of a standard weight and a standard reservoir of the same temperature as the gas, namely, T. By keeping the gas in thermal contact with the reservoir it may be compressed isothermally by means of the weight.

TABLE 2-1
Entropy changes during free expansion of a perfect gas

	ΔS of gas	ΔS of weight–reservoir system	Entropy production; total ΔS
Irreversible expansion of perfect gas into vacuum	$+nR\left(\dfrac{T}{\theta}\right)\ln\left(\dfrac{V_B}{V_A}\right)$	0	$+nR\left(\dfrac{T}{\theta}\right)\ln\left(\dfrac{V_B}{V_A}\right)$
Reversible compression	$-nR\left(\dfrac{T}{\theta}\right)\ln\left(\dfrac{V_B}{V_A}\right)$	$+nR\left(\dfrac{T}{\theta}\right)\ln\left(\dfrac{V_B}{V_A}\right)$	0
Total for both steps	0	$+nR\left(\dfrac{T}{\theta}\right)\ln\left(\dfrac{V_B}{V_A}\right)$	$+nR\left(\dfrac{T}{\theta}\right)\ln\left(\dfrac{V_B}{V_A}\right)$

The work done by the weight and the heat absorbed by the reservoir, if the compression is carried on in a reversible manner, are given by Eq. (2-10)*,

$$q = -w = nRT\ln\left(\dfrac{V_B}{V_A}\right)$$

Now q/θ is the increase in the entropy of the weight–reservoir system during the restoration step, and $-q/\theta$ must be the increase of entropy of the gas during this reversible step. During the free expansion of the gas, the entropy of the gas must have increased by q/θ. If we write S_A as the entropy of the gas before the expansion and S_B as the entropy of the gas after expansion, we find

$$S_B - S_A = nR\left(\dfrac{T}{\theta}\right)\ln\left(\dfrac{V_B}{V_A}\right) \qquad (2\text{-}12)*$$

Table 2-1 reviews the entropy changes in the successive steps that we have been considering. We see that there has been an overall entropy production of $nR(T/\theta)\ln(V_B/V_A)$. As the gas has been returned to its initial state, there is no net entropy change in the gas and the total entropy increase is found in the standard reservoir. As no net entropy increase took place during the reversible compression when entropy was merely transferred from the gas to the reservoir, the irreversible process that produced entropy took place during the free expansion of the gas.

Entropy as an Extensive Property

In expressing the entropy change during an irreversible process as the difference between the entropy at the end and the entropy at the beginning, we have implied that entropy is a *property* and therefore that the entropy change depends solely upon the initial and final states. Indeed, this follows directly from our definition, for, by whatever irreversible path we proceed from state A to state B, the minimum degradation of the weight–reservoir system necessary for the return from state B to state A is the same. It is true that we have not shown how to obtain the absolute value of S_B or S_A, but only their difference, nor shall we need to discuss this question until a later chapter. In the meantime, we shall regard the entropy, like the energy and heat content, as a quantity of which the absolute magnitude is undetermined.

Moreover, entropy is an extensive property, for we may consider two systems that are exactly alike and each of which undergoes the same irreversible process; evidently the change in the standard weight–reservoir system necessary for their restoration is twice as great as it would be for one of them alone.

Since entropy is extensive, we may regard the entropy of a system as equal to the sum of the entropies of its parts. It is therefore important to ascertain how to determine the localization of entropies in the various parts of a system. Owing to the special properties of the standard weight–reservoir system that we assumed at the outset, it will be convenient to postulate that in any operation of the weight–reservoir system the entropy changes occur in the reservoir alone, so that, if the standard reservoir gains heat from any source by the amount q, the reservoir changes in entropy by q/θ.

An Important Criterion for Reversible Process

We have seen that the total entropy change in a reversible process is zero. For such a process entropy is merely transferred with no entropy production. It follows that in such a process, the entropy change in any system must be equal and opposite in sign to the entropy change in all other systems involved. In order to study this case further, let us consider the energy changes that occur in a reversible process between some system and the standard weight–reservoir system. For the sake of simplicity, we shall choose an infinitesimal process. Letting S denote the entropy of the system and S_{st} that of the standard weight–reservoir system, the condition of reversibility, since the total entropy must remain constant, is

$$dS = -dS_{st} \qquad (2\text{-}13)$$

Bearing in mind the condition that the process is to be reversible, it is possible to conduct it so that the system and the standard weight merely exchange mechanical energy and the system and the reservoir merely exchange heat. There must, moreover, be a state of balance between the mechanical forces exerted by the system and by the weight; and the temperature of the system and the reservoir must not differ more than infinitesimally from one another. The total energy gained by the system is equal to the total energy lost by the weight–reservoir system, and, owing to the state of balance, the work done by the weight must equal the work done upon the system. Therefore, by the conservation law, the heat lost by the system is equal to the heat gained by the reservoir.

Algebraically speaking, if the heats absorbed by system and reservoir are δq and δq_{st}, then $\delta q = -\delta q_{st}$ or, since the temperatures are the same, $\delta q/\theta = -\delta q_{st}/\theta$. But by definition of the standard weight–reservoir system, $\delta q_{st}/\theta = dS_{st}$; and therefore, by Eq. (2-13),

$$dS = \frac{\delta q_{rev}}{\theta} \qquad (2\text{-}14)$$

where we have written δq_{rev} to emphasize that it is heat absorbed reversibly that determines the entropy change of the system.

Perfect Differential

We have shown that entropy is a state function. In order to further define the temperature θ, we note an important mathematical property of a function of two variables. Consider an expression of the type

$$\delta Z = L(x,y)\,dx + M(x,y)\,dy \qquad (2\text{-}15)$$

where δZ is an infinitesimal quantity and L and M are functions of the independent variables x and y as indicated. This type of expression may or may not be the total differential of a state function $Z(x,y)$. If there is a function $Z(x,y)$, then

$$L(x,y) = \left(\frac{\partial Z}{\partial x}\right)_y \quad \text{and} \quad M(x,y) = \left(\frac{\partial Z}{\partial y}\right)_x$$

and

$$\frac{\partial L}{\partial y} = \frac{\partial^2 Z}{\partial x\,\partial y} = \frac{\partial M}{\partial x} \qquad (2\text{-}16)$$

The equality $\partial L/\partial y = \partial M/\partial x$ is the necessary and sufficient condition that an expression of the type (2-15) is a perfect differential and can be integrated to a function $Z(x,y)$. Then $\Delta Z = Z(x_2,y_2) - Z(x_1,y_1)$ and Z is a state function. If Eq. (2-16) does not hold, however, it is impossible to integrate (2-15) unless the path from x_1,y_1 to x_2,y_2 is specified. Also the result will depend on the path chosen.

The Thermodynamic Temperature

At this point let us determine the relationship between the thermodynamic temperature θ and that of the perfect gas T. We have shown that entropy is a property; i.e., its change between a given pair of states is independent of the path followed. Thus the differential of the entropy must be a perfect differential, and, if we have an expression of the type

$$\delta S = L\,dT + M\,dV$$

the functions L and M must fulfill the condition

$$\frac{\partial L}{\partial V} = \frac{\partial M}{\partial T}$$

The properties of a perfect gas were defined by Eqs. (2-8)* and (2-9)*. The heat absorbed in a given change is

$$\delta q = dU - \delta w \qquad (2\text{-}5)$$

and for reversible expansion of a perfect gas this becomes

$$\delta q_{\text{rev}} = C_V dT + \frac{nRT}{V} dV \qquad (2\text{-}17)*$$

Although we know that δq is not a perfect differential, it is interesting to verify the point. The energy of a perfect gas depends only on the temperature and not on the volume; hence

$$\frac{\partial C_V}{\partial V} = \frac{\partial^2 U}{\partial V \partial T} = 0$$

But

$$\frac{\partial}{\partial T}\left(\frac{nRT}{V}\right) = \frac{nR}{V} \neq 0$$

which confirms that δq_{rev} is not a perfect differential.

Division of δq_{rev} by θ yields the differential of the entropy of a perfect gas,

$$dS = \frac{C_V}{\theta} dT + \frac{nRT}{V\theta} dV \qquad (2\text{-}18)*$$

The criterion of a perfect differential gives

$$\left[\frac{\partial}{\partial V}\left(\frac{C_V}{\theta}\right)\right]_T = \left[\frac{\partial}{\partial T}\left(\frac{nRT}{V\theta}\right)\right]_V \qquad (2\text{-}19)*$$

But we have already indicated that the left side is zero because the energy of a perfect gas does not depend on the volume. Also, the derivative on the right is at constant V, while n and R have no temperature derivative; consequently,

$$\left(\frac{nR}{V}\right)\left[\frac{\partial}{\partial T}\left(\frac{T}{\theta}\right)\right] = 0 \qquad (2\text{-}20)*$$

Since θ and T both have the qualitative properties of a temperature, their ratio must be constant to satisfy Eq. (2-20)*. But the size of the degree is arbitrary; consequently, we may define this ratio to be unity and make the thermodynamic temperature identical to the perfect-gas temperature.

Now Eq. (2-12)* for the entropy change on isothermal expansion of a perfect gas may be simplified,

$$S_B - S_A = nR \ln\left(\frac{V_B}{V_A}\right) = nR \ln\left(\frac{P_A}{P_B}\right) \qquad (2\text{-}21)*$$

General Equation for Entropy

Thus we write for the entropy

$$dS = \frac{\delta q_{rev}}{T} \tag{2-22}$$

From this important equation, we conclude that in any *reversible* process the increase in entropy of any system, or part of a system, is equal to the heat it absorbs divided by the absolute temperature. Indeed, it is this fundamental equation that Clausius used for his original definition of entropy. This result does not depend in any way upon the specific weight–reservoir system that we have used for measuring the degree of degradation of the system under examination. It is clear that any source of mechanical and thermal energies that can operate reversibly upon the system to return it to its initial state would yield the same results.

A more general equation is

$$dS = \frac{\delta q}{T} + dS_{irr} \tag{2-23}$$

where dS, the increase in entropy of the system, is given by $\delta q/T$, the entropy transferred from the surroundings, plus dS_{irr}, the entropy produced as a result of irreversible processes within the system. For a reversible process $dS_{irr} = 0$, and Eq. (2-23) reduces to (2-22).

Let us consider the entropy increase dS_{irr} caused by a few simple irreversible processes. These examples will illustrate the possible use of Eq. (2-23) with this term included.

Consider first the flow of an amount of heat δq from a body at temperature T_1 to a body at lower temperature T_2. If we imagine reversible transfer of heat at each temperature, the two entropy terms are readily computed and the net entropy increase is

$$dS_{irr} = \delta q \left(\frac{1}{T_2} - \frac{1}{T_1} \right) = \frac{\delta q (T_1 - T_2)}{T_1 T_2} \tag{2-24}$$

The result for the degradation of some form of work to heat follows directly from our discussion of the weight and heat reservoir. If an amount of work δw is degraded to heat at temperature T, the increase in entropy is

$$dS_{irr} = \frac{\delta w}{T} \tag{2-25}$$

Equation of State

We now consider the case of a substance in internal equilibrium and subject only to changes brought about by reversible exchange of heat with an external reservoir and reversible expansion work against an external restraining pressure. In this case the heat absorbed is $T\,dS$ and the work absorbed is $-P\,dV$; whereupon by the first law

$$dU = T\,dS - P\,dV \tag{2-26}$$

If $P_{\text{system}} > P_{\text{ex}}$, the process is irreversible. There is entropy production, and $\delta w = -P_{\text{ex}} dV = -P\, dV + T\, dS_{\text{irr}}$. Also, from Eq. (2-23), $\delta q = T\, dS - T\, dS_{\text{irr}}$, where dS_{irr} is the same as in the previous term if there is no other source of irreversibility such as incomplete chemical equilibrium within the system. The combination of these two equations yields again Eq. (2-26), which is now seen to be general for any closed system in internal equilibrium.

In summary, we may note that entropy has been defined such that in any irreversible process the total entropy of all systems concerned is increased, whereas in a reversible process the total increase in entropy of all systems is zero. The entropy increase of a system undergoing an irreversible process is the sum of entropy transferred from other systems and entropy produced by the irreversible processes within the system. For a reversible process, which involves only entropy transfers, the increase in the entropy of any individual system, or part of a system, is equal to the heat it absorbs divided by its absolute temperature. The definition of entropy together with the definition of the thermodynamic temperature scale makes entropy an extensive property of a system. It is important to see clearly that the idea of entropy is necessitated by the existence of irreversible processes; it is only for the purpose of convenient measurement of entropy changes that we have discussed reversible processes here.

ENTROPY AND PROBABILITY

We are now familiar with the idea that matter consists of a very large number of atoms and molecules and that their behavior follows statistical factors. Thus, a spontaneous process leads from a state of smaller probability to one of greater probability. And an equilibrium state is one of maximum probability subject to the criteria that define that state. This pattern also implies the possibility of fluctuations, of a momentary change to a state of less than maximum probability.

The science of statistical mechanics is a sister to thermodynamics, but it is built up from postulates at the atomic level. In statistical mechanics, the second law appears as the natural conclusion that changes should take place in the direction of increasing probability. Anyone who wishes to use thermodynamics most effectively should study statistical mechanics and make full use of its methods where appropriate. But many, indeed most, real systems of practical interest are too complex for accurate and rigorous treatment by statistical mechanics. Thus the thermodynamic relationships between and predictions of macroscopic properties are still very important. At the same time, one uses statistical methods wherever appropriate. In some cases, where the atomic and molecular data are adequate, the statistical results are rigorous and very precise. Examples of this type are given in Chapter 5 and certain later chapters. In other cases, statistical mechanics suggests the form of an equation but does not provide accurate values of certain parameters. These parameters can then be evaluated from experimental thermodynamic data. Examples of this type are given in several chapters.

At this point we shall only consider a case where the statistics are very simple. Assume two equal boxes connected by an opening. Place N identical balls inside and shake the system so that any one ball is equally likely to be at any particular point inside. The chance that all the N balls will be in a specified box at a given time is $(1/2)^N$. Likewise, if in a pair of similar flasks connected by a stopcock we have N molecules of a certain gas, then if the stopcock is closed at a certain instant the chance that all the molecules will be in one specified flask is $(1/2)^N$. Thus if $N = 20$, the chance in question is about one in a million, and this chance obviously diminishes enormously as we proceed to the large number of molecules such as we deal with in practice. The most recent determinations of the number of molecules in 1 mole give 6.02×10^{23}; in dealing with numbers so vast, the laws of chance lead inexorably to results of an accuracy far exceeding that which is possible even in the most refined physical measurements.

Thus, in the case before us, if 1 mole of gas is distributed between the two flasks, the very randomness of the molecular motions makes it logically certain that minute temporary changes in concentration will from time to time occur. Nevertheless, the relative deviations from complete uniformity of distribution between the two flasks must be so exceedingly small that it seems inconceivable that they could ever be detected experimentally. In other words, the chance that, within the limits of accuracy of our own observation, the gas will be equally distributed between the two flasks is, to all intents and purposes, unity. Expressing this mathematical chance or probability by the symbol $\mathcal{P}$, we can write, as a very close approximation, $\mathcal{P} = 1$. On the other hand, we have found that the probability of finding all the molecules in one flask is almost zero, namely, $\mathcal{P} = (1/2)^{N_A}$, where N_A is now the number of molecules in 1 mole.

When therefore the gas is at first enclosed in one of the flasks and the stopcock is then opened to allow it to distribute itself between the two flasks, it is legitimate to say that immediately after opening the cock the system passes from a state of very small probability to a state of very large probability, i.e., from $\mathcal{P}_A = (1/2)^{N_A}$ to $\mathcal{P}_B = 1$.

In order to obtain a relation that we are about to exhibit, let us in a purely arbitrary manner define a new quantity σ by the equation

$$\sigma = k \ln \mathcal{P} \tag{2-27a}$$

where k is the gas constant per molecule R/N_A. This is the Boltzmann constant with the value $1.38066 \times 10^{-23}\,\mathrm{J\,K^{-1}}$. Also,

$$\sigma_B - \sigma_A = k \ln\left(\frac{\mathcal{P}_B}{\mathcal{P}_A}\right) = \frac{R}{N_A} \ln\left(\frac{\mathcal{P}_B}{\mathcal{P}_A}\right) \tag{2-27b}$$

Using the above relationships, we obtain

$$\sigma_B - \sigma_A = \frac{R}{N_A} \ln 2^{N_A} = R \ln 2$$

If, instead of using two flasks of equal size, we had allowed the mole of gas to expand from any volume V_A to any other volume V_B, we should have found by precisely similar reasoning

$$\frac{\mathcal{P}_B}{\mathcal{P}_A} = \left(\frac{V_B}{V_A}\right)^{N_A}$$

and

$$\sigma_B - \sigma_A = R \ln\left(\frac{V_B}{V_A}\right) \qquad (2\text{-}28)$$

This equation is of very great interest since we have obtained in Eq. (2–21)* an identical expression for the change in entropy in the expansion of one mole of ideal gas, namely,

$$S_B - S_A = R \ln\left(\frac{V_B}{V_A}\right)$$

Hence in this simple case we find a very simple relation between the entropy and the logarithm of the probability, namely,

$$S_B - S_A = k(\ln \mathcal{P}_B - \ln \mathcal{P}_A) \qquad (2\text{-}29)$$

We shall make use of this last relationship in statistical calculations for other systems in Chapter 5 and thereafter.

A formal statement of the second law must be in terms of probability; an appropriate statement is: *Every system that is left to itself will, on the average, change toward a condition of maximum probability.* This law, which is true for *average* changes in any system, is also true for *any* changes in a system of very many molecules.

REFERENCES

1. A. Einstein, *Ann. Physik*, **18**, 639 (1905).
2. W. F. Giauque, *Nature*, **143**, 623 (1939).
3. B. W. Mangum and G. T. Furukawa, "Guidelines for Realizing the International Temperature Scale of 1990 (ITS-90)," NIST Technical Note 1265.

PROBLEMS

2-1. Calculate the net entropy increase of all parts of the system, which measures the total irreversibility, when 1 kg of ice of 0°C drops 1 m into an ice–water bath.

2-2. A galvanic cell thermostated at 20°C that produces a reversible emf of 1.07 V is short-circuited until 1000 C of electricity has passed. What is the net entropy increase of the cell and thermostat if it is assumed that the chemical changes in the cell are the same as would have occurred on reversible discharge by the same amount? What additional information would be needed to divide this entropy increase between the cell and the thermostat?

2-3. A flask initially containing benzene at its freezing point of 5.5°C is brought into contact with an ice–water bath until 1 mole of benzene has frozen. The heat of fusion of benzene is 126.8 J g^{-1}. Calculate the decrease in the entropy of the benzene; the net increase for the combined system.

2-4. What is the net increase in total entropy (because of irreversibility) when 1 kg of water at 30°C is added to a large ice–water bath at 0°C? Assume $C_P = 4.2$ J g^{-1} K^{-1} for water.

2-5. For which of the following changes of state is Eq. (2-26) applicable?
  (a) A sample of NO_2 gas is expanded slowly so that the equilibrium is maintained with respect to $NO + \tfrac{1}{2}O_2$.
  (b) A sample of NO_2 gas is expanded at a rate such that the dissociation to $NO + \tfrac{1}{2}O_2$ lags behind the equilibrium composition.
  (c) A sample of SO_3 gas is expanded under conditions (absence of catalyst) such that there is no dissociation into SO_2 and O_2.
  (d) A sample of water is frozen isothermally at $-10°C$.

CHAPTER 3

OTHER STATE FUNCTIONS AND THEIR RELATIONSHIPS

In this chapter we introduce a few new functions and derive a large number of very useful equations relating one thermodynamic property to another. The basis throughout is a constant amount of a single substance† in internal equilibrium. Often we will also consider cases where the only type of work is that of volume changes against an external pressure. The basic equation for this type of system was derived above:

$$dU = T\,dS - P\,dV \qquad (2\text{-}26)$$

This is the appropriate equation to use if entropy and volume are the independent variables of the problem, but this is rarely the case. More often, pressure is the appropriate variable instead of volume and temperature instead of entropy. In order to

---

† Relationships when there is more than one component with variation of composition will be considered in Chapter 10.

deal conveniently with these situations, Gibbs[1] in 1875 defined three new functions as follows:

$$H \equiv U + PV \tag{3-1}$$

$$A \equiv U - TS \tag{3-2}$$

$$G \equiv U - TS + PV = H - TS = A + PV \tag{3-3}$$

Note that H, A, and G are *state functions* or *properties* since all of the quantities entering their definitions have that characteristic. The process of their formation is sometimes called a Legendre transform.

It should be noted that Massieu[2] had already in 1869 defined two functions, which are $-A/T$ and $-G/T$ in terms of those defined above, and had developed some of the relationships of the type given below for A and G. Also Helmholtz[3] proposed in 1882 the function U-TS and applied it to certain chemical processes; this work was published after that of Gibbs and Massieu but was presumably independent. It was Gibbs,[1] however, who demonstrated the great utility of all of these functions in the interpretation of diverse physical–chemical phenomena and, in particular, for multicomponent systems.

Gibbs used the Greek letters ε, χ, ψ, and ζ, respectively, for U, H, A, and G, but these symbols have not been in general use for many years. After some differences concerning A and G, with accompanying confusion, the symbols shown above have been widely agreed upon and are recommended by IUPAC.[4]† We also follow the IUPAC[4] recommendations for names: enthalpy for H, Helmholtz energy for A, and Gibbs energy for G. Alternates in the literature are heat content for H; Helmholtz function, Helmholtz free energy, or work content for A; and Gibbs function or Gibbs free energy for G. The reader should always take care to ascertain the names and symbols in use for $U - TS$ and $H - TS$ in a particular book, article, or table.

We now take the differential of H,

$$dH = dU + P\,dV + V\,dP$$

and substitute Eq. (2-24) to obtain

$$dH = T\,dS + V\,dP \tag{3-4}$$

Similarly, one obtains

$$dA = -S\,dT - P\,dV \tag{3-5}$$

$$dG = -S\,dT + V\,dP \tag{3-6}$$

---

† The principal difference has been the use of F instead of either A or G. The literature in physics commonly used and sometimes still uses F instead of A for $U - TS$. The use of F instead of G for $H - TS$ was common among American physical chemists until recently; that choice was adopted by Lewis and Randall in 1923. In our revision in 1961, we retained F for $H - TS$ but recognized the need for some change. We also advocated the abandonment of F for either $U - TS$ or $H - TS$, and urged the adoption of A and G, respectively. We are pleased that this was the decision of IUPAC.

A number of very useful relationships now follow directly. At constant T or S,

$$\left(\frac{\partial A}{\partial V}\right)_T = -P = \left(\frac{\partial U}{\partial V}\right)_S \tag{3-7}$$

$$\left(\frac{\partial G}{\partial P}\right)_T = V = \left(\frac{\partial H}{\partial P}\right)_S \tag{3-8}$$

while at constant V or P,

$$\left(\frac{\partial A}{\partial T}\right)_V = -S = \left(\frac{\partial G}{\partial T}\right)_P \tag{3-9}$$

$$\left(\frac{\partial U}{\partial S}\right)_V = T = \left(\frac{\partial H}{\partial S}\right)_P \tag{3-10}$$

From further differentiation, we obtain

$$\frac{\partial^2 A}{\partial T \partial V} = -\left(\frac{\partial S}{\partial V}\right)_T = -\left(\frac{\partial P}{\partial T}\right)_V$$

hence,

$$\left(\frac{\partial S}{\partial V}\right)_T = \left(\frac{\partial P}{\partial T}\right)_V \tag{3-11}$$

and, similarly, from $\partial^2 G/\partial T \partial P$,

$$\left(\frac{\partial S}{\partial P}\right)_T = -\left(\frac{\partial V}{\partial T}\right)_P \tag{3-12}$$

from $\partial^2 H/\partial S \partial P$

$$\left(\frac{\partial T}{\partial P}\right)_S = \left(\frac{\partial V}{\partial S}\right)_P \tag{3-13}$$

and from $\partial^2 U/\partial S \partial V$

$$\left(\frac{\partial T}{\partial V}\right)_S = -\left(\frac{\partial P}{\partial S}\right)_V \tag{3-14}$$

The last four equations are known as the Maxwell relations.

For the effect of change in pressure on enthalpy at constant temperature, one combines Eqs. (3-3), (3-8), and (3-12) to obtain

$$\left(\frac{\partial H}{\partial P}\right)_T = \left(\frac{\partial G}{\partial P}\right)_T + T\left(\frac{\partial S}{\partial P}\right)_T = V - T\left(\frac{\partial V}{\partial T}\right)_P \tag{3-15}$$

If we now combine Eq. (3-9) with the original definitions (3-2) and (3-3) we obtain valuable relationships that are appropriately known as the Gibbs–Helmholtz equations:

$$A = U + T\left(\frac{\partial A}{\partial T}\right)_V \tag{3-16a}$$

$$G = H + T\left(\frac{\partial G}{\partial T}\right)_P \tag{3-16b}$$

These equations can be rearranged to more compact and useful forms as follows:

$$\left[\frac{\partial (A/T)}{\partial T}\right]_V = -\frac{U}{T^2} \tag{3-17}$$

$$\left[\frac{\partial (G/T)}{\partial T}\right]_P = -\frac{H}{T^2} \tag{3-18}$$

Alternate forms are sometimes more convenient:

$$\left[\frac{\partial (A/T)}{\partial (1/T)}\right]_V = U \tag{3-19}$$

$$\left[\frac{\partial (G/T)}{\partial (1/T)}\right]_P = H \tag{3-20}$$

Equations (3-18) and (3-20) are particularly important in the treatment of the temperature dependency of chemical equilibria under constant pressure.

Change of Variable

In a number of problems it is necessary to transform an expression from one set of independent variables to another. The most frequent case in thermodynamics is a change from T and V to T and P. We will use that transformation as an example, but the mathematics is general, of course. Suppose that the property Y is a function of T and V. Its differential is

$$dY = \left(\frac{\partial Y}{\partial T}\right)_V dT + \left(\frac{\partial Y}{\partial V}\right)_T dV$$

Therefore, along a constant-pressure path the temperature derivative is

$$\left(\frac{\partial Y}{\partial T}\right)_P = \left(\frac{\partial Y}{\partial T}\right)_V + \left(\frac{\partial Y}{\partial V}\right)_T \left(\frac{\partial V}{\partial T}\right)_P \tag{3-21}$$

It is also useful to consider an increment in T along the path of constant Y, i.e., $dY = 0$:

$$0 = \left(\frac{\partial Y}{\partial T}\right)_V + \left(\frac{\partial Y}{\partial V}\right)_T \left(\frac{\partial V}{\partial T}\right)_Y \tag{3-22}$$

This may be arranged to

$$\left(\frac{\partial V}{\partial T}\right)_Y = -\frac{(\partial Y/\partial T)_V}{(\partial Y/\partial V)_T} \quad (3\text{-}23)$$

Another rearrangement has an interesting symmetry:

$$\left(\frac{\partial Y}{\partial V}\right)_T \left(\frac{\partial V}{\partial T}\right)_Y \left(\frac{\partial T}{\partial Y}\right)_V = -1 \quad (3\text{-}24)$$

Note the cyclic order of the three variables; each occurs once "above," once "below," and once "outside."

Thermal Processes at Constant Pressure

In many calorimetric measurements and other processes involving heat transfer, the pressure rather than the volume is held constant. Then there is a work term $-P\,\Delta V$ and the first law yields

$$\Delta U = q - P\,\Delta V$$

But at constant pressure,

$$\Delta H = \Delta U + P\,\Delta V = q \quad (3\text{-}25)$$

Hence for any constant-pressure process with no work other than that of volume change, the enthalpy increase is exactly the heat absorbed. For this reason H is sometimes called the heat content.

Heat Capacity

In the previous chapter we obtained for the heat capacity at constant volume

$$C_V = \left(\frac{\partial U}{\partial T}\right)_V \quad (2\text{-}7)$$

From the definition of entropy, we now have for a process of heating at constant volume,

$$dS = \frac{C_V}{T} dT = C_V\,d\ln T \quad (3\text{-}26)$$

For a constant-pressure process, we showed in Eq. (3-25) that the heat absorbed is just ΔH; hence, the heat capacity at constant pressure is

$$C_P = \left(\frac{\partial H}{\partial T}\right)_P \quad (3\text{-}27)$$

and the entropy change with temperature at constant pressure is

$$dS = \frac{C_P}{T} dT = C_P\,d\ln T \quad (3\text{-}28)$$

It is often of interest to relate C_P to C_V. From the general relationship for a change of variable, Eq. (3-21), one has

$$\left(\frac{\partial S}{\partial T}\right)_P = \left(\frac{\partial S}{\partial T}\right)_V + \left(\frac{\partial S}{\partial V}\right)_T \left(\frac{\partial V}{\partial T}\right)_P$$

Then, from (3-11), (3-26), and (3-28), one obtains

$$\frac{C_P}{T} = \frac{C_V}{T} + \left(\frac{\partial P}{\partial T}\right)_V \left(\frac{\partial V}{\partial T}\right)_P$$

and

$$C_P = C_V + T\left(\frac{\partial P}{\partial T}\right)_V \left(\frac{\partial V}{\partial T}\right)_P \tag{3-29}$$

If, as is usually the case, one is using P and T as the independent variables, one uses the transformation, Eq. (3-23),

$$\left(\frac{\partial P}{\partial T}\right)_V = -\frac{(\partial V/\partial T)_P}{(\partial V/\partial P)_T}$$

to obtain

$$C_P = C_V - T\frac{(\partial V/\partial T)_P^2}{(\partial V/\partial P)_T} \tag{3-30}$$

Finally, one may prefer an expression in terms of the coefficient of expansion

$$\alpha = \frac{1}{V}\left(\frac{\partial V}{\partial T}\right)_P \tag{3-31}$$

and the compressibility

$$\kappa = -\frac{1}{V}\left(\frac{\partial V}{\partial P}\right)_T \tag{3-32}$$

The result is

$$C_P = C_V + \frac{TV\alpha^2}{\kappa} \tag{3-33}$$

Occasionally, one prefers to use the pair V, T as independent variables. Then a different transformation,

$$\left(\frac{\partial V}{\partial T}\right)_P = -\frac{(\partial P/\partial T)_V}{(\partial P/\partial V)_T}$$

is inserted into (3-29) to obtain

$$C_P = C_V - T\frac{(\partial P/\partial T)_V^2}{(\partial P/\partial V)_T} \tag{3-34}$$

Also of interest is the effect on C_P of a change in pressure; from Eqs. (3-15) and (3-27) one obtains

$$\left(\frac{\partial C_P}{\partial P}\right)_T = \frac{\partial^2 H}{\partial P \partial T} = -T\left(\frac{\partial^2 V}{\partial T^2}\right)_P \qquad (3\text{-}35)$$

If the substance is an ideal gas with $P = nRT/V$, one has

$$\left(\frac{\partial P}{\partial T}\right)_V = \frac{nR}{V}$$

and

$$\left(\frac{\partial P}{\partial V}\right)_T = -\frac{nRT}{V^2}$$

whereupon

$$C_P = C_V + nR \qquad (3\text{-}36)^*$$

Maximum Work

In order to more fully understand the meaning of the Helmholtz energy, A, let us consider the maximum work that can be obtained from a change at constant temperature. The work done by the system is, from the first law and the definition of A,

$$-w = q - \Delta U$$
$$= q - \Delta A - T\Delta S \qquad (3\text{-}37)$$

We know, however, from the second law, Eq. (2-23), that

$$q \leq T\Delta S$$

where the equality corresponds to a reversible process and the inequality to a dissipative process. Then

$$-w \leq -\Delta A \qquad (3\text{-}38)$$

which shows that the decrease in A gives the maximum limit to the amount of work done by the system. Also, if the process is reversible, the equality sign applies and the work done is $-\Delta A$.

In view of these relationships A can be called the "isothermal work content." It has, less precisely, been called the "work content" and first received the symbol A after the German word Arbeit.

Maximum Net (non-PV) Work

Let us now separate the work term into the work done against a constant-pressure atmosphere, $P\Delta V$, and any other work, which we designate w'. This quantity is sometimes called the useful work, but that is misleading because $P\Delta V$ can in some circumstances be useful work. It is only expansion against the natural atmosphere that is intrinsically useless; expansion against an artificial high-pressure system can be useful work of compression. Thus, we prefer the terms "net work" or "non-PV work" for w'.

From Eq. (3-38) and the definitions of A and G, we have

$$-w' = -w - P\Delta V$$
$$\leq -\Delta A - P\Delta V$$

and

$$-w' \leq -\Delta G \tag{3-39}$$

Thus, the change in Gibbs energy gives the maximum for the "net work" that can be obtained from an isothermal process. Again, the equality sign applies if the process is reversible.

If no work is involved other than PV work, then one has the very simple and important result

$$\Delta G \leq 0 \tag{3-40}$$

for an isothermal process. And the condition for equilibrium at constant T and P (and with no work other than $P\Delta V$) is

$$\Delta G = 0 \tag{3-41}$$

Example with Net Work

As an example where there is work of a type other than $P\Delta V$, consider an electrochemical system. Specifically, take a system in which zinc acts upon aqueous sulfuric acid to form aqueous zinc sulfate and hydrogen, under constant atmospheric pressure, and in a thermostat. Evidently this process can occur in such a way as to perform no work except the small amount done against the atmosphere by the evolved hydrogen. This, in fact, is just what will occur if impure zinc is added directly to the acid. This process is highly irreversible. On the other hand, if we place in the thermostat the same substances arranged as a galvanic cell, with zinc as one electrode and another electrode of hydrogen in contact with a platinized electrode, and if these two electrodes are connected to a motor or other electrical system in such a way as to utilize the electrical energy that is now available, an amount of work will be done that will depend upon the efficiency of our arrangements.

The maximum work would be obtained if at every instant the external electrical system were arranged to exert so large a counter emf that, when infinitesimally increased, it would force the current in the opposite direction, thus causing hydrogen to be consumed and zinc to precipitate. The process would then be reversible and the net work equal to $-\Delta G$.

In both the processes we have described, the system passes from the same initial to the same final states. Therefore $-\Delta G$ is the same in both cases. This difference $-\Delta G$ is, in the reversible process, the work that the system has performed, while in the irreversible process it is the maximum work that *might* have been performed.

As a corollary we may note that, while $T \Delta S$ is also the same for both processes, it represents in the reversible process the amount of heat actually absorbed from the thermostat, while in the irreversible process it merely shows the maximum amount of heat that might have been absorbed; for ΔU is the same in both cases, and therefore the heat absorbed will be greater, the greater the amount of work performed.

Equilibrium Between a Substance and Its Vapor

Let us apply Eq. (3-11) to a system composed of a substance and its vapor, the two being in equilibrium with one another at the vapor pressure P, which is moreover equal to the applied external pressure. In this case $\partial S/\partial V$ is the same as $\Delta S/\Delta V$, where ΔS is the increase in entropy and ΔV the increase in volume when an amount of the substance vaporizes. As we are dealing with an equilibrium, and therefore with a reversible process, $\Delta S = q/T$ and q, which we may also write as ΔH, is the ordinary heat of vaporization. Moreover, since the vapor pressure does not depend upon the volume of the system, we may omit the restriction of constancy of volume and thus Eq. (3-11) becomes

$$\frac{dP}{dT} = \frac{\Delta H}{T \Delta V} \tag{3-42}$$

This is the famous equation which Clapeyron[5] obtained essentially in this form in 1834. It was the first physicochemical application of what we now call the second law of thermodynamics. This equation is also valid for solid–solid and solid–liquid equilibria.

At low pressures the vapor can be approximated as an ideal gas, and the volume of the liquid or solid is negligible in comparison. Then, on a molar basis,

$$\Delta V_m \cong \frac{RT}{P}$$

and on substitution into Eq. (3-42) one obtains

$$\frac{d\ln P}{dT} = \frac{\Delta H_m}{RT^2} \tag{3-42a}*$$

This equation, which is sometimes called the approximate Clapeyron equation, is often adequate for approximate calculations of the sublimation pressure of a solid or the vapor pressure of a liquid below one atmosphere.

Changes in H, A, and G for Isothermal Expansion of an Ideal Gas

The changes in U and S for an isothermal expansion of an ideal (perfect) gas were given by Eqs. (2-9)* and (2-21)*

$$\left(\frac{\partial U}{\partial V}\right)_T = 0 \quad \text{ideal gas} \tag{2-9}*$$

$$(S_B - S_A)_T = nR \ln\left(\frac{V_B}{V_A}\right) = nR \ln\left(\frac{P_A}{P_B}\right) \quad \text{ideal gas} \quad (2\text{-}21)^*$$

while we have the assumed equation of state

$$PV = nRT \quad \text{ideal gas} \quad (2\text{-}8)^*$$

We may now substitute these results into the equations defining H, A, and G. For H we first note that for the ideal gas

$$\left[\frac{\partial (PV)}{\partial P}\right]_T = \left[\frac{\partial (PV)}{\partial V}\right]_T = 0 \quad (3\text{-}43)^*$$

This result in combination with (2-9)* yields for H,

$$\left(\frac{\partial H}{\partial V}\right)_T = \left(\frac{\partial H}{\partial P}\right)_T = 0 \quad \text{ideal gas} \quad (3\text{-}44)^*$$

Since $A = U - TS$ and with Eqs. (2-9)* and (2-19)*, one has for a perfect gas

$$(A_B - A_A)_T = nRT \ln\left(\frac{V_A}{V_B}\right) = nRT \ln\left(\frac{P_B}{P_A}\right) \quad (3\text{-}45)^*$$

Since the change A to B is an isothermal process, the value $A_B - A_A$ should be w, the work done on the system. Equation (2-10)* gave w for the compression of an ideal gas, and we note that there is complete agreement.

In view of Eq. (3-43)* it is clear that the behavior of G will be the same as A, just as that of H was the same as U, for the isothermal expansion of a perfect gas. Thus

$$(G_B - G_A)_T = nRT \ln\left(\frac{P_B}{P_A}\right) = nRT \ln\left(\frac{V_A}{V_B}\right) \quad (3\text{-}46)^*$$

These results will be of interest in a number of cases to be considered in later chapters.

Adiabatic Reversible Processes

We now consider processes where there is no heat transfer; these are said to be *adiabatic*. For our present discussion, we retain the basis of this chapter, that of a substance in internal equilibrium, and further assume that the only work is that of volume change. Thus, these processes are also reversible and the requirement that $\delta q = 0$ implies also that $dS = 0$; in other words, we consider *isentropic* processes.

In such an isentropic compression, there will ordinarily be a change in temperature, and by measuring the ratio between this temperature change and the pressure change we obtain $(\partial T/\partial P)_S$. By following the method of Eq. (3-23) and by employing also Eqs. (3-12) and (3-28), we find

$$\left(\frac{\partial T}{\partial P}\right)_S = -\frac{(\partial S/\partial P)_T}{(\partial S/\partial T)_P} = \frac{T}{C_P}\left(\frac{\partial V}{\partial T}\right)_P \quad (3\text{-}47)$$

By the corresponding method we obtain

$$\left(\frac{\partial T}{\partial V}\right)_S = -\frac{(\partial S/\partial V)_T}{(\partial S/\partial T)_V} = -\frac{T}{C_V}\left(\frac{\partial P}{\partial T}\right)_V \quad (3\text{-}48)$$

These equations have several important applications; one is discussed in the following section and others are considered in subsequent chapters.

If the fluid is an ideal gas, Eq. (3-47) simplifies to

$$\left(\frac{\partial T}{\partial P}\right)_S = \frac{RT}{PC_P} \quad (3\text{-}49\text{a})*$$

or

$$d\ln T = \frac{R}{C_P} d\ln P \quad \text{constant } S \quad (3\text{-}49\text{b})*$$

where C_P is the molar heat capacity. Similarly, by substitution of the ideal gas equation, one obtains from Eq. (3-48),

$$\left(\frac{\partial T}{\partial V}\right)_S = -\frac{RT}{VC_V} \quad (3\text{-}50\text{a})$$

or

$$d\ln T = -\frac{R}{C_V} d\ln V \quad \text{constant } S \quad (3\text{-}50\text{b})$$

These equations cannot be integrated without knowledge of the temperature dependence of C_V or C_P. But if the heat capacity is constant, which is often a good approximation, the integration is simple, and

$$\frac{T}{P^{R/C_P}} = \text{constant} \quad (3\text{-}51)*$$

and

$$TV^{R/C_V} = \text{constant} \quad (3\text{-}52)*$$

Equation (3-52)* will be used in connection with the Carnot cycle in Chapter 4. We next turn to the application of these equations for isentropic expansion to the measurement of the heat capacities of gases.

An Indirect Method of Measuring the Heat Capacity of Gases

There is a method whereby Eq. (3-47) yields the heat capacity of a gas from other measurements. The P–V–T behavior of the gas must be known. While assumption of the perfect-gas law yields approximate values of $(\partial V/\partial T)_P$ or $(\partial P/\partial T)_V$ for any gas at low pressure, accurate work requires knowledge of the equation of state for that gas (see Chapter 9).

In the method proposed by Lummer and Pringsheim,[6] the value of $(\partial T/\partial P)_S$ is measured directly by a thermometer of very low heat capacity suspended in the center of a large volume of gas. Kistiakowsky and Rice[7] refined earlier techniques and obtained excellent results. They used platinum wire of 0.0075 mm diameter as a thermometer and a 12-liter gas volume. It proved to be necessary to correct for direct heat exchange by radiation from the thermometer to the wall of the vessel.

Let us assume the simple equation of state that is valid for low pressure in most cases,

$$V = \frac{RT}{P} + B \tag{3-53*}$$

where the second virial coefficient B is a separate function of temperature for each substance. Then

$$\left(\frac{\partial V}{\partial T}\right)_P = \frac{R}{P} + \frac{dB}{dT}$$

and with Eq. (3-47)

$$C_P \frac{dT}{T} = \left(\frac{R}{P} + \frac{dB}{dT}\right) dP \quad \text{at constant } S \tag{3-54*}$$

If we take C_P and dB/dT to be independent of T and P for small changes, Eq. (3-54)* integrates to

$$\frac{C_P}{R} = \frac{\ln(P_i/P_f)}{\ln(T_i/T_f)} + \left(\frac{dB}{dT}\right)\frac{P_i - P_f}{R\ln(T_i/T_f)} \tag{3-55*}$$

where P_i, P_f are the initial and final pressures, respectively, and T_i, T_f are the corresponding temperatures. For carbon dioxide at 300.06 K and approximately 1 bar mean pressure, Kistiakowsky and Rice found the two terms in Eq. (3-55)* to be 4.463 and 0.054, respectively, yielding C_P/R = 4.517. This result agrees very well with the value calculated from the spectroscopic energy levels of carbon dioxide and corrected for gas imperfection at 1 bar (see Chapter 5).

UNITS OF ENERGY AND ENTROPY

We first state the units in which these quantities are being reported in current work of high quality with recommendations of one system that seems to us by far the best. Subsequently, we will give a brief account of past practices; this knowledge is of historical interest and is needed if one is to correctly interpret the older literature, which contains many results of current importance.

Entropy

We have seen that entropy is a logarithmic measure of probability. Thus, it is intrinsically dimensionless. In our view,[8] it should be expressed that way in practical work. Equation (2-29), when divided by R, becomes

$$\frac{(S_B - S_A)}{R} = N_A^{-1} \ln\left(\frac{\mathcal{P}_B}{\mathcal{P}_A}\right)$$

Statistical entropies should not be multiplied by R; rather they should be reported as values of S/R. Calorimetrically measured entropies should be divided by R and, likewise, reported as dimensionless quantities.

In the literature, one finds many values reported in the dimensions of R; currently, this is usually $J K^{-1} mol^{-1}$. This is, of course, a correct and generally satisfactory system, but it has no advantage and certain disadvantages. A trivial disadvantage is the nuisance of writing out dimensions instead of using a dimensionless quantity. More significantly, the most precise quantities in chemical thermodynamics are those calculated statistically from accurate atomic and molecular data. They should be reported as calculated without further unnecessary manipulation. The value of R is now known with a precision so high that no uncertainty will be introduced for thermal measurements by division by R.

Also, we shall see that the relationship of entropies to equilibrium constants or their equivalent always involves $\Delta S/R$. Thus, it is $\Delta S/R$ that one wants to use, and it saves division by R if S/R has been tabulated.

Energy

Measurements of mechanical or electrical energy yield results in joules. Also, thermal measurements usually involve either heat introduced electrically or calibration by comparison with electrically introduced heat; in either case the results are in joules. Thus, it is entirely appropriate to report and tabulate experimental energies and enthalpies in joules per mole.

If entropies are being reported and tabulated as dimensionless quantities, however, there are advantages in dividing energies by R, whereupon the values have the dimension temperature and the units are K. Again, if these energies are used to calculate equilibrium constants they will be divided by RT. Thus, values of $\Delta U/R$ or $\Delta H/R$ in K are divided by T in K and combined with dimensionless values of $\Delta S/R$. Also, we believe that an energy in kelvins is more easily interpreted than a value in joules. It is easy to compare values of $\Delta H/R$ with the temperature of interest to determine the significance of the enthalpy term.

Calories

Until about 1920 most measurements of heat were made in such a manner that the result was to increase the temperature of a known amount of liquid water or its equivalent. The results were reported in calories, with one calorie being the heat required to raise the temperature of one gram of water by one degree Celsius. Experiments were done at room temperature, which ranged from 12 to 25°C. The heat capacity of water is approximately but not exactly constant in this range.

Later an improved method was adopted in which the basic measurement was the amount of electrical energy that yielded the same heat as that observed in the process being investigated. Then a "thermochemical calorie" was defined in terms of a joule of a magnitude equal as nearly as possible to that corresponding to the earlier data.

For a period roughly 1920–1948 the electrical measurements were made in terms of "International" volts, amperes, etc., which were established by the various national standards laboratories to be accurately transferable among laboratories but only approximately equal to the absolute volt, ampere, etc. The thermochemical calorie was 4.1833 "International Joules." After 1948 the electrical calibrations were in absolute units and the equivalence became 1 cal = 4.1840 J. This was adopted as a defined value for the thermochemical calorie.

In the thermochemical calorie $R = 1.9872 \text{ cal K}^{-1} \text{ mol}^{-1}$. Since this value is close to 2.0, one could make an approximate mental division of entropies or energies by R very easily for estimates of equilibrium constants. Division by $8.3145 \text{ J K}^{-1} \text{ mol}^{-1}$ is more difficult, even if approximated to 8.3. While in one sense this is a trivial matter, it is a reason why there was less incentive to convert tables of entropies and energies from $\text{cal K}^{-1} \text{ mol}^{-1}$ and cal mol^{-1} to dimensionless and K at that time than there is now to convert from joules.

PV Units

For work associated with compression or expansion, the measurements involve pressure and volume. The SI units are pascals for pressure and cubic meters for volume and their product yields joules. From the viewpoint of human experience, however, we noted earlier that the pascal carries little meaning. Thus we prefer the unit bar, which approximates atmospheric pressure. Also, a cubic centimeter is a more realistic laboratory unit of volume than a cubic meter. Since 1 bar = 10^5 Pa and 1 cm^3 = 10^{-6} m^3, the energy unit bar cm^3 is equal to 0.1 J. Then the gas constant $R = 83.1451 \text{ bar cm}^3$.

REFERENCES

1. J. W. Gibbs, *Trans. Conn. Acad. Sci.*, **3**, 108, 343 (1875, 6); *The Collected Works of J. Willard Gibbs*, Yale University Press, New Haven, 1948, Chapter III; see p. 142 of the original or p. 87 of the republication.
2. M. F. Massieu, *Compt. rend.*, **69**, 858, 1057 (1869).
3. H. von Helmholtz, *Sitzungsber. Akad. Wiss. Berlin*, **1**, 22 (1882).
4. I. Mills, T. Evitas, K. Holmann, N. Kalloy, and K. Kuchitsu, *Quantities, Units, and Symbols in Physical Chemistry*, a Report of the International Union of Pure and Applied Chemistry, Blackwell Scientific, Oxford, 1988.
5. E. Clapeyron, *J. ecole polytech. (Paris)*, **14**(23), 153 (1834).
6. O. Lummer and E. Pringsheim, *Ann. Physik*, (3)**64**, 555 (1898).
7. G. B. Kistiakowsky and W. W. Rice, *J. Chem. Phys.*, **7**, 281 (1939).
8. K. S. Pitzer and L. Brewer, *J. Phys. Chem. Ref. Data*, **8**, 917 (1949); *High Temp. Sci.*, **11**, 49 (1979).

PROBLEMS

3-1. All gases at relatively low pressures approach the equation of state $PV = RT + BP$ for 1 mole, where B is independent of P or V. Derive an expression for the difference $(\partial H/\partial V)_T - (\partial U/\partial V)_T$ for 1 mole of such a gas.

52 THERMODYNAMICS

3-2. Show that Eq. (3-52)*, for adiabatic change, can be put into the equivalent form

$$PV^{C_P/C_V} = \text{constant}$$

3-3. Show that for an ideal gas

$$\left(\frac{\partial C_V}{\partial V}\right)_T = 0 \quad \left(\frac{\partial C_P}{\partial V}\right)_T = 0$$

3-4. The relation between the pressure, the temperature, and the volume of 1 mole of hydrogen gas may be expressed over a limited range by the equation of state $P = RT/(V - B)$, where B is independent of P but is a function of T.
  (a) Find $(\partial V/\partial T)_P$ and $(\partial V/\partial P)_T$, and express dV as a function of T and P.
  (b) Apply Eq. (2-16) to confirm that dV is a perfect differential.
  (c) Confirm Eq. (3-23) by using it to obtain $(\partial V/\partial T)_P$ for the above equation of state and comparison with the result obtained by rearrangement to $V = B + RT/P$ followed by differentiation with P constant. For a more complex equation of state, the second method may not be practical.

3-5. Prove that

$$C_P = C_V + \left[V - \left(\frac{\partial H}{\partial P}\right)_T\right]\left(\frac{\partial P}{\partial T}\right)_V$$

3-6. (a) 18.02 g of liquid water is enclosed under a frictionless weightless piston at 100°C and 1 atm pressure. The pressure above the piston is lowered slightly below 1 atm and the water is allowed to vaporize isothermally until all vaporized. For this process, $q = 40.71$ kJ. The specific volume of water at 100°C is 1.043 cm³, and the specific volume of steam is 1677 cm³ at 100°C and 1 atm. Calculate the work w attending this vaporization and ΔU and ΔH for the process. (b) Find the values of ΔU, ΔH, and q for the process where the piston is just removed and the water is allowed to vaporize freely isothermally into an evacuated space of such volume that the pressure has built up to 1 atm with all the water vaporized. (c) Calculate the work in joules attending the fusion of 18.02 g of ice at 0°C and 1 atm. The density of ice is 0.917 g cm⁻³ and the density of water is 1.000 g cm⁻³. For the fusion process, $q = 6.01$ kJ. Calculate the ΔU and ΔH.

3-7. Near the triple point the vapor pressure of liquid ammonia is given by (P in bars)

$$\ln P = 15.16 - 3063/T$$

This is the equation of the liquid–vapor boundary curve on a phase diagram. The vapor pressure of the solid is given by

$$\ln P = 18.70 - 3754/T$$

  (a) What is the temperature and pressure of the triple point?
  (b) What is the heat of vaporization (liquid→gas) and the heat of sublimation (solid→gas)?
  (c) What is the heat of fusion? (solid→liquid)
  (d) What is the change in Gibbs energy for the process solid→liquid at the triple point?

3-8. Calculate ΔG, ΔH, and ΔS for the process of freezing 1 kg of supercooled water isothermally at $-10°C$. Assume that $\Delta H = 80$ cal g⁻¹ for fusion of ice at 0°C and that the heat capacities of water and ice are 1.0 and 0.5 cal g⁻¹ K⁻¹, respectively.

3-9. The vapor pressure of liquid ammonia is 7.68 bar at 17°C and increases at the rate of 0.25 bar deg⁻¹. The specific volumes of vapor and liquid are, respectively, 165 and

$2\,\text{cm}^3\,\text{g}^{-1}$. Calculate the heat of vaporization per gram of ammonia, and compare with the more precisely measured value of $1238\,\text{J}\,\text{mol}^{-1}$.

3-10. A gram of ammonia liquid at 7.68 bar and 17°C is vaporized isothermally into an evacuated space to a final volume of $165\,\text{cm}^3$. Calculate q, ΔU, ΔH, ΔS, and ΔG for this process, using the data of Problem 3-9.

3-11. Calculate the percentage deviation of the volume of isopentane saturated vapor at 298 K from the ideal-gas volume. Use the heat of vaporization and the Clapeyron equation. $\Delta H_{\text{vap}} = 24.594\,\text{kJ}\,\text{mol}^{-1}$, density of liquid $= 0.6146\,\text{g}\,\text{cm}^{-3}$, both at 298 K; and the vapor pressure in millimeters of mercury is given by

$$\log P = 6.7897 - \frac{1020.0}{T - 40.0}$$

CHAPTER 4

HEAT ENGINES, HEAT PUMPS, CYCLES, AND FLOW PROCESSES

The Heat Engine

Any system that is not in equilibrium can be made to do useful work. However, in every irreversible process there is always some waste of opportunity in this regard. Let us consider the flow of heat between two reservoirs at different temperatures. Instead of allowing heat to flow directly from one to the other, we may obtain work by means of a steam engine, or a hot-air engine, or any one of the various inventions that are known generically as heat engines. These are characterized by operating in such manner that they themselves undergo no permanent change but do work at the expense of a part of the energy taken from a hot reservoir, while the rest of the energy passes into a cold reservoir. The operation of the heat engine is achieved by subjecting the working substance of the engine to a sequence of processes that form a cycle.

The ratio between the work done and the heat taken from the hot reservoir is known as the conversion factor, or the thermal efficiency, of the engine. The problem of determining the maximum value of this ratio is the one that occupied Carnot[1] in the great monograph that laid the foundations of the second law of thermodynamics.

Every actual heat engine is inefficient because of friction or imperfect design; but even if all sources of degradation are eliminated, it is evident that no engine could be constructed to give a conversion factor of 100 percent. For if all the heat taken from the hot reservoir were converted into work, the cold reservoir might be removed altogether.

The energy taken from the hot reservoir would then be converted into mechanical work without any degradation in other systems. This work could, by friction, be returned as heat to the reservoir, and we should thus find an irreversible process, bringing the whole system back to its original state, which would violate the second law of thermodynamics.

In order to obtain the maximum possible work from a heat engine, it would be necessary to eliminate friction, to prevent direct flow of heat from hot to cold portions of the system, and to maintain a state of balance with respect to the mechanical forces. In other words, the process must be reversible. Under given conditions, therefore, the maximum conversion factor is that of a heat engine that operates reversibly in all its stages, and if we find the thermal efficiency of such an engine, we know the limit that may be approached by any actual engine as its design and construction are improved.

If a heat engine operates reversibly and passes through a whole number of complete cycles, so that it is in the same state at the end of the operation as at the beginning, it will itself suffer no change of entropy. Hence all the entropy changes are in the rest of the system, and these must sum up to zero in a reversible process. These entropy changes are immediately obtained from Eq. (2-22). If q_h is the heat *taken from* the hot reservoir at T_h and q_c is the heat *given to* the cold reservoir at T_c, then the increase of entropy of the hot reservoir is $-q_h/T_h$, and that of the cold reservoir is q_c/T_c. Equating the sum to zero,

$$-\frac{q_h}{T_h} + \frac{q_c}{T_c} = 0 \qquad (4\text{-}1)$$

By the conservation law, the net work done by the engine, w, is given by

$$w = q_h - q_c \qquad (4\text{-}2)$$

and combining these two equations, we find

$$\frac{w}{q_h} = \frac{T_h - T_c}{T_h} \qquad (4\text{-}3)$$

This important equation gives the conversion factor, or thermal efficiency, of a perfectly efficient engine operating between any two temperatures T_h and T_c. Any actual engine operating between these temperatures has a lower efficiency, but one that may approach Eq. (4-3) as a limit. Thus, for a steam engine with a condenser at 27°C, or 300 K, and with a boiler at 327°C, or 600 K, the maximum work obtainable is equal to one-half the heat taken from the boiler.

The reader should note carefully the method used to obtain Eq. (4-3) because it is applicable to all types of more complex heat-engine and heat-pump problems. In the first step all entropy changes are calculated and their sum is set to zero, in accordance with the second law, since the process is assumed to be reversible. In the second step all energy changes are expressed, and their sum also is set to zero by the first law. The work term does not appear in the first equation, since it involves no entropy, but it does appear in the second equation. These two equations are then solved for the unknown quantities.

The Heat Pump

By reversing a heat engine, it is possible, through the expenditure of work, to transport heat from a cold to a hot reservoir. This is the method employed in a refrigerating machine. The performance of a heat pump for refrigerating purposes is expressed as the coefficient of performance, q_c/w, where w is now the net work done on the working material and q_c is the heat withdrawn from the cold reservoir.

$$\frac{q_c}{w} = \frac{T_c}{T_h - T_c} \tag{4-4}$$

Let us calculate, as an example, the minimum amount of work required to convert 1 kg of water at 0°C into ice at 0°C by an engine operating in a room at 30°C. In the operation heat will be given up to the room in an amount equal to the heat absorbed from the water, together with the work done on the working material. If the heat pump is reversible, we employ Eq. (4-4). The coefficient of performance is 273/30, q_c is 333.5 kJ, and w is 333.5 × 30/273 = 36.6 kJ.

The theory of the heat pump leads to an interesting consequence, long recognized theoretically, that has now acquired practical importance. In some localities buildings are heated by electricity, the electricity passing through some form of heater whose resistance converts electrical into thermal energy. At first sight it would appear, from the law of conservation of energy, that the maximum heating effect would be produced when a certain amount of electrical energy is completely converted into thermal energy. But this is very far from the truth.

If a heat pump were constructed with the inside of the building serving as the hot reservoir and the outdoor air as the cold reservoir, and if by means of a motor the electrical energy were used to operate this pump so that the heat would be taken from without and given up inside the building, the amount of heating thus produced would, in the limiting case of ideal efficiency, be given by Eq. (4-3), where w represents the electrical energy expended, q_h the heating effect in the building, and T_h and T_c the temperatures within and without. The amount of heating thus produced per unit of electrical energy expended, or the coefficient of performance of a heat pump for heating purposes, would, in the limiting case of ideal efficiency, be given by

$$\frac{q_h}{w} = \frac{T_h}{T_h - T_c} \tag{4-5}$$

If the internal temperature T_h were 18°C, or 291 K, and the external temperature T_c were 0°C, or 273 K, the coefficient of performance at the maximum would be 291/18, or more than 16.

Initially, the use of heat pumps for heating purposes was handicapped by the high capital-investment cost and high maintenance costs. However, the development of cheap and dependable refrigerating equipment has now made the heat pump quite feasible for even relatively small heating applications, including individual household heating.

Electrons can be used as the working material of heat engines or heat pumps. An electric current flowing through a junction of one conductor with another absorbs or evolves heat depending on the direction of current flow. This effect, the Peltier heat, is related by thermodynamic principles to the Seebeck effect, which is the potential of the complete thermocouple (see Chapter 26). Semiconductors show larger thermoelectric effects than metals; consequently fewer junctions are needed for a given function. To date, these devices have attained only a fraction of the thermal efficiency, or coefficient of performance, that would be expected from Eqs. (4-3), (4-4), or (4-5) because of direct thermal conduction from hot to cold junctions. Nevertheless, the lack of moving parts and simplicity of design make even presently available thermoelectric devices of considerable interest for small power or refrigerating applications, particularly at remote sites.

Carnot Cycle

The Carnot cycle is a reversible cycle that has played an important historical role in the development of the concept of entropy. A Carnot cycle, which is illustrated in Fig. 4-1, consists of an isothermal expansion, an adiabatic expansion, an isothermal compression, and, finally, an adiabatic compression to return the working material to its original state. The net result of such a cycle is the transfer of heat from a high-temperature reservoir to a low-temperature reservoir with partial conversion into work. For a perfect gas of constant heat capacity as the working material, one can readily calculate that the net work obtained by carrying out the Carnot cycle is equal to $RT_h \times \ln(V_2/V_1) + RT_c \ln(V_4/V_3)$. The two adiabatic processes absorb no heat, and one finds that their work terms cancel (Problem 4-8). Thus, the heat absorbed from the hot reservoir is given by $q_h = RT_h \ln(V_2/V_1)$. From Eq. (3-52)* one can relate the temperatures and volumes for the adiabatic steps, obtaining $V_3/V_2 = (T_h/T_c)^{C_V/R}$ and $V_4/V_1 = (T_h/T_c)^{C_V/R}$. From these results we see that $V_1/V_2 = V_4/V_3$ so that the net work reduces to $R(T_h - T_c) \times \ln(V_2/V_1)$ and the work obtained divided by the heat absorbed from the hot reservoir reduces to $w/q_h = (T_h - T_c)/T_h$ in agreement with Eq. (4-3). The Carnot cycle can be taken in the reverse direction and can thus operate as a heat pump, and one can confirm the results of Eq. (4-4).

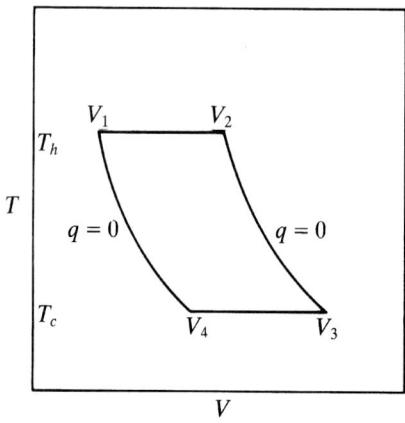

FIGURE 4-1
Carnot cycle.

Historically,[2-4] the Carnot cycle was used to develop a statement of the second law of thermodynamics and to develop the concept of entropy. Starting with the experimental observation that it is impossible to transfer heat from a low temperature to a higher temperature without the application of work by some outside system, it is possible to show that a function, entropy, as defined in Eq. (2-22) does exist and that dS is an exact differential for any substance carried through a Carnot cycle or any other reversible cycle and thus that entropy is a state function.

An alternate procedure[5-9] for the development of the second law and the concept of entropy is that of Born and Caratheodory, which does not depend upon heat engines but recognizes the relationship between the behavior of real systems and the properties of Pfaffian differential equations. In this treatment, the existence of a state function, entropy, is established by demonstrating the existence of an integrating factor that can convert the differential δq to a perfect differential dS. The integrating factor is defined as the thermodynamic temperature, which can be shown, as has been done in Chapter 2, to be identical with the perfect-gas temperature. These alternate procedures,[10] including the one presented in Chapter 2, are entirely equivalent. But these different approaches are useful for examining the concept of entropy from different points of view.

FLOW PROCESSES

The increasing importance of flow processes in recent years makes it desirable to present briefly the essential features of thermodynamics under these conditions. Examples include flow through jet and rocket engine nozzles as well as flow through all sorts of industrial equipment. Also, for laboratory measurements of heat capacities and other properties of fluids, flow methods have many advantages, especially at high temperatures and pressures. Initially, we consider cases where the kinetic energy of the flow can be neglected; later that term is added for particular examples.

In a flow system, work in the amount P_1V_1 is done on the fluid flowing into the region of interest, while the work P_2V_2 is done by the fluid leaving. Thus the fluid in any flow system absorbs an amount of energy $-\Delta(PV)$ from the surroundings. The first-law equation may be written in the form

$$\Delta U = q + w = q - \Delta(PV) + w'$$

where w' is the total of all work absorbed except pressure–volume work associated with the flow. In view of the definition of enthalpy, this becomes

$$\Delta H = q + w' \qquad (4\text{-}6)$$

One result that may be readily obtained from Eq. (4-6) is the mechanical work required to operate an adiabatic gas compressor. The gas flows into the compressor at a low pressure and is discharged at a higher pressure. Frequently it is a good approximation to assume that the gas is compressed adiabatically; that is, $q = 0$. Also, if flow rates are slow enough that kinetic energy effects are negligible, Eq. (4-6) yields the simple result

$$w' = \Delta H$$

While one can analyze any specific compression cycle and obtain this same result, thermodynamics yields it on a perfectly general basis without regard for the type of machine used.

Free Expansion

When flow rates are low so that the kinetic energy of flow is negligible and no work is extracted other than the $\Delta(PV)$ effect, w' in Eq. (4-6) is zero. And if there is no heat transfer, q is also zero, and $\Delta H = 0$. Thus, an adiabatic free expansion at low flow rate is a constant-enthalpy process. This experiment was first performed by Joule and Thomson (later Lord Kelvin)[11] in 1853 and was widely used by others in the determination of the properties of gases. This experiment consisted in allowing a gas to escape through a porous plug from a high pressure on one side to a low pressure on the other, the resistance of the plug being great enough to ensure a nearly constant pressure in the incoming and also in the outgoing gas. Thermometers placed on either side of the plug showed a temperature difference $T_2 - T_1$ depending upon the pressure difference $P_2 - P_1$. The apparatus was constructed of such poor thermal conductors that no appreciable amount of heat could pass into or out of the system.

At ordinary temperatures and pressures, all gases, except hydrogen and helium, show a cooling effect in such free expansion. This amounts, in the case of carbon dioxide, to more than 1 degree for a difference of pressure of 1 bar, while the cooling for air, which approaches more nearly to the perfect gas, is only about one-fifth as great. However, this effect in air is large enough to be of practical importance, and an apparatus for the production of liquid air makes use of this phenomenon. Thus, if a certain portion of compressed gas undergoes free expansion and the cooling effect is used to precool another portion, that portion upon expansion will fall to a still lower temperature. By continuing this process a certain fraction of the original compressed gas can be liquefied.

The experiments of Joule and Thomson and of others who have used the same method show that the cooling produced by a given pressure drop is nearly independent of the pressure (although it is noticeably smaller at the highest pressures that have been studied). On the other hand, it increases rapidly with diminishing temperature. Thus, a difference of 200 atm causes a lowering of 45 degrees in air initially at 0°C and of 100 degrees if the air is initially at −90°C. Indeed hydrogen and helium, which at ordinary temperatures are not cooled but are heated by free expansion, behave like other gases at low temperatures. If, therefore, hydrogen is precooled by liquid air, it can be further cooled and eventually liquefied by its own free expansion.

The ratio $(T_2 - T_1)/(P_2 - P_1)$ having been obtained at several values of $P_2 - P_1$, we may find its value in the limiting case as $P_2 - P_1$ approaches zero. This value is called the Joule–Thomson coefficient μ. It is formally defined by the equation

$$\mu = \left(\frac{\partial T}{\partial P}\right)_H \tag{4-7}$$

The temperature at which $\mu = 0$, i.e., where a gas is neither heated nor cooled by free expansion, is called its inversion temperature. Gases that at ordinary temperatures are cooled by free expansion show such an inversion at higher temperatures.

It is interesting to note the relationship of the Joule–Thomson coefficient to other properties. One starts with the basic mathematical expression Eq. (3-23), which for this case is

$$\left(\frac{\partial T}{\partial P}\right)_H = -\frac{(\partial H/\partial P)_T}{(\partial H/\partial T)_P}$$

But $(\partial H/\partial T)_P$ is just the heat capacity at constant pressure, and then

$$\left(\frac{\partial H}{\partial P}\right)_T = -\mu C_P \tag{4-8}$$

Figure 4-2 shows the results of Roebuck and Osterberg's[12] measurements of the Joule–Thomson coefficient of nitrogen. It is evident that μ approaches a finite value at zero pressure, and this result is characteristic of all gases. The experimental evidence

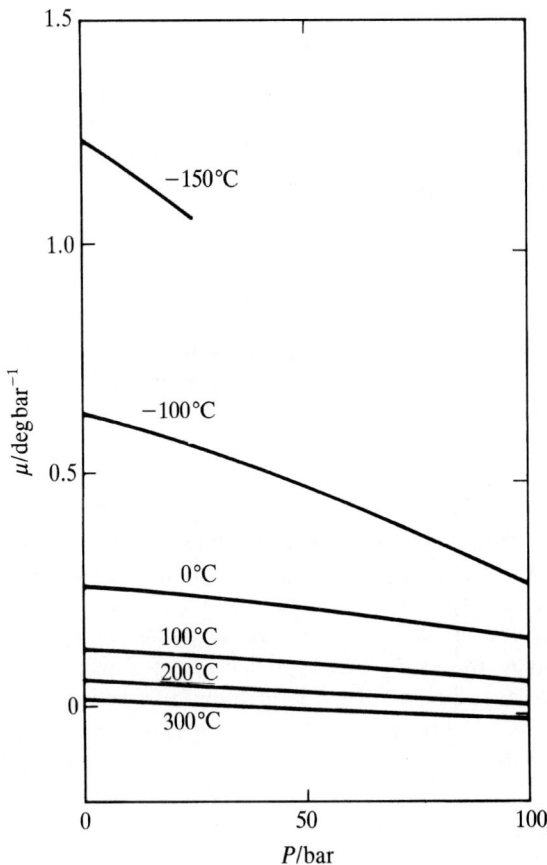

FIGURE 4-2
The Joule–Thomson coefficient of nitrogen gas. At the lowest temperature, –150°C, nitrogen liquefies; hence the curve for the gas terminates at the vapor pressure.

shows also that C_P approaches a finite value at zero pressure. From these conclusions we may prove that real gases follow the equation $(\partial U/\partial V)_T = 0$ in the limit of zero pressure.

$$\left(\frac{\partial U}{\partial V}\right)_T = \left(\frac{\partial H}{\partial V}\right)_T - P - V\left(\frac{\partial P}{\partial V}\right)_T = \left(\frac{\partial P}{\partial V}\right)_T\left[\left(\frac{\partial H}{\partial P}\right)_T - V\right] - P$$

But we have shown that $(\partial H/\partial P)_T = -\mu C_P$; hence

$$\left(\frac{\partial U}{\partial V}\right)_T = -\left(\frac{\partial P}{\partial V}\right)_T (\mu C_P + V) - P \qquad (4\text{-}9)$$

However, in the limit of zero pressure, Eq. (2-8)* applies, and

$$\left(\frac{\partial P}{\partial V}\right)_T = -\frac{nRT}{V^2} = -\frac{P}{V}$$

Hence

$$\lim_{P\to 0}\left(\frac{\partial U}{\partial V}\right)_T = \lim_{P\to 0}\frac{P\mu C_P}{V} = 0 \qquad (4\text{-}10)$$

We recall that temperature was defined in Chapter 2 in a fashion that makes Eq. (2-8)* true in the limit at zero pressure. Since we have now shown that Eq. (2-9)* is also true in this limit, we may conclude that real gases do conform in the limit of zero pressure with our criteria of the perfect gas.

Flow Calorimeters

While the free or Joule–Thomson expansion was important in the earlier exploration of the precise thermodynamics of gases, flow calorimeters have played a greater role recently. In such instruments an electrical heater introduces heat at an accurately measured rate into the flowing fluid, which may be either gas or liquid.

For heat-capacity measurements, the flow is slow enough and the pressure drop small enough that kinetic energy and pressure decrease effects are negligible. If a volatile liquid is being investigated, the pressure is kept high enough that no vapor is present, and there is no correction for change in vapor pressure with temperature. From the power input and the flow rate, the calculation of the apparent heat capacity is trivial. There is a problem with heat loss, however. Some of the heat is lost to the surroundings and this must be subtracted.

A very interesting advance was made by Picker et al.,[13] who in 1971 developed a twin flow calorimeter in which the heat capacity of the fluid being measured is compared with that of a very similar fluid of accurately known heat capacity in an apparatus with two identical units within the same surroundings. Usually it is an aqueous solution compared with pure water. By clever design, the heat losses can be made accurately equal and the heat capacity of the solution can be precisely related to that of pure water.

This type of flow calorimetry was extended to high temperatures and pressures by Smith-Magowan and Wood[14] and by Rogers and Pitzer[15] in 1981, and has been further refined by Carter and Wood.[16] Results from this source are presented and discussed in Chapter 18.

Another type of flow calorimeter is one measuring the "isothermal Joule–Thomson coefficient." In this case electrical heat is introduced to compensate for the cooling caused by the pressure decrease; then the entire unit is isothermal, which avoids the heat loss or gain corrections that necessarily arise in the original Joule–Thomson experiment. This procedure yields $(\partial H/\partial P)_T$ directly instead of indirectly via Eq. (4-8). Al-Bizreh and Wormald[17] describe a calorimeter of this type and report measurements on the vapors of benzene and cyclohexane. The results show a substantial trend with pressure similar to that in Fig. 4-2 for the Joule–Thomson effect itself.

The heat absorbed or released on isothermal mixing of two fluids can be measured accurately by a flow calorimeter in which the two fluid streams mix with precisely measured heat flow to or from the isolated unit where the mixing takes place. There are many good examples of heat of dilution or heat of mixing measurements by this method. Busey, Holmes, and Mesmer[18] describe an excellent calorimeter of this type for liquids at high temperatures and report heat of dilution measurements for $NaCl-H_2O$ extending to 673 K. Wormald[19] describes a mixing calorimeter for gases and vapors. Since there is some pressure drop in the mixing calorimeter, a Joule–Thomson effect is present. Wormald places in the exit stream a Joule–Thomson unit carefully designed to have an equal pressure drop. Thus, he is able to cancel out the Joule–Thomson cooling and report values for just the mixing effect at constant T and P.

General Equation for Flow Processes

While it is beyond the scope of this book to discuss all possible complexities in flow processes, it is important in some cases to include the kinetic and gravitational energy terms. The familiar equation of mechanics gives the kinetic energy, which is $mu^2/2$ per mass m of fluid, where u is the linear velocity. Also, if the fluid changes level, there is a change in gravitational potential energy. It is inconvenient to include gravitational energy on an absolute basis, but its change is $mg(h_2 - h_1)$, where h is the vertical height.

While the internal energy of a substance may be defined to include the kinetic and gravitational energies,† it is customary in dealing with flow problems to exclude these terms from U and to include them separately in energy-balance equations. The symbols U, H, etc., thus represent the energy content, enthalpy, etc., of the substance

---

† Indeed in Chapter 24 the gravitational energy is included in the energy or enthalpy of a substance.

when stationary and at the base height. A general equation for the first law of thermodynamics as applied to a mass m in a flow system is

$$\Delta H + m\frac{\Delta u^2}{2} + mg\,\Delta h = q + w' \qquad (4\text{-}11)$$

where q is the heat absorbed by the system and w' is work absorbed other than pressure–volume work.

Another interesting application is flow through a jet engine or rocket nozzle. In good approximation Δh, q, and w' are zero in this case, but Δu^2 is large. Equation (4-7) yields at once $\frac{1}{2}\Delta u^2 = -\Delta H/m$. Usually the gas velocity in the combustion chamber is small; hence the exit velocity u_e is given by the expression

$$\tfrac{1}{2}u_e^2 = -\frac{\Delta H}{m} \qquad (4\text{-}12)$$

where ΔH is the enthalpy change from the combustion chamber in the nozzle exit.

The second law is also applicable to flow systems. If no heat is transferred and irreversible effects are negligible, the condition $\Delta S = 0$ may be applied. Equation (3-51) gives the P–T relationship for an isentropic expansion of a perfect gas. Thus, if the initial pressure and temperature and the final pressure are fixed for an adiabatic and reversible flow process, the isentropic condition determines the final temperature. Hence the enthalpy change ΔH is determined and may be used in Eq. (4-11) or (4-12) to determine the other quantities of interest.

In some flow systems such as jet or rocket engine nozzles there may be chemical changes occurring in the fluid. If such changes maintain chemical equilibrium, then no irreversibility is introduced. Also, if a given chemical reaction does not occur at all, then it contributes no irreversible entropy increase. However, if reaction occurs but lags significantly behind equilibrium, then an irreversible entropy increase does occur and must be included in the final entropy of the fluid.

We may also obtain the Bernoulli equation, which was first discovered from reasoning based upon mechanical work long before the development of thermodynamics. We consider an incompressible fluid in frictionless adiabatic flow; hence the internal energy of the fluid is constant, and $\Delta H = V\,\Delta P$, while $q = 0$. Also we assume that no work is transferred except for that due to the volume of fluid flow; that is, $w' = 0$. Now Eq. (4-7) becomes

$$m\frac{\Delta u^2}{2} + mg\,\Delta h + V\,\Delta P = 0 \qquad (4\text{-}13)$$

The most interesting feature of this equation is the pressure decrease that accompanies an increase in velocity when $\Delta h = 0$.

In the chapters to follow we shall not usually consider flow processes explicitly. But it is hoped that this brief discussion will indicate the additional features that flow introduces into a problem and will assist those readers who wish to apply to flow systems the thermodynamic relations to be presented in later chapters. We note also that certain types of flow systems are discussed extensively in texts[20] directed primarily to chemical engineers and that very general consideration of hydrodynamic phenomena in fluids is given by Hirschfelder, Curtiss, and Bird.[21]

REFERENCES

1. S. Carnot, *Reflections sur la puissance motrice du feu*, Chez Bachelier Libraire, Paris, 1824.
2. R. J. E. Clausius, *The Mechanical Theory of Heat* (trans. by Hirst), Van Voorst, London, 1867.
3. H. Poincaré, *Thermodynamique*, Carré, Paris, 1892.
4. J. H. Keenan, *Thermodynamics*, Wiley, New York, 1941.
5. M. Born, *Physik. Z.*, **22**, 218, 249, 282 (1921).
6. S. Chandrasekhar, *An Introduction to the Study of Stellar Structures*, University of Chicago Press, Chicago, 1939.
7. H. A. Buckdahl, *Am. J. Phys.*, **17**, 41, 44, 212 (1949).
8. M. Born, *Natural Philosophy of Cause and Chance*, Oxford University Press, New York, 1949.
9. R. Eisenschitz, *Sci. Progr.*, **43**, 246 (1955).
10. M. Planck, *Treatise on Thermodynamics*, 3d English edn. (trans. by A. Ogg), Longmans, Green & Co., New York, 1927 (reprinted by Dover Publications, New York, 1945).
11. J. P. Joule and W. Thomson, *Proc. Roy. Soc. (London)*, **143**, 357 (1853).
12. J. R. Roebuck and H. Osterberg, *Phys. Rev.*, **48**, 450 (1935).
13. P. Picker, P.-A. Leduc, P. R. Philip, and J. E. Desnoyers, *J. Chem. Thermodynamics*, **3**, 631 (1971).
14. D. Smith-Magowan and R. H. Wood, *J. Chem. Thermodynamics*, **13**, 1047 (1981).
15. P. S. Z. Rogers and K. S. Pitzer, *J. Phys. Chem.*, **85**, 2886 (1981).
16. R. W. Carter and R. H. Wood, *J. Chem. Thermodynamics*, **23**, 1037 (1991).
17. N. Al-Bizreh and C. J. Wormald, *J. Chem. Thermodynamics*, **9**, 749 (1977).
18. R. H. Busey, H. F. Holmes, and R. E. Mesmer, *J. Chem. Thermodynamics*, **16**, 343 (1984).
19. C. J. Wormald, *J. Chem. Thermodynamics*, **9**, 901 (1977).
20. See, for example, J. M. Smith and H. C. Van Ness, *Introduction to Chemical Engineering Thermodynamics*, 4th edn., McGraw-Hill, New York, 1987; or T. E. Daubert, *Chemical Engineering Thermodynamics*, McGraw-Hill, New York, 1985.
21. J. O. Hirschfelder, C. F. Curtiss, and R. B. Bird, *Molecular Theory of Gases and Liquids*, Wiley, New York, 1954, chapter 11.

PROBLEMS

4-1. What is the maximum conversion factor of a steam engine operating with a condenser at 30°C and a boiler at 250°C?

4-2. A refrigerating machine operating in a room at 30°C is employed to maintain a cold storage tank at −10°C. What is the minimum amount of work required to withdraw 1000 cal from the tank?

4-3. A thermally powered refrigerator receives heat from a flame at 400°C and withdraws heat from a cold box at −40°C. If the machine is reversible and all heat is discharged at 30°C, calculate the ratio of the heat required at 400°C to that withdrawn at −40°C.

4-4. Assume that the total cost of operation of a heat pump is four times the theoretical power cost for perfect efficiency, whereas the cost of direct electrical heating is just the power cost. If room temperature is 27°C, at what outdoor temperature would the two systems yield equal total cost?

4-5. Calculate the work required per mole to compress an ideal monatomic gas at 1 bar and 300 K to 10 bar in an adiabatic reversible flow system. Note that the temperature change on reversible adiabatic compression of an ideal gas of constant heat capacity can be obtained from Eq. (3-51)*.

4-6. What is the exit velocity of perfect monatomic helium gas that is expanding from 3000 K and 50 bar through a jet nozzle to 1 bar? Assume reversible adiabatic flow with negligible initial velocity.

4-7. What is the minimum amount of work required to cool 1 kg of water from 25°C to 0°C if all heat is rejected at 25°C? Assume $C_P = 1.0 \, \text{cal g}^{-1} \, \text{K}^{-1}$ for water and that there is an ideal refrigeration device for this process.

4-8. Derive the work of an adiabatic expansion or compression of an ideal gas of constant heat capacity and show that the two terms of that type in the Carnot cycle cancel.

4-9. In a high-temperature gas turbine the gas flowing into the turbine (a mixture of N_2, CO_2, H_2O, etc.) can be approximated as an ideal gas with a constant $C_P = 4R$. If the input is at 1000 K and 5 bar with adiabatic expansion to 1 bar, calculate (a) the final temperature and (b) the useful work per mole of gas in a flow system. (c) Since the final temperature is relatively high, make suggestions about practical methods of extracting additional work from this system.

4-10. A thermally powered refrigerator receives heat from a hot spring at 95°C and withdraws heat from a cold box at −25°C. If the machine is reversible and all heat is discharged at 40°C, calculate the heat required at 95°C to transfer 10^4 cal from the cold box.

4-11. During a sunny day a solar heating unit on a house roof can heat 20 kg of water per hour from 30°C to 80°C. If this water is used in an air conditioning heat pump, how many kcal per hour can be transferred from inside the house at 25°C to the outside at 30°C? Any surplus heat is also discharged at 30°C. Assume reversible processes in the heat pump.

CHAPTER 5

ABSOLUTE ENTROPIES AND STATISTICAL MECHANICS; CALCULATIONS FOR SIMPLE MOLECULES

While thermodynamics deals with the properties of amounts of matter containing a very large number of atoms, it is interesting to examine the relationships between atomic and molecular properties and the corresponding thermodynamic properties. For the First Law the relationship is one of identity; the principle of conservation of energy applies at the atomic level as well as the macroscopic level. We saw in Chapter 2 that the Second Law is related to probability. In particular, Eq. (2-29) gave a relationship between differences in entropy and ratios of probabilities:

$$S_B - S_A = k \ln\left(\frac{\mathcal{P}_B}{\mathcal{P}_A}\right) \qquad (2\text{-}29)$$

Here, k is the Boltzmann constant which is equal to R/N_A. Note that there is no requirement in this relationship for an absolute value for entropy or an absolute unit of probability.

With the development of quantum theory, new possibilities arose. One could postulate that a single quantum state represented an absolute unit of probability. Then the probability of a thermodynamic state is the number of single quantum states in which the system might exist under the specified conditions of volume and energy (or of pressure and temperature ...) that define a thermodynamic state. This number of quantum states is commonly designated W. Then Eq. (2-29) becomes

$$S_B - S_A = k \ln\left(\frac{W_B}{W_A}\right) \tag{5-1}$$

The possibility now exists that the macroscopic system of many atoms might be in a single quantum state with $W = 1$. Then one could define this state as the zero of entropy and obtain entropies of other states on an absolute basis:

$$S_B = k \ln W_B \tag{5-2}$$

In this chapter, we explore this general concept by considering several simple examples. These examples involve relatively simple systems, ideal gases and idealized crystals, and their quantum energy patterns are further simplified here, but the results are good approximations to real substances. More precise treatments eliminating the approximations are given in later chapters.

If the present chapter were to be independent of all other knowledge, we would now have to make an extensive excursion into quantum mechanics, spectroscopy, and statistical mechanics. But most readers will be familiar with the elementary aspects of these subjects; hence, we shall now assume certain simple relationships. Appendix 1 gives a very brief derivation of some of these relationships. Indeed, for many readers, the entire content of this chapter may be familiar. But it is desirable to have this material in mind when considering the third law of thermodynamics in the following chapter. Also, one should realize that the statistical methods here described are the best source of entropies, heat capacities, and related properties for many simple gases. Indeed, a thorough study of statistical mechanics is recommended to anyone seriously interested in thermodynamic research.

Before proceeding further, it is interesting to ask how small W_A must be in Eq. (5-1) in order to make S_A negligible, i.e., to make the state A a practical zero of entropy. An amount 0.0001 in S/R is well within the uncertainty of the most precise macroscopic thermodynamic measurements. Hence, for one mole of particles,

$$\begin{aligned} \ln W_A &\leq 0.0001 \times 6 \times 10^{23} = 6 \times 10^{19} \\ W_A &\leq 10^{2.6 \times 10^{19}} \end{aligned} \tag{5-3}$$

Thus, we see that a practical state of zero entropy can contain an enormous number of individual quantum states.

We turn now to simple examples, first of ideal gases and then of very simple crystalline solids.

IDEAL GASES

We now know that an ideal gas is an idealized state of matter in which the individual gas molecules have no interaction upon one another. Thus, if we can calculate the properties of individual molecules, we should be able to calculate the properties of the ideal gas comprising these molecules.

It proves convenient for a number of reasons to consider the translational motion of entire molecules separately from all other motions that are internal within a given molecule. The quantized energy levels of the latter motions can be obtained by appropriate spectroscopic studies. Indeed, there is a very extensive body of spectroscopic data on molecular energy levels. There is no coupling between translation and the internal motions; the overall numbers of quantum states are multiplicative $W(\text{overall}) = W(\text{translational}) \times W(\text{internal})$, which yields entropies that are additive, as are the energies.

Translation

In contrast to the internal motions, where spectroscopic data are available, the translational motion of a molecule must be treated by other methods.† The heat capacity at constant volume of translation, $\frac{3}{2}R$ per mole, is calculated by elementary kinetic theory. Likewise the molar entropy change on expansion was calculated in Eq. (2-21)*:

$$S_{m,2} - S_{m,1} = R \ln\left(\frac{V_2}{V_1}\right) \qquad (2\text{-}21)^*$$

Since all of the equations in this chapter are on a molar basis, the subscript m is omitted hereafter. These two terms may be combined to yield an equation for the molar entropy of translation,

$$\frac{S_{tr}}{R} = \ln V_m + \tfrac{3}{2}\ln T + \text{constant} \qquad (5\text{-}4)\ddagger$$

with V_m the molar volume. This equation gives everything needed for the application of the first and second laws of thermodynamics. For use with the third law, however, we need the absolute value of S.

Sackur[1] showed that the constant in Eq. (5-4) for a number of substances could be expressed as the sum of a universal constant and a term $\tfrac{3}{2}R \ln M$, where M is the molecular mass of the molecule. The subsequent development of quantum mechanics has given full theoretical confirmation of Sackur's empirical relationship and in addition yields the value of the universal constant, which was first found by Tetrode[2] (see Appendix 1). The complete equation for the molar entropy of translation is

$$\frac{S_{tr}}{R} = \ln V_m + \tfrac{3}{2}\ln T + \tfrac{3}{2}\ln M + C \qquad (5\text{-}5)$$

---

† In principle, the translation of a molecule in a fixed volume is quantized, but the energy levels have not yet been measured by spectroscopy.

‡ All equations in this chapter might be designated by * since they pertain to the ideal gas state. The * is omitted throughout this chapter and is used hereafter only in circumstances where we wish to draw specific attention to the limitations on a given equation.

$$C = \frac{5}{2} + \frac{3}{2} \ln \left[\frac{2\pi k}{h^2}\right] - \tfrac{5}{2} \ln N_A = -5.57228 \qquad (5\text{-}6)$$

where M is the usual molecular mass and V_m is in cubic centimeters per mole.

Values calculated from Eq. (5-5) may be compared with the entropies calculated from the third law of thermodynamics by using low-temperature heat-capacity measurements for the solid and entropies of vaporization for monatomic gases that have no internal energy levels within the thermal range. These results are given in Table 6-4 in the next chapter.

The other thermodynamic functions are readily obtained from Eq. (5-5), and the value $\tfrac{3}{2}RT$ for the molar internal energy $U° - U_0°$ of translation derived by elementary kinetic theory [and Eq. (A1-11)]. Also it is convenient to convert to the standard state of ideal gas at 1 bar. The equations for the translation contributions are as follows:

$$\frac{S_{tr}°}{R} = \tfrac{3}{2} \ln M + \tfrac{5}{2} \ln T - 1.151\,69 \qquad (5\text{-}7)$$

$$-\frac{(G° - H_0°)_{tr}}{RT} = \tfrac{3}{2} \ln M + \tfrac{5}{2} \ln T - 3.651\,69 \qquad (5\text{-}8)$$

$$\frac{(H° - H_0°)_{tr}}{RT} = \frac{5}{2} \qquad (5\text{-}9)$$

$$\frac{(C_P°)_{tr}}{RT} = \frac{5}{2} \qquad (5\text{-}10)$$

Internal Energy Levels[3]

We now assume that we have the complete list of internal energy levels $\varepsilon_0, \varepsilon_1, \varepsilon_2, \ldots, \varepsilon_i, \ldots$ for the molecule (or atom). These values usually come from spectroscopic measurements. Quantum mechanics is of aid in interpreting the spectral data, and in some cases it is also possible to calculate the energy levels from theory alone.

We wish to calculate the properties of an ideal gas composed of these molecules in their equilibrium distribution among these various energy levels at a given temperature. For the moment let us consider a microscopic state to be defined by one of these internal energy levels. Each such microscopic state corresponds to certain definite quantum specifications not possessed in every particular by any other state. We assume that each microscopic state has the same a priori probability. By this we mean that, given equal opportunity to possess the energies necessary for their existence, all states are equally probable.

The Boltzmann distribution law gives the most probable distribution of particles among the quantum states. Many readers may be familiar with this expression; for those who are not, a simple statistical derivation is given in Appendix 1. The result, Eq. (A1-5), is

$$N_i = e^{-\alpha} g_i\, e^{-\varepsilon_i/kT} \qquad (5\text{-}11)$$

where N_i is the most probable number of particles in a group of g_i quantum states, all of which have the same energy ε_i. The factor $e^{-\alpha}$ is to be evaluated from the total number of particles present.

It is convenient to assume that the lowest energy level has zero energy ($\varepsilon_0 = 0$), which we may always arrange by an arbitrary shift in the energy scale. We may then substitute for the factor $e^{-\alpha}$ the number of particles N_0 in a single quantum state of zero energy. Then the number in any other state of multiplicity g_i and an energy ε_i above zero will be

$$N_i = N_0 g_i \exp(-\varepsilon_i/kT) \tag{5-12}$$

and the total number of molecules is

$$N = N_0 \sum_i g_i \exp(-\varepsilon_i/kT) \tag{5-13}$$

where the sum covers all the microscopic states.

The sum in Eq. (5-13) is given the name *partition function* (or sum over states) and the symbol Q, thus:

$$Q = \sum_i g_i \exp(-\varepsilon_i/kT) \tag{5-14}$$

We may now calculate the contribution of these internal energy levels to the enthalpy or the internal energy of all the molecules since we have the number of molecules in each state and the energy of each state.

$$(H - H_0)_{\text{int}} = (U - U_0)_{\text{int}} = \sum_i N_i \varepsilon_i$$

$$= N_0 \sum_i \varepsilon_i g_i \exp(-\varepsilon_i/kT)$$

$$= N_0 kT^2 \frac{dQ}{dT} \tag{5-15}$$

Since we usually do not know N_0, we eliminate this factor by Eqs. (5-13) and (5-14) and obtain for the enthalpy or the internal energy per mole,

$$(H - H_0)_{\text{int}} = \frac{N_A kT^2}{Q} \frac{dQ}{dT}$$

$$= RT^2 \, d\ln Q/dT \tag{5-16}$$

The heat-capacity contribution follows directly by differentiation:

$$\frac{C_{\text{int}}}{R} = T^2 \frac{d^2 \ln Q}{dT^2} + 2T \frac{d \ln Q}{dT} \tag{5-17}$$

Finally one may obtain the entropy contribution by integration:

$$\left[\frac{S - S_0}{R}\right]_{\text{int}} = \int_0^T \left[T \frac{d^2 \ln Q}{dT^2} + 2 \frac{d \ln Q}{dT}\right] dT$$

$$= \left[\ln Q + T\frac{d\ln Q}{dT}\right]_0^T$$

$$= \ln Q_T - \ln g_0 + T\frac{d\ln Q}{dT} \tag{5-18}$$

At 0 K the partition function becomes just g_0, which has been substituted for Q_0 in the last step of the derivation above. In ordinary cases it is found that $g_0 = 1$, even if other g's have multiple values. In that case $\ln g_0 = 0$. Even in a case where g_0 is not unity, we would take $S_0 = R \ln g_0$ in accordance with the general equation for the entropy in terms of probability. Thus in either event we obtain for our final result for the entropy contribution from internal energy levels the expression

$$\frac{S_{int}}{R} = \ln Q + T\frac{d\ln Q}{dT} \tag{5-19}$$

For practical computations it is frequently necessary to sum the partition function by direct numerical methods. Then the derivatives are not readily obtained, and it is best to work instead with the following functions:

$$Q' = \sum_i g_i(\varepsilon_i/kT)\exp(-\varepsilon_i/kT) \tag{5-20}$$

$$Q'' = \sum_i g_i(\varepsilon_i/kT)^2 \exp(-\varepsilon_i/kT) \tag{5-21}$$

These functions are readily computed at the same time that Q itself is calculated. It is easily shown that in these terms the thermodynamic functions are given by the following equations:

$$\left(\frac{C}{R}\right)_{int} = \frac{Q''}{Q} - \left(\frac{Q'}{Q}\right)^2 \tag{5-22}$$

$$\left(\frac{H - H_0}{RT}\right)_{int} = \left(\frac{U - U_0}{RT}\right)_{int} = \frac{Q'}{Q} \tag{5-23}$$

$$\left(\frac{S}{R}\right)_{int} = \ln Q + \frac{Q'}{Q} \tag{5-24}$$

Because the volume affects only the translational contributions, the internal contribution to U is also the value for H as indicated above, and the value for G is that for A. Since G and H are used more widely than A and U, we shall label the internal contributions $H - H_0$ and $G - H_0$, with the realization that the same expressions pertain to $U - U_0$ and $A - U_0$, respectively.

The function $(G_T^\circ - H_0^\circ)/RT$ will be seen in Chapter 8 and later chapters to be of great utility and convenience; it is often called the Gibbs energy function. Thus, we obtain by combination of Eqs. (5-23) and (5-24) the very simple result

$$-\left(\frac{G - H_0}{RT}\right)_{int} = \ln Q \tag{5-25}$$

Monatomic Gases at High Temperatures

The equations of the preceding section are appropriate for the calculation of the thermodynamic properties of monatomic gases at high temperatures, where excited electronic states become important. The energy levels are obtained by atomic spectroscopy, and the partition function is obtained by direct numerical summation. Table 5-1 gives a sample calculation for monatomic silicon gas at 5000 K. The energy-level values are taken from Moore.[4]

Since the precision of spectroscopic measurement of energies is high and the fundamental constants R and k are very accurately known, the resulting thermodynamic functions for a monatomic gas are very accurate provided the list of excited states is complete. In effect, one requires only that this list is complete up to an energy where the contributions to Q'' become negligible. For low or moderate temperatures, the existing tables of spectroscopic states usually suffice, although it is important to consider the quantum theory for the particular atom to be sure that significant states have not been missed because of low transition probability (forbidden transitions). For very high temperatures, however, it is often found that the published lists are incomplete and that estimates must be made for additional levels (see Brewer[5]).

These equations, when combined with those for the translational contributions, suffice for the calculation of the thermodynamic properties of any ideal gas when the molecular energy levels are known. In some cases further rearrangements of the equations are convenient, and these will be described in later sections.

TABLE 5-1
Calculation of thermodynamic properties of silicon gas at 5000 K

State	$\varepsilon_i/\text{cm}^{-1}$	g_i	$\dfrac{\varepsilon_i}{kT}$	$e^{-\varepsilon_i/kT}$	$\dfrac{\varepsilon_i}{kT} e^{-\varepsilon_i/kT}$	$\left(\dfrac{\varepsilon_i}{kT}\right)^2 e^{-\varepsilon_i/kT}$
$^3P_0$	0	1	0.0	1.00	0.0	0.0
$^3P_1$	77.15	3	0.022	0.978	0.022	0.000
$^3P_2$	223.31	5	0.064	0.938	0.60	0.004
$^1D_2$	6 298.81	5	1.812	0.163	0.295	0.535
$^1S_0$	15 394.24	1	4.430	0.011 9	0.053	0.234
$^3P_0^\circ$	39 683.10	1	11.419	0.000 01	0.000 1	0.001

$Q = 9.451 \quad Q' = 1.894 \quad Q'' = 2.930$

$-\left(\dfrac{G^\circ - H_0^\circ}{RT}\right)_{\text{int}} = \ln 9.451 = 2.246$

$(H^\circ - H_0^\circ)_{\text{int}} = RT(1.894/9.451) = 8.331 \text{ kJ mol}^{-1}$
$C_{\text{int}}/R = 2.930/9.451 - (1.894/9.451)^2 = 0.270$

Rotational Contributions for Linear Molecules

Even for relatively simple molecules the complete array of energy levels is so extensive that direct summation of the partition function becomes burdensome. Fortunately the energy levels follow such a systematic pattern that new methods become feasible. It is found that each molecule has a characteristic pattern of relatively closely spaced energy levels, which are attributed to molecular rotation. This pattern is repeated at various higher energies. The larger energy differences are usually those of molecular vibrations, although excited electronic states may also arise. Let us now consider just the contribution of the rotational levels.

In the case of linear molecules, the rotational levels are given in good approximation by the simple formula

$$\varepsilon_J = hcBJ(J + 1) \tag{5-26}$$

where B is a constant characteristic of the molecule and J is a quantum number taking integral values. Also, h is Planck's constant and c is the velocity of light. B is in the spectroscopist's unit, the reciprocal centimeter, and must be multiplied by hc to yield energy. It is found that for symmetrical molecules such as N–N, O–C–O, etc., J may have only alternate values. In some cases[6,7] the even values 0, 2, 4, ... are allowed and in other cases the odd values 1, 3, 5, ...†, but for most of our purposes this difference will be of no consequence. For unsymmetrical molecules all integral values of J are permitted. All levels with J greater than zero are multiple with $g_J = 2J + 1$. This multiplicity may be verified by the application of electric or magnetic fields, which separate the energy levels and hence split the lines in the spectrum.

Quantum theory[6] yields results in complete accord with spectroscopy and in addition gives for the constant B the formula

$$B = \frac{h}{8\pi^2 cI} \tag{5-27}$$

where I is the moment of inertia of the molecule about an axis through the center of gravity and perpendicular to the axis on which the atoms lie. Thus, if the molecular dimensions are known, B may be calculated even though it has not been observed spectroscopically.

Spectroscopists usually report B in units of reciprocal centimeters, but in recent years values are also being given in MHz. In that case the factor hc in Eq. (5-26) and in later equations is replaced by $10^6 h$.

Let us now define a quantity with the dimension temperature as follows,

$$\theta_r = \frac{hcB}{k} = \frac{h^2}{8\pi^2 Ik} \tag{5-28}$$

whereupon the partition function for rotation becomes

$$Q_r = \sum_J (2J + 1) \exp[-J(J + 1)\theta_r/T] \tag{5-29}$$

---

† This assignment also depends on the nuclear spin, see Chapter 22 for the details for H_2.

with the sum covering either all positive integral values of J or only alternate values as indicated above. For most substances the value of θ_r is so small that the sum in Eq. (5-29) may be replaced by an integral:

$$Q_r = \frac{1}{\sigma} \int_0^\infty e^{-J(J+1)\theta_r/T}(2J+1)\,dJ = \frac{T}{\sigma\theta_r} \tag{5-30}$$

Here σ, the symmetry number, is 1 for the unsymmetrical molecules, where all values of J are allowed, and is 2 for the symmetrical molecules, where only alternate values are permitted.

The simple result of Eq. (5-30) is adequate, provided θ_r is much smaller than T. At 300 K, θ_r/T is 0.0012 for Cl_2, 0.0069 for O_2, 0.0095 for N_2, 0.050 for HCl, and 0.283 for H_2. Thus, the simple approximation is adequate for many molecules and nearly adequate for all others except for the various isotopes of hydrogen. Hydrogen is best handled by direct numerical summation, and the results are available. By more elaborate mathematical methods one may show that the partition function[7] is

$$Q_r = \frac{T}{\sigma\theta_r}\left(1 + \frac{\theta_r}{3T} + \frac{\theta_r^2}{15T^2} + \cdots\right) \tag{5-31}$$

This result will be adequate for all practical cases except for the hydrogen isotopes.

We now introduce the result just obtained into Eqs. (5-16) to (5-19). We also note that $\ln(1+x) = x - (x^2/2) + \ldots$ when x is small. The final results for the rotation of linear molecules are then as follows:

$$-\left(\frac{G-H_0}{RT}\right)_r = \ln T - \ln \sigma - \ln \theta_r + \frac{\theta_r}{3T} + \frac{\theta_r^2}{90T^2} + \cdots \tag{5-32}$$

$$\left(\frac{H-H_0}{RT}\right)_r = 1 - \frac{\theta_r}{3T} - \frac{\theta_r^2}{45T^2} \cdots \tag{5-33}$$

$$\left(\frac{S}{R}\right)_r = 1 + \ln T - \ln \sigma - \ln \theta_r - \frac{\theta_r^2}{90T^2} \cdots \tag{5-34}$$

$$\left(\frac{C}{R}\right)_r = 1 + \frac{\theta_r^2}{45T^2} \cdots \tag{5-35}$$

It is interesting to note that the term $\theta_r/3T$ drops out of the expressions for the entropy and heat capacity.

The treatment for rotation of nonlinear molecules is similar to that for linear molecules but more complex; hence, its presentation is postponed until Chapter 20.

Vibrational Contributions

The energy levels for a vibrating system of atoms, whether in a molecule or a crystal, are given by the equation[6]

$$\varepsilon - \varepsilon_0 = hc\omega v \tag{5-36}$$

where ω is the frequency of the vibration in units of reciprocal centimeters and v is a quantum number that takes the integral values 0, 1, 2, ... The energy ε_0 of the lowest vibrational quantum state is $hc\omega/2$ above that of the "vibrationless" molecule. But we may define $\varepsilon_0 = 0$ in this case as was done earlier, and we then omit ε_0 hereafter. The true frequency is $\nu = c\omega$. This formula applies to harmonic vibration, i.e., where the restoring force is strictly proportional to the displacement. Molecules usually deviate slightly from the harmonic condition; consequently corrections for anharmonicity must be considered. It is found, however, that the harmonic formula always yields a good first approximation to the thermodynamic properties.

If the molecule has more than two atoms, it will have more than one vibrational motion. A complete discussion of this situation would involve the classical theory of normal modes of vibration and the quantum theory of such systems. The result, however, is quite simple. For each mode of vibration there is a frequency ω_i, and the vibrational energy is just the sum of terms for the separate modes of vibration

$$\varepsilon = hc \sum_i \omega_i v_i \qquad (5\text{-}37)$$

The frequencies observed in the infrared or Raman spectra are, of course, the frequencies that appear in this equation.

Since in the approximation of Eq. (5-37) there are no cross terms between different vibrations, we shall find that their contributions separate conveniently. We define $u_i = hc\omega_i/kT = 1.4388\omega_i/T$. Then

$$Q_{\text{vib}} = \sum_{v_1} \sum_{v_2} \cdots \sum_{v_n} \exp[-(u_1 v_1 + u_2 v_2 + \cdots u_n v_n)]$$

where each of the v's independently has the values 0, 1, 2, ... Since $\exp[-(u_1 v_1 + u_2 v_2 + \cdots + u_n v_n)] = [\exp(-u_1 v_1)][\exp(-u_2 v_2)] \cdots [\exp(-u_n v_n)]$, the multiple sum may be factored:

$$Q_{\text{vib}} = \left[\sum_{v_1} e^{-u_1 v_1}\right]\left[\sum_{v_2} e^{-u_2 v_2}\right] \cdots \left[\sum_{v_n} e^{-u_n v_n}\right]$$

$$= \prod_i \left[\sum_{v_i} e^{-u_i v_i}\right] \qquad (5\text{-}38)$$

The thermodynamic functions all depend on $\ln Q$, and the product in Q becomes a sum.

$$\ln Q_{\text{vib}} = \sum_{v_i} \ln \left[\sum_{v_i} e^{-u_i v_i}\right] \qquad (5\text{-}39)$$

Parenthetically, we may remark that this separability of thermodynamic contributions into a sum is generally valid whenever the energy is a sum of independent terms. By an analogous argument one may prove the separability of rotational, translational, vibrational, and other contributions to thermodynamic functions if the corresponding energy terms are independent in the energy-level formula.

We return now to the calculation of the partition function for a single vibration.

$$Q_{vib} = \sum_v e^{-uv}$$

$$= 1 + e^{-u} + (e^{-u})^2 + (e^{-u})^3 + \cdots$$

$$= \frac{1}{1 - e^{-u}} \tag{5-40}$$

It is very fortunate that this infinite sum can be written in a closed, simple form. The thermodynamic functions are as follows:

$$-\left[\frac{G - H_0}{RT}\right]_{vib} = -\ln(1 - e^{-u}) \tag{5-41}$$

$$\left[\frac{H - H_0}{RT}\right]_{vib} = \frac{ue^{-u}}{1 - e^{-u}} = \frac{u}{e^u - 1} \tag{5-42}$$

$$\left(\frac{S}{R}\right)_{vib} = \frac{u}{e^u - 1} - \ln(1 - e^{-u}) \tag{5-43}$$

$$\left(\frac{C}{R}\right)_{vib} = \frac{u^2 e^u}{(e^u - 1)^2} \tag{5-44}$$

where

$$u = \frac{h\nu}{kT} = \frac{hc\omega}{kT} = 1.4388 \frac{\omega}{T}$$

As $T \to \infty$ and $u \to 0$, both $(C/R)_{vib}$ and $[(H - H_0)/RT]$ approach the limiting value 1.0, but the S and G functions increase indefinitely. These are the formulas first obtained by Einstein in his treatment of a quantized vibration.

Higher Approximations

The observed molecular spectra do not follow exactly Eq. (5-26) for rotation and Eq. (5-36) for vibration. The differences are small for the energy levels important at moderate temperatures but can be very important at high temperatures where the molecules begin to dissociate. Higher approximations have been developed in the form of additional terms for the thermodynamic properties; the terms given above remain unchanged as the leading terms. These more complex expressions are presented in Chapter 20. At various points, where we list calculated values for entropies, heat capacities, etc., we give the exact values including these additional terms.

CRYSTALLINE SOLIDS

One can think of a solid as a giant molecule with a great many vibrations. Translations and rotations can be ignored except for extremely small particles.[8] In three dimensions there will be $3N$ vibrations where N is the number of atoms. While the heat capacity over the full range of temperature is of importance, we are particularly interested in the behavior at very low temperatures where a crystal might fall into a single quantum state.

The idea that the lattice vibrations of a crystal might be quantized was proposed by Einstein[9] soon after Planck[10] had shown that there were quanta of electromagnetic radiation. Einstein assumed that the frequencies of all atomic vibrations were the same and he adopted Planck's expression for the energy $\varepsilon = nh\nu$. We now know that this energy expression is correct for a harmonic oscillator and the resulting heat capacity follows directly from Eq. (5-44). For three directions of vibration, this yields $3N_A$ vibrations per gram-atom and

$$\frac{C}{R} = \frac{3u^2 e^u}{(e^u - 1)^2} \qquad u = \frac{h\nu}{kT} \qquad (5\text{-}45)*$$

At high temperatures, this equation gives the value $C = 3R$, which is the magnitude noted many years earlier by Dulong and Petit[11] from experiment and explained in classical statistical mechanics by Boltzmann.[12]

Equation (5-45) yields a heat capacity that decreases with decrease in temperature. At that time, this behavior had already been noted for diamond, and Einstein showed that there was reasonable agreement between his calculated values and the experimental measurements, which then extended from 1200 K down to 220 K. About this same time, refrigeration methods were being developed for lower temperatures, and heat capacity measurements were extended below 200 K, first by Dewar,[13] and soon by others. It then became apparent that the assumption of a single frequency for all vibrations in a solid was not correct.

A crystal has many different modes of vibration and each mode has its own frequency. Most of these modes depend in detail on the masses and interatomic forces in a complex manner. But for long-wavelength motions where several atoms move together, the modes and frequencies of a real crystal approach those of a continuous material. This pattern was well known, and Debye[14] recognized that it would be the dominant contribution at low temperature. In particular, the number of modes of frequency from ν to $\nu + d\nu$ is proportional to $\nu^2 d\nu$. Debye adopted this distribution, known to be correct for the low frequencies, and then extended it to higher frequencies until the total number $3N_A$ were included for N_A atoms in three-dimensional space.

It is convenient to define a temperature θ_D by the maximum frequency of the Debye distribution,

$$\theta_D = \frac{h\nu_{\max}}{k} \qquad (5\text{-}46)$$

Each substance will have a different θ_D. Then the heat capacity is

$$\frac{C}{R} = f(T/\theta_D)$$
$$= 9\left(\frac{T}{\theta_D}\right)^3 \int_0^{\theta_D/T} (e^u - 1)^{-2} e^u u^4 \, du \qquad (5\text{-}47)^*$$

At high temperatures, the integral approaches $\frac{1}{3}(\theta_D/T)^3$ and C/R becomes 3 as expected from the law of Dulong and Petit. At low temperatures, the integral approaches a constant value and

$$\frac{C}{R} \cong \left(\frac{12\pi^4}{5}\right)\left(\frac{T}{\theta_D}\right)^3 = \text{constant} \times T^3 \qquad (5\text{-}48)^*$$

This proportionality of C to T^3 is verified approximately for many substances and there is every reason to believe it is exact for crystal vibrations (phonons) in the limit as T approaches zero. At intermediate temperatures, the Debye function is complex to express analytically but readily calculated numerically; tables are given in Appendix 2 for the heat capacity, the entropy, and other functions.

One important feature of the Debye formula, which arises from the form of the equation, is the property that C_V is a general function of $\log T - \log \theta_D$. Consequently, if we plot C_V for various elements as a function of $\log T$, we should obtain curves of the same shape but displaced from one another horizontally by the differences in $\log \theta_D$ for the various substances. Figure 5-1 presents such a plot for several solid elements. The quantity θ_D is proportional to the frequency of atomic vibration which in turn increases with the strength of interatomic forces and decreases with atomic mass. One notes the large increase in θ_D from lead with softly bound heavy atoms to C (diamond) with tightly bound light atoms.

In addition to the vibrations of the atomic nuclei in a crystal, there may be other contributions to the heat capacity. One of rather general interest is that of free

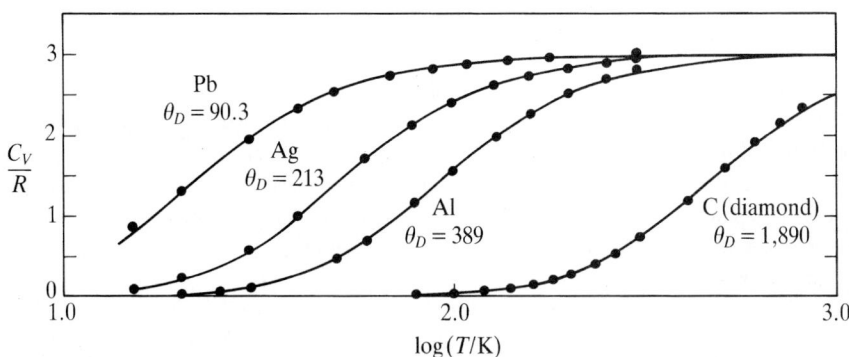

FIGURE 5-1
The heat capacity, C_V/R, for several solid elements. The curves are from the Debye function with the θ_D values given.

conduction electrons in a metal. Although electrons are free to move through the metal to conduct electricity, their contribution to the heat capacity is so small that it was unnoticed for many years. This peculiar situation is now explained by the quantum theory of a degenerate Fermi–Dirac gas. This theory predicts that at low temperatures the electronic heat capacity will rise with the first power of the absolute temperature.

$$\frac{C}{R} = \gamma T \qquad (5\text{-}49)*$$

The proportionality constant γ is a property of each metal. It is interesting to note that the electronic heat capacity can best be detected at either high or very low temperatures. At high temperatures [although Eq. (5-49)* is no longer strictly valid], one finds a small linear increment over the theoretical value of $3R$ for C, due to anharmonicity of lattice vibrations as well as the electronic effect, while at very low temperatures the normal atomic vibrational heat capacity drops off more rapidly (as T^3) than the electronic term. Table 5-2 indicates the relative magnitude of the electronic term at 2 K and 30 K.

The discussion of solids to this point has emphasized the heat capacity. The integration to the entropy and other thermodynamic functions is straightforward. If in the region of very low temperatures, only the lattice vibrations (phonons) and conduction electrons, if present, contribute, one has

$$C = aT^3 + \gamma T \qquad (5\text{-}50)*$$

where a is the coefficient of the Debye contribution and γ that for metallic electrons. If this expression is integrated to yield the entropy, the result is

$$S = \left(\frac{a}{3}\right)T^3 + \gamma T \qquad (5\text{-}51)*$$

In the case that the electronic term is absent or negligible, one has the simple result that, at a given low temperature, $S = C/3$. We shall use the results just derived to estimate the entropy increment from 0 K up to the lowest temperature of measurement.

There are a number of other phenomena that yield heat-capacity contributions in certain solids. Examples include the change from an ordered to a disordered arrangement of certain atoms or the beginning of rotation of a group of atoms. If any atom or ion in the solid has a magnetic moment, then heat-capacity effects can arise

TABLE 5-2
Electronic contribution to heat capacity, C/R

	$\gamma \times 10^4$	At 2 K		At 30 K	
		C_{el}	C_{total}	C_{el}	C_{total}
Copper	0.90	0.000 18	0.000 23	0.002 7	0.206
Aluminum	1.73	0.000 35	0.000 38	0.005 2	0.102

from the reorientation of these magnetic moments. Nonzero nuclear spin yields significant effects for most substances only at extremely low temperatures, but for H_2 this temperature is much higher, as discussed in Chapter 22. Similarly, special effects occasionally arise from reorientation of molecules with electric-dipole moments. Several of these effects will be considered in later chapters.

PRACTICAL ZERO OF ENTROPY

In addition to the various contributions to the multiplicity W and to the entropy discussed above, there may be other multiplicities that have no practical effect and are best ignored for practical chemical thermodynamics. The first is nuclear spin, which is nonzero for some isotopes. Except for hydrogen itself, methane, and possibly a few other hydrogen compounds, any thermal effects of nuclear spin occur at such extremely low temperatures (microkelvins) that there is no overlap with the thermal effects of present interest. Furthermore, the spin multiplicity of a given isotope remains the same in the gas, liquid, or solid element and in any chemical compound. Thus, the entropy associated with nuclear spin multiplicity would cancel for any physical or chemical process, and it is best simply to omit that term.

Although various isotopes of hydrogen and methane show effects due to nuclear spin in the solid at temperatures where the lattice heat capacity is still significant, the two contributions can be separately evaluated. Also, the molecular properties are very well known; hence, we can calculate entropies on a basis excluding nuclear spin multiplicity.

A second phenomenon to be discussed is the mixing of isotopes. Here, one may wish specifically to consider isotopic separation processes. Then, one wishes thermodynamic properties for pure isotopes or molecules of defined isotopic structure. Such values are available for H_2, HD, D_2, HT, and T_2, and can readily be calculated whenever the structural data are known.

With the exception again of hydrogen, for many purposes the isotopes of each element remain randomly mixed. Thus, the inclusion of the entropy of mixing would serve no useful purpose; it would cancel in chemical reactions and in vaporization and other processes. In principle, we note that there may also be undiscovered multiplicities that do not show any thermal effects even in the millikelvin range of temperatures. If so, they will cancel for all of our purposes and would be omitted even if known.

In summary, the multiplicity due to nuclear spin is always omitted in the calculation of thermodynamic properties for practical purposes and, correspondingly, the entropy of nuclear spin is likewise omitted from the practical zero. Thermodynamic functions are calculated both for pure isotopes and for the natural mixture of isotopes, but in the latter case the entropy of isotopic mixing is omitted.

REFERENCES

1. O. Sackur, *Ann. Physik*, (4) **36**, 598 (1911); **40**, 67 (1913).
2. H. Tetrode, *Ann. Physik*, (4) **38**, 434 (1912).
3. This treatment follows closely that of W. F. Giauque, *J. Am. Chem. Soc.*, **52**, 4808 (1930).

4. C. E. Moore, Atomic Energy Levels, *Natl. Bur. Standards (U.S.) Circ. 467*, 1949.
5. L. Brewer, *High Temperature Science*, **17**, 1 (1984).
6. L. Pauling and E. B. Wilson, Jr., *Introduction to Quantum Mechanics*, McGraw-Hill, New York, 1935.
7. J. E. Mayer and M. G. Mayer, *Statistical Mechanics*, 2d edn., Wiley, New York, 1977, p. 184.
8. G. Jura and K. S. Pitzer, *J. Am. Chem. Soc.*, **74**, 6030 (1952).
9. A. Einstein, *Ann. Physik*, (4) **22**, 180 (1907).
10. M. Planck, *Ann. Physik*, (4) **1**, 99 (1900); **4** (4), 553 (1901).
11. P. L. Dulong and A. T. Petit, *Ann. Chim. Phys.*, **10**, 395 (1819).
12. L. Boltzmann, *Sitzber, kgl. Akad. Wiss. Wien*, **63** (2), 679 (1871).
13. J. Dewar, *Proc. Roy. Soc. (London)*, **A76**, 330 (1905).
14. P. Debye, *Ann. Physik.*, (4) **39**, 789 (1912).

PROBLEMS

5-1. Calculate the Gibbs energy function $-(G° - H_0°)/RT$ for CO at 2000 K for the harmonic oscillator rigid rotator approximation. The rotational constant $B = 1.931\,\text{cm}^{-1}$, and the vibration frequency is $\omega = 2167\,\text{cm}^{-1}$. Compare this result with the more accurate value in Appendix 6, and note the magnitude of the effects of anharmonicity, stretching, etc., at this temperature.

5-2. Calculate $C_P°$ for $Br_2(g)$ at 600 K. The vibration frequency is $\omega = 322\,\text{cm}^{-1}$.

5-3. Calculate $C_V°$ for aluminum at 25 K and 100 K from Eq. (5-48)* with the value 389 K for θ_D. Compare with the values interpolated from the table in Appendix 2. What do you conclude about the range of applicability of Eq. (5-48)*?

5-4. Estimate the value of $C_P°$ for rhodium at 300 K. Give your reasoning.

CHAPTER 6

THE THIRD LAW OF THERMODYNAMICS; ENTROPY CALCULATIONS

For the application of thermodynamics to various substances and processes, a database of enthalpies and chemical potentials is needed. These quantities are usually given for the process of formation from the component elements in defined standard states. The required measurements for enthalpies are straightforward; the heat evolved or absorbed is measured for one or a chain of spontaneous processes connecting the substance of interest with its elements. For example, any hydrocarbon, alcohol (or similar organic compound) can be burned in oxygen. Then this heat of combustion is subtracted from the heats of combustion of the corresponding amounts of hydrogen and carbon to obtain the desired enthalpy of formation. Corresponding methods are available for other types of compounds. Indeed, a comprehensive array of enthalpy data had been measured by Thomsen, Berthelot, and others before the end of the nineteenth century.

For the prediction of the direction of chemical reactions and their points of equilibrium, however, it is the chemical potential—the molar (or partial molar) Gibbs energy—that is required. This can be obtained at ordinary temperatures only from processes that come to a measurable equilibrium either directly or in an electrical cell or similar situation. It is not always possible to find such processes. Their discovery and measurement, even if successful, is a difficult task that will differ from one compound to another and may require the development of new catalysts, etc.

An alternate approach to Gibbs energies arises if entropy values can be obtained. Thus, when it became apparent that for a single substance the entropy difference above the value at 0 K was finite and readily measurable, there was an immediate interest in the possibility that entropy changes for formation from the elements might be zero at $T = 0$. This would correspond, of course, to unit probability on a quantum basis for the compound and for its constituent elements at 0 K. Before testing this principle experimentally, we discuss practical calculations of entropy.

NUMERICAL CALCULATIONS OF ENTROPY

As a starting point we may recall that in any *reversible* process a system, or any part of a system, undergoes an increase of entropy just insofar as it absorbs heat from the surroundings, resulting in an equal decrease of the entropy of the surroundings, and that the increase in entropy is equal to the heat so absorbed divided by the absolute temperature,

$$dS = \frac{\delta q_{rev}}{T} \tag{2-22}$$

Heat-Capacity and Enthalpy Measurements

Since it is heat absorbed that yields an entropy increase, calorimetric measurements are important. Flow calorimeters for fluids were discussed briefly in Chapter 4. For heat-capacity measurements for solids, the sample is normally enclosed in a container carrying an electrical heater and a thermometer. Electrical resistance thermometers are excellent. A small electrical energy input q yields a small temperature increase ΔT, whereupon the heat capacity is obtained as $q/\Delta T$. The heat capacity of the empty container must be subtracted, of course. The difficulty lies primarily in avoiding or measuring extraneous heat transfer by radiation, through lead-in wires, etc. One method is to surround the measuring unit with a shield that is maintained by heaters and thermometric controls at exactly the same temperature. Such adiabatic calorimeters are very successful.

For isothermal processes, melting, solid–solid transitions, and vaporization, the problem is simplified by the absence of a temperature change but may be complicated by other aspects such as sluggishness for a transition or transfer and control of the vapor for vaporization.

An excellent book is available that discusses calorimeter design and operation in detail and presents examples.[1]

Entropy Change in Fusion

Let us use this basic equation, Eq. (2-22), to calculate the change of entropy when a substance changes from one phase to another at constant temperature and under conditions of equilibrium. We shall consider the fusion of 1 mole of solid mercury at its melting point, which (at atmospheric pressure) is 234.29 K. At the melting point the two phases are in equilibrium. That is to say, there is a state of balance such that if the external temperature is raised by an infinitesimal amount the solid will melt and if it is diminished by an infinitesimal amount the liquid will freeze. So also, at constant temperature, if the pressure is lowered or raised by any amount, the process will occur in the one direction or the other. Hence the process of fusion at the melting point is a reversible one.

In the case of a pure substance like mercury the temperature remains constant during fusion, and we have from Eq. (2-22)

$$\Delta S = \frac{\Delta H}{T} \qquad (6\text{-}1)$$

If $\Delta_{fus}H$ is the heat of fusion of 1 mole, namely 2295 J, and T is 234.29 K, we may write

$$Hg(s) = Hg(l) \qquad \frac{\Delta_{fus}S}{R} = \frac{2295}{8.3145 \times 234.29} = 1.178$$

It is convenient to divide molar entropies and heat capacities by the gas constant R since the resulting ratio is dimensionless. This follows the pattern in Chapter 5.

It must be borne in mind that such a calculation is based upon the fact that we have a state of equilibrium in which every process is reversible. If, on the other hand, we consider the difference in entropy between ice at $-10°C$ and supercooled water at the same temperature, that difference cannot be obtained by dividing the difference in enthalpy by the absolute temperature, 263 K.

Entropy Change in Vaporization

In vaporization the change of entropy is usually much larger than in fusion. Ethyl ether at its boiling point, 307.7 K, absorbs 27 200 J mol^{-1} by evaporation. Hence we write for the vaporization,

$$(C_2H_5)_2O(l, 1\text{ atm}) = (C_2H_5)_2O(g, 1\text{ atm})$$

$$\frac{\Delta_{vap}S}{R} = \frac{27200}{8.3145 \times 307.7} = 10.63$$

Similarly, for benzene, cyclohexane, and toluene at their respective boiling points, the corresponding values of $\Delta_{vap}S/R$ are 10.47, 10.22, and 10.47.

It will be noted that the four values are nearly equal. An empirical principle, known as *Trouton's rule*,[2] states that the entropy increase per mole is the same for all so-called normal, or nonpolar, liquids at their boiling points. The constant of Trouton's rule is usually given as about 10.5 for $\Delta_{vap}S/R$.

Entropy Change with Pressure or Volume

The general equations for the change of entropy with pressure or volume at constant temperature were derived in Chapter 3:

$$\left(\frac{\partial S}{\partial V}\right)_T = -\left(\frac{\partial P}{\partial T}\right)_V \quad (3\text{-}11)$$

$$\left(\frac{\partial S}{\partial P}\right)_T = -\left(\frac{\partial V}{\partial T}\right)_P \quad (3\text{-}12)$$

For a perfect gas these integrate to the equations from Chapter 2:

$$S_B - S_A = nR \ln\left(\frac{V_B}{V_A}\right) = nR \ln\left(\frac{P_A}{P_B}\right) \quad (2\text{-}21)^*$$

In the case of solids or dense liquids this entropy change is very small unless the pressure change is large. But for gases the entropy change is large. At low pressures a gas is nearly perfect and Eq. (2-21)* is often a good approximation. Corrections for gas imperfection are considered in Chapter 9.

Change of Entropy with Temperature

We have seen in Eqs. (3-26) and (3-28) how the entropy of a substance changes with the temperature. The equations read

$$\left(\frac{\partial S}{\partial T}\right)_V = \frac{C_V}{T} \qquad \left(\frac{\partial S}{\partial T}\right)_P = \frac{C_P}{T}$$

Thus, if a substance is heated at constant pressure, the change of entropy is given by the equation

$$\int dS = \int \frac{C_P}{T} dT = \int C_P \, d\ln T \quad (6\text{-}2)$$

If, therefore, C_P is known at various temperatures, we may perform the integration by analytical or graphical methods and find the change in entropy of a substance between two temperatures.

The simplest case is the one in which C_P is constant, when we find, between the temperatures T and T',

$$S' - S = C_P \ln\left(\frac{T'}{T}\right) \quad (6\text{-}3)^*$$

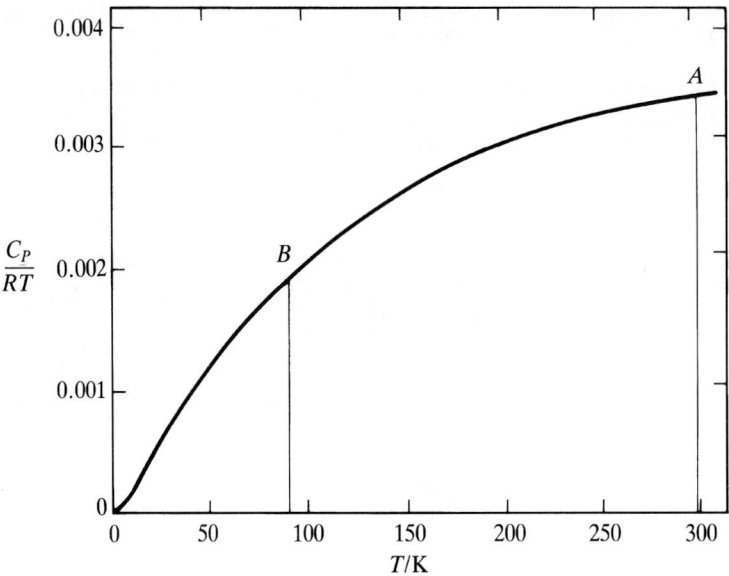

FIGURE 6-1
C_P/RT for graphite.

For liquid mercury in the small temperature range between $T' = 298.15$ K and the freezing point $T = 234.29$ K, we may regard the molar heat capacity as approximately constant and with $C_P/R = 3.40$. Hence

$$\frac{S' - S}{R} = 3.40 \ln \frac{298.15}{234.29} = 0.820$$

At moderate and high temperatures, the heat capacity can often be represented by a simple empirical equation, and this can be integrated to yield entropy, enthalpy, etc. This will be illustrated in Chapter 8. For our consideration of entropies at 0 K, however, we must evaluate Eq. (6-2) over the full range of temperature from zero upward. In this range the variation of C_P is very complex, and graphical methods have been widely used. Alternatively, one can use numerical integration with a large number of smoothed values of C_P. Either method can be used for the actual evaluations in the following examples.

Figure 6-1 shows that the atomic heat capacity of graphite[3] plotted as C_P/RT vs T.

It is evident from Fig. 6-1 that the total entropy above 0 K is a finite quantity. From the total area under the curve up to the point A one obtains $(S_{298} - S_0)/R = 0.690$, where S_0 represents the entropy at absolute zero.

Extrapolation of Heat Capacity to 0 K

While the heat capacity of graphite was so small at the lowest measured temperature that the extrapolation to 0 K was trivial, for many cases this extrapolation must be carefully considered. The pertinent theory was presented in Chapter 5 together with simple examples. One often finds that the measured points at the lowest temperatures fit the T^3 law, Eq. (5-48)*, or, for metals, the slightly more complex equation

$$C = aT^3 + \gamma T \qquad (5\text{-}50)*$$

which integrates to

$$S = \frac{a}{3}T^3 + \gamma T \qquad (5\text{-}51)*$$

A graph of C_P/RT vs. T^2 is very convenient since the function of Eq. (5-50)* becomes a straight line with a slope a/R and intercept γ/R on the $T^2 = 0$ axis. Figures 6-2 to 6-4 present data for a number of substances on this basis, and in all cases the curves approach the behavior predicted by Eq. (5-50)* at the lowest temperatures. Also the intercept is at the origin ($\gamma = 0$) except for metals.

In our opinion a plot of the heat-capacity data on the C_P/RT versus T^2 basis along with the available data extending to lower temperatures for similar substances provides the best basis for extrapolation to 0 K. The already substantial and rapidly growing body of data extending to very low values of C_P makes this a feasible procedure. One then draws an extrapolated curve similar to the observed curves for similar substances and in a manner consistent with Eq. (5-51)* in the limit at 0 K. The two curves for $(CH_2)_3S$ in Fig. 6-3 illustrate the probable limits within which the correct extrapolation should lie. Finally, the extrapolated values of C_P may be transferred to a plot of C_P/T versus T for the calculation of the entropy by graphical integration.

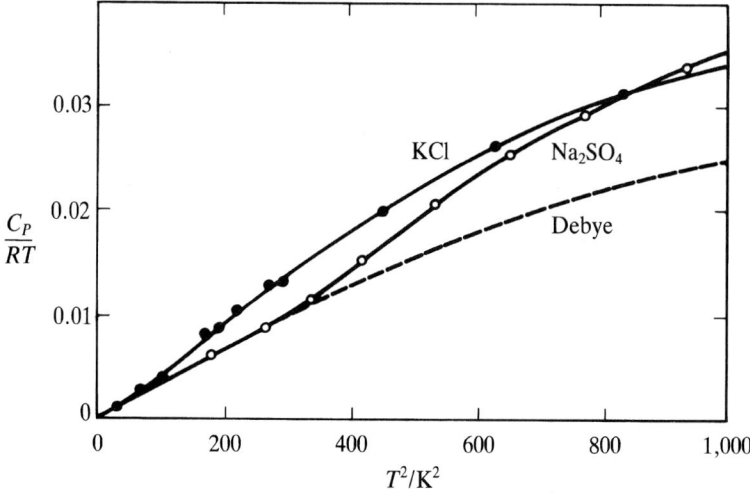

FIGURE 6-2
Heat-capacity curves for KCl and Na_2SO_4, together with a Debye curve fitted to the latter substance in the T^3 region. [KCl data from W. H. Keesom and C. W. Clark, *Physica*, **2**, 698 (1935); J. C. Southard and R. A. Nelson, *J. Am. Chem. Soc.*, **55**, 4865 (1933). Na_2SO_4 data from K. S. Pitzer and L. V. Coulter, *J. Am. Chem. Soc.*, **60**, 1310 (1938).]

88 THERMODYNAMICS

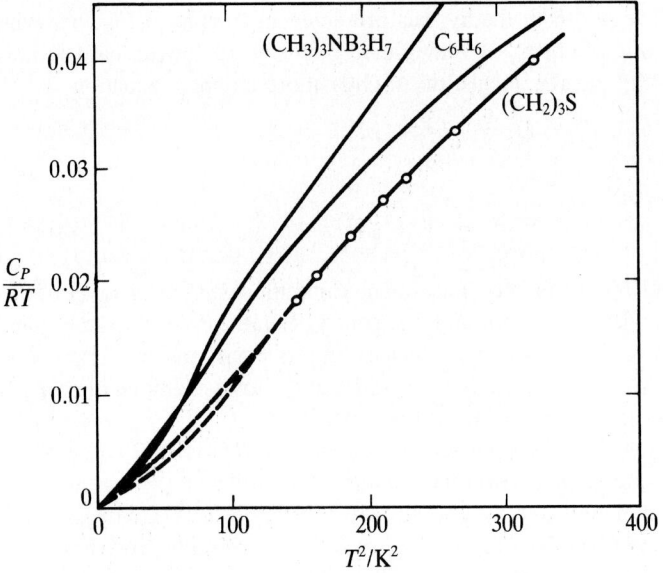

FIGURE 6-3
Heat-capacity curves for three molecular crystals. The pair of dashed lines for $(CH_2)_3S$ indicate the range of reasonable extrapolation curves. [J. E. Ahlberg et al., *J. Chem. Phys.*, **5**, 539 (1937), and G. D. Oliver et al., *J. Am. Chem. Soc.*, **70**, 1502 (1948), for C_6H_6; D. W. Scott et al., *ibid.*, **75**, 2795 (1953), for $(CH_2)_3S$; N. E. Levitin et al., *ibid.*, **81**, 3547 (1959) for $(CH_3)_3NB_3H_7$.]

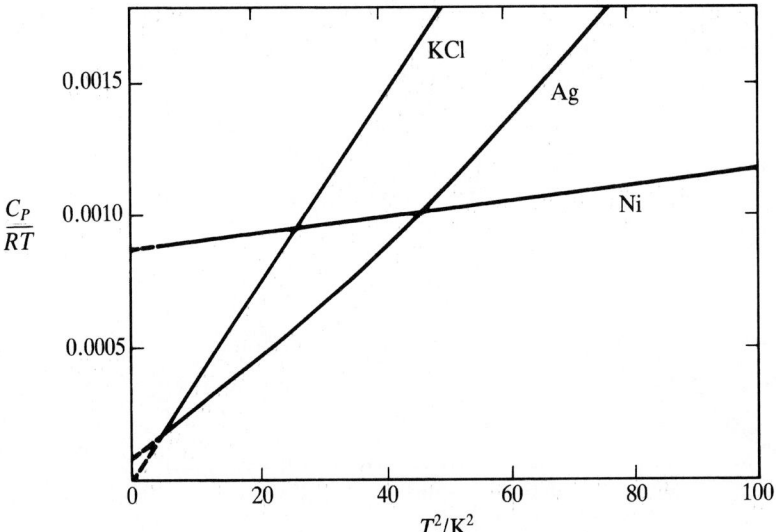

FIGURE 6-4
Heat-capacity curves for KCl, Ag, and Ni in the range below 10 K. Note that the metals have nonzero intercepts for C_P/RT at 0 K. [KCl data from W. H. Keesom and C. W. Clark, *Physica*, **2**, 698 (1935). Ag data from W. S. Corak et al., *Phys. Rev.*, **98**, 1699 (1955); W. H. Keesom and J. A. Kok, *Proc. Acad. Sci. Amsterdam*, **35**, 301 (1932). Ni data from W. H. Keesom and C. W. Clark, *Physica*, **2**, 513 (1935).]

Absolute Entropies

We now assume that the entropy at 0 K, S_0, is zero and proceed to calculate the total entropy of substances at the standard temperature 298.15 K, the boiling point of a liquid, or another desired temperature. If there are solid–solid transitions, each entropy contribution is exactly like that for fusion. Liquid and gas heat capacities have simple temperature dependencies that are readily handled.

Oxygen constitutes a relatively complex example for which there are the measurements of Giauque and Johnston.[4] The results are presented in Fig. 6-5 and Table 6-1. In this case there are three solid forms as well as the liquid and gas.

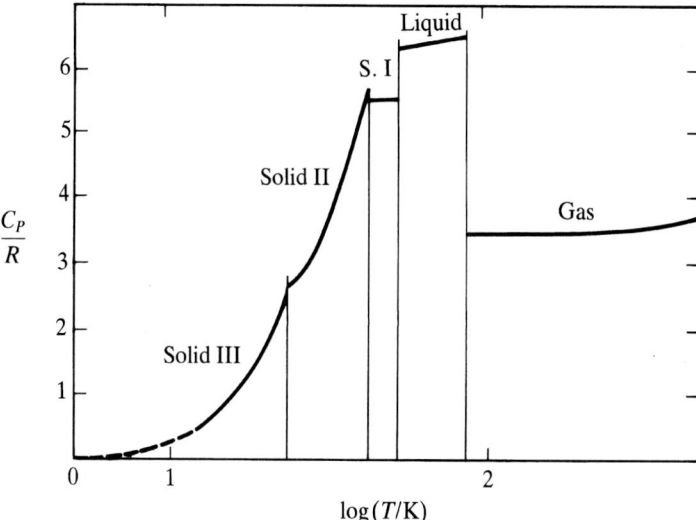

FIGURE 6-5
The molal heat capacity of oxygen. The areas in the graph are terms in the entropy calculation.

TABLE 6-1
The entropy of oxygen
Values of S/R and $(\Delta H/R)/K$

0–14 K, extrapolation	0.27
14–23.66 K, solid III, graphical	0.755
Transition, 11.28/23.66	0.477
23.66–43.76 K, solid II, graphical	2.345
Transition, 89.36/43.76	2.042
43.76–54.39 K, solid I, graphical	1.206
Fusion, 53.48/54.39	0.983
54.39–90.13 K, liquid, graphical	3.252
Vaporization 819.6/90.13	9.094
Total, $S_{90.13}$ (gas, 1 atm.)	20.42

Tests of Entropy Equality at 0 K

We now calculate the absolute entropies of a given substance by different routes and thereby test the validity of the postulate that entropies at 0 K (S_0) may be taken to be zero. For the present, we consider only simple cases where there are no special disorders or multiplicities present. Then the calculations for measured properties follow the methods just described.

As a first example, we consider the entropy of formation of solid AgCl:

$$\text{Ag(cr)} + \tfrac{1}{2}\text{Cl}_2\text{(g)} = \text{AgCl(cr)}$$

The Gibbs energy, entropy, and enthalpy changes for this reaction were measured by Gerke[5] using a Galvanic cell with Ag, AgCl, and Cl$_2$ electrodes, and aqueous HCl electrolyte. (Galvanic cells are discussed briefly in Chapter 8 and in detail in Chapter 19.) Cell potentials were measured very precisely at 15°C, 25°C, and 35°C, and the value $\Delta_f S/R = -6.91 \pm 0.1$ was determined from the temperature coefficient. The entropy of Cl$_2$ (g) is accurately calculated from molecular properties. Values for Ag(cr)[6] and AgCl(cr)[7] are available from heat-capacity measurements extending down to 15 K or below, from which point the extrapolation to 0 K is unambiguous. Table 6-2 shows the results, which yield $\Delta_f S/R = -6.98 \pm 0.06$, in good agreement.

The entropy of chlorine is also available from measurements of the heat capacity, heat of fusion, and heat of vaporization by Giauque and Powell.[8] Their results are given in Table 6-3 with a comparison between calorimetric and calculated values at the

TABLE 6-2
Entropy of formation of AgCl(cr) at 298.15 K

	S/R
AgCl	11.57 ± 0.05
−Ag	−5.13 ± 0.02
$\tfrac{1}{2}$Cl$_2$	−13.42 ± 0.01
	−6.98 ± 0.06

TABLE 6-3
Entropy of Cl$_2$ at 239.05 K from calorimetric measurements and from statistical calculations

	S/R
0–15 K, Debye function $\theta_D = 115$	0.167
15–172.15 K integral of C_P	8.340
Fusion at 172.12 K	4.476
172.12–239.05 K integral of C_P	2.632
Vaporization at 239.05 K	10.269
Correction to ideal gas	0.06
Ideal gas, calorimetric	25.94
Ideal gas, calculated statistically	25.94

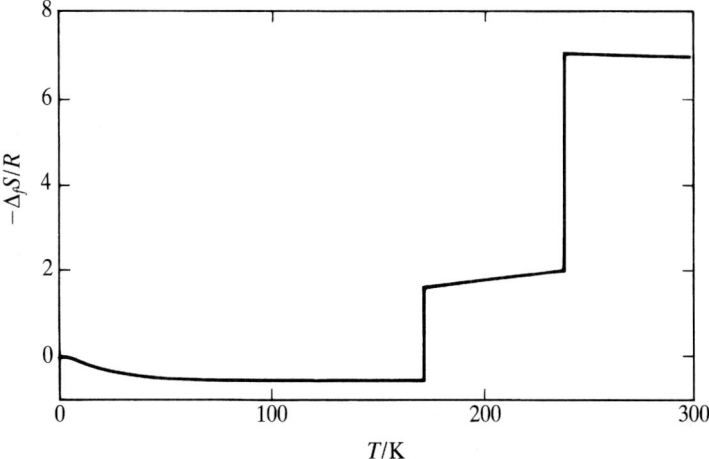

FIGURE 6-6
The molar entropy of formation of AgCl(cr) as a function of temperature.

boiling point. The agreement is exact, well within an uncertainty of about 0.05 for the calorimetric value.

It is interesting to see from Figure 6-6 how the ΔS of formation of AgCl from the elements changes as the temperature is reduced. The large changes occur on condensation and freezing of the chlorine; indeed, $\Delta_f S$ happens to change sign at the freezing point. Finally, $\Delta_f S$ approaches zero with zero slope as $T \to 0$.

Another type of test of the equality of entropies at 0 K is a comparison of two crystalline forms of the same substance, one of which can be supercooled. Phosphine provides an excellent example of this type, with precise data from Stephenson and Giauque[9] shown in Table 6-4.

The comparison of the difference in entropies of rhombic and monoclinic sulfur obtained by integration of C_P/T from 0 K with that for their direct transition at 368.5 K provides another check of the same type as that of phosphine. Montgomery[10] reports that these two $\Delta S/R$ values agree exactly with an uncertainty of only 0.01.

TABLE 6-4
Entropy of phosphine, values of S/R

	Form stable above 49.43 K		Form stable below 49.43 K	
0–15 K, Debye extrapolation		0.249		0.170
Integration of C_P	15–30.29 K	1.100	15–49.43 K	2.034
Transitions	at 30.29 K	0.326	at 49.43 K	1.891
Integration of C_P	30.29–49.43 K	2.415		
Entropy, S/R at 49.43 K		4.090		4.095

Third Law of Thermodynamics

There are many examples of the types just illustrated, although the uncertainties are somewhat larger in most cases. This body of information about the macroscopic properties, heat capacities and entropies, is the basis for the *third law of thermodynamics*, which states that, *for any substance in a perfect crystalline state, the entropy is zero in the limit $T \to 0$*. This is interpreted to be the practical zero of entropy defined in the last chapter, which excludes effects of nuclear spin and of isotopic mixing whenever the consideration is of a natural mixture of isotopes.

There remains the matter of definition of a perfect crystalline state. Some solids are obviously disordered, such as glasses, and can be excluded. In other cases, the substance is clearly crystalline but might have an internal disorder such as random magnetic moments or orientations of molecules. There is no simple test of crystalline perfection; one must consider the possibilities of residual disorder for each case.

There are many crystals, however, for which there is no plausible disorder, and one can adopt the entropy values based on the third law with full confidence. The monatomic gases neon, argon, krypton and xenon have a very simple structure in the solid state, and no rotational disorder is possible. Their entropies from low-temperature heat capacities[11] agree accurately with the statistical values as shown in Table 6-5. The possibility of disorder in solid Cl_2 is most unlikely, and the agreement between statistical and calorimetric entropies in Table 6-3 is excellent. Whenever the structural pattern is very simple and the interparticle forces are strong, randomness in structure is most improbable. Entropies based on the third law have been widely adopted for such substances without any indication of discrepancy.

There are several types of crystals, however, where randomness or disorder are known to occur at room temperature. In some cases there is a transition or other thermal anomaly where the disorder is removed and the crystal then meets the perfection requirement as $T \to 0$. But in other cases, the disorder is removed only at extremely low temperature or becomes "frozen" and remains as $T \to 0$. The disorder may relate to orientation of molecules, to orientation of electronic magnetic moments, or to other phenomena. Chapter 21 describes several cases of frozen disorder.

For small molecules and some larger ones, there is adequate information about molecular properties; hence statistically calculated entropies for the ideal gas are available and should be adopted. Calorimetric measurements on the condensed phases and the vaporization process then yield entropies related to the practical zero without

TABLE 6-5
Entropies of monatomic gases, $S°/R$ (ideal gas)

	T/K	P/bar	Calorimetric	Statistical
Ne	24.55	0.4335	12.19	12.20
Ar	83.78	0.6870	15.83	15.80
Kr	115.76	0.7319	17.68	17.69
Xe	161.36	0.8160	19.08	19.09

ambiguity. The possibility of residual disorder at $T = 0$ is an interesting structural question, but it does not affect thermodynamic properties at higher temperatures when calculated by this procedure. Some examples of this type are considered in Chapter 21. For H_2 and the other isotopic forms of hydrogen, there are interesting complications related to nuclear spin that are described in Chapter 22. Again, the properties for these species as gases are given accurately and reliably by the statistical calculations, and those values should be used wherever applicable.

Entropies from Magnetic Effects

There are many crystals that contain atoms or ions with incomplete d or f electron shells. These atoms or ions have magnetic moments with quantized orientations. In their high-temperature, paramagnetic state the magnetic moments are at least partially randomly oriented in the absence of a very strong magnetic field. Thus, they have substantial magnetic entropy. It is interesting to examine whether these substances become perfect crystals in the sense of the third law as $T \to 0$. Now one can observe not only the thermal properties but also the magnetic properties, and the latter may indicate clearly whether the magnetic moments approach a completely ordered state as $T \to 0$.

As a first example, we consider $NiCl_2$, which was studied by Busey and Giauque.[12] Their heat capacities (Figure 6-7) show an anomaly or transition at 52 K, below which the substance becomes antiferromagnetic. Above 52 K, the magnetic

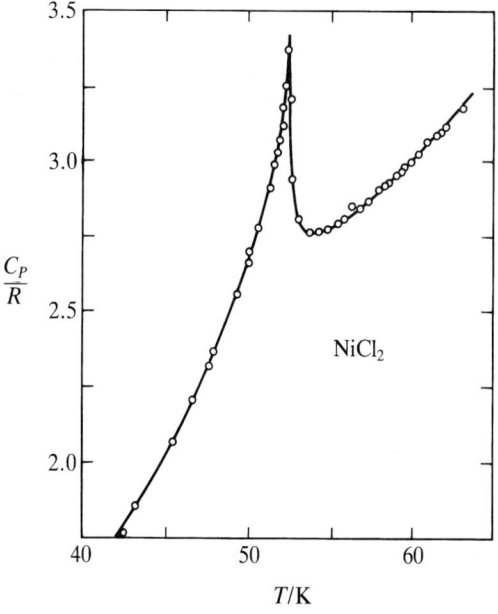

FIGURE 6-7
The heat capacity of $NiCl_2$(cr), showing the magnetic anomaly.

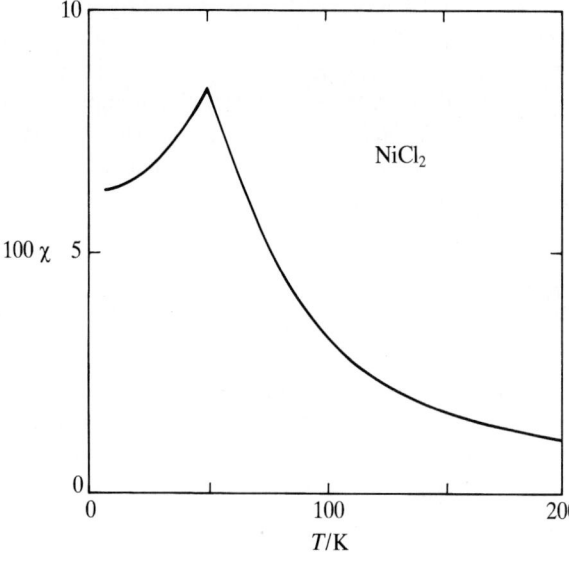

FIGURE 6-8
The magnetic susceptibility of $NiCl_2(cr)$.

susceptibility[13] (Figure 6-8) shows the typical Curie's law behavior with χ approximately proportional to $1/T$. Below the transition, the susceptibility decreases with decrease in temperature and then becomes approximately constant. This indicates that all of the magnetic moments are ordered antiferromagnetically, i.e., in an antiparallel pattern with zero net magnetization except for thermal excitations. Thus, we expect that the entropy of $NiCl_2$ approaches the practical zero as $T \to 0$.

There is the alternative of ferromagnetic behavior below the transition with the magnetic moments adopting a parallel pattern. Again, this is an ordered pattern that should yield zero magnetic entropy as $T \to 0$. Metallic iron, cobalt, and nickel provide examples of ferromagnetism; their transition temperatures are above room temperature at 1033, 1400, and 633 K, respectively.

There are many salts that are similar to $NiCl_2$ with paramagnetic behavior above a transition and antiferromagnetic properties below it. Presumably, the magnetic entropy approaches zero as $T \to 0$ in all such cases. There are few thermodynamic checks for these magnetic salts, but there is a verification for $NiCl_2$. Busey and Giauque[14] measured the equilibrium for the reaction

$$NiCl_2(cr) + H_2(g) = 2HCl(g) + Ni(cr)$$

over the range 630–737 K. The entropies of H_2 and HCl are accurately known from statistical calculations and those of $NiCl_2$ and Ni from experimental heat-capacity measurements. The entropy change calculated on the basis of the third law agrees with that measured for the chemical reaction to $0.05R$.

As the magnetic ions are separated more widely by nonmagnetic material, their interaction decreases and the temperature of the ordering transition is lowered.

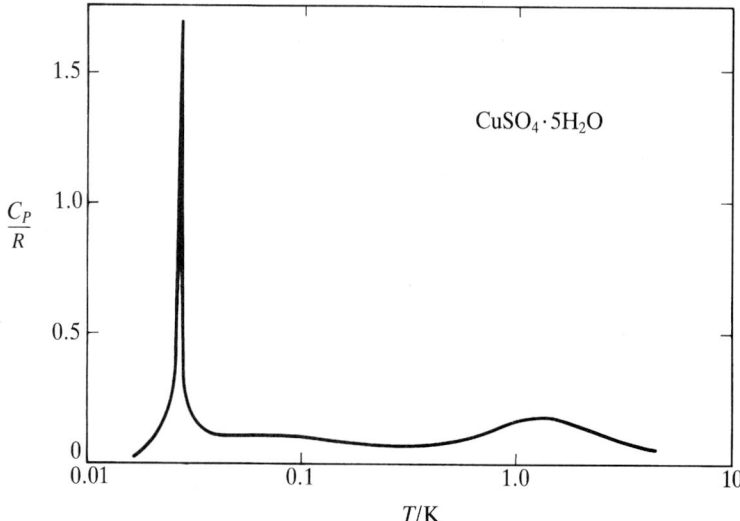

FIGURE 6-9
The heat capacity of $CuSO_4 \cdot 5H_2O(cr)$ at very low temperatures, showing two anomalies.

Indeed, if these interactions are very weak, the most effective method of removing the magnetic disorder is a strong external magnetic field. One can take the double limit $T \to 0$, $(1/H) \to 0$ as the method of extrapolation to the practical zero of entropy.

An important method of attainment of very low temperatures is that of adiabatic demagnetization. This is described in Chapter 24; it requires a magnetic material with extremely weak interactions.

We conclude this section with a consideration of $CuSO_4 \cdot 5H_2O$ as an example of a case with weak but not extremely weak magnetic interactions. It is interesting in that it shows two thermal anomalies; one has a very broad maximum in the range 1–2 K, and the other is very sharp at 0.035 K (see Figure 6-9). The magnetic susceptibility increases steadily with reduction in temperature through the 1–2 K range, which indicates that much of the magnetic disorder remains below 1 K. The susceptibility reaches a peak at 0.035 K and then decreases with further decrease in T. This is consistent with full ordering as $T \to 0$.

The crystal structure of $CuSO_4 \cdot 5H_2O$ is triclinic, i.e., without any axes or planes of symmetry, and has two nonequivalent cupric ions per unit cell. The presence of two magnetic anomalies is undoubtedly related to this nonequivalence and the absence of symmetry. It appears that half of the magnetic entropy is associated with the transition at 0.035 K and the other half with the region of high heat capacity that extends to 5 K. However, we are not aware of a detailed theory yielding quantitatively the properties of $CuSO_4 \cdot 5H_2O$. The very thorough experimental investigations of copper sulfate pentahydrate are summarized by Giauque et al.,[15] who discuss many additional features of this interesting system.

The Entropies of Particular Substances

The heat capacity has been measured down to very low temperatures for most of the chemical elements and for a substantial number of compounds. While various types of possible disorder have been discussed in the sections above, they represent special cases that can readily be excluded for most particular substances. Then the entropy above the practical zero defined in Chapter 5 is given by

$$S_T = \int_0^T C_P \, d\ln T' \tag{6-4}$$

Many entropy values have been calculated by the method described above. It is convenient to tabulate the entropy for a substance in a standard state. For solids and liquids, the standard state is taken at 1 bar. Since the entropy of a gas does not approach a constant value as the pressure is reduced, one cannot choose a low-pressure perfect-gas standard state for the tabulation of the entropy of a gas. The standard state for the tabulation of the entropy of a gas will be the *hypothetical perfect gas at a designated pressure of 1 bar*. Standard-state entropies are indicated by a superscript, $S°$.

To obtain this standard entropy of a gas, we first note the entropy difference between the entropy of a perfect gas at some low pressure P and at 1 bar by Eq. (2-21),

$$S° = S_P + R \ln\left(\frac{P}{1}\right) \tag{6-5}*$$

TABLE 6-6
Entropies of some common substances at 298.15 K, values of $S°_{298}/R$[a]

Solids		Liquids		Gases	
Al	3.401	Br_2	18.306	CH_4	22.401
C(gr)	0.690	H_2O	8.413	CO	23.772
Ca	5.002	Hg	9.144	CO_2	25.714
CaO	4.596			Cl_2	26.830
Cr	2.840			F_2	24.390
Cu	3.989			H_2	15.717
Fe	3.286			HCl	22.479
I_2	13.969			H_2S	24.747
Mg	3.929			N_2	23.045
MgO	3.238			NO	25.348
NaCl	8.673			O_2	24.673
Ni	3.925				
S(rh)	3.855				
Si	2.264				
SiO_2(qu)	4.987				

[a] M. W. Chase, Jr., C. A. Davies, J. R. Downey, Jr., D. J. Frurip, R. P. MacDonald, and A. N. Syverud, "JANAF Thermochemical Tables," 3d edn., *J. Phys. Chem. Ref. Data*, **14**, supplement no. 1, 1985.

If one now takes the limit $P \to 0$, the real gas approaches the ideal-gas state and the quantity $S_P + R \ln P$ will approach a constant independent of pressure. Thus one has the equation for the standard entropy,

$$S° = \lim_{P \to 0} (S_P + R \ln P) \qquad (6\text{-}6)$$

Convenient methods of dealing with gas imperfection effects are discussed in Chapter 9.

Table 6-6 gives a brief list of entropy values at 298.15 K (= 25°C); a more extensive table is given in Appendix 6. Note that the values of $-(G_T° - H_{298}°)/RT$, which are tabulated in Appendix 6 at 298.15 K, are values of $S_{298}°/R$. For most gases the values calculated by the methods described in Chapters 5 and 20 are more precise than those obtained by calorimetry and the third law. The values in Table 6-6 are those given by the most precise method in each case.

REFERENCES

1. J. P. McCullough and D. W. Scott, Eds., *Experimental Thermodynamics*, Volume I. *Calorimetry of Nonreacting Systems*, Butterworths, London, 1968.
2. F. Trouton, *Phil. Mag.*, **18** (5), 54 (1884).
3. W. DeSorbo and W. W. Tyler, *J. Chem. Phys.*, **21**, 1660 (1953).
4. W. F. Giauque and H. L. Johnston, *J. Am. Chem. Soc.*, **51**, 2300 (1929).
5. R. H. Gerke, *J. Am. Chem. Soc.*, **44**, 1684 (1922).
6. P. F. Meads, W. R. Forsythe, and W. F. Giauque, *J. Am. Chem. Soc.*, **63**, 1902 (1941).
7. E. D. Eastman and R. T. Milner, *J. Chem. Phys.*, **1**, 444 (1933).
8. W. F. Giauque and T. M. Powell, *J. Am. Chem. Soc.*, **61**, 1970 (1939).
9. C. C. Stephenson and W. F. Giauque, *J. Chem. Phys.*, **5**, 149 (1937).
10. R. L. Montgomery, *Science*, **184**, 562 (1974).
11. V. A. Rabinovich, A. A. Vasserman, V. I. Nedostup, and L. S. Vekslu, *Thermophysical Properties of Neon, Argon, Krypton, and Xenon*, Hemisphere, Washington, D.C., 1988 (English translation).
12. R. H. Busey and W. F. Giauque, *J. Am. Chem. Soc.*, **74**, 4443 (1952).
13. C. Starr, F. Bitter, and A. R. Kaufmann, *Phys. Rev.*, **58**, 977 (1940).
14. R. H. Busey and W. F. Giauque, *J. Am. Chem. Soc.*, **75**, 1791 (1953).
15. W. F. Giauque, R. A. Fisher, E. W. Hornung, and G. E. Brodale, *J. Chem. Phys.*, **53**, 3733 (1970).

PROBLEMS

6-1. Taking $C_P/R = 3.50$ for gaseous oxygen and the entropy of vaporization from Table 6-1, show that for the process $O_2(g, 298.15\,K) = O_2(l, 90.13\,K)$, $\Delta S/R = -13.28$.

6-2. Calculate the minimum amount of work required to convert 1 mole of oxygen gas at 298.15 K and 1 bar to the liquid state at 90.13 K in a reversible process in which heat is transferred to a heat reservoir at 298.15 K. Note that the entropy decrease of the oxygen was calculated in Problem 6-1 and that the entropy increase of the heat reservoir must be equal so that the net entropy change will be zero. The first law may then be used to calculate the work.

CHAPTER 7

CHEMICAL POTENTIAL, FUGACITY, CONDITIONS FOR EQUILIBRIUM

One of the most important quantities in the thermodynamics of multicomponent systems is the potential that indicates the tendency of a substance to transfer from one phase to another or to enter into a chemical reaction. In qualitative terms we shall call this the "escaping tendency" while the quantitative potential is called the "chemical potential." In Chapter 10, we will derive the general equations for the chemical potentials of the various components in any multicomponent system and their interrelationships to various other properties. At this point, we shall examine several simple situations that will illustrate the role of the chemical potential. Also, we will introduce an alternate measure of escaping tendency, the fugacity, which becomes equal to the pressure for ideal gases.

These simple examples will illustrate the criteria determining the equilibrium of a single substance between different phases. We also consider the criterion of equilibrium in a chemical reaction, the equilibrium constant, for ideal gases and for other systems where the chemical potential of each substance is easily determined.

We first consider a pure liquid substance X in equilibrium with its vapor. Since the liquid and vapor are in direct contact, they are at the same temperature and pressure.

Thus, we have a constant-temperature, constant-pressure situation and Eq. (3-40) applies:

$$\Delta G \leq 0 \qquad (3\text{-}40)$$

Here the equality applies at equilibrium and the inequality relates to a change toward equilibrium. Now G pertains to the entire system:

$$n_{\text{liq}} X_{\text{liq}} + n_{\text{vap}} X_{\text{vap}}$$

$$n_{\text{liq}} + n_{\text{vap}} = \text{constant} \qquad (7\text{-}1)$$

$$G = n_{\text{liq}} G_{X,\text{liq}} + n_{\text{vap}} G_{X,\text{vap}}$$

where $G_{X,\text{liq}}$ and $G_{X,\text{vap}}$ are the molar Gibbs energies of the liquid and vapor, respectively, of which there are n_{liq} and n_{vap} moles present. The subscript X implies a molar quantity; hence the subscript m is omitted.

If Δn moles of liquid changes to vapor, the change in Gibbs energy of the combined system is

$$\Delta G = \Delta n (G_{X,\text{vap}} - G_{X,\text{liq}})$$

If $G_{X,\text{vap}}$ is less than $G_{X,\text{liq}}$, ΔG is less than zero and spontaneous evaporation will occur. Conversely, if $G_{X,\text{vap}}$ is greater than $G_{X,\text{liq}}$, then ΔG can be negative only if Δn is negative, and this means that condensation will occur.

From this simple example, we see that the molar Gibbs energy of the substance is the potential that determines the direction of a change from one phase to another. Also, equality of molar Gibbs energy is the requirement for equilibrium.

CHEMICAL POTENTIAL

The chemical potential is a very important quantity, and it was given a separate symbol, μ, by Gibbs. For the case of a pure substance, we saw above that the chemical potential is the molar Gibbs energy; thus

$$\mu_X = G_X \qquad (7\text{-}2)$$

We shall use these two symbols interchangeably, but prefer μ_X in most cases. The reader should remember their equality since future equations in μ will be derived on the basis of relationships previously given for the Gibbs energy.

THE EFFECT OF PRESSURE AND TEMPERATURE UPON SIMPLE TYPES OF EQUILIBRIUM

If two phases of a pure substance are in equilibrium with one another and the pressure is increased, the chemical potential of the substance will be increased in each phase, and to a greater extent in the phase of larger molar volume [Eq. (3-8)]. That phase will therefore disappear. Thus at 1 bar ice and water are in equilibrium at 0°C; if the pressure exerted upon the two phases is now increased, the chemical potential of the ice, which is more voluminous, is increased more than the chemical potential of the water and the ice will disappear if the temperature remains constant. So also, if the two phases are in equilibrium and the temperature is changed, the two chemical potentials will change differently in accordance with Eq. (3-9) and equilibrium will no longer exist.

We may, however, change both temperature and pressure in such a way as to keep the two chemical potentials equal to one another. This is the condition for the maintenance of equilibrium, which now we need only translate into mathematical form.

Change of Equilibrium Pressure with Temperature

Suppose that we have two phases of the same pure substance, in equilibrium, and that μ_A, V_A, S_A, H_A, and μ_B, V_B, S_B, H_B are the chemical potentials, molar volumes, etc., in the two phases. Then, with $\mu_i = G_i$, our condition for equilibrium is

$$G_A = G_B \tag{7-3}$$

Moreover, if any change occurs and equilibrium is to be maintained, it is necessary that

$$dG_A = dG_B \tag{7-4}$$

But the two states are completely determined by the two variables P and T. Hence, we have

$$dG_A = \left(\frac{\partial G_A}{\partial P}\right)_T dP + \left(\frac{\partial G_A}{\partial T}\right)_P dT \qquad dG_B = \left(\frac{\partial G_B}{\partial P}\right)_T dP + \left(\frac{\partial G_B}{\partial T}\right)_P dT$$
$$\tag{7-5}$$

By combining this with Eq. (7-3) and substituting the values of the differential coefficients from Eqs. (3-8) and (3-9), we find (since $\Delta G = 0$)

$$\frac{dP}{dT} = \frac{S_B - S_A}{V_B - V_A} = \frac{\Delta S}{\Delta V} = \frac{\Delta H}{T \Delta V} \tag{7-6}$$

This equation shows how the equilibrium pressure must change with the temperature in any two-phase system such as vapor and liquid, liquid and solid, or two solid forms like rhombic and monoclinic sulfur. Equation (7-6) is evidently identical with Eq. (3-46), the Clapeyron equation, which we obtained before by using the fundamental equations of entropy. Indeed, in this simple case that method seems a little less cumbrous, but the method that we have just described is of such general applicability that it will be used frequently.

FUGACITY

While all of the relationships related to the tendency of a substance to react or to change phase could be expressed by the chemical potential as defined above, an alternate function is more convenient in many cases and, in addition, conveys a simple physical meaning. This is the *fugacity*, which was first defined and used by G. N. Lewis[1] in 1901.

Before defining the fugacity, however, let us consider a familiar property, the vapor pressure, which also gives a measure of the escaping tendency. Thus, the vapor pressure of ice is equal to that of liquid water at the melting point, but less than that of supercooled water at all lower temperatures. And, if one has two aqueous salt solutions with a gas-phase contact between them, water will evaporate from the solution with the higher vapor pressure and condense in the other until the two vapor pressures become equal.

Indeed we could with entire correctness use the vapor pressure also as a quantitative measure of escaping tendency, and this would be a very satisfactory procedure if every vapor behaved as a perfect gas. In the fugacity we are going to define a measure of escaping tendency that bears to the vapor pressure a relation analogous to the relation between the perfect-gas thermometer and a thermometer of some actual gas. The fugacity will be equal to the vapor pressure when the vapor is a perfect gas, and in general it may be regarded as an "ideal" or "corrected" vapor pressure.

We have found in Eq. (3-45)* the difference in Gibbs energy of a perfect gas between two pressures, at the same temperature. We also defined the chemical potential to be equal to the molar Gibbs energy; thus,

$$\mu_B - \mu_A = G_B - G_A = RT \ln\left(\frac{P_B}{P_A}\right) \qquad (7\text{-}7)*$$

We wish the same functional form to apply exactly for the difference in fugacity between states A and B, thus:

$$\ln f_B - \ln f_A = \frac{\mu_B - \mu_A}{RT} \qquad (7\text{-}8a)$$

or

$$\ln f_B = \frac{\mu_B}{RT} + \left[\ln f_A - \frac{\mu_A}{RT}\right] \qquad (7\text{-}8b)$$

But in the limit of a gas at low pressure we wish to make $f = P$; hence *we define fugacity by the equation*

$$\ln f_B = \frac{\mu_B}{RT} + \lim_{P_A \to 0}\left[\ln P_A - \frac{\mu_A}{RT}\right] \qquad (7\text{-}9)$$

The second term on the right is a constant at a given temperature, and one obtains the simple differential form

$$d\ln f = \frac{d\mu}{RT} \quad \text{at constant } T \qquad (7\text{-}10)$$

This definition of the fugacity applies not only to real gases but to liquids and solids as well. For if we admit that every substance at a finite temperature has a finite vapor pressure, then if the pressure upon a substance is decreased without limit the substance will eventually vaporize and with further diminution in pressure of vapor will approach nearer and nearer to the condition of a perfect gas. Fugacity will be regarded as having the same dimensions as pressure, and we will take the unit of fugacity to be the bar. The standard atmosphere is often the unit of fugacity in the older literatures; also one may use the pascal or any other unit of pressure.

Change of Fugacity with Pressure

We have seen by Eq. (3-8) that

$$\left(\frac{\partial \mu}{\partial P}\right)_T = \left(\frac{\partial G_m}{\partial P}\right)_T = V_m$$

and that by Eq. (7-10)

$$\left(\frac{\partial \ln f}{\partial \mu}\right)_T = \frac{1}{RT} \tag{7-11}$$

whence

$$\left(\frac{\partial \ln f}{\partial P}\right)_T = \frac{V_m}{RT} \tag{7-12}$$

where G_m and V_m are molar properties. By integrating this equation at constant temperature, we may ascertain the fugacity of a substance at any pressure if it is known at some other pressure and if V is known as a function of P. In Chapter 9 we shall discuss the application of this method to imperfect gases and to the fluid state.

Since the fugacity is defined with reference to the gaseous state at low pressure, it is the molecular mass in that state that we must use consistently in our fugacity equations. Thus liquid nitrogen dioxide presumably is composed principally of N_2O_4 molecules, but the vapor dissociates at low pressure to NO_2 molecules. Consequently one would adopt the mole of NO_2 as the basis and use the volume of 46 g in all calculations of fugacity. In the case of a substance such as chlorine, however, the dissociation from Cl_2 to $2Cl$ occurs only at very high temperatures or extremely low pressures. At ordinary temperatures we have no difficulty in obtaining the fugacity of chlorine on the basis of a perfect gas of Cl_2 molecules. Thus the usual basis of fugacity of chlorine is the diatomic molecule, but at high temperatures an ambiguity would arise, and it would be necessary to specify whether the basis chosen was Cl or Cl_2.

Change of Fugacity with Temperature

We have seen that at least theoretically any substance may be converted isothermally into a vapor and that this vapor may be made to approach the perfect gas, without limit, as the pressure is indefinitely diminished. Suppose now that we have a substance in a given state and compare its chemical potential μ with the chemical potential μ^* in the vapor state, at some very low pressure and at the same temperature. We are to consider the change in chemical potential when 1 mole of the given substance is converted into the highly attenuated vapor. By Eq. (7-8)

$$\mu^* - \mu = RT \ln \left(\frac{f^*}{f}\right) \tag{7-13}$$

Differentiating this equation with respect to temperature, while the pressure upon each of the two states remains constant, we obtain

$$\left(\frac{\partial \mu^*}{\partial \mu T}\right)_P - \left(\frac{\partial \mu}{\partial T}\right)_P = R \ln \left(\frac{f^*}{f}\right) + RT \left[\frac{\partial \ln f^*}{\partial T}\right]_P - RT \left[\frac{\partial \ln f}{\partial T}\right]_P \tag{7-14}$$

In the gas at very low pressure the fugacity is equal to the pressure. Hence f^* does not change with the temperature at constant pressure, and the next to the last term disappears. Also by Eq. (7-13),

$$R \ln \left(\frac{f^*}{f}\right) = \frac{\mu^* - \mu}{T}$$

whence

$$\left(\frac{\partial \mu^*}{\partial T}\right)_P - \left(\frac{\partial \mu}{\partial T}\right)_P = \frac{\mu^*}{T} - \frac{\mu}{T} - RT \left[\frac{\partial \ln f}{\partial T}\right]_P \tag{7-15}$$

Now, employing Eqs. (3-16) and (7-2), we obtain the simple equation

$$\left[\frac{\partial \ln f}{\partial T}\right]_P = \frac{H_m^* - H_m}{RT^2} \tag{7-16}$$

where H_m and H_m^* are molar enthalpies. H_m pertains to the real gas at pressure P, while H^* is the enthalpy in the limit as $P \to 0$.

Standard States and Fugacity

For the tabulation of the thermodynamic properties of various substances, standard states are adopted in order to make the subsequent use of the information unambiguous and convenient. While one can refer to a standard state at any temperature, the primary standard for temperature is 25°C = 298.15 K. For pressure at any temperature, 1 bar is the current standard, although it was formerly 1 atmosphere = 1.013 bar. This implies for a gas a fugacity of 1 bar; in other words, a hypothetical ideal-gas state with a pressure of 1 bar. Then Eq. (7-13) can be rewritten as

$$\ln f = \frac{\mu - \mu^\circ}{RT} \tag{7-17}$$

with μ° the chemical potential in the standard state.

Since fugacity has the dimension of pressure, it is not strictly correct to take its logarithm. Rather, one should replace $\ln f$ by $\ln(f/1)$, the logarithm of the dimensionless ratio of f to 1 bar. This change is implied throughout this and later chapters. The symbol for the standard state is usually a small superscript zero or degree symbol; thus $G°$, $\mu°$, or $H°$. The symbol $^\varnothing$ is also used. The standard-state enthalpy or heat capacity for a gas is the limiting value at zero pressure which is, of course, the ideal-gas value, i.e., $H° = H_m^*$ in Eq. (7-16). The standard state for a pure liquid or solid is the real liquid or solid under a pressure of 1 bar.

In using values from older tables, one must keep in mind that the standard pressure in effect may be 1 atm. instead of 1 bar. For a gas the correction is clearly a factor of 1.013, or slightly more than 1% in fugacity. In chemical potential, it is $298 \ln(1.013)$ or 3.85 K in $\mu°/R$ or 32 J mol^{-1} at 298 K.

The effect of this change in standard pressure is much smaller for solids or liquids. We see this from Eq. (7-12), which integrates to

$$\delta \ln f = \frac{(\delta P)V_m}{RT} \qquad (7\text{-}18)$$

For the present problem, this becomes

$$\delta \ln f = \frac{0.013 V_m}{298 \times 83.1}$$

and for typical molar volumes of 20 to 50 cm^3 this yields from 1×10^{-5} to 3×10^{-5} for $\delta \ln f$. This change is negligible in almost all cases.

CHEMICAL EQUILIBRIA; EQUILIBRIUM CONSTANTS

Let us now consider a chemical reaction in an ideal or nearly ideal gaseous state. The chemical potentials of the various substances are given exactly by their fugacities, which are, in turn, approximately the various partial pressures. In a later chapter, we present the general methods for determining fugacities in mixed, nonideal gases; for the present, we assume that the approximation by partial pressures will suffice or that experiments are carried out at various total pressures and the results extrapolated to zero pressure.

We represent the chemical reaction by the equation

$$l\text{L} + m\text{M} + \cdots = q\text{Q} + r\text{R} + \cdots$$

which states that l moles of substance L react with m moles of M, etc., to give q moles of Q, r moles of R, etc. The change in Gibbs energy for this reaction is

$$\Delta G = (q\mu_Q + r\mu_R + \cdots) - (l\mu_L + m\mu_M + \cdots) \qquad (7\text{-}19)$$

where the various μ_i are the chemical potentials of those substances in their actual nonstandard states. We also consider the change in Gibbs energy when all substances are in their standard states:

$$\Delta G° = (q\mu_Q° + r\mu_R° + \cdots) - (l\mu_L° + m\mu_M° + \cdots) \qquad (7\text{-}20)$$

Next we shift from chemical potentials to fugacities by use of Eq. (7-17), together with the value $f = 1$ for the standard state of a gas. Thus,

$$q(\mu_Q - \mu_Q^\circ) = qRT \ln f_Q = RT \ln f_Q^q \qquad (7\text{-}21)$$

and so on for M, etc.

Combining the several equations, we find

$$\Delta G - \Delta G^\circ = RT \ln \left[\frac{f_Q^q f_R^r \cdots}{f_L^l f_M^m \cdots} \right] \qquad (7\text{-}22)$$

The important quotient appearing on the right we may call the fugacity quotient for the reaction. To the degree that partial pressures approximate fugacities, p_Q, etc., can be substituted for f_Q, etc., in Eq. (7-22); this approximation becomes excellent for gases at very low pressures.

Equilibrium Constant

Since this is a process at constant temperature and pressure, the criterion for equilibrium is

$$\Delta G = 0 \qquad (3\text{-}40)$$

Then we have the very important relationship

$$\frac{\Delta G^\circ}{RT} = -\ln \left[\frac{f_Q^q f_R^r \cdots}{f_L^l f_M^m \cdots} \right] \qquad (7\text{-}23)$$

At a given temperature, ΔG° is a constant; hence, the requirement for equilibrium is that the fugacity quotient shall also be a constant. The value of this quotient, when the system is at equilibrium, we shall call the equilibrium constant K, and write

$$\frac{\Delta G^\circ}{RT} = -\ln K \qquad (7\text{-}24)$$

If one or more of the reactants or products in the chemical reaction are pure liquids or solids, Eq. (7-19) still applies. But now there is no need to introduce Eq. (7-21) since the fugacity of the substance in the gas is just the fugacity of the pure solid or liquid. If the total pressure of the gas is small, that fugacity is equal to that for the solid or liquid in its standard state to high accuracy. Consequently, the term f_J^j is omitted in the fugacity quotient in Eqs. (7-22 and 7-23) for any substance present as a pure solid or liquid, and its standard chemical potential in Eq. (7-20) is that for the condensed state.

Equation (7-23) and those preceding it are generally valid and are not limited to low-pressure gases or pure solids or liquids. However, these equations are only formalities for highly nonideal gases or liquid or solid solutions until the fugacity relationships are developed for such systems. Thus, the equations are presented here in the context of the ideal gas with or without pure solids or liquids. The fugacity relationships for more complex systems are the subject of several later chapters.

Change of Equilibrium Constant with Temperature

The change of Gibbs energy with temperature is determined by the enthalpy as given by Eqs. (3-18) or (3-20):

$$\left[\frac{\partial (G/T)}{\partial T}\right]_P = -\frac{H}{T^2} \qquad (3\text{-}18)$$

$$\left[\frac{\partial (G/T)}{\partial (1/T)}\right]_P = H \qquad (3\text{-}20)$$

Since we are concerned at present with changes in the standard-state Gibbs energy, the condition of constancy of pressure is satisfied by the definition of the standard state. Then one can take the sums and differences of Eq. (7-20), and one obtains

$$\frac{d(\Delta G°/T)}{d(1/T)} = \Delta H° \qquad (7\text{-}25)$$

and from Eq. (7-24),

$$\frac{d\ln K}{d(1/T)} = -\frac{\Delta H°}{R} \qquad (7\text{-}26)$$

Thus, if the change in heat capacity, $\Delta C_P°$, is known, one has $\Delta H°$ as a function of temperature, and Eq. (7-26) can be integrated to give the equilibrium constant at various temperatures. Also, if $\Delta C_P°$ is very small, as it often is, a plot of $\ln K$ vs. $1/T$ gives a straight line of slope $-\Delta H°/R$. Examples illustrating these statements are given in Chapter 8.

THE PHASE RULE

When the state of a system cannot be completely determined until at least r data are given, we say that it possesses r degrees of freedom. This is only another way of saying that the state of the system depends upon r independent variables. The state of a single phase of a pure substance depends usually upon only two variables, let us say temperature and pressure. If we impose the further condition that two such pure phases are to exist together in equilibrium, the number of degrees of freedom is reduced to one. With three coexisting phases we have what is known as a nonvariant system. Thus, if water and water vapor are to be coexistent, we may arbitrarily fix the pressure or the temperature, but not both, while the condition that ice, water, and water vapor exist together completely determines the state of the system at a so-called triple point.

There may be other variables besides temperature and pressure that are requisite to determine the state of a phase. Thus we may have to consider the presence of an electric or of a magnetic field or the size of particles. Especially if the phase in question is a solution, the number of degrees of freedom is increased by one for each component beyond the first.

Whatever the number of variables, it still remains true that the number of degrees of freedom of a system as a whole is equal to the number of variables requisite to determine the state of the individual phases, less the number of phases in the system, beyond the first. This is the celebrated phase rule of Gibbs.

Various examples in later chapters will illustrate the phase rule. But in most cases we will be dealing quantitatively with the chemical potential, the fugacity, or the equivalent and not merely noting the number of degrees of freedom.

REFERENCE

1. G. N. Lewis, *Proc. Am. Acad.*, **37**, 49 (1901); *Z. physik. Chem.*, **38**, 205 (1901).

PROBLEMS

7-1. How many degrees of freedom has the three-phase system water vapor, ice, liquid solution of NaCl of molality m? Would this conclusion be the same or different if the liquid solution was of HCl that was also present in the vapor?

7-2. For the reaction $Ca(OH)_2(s) + CO_2(g) = CaCO_3(s) + H_2O(g)$, write the equilibrium constant expression, Eq. (7-23). At equilibrium at a given temperature, how will the composition of the gas vary with the pressure? Interpret this situation in terms of the phase rule.

CHAPTER 8

PROPERTIES OF SIMPLE SYSTEMS; TABLES OF THERMODYNAMIC PROPERTIES

To devise various methods of calculating the change of chemical potential in chemical reactions is a task that will occupy us throughout a major part of this book. Already, however, we have become acquainted with one or two of the important methods. Thus we know that, whenever a condition of equilibrium is reached in a chemical reaction, the Gibbs-energy change of the reaction is zero. For example, there is a transition point between rhombic and monoclinic sulfur at 1 bar and 95.4°C = 368.5 K. Hence we may write

$$S(\text{rhombic}) = S(\text{monoclinic}) \quad \Delta\mu_{368.5} = 0$$

So also we have seen that $\Delta\mu$ for a reaction that occurs in a galvanic cell is determined by the maximum electrical work that the cell is capable of performing. A further study here of such a cell will be profitable.

Gibbs Energy and the Electromotive Force of a Galvanic Cell

Let us consider a cell composed of a lead electrode in contact with solid lead chloride, a mercury electrode in contact with mercurous chloride, and a solution of potassium chloride as electrolyte. When electrical contact is established between the two electrodes, a current will pass through the cell so that metallic lead is used up, metallic mercury is precipitated, and at the same time the lead chloride increases and the mercurous chloride diminishes in amount according to the chemical equation,

$$Pb(s) + Hg_2Cl_2(s) = PbCl_2(s) + 2\,Hg(l)$$

In such a case, by Eq. (3-39)

$$\Delta G = w'$$

where $-w'$ is the electrical work capable of being obtained under conditions of maximum efficiency. Under such conditions the counter-emf of the storage battery, motor, or other apparatus upon which the electrical work is done must differ only infinitesimally from the maximum or reversible emf of the cell. This emf E, multiplied by the amount of electricity flowing through the cell, measures the maximum output of electrical work. If F is the Faraday equivalent and n is the number of such equivalents passing through the cell when the above reaction occurs, we may write,

$$\Delta G = w' = -nFE \tag{8-1}$$

where E is positive if the reaction as written is a spontaneous one.

In accordance with the chemical equation we have written, $n = 2$ and

$$\Delta G = -2 \times 96\,485 E$$

The measured value of the emf at 25°C is 0.5359 V, whence

$$Pb(s) + Hg_2Cl_2(s) = PbCl_2(s) + 2\,Hg(l)$$

$$\Delta G_{298} = -103\,413\,\text{J}\,\text{mol}^{-1}$$

This is the change of Gibbs energy that results from the above reaction, whether the reaction occurs in the cell or lead is added to mercurous chloride and the process takes place in an irreversible way. Any other reversible cell in which this reaction, and only this reaction, occurs can equally well be used to measure the Gibbs-energy change and, since n would be the same, E would be the same. Thus the cell that we have described must give the same emf independently of the particular chloride used as electrolyte, and of its concentration, and of the solvent, provided always that when the current passes through the cell no other process occurs than the one stated. If, for example, we used a very dilute solution of potassium chloride as an electrolyte, the solubility of the two chlorides no longer being negligible, the cell process would not correspond exactly to this chemical equation and the emf would be slightly different.

The Gibbs–Helmholtz Equation

This relationship, Eq. (3-16), yields a valuable formula for the temperature coefficient of the emf of a reversible galvanic cell:

$$\left(\frac{\partial G}{\partial T}\right)_P = -\Delta S = \frac{G - H}{T} \tag{3-16}$$

Substitution of Eq. (8-1) then yields

$$-nF\left(\frac{\partial E}{\partial T}\right)_P = -\Delta S = \frac{-nFE - \Delta H}{T} \tag{8-2a}$$

or in simpler form

$$T\left(\frac{\partial E}{\partial T}\right)_P = E + \frac{\Delta H}{nF} \tag{8-2b}$$

Thus, for the cell described above, with $E_{298} = 0.5359$ V, the emf increases $0.000\,145$ V K^{-1}. Hence, we may calculate the heat of this reaction, namely

$$\Delta H^{\circ}_{298} = 2 \times 96\,485\,(298.15 \times 0.000\,145 - 0.5359) = -95\,070\,J$$

This is probably more accurate than a value from calorimetry.

Gibbs Energy and Heat of Reaction

In the early days of the first law of thermodynamics, before the second law was fully understood, it was assumed as a matter of course that the most efficient utilization of a chemical reaction for the production of work would consist in converting all the heat of that reaction into work. In other words, the quantity $-\Delta H$ was assumed to represent the limiting quantity of work that could be obtained under the conditions of maximum efficiency. The work of Thomsen, Berthelot, and others has given us a great mass of thermochemical data of all grades of accuracy. The hope of these investigators that the results of their labors would give a direct measure of chemical affinity proved to be a vain one. Nevertheless, their data are of great utility in the calculations that lead to the true measure of chemical affinity and will be often employed in our later calculations.

However, we have just seen that it is not $-\Delta H$ but $-\Delta G$ that measures this maximum capacity for performing useful work. And these two quantities are not equal unless the entropy of the system in question is the same at the beginning and end of the isothermal reaction under consideration. This is shown by applying Eq. (3-3) to an isothermal process, which gives

$$\Delta G - \Delta H = -T\Delta S \tag{8-2}$$

According to the sign of ΔS, the work obtainable in a given reversible process may be greater or less than the heat of the reaction.

It is true that, according to the first law, the external work performed must be equal to the loss in enthalpy of a system, unless some heat is given to or taken from the surroundings, but this is precisely the point first clearly seen by Gibbs. When an isothermal reaction runs reversibly, $T\Delta S$ is the heat absorbed from the surroundings, and if this is positive, the work done will be even greater than the heat of reaction.

In the specific reaction which we have just been considering

$$\Delta H = -95\,070\,\text{J mol}^{-1}$$

Hence, if 1 mole of lead reacts irreversibly with mercurous chloride, as in the calorimeter, 95 070 J mol^{-1} is given up. But we have seen that

$$\Delta G = -103\,413\,\text{J}\,\text{mol}^{-1}$$

so that, in the reversible isothermal process, the work done is greater than the heat of reaction. Therefore, when this galvanic cell operates reversibly in a thermostat, heat is not given to, but taken from, the thermostat, to the extent of 8343 J for each mole of lead consumed.

THE CHANGE OF EQUILIBRIUM CONSTANT WITH THE TEMPERATURE

In Chapter 7, the change with temperature of an equilibrium constant was considered and Eq. (7-26) was derived.

$$\frac{d\ln K}{d(1/T)} = -\frac{\Delta H^\circ}{R} \tag{7-26}$$

Since ΔH° is often nearly constant over the limited temperature range of the equilibrium measurements, a graph of $\ln K$ vs. $1/T$ is nearly a straight line and its slope is readily determined. The resulting ΔH° can be assigned the mean temperature of the set of K values.

As an example, we may employ the measurements of Preuner and Schupp[1] and of Randall and Bichowsky[2] on the equilibrium between hydrogen, sulfur, and hydrogen sulfide, at temperatures above 1000 K. Now at these temperatures sulfur vapor is in the form of S_2, a species that exists to no measurable extent at ordinary temperatures and is therefore not subject to the usual kind of calorimetric investigation. The reaction is written

$$H_2(g) + \tfrac{1}{2}S_2(g) = H_2S(g) \qquad K = \frac{[H_2S]}{[H_2][S_2]^{1/2}}$$

where $[H_2S]$ represents f_{H_2S} or approximately P_{H_2S} at low to moderate pressures. The results are given in Table 8-1 and the data of the two middle columns are plotted in

TABLE 8-1
Equilibrium between H_2, S_2, and H_2S

T	1/T	ln K	($\Delta H_0^\circ/R$)/K
1023	0.000 9775	4.663	−9810
1103	0.000 9066	3.937	−9850
1218	0.000 8210	3.005	−9850
1338	0.000 7474	2.220	−9880
1362	0.000 7342	2.077	−9880
1405	0.000 7117	1.826	−9880
1473	0.000 6789	1.481	−9900
1537	0.000 6506	1.128	−9830
1667	0.000 5999	0.592	−9850
		Average	−9859

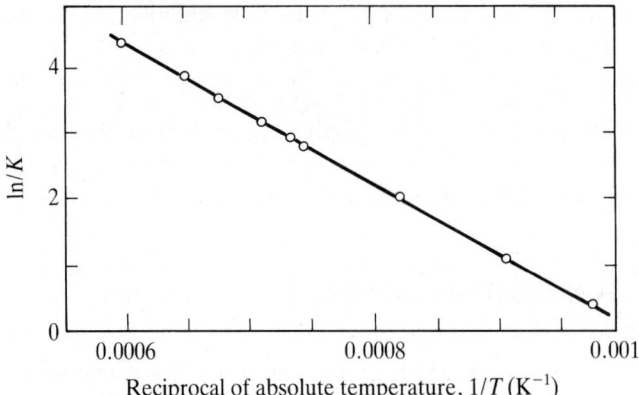

FIGURE 8-1
Equilibrium between hydrogen, sulfur, and hydrogen sulfide.

Fig. 8-1. The individual points fall beautifully upon a smooth curve that is nearly a straight line. The slope is nearly constant with a mean value of $-10\,700$ K, which may be taken for $\Delta H°$ at 1300 K.

This treatment is often called the *second-law* method and the resulting $\Delta H°$ the *second-law* $\Delta H_T°$. If the heat capacities of the substances involved are known over a given range of temperature, this $\Delta H_T°$ can be converted to another temperature such as 298 K or even 0 K. In the present example, this would be possible only with statistically calculated heat capacities because the species $S_2(g)$ is not the important species of sulfur at lower temperatures. With statistically calculated values of $(H_{1300}° - H_0°)$ for H_2S, H_2, and S_2, one obtains $\Delta(H_{1300}° - H_0°)/R = -370$ K and subtraction from $-10\,700$ K yields $\Delta H_0°/R = -10\,330$ K for this reaction. But, however calculated, values are reported for $\Delta H_{298}°$ or $\Delta H_0°$ that were determined from equilibria at much higher temperatures by this second-law method and then converted to a standard temperature for tabulation and further use.

Where heat capacities and often entropies are known over a wide range of temperature, possibly even to 0 K, there are several other methods of treating (or predicting) equilibrium constants that have many advantages; we consider them next.

Gibbs Energies from Third-law Entropies and Enthalpy Data

In addition to the various methods that measure Gibbs energy directly, the third law of thermodynamics provides a very important indirect method through the equation given earlier:

$$\Delta G = \Delta H - T\Delta S \qquad (8\text{-}3)$$

We noted before that even by the beginning of the present century there were extensive data available on the ΔH of various chemical reactions. Until the development of the third law, however, there was no source of ΔS values that did not essentially require a measurement of ΔG. The third law provided a means of obtaining from heat-capacity measurements at low temperatures the absolute entropy of each substance in a reaction. If such entropy values are available for each reactant and each product, then ΔS may be calculated and combined with a calorimetrically determined ΔH according to Eq. (8-3) to obtain ΔG. Thus, the third law opened a great new opportunity for the prediction of the chemical potential changes of reactions that are too slow to be measured by direct methods. Furthermore, it has been found that for many types of substances the entropy values show regular trends. This allows one to estimate entropy values as accurately as are needed for use with many experimental heat-of-reaction data. Latimer made particularly effective use of these methods in his work on the free energies of inorganic substances.[3] Similar methods are equally useful for organic compounds. Some of the systems for estimating entropies are discussed in Appendix 13; others are mentioned at various points throughout the book. In addition to experimental entropy measurements, the third law also provides the basis for the calculation of these statistically calculated entropies of gases when enough information is available about the individual quantum energy levels of their molecules. These methods are given in Chapter 5 for simple cases and in Chapter 20 for more complex examples.

An interesting application of the third-law method concerns the reaction

$$C(\text{diamond}) = C(\text{graphite})$$

This reaction was known to take place in the direction diamond to graphite at high temperature and low pressure, but the first and, to the present date, the only precise information about ΔG comes from enthalpy and entropy data. The heats of combustion of diamond and of graphite have been measured many times, and the difference is found to be $\Delta H^\circ_{298} = -1895 \text{ J mol}^{-1}$. The calculation of the entropy of graphite from heat capacities was discussed in Chapter 6, and the result was $5.74 \text{ J K}^{-1} \text{ mol}^{-1}$ at 298.15 K. A similar calculation for diamond yields $2.38 \text{ J K}^{-1} \text{ mol}^{-1}$. Consequently,

$$\Delta G^\circ_{298} = -1895 - 298 \times 3.36 = -2896 \text{ J mol}^{-1}$$

From this result, and ΔV derived from the densities of diamond and graphite, one may calculate by Eq. (3-8) that ΔG would be zero at approximately 15 000 bar pressure at room temperature and that at higher pressures diamond would become the stable form of carbon.

Another example is the reaction of isomerization of 1-butene to cyclobutane, which has not been observed directly as yet. The heat of combustion of each hydrocarbon has been measured[4,5] and subtraction of the results yields

$$\text{1-butene (g)} = \text{cyclobutane (g)} \qquad \Delta H_{298} = 26.8 \text{ kJ mol}^{-1}$$

The entropy of each substance[4,6] has also been obtained by the methods of Chapter 6 from low-temperature heat-capacity measurements on the solid and liquid together with heat-of-fusion and heat-of-vaporization measurements. There results

$$\Delta S_{298} = -40.2 \text{ J K}^{-1} \text{ mol}^{-1}$$

$$\Delta G_{298} = 26.8 - 298.15(-40.2) \times 10^{-3} = 38.8 \text{ kJ mol}^{-1}$$

Consequently, we may predict that at 298 K cyclobutane would react almost completely to 1-butene.

Gibbs Energy Functions and Heat Capacity Equations

The equation for the change of $\Delta G°$ with temperature was derived in Chapter 3 along with the relationship of $\Delta G°$ to the equilibrium constant for a reaction. We now consider several efficient procedures involving the integration of this equation. The situation where the change in heat capacity can be given with a simple expression will be considered as one case. We first rewrite the basic equations in a convenient form for the case where there are no transitions or other thermal anomalies in the temperature range considered. If such transitions occur, the required terms must be added.

$$\Delta H_T° = \Delta H_{Tr}° + \Delta(H_T° - H_{Tr}°) = \Delta H_{Tr}° + \int_{Tr}^{T} \Delta C_P \, dT' \tag{8-4}$$

$$\Delta S_T° = \Delta S_{Tr}° + \Delta(S_T° - S_{Tr}°) = \Delta S_{Tr}° + \int_{Tr}^{T} \left(\frac{\Delta C_P}{T'}\right) dT' \tag{8-5}$$

$$\ln K_T = -\frac{\Delta G_T°}{RT} = \frac{\Delta S_T°}{R} - \frac{\Delta H_T°}{RT} \tag{8-6}$$

Here Tr is a convenient reference temperature; either 0 K or 298.15 K is usually chosen, but it can be any temperature. If $Tr = 0$ and there is no frozen-in entropy, then $\Delta S_{Tr}° = \Delta S_0° = 0$, and Eq. (8-5) represents just the usual third-law entropy calculation.

In situations where the necessary spectroscopic and molecular data are available and the statistical mechanical equations are used, it is more convenient to take $Tr = 0$, whereupon

$$\ln K_T = -\frac{\Delta H_0°}{RT} - \Delta\left(\frac{G_T° - H_0°}{RT}\right) \tag{8-7}$$

with the function $-(G_T° - H_0°)/RT$, often called the *Gibbs energy function*, calculated from Eqs. (5-8) and (5-25) or the methods of Chapter 20 for more complex cases. The change in enthalpy at 0 K must still be determined.

If there are several accurate values of K_T, a value of $\Delta H_0°$ can be calculated from each, and the equality of these values is a test of the validity of various parts of the treatment. The last column of Table 8-1 shows the result for this method for the reaction of formation of $H_2S(g)$ from $S_2(g)$ and $H_2(g)$. The values of $-(G_T° - H_0°)/RT$ for each substance are given in Appendix 6, or with closer spacing in T in sources cited at the end of this chapter. The average value of $\Delta H_0°$ from this method is reported, $-9859R$ in this case. It is often called the *third-law* $\Delta H_0°$ because $\Delta S_0° = 0$ was implicit in this treatment.

In some cases there may be errors in the molecular properties used in the calculation of the Gibbs energy function. This can be detected by a trend with T in the

ΔH_0° values or by an unacceptable difference between the second-law and third-law values of ΔH_0°. For the reaction forming $H_2S(g)$ there was no trend in the ΔH_0° values in Table 8-1, and the second-law value agrees within its larger uncertainty of at least 500 K.

The equilibrium between Pb and Pb_2 in lead vapor near 1000 K provides an example of a serious disagreement between second-law and third-law values. As shown in Table 8-2 the originally reported treatment yielded second-law and third-law values of ΔH_0° that differed by several times their total uncertainties. But in the calculation of $-(G_T^\circ - H_0^\circ)/T$ for Pb_2 the original authors[7] had made the assumption that the ground electronic state was $^3\Sigma^-$ by analogy with Si_2. This yields $g_0 = 3$ in Eq. (5-14). Subsequently, with the development of relativistic quantum theory for molecules, it was shown that the three states of $^3\Sigma^-$ were widely split for Pb_2 with a single 0_g^+ state very much lower in energy than the others. With only a single electronic state populated at 1000 K, the entropy and Gibbs energy function were smaller by $R \ln 3$ than had been assumed. With this correction the second-law and third-law values of ΔH_0° agree very satisfactorily and one can have confidence in the result.

For many reactions the enthalpy change has been measured calorimetrically and ΔH° is available at 298.15 K. Also, where statistically calculated Gibbs energy functions are not available, one often has third-law entropies for 298.15 K. Then it is convenient to adopt $Tr = 298.15$ K in Eqs. (8-4) and (8-5) and to combine these equations to yield

$$-\left(\frac{G_T^\circ - H_{298}^\circ}{RT}\right) = \frac{S_{298}^\circ}{T} + \int_{298}^T \frac{C_P^\circ}{R}\left(\frac{1}{T'} - \frac{1}{T}\right) dT' \qquad (8\text{-}8)$$

Here as elsewhere 298 implies the more precise 298.15.

A value for the Gibbs energy function with $Tr = 0$ is readily converted to $Tr = 298.15$ K as follows:

$$-\left(\frac{G_T^\circ - H_{298}^\circ}{RT}\right) = -\left(\frac{G_T^\circ - H_0^\circ}{RT}\right) + \frac{H_{298}^\circ - H_0^\circ}{RT} \qquad (8\text{-}9)$$

The value of $(H_{298}^\circ - H_0^\circ)$ is always available from the same basic information that yielded the Gibbs energy function. It is also interesting to note explicitly that at 298.15 K

$$-\left(\frac{G_T^\circ - H_{298}^\circ}{RT}\right)_{298.15} = \frac{S_{298.15}^\circ}{R} \qquad (8\text{-}10)$$

TABLE 8-2
Dissociation equilibrium of Pb_2 near 1000 K

Original third-law ΔH_0°/kJ mol^{-1}	74 ± 1
Second-law ΔH_0°/kJ mol^{-1}	84 ± 2
Revised third-law ΔH_0°/kJ mol^{-1}	82.5 ± 1
Recommended value ΔH_0°/kJ mol^{-1}	83 ± 1

Let us now assume a simple form of equation for ΔC_P° and proceed to evaluate the integral in Eq. (8-8). A very widely used expression is

$$\Delta C_P = \Delta a + \Delta b T + \Delta c T^{-2} \qquad (8\text{-}11)*$$

whereupon

$$-\Delta\left(\frac{G_T^\circ - H_{Tr}^\circ}{RT}\right) = \frac{\Delta S_{Tr}^\circ}{R} + \frac{\Delta a}{R}\left[\ln\left(\frac{T}{Tr}\right) - \frac{T - Tr}{T}\right] + \frac{\Delta b}{2R}\frac{(T - Tr)^2}{T}$$

$$+ \frac{\Delta c}{2R}\frac{(T - Tr)^2}{T^2 Tr^2} \qquad (8\text{-}12)*$$

and

$$\ln K_T = -\frac{\Delta H_{Tr}^\circ}{RT} - \Delta\left(\frac{G_T^\circ - H_{Tr}^\circ}{RT}\right) \qquad (8\text{-}13)$$

For an example we take $Tr = 298.15$ and the reaction

$$\mathrm{CaSO_4 \cdot 2H_2O(s)} = \mathrm{CaSO_4(s)} + \mathrm{2H_2O(l)}$$

Robie et al.[9] measured the heat capacity from 8 K for each solid and calculated the entropy for 298.15 K with the results for S_{298}°/R of 23.31 and 12.92 for the dihydrate and anhydrous solids, respectively. The value for $\mathrm{2H_2O(l)}$ from Chase et al.[10] is 16.83, which yields $\Delta S_{298}^\circ/R = 6.44$. For temperatures near 298 K Robie et al. adopt a simple linear equation for the change in heat capacity:

$$\Delta C_P^\circ/R = 14.88 - 0.024T$$

Thus, $\Delta a/R = 14.88$, $\Delta b/R = -0.024\,\mathrm{K}^{-1}$ and $\Delta c = 0$, whereupon Eq. (8-12)* yields the change in the Gibbs energy function near 298 K:

$$-\Delta\left(\frac{G_T^\circ - H_{298}^\circ}{RT}\right) = \frac{3370}{T} - 86.07 + 14.88\ln T - 0.012T$$

Robie et al.[9] next adopt the value 2028 ± 10 K for $\Delta H_{298}^\circ/R$, which was determined from heat of solution measurements by Kelley et al.[11] The dissolution was into aqueous HCl rather than pure water because the solubility in water is so small that accuracy is not attainable. Then from Eq. (8-13) Robie et al. calculate that $K_T = 1$, $\Delta G^\circ = 0$ at 314.7 ± 3.5 K. Posnjak[12] measured the solubility of gypsum (the dihydrate) and anhydrite (anhydrous $\mathrm{CaSO_4}$) in water and found them to be equal at 42°C, 315 K. Since the solubility of $\mathrm{CaSO_4}$ is small, the chemical potential of water in the saturated solution is only slightly less than that of pure water, and the agreement is excellent. The chemical potential in the saturated solution could be measured or calculated by methods to be described in later chapters, but such refinement would be meaningless in this case with the uncertainties in various measurements. Indeed, other investigators cited by Robie et al.[9] find appreciably different values for the solubility of anhydrite and for the temperature at which $\Delta G^\circ = 0$.

One could, of course, regard the equilibrium constant as the known quantity and solve Eq. (8-13) for ΔH_{298}°, but this would not be appropriate in view of the uncertainty in K_T. Rather one concludes that the Robie et al.[9] calculations offer strong support for the solubilities of Posnjak.[12]

It is possible to rearrange Eqs. (8-12)* and (8-13) so that a series of equilibrium-constant values can be adjusted so that a plot like Fig. 8-1 is strictly linear. One does this by transposing all terms other than constants or those depending on $1/T$ to yield a combination with $\ln K_T$ as follows:

$$\Sigma = -R \ln K_T + \Delta a \ln T + \frac{\Delta b T}{2} + \frac{\Delta c}{2T^2} \quad (8\text{-}14a)^*$$

$$= I + \frac{\Delta H_I}{T} \quad (8\text{-}14b)^*$$

Thus a plot of Σ vs. $1/T$ gives a straight line if the ΔC_P equation and the K_T values are accurate. This method is useful if ΔS_T° is not known independently but ΔC_P is known accurately in the range of the equilibrium measurements. After I and ΔH_I are determined from the intercept and slope, respectively, one can find ΔS° and ΔH° at a temperature within the range of validity of the equation for ΔC_P. Since this method is now rarely used, no example will be given. This description should be useful, however, when reading and evaluating the older literature where it was often used.

In certain recent studies of mineral reaction equilibria at high temperatures, heat-capacity equations more complex than Eq. (8-11)* have been used. Berman[13] and Holland and Powell[14] use different four-term combinations of the following terms:

$$C_1 + C_2 T + C_3 T^2 + C_4 T^{-1/2} + C_5 T^{-2} + C_6 T^{-3}$$

These more complex equations give greater accuracy when required over a wide range of temperature.

BIBLIOGRAPHY OF COMPILATIONS OF THERMODYNAMIC PROPERTIES OF PURE SUBSTANCES

In Appendix 6, tables of Gibbs-energy functions from 298 to 2000 K and values of heats of formation at 298 K are given for many substances, which allow calculation of ΔG° values over a range of temperature. These tables are restricted to rather common substances and generally to data of high accuracy. In addition, the following list presents a number of comprehensive compilations of Gibbs-energy functions and heats of formation that will be useful sources of information. Most of these compilations give values at smaller temperature intervals than Appendix 6 and therefore may allow easier and more accurate interpolation. After listing compilations that include Gibbs energy functions at various temperatures, several other compilations are added that present Gibbs energies, enthalpies, and entropies for 298.15 K only but include substances not present in the other tables. In some cases high-temperature heat-capacity expressions are included.

1. I. Barin et al., *Thermochemical Data of Pure Substances*, V.H.S. Publishers, New York, 1989, presents Gibbs energy functions and enthalpies referenced to

298.15 K together with entropies, heat capacities, heats and Gibbs energies of formation and values of $\log K_f$, the equilibrium constant of formation, all at 100-degree intervals from 298.15 K upward. This is essentially a compilation combining other compilations and covers an extremely wide range of substances. References to the source compilations are given, but the source compilation must be consulted for information on the accuracy and basis of the information. Earlier editions have the title, *Thermochemical Properties of Inorganic Substances*.

2. M. W. Chase, Jr., et al., "JANAF Thermochemical Tables," 3d edn., *J. Phys. Chem. Ref. Data*, **14**, supplement no. 1, 1985, presents Gibbs energy functions and enthalpies referenced to 298.15 K, together with entropies, heat capacities, heats and Gibbs energies of formation and values of $\log K_f$, the equilibrium constant of formation, all at 100-degree intervals extending from 0 K to as high a temperature as the underlying data allow. Included are values for crystalline solids, liquids, and gases as well as brief statements about the sources of the basic data and their accuracy and reliability. A large number of elements and inorganic compounds are included as well as some small organic molecules, but there are surprising omissions; for example, neither Ag nor any silver compound is included.

3. J. D. Cox, D. D. Wagman, and V. A. Medvedev, *CODATA Key Values for Thermodynamics*, Hemisphere, New York, 1989. CODATA was an international program for the selection of values for key substances and this is the final report. Gibbs energy and enthalpy functions based on 0 K, together with entropies and heat capacities, are given at 100-degree intervals to 4000 K (or less as appropriate) and at 298.15 K for many elements and common compounds. In separate tables are values at 298.15 K for the entropy and the $\Delta H°$ of formation from the elements; some aqueous ions are included in these last tables.

4. *TRC Thermodynamic Tables*: (a) *Hydrocarbons* and (b) *Nonhydrocarbons*, Thermodynamics Research Center, K. N. Marsh, Director, Texas A&M University, College Station, Texas, are loose-leaf, frequently updated tables including all thermodynamic properties for organic compounds and reference elements and compounds. The Gibbs energy functions are referenced to 0 K. Currently, there are 12 volumes for hydrocarbons and 9 volumes for nonhydrocarbons. These tables are successors of a single volume of 1953 that is still useful and more convenient when adequate: *Selected Values of Physical and Thermodynamic Properties of Hydrocarbons and Related Compounds*, F. D. Rossini, K. S. Pitzer, R. L. Arnett, R. M. Braun, and G. C. Pimentel, Carnegie Press, Pittsburgh, Pennsylvania.

5. R. A. Robie, B. S. Hemingway, and J. R. Fisher, *Thermodynamic Properties of Minerals and Related Substances at 298.15 K and 1 Bar (10^5 Pascals) Pressure and at Higher Temperatures*, U.S. Geological Survey Bulletin 1452, U.S. Government Printing Office, Washington, 1978, gives Gibbs energies and enthalpies referenced to 298.15 K together with entropies, heat capacities, and $\Delta_f H°$, $\Delta_f G°$, and K_f of formation at 100-degree intervals to 1800 K or a lower temperature as appropriate for minerals and reference substances. A separate table gives additional information for 298.15 K only.

6. L. V. Gurvich, I. V. Veyts, and C. B. Alcock, *Thermodynamic Properties of Individual Substances*, 4th edn., Hemisphere, New York, 1989 (for vol. 1). This is an English translation, with some revision, of a Russian edition of 1978; it is appearing as a series of pairs of volumes with the extensive tables in the second volume of each pair. Note that the symbol $\Phi°$ is used for $-(G°-H_0°)/T$. Included are $C_P°$, Φ, $S°$, $(H°-H_0°)$, and $\log K_f°$ at 100-degree intervals, together with an equation for $\Phi°(T)$ and values of $\Delta_f H_0°$. It is ironic that the principal uncertainty in some tables for gases lies in the value of R at the time of initial calculation; if $\Phi°/R$, $S°/R$, etc., had been tabulated, this uncertainty would have been avoided.

7. R. Hultgren et al., *Selected Values of the Thermodynamic Properties of the Elements*, American Society for Metals, Metals Park, Ohio, 1973. The section on each element includes tables of the information available including Gibbs energy, entropy, enthalpy, etc., also vapor pressures and boiling and melting points. The section editor is identified with date and his comments on the selection of best values is included. Although the dates are 1973 and earlier, this is still valuable.

8. D. D. Wagman et al., The NBS Tables of Chemical Thermodynamic Properties. Selected Values for Inorganic and C_1 and C_2 Organic Substances in SI Units, *J. Phys. Chem. Ref. Data*, **11**, supplement no. 2, 1982, gives values of various functions at 298.15 K and of $\Delta_f H_0°$ for a very large array of substances. Although the volume is dated 1982, most tables are from earlier NBS publications after conversion to SI units from calories. Individual tables carry original dates in the range 1963–1968.

9. O. Kubaschewski and C. B. Alcock, *Metallurgical Thermochemistry*, 5th edn., Pergamon Press, New York, 1979, gives heats of formation and entropies at 298 K, heat capacity equations, vapor pressures, and Gibbs energies for metals and intermetallic compounds.

10. K. K. Kelley, "Contributions to the Data on Theoretical Metallurgy: XIII, High-Temperature Heat-content, Heat-capacity, and Entropy Data for Inorganic Compounds," *U.S. Bur. Mines Bull. 584*, 1960, and XIV, Entropies of Inorganic Substances, by K. K. Kelley and E. G. King, *U.S. Bur. Mines Bull. 592*, 1961, present a critical summary of $S_{298}°$, $H_T° - H_{298}°$, and $S_T° - S_{298}°$ data at 100-degree intervals from which $(G_T° - H_{298}°)/T$ values can readily be calculated. Theoretical Metallurgy: XII, Heats and Free Energies of Formation of Inorganic Oxides, *U.S. Bur. Mines Bull. 542*, 1954, by J. P. Coughlin, presents a critical summary of $\Delta H°$ and $\Delta G°$ data and estimated values up to 2000 K. Although not updated for recent contributions, these bulletins are still useful and for some quantities give more background than other tables.

11. F. L. Oetting, executive editor, *The Chemical Thermodynamics of Actinide Elements and Compounds*, International Atomic Energy Agency, Vienna. This is a series of separately authored and titled volumes. An example is vol. 8, J. Fuger et al., "The Actinide Halides," 1983.

12. *Atomic Energy Review*; Special Issues, International Atomic Energy Agency, Vienna. This is a series of volumes under different editors and authors; each gives thermodynamic properties as well as phase diagrams for compounds of a particular element. Examples are vol. 5, "Thorium," O. Kubaschewski (ed.), and vol. 7, "Molybdenum," L. Brewer (ed.).

13. L. Brewer, L. A. Bromley, P. W. Gilles, and N. L. Lofgren, Papers 6 and 8, in L. L. Quill (ed.), *The Chemistry and Metallurgy of Miscellaneous Materials—Thermodynamics*, McGraw-Hill, New York, 1950. Experimental and estimated Gibbs-energy functions based on 298 K are tabulated for virtually all known halides at 298.15, 500, 1000, and 1500 K, together with $\Delta_f H_{298}$ for formation from the elements.

14. R. G. Berman, "Internally Consistent Thermodynamic Data for Minerals in the System $Na_2O-K_2O-CaO-MgO-FeO-Fe_2O_3-Al_2O_3-SiO_2-TiO_2-H_2O-CO_2$," *J. Petrology*, **29**, Part 2, 445 (1988) presents for 67 minerals values of $\Delta_f G°$, $\Delta_f H°$, $S°$, $V°$, at 298.15 K, together with parameters for a simple expression for the heat capacity in the range above 298 K.

15. T. J. B. Holland and R. Powell, "An Enlarged and Updated Internally Consistent Thermodynamic Dataset with Uncertainties and Correlations: the System $K_2O-Na_2O-CaO-MgO-MnO-FeO-Fe_2O_3-Al_2O_3-TiO_2-SiO_2-C-H_2-O_2$," *J. Metamorphic Geol.*, **8**, 89 (1990) presents for 123 mineral and fluid end-members similar information to that in the Berman compilation above.

16. Finally, we note that the *Journal of Physical and Chemical Reference Data* is a rich source of values of thermodynamic properties. For example, volume 22 (1993) includes extensive tables for organic nitrogen compounds from K. N. Marsh and associates[15] (see entry (4) above), values from Steele and Chirico[16] for mono-olefins larger than C_4, and papers[17,18] from Domalski and associates on the estimation of properties for various groups of compounds.

REFERENCES

1. G. Preuner and W. Schupp, *Z. physik. Chem.*, **68**, 157 (1909).
2. M. Randall and F. R. Bichowsky, *J. Am. Chem. Soc.*, **40**, 368 (1918).
3. W. M. Latimer, *The Oxidation States of the Elements and Their Potentials in Aqueous Solutions*, 2d edn., Prentice-Hall, Englewood, Cliffs, N.J., 1952.
4. Values for 1-butene are summarized in F. D. Rossini et al., *Selected Values of Physical and Thermodynamic Properties of Hydrocarbons*, Carnegie Press, Pittsburgh, 1953 edition and subsequent revisions.
5. S. Kaarsemaker and J. Coops, *Rec. trav. chim.*, **71**, 261 (1952).
6. G. W. Rathjens, *J. Am. Chem. Soc.*, **75**, 5629, 5634 (1953).
7. K. A. Gingerich. D. L. Cooke, and F. Miller, *J. Chem. Phys.*, **64**, 4027 (1976).
8. K. S. Pitzer, *J. Chem. Phys.*, **74**, 3078 (1981).
9. R. A. Robie, S. Russell-Robinson, and B. S. Hemingway, *Thermochemica Acta*, **139**, 67 (1989).
10. M. W. Chase, Jr., C. A. Davies, J. R. Downey, Jr., D. J. Frurip, R. A. McDonald, and A. N. Syverud, *J. Phys. Chem. Ref. Data*, **14**, supplement 1 (1985).
11. K. K. Kelley, J. C. Southard, and C. T. Anderson, "Thermodynamic Properties of Gypsum and Its Dehydration Products," *U.S. Bur. Mines Tech. Paper, 625* (1941).
12. E. Posnjak, *Am. J. Sci.*, **35A**, 247 (1938).
13. R. G. Berman, *J. Petrology*, **29**, part 2, 445 (1988).
14. T. J. B. Holland and R. Powell, *J. Metamorphic Geol.*, **8**, 89 (1990).

15. A. Das, M. Frenkel, N. A. M. Gadalla, S. Kudchadker, K. N. Marsh, A. S. Rodgers, and R. C. Wilhoit, *J. Phys. Chem. Ref. Data*, **22**, 659 (1993).
16. W. V. Steele and R. D. Chirico, *J. Phys. Chem. Ref. Data*, **22**, 377 (1993).
17. V. Ruzicka, Jr., and E. S. Domalski, *J. Phys. Chem. Ref. Data*, **22**, 597, 619 (1993).
18. E. S. Domalski and E. D. Hearing, *J. Phys. Chem. Ref. Data*, **22**, 805 (1993).

PROBLEMS

8-1. (a) From the value 1890 K for θ_D for diamond and the table and equation in Appendix 2, calculate $-(G_T^\circ - H_0^\circ)/RT$ at 298.15 K, 500 K, and 1000 K; also $(H_{298}^\circ - H_0^\circ)$; and combine these values to yield values of $-(G_T^\circ - H_{298}^\circ)/RT$. (b) With the value in Appendix 6 for $-(G_T^\circ - H_{298}^\circ)/RT$ for graphite and the value $\Delta H_{298}^\circ = -1895$ J mol^{-1} for the reaction diamond = graphite, calculate ΔG° at 298.15 K, 500 K, and 1000 K, and compare with the value -2896 J mol^{-1} for 298.15 K given in the text.

8-2. Use the Gibbs-energy functions from Appendix 6 to calculate $(\Delta G_{2000}^\circ - \Delta H_0^\circ)/T$ and ΔG_{2000}° for

$$CO(g) + \tfrac{1}{2}O_2(g) = CO_2(g)$$

Calculate the equilibrium constant and the percentage dissociation of CO_2 at this temperature and a total pressure of 1 bar.

8-3. For the "water-gas reaction" $CO_2 + H_2 = CO + H_2O(g)$, it has been reported that equilibrium is reached at 1538 K when the four gases have, respectively, the partial pressures 0.10, 0.10, 0.10, and 0.24 bar. (a) Calculate K and ΔG_{1538}°. (b) From the Gibbs-energy functions and $\Delta_f H_{298}^\circ$ values in Appendix 6, calculate ΔG_{1538}° for comparison.

8-4. For the reaction $I_2(g) = 2I(g)$, M. L. Perlman and G. K. Rollefson [*J. Chem. Phys.*, **9**, 362 (1941)] reported the following equilibrium constant values:

T/K	872	973	1073	1173	1274
K	1.81×10^{-4}	1.80×10^{-3}	0.0108	0.0480	0.167

Combine these with Gibbs-energy functions from Appendix 6 to calculate ΔH_{298}° at each temperature. Are these values of ΔH_{298}° the same within reasonable uncertainties in your opinion?

8-5. From the values in Appendix 6, calculate the pressures of $KCl(g)$ and $K_2Cl_2(g)$ in equilibrium with liquid KCl at 1500 K.

8-6. (a) Use the data of Appendix 6 to calculate the equilibrium constant at 500 K for the reaction $Hg(g) + Cl_2(g) = HgCl_2(g)$. (b) Calculate the percent of $HgCl_2$ gas at 0.01 bar decomposed to the elements at 500 K. (c) Of the halogen gas that results from the decomposition of the $HgCl_2$ at 0.01 bar, what fraction is monatomic chlorine?

8-7. From the Gibbs-energy functions and other properties given in Appendix 6, calculate ΔG° for the reaction $4Fe_{0.947}O(s) = Fe_3O_4(s) + 0.788Fe(s)$ at 298.15 K and at 1000 K. What do you conclude about the stability of $Fe_{0.947}O(s)$ at these temperatures?

CHAPTER 9

REAL GASES AND THE FLUID STATE

The ideal gas law is a satisfactory approximation for the $P-V-T$ properties of gases at low pressures with molar volumes large compared to the volumes of the molecules and where the attractive intermolecular potentials are small in comparison to thermal energy. The initial deviations of real gases from ideal behavior are readily represented by simple equations. In the separate range of low temperatures, the properties of liquids can be represented by simple empirical equations giving the linear dependence of density on temperature and pressure. The behavior of a fluid outside of these two regions, however, is complex and no single equation of state has been found that is fully satisfactory for representing the properties of fluids over the full range of temperature and pressure.

In this chapter, we first describe the general pattern of properties of the fluid state ranging from nearly ideal gases to dense liquids. At temperatures near critical or above, the appropriate reference state is the ideal gas. At lower temperatures one uses both the ideal gas and the liquid reference states. General thermodynamic relationships are next illustrated with relatively simple equations of state that further simplify for the low-pressure range. Then, in later sections, other empirical equations of state that describe the entire P–V–T range from ideal gas to dense liquid are described. Finally, we discuss the problem of estimation of fluid properties for substances for which there are only a few measurements. Only pure fluids are considered in this chapter; fluids with two or more components are treated in later chapters.

Throughout this chapter it will be convenient to consider the quantity z defined as

$$z = PV/nRT = P/\rho RT \qquad (9\text{-}1)$$

which is called the *compression factor*, or more frequently but less appropriately, the *compressibility factor*. Here $\rho = n/V$ is the molar density. For an ideal gas, $z = 1.0$, and for real gases z may be either greater or smaller than unity. Figure 9-1 shows the typical pattern of the compression factor in the range of low and moderate pressures. At low pressures the isotherms are nearly straight lines with slope negative at low temperature but becoming positive at high temperature. At somewhat higher pressures the isotherms show some curvature downward at low temperatures and upward at higher temperatures. A plot of z vs. ρ shows a similar pattern. This behavior is represented effectively by a power series in either P or ρ.

Throughout this and subsequent chapters and appendices on fluids, we ordinarily use the density rather than the volume as a variable. We believe this has several advantages but note that much of the literature on this field uses the molar volume $V_m = 1/\rho$ instead.

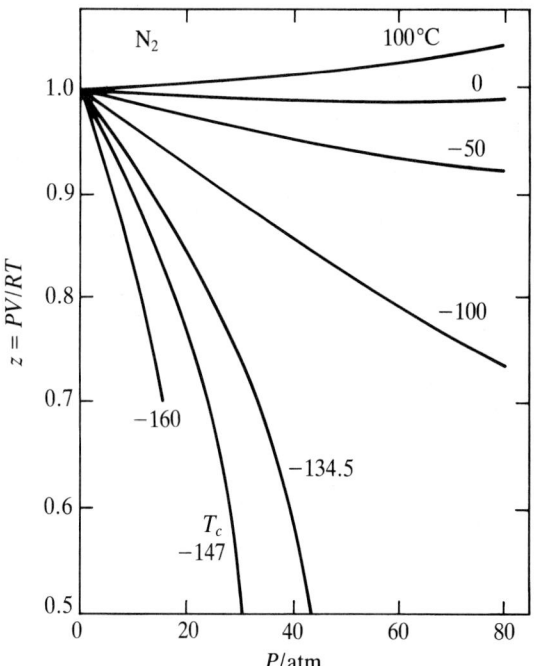

FIGURE 9-1
The compression factor for nitrogen.

Virial Equations of State

In 1901 Onnes first used a power series to express the P–V–T properties of a gas. Such an expression is known as a virial equation of state, and a convenient form is

$$\frac{P}{RT} = \rho + B\rho^2 + C\rho^3 + D\rho^4 + \cdots \qquad (9\text{-}2)$$

The unit coefficient for the first term on the right is the first virial coefficient and represents the ideal gas law. B is the second virial coefficient, C the third, etc.; each is a function of temperature, and is specific to a particular gas or vapor. The dimension of B is volume per mole, of C is $\text{vol}^2\,\text{mol}^{-2}$, etc. In terms of the compression factor, this series becomes

$$z = 1 + B\rho + C\rho^2 + D\rho^3 + \cdots \qquad (9\text{-}3)$$

Often it is more convenient to take P and T as the independent variables (instead of ρ and T). Then one adopts the series

$$z = 1 + B'P + C'P^2 + D'P^3 + \cdots \qquad (9\text{-}4)$$

By comparison of Eqs. (9-3) and (9-4) in the limit of small ρ and P, one can show for the second virial coefficient that $B' = B/RT$. The higher coefficients are related, $C' = (C-B^2)/(RT)^2$, but in an increasingly complex manner. In view of their approximate nature, one should use caution in converting the higher coefficients from one series for use in the other.

The coefficients B, C, etc., of Eqs. (9-2) and (9-3) are related in statistical mechanics to molecular properties and, in a strict sense, only they and not B', C', etc., should be called virial coefficients.

Fluids Over Wide Ranges of T and P

Figure 9-2 shows a series of isotherms of pressure as a function of density for a fluid. The example is methane but the pattern is universal. At high temperatures, P increases with ρ in a simple, nearly linear manner that is readily expressed by a virial series, Eq. (9-4). With decreasing temperature, the pattern becomes more complex with an intermediate point of minimum slope. The point where this minimum slope becomes zero is designated the critical point; it is the unique point for the entire diagram. The critical conditions are

$$\left(\frac{\partial P}{\partial \rho}\right)_T = 0 \qquad \left(\frac{\partial^2 P}{\partial \rho^2}\right)_T = 0 \qquad (9\text{-}5)$$

or

$$\left(\frac{\partial P}{\partial V}\right)_T = 0 \qquad \left(\frac{\partial^2 P}{\partial V^2}\right)_T = 0 \qquad (9\text{-}6)$$

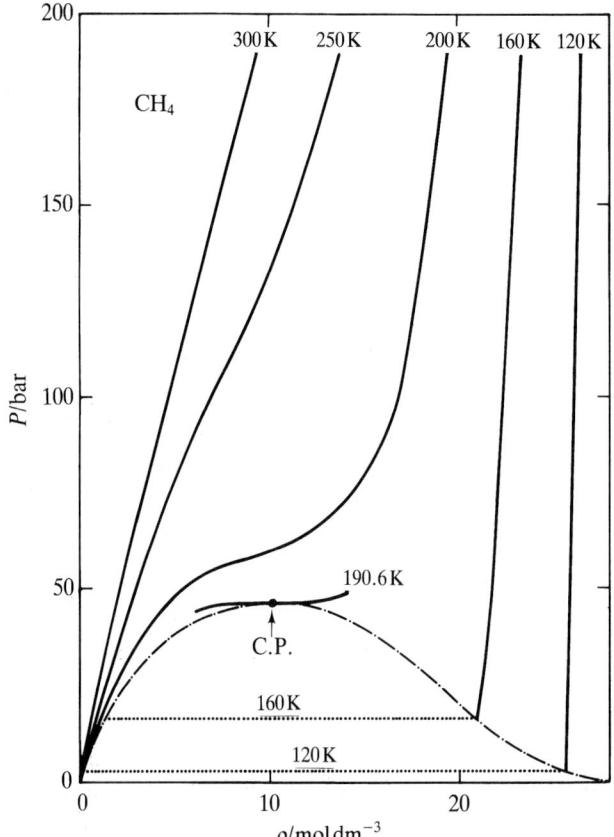

FIGURE 9-2
$P-\rho$ isotherms illustrating the critical region and differences at other temperatures.

Below the critical temperature one has a range of density in which the homogeneous fluid is unstable and the equilibrium state is that of the coexisting saturated vapor and saturated liquid. The conditions are

$$T_{vap} = T_{liq} \quad P_{vap} = P_{liq} \quad \mu_{vap} = \mu_{liq} \tag{9-7}$$

The pressure equality is obvious; the equality of chemical potential requires a more complex calculation, but we shall see that it is implicit in a comprehensive equation of state.

From Figure 9-2, we see that the vapor isotherms are simple (nearly linear) at temperatures well below critical and have a simple shape below a density one half that of the critical point at any temperature. This is the region in which a simple virial representation is always satisfactory; indeed, the use of just the second virial coefficient frequently suffices. Also, as noted at the beginning of this chapter, the isotherms become simple for small ranges of density near that of the saturated liquid at

Reduced Variables and Equations

The critical point is the unique point in the P–V–T diagram, and van der Waals proposed in 1873 that the properties of various fluids might be the same if compared on a reduced basis, i.e., as functions of $T_r = T/T_c$, $P_r = P/P_c$, $V_r = V/V_c$, $\rho_r = \rho/\rho_c$. Figure 9-3 shows isotherms of the compression factor on the reduced basis. It is clear that N_2, CO_2, and H_2O have similar reduced isotherms, but that they are not exactly the same. Nevertheless, it has been found to be very useful to compare fluid properties on a reduced basis and to develop equations in terms of reduced properties. It is found that there are small groups of fluids that have quite accurately the same properties on a reduced basis. The simplest and most important group comprises the heavier noble gases (Ar, Kr, Xe) and methane; they are sometimes denoted simple fluids.

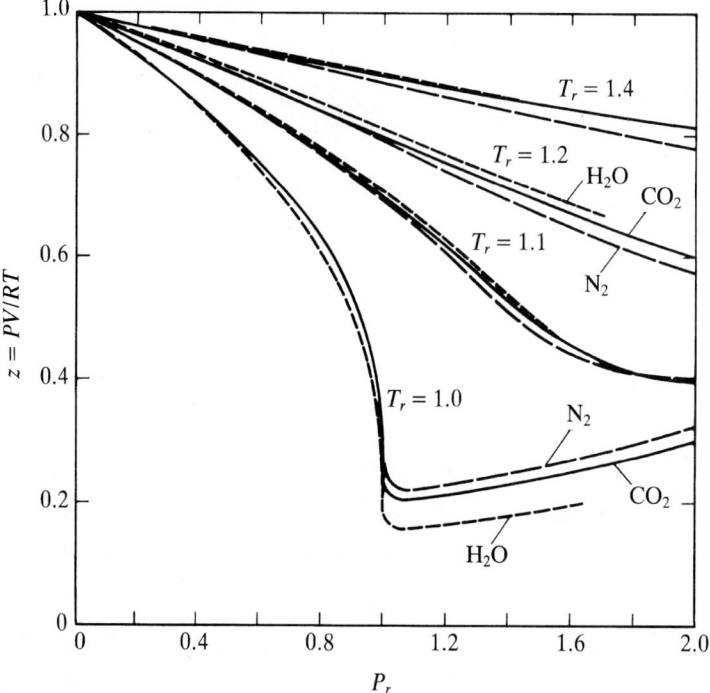

FIGURE 9-3
A test of the hypothesis of corresponding states. Isotherms of nitrogen (long-dashed lines), carbon dioxide (solid lines), and steam (short-dashed lines) as functions of reduced pressure.

Evaluation of Virial Coefficients

Various methods may be used to determine the second virial coefficient or an entire series of virial coefficients from a sequence of pressure–volume measurements at constant temperature. One may select one or another of Eqs. (9-2) to (9-4), retaining an appropriate number of terms, and then carry out a least-squares regression of the data to evaluate the virial coefficients retained, B, C, ... It is important to consider the experimental uncertainties and to assign appropriate weights to the measured values. Also, in most cases, one should recalculate with a different number of coefficients in order to ascertain the degree to which the individual coefficients are unambiguously determined.

An alternate procedure is to rearrange Eq. (9-3) to

$$\frac{z-1}{\rho} = B + C\rho + D\rho^2 + \cdots \tag{9-8}$$

Then a graph of $(z-1)/\rho$ vs. ρ will give B as the intercept at $\rho = 0$. This shows visually the uncertainty and the possible redundancy of B with C.

One may carry this procedure another step and write

$$\frac{z-1}{\rho^2} - \frac{B}{\rho} = C + D\rho + \cdots \tag{9-9}$$

whereupon extrapolation of the left-hand side gives C as the intercept at $\rho = 0$. The work of Douslin[1] on methane is a good example of the use of Eqs. (9-8) and (9-9).

Most investigators, however, have carried out direct regressions of Eq. (9-2) with varying numbers of terms. The work of Michels et al.[2] on argon is an excellent example that shows clearly that the second virial coefficient is relatively insensitive to the number of terms. The values of the third virial coefficient, however, vary significantly between different treatments of the same data. For example, at 188.15 K, they used a three-term equation (ending in C) fitted to the data up to 50 bar and obtained $B = -54.83 \text{ cm}^3 \text{mol}^{-1}$ and $C = 1791 \text{ cm}^6 \text{mol}^{-2}$. They also fitted a seven-term equation to their full array of data extending to 1000 bar. In the latter case, B is almost unchanged at -54.27 while C is greatly decreased to 1416; also the next term D is large and positive and is expressing some of the effect included in C for the shorter equation.

At temperatures well below critical, where the saturated vapor pressure is small, only the second virial coefficient is readily determined. Surface adsorption effects can be more important than those of the third virial coefficient. The study of argon in this range by Fender and Halsey[3] indicates how these problems can be handled.

There are now accurate data for the second virial coefficient for many gases; these have been evaluated and summarized by Dymond and Smith,[4] who also give third virial coefficient values when available.

Second Virial Coefficients: Temperature Dependence and Other Properties

The second virial coefficients for several gases are shown in Fig. 9-4 on a reduced basis. The reduced virial coefficient is

$$B_r = \frac{nB}{V_c} = B\rho_c \qquad (9\text{-}10)$$

while the reduced temperature is $T_r = T/T_c$. At reduced temperatures above about 2.7, the virial coefficient is positive, reflecting a dominant effect of repulsive forces. The temperature at which $B = 0$ is known as the Boyle point. At lower temperatures, the relative importance of intermolecular attractive forces increases. The pattern of negative values below $T_r \cong 0.8$ represents the increasing proportion of double molecules in the vapor. For normal fluids, this dimerization does not proceed very far

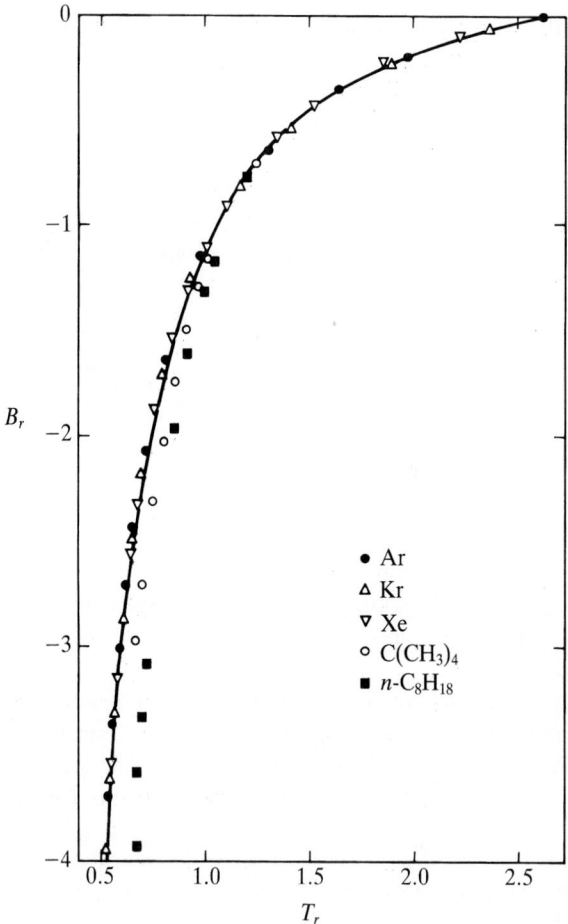

FIGURE 9-4
The reduced second virial coefficient for several gases together with the curve of Eq. (9-12)*, which represents the behavior of the inert gases Ar, Kr, and Xe.

before the vapor condenses to the liquid. Thus, the virial coefficient expression is usually more satisfactory than a dimerization reaction equilibrium treatment because the repulsive-force effects are naturally included in the former but not in the latter.

There are special cases, usually involving hydrogen bonding, where the dimerization or polymerization proceeds to a large extent before liquefaction. Such examples and the appropriate methods of treatment are described in a separate section below where the statistical mechanics of the second virial coefficient is also given.

It is not surprising that the points for Ar, Kr, and Xe fall close to a single curve on Fig. 9-4. For other normal fluids the reduced virial coefficient is even more negative at low T_r, as illustrated by the points for $C(CH_3)_4$ and n-octane. This pattern of behavior can be expressed in a general equation described in a later section and in Appendix 4. For the present, we take as an example the equation[5] fitting the data for the simple fluids Ar, Kr, and Xe on a reduced basis:

$$B_r = 0.442\,26 - \frac{0.980\,97}{T_r} - \frac{0.611\,14}{T_r^2} - \frac{0.005\,156}{T_r^6} \qquad (9\text{-}11)*$$

If we insert the critical properties, $T_c = 150.86$ K and $\rho_c = 13.41$ mol dm^{-3}, the second virial coefficient of argon is

$$B = 32.98 - \frac{1.1036 \times 10^4}{T} - \frac{1.0372 \times 10^6}{T^2} - \frac{4.53 \times 10^{12}}{T^6} \quad \text{cm}^3\,\text{mol}^{-1}$$

$$(9\text{-}12)*$$

This equation fits the experimental values[4] over the entire range from 80 to 1000 K within experimental uncertainty.

Expansion and Compression Coefficients

The coefficient of expansion, the compressibility and related quantities are obtained by differentiation after appropriate rearrangement of the virial equations. From Eq. (9-4) after multiplication by RT/P, one obtains

$$\left(\frac{\partial V_m}{\partial T}\right)_P = \frac{R}{P} + RT\left(\frac{dB'}{dT} + P\frac{dC'}{dT} + P^2\frac{dD'}{dT} + \cdots\right) \qquad (9\text{-}13)$$

$$\left(\frac{\partial V_m}{\partial P}\right)_T = RT\left(-\frac{1}{P^2} + C' + 2D'P + \cdots\right) \qquad (9\text{-}14)$$

As usually defined, the coefficient of expansion α is

$$\alpha = \frac{1}{V}\left(\frac{\partial V}{\partial T}\right)_P = \frac{1}{zT} + RT\rho\left(\frac{dB'}{dT} + P\frac{dC'}{dT} + P^2\frac{dD'}{dT} \cdots\right) \qquad (9\text{-}15)$$

while the compressibility is

$$\kappa = -\frac{1}{V}\left(\frac{\partial V}{\partial P}\right)_T = RT\rho\left(\frac{1}{P^2} - C' - 2D'P + \cdots\right) \qquad (9\text{-}16)$$

In a similar manner, Eq. (9-2) yields

$$\left(\frac{\partial P}{\partial T}\right)_\rho = R\left[\rho + \left(B + T\frac{dB}{dT}\right)\rho^2 + \left(C + T\frac{dC}{dT}\right)\rho^3 + \left(D + T\frac{dD}{dT}\right)\rho^4 + \cdots\right] \quad (9\text{-}17)$$

$$\left(\frac{\partial P}{\partial \rho}\right)_T = RT[1 + 2B\rho + 3C\rho^2 + 4D\rho^3 + \cdots] \quad (9\text{-}18)$$

The last two results can be combined to obtain $(\partial V/\partial T)_P$ as a function of density:

$$\left(\frac{\partial V_m}{\partial T}\right)_P = \frac{1}{\rho^2}\frac{(\partial P/\partial T)_\rho}{(\partial P/\partial \rho)_T} \quad (9\text{-}19)$$

THERMODYNAMIC RELATIONSHIPS

We now apply the general thermodynamic equations to fluids, both on a general basis and then with the virial equations as examples.

Gibbs and Helmholtz Energies

All of the other thermodynamic properties of a fluid can be expressed very conveniently in terms of the Gibbs energy, if the independent variables are T and P, or in terms of the Helmholtz energy, if the independent variables are T and ρ (or V). For the region of low and moderate pressures shown in Fig. 9-1, either set of variables is fully satisfactory and both methods will be described. When the two-phase, vapor–liquid region is involved, however, the Helmholtz energy is a much more satisfactory basis since A and P remain single-valued functions of T and ρ. In this region ρ (or V) is a multiple-valued function of T and P. One can tabulate or write separate equations for vapor and liquid properties, including the Gibbs energy, as functions of P. But if one wishes a single comprehensive equation of state, one uses T and ρ (or V) as variables and the Helmholtz energy as the parent function.

The parent function in either case is the sum of two functions, one for the ideal-gas or standard-state properties and the other for the departure of the fluid from ideal-gas behavior. Thus, on a molar basis and considering first the ideal gas,

$$G^{\text{id}}(T,P) = G^\circ(T) + RT \ln\left(\frac{P}{P^\circ}\right) \quad (9\text{-}20)$$

where P° is the pressure of the standard state, normally 1 bar, and $G^\circ(T)$ is the Gibbs energy in the standard state, which is a function of T. Then, at constant T,

$$\left[\frac{\partial (G - G^{\text{id}})}{\partial P}\right]_T = V_m - \frac{RT}{P} \quad (9\text{-}21)$$

$$G(T,P) - G^{\text{id}}(T,P) = \int_0^P \left(V_m - \frac{RT}{P'}\right)dP' \quad (9\text{-}22)$$

$$G(T,P) = G°(T) + RT\left[\ln\left(\frac{P}{P°}\right) + \int_0^P (z-1)\,d\ln P'\right] \quad (9\text{-}23)$$

Insertion of the virial equation for z yields

$$G(T,P) = G°(T) + RT\left[\ln\left(\frac{P}{P°}\right) + B'P + \frac{C'P^2}{2} + \frac{D'P^3}{3} + \cdots\right] \quad (9\text{-}24)$$

The corresponding treatment for the molar Helmholtz energy also begins with the ideal gas,

$$A^{id}(T,\rho) = A°(T) + RT\ln\left(\frac{\rho}{\rho°}\right)$$

$$= G°(T) + RT[-1 + \ln(\rho RT)] \quad (9\text{-}25)$$

since $\rho° = P°/RT$, $P° = 1$ bar, and $G° = A° + RT$.

The initial formulation of an equation of state is often given for the Helmholtz energy instead of the compression factor. In particular, one first divides A into an ideal and a residual term,

$$A = A^{id} + A^{res} \quad (9\text{-}26)$$

where A^{id} was given in Eq. (9-25). Then

$$z = 1 + \rho\left[\frac{\partial(A^{res}/RT)}{\partial\rho}\right]_T \quad (9\text{-}27a)$$

If z is initially defined, one has the converse relationship

$$\frac{A^{res}}{RT} = \int_0^\rho (z-1)\,d\ln\rho' \quad (9\text{-}27b)$$

This yields A^{res} if the equation was first formulated for z, and the complete equation for the Helmholtz energy is

$$A(T,\rho) = G°(T) + RT[-1 + \ln(\rho RT)] + A^{res} \quad (9\text{-}28)$$

If we substitute the virial equation, we obtain from Eqs. (9-27b) and (9-28)

$$A(T,\rho) = G°(T) + RT\left[-1 + \ln(\rho RT) + B\rho + \frac{C\rho^2}{2} + \frac{D\rho^3}{3} + \cdots\right] \quad (9\text{-}29)$$

Note that, in the last two equations, R must be in bars for the quantity ρRT, while in the factor outside the brackets R has the units used for $G°$ and A.

Other Thermodynamic Functions

The chemical potential, fugacity, entropy, enthalpy and other functions are readily derived from the expressions for the molar Gibbs and Helmholtz energies. All extensive quantities in this section are for one mole. The results with pressure as the independent variable are

$$\frac{\mu}{RT} = \frac{G}{RT} = \frac{G°}{RT} + \ln\left(\frac{P}{P°}\right) + \int_0^P (z-1)\, d\ln P' \qquad (9\text{-}23)$$

$$\ln\left(\frac{f}{P}\right) = \int_0^P (z-1)\, d\ln P' \qquad (9\text{-}30)$$

$$\frac{H - H°}{R} = -T^2 \int_0^P \left(\frac{\partial z}{\partial T}\right)_P d\ln P' \qquad (9\text{-}31)$$

$$\frac{S - S°}{R} = -\ln\left(\frac{P}{P°}\right) + \int_0^P \left[1 - z - T\left(\frac{\partial z}{\partial T}\right)_P\right] d\ln P' \qquad (9\text{-}32)$$

$$\frac{C_P - C_P°}{R} = -T \int_0^P \left[2\left(\frac{\partial z}{\partial T}\right)_P + T\left(\frac{\partial^2 z}{\partial T^2}\right)_P\right] d\ln P' \qquad (9\text{-}33)$$

If the virial series in pressure, Eq. (9-5), is substituted, one obtains after integration,

$$\frac{\mu - G°}{RT} = \ln\left(\frac{P}{P°}\right) + B'P + \frac{C'P^2}{2} + \frac{D'P^3}{3} + \cdots \qquad (9\text{-}24)$$

$$\ln\left(\frac{f}{P}\right) = PB' + \frac{P^2 C'}{2} + \frac{P^3 D'}{3} + \cdots \qquad (9\text{-}34)$$

For calculations with density as the independent variable, we first recall Eqs. (9-25)–(9-27) for the Helmholtz energy. Then the standard relationships yield expressions for other molar functions as follows.

$$\frac{A - G°}{RT} = -1 + \ln(\rho RT) + \int_0^\rho (z-1)\, d\ln \rho' \qquad (9\text{-}28)$$

$$\frac{\mu - G°}{RT} = \ln(\rho RT) + z - 1 + \int_0^\rho (z-1)\, d\ln \rho' \qquad (9\text{-}35)$$

$$\ln\left(\frac{f}{P}\right) = z - 1 - \ln z + \int_0^\rho (z-1)\, d\ln \rho' \qquad (9\text{-}36)$$

$$\frac{U - U°}{R} = -T^2 \int_0^\rho \left(\frac{\partial z}{\partial T}\right)_\rho d\ln \rho' \qquad (9\text{-}37)$$

$$\frac{H - H°}{R} = T(z-1) - T^2 \int_0^\rho \left(\frac{\partial z}{\partial T}\right)_\rho d\ln \rho' \qquad (9\text{-}38)$$

$$\frac{S - S^\circ}{R} = -\ln(\rho RT) + \int_0^\rho \left[1 - z - T\left(\frac{\partial z}{\partial T}\right)_\rho\right] d\ln\rho' \quad (9\text{-}39)$$

$$\frac{C_V - C_V^\circ}{R} = -T \int_0^\rho \left[2\left(\frac{\partial z}{\partial T}\right)_\rho + T\left(\frac{\partial^2 z}{\partial T^2}\right)_\rho\right] d\ln\rho' \quad (9\text{-}40)$$

Substitution of Eq. (9-27b) yields alternate expressions involving A^{res}/RT.

Now any equation of state may be substituted for z (and A^{res}) and the desired function obtained. For example, if the virial series in density, Eq. (9-3), is chosen, one obtains the following.

$$\frac{A - G^\circ}{RT} = -1 + \ln(\rho RT) + B\rho + \frac{C\rho^2}{2} + \frac{D\rho^3}{3} + \cdots \quad (9\text{-}29)$$

$$\frac{\mu - G^\circ}{RT} = \ln(\rho RT) + 2B\rho + \frac{3C\rho^2}{2} + \frac{4D\rho^3}{3} + \cdots \quad (9\text{-}41)$$

$$\ln f = \ln(\rho RT) + 2B\rho + \frac{3C\rho^2}{2} + \frac{4D\rho^3}{3} + \cdots \quad (9\text{-}42)$$

$$\ln\left(\frac{f}{P}\right) = B\rho + \frac{(C + B^2)\rho^2}{2} + \left(BC + \frac{D}{3}\right)\rho^3 + \cdots \quad (9\text{-}43)$$

$$U - U^\circ = -RT^2 \left[\rho\left(\frac{dB}{dT}\right) + \frac{\rho^2}{2}\left(\frac{dC}{dT}\right) + \cdots\right] \quad (9\text{-}44)$$

$$H - H^\circ = RT\left[\rho\left(B - T\frac{dB}{dT}\right) + \frac{\rho^2}{2}\left(C - T\frac{dC}{dT}\right) + \cdots\right] \quad (9\text{-}45)$$

$$S - S^\circ = -R\left[\ln(\rho RT) + \rho\left(B + T\frac{dB}{dT}\right) + \frac{\rho^2}{2}\left(C + T\frac{dC}{dT}\right) + \cdots\right] \quad (9\text{-}46)$$

$$C_V - C_V^\circ = -RT\left[\rho\left(2\frac{dB}{dT} + T\frac{d^2B}{dT^2}\right) + \rho^2\left(\frac{dC}{dT} + \frac{T}{2}\frac{d^2C}{dT^2}\right) + \cdots\right] \quad (9\text{-}47)$$

The fugacity coefficient $\phi = f/P$ is directly available from Eqs. (9-30), (9-34), (9-36), and (9-43). Also, as noted above, if the standard state is the perfect gas at 1 bar, then R must be in bars for the quantity (ρRT) in these equations. Where R appears elsewhere, it has the same units as the function calculated: A, G, etc.

REAL GASES AT LOW PRESSURE; THE SECOND VIRIAL COEFFICIENT

In many practical situations with gases at or near atmospheric pressure, the deviation from the ideal gas law is small but is significant for accurate work. Under these conditions the higher virial coefficients can be ignored and attention concentrated on the second virial coefficient. The simplification of equations through (9-47) is obvious; in particular, Eq. (9-43) becomes

$$\frac{f}{P} = \exp(B\rho) \cong 1 + B\rho \qquad (9\text{-}43)^*$$

where the second equality arises from the expansion of the exponential. But from Eq. (9-3) we see that in this approximation

$$\frac{f}{P} = z = \frac{P}{\rho RT} \qquad (9\text{-}48)^*$$

and, if we note that the ideal pressure at ρ and T is $P_i = \rho RT$, we find

$$\frac{f}{P} = \frac{P}{P_i} \qquad (9\text{-}49)^*$$

Thus, the actual pressure lies between the fugacity and the pressure calculated from the ideal-gas law and is the geometrical mean of the two.

Very simple and useful equations are obtained by making the substitution $\rho = P/RT$, which is valid at the second virial level, into Eq. (9-45) and then differentiating:

$$H - H° = P\left(B - T\frac{dB}{dT}\right) \qquad (9\text{-}50)^*$$

$$C_P - C_P° = -PT\frac{d^2B}{dT^2} \qquad (9\text{-}51)^*$$

It is interesting to test the error caused by the omission of the third virial coefficient. In an earlier section, the values $B = -54.8 \text{ cm}^3\text{ mol}^{-1}$ and $C = 1791 \text{ cm}^6\text{ mol}^{-2}$ were given for argon at 188.15 K, which is somewhat above the critical temperature at 150.86 K. At 188 K and 10 bar, the density is about $6.4 \times 10^{-4} \text{ mol cm}^{-3}$ and the second virial term reduces z by 3.5%. For these conditions, the third virial term raises z by 0.07%, which is negligible for many purposes.

An example of application of the equation in pressure is obtained from the work of Otto, Michels, and Wouters[6] for nitrogen to 200 bar. Their expression for 0°C may be converted to the basis of pressure in bars and density in mol cm^{-3}.

$$\frac{P}{\rho} = 22\,712 - 10.281P + 0.064\,337P^2 + 4.9943 \times 10^{-7}P^4$$

$$- 1.2318 \times 10^{-11}P^6 + 9.20 \times 10^{-17}P^8 \qquad (9\text{-}52)^*$$

From Eqs. (9-4), and (9-34)*, we then obtain

$$\ln\left(\frac{f}{P}\right) = \ln\phi = -4.527 \times 10^{-4}P + 1.4164 \times 10^{-6}P^2 + 5.497 \times 10^{-12}P^4$$
$$- 9.039 \times 10^{-17}P^6 + 5.06 \times 10^{-22}P^8 \qquad (9\text{-}53)*$$

The change of heat capacity with pressure in the low-pressure range is sometimes of interest and its calculation will serve as an example of other derived quantities. If, for example, we substitute Eq. (9-12)* for argon into Eq. (9-51)* and express the results as C_P/R, we obtain in bar^{-1}

$$\left(\frac{\partial C_P/R}{\partial P}\right)_T = \frac{265.5}{T^2} + \frac{7.485 \times 10^4}{T^3} + \frac{2.29 \times 10^{12}}{T^7} \qquad (9\text{-}54)*$$

This effect for argon is only 0.1 bar^{-1} at 100 K and decreases rapidly with increase in temperature. For larger and more complex molecules, the effect is larger, as we show in a later section.

EQUATIONS OF STATE VALID OVER THE FULL RANGE OF T AND P

The simplest equation that yields all of the qualitative features of real fluids is that of van der Waals. In terms of density, it is

$$P = \frac{RT\rho}{1 - b\rho} - a\rho^2 \qquad (9\text{-}55)*$$

Here a and b are parameters to be adjusted for each substance. The first term involving b includes the repulsive interactions, while the second term represents the intermolecular attractive forces.

It is interesting to apply certain transformations to this simple case. The conditions for the critical point Eq. (9-5) yield values for the parameters in terms of the critical temperature T_c and density ρ_c:

$$b = \frac{1}{3\rho_c} \qquad a = \frac{9RT_c}{8\rho_c} \qquad \text{also} \qquad \rho_c = \frac{8P_c}{3RT_c}$$

The molar Helmholtz energy is readily obtained from Eqs. (9-27b) and (9-28):

$$\frac{A^{\text{res}}}{RT} = -\ln(1 - b\rho) - \frac{a\rho}{RT} \qquad (9\text{-}56)*$$

$$A = G° + RT[-1 + \ln(\rho RT) - \ln(1 - b\rho)] - a\rho \qquad (9\text{-}57a)*$$

The chemical potential or molar Gibbs energy is then

$$\mu = G = A + \frac{P}{\rho}$$

$$P = \mu° + RT\left(\ln(\rho RT) - \ln(1 - b\rho) + \frac{b\rho}{1 - b\rho}\right) - 2a\rho \qquad (9\text{-}57b)*$$

Other properties can now be calculated as desired using Eq. (9-56)*.

If T and P are the independent variables, the van der Waals equation can be solved noniteratively for the density as a cubic:

$$ab\rho^3 - a\rho^2 + (RT + bP)\rho - P = 0 \qquad (9\text{-}55a)*$$

Unfortunately, the van der Waals equation is not useful as an accurate representation of real fluid properties. For example, the compression factor at the critical point $z_c = P_c/RT_c\rho_c$ is 0.375, while for simple fluids such as Ar, it is 0.291 and it is still smaller for most other real fluids. This 30% error is typical for the van der Waals equation, except at very low densities, where it reduces to the ideal gas law.

Calculation of the Pressure and the Densities of Coexisting Vapor and Liquid

One can simultaneously apply the conditions of equality of pressure and chemical potential, Eq. (9-7), in an iterative calculation. With modern computers there is no difficulty, but it is useful to know about a graphical method that yields the desired vapor pressure and the densities of both liquid and vapor in a single procedure. One plots the Helmholtz energy as a function of molar volume $(1/\rho)$ at constant temperature, as shown in Figure 9-5. At the two points connected by a common tangent, the slope is the same, which demonstrates the equality of pressure and its value. Also, the molar Gibbs energy is the same because $\Delta G = \Delta A + \Delta(PV) = 0$. Thus, the densities corresponding to the two points of common tangency are the densities of the saturated vapor and liquid. The curve in Figure 9-5 was calculated for the van der Waals equation with $T = 0.8\ T_c$.

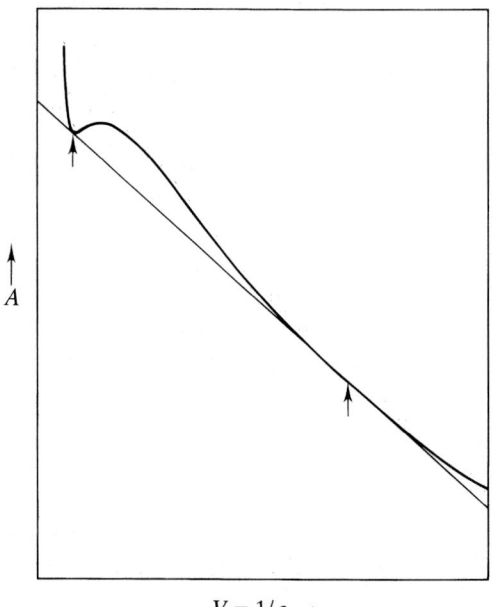

FIGURE 9-5
Example of the common tangent method of determining the densities of the coexisting vapor and liquid.

Improved Equations of State

A very large number of forms have been proposed for equations of state for fluids. In 1949 Redlich and Kwong[7] identified the most serious deficiencies of the van der Waals equation and proposed very simple remedies. The *Redlich–Kwong* (RK) equation is

$$P = \frac{RT\rho}{1 - b\rho} - \frac{a\rho^2}{(1 + b\rho)T^{1/2}} \qquad (9\text{-}58)*$$

The introduction of $T^{1/2}$ in the denominator of the second term greatly improves the temperature dependence of the second virial coefficient and the vapor pressure, while the factor $(1 + b\rho)$ improves the density dependence of the pressure and other properties. This equation is essentially as simple to use as the van der Waals equation with only two parameters, a and b, and a cubic form in density. It also integrates as easily to the Helmholtz energy or other functions via A^{res}.

$$\frac{A^{res}}{RT} = -\ln(1 - b\rho) - \left(\frac{a}{bRT^{3/2}}\right)\ln(1 + b\rho) \qquad (9\text{-}59)*$$

$$A = G° + RT[-1 + \ln(\rho RT) - \ln(1 - b\rho)] - \left(\frac{a}{bT^{1/2}}\right)\ln(1 + b\rho) \qquad (9\text{-}60)*$$

In terms of the critical properties, the parameters are

$$a = \frac{0.427\,48R^2T_c^{2.5}}{P_c} \qquad b = \frac{0.086\,64RT_c}{P_c} \qquad z_c = \frac{1}{3}$$

The Redlich–Kwong equation is accurate enough to be useful for many purposes, especially if its parameters are adjusted to fit the properties of particular interest. While values of a and b can be obtained from the critical temperature and density, the RK equation departs substantially from real fluid properties in the critical region. Thus, better values of a and b can often be determined from other data (see Appendix 4). It is for simple gases such as N_2, O_2, CH_4, and Ar that the RK equation is relatively satisfactory; larger deviations occur for fluids with molecules of larger size or significant polarity. In the years after its proposal, many modifications were suggested with more complex dependencies on T and ρ in the second term. These equations retain the first term and the overall cubic dependency on ρ but do involve a third parameter. Usually, this is the acentric factor, which is described below. Several of these equations are listed in Appendix 4. These and other similar equations are also discussed in reviews listed at the end of this chapter.

In order to represent the best measurements within experimental accuracy, very complex equations are required. While modern computers handle such equations without difficulty for pure fluids, equations with many parameters can give grossly erroneous results if extrapolated beyond their database. Also, there are severe theoretical as well as computational difficulties in their use for mixed fluids.

Since the precision of measurements for mixtures is generally less than that for pure fluids, one seeks an equation that is both adequate in accuracy for a mixed system yet simple enough to be used in a theoretically rigorous manner. An example of such an equation is that of Anderko and Pitzer.[8] It combines accurate expressions for the temperature dependency of the second, third, and fourth virial coefficients with a simple term for all higher-order repulsive effects. This equation is

$$z = \frac{1 + c\rho}{1 - b\rho} + \alpha\rho + \beta\rho^2 + \gamma\rho^3 \qquad (9\text{-}61)*$$

This form of repulsive term yields good agreement at high densities with minimal complexity. Integration to the Helmholtz energy is straightforward. Here b is a single constant, but there are temperature-dependent expressions for the other terms:

$$c = c_0 + \frac{c_1}{T} + \frac{c_2}{T^2} \qquad (9\text{-}61a)*$$

$$\alpha = \alpha_0 + \frac{\alpha_1}{T} + \frac{\alpha_2}{T^2} + \frac{\alpha_3}{T^6} \qquad (9\text{-}61b)*$$

$$\beta = \beta_0 + \frac{\beta_1}{T} + \frac{\beta_2}{T^2} + \frac{\beta_3}{T^6} \qquad (9\text{-}61c)*$$

$$\gamma = \gamma_0 + \frac{\gamma_1}{T} + \frac{\gamma_2}{T^2} + \frac{\gamma_3}{T^6} \qquad (9\text{-}61d)*$$

This equation has been shown to fit accurately various normal fluids with parameters predictable from the critical properties and the acentric factor. Also, it fits H_2O but with parameters specific to that substance. Numerical values of parameters are given in Appendix 4 for normal fluids.

A different approach for an equation of intermediate accuracy is that proposed in 1940 by Benedict, Webb, and Rubin (BWR).[9] It was a great advance at that time and constitutes the starting point for a variety of extensions (eBWR). Here the van der Waals type of repulsive term is discarded and in its place there are both high-order virial coefficients and an exponential expression. The original BWR equation is

$$z = 1 + \left((c_1 + \frac{c_2}{T} + \frac{c_3}{T^3} \right)\rho + \left[c_4 + \frac{c_5}{T} + \frac{c_6(1 + \gamma\rho^2)}{T^3} \exp(-\gamma\rho^2) \right]\rho^2$$

$$+ \frac{c_7}{T}\rho^5 \qquad (9\text{-}62)*$$

With eight parameters, it was very successful for normal fluids up to a density about $1.8\rho_c$. The exponential term was chosen to be integrable to a simple form for the Helmholtz energy.

In order to maintain accuracy at higher density (above $1.8\rho_c$), additional terms of the same type have been added by various investigators. A 33-parameter eBWR equation was used by Younglove and Ely[10] to treat the extensive data for methane, ethane, propane, and both butanes. Still more elaborate equations have been developed by Hill[11] and by Saul and Wagner[12] for H_2O.

For mixed fluids, the original BWR equation was applied in a rigorous manner and gave good results up to densities about 1.8 times critical. But the eBWR equations with 30 or more terms, which are required for accuracy over the full density range, yield such complex equations, if rigorously extended for mixtures, that they are seldom used in that form. Instead, an approximate extension to mixtures is often used that violates the rigorously derived composition dependency of the second virial coefficient. Thus the Anderko–Pitzer equation, or its equivalent, is recommended for mixed fluids.

The most difficult region of ρ–T space to represent accurately is that near the critical point. Indeed, there are special effects related to fluctuations that cannot be represented rigorously by a mathematically analytic equation; these are considered in Appendix 5. For practical purposes, however, an analytic equation is more convenient and, with enough terms, can be made as accurate as desired. Schmidt and Wagner[13] examined this problem thoroughly and recommend a system for the choice of terms for an eBWR equation.

Corresponding States; Reduced Equations of State

From the complexity of fluid behavior and the large number of parameters in an accurate equation of state, it is apparent that a very large number of measurements must be made to fully define the properties of a single fluid. But various fluids show similarities; hence, one seeks to use those aspects to obtain good estimates for additional fluids from only a few measurements.

We saw in Fig. 9-3 that curves for the compression factor for various fluids have exactly the same pattern but differ somewhat quantitatively when presented in terms of the reduced variables $T_r = T/T_c$, $P_r = P/P_c$, and $\rho_r = \rho/\rho_c$. If these curves agree exactly, the properties are said to follow *corresponding states*. Only small groups of fluids, such as Ar, Kr, Xe, CH_4 actually follow corresponding states to nearly full experimental accuracy. The similarity of the reduced isotherms for very different fluids, however, encourages one to seek an extension of the corresponding-states concept that yields more accurate correlations. Such a method would then provide predictions of great value in the absence of extensive and laborious measurements on each fluid.

One can show by statistical mechanics[14,15] that certain molecular properties will yield exact corresponding-states behavior. The simplest category of this type comprises spherical particles (atoms), whose interparticle potentials are given by a universal function with individual energy and distance scaling factors. The particle mass must also be large enough that quantum effects on translational motion are negligible. The heavier noble gases Ar, Kr, Xe conform quite closely to these criteria and do follow corresponding states quite accurately. To this group CH_4 can be added; its departure from spherical symmetry is not significant for fluid properties. These four, Ar, Kr, Xe, CH_4, are termed "simple fluids" and their behavior is often used as a reference in studies of more complex fluids.

Various types of molecular shapes and molecular dipole moments might be expected to cause different deviations from the macroscopic properties of simple fluids. It was found,[16] however, that the theoretical reduced second virial coefficients for a wide variety of molecular types fall into a single family of curves that may be characterized by a single parameter. The only exception noted was that of molecules with large dipole moments, although there are unusual types of abnormality that were not tested. The molecules falling into this single family are just those commonly called normal liquids or fluids.

Before proceeding to more detailed consideration of general equations for all normal fluids, we should emphasize that there are very important fluids such as H_2O that do not fall in the "normal" category. Excluded from the normal category are metallic fluids as well as those where there are strong hydrogen bonds or large dipole moments, as well as He, H_2, and to a lesser degree Ne, because of quantum effects. Equations of the eBWR type, as well as Eq. (9-61)*, are appropriate for such cases, but the parameters must be evaluated for each particular fluid. Simpler equations are also appropriate when the highest accuracy is not required.

Normal Fluids; Acentric Factor

The theory thus suggests an extension of the corresponding-states correlation involving a third parameter.[17-20] The slope of the reduced vapor pressure curve is the most sensitive property upon which to base the third parameter, and it has the additional advantage that vapor pressures are readily measured with high accuracy. An arbitrary but convenient definition[19] is based upon the reduced vapor pressure at a point well removed from the critical point and takes the form

$$\omega = -\log\left(\frac{P_s}{P_c}\right) - 1.000 \tag{9-63}$$

where

$$P_s = \text{vapor pressure at } T_r = 0.700.$$

The form is chosen to make $\omega = 0$ for the simple fluids, Ar, Kr, and Xe, with simple spherical molecules. Other normal fluids have small positive values of ω. The name *acentric factor* was adopted to indicate that the factor measures the deviation of the intermolecular potential function from that of the simple spherical molecules. Riedel[18] chose a different definition for his parameter α_k, but the results are equivalent with $\alpha_k = 5.808 + 4.93\omega$.

Any property of the fluid, in reduced or dimensionless form, is assumed to be given by a function of the three variables: reduced temperature, reduced density or reduced pressure, and acentric factor. For example, the compression factor may be written as either

$$z = z(T_r, \rho_r, \omega) \tag{9-64}$$

or

$$z = z(T_r, P_r, \omega) \tag{9-65}$$

In its original development and testing, the T_r, P_r, ω basis was used with graphical methods and the results were presented in tabular form.[19,20] These results are still useful for simple estimation without elaborate computation and are presented in Appendix 3. Equations of state, such as Eqs. (9-61)* and (9-62)*, can be used with each parameter dependent on the acentric factor and with the reduced temperature and density as the variables. Examples of this type including details for the Anderko–Pitzer equation, Eq. (9-61)*, are given in Appendix 4.

Equation for the Second Virial Coefficient

As noted earlier in this chapter, there are many situations where a gas is nearly ideal but a small correction is needed that is given by the second virial coefficient. Reduced equations with the acentric factor as a third characterizing parameter have been given by several investigators.[5,20,21] A recent equation of Schreiber and Pitzer[5] was based upon evaluated experimental data of Dymond and Smith;[4] it is

$$B_r = 0.442\,259 + 0.725\,650\,\omega - (0.980\,970 - 0.218\,714\,\omega)/T_r$$
$$- (0.611\,142 + 1.249\,76\,\omega)/T_r^2 - (0.005\,156\,24 + 0.189\,187\,\omega)/T_r^6$$

(9-66a)*

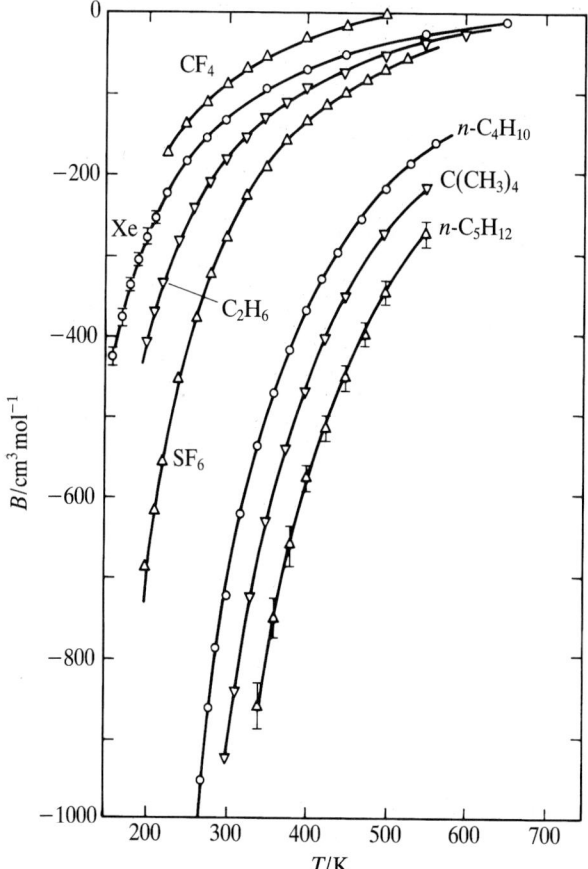

FIGURE 9-6
Comparison of experimental second virial coefficients with curves given by Eq. (9-66)* in the acentric factor system.

TABLE 9-1
Pressure effect on the molar heat capacity of gaseous *n*-heptane

T	$(\partial C_P/\partial P)/R$	
K	Exp.[22]	Eq. (9-66)
357.1	0.72	0.77
373.2	0.57	0.60
400.4	0.35	0.41
434.4	0.22	0.28
466.1	0.13	0.20

with

$$B = B_r V_c \quad (9\text{-}66b)^*$$

or, more precisely, since values of V_c are often less accurate than those of P_c,

$$B = \frac{RT_c z_c^* B_r}{P_c} \quad (9\text{-}66c)^*$$

$$z_c^* = 0.2905 - 0.0787\,\omega \quad (9\text{-}66d)^*$$

Figure 9-6 shows the excellent agreement between the experimental second virial coefficients for several gases with curves calculated from Eq. (9-66)*. The special case of Eq. (9-66)* with $\omega = 0$ was given above as an equation for Ar, Kr, and Xe.

The effect of pressure on the heat capacity provides a very severe test of an equation for the second virial coefficient since it requires the second temperature derivative, Eq. (9-51)*. Table 9-1 compares experimental measurements of $(\partial C_P/\partial P)$ for *n*-heptane from Waddington et al.[22] with the predictions from Eq. (9-66)*. While the agreement is not perfect, it does not seriously exceed the experimental uncertainties.

Appropriate manipulations, as indicated by equations given earlier in this chapter, when applied to Eq. (9-66)* yield fugacities, entropies, enthalpies, etc., for the low-pressure range where the second virial coefficient is an adequate approximation.

MOLECULAR MODELS FOR FLUID PROPERTIES

In this section, we consider first those aspects of the statistical theory of fluids that are simple and useful for thermodynamic applications and later those special situations where one or more association equilibria constitute a useful model.

There is an extensive structure of statistical mechanics providing relationships between the intermolecular potentials and fluid properties. There are two general methods. One involves very extensive calculations in which a substantial number of molecules are considered simultaneously with statistically guided changes of location (or removal or addition) of individual molecules. Both Monte Carlo (MC) and molecular dynamics (md) methods are of this type and are called "simulations." Many very interesting results have been obtained for very simple molecular models. The computational requirements are very great, however, and exceed reasonable limits at present for realistic models for most fluids with polyatomic molecules. With further increases in computer power, this situation may change, but we will not consider these methods further at this point.

The second general method is that of the virial expansion, in which binary interactions of molecules determine the second virial coefficient, binary and ternary interactions the third coefficient, etc. The equation for the second virial coefficient for spherical molecules is very simple:[23]

$$B = 2\pi N_A \int_0^\infty (1 - e^{-u(r)/kT}) r^2 \, dr \qquad (9\text{-}67)$$

Here N_A is the Avogadro number, k is the Boltzmann constant, and $u(r)$ is the intermolecular potential. For the third virial coefficient the expression is much more complex, even if the potential is assumed to be a sum of binary terms:

$$C = -\frac{8\pi^2 N_A^2}{3} \int_0^\infty \int_0^\infty \int_{|r_{12}-r_{13}|}^{r_{12}+r_{13}} f_{12} f_{13} f_{23} r_{12} r_{13} r_{23} \, dr_{12} \, dr_{13} \, dr_{23}$$

$$(9\text{-}68a)$$

with

$$f_{ij} = \exp[-u_{ij}(r)/kT] - 1 \qquad (9\text{-}68b)$$

If there is a triple interaction term in the potential, the expression is complicated further.

While the much more complex expressions can be written for the fourth and higher virial coefficients, they are not very useful. Indeed, a MC simulation is a more practical method than analytical integration for these coefficients for any realistic potentials.

Theoretical results are available for the second virial coefficient for a great many realistic intermolecular potentials.[21] The extension to nonspherical models is straightforward; the angular coordinates are introduced both in the potential and for the integration over all space. A potential that yields a very simple expression for B, and which, nevertheless, yields excellent agreement with experimental data, is the square well:

$$u(r) = \infty \quad \text{for } 0 < r < \sigma \qquad (9\text{-}69)$$
$$u(r) = -\varepsilon \quad \text{for } \sigma < r < R\sigma$$
$$u(r) = 0 \quad \text{for } r > R\sigma$$

Substitution of this potential into Eq. (9-67) yields

$$B = b_0 \{1 - (R^3 - 1)[\exp(\varepsilon/kT) - 1]\} \qquad (9\text{-}70a)*$$

$$b_0 = \frac{2\pi N_A \sigma^3}{3} \qquad (9\text{-}70b)*$$

The expression for the third virial coefficient is somewhat more complex but is also a closed analytic expression.[23]

For some purposes, a rearrangement of Eq. (9-70) is convenient:

$$B = b - c \exp(\varepsilon/kT) \qquad (9\text{-}71)*$$

with

$$b = b_0 R^3 \quad \text{and} \quad c = b_0(R^3 - 1)$$

If the three parameters b, c, and ε are adjusted, this equation fits second virial coefficient data very well. As we shall see below, the second term of Eq. (9-71)* corresponds to an association equilibrium between single and double molecules, with $-\varepsilon$ the energy change on dimerization, while c is related to the corresponding entropy change.

If the exponential in Eq. (9-71)* is expanded, one obtains

$$B = (b - c) - \frac{c\varepsilon}{kT} - \frac{c\varepsilon^2}{2k^2T^2} \cdots \qquad (9\text{-}71\text{a})*$$

One notes that this expansion yields terms of exactly the same form as the first three terms of Eq. (9-12)* and that the final term of (9-12)* serves to approximate the higher terms of Eq. (9-71a)*, which are small for all conditions of interest. Thus, the empirical success of both equations is understandable.

The effectiveness of the obviously incorrect square well potential in representing second virial coefficient data indicates clearly that additional information is required to define the true intermolecular potential. This has now been accomplished. Quantum theory shows that the potential at large distances is given by $-(\text{constant})r^{-6}$ with further terms in r^{-8}, r^{-10}, etc., all with negative coefficients. Theory also indicates that the potential will rise sharply at short distances. For the exact shape of the potential near its minimum, the most detailed information comes from emission spectra; the frequencies of over 100 lines for Ar_2 closely define the exact potential in this region. Molecular beam scattering data are also important. Thermal conductivity and viscosity data must also be fitted but they, like the second virial coefficient, are less sensitive to the exact shape of the potential. The solid curve in Fig. 9-7 shows the accurate Ar–Ar potential,[24] for which the mathematical expression is rather complex. There are a number of less complex but still quite accurate expressions[25] that would be indistinguishable in Fig. 9-7.

A very widely used potential with the correct qualitative characteristics including the r^{-6} behavior at large r is that proposed by Lennard-Jones (LJ),

$$u(r) = 4\varepsilon\left[\left(\frac{\sigma}{r}\right)^{12} - \left(\frac{\sigma}{r}\right)^6\right] \qquad (9\text{-}72\text{a})$$

in which ε is the depth of the potential well and σ the collision diameter. The potential minimum lies at $r_0 = \sigma 2^{1/6}$ and in terms of that distance

$$u(r) = \varepsilon\left[\left(\frac{r_0}{r}\right)^{12} - 2\left(\frac{r_0}{r}\right)^6\right] \qquad (9\text{-}72\text{b})$$

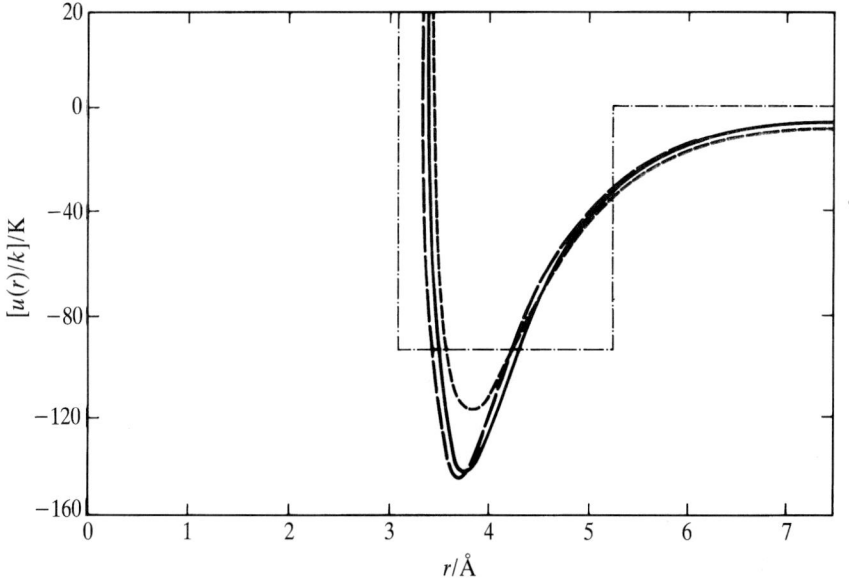

FIGURE 9-7
Binary interaction potentials for argon. The solid line is the accurate potential based on many properties. The long-dashed and the dot-dashed lines are for the Kihara core potential and the square well, respectively; both fit the second virial coefficient accurately. The short-dashed curve is the Lennard-Jones 6–12 potential, which gives a less accurate virial coefficient.

This potential, fitted to data for Ar, is shown as the short-dashed curve in Fig. 9-7. This potential, with only two adjustable parameters, does not yield really good agreement with the measured second virial coefficients. Indeed, the agreement for the square well with three parameters is very much better than for the LJ potential.

Kihara[26] made a very useful generalization of the LJ potential that includes nonspherical models. He assumed cores of various shapes and that the LJ 6–12 potential acted for the shortest distance between the cores. The results obtained by Kihara were important in the basis for the acentric factor theory for fluid properties.[14] For argon[27] the Kihara potential with a small spherical core of radius 0.179 Å, together with the values $r_0 = 3.321$ Å and $\varepsilon/k = 146.52$ K, yields the long-dashed curve in Fig. 9-7 which agrees very well with the accurate potential. This Kihara potential also yields second virial coefficients in essentially perfect agreement with experiment.

Danon and Pitzer[28] showed that the radius a of a spherical core correlates closely with the acentric factor according to the equation

$$\frac{a}{r_0} = 0.04 + 1.17\omega \qquad (9\text{-}73)*$$

They also report[28] calculations for tetrahedral cores for CH_4 and CF_4, a hexagonal core for C_6H_6, and thin rod or cylindrical cores for N_2 and CO_2. In addition, the effect of a quadrupole moment is considered for CO_2. For CF_4, C_6H_6, and N_2, the core sizes

TABLE 9-2
Kihara core potentials[28]

Molecule	$(\varepsilon/k)/K$	$r_0/\text{Å}$	$l/\text{Å}$
	(Tetrahedral: l is center to vertex distance)		
CH_4	175	3.75	0.28
CF_4	350	3.08	1.01
	(Hexagonal: l is length of one side)		
C_6C_6	990	3.00	1.16
	(Thin rod of length l)		
N_2	132	3.40	1.19

relate rather closely to the C–F, C–C, and N–N bond lengths, while for CH_4 the core is very small, as is shown in Table 9-2. For CO_2 the effect of the quadrupole moment is substantial. Other calculations for the core model have been reported by Tee et al.[27] and Prausnitz et al.[29]

While these results for the Kihara core model are very interesting, the use of an empirical equation with the acentric factor such as Eq. (9-70) remains the best and most convenient source of values for second virial coefficients for normal fluids. Conversely, where spectroscopic, molecular beam, and other molecular-level experiments are available, they yield the best intermolecular potentials.[24]

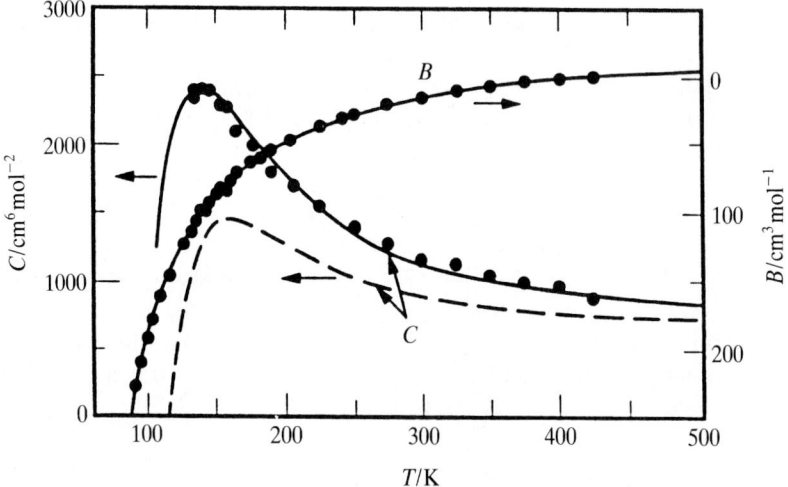

FIGURE 9-8
Second and third virial coefficients for argon with experimental points and curves calculated from the accurate binary potential. For C the dashed curve is for the binary potential only, while the solid curve includes the three-body potential.

Given an accurate two-body potential for a gas such as argon, one can calculate the third virial coefficient on the assumption that only binary forces are present from Eq. (9-68). The agreement with experimental third virial coefficients is not good (Fig. 9-8), and indicates that three-body forces are significant. But when the three-body terms predicted from quantum theory are introduced, reasonable agreement is obtained (Fig. 9-8). The more complex equations for third virial coefficients with three-body forces and related matters are considered by Hirschfelder et al.[23] and Barker and Henderson.[30]

Association Reaction Models

The attractive forces between molecules lead to their association, especially at low temperatures, into dimers, trimers, and larger clusters. Thus, one can express gas nonideality by the relationships for chemical equilibria for one or more reactions of the type

$$2A = A_2 \quad 3A = A_3 \quad nA = A_n$$

with

$$K_2 = \frac{P_2}{P_1^2}, \text{ etc.} \tag{9-74a}$$

and

$$\ln K_2 = \frac{\Delta S_2}{R} - \frac{\Delta H_2}{RT}, \text{ etc.} \tag{9-74b}*$$

where P_1, P_2, etc., are partial pressures, ΔS and ΔH are the entropy and enthalpy of association, respectively, and the change in heat capacity ΔC_P is neglected in the last equation. One may assume ideal gas behavior for the mixture of A, A_2, etc., or one may make a small empirical correction for the repulsive interactions at short distances. Similar equations can be written on a concentration basis with ΔC_V and ΔU replacing ΔC_P and ΔH. It seems probable that the approximation $\Delta C_P = 0$ is better than $\Delta C_V = 0$; consequently, we have shown the more familiar equations on the partial-pressure basis.

These association methods are particularly valuable where there are special bonding interactions that guide the selection of the species thought to be important. Hydrogen bonding is an important effect of this type (and will be the basis of two examples).

There is an extensive literature on equations of state including association. Some of these equations are readily applied to mixtures while others have been used only for pure fluids. We give first two general models that yield closed expressions for z and the Helmholtz energy and then two specific examples limited to low-pressure gases but illustrating different patterns of association.

One general assumption is that of a fixed association constant K for each step in successive association:

$$A_n + A = A_{n+1} \quad K = \frac{C_{n+1}}{C_1 C_n} \tag{9-75}*$$

with C_i the concentration of the ith species. With one or another of two approximations, a simple equation can be derived for z or the Helmholtz energy. The first basis involves separate repulsive and attractive terms as proposed by Heidemann and Prausnitz:[31]

$$z = z_{rep} + z_{attr} - 1 \qquad (9\text{-}76)*$$

Various detailed assumptions can be made for each term.

The second general basis, proposed by Anderko,[32] involves separate "chemical" and "physical" terms with the former just the ratio of the actual number of separate molecules to the total number of monomer units. This ratio is readily evaluated and found to be

$$z_{ch} = \frac{\Sigma C_n}{\Sigma n C_n}$$

$$= \frac{2}{1 + (1 + 4RTK/V)^{1/2}}$$

A simple equation, such as that of Redlich and Kwong, is chosen for the "physical" term, and the complete equation becomes

$$z = z_{ph} + \frac{2}{1 + (1 + 4RTK/V)^{1/2}} - 1 \qquad (9\text{-}77)*$$

Alcohols. The O—H···O hydrogen bond has an energy in the range 3000 K or 6 kcal mol^{-1} and is much larger than other attractive interactions for alcohol molecules. The hydrogen bond is strongest in linear geometry. Thus, the alcohol dimer can either form just one strong hydrogen bond or two weak bonds for a net bonding of about 1500 K per monomer unit. A larger cluster can form hydrogen bonds around a ring, as illustrated for the tetramer in Fig. 9-9a. The bonding energy per monomer is now about twice as large, or 3000 K.

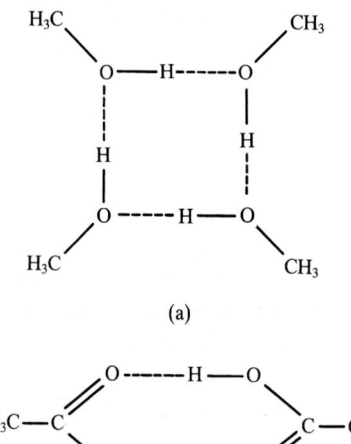

FIGURE 9-9
Structures for strong hydrogen bonding in (a) a tetramer of methanol and (b) a dimer of acetic acid.

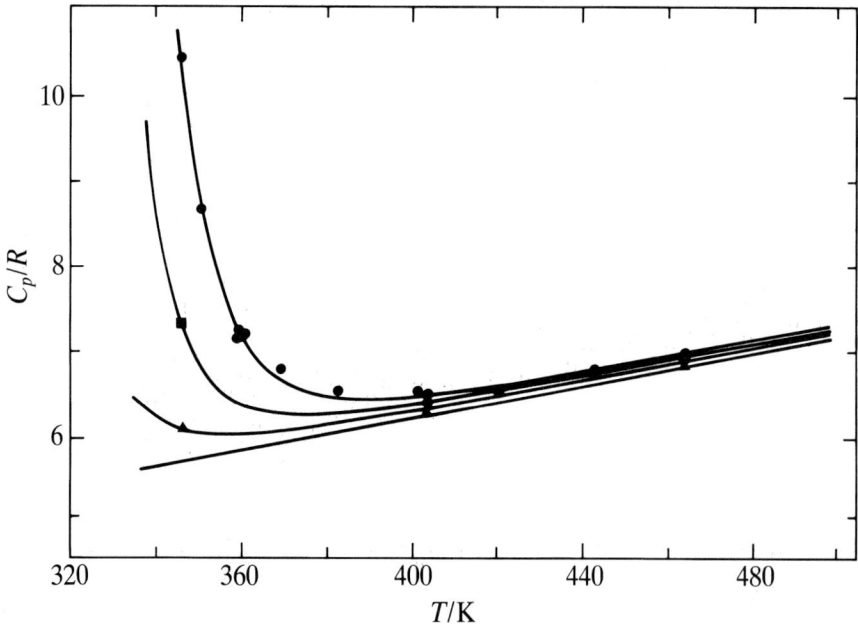

FIGURE 9-10
The heat capacity of methanol vapor, showing the anomalously high values arising from the heat absorbed in dissociation of the hydrogen-bonded tetramer.[31]

The volumetric properties, including the second virial coefficients, of fluid methanol or ethanol do not conform quantitatively to the pattern of normal fluids, but they do not depart grossly. The pressure dependence of the heat capacities of the vapors, however, has an anomalous pattern, as shown in Fig. 9-10 for methanol.

Weltner and Pitzer[33] treated the volumetric and heat capacity data for methanol on the basis of a virial expansion including a second virial coefficient and an nth virial coefficient based in each case on an entropy and enthalpy of association. This yields the following equations:

$$z = 1 + B_2'P + B_n'P^{n-1} \qquad (9\text{-}78a)*$$

where B_n' is the nth virial coefficient in the pressure series, Eq. (9-4), and

$$B_2' = \frac{b}{RT} - K_2 = \frac{b}{RT} - \exp(\Delta S_2/R)\exp(-\Delta H_2/RT) \qquad (9\text{-}78b)*$$

$$B_n' = -(n-1)K_n = -(n-1)\exp(\Delta S_n/R)\exp(-\Delta H_n/RT) \qquad (9\text{-}78c)*$$

Here b is a volume related to the repulsive forces. For the heat capacity, application of Eq. (9-33) yields

$$C_P = C_P^\circ + aP + cP^{n-1} \qquad (9\text{-}79a)*$$

$$\frac{a}{R} = \left(\frac{\Delta H_2}{RT}\right)^2 \exp(\Delta S_2/R) \exp(-\Delta H_2/RT) \tag{9-79b}*$$

$$\frac{c}{R} = \left(\frac{\Delta H_n}{RT}\right)^2 \exp(\Delta S_n/R) \exp(-\Delta H_n/RT) \tag{9-79c}*$$

Note the factor ΔH_n^2 in the c/R term for the heat capacity, which makes that quantity very sensitive to an association with a large ΔH. This factor arises from the shift with temperature of the amount of polymeric species and the enthalpy change for that amount of reaction.

Test calculations for various integral values of n indicated a strong preference for $n = 4$, whereupon the other parameters have the values

$$\frac{\Delta H_2}{R} = -1620\,\text{K} \qquad \frac{\Delta S_2}{R} = -8.2_6$$

$$\frac{\Delta H_4}{R} = -12\,180\,\text{K} \qquad \frac{\Delta S_4}{R} = -40.9$$

with $b = 80\,\text{cm}^3\,\text{mol}^{-1}$. The agreement of calculated values with the heat capacity measurements is shown in Fig. 9-10; the agreement for volumetric data is also excellent.[33] The value of $\Delta H_4/R$, approximately 12 000 K, is about that expected for four strong O—H$\cdots$O hydrogen bonds. This same pattern has been reported for the heat capacity of other alcohol gases and is included in a more elaborate treatment[34] of methanol vapor, which considers enthalpy measurements in addition to heat capacities. At higher temperatures and pressures the cyclic tetramer is no longer important, and alcohols are successfully modeled with a fixed association constant as discussed above.[32]

Carboxylic acids. In contrast to alcohols, acetic acid and other carboxylic acids form dimers with two strong hydrogen bonds (Fig. 9-9b). Thus, the dimerization equilibrium dominates and there is no need to consider higher polymers at low pressures. But the virial treatment is now inappropriate because the vapor can become largely dimer with only a little monomer.

We assume ideal behavior of the gas comprising the two species, and let the fraction of monomer associated be α. Thus, starting with one mole of monomer, we have $n_1 = 1-\alpha$ and $n_2 = \alpha/2$ at equilibrium, and

$$P_1 = (1 - \alpha)\frac{RT}{V} \tag{9-80a}*$$

$$P_2 = \left(\frac{\alpha}{2}\right)\frac{RT}{V} \tag{9-80b}*$$

$$P = P_1 + P_2 = \left(\frac{1-\alpha}{2}\right)\frac{RT}{V} \tag{9-81}*$$

$$K_2 = \frac{P_2}{P_1^2} = \exp(\Delta S_2/R) \exp(-\Delta H_2/RT) \qquad (9\text{-}74)*$$

From the measured pressure and Eq. (9-81)*, one can calculate α and eventually K_2 for that temperature. Then from measurements at a series of temperatures, the values of ΔS_2 and ΔH_2 can be determined. From measurements on acetic acid at very low pressure, where the ideal gas assumption is most accurate, Taylor[35] found $\Delta H_2/R = -7707$ K and $\Delta S_2/R = -25.2$. At 51.2°C and a pressure of only 0.035 bar, he found 82% of acetic acid to be dimeric. With two hydrogen bonds in the dimer, this result indicates a hydrogen bond energy of 3854 K for acetic acid, which is somewhat larger than the value from the tetramer of methanol, 3045 K.

At higher pressures there are further gas-imperfection effects in acetic acid vapor that arise from the longer-range attractive forces between dimeric molecules and, possibly, from hydrogen bonded trimers or larger species. It is difficult to sort out these effects with only volumetric data. Weltner[36] measured heat capacities and made spectral measurements for acetic acid, however, and from a very careful analysis concluded that Taylor's equation for the dimerization equilibrium was quite accurate and that there was no need to assume larger hydrogen-bonded polymers. Similar properties have been reported from less extensive studies of formic, propionic, and other carboxylic acids.[37] Very recently, Twu et al.[38] presented an alternate treatment for carboxylic acids. The values of ΔH_2 and ΔS_2 remain about the same for these other acids. Information is also available for dimerization in other polar gases.[39]

Experimental Methods

The methods of measurements are described in various papers cited above; also, recently developed flow methods are discussed in Chapter 4. We also note a comprehensive volume on experimental measurements for fluids that was edited by Le Neindre and Vodar;[40] here pressure measurements are described in detail along with discussions of temperature, velocity of sound, etc.

References

1. D. R. Douslin, *Progress in International Research on Thermodynamic and Transport Properties*, ASME 1962, p. 135.
2. A. Michels, J. M. Levelt, and W. DeGraff, *Physica*, **24**, 659 (1958); this paper uses the Amagat unit which is the volume of 1 mol at 0°C and 1 atm.
3. B. E. F. Fender and G. D. Halsey, Jr., *J. Chem. Phys.*, **36**, 1881 (1962).
4. J. C. Dymond and E. B. Smith, *The Second Virial Coefficients of Gases: A Critical Compilation*, Clarendon Press, Oxford, 1980.
5. D. R. Schreiber and K. S. Pitzer, *Fluid Phase Equil.*, **46**, 113 (1989).
6. J. Otto, A. Michels, and H. Wouters, *Physik Z.*, **35**, 97 (1934); similar equations are given for other temperatures up to 150°C.
7. O. Redlich and J. N. S. Kwong, *Chem. Rev.*, **44**, 233 (1949).

8. A. Anderko and K. S. Pitzer, *Am. Inst. Chem. Eng. J.*, **37**, 1379 (1991); *Fluid Phase Equil.*, **79**, 125 (1992).
9. M. Benedict, G. B. Webb, and L. C. Rubin, *J. Chem. Phys.*, **8**, 334 (1940); **10**, 747 (1942).
10. B. A. Younglove and J. F. Ely, *J. Phys. Chem. Ref. Data*, **16**, 577 (1987).
11. P. G. Hill, *J. Phys. Chem. Ref. Data*, **19**, 1233 (1990).
12. A. Saul and W. Wagner, *J. Phys. Chem. Ref. Data*, **18**, 1537 (1989).
13. R. Schmidt and W. Wagner, *Fluid Phase Equil.*, **19**, 175 (1985).
14. J. de Boer and A. Michels, *Physica*, **5**, 945 (1938).
15. K. S. Pitzer, *J. Chem. Phys.*, **7**, 583 (1939).
16. K. S. Pitzer, *J. Am. Chem. Soc.*, **77**, 3427 (1955).
17. A third parameter was first suggested by Nernst in 1907 and by various authors since. See papers by Riedel and by Pitzer et al. below for references.
18. L. Riedel, *Chem. -Ing. Tech.*, **26**, 83, 259, 679 (1954); **27**, 209, 475 (1955); **28**, 557 (1956).
19. K. S. Pitzer, D. Z. Lippmann, R. F. Curl, Jr., C. M. Huggins, and D. E. Petersen, *J. Am. Chem. Soc.*, **77**, 3433 (1955).
20. R. F. Curl, Jr., and K. S. Pitzer, *Ind. Eng. Chem.*, **50**, 265 (1958); *J. Am. Chem. Soc.*, **79**, 2369 (1957).
21. C. Tsonopoulos, *AIChE J.*, **20**, 263 (1974).
22. G. Waddington, S. S. Todd, and H. M. Huffman, *J. Am. Chem. Soc.*, **69**, 22 (1947).
23. Theory for virial coefficients is given in many books on statistical mechanics. For theory plus numerous examples, see J. O. Hirschfelder, C. F. Curtiss, and R. B. Bird, *Molecular Theory of Gases and Liquids*, Wiley, New York, 1954.
24. R. A. Aziz and M. J. Slaman, *Mol. Phys.*, **57**, 825 (1986); **58**, 679 (1986).
25. J. A. Barker, R. A. Fisher, and R. O. Watts, *Mol. Phys.*, **21**, 657 (1971).
26. T. Kihara, *Adv. Chem. Phys.*, **1**, 276 (1958); **5**, 147 (1963); *Rev. Mod. Phys.*, **25**, 831 (1953).
27. L. S. Tee, S. Gotoh, and W. E. Stewart, *Ind. Eng. Chem. Fundam.*, **5**, 363 (1966).
28. F. Danon and K. S. Pitzer, *J. Chem. Phys.*, **36**, 425 (1962).
29. J. M. Prausnitz, R. N. Lichtenthaler, and E. G. de Azevedo, *Molecular Thermodynamics of Fluid Phase Equilibria*, Prentice-Hall, Englewood Cliffs, N.J., 1986, pp. 112–115.
30. J. A. Barker and D. Henderson, *Rev. Mod. Phys.*, **48**, 587 (1976).
31. R. A. Heidemann and J. M. Prausnitz, *Proc. Natl. Acad. Sci. USA*, **73**, 1773 (1976).
32. A. Anderko, *J. Chem. Soc. Faraday Trans.*, **86**, 2823 (1990).
33. W. Weltner, Jr., and K. S. Pitzer, *J. Am. Chem. Soc.*, **73**, 2606 (1951).
34. M. Massucci, A. P. du'Gay, A. M. Diaz-Laviada, and C. J. Wormald, *J. Chem. Soc. Faraday Trans.*, **88**, 427 (1992).
35. M. D. Taylor, *J. Am. Chem. Soc.*, **73**, 315 (1951).
36. W. Weltner, Jr., *J. Am. Chem. Soc.*, **77**, 3941 (1955).
37. C. Tsonopoulos and J. M. Prausnitz, *Chem. Eng. J.*, **1**, 273 (1970), and references there cited.
38. C. H. Twu, J. E. Coon, and J. R. Cunningham, *Fluid Phase Equil.*, **82**, 379 (1993).
39. A. Ksiazczak and A. Anderko, *Ber. Bunsenges. Phys. Chem.*, **91**, 1048 (1987).
40. B. Le Neindre and B. Vodar, *Experimental Thermodynamics*, Volume II, *Experimental Thermodynamics of Non-reacting Fluids*, Butterworths, London, 1975 (Vol. I was cited as Ref. 1 of Chapter 6).

Additional references for equations of state

Books on equations of state

J. M. Prausnitz, R. N. Lichtenthaler, and E. G. de Azevedo, *Molecular Thermodynamics of Fluid Phase Equilibria*, 2d edn., Prentice-Hall, Englewood Cliffs, N.J., 1986.

K. C. Chao and R. L. Robinson, Jr., eds., Equations of State, *ACS Symposium Series No. 300*, American Chemical Society, Washington, D.C., 1986.

K. C. Chao and R. L. Robinson, Jr., eds., Equations of State, *Advances in Chemistry Series No. 182*, American Chemical Society, Washington, D.C., 1979.

Reviews of cubic equations of state

C. Tsonopoulos and J. L. Heidman, *Fluid Phase Equil.*, **24**, 1 (1985).
J. J. Martin, *Ind. Eng. Chem. Fundam.*, **18**, 81 (1979).
M. M. Abbot, *A.C.S. Adv. Chem. Ser.*, **182**, 47 (1979).
G. Soave, *Fluid Phase Equil.*, **82**, 345 (1993).

PROBLEMS

9-1. Calculate the second virial coefficient of argon at its normal boiling point of 87.45 K from Eq. (9-12)*. What is the percentage deviation of its molar volume from that of an ideal gas?

9-2. (a) Derive the relationships stated in the text for the van der Waals parameters in terms of the critical properties. (b) Derive the corresponding relationships for the original Redlich–Kwong equation, Eq. (9-59)*. (c) Derive these relationships for the Soave–Redlich–Kwong equation as given in Appendix 4.

9-3. Derive the expression $C' = (C-B^2)/RT$, where C' is the third virial coefficient in the pressure series, Eq. (9-4).

9-4. Find expressions for $(\partial S/\partial \rho)_T$, $(\partial S/\partial P)_T$, and $(\partial H/\partial P)_T$ for the Redlich–Kwong equation.

9-5. Use the tables of Appendix 3 to obtain an estimate for the molar volume and the value of f/P for propane at 400 K and 40 bar.

9-6. Calculate the second virial coefficient for n-butane at its normal boiling point of 272.6 K from Eq. (9-66a)* and its parameters $\omega = 0.200$, $T_c = 425.16$ K. What is the percentage deviation of its molar volume from the ideal gas law?

9-7. Estimate the fugacity coefficient for n-butane at its normal boiling point using the second virial coefficient from Problem 9-6.

9-8. Compare Eqs. (9-11)* and (9-71)* for the second virial coefficient. Use a graph of $\ln B$ vs. $\ln T$ and shift the B_r/B and T_r/T ratios for best agreement; then note any detailed differences.

9-9. Calculate the values of z, V_m, and $\phi = f/P$ for methane at 50 bar and 300 K. Use the Soave–Redlich–Kwong equation (Appendix 4), the critical parameters in Table A3-11, and the answers to Problem 9-2c.

9-10. Calculate the same quantities as in Problem 9-9, but use the Acentric Factor tables and equations of Appendix 3.

CHAPTER 10

MULTICOMPONENT SYSTEMS; BASIC RELATIONSHIPS

We have now completed the presentation of the general principles of thermodynamics together with examples of application to one-component systems. Next we turn to systems of variable composition, i.e., gaseous mixtures and liquid or solid solutions. Most chemical processes involve mixed systems; hence their treatment is of the utmost importance. Indeed, most of the materials of the real world are not pure substances with all atoms or molecules identical but rather are mixtures of one type or another. And much of the interest in thermodynamics arises from its power to deal rigorously and quantitatively with the properties of such mixtures as a function of composition, and to predict or interrelate the conditions in which a change in composition occurs naturally or can be made to occur.

The pure substances from which a solution may be prepared are called the *components*, or constituents, of the solution. There is always something arbitrary in the choice of these components; thus a given aqueous sulfuric acid could be prepared equally well from H_2SO_4 and H_2O or from SO_3 and H_2O. We are at liberty to choose either of these pairs as the components of the solution in question, or indeed, we might even consider all three substances, SO_3, H_2SO_4, H_2O, as the components. However, it is always possible to state the *minimum* number of pure components from which the solution may be made, and this number plays an important role in thermodynamics. In the case we have just cited this number is two, and such a solution is called a binary solution, or a binary mixture.

Under ordinary circumstances, the extensive properties of a pure substance are determined by pressure, temperature, and amount, and its intensive properties by pressure and temperature alone. Likewise we shall assume, unless the contrary is especially stated, that the extensive properties of a solution are determined by pressure, temperature, and the amount of each component and its intensive properties by pressure, temperature, and the relative amounts of the several components, or, in other words, by pressure, temperature, and *composition*. The only important alternative is the substitution of volume or density for pressure as an independent variable for highly compressible systems. Thus, for the treatment of a mixed fluid in the critical region, the preferable variables are the temperature and the densities of the various components.

MEASUREMENTS OF COMPOSITION

The composition of a solution is often best expressed by the *mole fraction*, the ratio of the number of moles of each component to the total number of moles. Thus, the mole fraction of substance 1 is

$$x_1 = \frac{n_1}{\sum n_i} \tag{10-1}$$

where n_i is the number of moles of component i. It is evident that the sum of all mole fractions is unity:

$$\sum x_i = 1 \tag{10-2}$$

In the case of a binary solution we note that

$$x_1 + x_2 = 1 \qquad dx_1 = -dx_2 \tag{10-3}$$

Other Measures of Composition

Several other quantities are used to express the composition. The mass fraction or the percentage by weight is directly related to experimental measurements and is preferable in those cases where the definition of the molecular mass is ambiguous. This situation arises, for example, for a solute comprising polymer molecules of the same type but with a distribution of molecular masses.

Sometimes it seems convenient to think of a solvent as a background material in which solute molecules move much like gas molecules. In that case the common measures include the concentration (amount of solute per unit volume). If the unit is moles per liter (or per dm^3) of solution, it is also called the *molarity*. Another measure is the *molality*, moles per kilogram of solvent. The molality is usually to be preferred, since it does not depend on temperature or pressure, whereas any concentration unit is so dependent. The use of molality is nearly universal for aqueous solutions. For multicomponent systems, equations in molality are usually simpler than equations in mole fraction.

Other measures of composition include the volume fraction, and for binary systems the mole ratio or the volume ratio. For the remainder of this chapter, however, we will express all compositions as mole fractions; the other measures will be used later when appropriate. Indeed, there are various definitions and working equations that have been found to be convenient for different types of solutions: neutral-molecule liquids or solids, electrolytes, high polymers, gases, etc. These are described with examples in subsequent chapters. At this point we extend the basic equations of Chapters 2, 3, and 7 to include the variation of composition.

ESCAPING TENDENCY IN SOLUTIONS

Let us now consider with some care a simple process in which one component is transferred to or from a solution. Suppose only component 1 of a liquid solution has appreciable volatility; then the gas phase above the solution will consist of pure gas. Thus at room temperature H_2SO_4 has practically zero vapor pressure, and the gas in equilibrium with aqueous sulfuric acid is pure water vapor. We may write

$$H_2O(g,P) = H_2O(\text{soln}, x_1)$$

where P is the gas pressure and x_1 the mole fraction of water in the solution.

Now, if this system is at equilibrium, we have seen that $dG = 0$ for an infinitesimal transfer of water from gas to solution. But the Gibbs energy of the entire system, gas and solution, is merely the sum of the Gibbs energies of the two phases. Thus

$$G = G(g) + G(\text{soln}) \tag{10-4}$$

where $G(g)$ and $G(\text{soln})$ are the Gibbs energies of gas and solution, respectively. From Eq. (7-2) we recall that the Gibbs energy of the water vapor is just the chemical potential $\mu_1(g)$; then $G(g) = n_g \mu_1(g)$. However, the Gibbs energy of the solution depends on the composition as well as the total amount. Hence we must regard $G(\text{soln})$ as a function of n_1 and n_2, the number of moles of water and sulfuric acid, respectively, as well as the temperature and pressure. The total differential is

$$dG(\text{soln}) = \left(\frac{\partial G}{\partial T}\right)_{P,n_1,n_2} dT + \left(\frac{\partial G}{\partial P}\right)_{T,n_1,n_2} dP$$
$$+ \left(\frac{\partial G}{\partial n_1}\right)_{P,T,n_2} dn_1 + \left(\frac{\partial G}{\partial n_2}\right)_{P,T,n_1} dn_2 \tag{10-5}$$

and if we hold P, T, and n_2 constant, we have

$$dG(\text{soln}) = \left(\frac{\partial G}{\partial n_1}\right)_{P,T,n_2} dn_1$$

The change in total free energy on transfer of dn_1 moles of water from gas to solution is then

$$dG = dG(g) + dG(\text{soln})$$
$$= -\mu_1(g)\, dn_1 + \left(\frac{\partial G}{\partial n_1}\right)_{P,T,n_2} dn_1$$

and if this is at equilibrium with $dG = 0$, we have

$$\mu_1 = \left[\frac{\partial G(\text{soln})}{\partial n_1}\right]_{P,T,n_2} \tag{10-6}$$

This shows that the chemical potential of one component in a multicomponent system is given by the partial derivative of the Gibbs energy, as indicated.

Equations for the Chemical Potential

We now restate in more formal and general terms the results of the preceding section. As in Chapter 3, we assume a system in internal equilibrium and that the only work is that of volume change against an external pressure. Also, we consider first the situation with pressure and temperature as independent variables since this pattern is both the easiest to comprehend and the most realistic experimentally. Thus, the Gibbs energy is the appropriate function, and it is now a function not only of T and P, but also of n_1, n_2, etc., the amounts of component 1, component 2, etc., contained in the system. We shall take the n_1, n_2, etc., to be numbers of moles, although amounts in mass or other units can be used. Then the differential of G becomes

$$dG = \left(\frac{\partial G}{\partial T}\right)_{P,n_i} dT + \left(\frac{\partial G}{\partial P}\right)_{T,n_i} dP + \sum_i \left(\frac{\partial G}{\partial n_i}\right)_{P,T,n_{j\neq i}} dn_i \tag{10-7}$$

If the composition is fixed, i.e., all dn_i are zero, then this equation reduces to (3-6), and we obtain

$$-S = \left(\frac{\partial G}{\partial T}\right)_{P,n_i} \tag{10-8}$$

and

$$V = \left(\frac{\partial G}{\partial P}\right)_{T,n_i} \tag{10-9}$$

Now we show explicitly the requirement that all n_i are constant for Eqs. (10-8) and (10-9), whereas a constant amount of substance was an implicit condition for the corresponding equations of Chapter 3.

If there is only one component, $(\partial G/\partial n_i)_{P,T}$ is just the molar Gibbs energy for that substance i. We saw in Chapter 7, and in the preceding section, that this was the measure of the tendency of a substance to move from one phase to another, i.e., the escaping tendency, and for that reason it was called the *chemical potential* with the symbol μ_i. Thus, we now rewrite Eq. (10-7) as

$$dG = -S\,dT + V\,dP + \sum_i \mu_i\,dn_i \tag{10-10}$$

with

$$\mu_i = \left(\frac{\partial G}{\partial n_i}\right)_{P,T,n_{j\neq i}} \tag{10-11}$$

Equation (10-10) may now be integrated at constant T and P by increasing each n_i but maintaining all ratios of n_i to n_j constant. This yields

$$G = \sum_i n_i \mu_i \tag{10-12}$$

Next, we verify that the μ_i measure the escaping tendencies of the various components in a multicomponent system. For this purpose, assume a system with two phases, α and β, in equilibrium at the same P and T, and consider a change in which an amount Δn_i of component i transfers from phase α to phase β. Then,

$$\Delta G = \Delta G^\alpha \, \Delta G^\beta$$
$$= \Delta n_i (\mu_i^\beta - \mu_i^\alpha)$$

Also, by the same argument as that given at the beginning of Chapter 7, for a spontaneous process $\Delta G < 0$. Thus, if $\mu_i^\beta < \mu_i^\alpha$, $\Delta n_i > 0$ and material moves from phase α to phase β, or vice versa if $\mu_i^\beta > \mu_i^\alpha$. And for equilibrium, $\Delta G = 0$ and $\mu_i^\alpha = \mu_i^\beta$, which establishes μ_i as a measure of the escaping tendency.

By the reverse of the transformations that yielded Eqs. (3-1) to (3-6), one obtains

$$dA = -S\,dT - P\,dV + \sum \mu_i dn_i \tag{10-13}$$

$$\mu_i = \left(\frac{\partial A}{\partial n_i}\right)_{T,V,n_{j \neq i}} \tag{10-14}$$

$$dH = T\,dS + V\,dP + \sum \mu_i dn_i \tag{10-15}$$

$$\mu_i = \left(\frac{\partial H}{\partial n_i}\right)_{S,P,n_{j \neq i}} \tag{10-16}$$

$$dU = T\,dS - P\,dV + \sum \mu_i dn_i \tag{10-17}$$

$$\mu_i = \left(\frac{\partial U}{\partial n_i}\right)_{S,V,n_{j \neq i}} \tag{10-18}$$

Indeed, Gibbs first introduced the chemical potential with Eqs. (10-17) and (10-18). We find the implication of Eq. (10-18), an addition of material at constant entropy and volume, difficult to picture, and therefore prefer to introduce μ_i initially for the constant pressure and temperature process with the Gibbs energy. But the method used by Gibbs is equally valid. Also, Gibbs used Eq. (10-17) as the basis for a rigorous mathematical demonstration that two phases are in equilibrium only if the pressures, temperatures, and all chemical potentials are equal.[1]

While we shall not ordinarily use Eqs. (10-16) or (10-18) for μ_i, it is important to note Eq. (10-14). It is a physically realistic process to add an amount of one component to a gas or compressible fluid at constant volume, with accompanying increase in pressure. Also, we saw in Chapter 9 that the Helmholtz function is a very valuable basis for an equation of state. Thus, Eq. (10-14) provides the very important route to the various chemical potentials in a mixed compressible fluid.

PARTIAL MOLAR QUANTITIES

In addition to the chemical potential, it is valuable to consider other quantities obtained from partial derivatives with respect to n_i with all other n_j constant. Physically, this corresponds to the addition to a large mixed system of a small amount of component i and observing the increase in volume or some other property. Mathematically, for an extensive property Y, the partial molar quantity for component i is

$$\bar{Y} = \left(\frac{\partial Y}{\partial n_i}\right)_{P,T,n_{j \neq i}} \tag{10-19}$$

Note that this derivative is taken at constant P and T; thus $\bar{G}_i = \mu_i$, but $\bar{A}_i$ is not μ_i, etc.

We shall use the bar above the symbol for the partial molar property to avoid ambiguity with the corresponding molar property $(Y/n)_i$ for pure substance i. The standard-state properties of substance i are written as Y_i°. These may be properties of one mole of pure i or partial molar properties of i in a solvent such as water; in the latter case we shall designate them as $\bar{Y}_i^\circ$, with the bar. On other occasions, symbols, such as Y_i^*, will be used for one mole of pure substance i in a particular state. Hence, we believe it desirable to use the bar for the partial molar quantity, although it is often omitted.

Relationships to the Chemical Potential

The pressure dependence of the chemical potential can be obtained from the second derivative of the Gibbs energy, i.e., from Eq. (10-10). Since the order of differentiation is immaterial,

$$\frac{\partial^2 G}{\partial P \partial n_i} = \frac{\partial V}{\partial n_i} = \frac{\partial \mu_i}{\partial P} \quad \text{at constant } T \text{ and } n_{j \neq i}$$

Thus,

$$\left(\frac{\partial \mu_i}{\partial P}\right)_T = \bar{V}_1 \tag{10-20}$$

and by similar methods one finds the following:

$$\left(\frac{\partial \mu_i}{\partial T}\right)_P = \bar{S}_i \tag{10-21}$$

$$\left[\frac{\partial (\mu_i/T)}{\partial T}\right]_P = -\frac{\bar{H}_i}{T^2} \tag{10-22}$$

$$\left[\frac{\partial (\mu_i/T)}{\partial (1/T)}\right]_P = \bar{H}_i \tag{10-23}$$

also

$$\left(\frac{\partial \bar{H}_i}{\partial T}\right)_P = \bar{C}_{P,i} \tag{10-24}$$

In each of the Eqs. (10-20) to (10-24), the composition is held constant; also recall that $\bar{G}_i$ is identical to μ_i. One notes that each of these relationships is the counterpart of one of Eqs. (3-8), (3-9), (3-18), (3-20), or (3-27).

General Partial Molar Equations

We may now consider certain characteristics and relationships common to all partial molar quantities. Let Y be any extensive property of a given solution, such as volume, heat capacity, or internal energy, that is a function of temperature, pressure, and the amounts of the several constituents. For the sake of clarity we shall assume for the present that temperature and pressure are constant, so that Y depends only upon $n_1, n_2, \ldots$.

We defined the partial molar values by the equations

$$\bar{Y}_1 = \left(\frac{\partial Y}{\partial n_1}\right)_{P,T,n_2,n_3,\ldots} \qquad \bar{Y}_2 = \left(\frac{\partial Y}{\partial n_2}\right)_{P,T,n_1,n_3,\ldots} \tag{10-19}$$

Now, by the chief equation of partial differentiation,

$$dY = \left(\frac{\partial Y}{\partial n_1}\right)_{P,T,n_2,n_3,\ldots} dn_1 + \left(\frac{\partial Y}{\partial n_2}\right)_{P,T,n_1,n_3,\ldots} dn_2 + \cdots \tag{10-25}$$

or

$$dY = \bar{Y}_1 dn_1 + \bar{Y}_2 dn_2 + \cdots \tag{10-26}$$

It is evident that these partial molar quantities such as $\bar{Y}_1$ are not extensive but intensive properties of the solution. They depend, therefore, not upon the total amount of each constituent, but only upon the composition, i.e., upon the **relative** amounts of the several constituents.

If, therefore, to a given solution at constant temperature and pressure we add the several constituents simultaneously, keeping their ratios constant, these partial molar quantities will remain constant. We may therefore integrate Eq. (10-26), keeping $n_1, n_2, \ldots$ in constant proportions, and find

$$dY = (\bar{Y}_1 x_1 + \bar{Y}_2 x_2 + \cdots) dn$$
$$Y = (\bar{Y}_1 x_1 + \bar{Y}_2 x_2 + \cdots) n$$

and

$$Y = n_1 \bar{Y}_1 + n_2 \bar{Y}_2 + \cdots \tag{10-27}$$

In deriving Eq. (10-27) we did not limit ourselves to any special values of $n_1, n_2, \ldots$. Hence this equation, being entirely general, can be differentiated with respect to any change of composition, however this change is produced (whether by addition or subtraction of infinitesimal amounts of any or all of the components). This general differentiation gives

$$dY = n_1 d\bar{Y}_1 + \bar{Y}_1 dn_1 + n_2 d\bar{Y}_2 + \bar{Y}_2 dn_2 + \cdots \tag{10-28}$$

and this equation combined with Eq. (10-26) gives

$$n_1 d\bar{Y}_1 + n_2 d\bar{Y}_2 + \cdots = 0 \tag{10-29}$$

which shows, for any infinitesimal alteration in composition, at constant temperature and pressure, the relation between the change in any one $\bar{Y}_i$ and the change in all the others. Equations (10-27) and (10-29) may for brevity be called the *partial molar equations*.

These equations assume a number of special forms, which are frequently useful. Thus, if we are dealing with 1 mole of solution, Eqs. (10-27) and (10-29) become

$$Y_m = x_1 \bar{Y}_1 + x_2 \bar{Y}_2 + \cdots \tag{10-30}$$

$$x_1 d\bar{Y}_1 + x_2 d\bar{Y}_2 + \cdots = 0 \tag{10-31}$$

We may regard the number of moles of one constituent, say n_1, as the main variable. Then if, as a reminder, we indicate also the constancy of P and T, Eq. (10-29) takes the form

$$n_1 \left(\frac{\partial \bar{Y}_1}{\partial n_1} \right)_{P,T} + n_2 \left(\frac{\partial \bar{Y}_2}{\partial n_1} \right)_{P,T} + \cdots = 0 \tag{10-32}$$

Similarly Eq. (10-31) becomes

$$x_1 \left(\frac{\partial \bar{Y}_1}{\partial x_1} \right)_{P,T} + x_2 \left(\frac{\partial \bar{Y}_2}{\partial x_1} \right)_{P,T} + \cdots = 0 \tag{10-33}$$

While the derivation just given is quite sufficient, it may be helpful to some readers to verify Eq. (10-32) by another method. We recall that each partial molar quantity is already a partial derivative. Hence

$$\frac{\partial \bar{Y}_2}{\partial n_1} = \frac{\partial}{\partial n_1}\left(\frac{\partial Y}{\partial n_2} \right) = \frac{\partial}{\partial n_2}\left(\frac{\partial Y}{\partial n_1} \right) = \frac{\partial \bar{Y}_1}{\partial n_2} \tag{10-34}$$

which follows because the order of differentiation is immaterial. Then Eq. (10-32) becomes

$$n_1 \left(\frac{\partial \bar{Y}_1}{\partial n_1} \right) + n_2 \left(\frac{\partial \bar{Y}_1}{\partial n_2} \right) + \cdots = 0 \tag{10-35}$$

Each of these derivatives gives the change in $\bar{Y}_1$ per mole addition of the specified component. But if we add these components in the same proportion in which they were originally present (n_1, n_2, etc.), then the composition of the solution has not changed and the value of $\bar{Y}_1$ must remain unchanged. Thus we see that the expression in Eq. (10-35) should be equal to zero.

These equations that were derived in the present section are of fundamental importance. It is interesting to note that they do not depend on thermodynamics but follow from mathematics for any extensive quantity that is a continuous and single-valued function of the composition of the solution.

It will be noted that, while n_1 can vary without any change in n_2, n_3, etc., we cannot change x_1 without some change in x_2, x_3, etc. So, in a binary solution,

$$dx_1 = -dx_2$$

It is also to be noted that, in spite of the similarity of Eqs. (10-32) and (10-33), $d\bar{Y}_1/dn_1$ has a very different meaning from $d\bar{Y}_1/dx_1$. Thus in a binary mixture, when n_2 is constant,

$$dx_1 = \frac{n_2}{(n_1 + n_2)^2} dn_1 \qquad (10\text{-}36)$$

Applications to Binary Solutions

There are several equations for partial molar quantities in systems with only two components that are very useful. Thus, if we apply Eq. (10-31),

$$(1 - x_2) d\bar{Y}_1 + x_2 d\bar{Y}_2 = 0$$

or

$$d\bar{Y}_1 = -\left(\frac{x_2}{1-x_2}\right) d\bar{Y}_2 \qquad (10\text{-}37)$$

Thus, if $\bar{Y}_2$ is known as a function of x_2, Eq. (10-37) can be integrated to yield $\bar{Y}_1$. If, however, it is the total quantity Y that is known initially, we consider the molar quantity $Y_m = Y/(n_1 + n_2)$ and differentiate with respect to x_2:

$$\left(\frac{\partial Y_m}{\partial x_2}\right)_{T,P} = \frac{\partial}{\partial x_2}[(1-x_2)\bar{Y}_1 + x_2\bar{Y}_2]$$

$$= \bar{Y}_2 - \bar{Y}_1 + x_1 \frac{\partial \bar{Y}_1}{\partial x_2} + x_2 \frac{\partial \bar{Y}_2}{\partial x_2}$$

From Eq. (10-31) or (10-37), the sum of the last two terms is zero. Then one has

$$\left(\frac{\partial Y_m}{\partial x_2}\right)_{T,P} = Y_2 - Y_1 \qquad (10\text{-}38)$$

and from Eq. (10-30)

$$Y_m = (1 - x_2)\bar{Y}_1 + x_2\bar{Y}_2 \qquad (10\text{-}30a)$$

From these two equations one can solve for $\bar{Y}_1$ and $\bar{Y}_2$:

$$\bar{Y}_1 = Y_m - x_2\left(\frac{\partial Y_m}{\partial x_2}\right) \qquad (10\text{-}39a)$$

$$\bar{Y}_1 = Y_m + (1 - x_2)\left(\frac{\partial Y_m}{\partial x_2}\right) \qquad (10\text{-}39b)$$

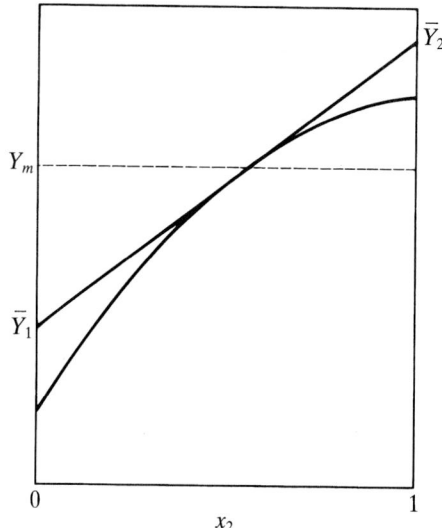

FIGURE 10-1
This illustrates Eqs. (10-39a) and (10-39b).

These relationships are very useful with a graph of Y_m vs x_2, which is illustrated in Fig. 10-1. It is apparent that the intercepts at $x_2 = 0$ and $x_2 = 1$ of the tangent to Y_m yield the values of $\bar{Y}_1$ and $\bar{Y}_2$. Note that if Y_m is the molar Gibbs energy G_m, the intercepts are the chemical potentials μ_1 and μ_2.

CHANGE OF CHEMICAL POTENTIAL AND FUGACITY WITH COMPOSITION

We now rewrite Eq. (10-33) specifically for the chemical potential:

$$x_1\left(\frac{\partial \mu_1}{\partial x_1}\right)_{P,T} + x_2\left(\frac{\partial \mu_2}{\partial x_1}\right) + \cdots = 0 \qquad (10\text{-}40)$$

This very important equation was derived by Gibbs[1] in 1875. Also, for a binary solution, Eq. (10-37) becomes

$$d\mu_1 = -\left(\frac{x_2}{1-x_2}\right)d\mu_2 \qquad (10\text{-}41)$$

These equations do not permit us to determine in any case how the escaping tendency of each constituent changes with its mole fraction, nor is this possible from thermodynamics alone. But if we have experimentally solved this problem for one constituent of a binary mixture, then it is solved for the other constituent.

The trend of the chemical potential of each constituent in a mixture of carbon disulfide and acetone is shown in Fig. 10-2. In any plot such as Fig. 10-2, Eq. (10-40) tells us that, when the two mole fractions are equal, the slopes of the two curves are equal and opposite in sign; when $x_1 = \frac{1}{4}$ and $x_2 = \frac{3}{4}$, $\partial \mu_2/\partial x_1 = -\frac{1}{3}\partial \mu_1/\partial x_1$, and so

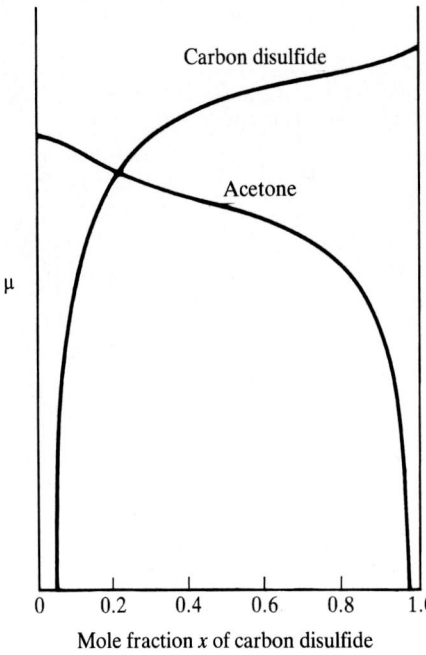

FIGURE 10-2
Chemical potentials in solutions of carbon disulfide and acetone.

on. In general, if one of the curves is known and we know a single point on the other curve, the slope of the second curve is determined at that point. Hence it is possible, by graphical or by analytical methods, to build up the second curve through the range of composition over which the first curve is given.

We note from Fig. 10-2 that the chemical potential decreases and approaches negative infinity as that component becomes more dilute. This is inconvenient, but the problem is readily solved by the use of the fugacity, which was defined by Eq. (7-21):

$$\mu_1 = RT \ln f_1 + \mu_1^\circ \tag{7-21}$$

where μ_1° is a constant, specifically the chemical potential in the standard state. One notes that, with μ_1° finite, if f_1 becomes zero, μ_1 becomes $-\infty$.

We have not explained how we obtained the values of μ plotted in Fig. 10-2. In fact, of the various methods that are employed in the determination of a chemical potential, there are several that we are not yet in a position to discuss; but the simplest and one of the most useful of these methods consists in determining the vapor pressure, and thence the fugacity, of the constituents of a liquid or solid solution.

If we use fugacity in place of Gibbs energy, Eq. (10-40) becomes

$$x_1 \left(\frac{\partial \ln f_1}{\partial x_1} \right)_{P,T} + x_2 \left(\frac{\partial \ln f_2}{\partial x_1} \right)_{P,T} + \cdots = 0 \tag{10-42a}$$

In a binary solution, where $dx_1 = -dx_2$, this equation becomes

$$x_1\left(\frac{\partial \ln f_1}{\partial x_1}\right)_{P,T} = x_2\left(\frac{\partial \ln f_2}{\partial x_2}\right)_{P,T} \quad (10\text{-}42b)$$

When the vapors are nearly perfect gases, we may substitute partial pressures[†] for fugacities and obtain the approximate equation

$$x_1\left(\frac{\partial \ln p_1}{\partial x_1}\right)_{P,T} = x_2\left(\frac{\partial \ln p_2}{\partial x_2}\right)_{P,T} \quad (10\text{-}43)*$$

This approximate equation was found first by Duhem and independently by Margules,[2] while the more rigorous Eq. (10-42b) follows from the work of Gibbs; hence, this relationship is often called the Gibbs–Duhem equation. These fundamental relations between escaping tendency and composition soon received extensive experimental verification, especially in the work of Zawidzki,[3] who determined, over the whole range of concentration, the partial vapor pressures for numerous binary liquid mixtures. We reproduce in Figs. 10-3 to 10-6 his curves for the four pairs, 1,2-dibromoethane–1,2-dibromopropane, carbon disulfide–acetone, acetone–chloroform, and pyridine–water. No discernible difference would be made in these figures if we plotted the actual

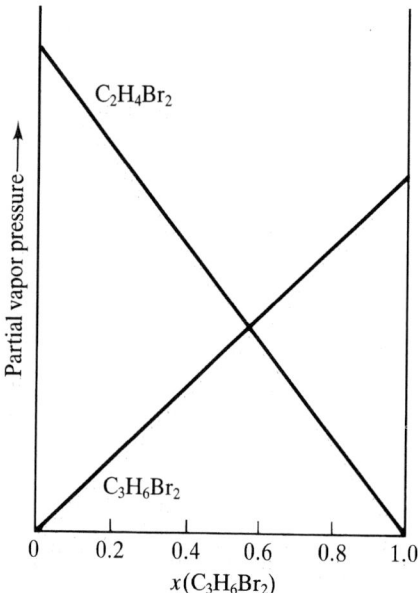

FIGURE 10-3

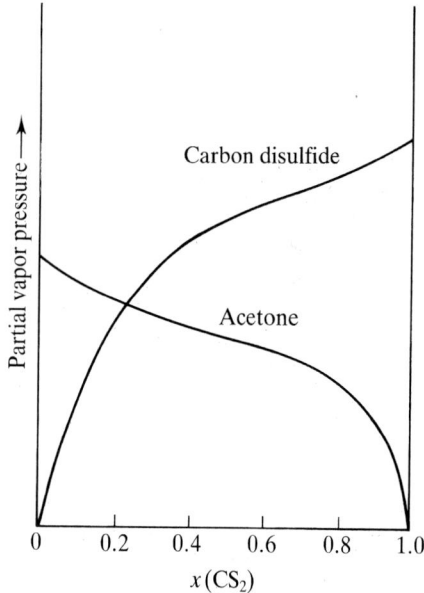

FIGURE 10-4

---

† The *partial pressure* is defined as the product of the total pressure of vapor times the mole fraction in the vapor, for example, $p_1 = y_1 P$, with y_1 the vapor mole fraction.

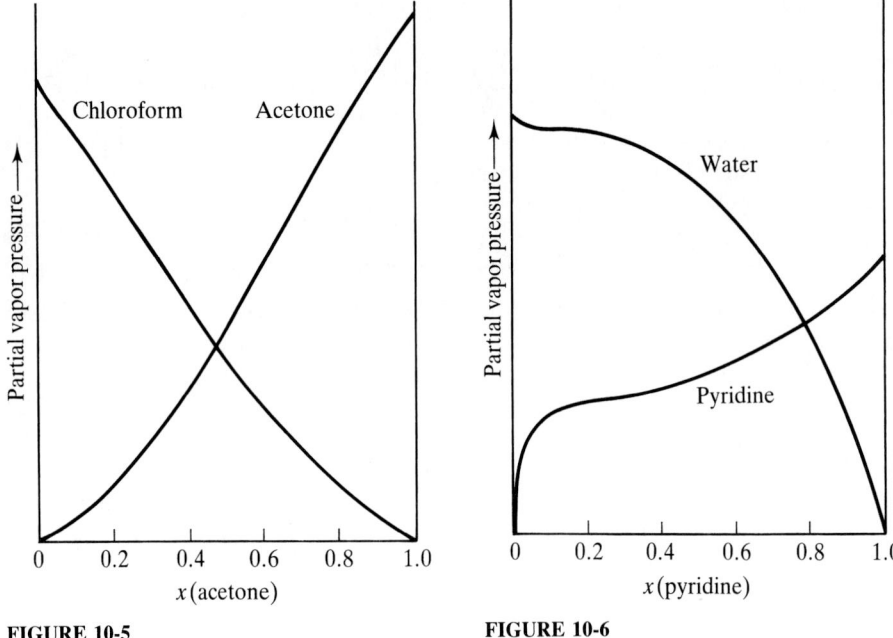

FIGURE 10-5

FIGURE 10-6

fugacity in place of the partial pressures. Hundreds of measurements of this type have subsequently been made for similar solutions.

An inspection of the curves in Figs. 10-3 to 10-6 shows that their slopes conform to Eq. (10-43)*, which may also be put in the alternative form

$$\frac{\partial p_1/\partial x_1}{\partial p_2/\partial x_2} = \frac{p_1/x_1}{p_2/x_2} \qquad (10\text{-}44)^*$$

One of the simplest corollaries of this equation is that, if one of the curves is a straight line across the entire range of composition, the other is also a straight line across the entire range. For if the first curve is straight, $\partial p_1/\partial x_1 = p_1/x_1$, whence $\partial p_2/\partial x_2 = p_2/x_2$. Such a case is illustrated in Figure 10-3.

The Critical Mixing Point

If we study, over a range of temperature, the fugacity–mole fraction curve of one of the components of such a mixture as that represented in Fig. 10-6, we often find that as the temperature is lowered the curvature becomes more pronounced. Finally, a temperature T' is reached, where the curve is horizontal at a point C, as illustrated in Fig. 10-7. If the solution at C is further cooled, it breaks into two phases. The whole phenomenon is closely analogous to the phenomenon of the critical point in the case of a pure

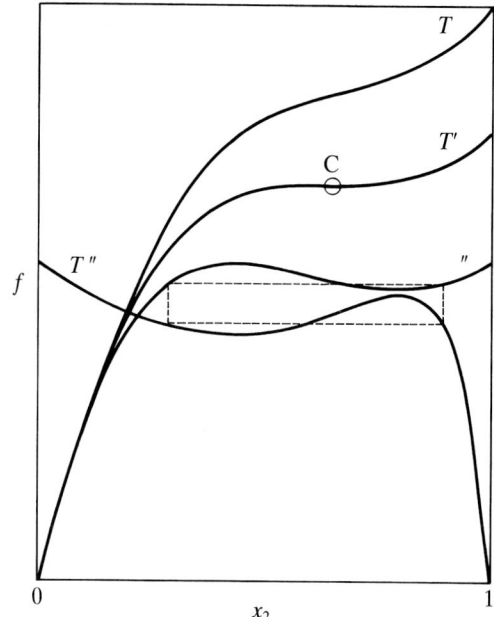

FIGURE 10-7

substance, and therefore the temperature at this point is known as the critical mixing temperature† and the composition as the critical composition.

At this critical point, since the curve is horizontal,

$$\frac{\partial f_1}{\partial x_1} = 0 \qquad \frac{\partial \mu_1}{\partial x_1} = 0 \qquad (10\text{-}45a)$$

Hence for the other component, by Eqs. (10-42) and (10-38),

$$\frac{\partial f_2}{\partial x_1} = 0 \qquad \frac{\partial \mu_2}{\partial x_1} = 0 \qquad (10\text{-}45b)$$

We note also that the critical point in Fig. 10-7 is not a simple maximum or minimum but rather is a point of inflection. Thus the second derivatives are also zero:

$$\frac{\partial^2 f_1}{\partial x_1^2} = 0 \qquad \frac{\partial^2 \mu_1}{\partial x_1^2} = 0 \qquad (10\text{-}46a)$$

$$\frac{\partial^2 f_2}{\partial x_1^2} = 0 \qquad \frac{\partial^2 \mu_2}{\partial x_1^2} = 0 \qquad (10\text{-}46b)$$

---

† In peculiar cases, we find two substances that are miscible in all proportions *below* a certain critical mixing temperature and form two phases above that temperature.

Below the critical mixing temperature phase separation occurs as indicated by the dotted lines. If we could prevent the separation into two phases, we should expect to obtain for both constituents something like the solid portions of the two curves T'', each having a maximum and a minimum, and a certain region in which the escaping tendency would diminish with an increase in mole fraction. In practice, however, except for a small degree of possible supersaturation, separation into two phases occurs. This separation must occur in such measure as to make the fugacity of either constituent the same in both phases.

EQUILIBRIUM BETWEEN PHASES THAT MAY BE SOLUTIONS

The types of equilibrium in systems containing two or more substances are so numerous and so complex that we must content ourselves here with a few simple illustrations.

The properties of a given amount of a binary solution are determined by temperature, pressure, and composition, the latter being fixed when the mole fraction of either constituent is given. When two phases in equilibrium are present, for example, solid water and brine, or the two liquid phases obtained by mixing ether and water, or aqueous hydrochloric acid and its vapor, the system is restricted to two degrees of freedom and in this respect resembles a single phase of a pure substance.

Equilibrium Between Two Phases

The criteria for equilibrium between two phases α and β are the equalities

$$T^\alpha = T^\beta \qquad P^\alpha = P^\beta \qquad \mu_1^\alpha = \mu_1^\beta \qquad \mu_2^\alpha = \mu_2^\beta \qquad (10\text{-}47)$$

These criteria can be applied by iterative calculations that are readily handled with modern computers. Other methods are useful in special cases. The graphical presentation illustrated in Fig. 10-8 is particularly useful for cases involving two fluid phases (or two solid phases). Now the parent function is the molar Gibbs energy. The pattern shown in Fig. 10-8 indicates phase equilibrium, since the common tangent implies equality of both chemical potentials by Eq. (10-39) and the entire presentation is at constant temperature and pressure. Situations in which this method is useful will be noted in later chapters.

Change of Eutectic Temperature with Pressure

With three phases in equilibrium the case is analogous to two phases of a pure substance, with one degree of freedom. Thus the whole system is determined when we choose the temperature, or the pressure, or the composition of one phase that contains both constituents. As an example of such a system, we may consider a solid salt, ice, and a saturated solution at the eutectic point. If the temperature is changed, the pressure, as well as the composition of the solution, must change by a fixed amount in order to maintain equilibrium.

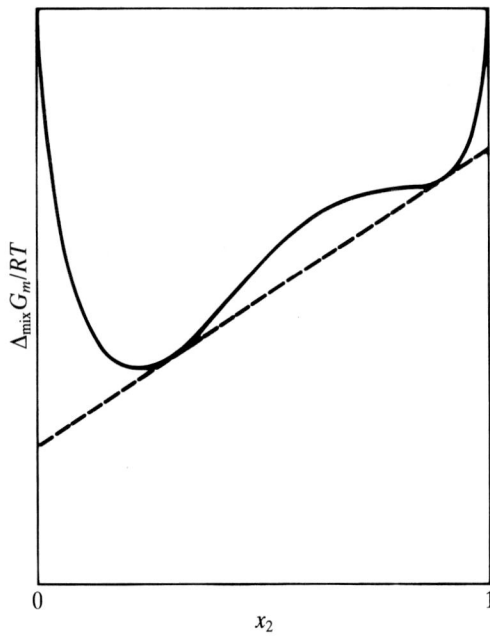

FIGURE 10-8
A graphical method of determining the compositions of coexisting phases.

In this very simple case the rate of change of the eutectic pressure with the temperature may be immediately ascertained by the methods of Chapter 3. But it will be instructive to go through the more complicated calculation in which we investigate the escaping tendency of each substance in the several phases. If we denote by μ_1^* the chemical potential of water in the solid state and by μ_1 its chemical potential in the solution, the first condition of sustained equilibrium is that $\mu_1^* = \mu_1$ and $d\mu_1^* = d\mu_1$. But we may write

$$d\mu_1 = \frac{\partial \mu_1}{\partial T} dT + \frac{\partial \mu_1}{\partial P} dP + \frac{\partial \mu_1}{\partial x_1} dx_1 \tag{10-48a}$$

and for ice, since it is a pure phase,

$$d\mu_1^* = \frac{\partial \mu_1^*}{\partial T} dT + \frac{\partial \mu_1^*}{\partial P} dP \tag{10-48b}$$

Equating these two expressions, and substituting the values of the temperature and pressure coefficients, by means of Eqs. (10-21) and (10-22) and the corresponding equations for pure substances, we find

$$(S_1^* - \bar{S}_1) dT + (\bar{V}_1 - V_1^*) dP + \frac{\partial \mu_1}{\partial x_1} dx_1 = 0 \tag{10-49}$$

and likewise for the second constituent, the salt,

$$(S_2^* - \bar{S}_2) dT + (\bar{V}_2 - V_2^*) dP + \frac{\partial \mu_2}{\partial x_1} dx_1 = 0 \tag{10-50}$$

Now the remaining differential coefficients are eliminated, in accordance with Eq. (10-40), if we multiply the first equation by x_1 and the second by x_2 and then add. Hence

$$(x_1 S_1^* - x_1 \overline{S}_1 + x_2 S_2^* - x_2 \overline{S}_2) dT = (x_1 V_1^* - x_1 \overline{V}_1 + x_2 V_2^* - x_2 \overline{V}_2) dP \quad (10\text{-}51)$$

But we recall that, by Eq. (10-30), $x_1 \overline{S}_1$ and $x_2 \overline{S}_2$ together equal S_m, the entropy of 1 mole of the solution, while $x_1 S_1^* + x_2 S_2^*$ gives the entropy of the corresponding amounts of the two solids. The algebraic sum of all these entropy terms is therefore ΔS, the increase in entropy when 1 mole of the solution is formed from the two solids. Similarly, the volume terms together give ΔV in the same process, and we obtain an equation that we might have taken directly from Eq. (3-42). It is identical with the equation for the change of pressure with temperature when only one component is present and only two phases, namely,

$$\frac{dP}{dT} = \frac{\Delta S}{\Delta V} = \frac{\Delta H}{T \Delta V} \quad (10\text{-}52)$$

The Solubility Curve of a Dissociable Solute

Let us consider one further case in which we deal simultaneously with an equilibrium between phases and an equilibrium in a chemical reaction. We may select for such a study the solubility curve of a hydrated salt such as $CaCl_2 \cdot 6H_2O$.

If we are dealing with a system of two components and one of our variables is fixed, as when the pressure is kept constant at 1 bar, a system of two phases has one remaining degree of freedom and either temperature or composition may be arbitrarily varied, but not both. Thus, if the two components are $CaCl_2$ and H_2O and the two phases are $CaCl_2 \cdot 6H_2O$ and a solution, the relation between the temperature and the composition of the solution may be represented by a continuous curve, as in Fig. 10-9, where temperature is the ordinate and the mole fraction x_2 of $CaCl_2$ in the solution is the abscissa.

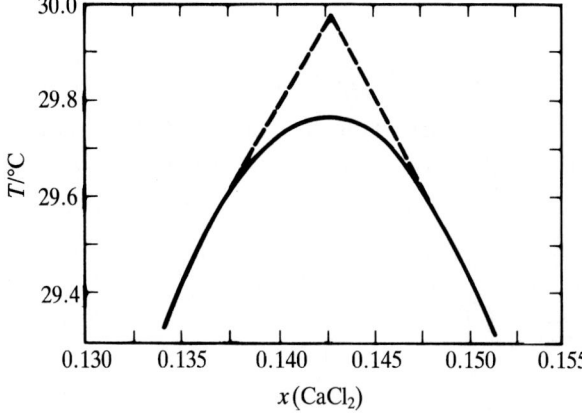

FIGURE 10-9
Solubility of $CaCl_2 \cdot 6H_2O$.

At the maximum point of the curve, where the composition of the solution is the same as that of the solid, any finite addition of either H_2O or $CaCl_2$ will lower the equilibrium temperature. But if in the solution there is any appreciable dissociation of $CaCl_2 \cdot 6H_2O$ into its constituents, an infinitesimal addition of either component does not change the equilibrium temperature. In other words, the point in question is a true maximum of a continuous curve, and not a cusp, or a point of intersection of two curves. This theorem, due to Lorentz and Stortenbeker,[4] may be demonstrated as follows.

Although this system is one of two *independent* constituents, we may, if we choose, consider three constituents present in the solution, namely, n_1 moles of H_2O, n_2 moles of $CaCl_2$, and n_3 moles of $CaCl_2 \cdot 6H_2O$, with chemical potentials μ_1, μ_2, and μ_3. At the melting point of the pure hydrate, its composition is that of the solution, and therefore $x_1 = 6x_2$ (however far dissociation may occur according to the equation $CaCl_2 \cdot 6H_2O = CaCl_2 + 6H_2O$).

Let us now find the effect of adding dn_1 moles of water. By Eq. (10-29), we have, at constant temperature and pressure,

$$n_1 \, d\mu_1 + n_2 \, d\mu_2 + n_3 \, d\mu_3 = 0 \tag{10-53}$$

or

$$n_2(6 \, d\mu_1 + d\mu_2) + n_3 \, d\mu_3 = 0 \tag{10-54}$$

Now $CaCl_2 \cdot 6H_2O$, $CaCl_2$, and H_2O are in equilibrium and must so remain. Therefore, by Eq. (3-41),

$$\Delta G = \mu_3 - 6\mu_1 - \mu_2 = 0$$

$$d\mu_3 = 6 \, d\mu_1 + d\mu_2$$

Finally, if we combine this with Eq. (10-54), we obtain

$$(n_2 + n_3) \, d\mu_3 = 0 \tag{10-55}$$

And, since $n_2 + n_3$ are not zero, $d\mu_3 = 0$. In other words, the escaping tendency of $CaCl_2 \cdot 6H_2O$ in the liquid phase is not changed by an infinitesimal addition of water, and it will therefore remain in equilibrium with the solid $CaCl_2 \cdot 6H_2O$ without change in the equilibrium temperature.

Of course, thermodynamics is unable to predict how flat such a curve as that of Fig. 10-9 will be. This must depend upon the extent to which the compound dissociates, for if there were no dissociation at all in the solution, adding $CaCl_2$ or H_2O would be like adding some foreign constituent and the continuous curve would be replaced by two curves intersecting at the melting point, as illustrated by the dotted lines in the figure.

REFERENCES

1. *Collected Works of J. W. Gibbs*, Yale University Press, New Haven, 1948, vol. 1, Chapter III, p. 65; J. W. Gibbs, *Trans. Conn. Acad. Sci.*, **3**, 118 (1875).
2. P. Duhem, "Dissolutions et Melanges III," *Trav. Univ. Lille*, **III** (13), 79 (1894); M. Margules, *Sitzber. Akad. Wiss. Wien. Math.-naturw. Kl., Abt. IIa*, **104**, 1243 (1895).
3. J. von Zawidzki, *Z. physik. Chem.*, **35**, 129 (1900).
4. W. Stortenbeker, *Z. physik. Chem.*, **10**, 194 (1892).

PROBLEMS

10-1. What is the mole fraction of the solute in a molal aqueous solution? What is its mole ratio?

10-2. At 15°C the density of aqueous sulfuric acid containing 5 mol l^{-1} is 1.2894. Calculate the molality of H_2SO_4 and its mole fraction.

10-3. Show that in the curves T'' of Fig. 10-7 the maximum in the curve for one constituent must come at the same composition as the minimum in the curve for the other.

10-4. Show diagrammatically the sort of fugacity–composition curves that would characterize a solution possessing two critical mixing temperatures between which separation into two phases occurs.

10-5. A constant-boiling solution is one in which evaporation causes no change in the composition of the liquid. In other words, the composition of liquid and gaseous phase must be identical. If the vapors may be assumed to be perfect gases, then the ratio of the two partial pressures is equal to the ratio of the mole fraction in the liquid. Hence show that

$$\frac{dp_1}{dx_1} + \frac{dp_2}{dx_1} = 0$$

and that the total vapor pressure $P = p_1 + p_2$ is a maximum or minimum. This is an important principle in the theory of distillation.

10-6. Show from the phase rule that, under constant atmospheric pressure, a sodium chloride solution in the presence of the solid salt has one degree of freedom. Let us calculate in this case the change in the fugacity of water in the saturated solution as the temperature varies. Let f_1 be the fugacity of water in the solution, f_2' that of solid salt, and f_2 that of the dissolved salt. We write $d\ln f_2 = d\ln f_2'$ and further

$$d\ln f_1 = \frac{\partial \ln f_1}{\partial T} dT + \frac{\partial \ln f_1}{\partial x_1} dx_1$$

$$d\ln f_2 = \frac{\partial \ln f_2}{\partial T} dT + \frac{\partial \ln f_2}{\partial x_1} dx_1$$

$$d\ln f_2' = \frac{\partial \ln f_1'}{\partial T} dT$$

From these equations show that

$$\frac{d\ln f_1}{dT} = \frac{H_{m,1}^* - \overline{H}_{m,1}}{RT^2} + \frac{x_2}{x_1} \frac{H_{m,2}' - \overline{H}_{m,2}}{RT^2}$$

where the numerators of the two last fractions are the negative of the differential molar heats of solution of attenuated water vapor and of solid salt.

CHAPTER 11

IDEAL SOLUTIONS

Even though most real gases depart appreciably in their behavior from the ideal or perfect gas law, that equation is very valuable both as an approximate representation of the properties of real gases at low pressure and as a leading term in more complex equations of state. In a somewhat similar manner, there is an *ideal solution* equation that is useful both as an approximation to the properties of many real solutions and as a leading term for more complex equations for solution properties. Also, both equations become exact under certain limiting conditions, but the practical importance of this characteristic is much greater for the perfect gas than for the ideal solution equation. In any event, the concept of the ideal solution is very important in the study of solutions. This pattern of solution behavior was first recognized by Raoult[1] in 1887 and is also known as *Raoult's law*.

In the early experiments on mixing to form solutions, it was observed that hydrocarbons of similar molecular mass (or other pairs of organic liquids similar to one another) mixed with very little, if any, heat effect or volume change. Also, the vapor pressures for their solutions followed the pattern of Fig. 10-3 with two straight lines.

In order to focus our attention upon the solution itself and not upon another phase in equilibrium with it, let us consider the fugacities rather than vapor pressures and define the ideal solution as one in which the fugacity of each constituent is proportional to the mole fraction of that constituent. Indeed, we go still further and require that this proportionality exists at every pressure and at every temperature.

In accordance with our definition, we therefore write for the first component of an ideal solution,†

$$f_1^{id} = f_1^\circ x_1 \tag{11-1}$$

where f_1° at a given temperature and pressure is a constant. If we can proceed over the whole range of concentration to $x_1 = 1$, f_1° appears as the fugacity of the pure constituent. In the ordinary case where the solution in question is a liquid, f_1° represents the fugacity of the pure component in the liquid state. Similar relationships hold for other components. Indeed, it was shown above in connection with Eq. (10-44) that, if Eq. (11-1) holds for a binary solution over the whole range 0–1 of mole fraction, the fugacity of the other component must follow an expression of the same form:

$$f_2^{id} = f_2^\circ x_2 \tag{11-1a}$$

The corresponding equation for either chemical potential has the form

$$\mu_i^{id} - \mu_i^\circ = RT \ln(f_i/f_i^\circ) = RT \ln x_i \tag{11-2}$$

Then, for the process of mixing n_1 moles of component 1 with n_2 moles of 2, the Gibbs energy of mixing is

$$\Delta_{mix} G^{id} = n_1(\mu_1^{id} - \mu_1^\circ) + n_2(\mu_2^{id} - \mu_2^\circ) = RT(n_1 \ln x_1 + n_2 \ln x_2) \tag{11-3}$$

In this derivation we used Eq. (10-27) and the fact that μ_i is the partial molar Gibbs energy $\overline{G}_i$.

The generalization of this equation to three or more components is straightforward:

$$\Delta_{mix} G^{id} = \sum_i n_i(\mu_i^{id} - \mu_i^\circ)$$
$$= RT \sum_i n_i \ln x_i \tag{11-4}$$

Before proceeding further we note explicitly that *a mixed perfect gas is an ideal solution*. In a mixed gas, the partial pressure is defined as $p_i = y_i P$ with y_i the mole fraction‡ and P the total pressure. Then, if this is a perfect gas,

$$V^{pg} = \left(\sum_i n_i\right) \frac{RT}{P} \tag{11-5}$$

and each partial molar volume is

$$\overline{V}_i^{pg} = \frac{RT}{P} \tag{11-6}$$

---

† Most of the equations of this chapter are approximate by themselves for real solutions but are exact as definitions of ideal behavior for use in later chapters. Hence, we use the "id" superscript but omit the asterisk for those equations cited later as definitions.

‡ The symbol y_i is commonly used for the mole fraction in the gas phase with x_i for the liquid (or solid) phase.

Then, from Eqs. (7-15) and (10-21),

$$d\ln f_i^{pg} = \left(\frac{\overline{V}_i}{RT}\right) dP$$

$$= d\ln P \tag{11-7}$$

This shows that f_i is proportional to P. Then we take the proportionality constant to be the mole fraction, whereupon

$$f_i^{pg} = y_i P = p_i \tag{11-8}$$

This confirms the result anticipated in Chapters 7 and 10 that the fugacity was equal to the partial pressure for an ideal gas; and also that the fugacity is proportional to the mole fraction at any given T and P, which is the characteristic of an ideal solution. Thus, the Gibbs energy of mixing of a mixed perfect gas is given by Eq. (11-4) and other properties by the equations to follow.

Equations for Ideal Mixing

We have, in Eq. (11-4) obtained the equation for the Gibbs energy of mixing for the ideal solution. For one mole of total solution this becomes

$$\Delta_{mix} G_m^{id} = RT\left(\sum_i x_i \ln x_i\right) \tag{11-9}$$

We now use the basic thermodynamic equations of Chapter 3 to obtain the expressions for the change in entropy, enthalpy, and other properties for the mixing process. In these derivations, we assume that the solution remains ideal as the temperature or pressure is changed. Then, from Eqs. (3-9) and (11-4),

$$\Delta_{mix} S^{id} = -\left[\frac{\partial (\Delta_{mix} G^{id})}{\partial T}\right]_P = -R\sum_i n_i \ln x_i \tag{11-10}$$

If there are only two components, this becomes

$$\Delta_{mix} S^{id} = -R(n_1 \ln x_1 + n_2 \ln x_2) \tag{11-11}$$

and for one mole of solution,

$$\Delta_{mix} S_m^{id} = -R(x_1 \ln x_1 + x_2 \ln x_2) \tag{11-12}$$

Similar methods yield, for the enthalpy and the volume of mixing,

$$\Delta_{mix} H^{id} = \left[\frac{\partial (\Delta_{mix} G^{id}/T)}{\partial (1/T)}\right]_P = 0 \tag{11-13}$$

$$\Delta_{mix} V^{id} = \left[\frac{\partial (\Delta_{mix} G^{id})}{\partial P}\right]_T = 0 \tag{11-14}$$

These results follow from the fact that all $n_i \ln x_i$ are independent of T and P. For properties given by derivatives of H or V, such as heat capacity or coefficient of expansion, the change on mixing is also clearly zero.

Statistics of Ideal Entropy of Mixing

The expression for the entropy of ideal mixing can also be obtained from a rather simple statistical derivation, which we give next. It offers further insight into and understanding of the nature of an ideal solution. Assume that there are N_1 and N_2 molecules of the two types, respectively, and that they are completely interchangeable without affecting the internal energy. Thus we may take the positions of all these molecules at some instant as constituting an array of $N_1 + N_2$ sites. We calculate the multiplicity of a random distribution. The first molecule can be placed on any of the $N_1 + N_2$ sites, the second on any of the $N_1 + N_2 - 1$ empty sites, the third on any of the $N_1 + N_2 - 2$ empty sites, and so on. The total number of possibilities is $(N_1 + N_2)(N_1 + N_2 - 1) \ldots = (N_1 + N_2)!$. However, the molecules of the same kind are not distinguishable from one another; consequently we must divide by $N_1!$, which is the number of possible interchanges of molecules of the first kind, and by $N_2!$ for the interchanges of molecules of the second kind. Thus the number of distinguishable arrangements of the $N_1 + N_2$ molecules is

$$W^{id}_{mixed} = \frac{(N_1 + N_2)!}{N_1! N_2!} \tag{11-15}$$

For the pure components before mixing one has the trivial values

$$W_1 = \frac{N_1!}{N_1!} = 1 \qquad W_2 = \frac{N_2!}{N_2!} = 1$$

The entropy of mixing is given by the multiplicity ratio

$$\Delta_{mix} S^{id} = k \ln \frac{W^{id}_{mix}}{W_1 W_2} = k \ln \frac{(N_1 + N_2)!}{N_1! N_2!} \tag{11-16}$$

Since these numbers are large, we may use Stirling's approximation† for the factorials:

$$\ln N! = N \ln N - N \tag{11-17}$$

---

† In case the reader is unfamiliar with Stirling's formula, we note that $\ln N! \cong \int_1^N \ln y \, dy = [y \ln y - y]_1^N = N \ln N - N + 1$. If N is large, the 1 may be neglected.

Then we find

$$\Delta_{mix}S^{id} = k[(N_1 + N_2)\ln(N_1 + N_2) - N_1 \ln N_1 - N_2 \ln N_2]$$
$$= -k\left[N_1 \ln\left(\frac{N_1}{N_1 + N_2}\right) + N_2 \ln\left(\frac{N_2}{N_1 + N_2}\right)\right]$$
$$= -k(N_1 \ln x_1 + N_2 \ln x_2) \qquad (11\text{-}18)$$

where the mole fractions x_1 and x_2 are introduced in the final step. This is the ideal, or Raoult's law, entropy of mixing. In terms of the numbers of moles n_1 and n_2, the result is

$$\Delta_{mix}S^{id} = -R(n_1 \ln x_1 + n_2 \ln x_2) \qquad (11\text{-}11)$$

which is exactly the equation obtained above.

This derivation is equally appropriate to liquid or solid solutions. In an earlier section we derived Raoult's law for mixtures of perfect gases, and in Chapter 13 we give the derivation for mixtures of real gases with approximately equal imperfection properties. In all cases except the perfect gas, the assumptions imply that the molecules have the same size and that the intermolecular forces between pairs of like molecules of each type, as well as between unlike molecules, are all the same. These are very stringent conditions. Molecules that differ only by the substitution of isotopes of heavy atoms provide the examples that fully conform to this model, and indeed they form ideal solutions. Quantum effects yield nonidealities for the substitution of isotopes of hydrogen and, to a rapidly decreasing extent with increase in atomic mass, for other light atoms.

Actually, if we allow even 1 or 2 percent deviation, we find that a considerably broader range of molecular properties still yields ideal behavior. Examples include (1) benzene and toluene, (2) 1,2-dibromoethane and 1,2-dibromopropane, and (3) methyl iodide and chloroform.

From the point of view of statistical theory, it is not clear at present just why these systems with greater differences in molecular properties still yield ideal solutions. The toluene molecule is substantially larger than that of benzene; consequently they cannot be regarded as interchangeable. It seems likely that small differences in molecular size and small differences in intermolecular force yield opposite deviations from Raoult's law and that there is a cancellation of opposing effects in some cases. These questions are discussed further in Chapter 12 after equations are developed for nonideal solutions.

Solubility of a Solid in an Ideal Liquid Solution

If one component of an ideal liquid solution has a high melting point, it may not be soluble to mole fraction unity. Instead, the solid may come to equilibrium with a saturated solution of the liquid. If this is component 2, then in the equation

$$f_2(T) = x_2(T)f_2^\circ(T) \qquad (11\text{-}1a)*$$

$f_2^\circ(T)$ is the fugacity of supercooled pure liquid 2, and the dependencies on temperature are indicated explicitly. Let $f_2^*(T)$ be the fugacity of the pure solid component 2 at T.

Consider first the situation at the melting point, T_m, where pure liquid and solid are in equilibrium; hence,

$$f_2^\circ(T_m) = f_2^*(T_m)$$

At a lower temperature, the equilibrium mole fraction is

$$x_2(T) = \frac{f^*(T)}{f^\circ(T)} \qquad (11\text{-}19)*$$

Next we consider the effect of temperature on the fugacities. Taking logarithms in Eq. (11-19)*, we obtain for the temperature derivative

$$\left(\frac{\partial \ln x_2}{\partial T}\right)_P = \left(\frac{\partial \ln f^*}{\partial T}\right)_P - \left(\frac{\partial \ln f_2^\circ}{\partial T}\right)_P \qquad (11\text{-}20)*$$

Next, we note that $d\ln f = R^{-1} d(G/T)$ and that

$$\left[\frac{\partial (G/T)}{\partial T}\right]_P = -\frac{H}{T^2} \qquad (3\text{-}18)$$

Therefore,

$$\left(\frac{\partial \ln x_2}{\partial T}\right)_P = \frac{H^\circ(\text{liq}) - H^*(\text{solid})}{RT^2} = \frac{\Delta H(\text{fusion})}{RT^2} \qquad (11\text{-}21)*$$

where ΔH(fusion) is the heat of melting the solid to the supercooled liquid. Over a small range of temperature ΔH may be taken as constant, in which case a simple integration gives

$$\ln\left(\frac{x_2''}{x_2'}\right) = \int_{T'}^{T''} \frac{\Delta H}{RT^2} dT = \frac{-\Delta H}{R}\left(\frac{1}{T''} - \frac{1}{T'}\right) = \frac{\Delta H}{R}\left(\frac{T'' - T'}{T''T'}\right) \qquad (11\text{-}22)*$$

This equation was first obtained by Schroder,[2] who also made a thorough experimental investigation of solubilities of substances of this type. For example, he found that naphthalene dissolves in benzene at 61°C until $x_2 = 0.689$. Naphthalene melts at 79.9°C, and the heat of fusion is 2295R. Substitution of these values in Eq. (11-22)* yields a calculated value of $x_2 = 0.692$, which agrees very satisfactorily with the measured value.

The ΔH of fusion was assumed to be a constant in Eq. (11-22)*, but the heat capacity of a liquid is frequently higher than that of the solid. Thus the ΔH of fusion should be expressed as a function of temperature before integration of Eq. (11-21)* if high accuracy is desired over a wide temperature range. A next approximation is to assume ΔC_P of fusion to be constant, which yields

$$\Delta H = \Delta H' + \Delta C_P(T - T') \qquad (11\text{-}23)^*$$

$$\ln\left(\frac{x_2''}{x_2'}\right) = \int_{T'}^{T''} \frac{(\Delta H' - T'\Delta C_P)}{RT^2} dT + \int_{T'}^{T''} \frac{\Delta C_P}{RT} dT$$

$$= \frac{(\Delta H' - T'\Delta C_P)(T'' - T')}{RT''T'} + \frac{\Delta C_P}{R}\ln\left(\frac{T''}{T'}\right) \qquad (11\text{-}24)^*$$

Here T' may be any fixed temperature, but we ordinarily take it to be the melting point, T_m. Then $\Delta H'$ is the heat of fusion at the melting point, and $x_2' = 1$.

Equation (11-24)* was applied by Pitzer and Scott[3] to solutions of xylene isomers in one another. If we choose p-xylene and m-xylene as an example, the pertinent data are given in Table 11-1. The values of $3R$ for ΔC_P for each xylene were obtained by extrapolation of the curves for solid and liquid to the melting point and are only approximate.

In this case the heat of fusion of each component is known, and we can calculate the solubility of each in the liquid solution, which we have assumed to be ideal. At the eutectic point both solids are in equilibrium with the same solution. This point is found from the pair of solubility equations; a graphical solution such as is shown in Fig. 11-1 is usually most convenient. The close agreement found in Table 11-1 for the xylene solution indicates close conformity to the ideal-solution law in that case. This result is reasonable in view of the close similarity of the two isomeric components.

It is interesting to note that the specific properties of the solvent (i.e., the other component) do not enter this calculation at all. Hence the solubility in mole fraction units at a given temperature and pressure for one substance is the same in all solvents that yield ideal solutions. Some results of Schroder[2] illustrate this situation very well. He measured the solubility of p-dibromobenzene in several solvents. The results are shown in Fig. 11-2. The curve of solubility with temperature is almost identical for benzene, bromobenzene, and carbon disulfide but is quite different for ethyl alcohol, which would not be expected to form an ideal solution.

TABLE 11-1
Solubility of *m*- and *p*-xylenes

	m-Xylene	*p*-Xylene
T_m(K)	225.27	286.39
$\Delta H(T_m)/R$/K	1391	2058
$\Delta C_P/R$	3	3

Eutectic temperature: Calculated 220.3 K, observed 220.2 K
Eutectic composition: Calculated 13.0% *para*, observed 12.9% *para*

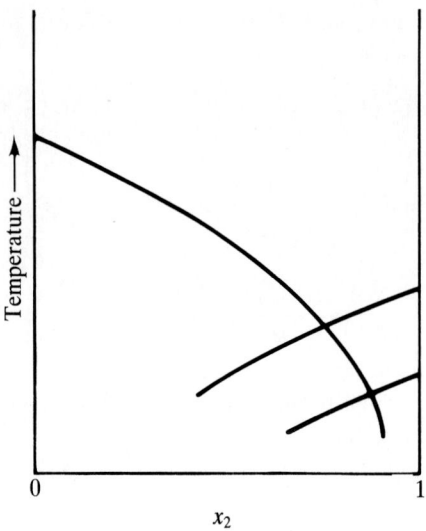

FIGURE 11-1
Curves for solid–ideal liquid solution equilibria for the three xylenes. The intersections represent the eutectic points for the pairs; extensions of curves below each eutectic temperature represent solutions supersaturated with respect to one of the two components.

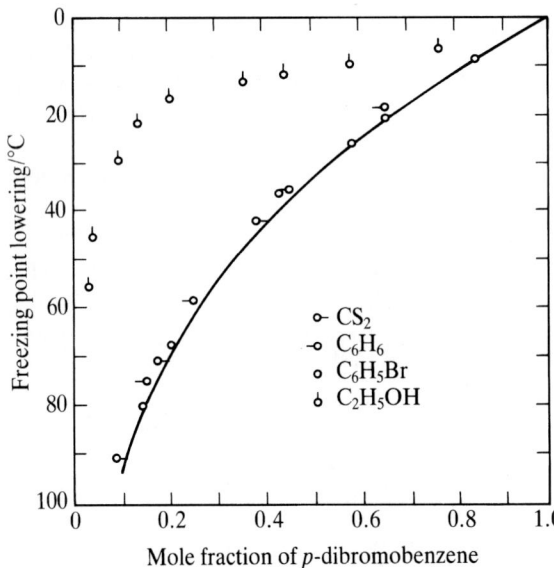

FIGURE 11-2
Solubility of p-dibromobenzene in several organic liquids presented as freezing point lowering.

The points in Fig. 11-2 may be interpreted as expressing the solubility of p-dibromobenzene in the several solvents, or the lowering of the freezing point of p-dibromobenzene by the same substances.

In this chapter we have gone rather fully into the theory of the ideal solution, notwithstanding the fact that in the ordinary calculations of thermodynamic chemistry we do not often meet with solutions that can be regarded as ideal. We have presented the properties of these solutions because the ideal solution is in some sense the norm with respect to which we may classify solutions in general. Indeed, most of the important methods of treating nonideal solutions proceed by dealing with the departure from ideal behavior as an activity coefficient, in the case of fugacity, or as an excess function in the case of the enthalpy or Gibbs energy.

REFERENCES

1. F. M. Raoult, *Compt. rend.*, **104**, 1430 (1887); *Z. physik. Chem.*, **2**, 353 (1888).
2. I. Schroder, *Z. physik. Chem.*, **11**, 449 (1893).
3. K. S. Pitzer and D. W. Scott, *J. Am. Chem. Soc.*, **65**, 803 (1943).

PROBLEMS

11-1. Show that, if a binary solution is ideal with respect to one component over the full range of mole fraction 0 to 1, it must be ideal also with respect to the other component.

11-2. Calculate the total ΔG of mixing to form an ideal liquid solution from 6 moles of liquid hexane, 3 moles of liquid pentane and 1 mole of butane gas, initially at 1 bar, all at 0°C. The vapor pressure of pure butane at 0°C is 5.00 bar. Assume the perfect gas law as well as the ideal solution equations.

11-3. Assuming an ideal solution, what is the ratio between the fugacity of pure ether and the fugacity of ether in a solution containing 10 g of ether, 10 g of benzene and 10 g of carbon disulfide?

11-4. Predict the freezing point of a 1 mol% solution of any ideal solute in benzene. The melting point of pure benzene is 5.53°C and the heat of fusion 9837 J mol^{-1}.

11-5. (a) Calculate the equilibrium vapor pressure at 273 K of butane above a liquid solution of butane and benzene saturated with, i.e., in equilibrium with, solid benzene. Assume the properties of benzene given in Problem 11-4, that the vapor pressure of pure butane is 5.0 bar at 273 K, and that the liquid solution is ideal. (b) What is the ΔG for the formation of 1 kg of this solution from solid benzene and butane gas at 2.0 bar, all at 273 K? Assume the ideal gas law.

11-6. Calculate the minimum work necessary to separate 1 mole of pure benzene from (a) a large volume of an equimolal solution of benzene and toluene; (b) a solution composed of exactly 1 mole of benzene and 1 mole of toluene.

11-7. Investigate the effect of the $\Delta \overline{C}_P$ term on the change of solubility of *p*-xylene with temperature. Compare the results calculated from Eq. (11-24)* with those from Eq. (11-22)* at 220.3 K. Note data in Table 11-1.

CHAPTER 12

NONIDEAL SOLUTIONS, EXCESS FUNCTIONS, ACTIVITIES, AND ACTIVITY COEFFICIENTS

While the basic thermodynamics of mixed systems was given in Chapter 10, there are additional working equations and definitions that are very valuable and in general use. The first step toward their presentation was the definition and description of the *ideal solution* in Chapter 11. While few real solutions are ideal within modern experimental accuracy, many depart only a little from ideality. Thus, it is very convenient to subtract the larger effects on fugacity of the ideal solution equations and, for each individual case, deal only with the small departures from ideality.

Excess Functions

The excess Gibbs energy is defined simply as the difference of the actual Gibbs energy at a given temperature, pressure, and composition from that of an ideal solution at the same conditions

$$G^E(T, P, x_i) = G(T, P, x_i) - G^{id}(T, P, x_i) \tag{12-1}$$

where the superscripts E and id denote excess and ideal, respectively. If we are concerned for the moment with the solution and its nonideality, we take the pure components at the same T and P as references and consider just the mixing process. Then at a given T, P, and x_i,

$$G^E = \Delta_{mix}G - \Delta_{mix}G^{id}$$

and from Eq. (11-4),

$$\Delta_{mix}G^{id} = RT \sum_i n_i \ln x_i \tag{12-2}$$

The familiar relationships of S, H, C_P, and V to G apply for excess functions, with the results of Chapter 11 for ideal behavior. Thus,

$$S^E = \Delta_{mix}S - \Delta_{mix}S^{id}$$

$$= \Delta_{mix}S + R \sum_i n_i \ln x_i \tag{12-3}$$

Furthermore, the volume, enthalpy, and heat capacity of ideal mixing are all zero; hence,

$$V^E = \Delta_{mix}V \tag{12-4}$$

$$H^E = \Delta_{mix}H \tag{12-5}$$

$$C_P^E = \Delta_{mix}C_P \tag{12-6}$$

The excess chemical potential follows from differentiation of Eqs. (12-1) and (12-2):

$$\mu_i^E = \left(\frac{\partial G^E}{\partial n_i}\right)_{T,P,n_{j \neq i}}$$

$$= \mu_i - RT \ln x_i \tag{12-7}$$

NONIDEALITY IN SIMPLE SYSTEMS

At this point, we consider the methods of treating nonideality in simple systems; this will illustrate many aspects of solution nonideality. In later sections these same aspects are revisited with more comprehensive consideration of various types of systems where alternate definitions must be adopted.

Activity and Activity Coefficients

The equations for the chemical potential μ_i of each component have, to this point, been given in terms either of that quantity or of the fugacity, which is logarithmically related and referenced to the pressure as a perfect gas. It is more convenient for liquid or solid solutions to take as the reference the pure component in the same state, liquid or solid, as the solution. Thus, we define for the activity

$$a_i = \frac{f_i}{f_i^*} \tag{12-8}$$

where f_i^* is the fugacity of the pure liquid or solid. Often, the symbol f_i^o is used instead of f_i^*; we make no distinction between them at this point. If a pure component does not exist in the same solid or liquid state as the solution at the given T and P, different definitions, to be given below, must be used.

The ideal activity follows from Eq. (11-1); it is just the mole fraction

$$a_i^{\text{id}} = x_i \tag{12-9}$$

It is convenient to define still another quantity as the ratio of the actual activity to the ideal activity. It is called the *activity coefficient* and given the symbol γ_i; thus,

$$\gamma_i = \frac{a_i}{x_i} \tag{12-10}$$

and from Eqs. (10-40) and (12-7),

$$\ln \gamma_i = \frac{\mu_i - \mu_i^{\text{id}}}{RT} = \frac{\mu_i^{\text{E}}}{RT} \tag{12-11a}$$

$$= \frac{1}{RT}\left(\frac{\partial G^E}{\partial n_i}\right)_{T,P,n_{j \neq i}} \tag{12-11b}$$

Quantities called activity coefficients are also defined on the basis of compositions given in molality or concentration, and various symbols are used. The term "rational activity coefficient" and the symbol f_i are often used for the quantity defined above, with "activity coefficient" and γ_i reserved for the corresponding quantity when composition is given in molality (see Chapter 14). We prefer to retain f_i for fugacity, and use an extra subscript x for mole fraction or m for molality when there is any ambiguity.

A Very Simple Example

With these definitions, the excess Gibbs energy must be zero for each pure component, i.e., if either x_1 or x_2 is zero. Then for a binary system containing 1 mole, the simplest expression is

$$\frac{G_m^E}{RT} = wx_1x_2 = wx_2(1 - x_2) \tag{12-12}*$$

with w a constant at a given T and P. For an indefinite amount of material, this becomes

$$\frac{G^E}{RT} = \frac{wn_1n_2}{n_1 + n_2} \tag{12-13}*$$

and the activity coefficients are

$$\ln \gamma_1 = wx_2^2 \tag{12-14a}*$$

$$\ln \gamma_2 = wx_1^2 \tag{12-14b}*$$

This pattern of behavior was first described by Margules[1] and his name is often associated with these equations. The term *regular solutions* is often used for this behavior, although this designation sometimes carries further implications about the value of the parameter w and/or a more complex composition dependence.

Here w is defined as a dimensionless quantity. In the literature it is often defined without the factor RT in Eq. (12-12)*, whereupon w has the dimension of energy. We prefer the present definition for general use and relate the product wRT to interaction energies in a later section.

From the equations relating $S, H, C_P,$ and V to G, one finds for one mole of binary solution,

$$H_m^E = -x_1 x_2 RT^2 \left(\frac{\partial w}{\partial T}\right)_P \qquad (12\text{-}15)*$$

$$S_m^E = -x_1 x_2 R \left[w + T\left(\frac{\partial w}{\partial T}\right)_P\right] \qquad (12\text{-}16)*$$

$$C_{P,m}^E = -x_1 x_2 RT \left[2\left(\frac{\partial w}{\partial T}\right)_P + T\left(\frac{\partial^2 w}{\partial T^2}\right)_P\right] \qquad (12\text{-}17)*$$

$$V_m^E = x_1 x_2 RT \left(\frac{\partial w}{\partial P}\right)_T \qquad (12\text{-}18)*$$

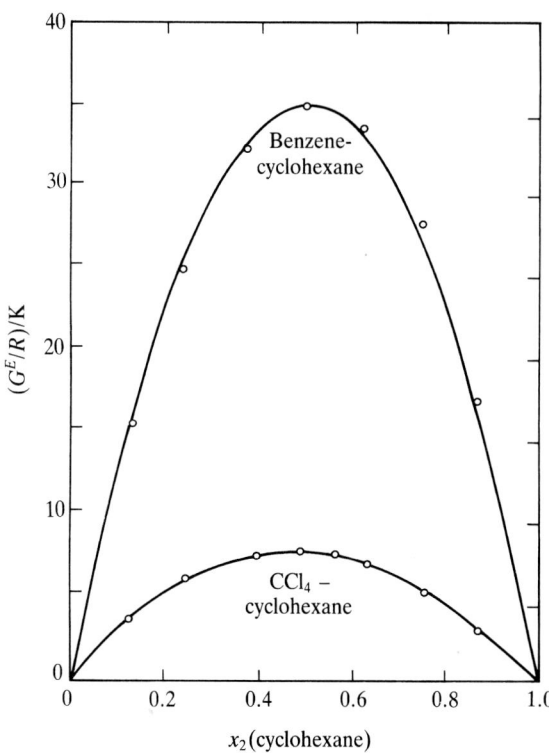

FIGURE 12-1
The excess Gibbs energy of mixing vs. mole fraction cyclohexane. The points are experimental; the curves are theoretical for the Margules equation (12-12)*.

TABLE 12-1
Properties of some simple solutions $n_1 = n_2 = x_1 = x_2 = 0.50$; dimensionless or in kelvins

	T (K)	$\ln \gamma = w/4$	$H^E/R =$ $-\dfrac{T^2}{4}\dfrac{\partial w}{\partial T}$ (K)	$S^E/R =$ $-\dfrac{1}{4}\left(w + T\dfrac{\partial w}{\partial T}\right)$	$C_P^E/R =$ $-\dfrac{T}{4}\left[T\dfrac{\partial^2 w}{\partial T^2} + 2\dfrac{\partial w}{\partial T}\right]$
CCl_4–benzene[a]	298	0.033	13.1	0.011	0.19
CCl_4–benzene	343	0.026	21.5	0.037	
CCl_4–cyclohexane[b]	313	0.026	17	0.028	
Benzene–cyclohexane[b]	293	0.13	100	0.21	−0.4
Benzene–cyclohexane	343	0.09	81	0.15	
Benzene–1,2-$C_2H_4Cl_2$[a]	298	0.009	8	0.015	0.0
Benzene–1,2-$C_2H_4Cl_2$[a]	343	0.006	7	0.015	
Benzene–toluene[c]	353	−0.004	6	0.019	−0.05
CS_2–acetone[a]	308	0.41	175	0.16	0.4
n-C_5F_{12}–n-C_5H_{12}[d]	277	0.53	186	0.14	

[a] L. A. K. Stavely, W. I. Tupman, and K. R. Hart, *Trans. Faraday Soc.*, **51**, 323 (1955).
[b] G. Scatchard, S. E. Wood, and J. M. Mochel, *J. Phys. Chem.* **43**, 119 (1939); *J. Am. Chem. Soc.*, **61**, 3206 (1939), **62**, 712 (1940).
[c] C. H. Cheesman and W. R. Ladner, *Proc. R. Soc. (London)*, **A229**, 387 (1955); A. P. Rollet, G. Elkaim, P. Toledano, and M. Senez, *Compt. rend.* **242**, 2560 (1956).
[d] J. H. Simons and R. D. Dunlap, *J. Chem. Phys.* **18**, 335 (1950).

Figure 12-1 shows the excess Gibbs energy for two examples that follow the Margules pattern accurately, while Table 12-1 gives numerical values for these and some additional examples. In most cases, both pressure and composition were measured for the vapor in equilibrium with the liquid solution. Corrections for nonideality of the vapor phase are small; these are considered in Chapter 13. Heats and volume changes on mixing can be measured directly or obtained from the temperature and pressure derivatives of the Margules parameter.†

Limiting Properties at Great Dilution

It is interesting to note that the behavior of this simple system as one mole fraction becomes very small. If $x_2 \to 0$,

$$\gamma_1 = 1 + wx_2^2 + \cdots \qquad (12\text{-}18\text{a})*$$

$$a_1 = x_1 + w(x_2^2 - x_2^3 + \cdots) \qquad (12\text{-}18\text{b})*$$

---

† In general the Margules equation is a series in powers of mole fraction, of which Eq. (12-12)* is just the first term. The full equation is given later as Eq. (12-73)*.

Thus the abundant component or solvent follows Raoult's law or is ideal to an excellent approximation with departure only as x_2^2.

For the solute or rare component, one obtains

$$\gamma_2 = \exp[w(1 - 2x_2 + x_2^2)]$$

$$= \exp(w)[1 - 2wx_2 + \cdots] \tag{12-19a}$$

$$a_2 = x_2 \exp(w)[1 - 2wx_2 + \cdots] \tag{12-19b}$$

Hence, we find that, in the limit of zero x_2, the solute activity is proportional to its mole fraction with a proportionality factor that is $\exp(w)$ for this simple model. This result is general (unless the solute dissociates on dilution) and is known as Henry's law. As compared to the case of the solvent, however, the departure from this limiting law is much more rapid, since it is given by $-2wx_2$ in comparison to unity. One should always anticipate departures from Henry's law if the measurements extend over any significant range of mole fraction.

This pattern of solute behavior also occurs if the solute is a gas rather than a liquid at the given T and P. Indeed, it is for that situation that the customary definition of Henry's constant is particularly appropriate; it is

$$H_{2,1} = \lim_{x_2 \to 0} \left(\frac{f_2}{x_2} \right) \tag{12-20}$$

Then the solute activity coefficient becomes

$$\gamma_2 = \frac{f_2}{x_2 H_{2,1}} \tag{12-21}$$

and other relationships can be derived for a liquid solution where one pure component is a gas at the given T and P.

STANDARD STATES AND ACTIVITIES

All of the equations given above through (12-11) are generally valid; indeed, several serve as definitions of excess functions, activities, and activity coefficients. But it was noted that there was need to consider the choice of standard states on a more comprehensive basis. This is done in the following paragraphs, which also include new definitions of activity and activity coefficient for the new situations as well as recapitulation in the familiar cases.

1. **FOR A GAS.** $a = f$: Thus, $a_i/y_i P = 1$ when $P = 0$ with y_i the mole fraction; also for a pure gas $a/P = 1$ when $P = 0$. Since the activity of a substance at a given temperature is always proportional to its fugacity, it is convenient in the case of a gas to make the activity of each component equal to its fugacity.

In Chapter 8 we have briefly referred to a 1-bar perfect-gas standard state for tabulation of entropies and Gibbs energies, and in Chapter 9 we have presented the procedure for calculating standard entropy and Gibbs energy values for pure gases from experimental data and the reverse procedure for calculating values for a real gas at any pressure from the standard values. At this time we can characterize this standard state more explicitly. For a perfect gas this is the same as making the activity unity at unit pressure; but in general the activity will not be exactly equal to the pressure. It is therefore to be borne in mind that, while the standard state of a liquid or solid is always taken at unit pressure, the standard state of a gas is one in which not the pressure but the fugacity is unity.

As a matter of fact we wish to go a little further than this. The standard state, as we shall use the term, implies not only that the fugacity is unity but that the heat capacity, enthalpy, etc., are those of the gas at infinite attenuation. Our standard state therefore is a hypothetical one and corresponds to no real state of the gas.

The characterization of the standard state for a pure gas is illustrated in Fig. 12-2, where the fugacity of a real gas is plotted against the pressure. For a mixed gas, one substitutes the partial pressure $p_i = y_i P$ for the abscissa, and Fig. 12-2 still pertains. At low pressures the fugacity vs. pressure curve approaches a limiting slope of unity, which is indicated by the dotted line and represents perfect-gas behavior. For purposes of tabulating heat capacities and enthalpies, it is sufficient to characterize the gaseous standard state as corresponding to a state of the gas at a low enough pressure that the difference between the curve for the real gas and the perfect gas is less than experimental error. To specify the entropy or the Gibbs energy, one must also give a specific pressure or fugacity. This is done by fixing the standard state for a pure gas at a value of fugacity equal to unity as indicated by the point of the dotted curve of

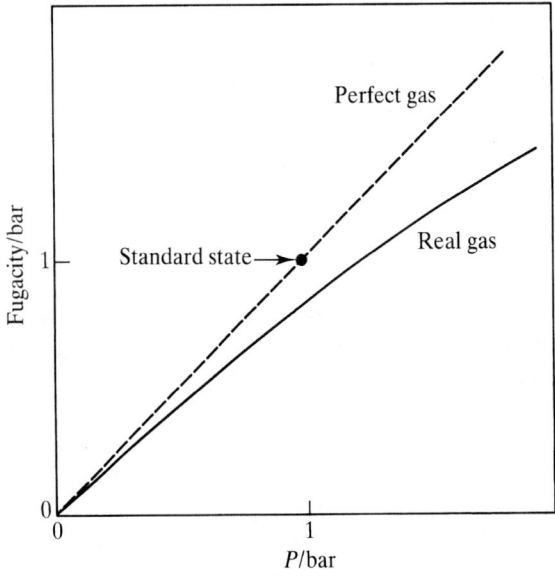

FIGURE 12-2

Fig. 12-2. To relate this hypothetical state to any real state of the gas, one must move along the dotted line that represents perfect-gas behavior until the difference between it and the real curve is negligible. Then one must move along the real curve by the methods given in Chapter 9 until one has reached the desired state. It may seem to be an unnecessary complication to use such a hypothetical state as a standard state. However, many of our calculations will be applied to systems where the difference between the perfect-gas behavior and the real-gas behavior is small and the difference is easily calculated as a second-order correction. Under those circumstances, the use of the perfect-gas standard state is quite advantageous. When we deal with actual equilibrium calculations in detail, it will be obvious that even when deviations from ideality are large there are additional advantages in choosing the perfect-gas standard state for which the dependence of properties upon pressure and temperature is readily separable.

For a mixed gas the assumption of perfect gas behavior also implies ideal solution properties—ideal entropy of mixing, etc. Thus, in real gases there can be deviations from perfect-gas properties with or without deviations from ideal-solution behavior. These and other aspects of the properties of mixed gases are considered in Chapter 13.

2. FOR A LIQUID OR SOLID THAT MAY ACT AS A SOLVENT. $a/x = 1$ when $x = 1$: We have noted previously that the pure liquid or pure solid at 1 bar pressure is chosen as the standard state for the tabulation of heat capacities, enthalpies, entropies, Gibbs energies, and fugacities. This same standard state is applicable to solutions also. It is always used for the component in largest amount, e.g., the solvent, and is frequently used for all components. The activity of a solvent is defined by

$$a_i = \frac{f_i}{f_i^\circ} \tag{12-22}$$

where, unlike the case of the gas, f_i° is not unity.

Now, if a solvent has a small but measurable vapor pressure, we may let P_i be the measured vapor pressure of the solvent from the solution and P_i° be the vapor pressure of the pure solvent. Then, if we may assume that the vapor behaves as a perfect gas, $P_i = f_i$ and $P_i^\circ = f_i^\circ$, so that

$$a_i = P_i/P_i^\circ \tag{12-22a}*$$

If more than one component has appreciable vapor pressure, P_i becomes the partial pressure given by the mole fraction times the total pressure. If the vapor is not a perfect gas, one follows the methods described in Chapters 9 and 13 to obtain the fugacities.

When the vapor pressure of the solvent is not measurable, numerous other methods of determining the activity are available. Thus, if we should measure the emf of some galvanic cell with a silver electrode and then substitute for the silver a solid solution of gold in silver, the emf would be found to change, owing to the fact that the gold lowers the chemical potential of the silver. A careful investigation of this sort would enable us, by methods that we shall illustrate later, to determine the activity of the silver in a series of gold–silver alloys.

3. FOR AN UNCHARGED SOLUTE†. $a_2/x_2 = 1$, when $x_2 = 0$:

Again, the pressure is assumed to be 1 bar. In the typical case of a solution at infinite dilution, we have Henry's law, which states that the fugacity of the solute, and therefore its activity, is proportional to its mole fraction. It is therefore convenient to choose the standard state of a solute A_2 so that at 1 bar pressure and at infinite dilution, the ratio of the activity to the measure of concentration is unity. Thus, one chooses $a_2/x_2 = 1$ when $x_2 = 0$.

This definition yields a hypothetical state as the standard state and in this respect is exactly like the definition for a gas. The standard state is not the real state of unit activity but rather a hypothetical ideal solution in which the partial molar enthalpy and heat capacity of the solute have the values of the infinitely dilute solution but the Gibbs energy and entropy correspond to unit activity. Figure 12-3 illustrates this definition of the solute standard state. At low mole fraction the plot approaches asymptotically the Henry's law slope, which is indicated by the dashed line. However, it is not sufficient just to refer to the infinitely dilute solution, since there the partial molar Gibbs energy

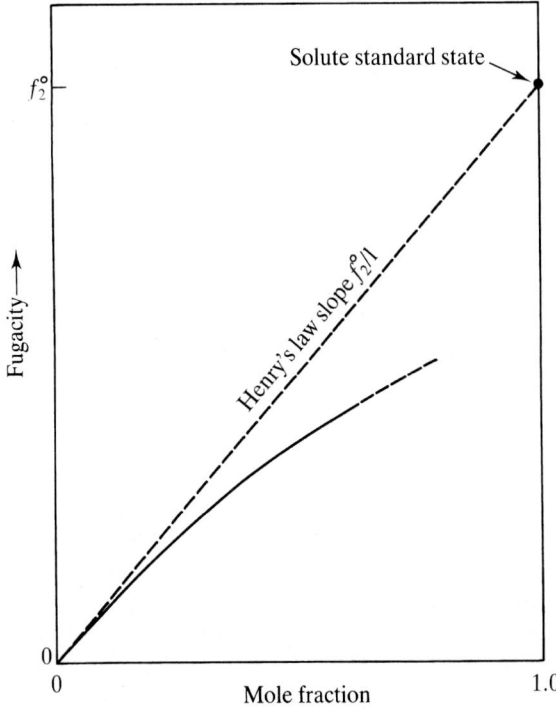

FIGURE 12-3

---

† The case of ionic solutes will be discussed in Chapter 15.

becomes minus infinity, as shown in Fig. 10-1. Hence one specifies unit activity as the standard state.

In mathematical terms the equations for the very dilute solution are

$$f_2^\circ = \lim_{x_2 \to 0} \left(\frac{f_2}{x_2}\right) \tag{12-23}$$

and for the standard chemical potential

$$\mu_2^\circ = \overline{G}_2^\circ = \lim_{x_2 \to 0} [\overline{G}_2(x_2) - RT \ln x_2] \tag{12-24}$$

For the determination of the solute standard state of a solute component, this limiting process must be carried out in the binary system of one solvent and one solute. Thus, a solute standard state is a property dependent on each of the two components. The situation with more than two components is considered below. Also, there are situations where the solute dissociates in the dilute range; this is discussed later in this chapter.

In order to avoid ambiguity between the solvent or solute definitions for activity coefficients, one often adds an asterisk in the solute case. Thus,

$$\gamma_2^* = \frac{f_2}{f_2^\circ} \tag{12-25}$$

with f_j° defined by the limiting process in Eq. (12-23). Then

$$\lim_{x_2 \to 0} \gamma_2^* = 1 \tag{12-26a}$$

whereas for the solvent standard state

$$\lim_{x_1 \to 1} \gamma_1 = 1 \tag{12-26b}$$

General Relationships for Activities, Fugacities, and Standard States

We can now summarize the preceding sections by noting that in all cases

$$a_i = \frac{f_i}{f_i^\circ} \tag{12-27a}$$

and

$$\mu_i - \mu_i^\circ = RT \ln a_i \tag{12-27b}$$

Most data on aqueous solutions are reported in terms of the molality, i.e., the number of moles per 1000 g of water rather than as mole fractions. Also, many aqueous solutions involve ionic solutes and in those cases there are additional factors to be considered. A satisfactory treatment can be given on the molality basis, but even for neutral solutes the simplest behavior pattern is different from that of ideal solutions on the mole fraction basis. Consequently, it seems best to postpone this presentation to Chapter 14. This will avoid both the complication and possible confusion of describing at this point the differences between ideality on the mole fraction basis and the pattern implied by unit activity coefficient on the molality basis.

In this chapter we emphasize binary solutions and leave to later chapters the full treatment of solutions with three or more components. But it is desirable to present certain equations in general, multicomponent form and to note when particular problems arise in generalization from two to three or more components. All equations through (12-11) are in multicomponent form; also, there is no problem with the preceding definitions for gases or for liquids or solids when the solvent (pure liquid or solid) standard state is used. For that reason, we have written the accompanying equations in general terms for component i. But the solute standard state, for either mole fraction or molality units, is a function of the solvent as well as the solute. If there is a single solvent, one can have multiple solutes without complication. But, if there is a mixed solvent, i.e., multiple solvent components with variable composition, the situation becomes much more complex and is not considered further at this point.

Let us make sure that there is no misunderstanding of the way in which we are to use the term activity. We have seen in a previous chapter that in a state of equilibrium the fugacity of a given substance is the same in every phase, or in every part of a system. Since the activity is defined as the relative fugacity, if we should choose, for the whole system, a single standard state of the given substance, its activity would be the same in every part of the system. On the other hand, since we have decided to use for the substance in question different standard states in different phases, its activities in the several phases that are in equilibrium will not be equal but will nevertheless remain proportional to one another as long as equilibrium persists, the factor of proportionality depending upon the choice of standard states.

Thus, in the case of a solvent in equilibrium with its vapor, the activity in the vapor phase being defined in the section on the activity of a gas and that in the liquid phase being defined in the section on the activity of a liquid solvent, the ratio of these two activities remains constant no matter how the concentration of solute changes (and this ratio is equal to f_l^o, the fugacity of the pure liquid).

EFFECT OF PRESSURE ON ACTIVITY

Since small pressure changes have very little effect on the chemical potential of liquid or solid phases, these pressure effects are frequently ignored. This is in contrast to the case for gases, where pressure effects are large and the fugacity of a component would never be regarded as independent of the pressure. However, accurate work requires the consideration of pressure effects on condensed-phase chemical potentials. It is best to consider the mixing of components to form a solution at constant pressure and temperature; hence we consider the pressure effect on each component in its reference state.

The term *standard state* and the symbol ° are reserved for the condition of 1 bar pressure† (or the ideal gas), and the term *reference state* will be used for states reached from standard states by a change of pressure. The ratio of activity in a reference state at a pressure P' to that in the standard state at 1 bar, for which we shall adopt the symbol Γ, is given by the integral of Eq. (3-8) or (10-20),

$$\ln \Gamma = \frac{\mu' - \mu°}{RT} = \int_1^{P'} \frac{V'}{RT} dP \tag{12-28}$$

where the lower limit of integration is 1 bar pressure. If a solute reference state is being used, V' is the partial molar volume $\overline{V}$ at infinite dilution; otherwise, it is the molar volume V_m^* of the pure liquid or solid.

The complete equations for activity and chemical potential with mole fraction as the composition measure now become

$$a_i = \gamma_i \Gamma_i x_i \quad \text{and} \quad \mu_i - \mu_i° = RT \ln x_i \gamma_i \Gamma_i \tag{12-29}$$

It is also useful to express γ_i in terms of fugacities,

$$\gamma_i = \frac{f_i}{f_i° \Gamma_i x_i} = \frac{f_i}{f_i' x_i} \tag{12-30}$$

where $f_i°$ is the fugacity of the standard state at 1 bar and f_i' is the fugacity in the reference state at the actual pressure P' of the solution.

In approximate calculations for low pressures one may frequently set $\Gamma_i \cong 1$ and ignore that factor in the equations. In precise work the Γ_i must be considered, but if the solutions are nearly ideal, the activity coefficients are nearly independent of pressure. The pressure derivative of Eq. (12-30) yields

$$\frac{\partial \ln \gamma_i}{\partial P} = \frac{\overline{V}_i - V'}{RT} \tag{12-31}$$

and $\overline{V} - V'$ is zero for an ideal solution and very much smaller than V' or $\overline{V}$ for nearly ideal solutions.

In most of our subsequent work we shall omit the factor Γ because the system is at 1 bar or so near to 1 bar that the difference is insignificant. At the other extreme, there may be special complications near the critical point of a gas where the compressibility becomes infinite; these are not considered at this point.

---

† Where the vapor pressure of a pure component exceeds 1 bar, the standard state is often taken as that for saturation pressure.

EXCESS FUNCTIONS AND ACTIVITY COEFFICIENTS FOR REAL SYSTEMS

The best and most convenient method of treating the nonideality of a real system is the use of an appropriate analytical equation defined initially for the excess Gibbs energy. Next, working equations are derived by differentiation for all of the measured properties and then the parameters in the equation are evaluated by least-squares procedures to fit the experimental data. Subsequently, any other thermodynamic quantity can be calculated. The choice of the initial equation may be based on a statistical model, or a simple power series expansion may be used provided it conforms to the known limits at mole fractions zero and unity. In practice, one may test alternate equations or determine how many terms are needed in a series to represent the data within experimental uncertainty. This general procedure assures that all equations for activity coefficients, enthalpies, and other properties are completely consistent.

Gibbs–Duhem Equation

One of the important consistency conditions is that of the Gibbs–Duhem equation, which was discussed for fugacity in Chapter 10 (Eq. 10-42). Since activity is directly proportional to fugacity, the corresponding equation has the same form, which for a binary system is

$$x_1 \left(\frac{\partial \ln a_1}{\partial x_1} \right)_{P,T} = x_2 \left(\frac{\partial \ln a_2}{\partial x_2} \right)_{P,T} \tag{12-32a}$$

Substitution of $a_i = \gamma_i x_i$ also yields the same form since terms arising from derivatives of x_i cancel:

$$x_1 \left(\frac{\partial \ln \gamma_1}{\partial x_1} \right)_{P,T} = x_2 \left(\frac{\partial \ln \gamma_2}{\partial x_2} \right)_{P,T} \tag{12-32b}$$

In an earlier period the excess function method had not come into common use, and graphical methods were often employed. Then, it was emphasized that the measured values of γ_1 and γ_2 should be tested for consistency by Eq. (12-32b). Alternatively, if γ_1 was known but γ_2 was not readily measurable, a transform of Eq. (12-32b) was used to calculate γ_2. This last situation arises frequently when only one component has an appreciable vapor pressure. Details and examples of such calculations are available from many sources[2,3] and will not be duplicated here.

If working equations derived from an expression for the excess Gibbs energy are used, consistency with Eq. (12-32b) is automatically attained. Also, it is often more convenient to give appropriate weights to various measurements by these procedures than by some of the older methods.

Simple Expansion for Excess Gibbs Energy

For solutions that depart from the simple Margules equation but do not involve ions or high polymers, a simple series extending Eq. (12-12)* is very satisfactory

$$\frac{G_m^E}{RT} = x_1 x_2 [A + B(x_1 - x_2) + C(x_1 - x_2)^2 + \cdots] \quad (12\text{-}33)$$

The second term in B provides for some asymmetry between the two components, while higher terms can represent more subtle or more complex aspects. This is commonly known as the Redlich–Kister[4,5] expansion in view of its extensive use by those investigators, although it had been proposed earlier by Guggenheim,[6] and it can be regarded as a Margules equation.

For an indefinite amount of substance, this becomes

$$\frac{G^E}{RT} = \frac{n_1 n_2}{(n_1 + n_2)} \left[A + B \left(\frac{n_1 - n_2}{n_1 + n_2} \right) + C \left(\frac{n_1 - n_2}{n_1 + n_2} \right)^2 + \cdots \right] \quad (12\text{-}34)$$

and differentiation yields the activity coefficients

$$\ln \gamma_1 = (A + 3B + 5C + \cdots)x_2^2 - 4(B + 4C + \cdots)x_2^3 + (12C + \cdots)x_2^4 + \cdots \quad (12\text{-}35)$$

$$\ln \gamma_2 = (A - 3B + 5C + \cdots)x_1^2 + 4(B - 4C + \cdots)x_1^3 + (12C + \cdots)x_1^4 + \cdots \quad (12\text{-}36)$$

The procedure for determination of the parameters $A, B, C, \ldots$ depends on the experimental database available. If there are measured values for both γ_1 and γ_2, then the best procedure is a simultaneous regression for $A, B, C, \ldots$; this is not difficult with modern computers. Redlich and Kister[4] describe alternate stepwise procedures in which only very simple calculations are required in each step.

In the case where only one activity coefficient γ_1 has been measured, a very simple regression yields the coefficients of the $x_2^2, x_2^3, x_2^4 \ldots$ terms in Eq. (12-35), and it is easy to determine how many terms are needed. Then a simple calculation yields $A, B, C, \ldots$ and their values can be substituted in Eq. (12-36) to yield γ_2.

Where the activity coefficients have been measured at a series of temperatures, one may solve for $A, B, C, \ldots$ at each temperature and then fit temperature-dependent expressions for each of these parameters. Alternatively, one may assume an expression such as $A_0/T + A_1 + A_2 T$ for parameter A and similar expressions for $B, C, \ldots$ and then carry out a comprehensive regression for $A_0, A_1, A_2, B_0, \ldots$. In an even more comprehensive treatment, there may be heat of mixing data that yield the excess enthalpies. The equation for H^E has the same form as Eq. (12-33) for G^E, except that RTA is replaced by $-RT^2 (\partial A/\partial T)_p$ and similarly for B, and so on. Thus, one has additional equations involving only the same parameters $A_0, A_1, A_2, B_0, \ldots$, and all can be regressed simultaneously after assigning appropriate weights or uncertainties.

Up to this point, we have assumed that values of the activity coefficient were available, and that requires knowledge of fugacities if the measurements were of the pressure and composition of the equilibrium vapor. This implies either that the vapor is a perfect gas to the approximation needed or that corrections can be made for gas imperfection. This last topic is considered in Chapter 13. But in many cases, the full information for conversion of partial pressure data to fugacities is not available. Many clever methods have been devised in which a missing gas-imperfection parameter is determined along with the nonideality parameters for the liquid solution. An early example is that of Scatchard and Raymond;[7] several are described by Prausnitz et al.[3]

Measurements of pressure are easier and more precise than those of vapor composition. Thus, a number of investigations report only the total vapor pressure, together with the liquid composition. If the vapor is a perfect gas, the total pressure is given by the simple equation

$$P = \gamma_1 x_1 P_1^* + \gamma_2 x_2 P_2^* \tag{12-37}*$$

with P_1^* and P_2^* the vapor pressure of the two pure liquids. Then it is straightforward to substitute expressions for γ_1 and γ_2 from Eqs. (12-35) and (12-36) and obtain the parameters A, B, etc., by regression from a series of values of P and x. In this situation it is especially important to correct for gas imperfections unless the vapor is accurately a perfect gas. Given enough information on the mixed gas, this can be done, but then the calculation becomes iterative. Several systems have been treated successfully by this method by Barker[8] and by others,[9,10] who give the details.

A large number of binary solutions have been investigated and their behavior expressed in terms of the parameters A, B, C, ... of Eq. (12-33) or their equivalent.

Table 12-2 lists the values of these parameters for a number of solutions.[6,10–12] One can get a picture of the ideality of these solutions from the function.

$$\frac{G_m^E}{RTx_1x_2} = A + B(x_1 - x_2) + C(x_1 - x_2)^2 + \cdots \tag{12-38}$$

TABLE 12-2
Parameters of Eq. (12-38) for several solutions

Components 1–2	T/K	A	B	C	Ref.
1. Ethanol–methylcyclohexane	305	2.118	−0.239	0.375[a]	[5]
2. Methylcyclohexane–acetone	318	1.6907	−0.0001	0.1832	[11]
3. Toluene–nitroethane	318	0.7637	0.0703	0.0619	[11]
4. Toluene–acetone	318	0.6637	−0.0048	0.0023	[11]
5. Benzene–cyclopentane	298	0.4560	−0.0182	–	[10]
6. Pyridine–acetone	303	0.1919	0.0050	0.0075	[12]
7. Chloroform–furan	303	−0.1083	−0.0177	0.0071	[12]
8. Dichloromethane–acetone	303	−0.6479	−0.0887	−0.0299	[12]
9. Pyridine–chloroform	303	−1.0271	0.2270	0.0930	[12]
10. Chloroform–1,4-dioxane	303	−1.2006	−0.4131	0.0318	[12]
11. Pyridine–dichloromethane	303	−0.5140	0.0652	0.0088	[12]

[a] Also $D = -0.173$.

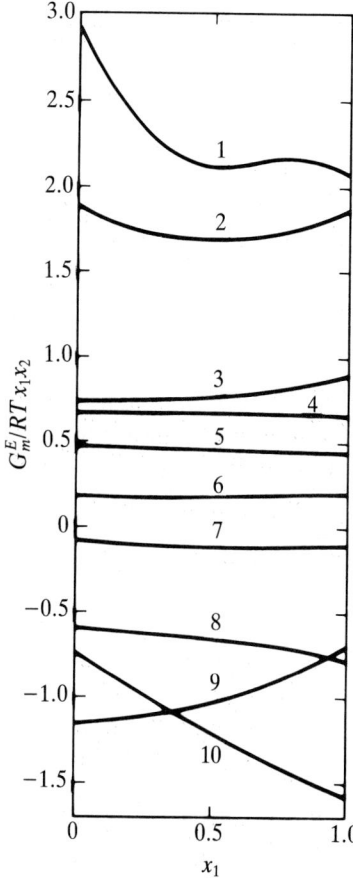

FIGURE 12-4
Excess Gibbs energies for the first ten systems listed in Table 12-2 with parameters for Eq. (12-38).

This is shown in Fig. 12-4 where a horizontal line at zero level indicates an ideal solution, a horizontal line away from zero level corresponds to zero $B, C, \ldots$ but nonzero A; this is the case of the very simple model of Eq. (12-12)* with $A = w$. A sloping but straight line indicates that C and higher terms are zero but that B is significant; indeed, the slope is $2B$. A curved line with constant curvature corresponds to a finite C but no higher terms. Usually, B is larger than C, which yields a slope but only slight curvature; an exception is number 2 (methylcyclohexane–acetone), where B is practically zero but C is not. Any more complex shape implies that terms beyond C are significant; the only example on Fig. 12-4 is number 1 (ethanol–methylcyclohexane).

Van Laar and Related Equations

Instead of adding terms to the simplest equation for nonideality, Eq. (12-12)*, one may retain its form but change the composition variable from the mole fraction. An obvious alternative is the volume fraction. This was suggested in 1906 by van Laar[13] as a part of a theory built upon the van der Waals equation. Scatchard[14] and Hildebrand and Wood[15] later proposed a theory that also uses the volume fraction; it is discussed more fully later in this chapter. At this point, we define a pair of composition variables

$$z_1 = \frac{n_1 b_1}{n_1 b_1 + n_2 b_2} \qquad z_2 = \frac{n_2 b_2}{n_1 b_1 + n_2 b_2} \qquad (12\text{-}39)$$

where the ratio b_1/b_2 may be adjusted arbitrarily or may be assigned the value of the ratio of molar volumes $V_1°/V_2°$. In the latter case the variable becomes the volume fraction usually designated by ϕ.

$$\phi_1 = \frac{n_1 V_1°}{n_1 V_1° + n_2 V_2°} \qquad \phi_2 = \frac{n_2 V_2°}{n_1 V_1° + n_2 V_2°} \qquad (12\text{-}40)$$

In proceeding, we shall write most equations only in terms of the generalized variables b_1, b_2, z_1, z_2, since the substitution of $V_1°, V_2°, \phi_1, \phi_2$ is obvious. The excess Gibbs energy is given by the equation

$$G^E = a_{12}\left(\frac{b_1 b_2 n_1 n_2}{n_1 b_1 + n_2 b_2}\right) \qquad (12\text{-}41)^*$$

whereupon the activity coefficients become

$$RT \ln \gamma_1 = a_{12} b_1 z_2^2 \qquad (12\text{-}42a)^*$$

$$RT \ln \gamma_2 = a_{12} b_2 z_1^2 \qquad (12\text{-}42b)^*$$

For solutions where the two components are similar in general character but differ substantially in volume, the van Laar equation is quite satisfactory. If the ratio b_1/b_2 is freely adjusted, it is usually found to be near the actual volume ratio. Thus, one frequently obtains good results by using volume fractions, whereupon there is only one parameter to adjust and the treatment is very simple.

An example of this general type is that of benzene–isooctane, which was studied by Weissman and Wood.[16] Prausnitz et al.[3] fitted the van Laar equation to their data at 45°C and report for b_1/b_2 the value 0.562, while the volume ratio is 0.536, which differs only slightly. The value 0.419 for $a_{12}b_1/RT$ was also reported. These results are shown in Fig. 12-5 where the asymmetry between the two components is clear.

Weissman and Wood originally represented their data for 35–75°C with Eq. (12-33) and the following temperature-dependent equations for the parameters (in cal mol^{-1}):

$$ART = 2428.10 - 10.9940T + 0.013\,961T^2$$

$$BRT = -1113.1 + 7.662T - 0.012\,07T^2$$

$$CRT = 33 - 0.027T$$

Although three parameters were used for the Redlich–Kister equation, C is very small. Thus, a two-parameter Redlich–Kister fit is, in this case, quite comparable to the two-parameter van Laar fit. If the ratio of the pure-liquid volumes is inserted into the van Laar equation, then only one parameter remains to be fitted from the solution data and reasonable agreement is still found.

From the temperature derivatives of A, B, and C, the enthalpy of mixing and the excess entropy were calculated; these are shown in Fig. 12-6. It is apparent that both enthalpy and excess entropy are positive and partially cancel to yield a smaller excess Gibbs energy. Also, all three properties decrease with increase in temperature. This type of system will be considered further in relation to theory in a later section.

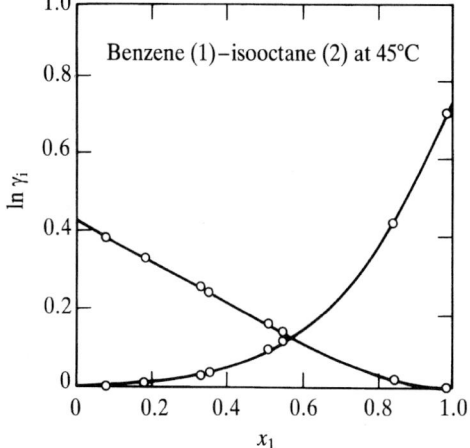

FIGURE 12-5
Application of van Laar's equations to a mixture whose components differ appreciably in molecular size. See text following Eq. (12-42)*. (From Prausnitz et al.[3])

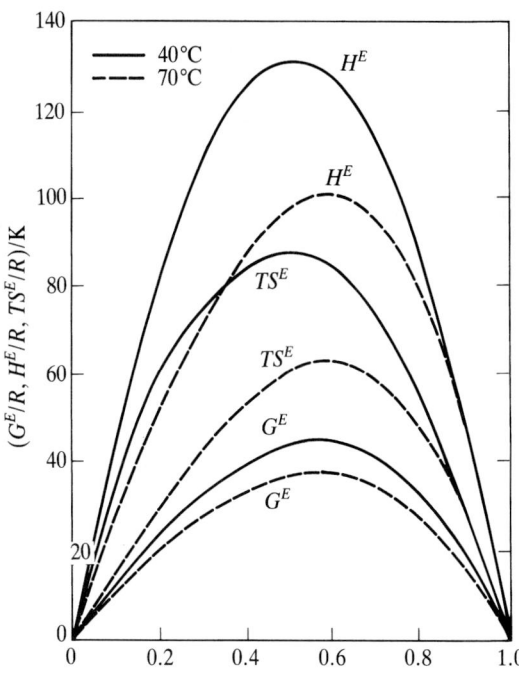

FIGURE 12-6
Excess properties for the system benzene–isooctane at two temperatures. See text following Eq. (12-42)*.

TABLE 12-3
Systems represented by the asymmetric equations (12-42a)*, (12-42b)*

Components 1–2	T/K	b_2/b_1	$a_{12}b_1$
Mercury–tin[a]	596	3.7	0.51
Mercury–cadmium[a]	595	0.53	−3.2
Carbon disulphide–isopentane[b]	298	1.94	0.58
n-Heptane–benzene[c]	343	1.50	0.33

[a] J. H. Hildebrand, A. H. Foster, and C. W. Beebe, *J. Am. Chem. Soc.*, **42**, 545 (1920).
[b] J. Hirshberg, *Bull. soc. chim. Belg.*, **41**, 163 (1932).
[c] I. Brown and A. H. Ewald, *Australian J. Sci Research*, **A4**, 198 (1951).

If in these equations b_2/b_1 is independent of temperature, then one may express the heat of mixing and excess entropy of mixing readily in terms of a_{12} and $\partial a_{12}/\partial T$. Where b_2/b_1 is equal to the ratio of molal volumes, it is likely that this situation will arise. However, in other cases there is every reason to assume that b_2/b_1 may be a function of temperature, and in that case the equation for the heat of mixing becomes much more complicated. There are too few examples to justify a general discussion at this time of the temperature dependence of b_2/b_1. All of the examples given above are for nonmetallic systems, but some aspects are equally applicable to metallic solutions. Indeed, there are other systems, particularly solutions of metals, whose properties fit the van Laar type of equations provided that the ratio of b_1 to b_2 is adjusted arbitrarily in each case to a value that has no relation to either the van der Waals constants or the molal volumes. A few examples are given in Table 12-3.

Other Equations

There are other equations that successfully represent the properties of nonideal solutions. A few are listed and described briefly in Appendix 11. Prausnitz et al.[3] discuss these equations more fully and comment on their relative merit for various types of systems.

The simplest of these equations is that of Wilson.[17] As reformulated by Orye and Prausnitz,[18] the molar excess Gibbs energy is taken to be

$$\frac{G_m^E}{RT} = -x_1 \ln(x_1 + \Lambda_{12}x_2) - x_2 \ln(x_2 + \Lambda_{21}x_1) \qquad (12\text{-}43)^*$$

It has only two parameters, Λ_{12} and Λ_{21}. Ideality arises, not from a zero value of these parameters, but rather when $\Lambda_{12} = \Lambda_{21} = 1.0$; thus the measure of nonideality is the departure of the parameters from 1.0. The activity coefficients are given in Appendix 11. Wilson parameters have been tabulated for a very large number of volatile, binary systems, together with vapor pressure equations for each pure component. The Wilson equation generalizes easily to more than two components with

$$\frac{G_m^E}{RT} = -\sum_{i=1}^{m} x_i \ln\left(\sum_{j=1}^{m} \Lambda_{ij}x_j\right) \qquad (12\text{-}43\text{a})^*$$

Now $\Lambda_{ii} = 1.0 = \Lambda_{jj}, \ldots$ and the other Λ_{ij} are just the binary parameters. The absence of any new interaction parameters for three different species seems attractive initially. But interactions of three different species do occur, and a provision is needed for their representation when they are significant.

OSMOTIC PRESSURE

For liquid solutions formed from simple molecules, the vapor pressure is the simplest measurement from which fugacity can be derived. But for very dilute solutions of nonvolatile solutes, the reduction of solvent vapor pressure becomes very small and difficult to measure. One method of overcoming this difficulty is that of isopiestic measurements in which the unknown solution is equilibrated with respect to solvent vapor pressure with another solution whose properties are known. Examples of this procedure are discussed in connection with electrolyte solutions, where it has been very widely used.

Another method that is valuable for some types of solutes in very dilute solution is that of osmotic pressure measurement. Here the solution is separated by a membrane from pure solvent. Certain membranes, known as semipermeable membranes, permit the free passage of solvent but do not permit the passage of certain substances dissolved in the solvent. If the pressure is the same on both sides of the membrane, the chemical potential of the solvent is less in the solution and the phenomenon of osmosis will occur; i.e., solvent will flow through the membrane into the solution. If the pressure on the solution, P, is raised above that on the solvent, $P°$, this will raise the chemical potential of the solvent in the solution and equilibrium can be attained. This pressure difference at equilibrium is the osmotic pressure $P - P°$ or Π.

This effect was first discovered for aqueous solutions, and the botanist Pfeffer[19] was the first to obtain reproducible osmotic pressure measurements in 1877. The osmotic pressure must restore the chemical potential of the solvent μ_1 in the solution to that of the pure solvent. Hence we write

$$d\mu_1 = 0 = \frac{\partial \mu_1}{\partial P} dP + \frac{\partial \mu_1}{\partial x_2} dx_2 \qquad (12\text{-}44)$$

and on substitution of Eq. (10-20) for the pressure effect and transformation from chemical potential to fugacity, we have

$$0 = \frac{\overline{V}_1}{RT} dP + \frac{\partial \ln f_1}{dx_2} dx_2 \qquad (12\text{-}45)$$

For a dilute solution we can take the partial molar volume to be the molar volume of the solvent $V_1°$ and integrate to obtain

$$\ln a_1 = \ln\left(\frac{f_1}{f_1°}\right) = \frac{(P° - P)V_1°}{RT} \qquad (12\text{-}46)*$$

Van't Hoff made the further substitution of the Gibbs–Duhem relationship between solute and solvent fugacity for a dilute solution and obtained

$$P - P^\circ = \frac{n_2 RT}{V} \qquad (12\text{-}47)*$$

which has the same form as the perfect gas law with n_2 moles of solute in a total volume V.

In our view, the important applications of osmotic pressure measurements are best discussed in terms of solvent activity or fugacity with the exact Eq. (12-45) or, in most cases, the excellent approximation, Eq. (12-46)*. Osmotic pressure measurements have been important in studies of solutions of large polymer molecules; examples will be discussed below.

THE EQUILIBRIUM CONSTANT

The equilibrium constant and its temperature dependence have been considered in several preceding chapters for systems comprising pure substances and perfect gases. At this point, we could just introduce the appropriate expression for the fugacity in relation to the activity in Eq. (7-26) for each component; this would yield the correct expression for solutions. It seems desirable, however, to review the fundamental relationships for this most fundamental expression in chemical thermodynamics. As before, we consider a general reaction.

$$l\mathrm{L} + m\mathrm{M} + \cdots = q\mathrm{Q} + r\mathrm{R} + \cdots$$

Let ΔG be the molar Gibbs energy or chemical potential† in this reaction when the substances are in any given states,

$$\Delta G = (q\mu_Q + r\mu_R + \cdots) - (l\mu_L + m\mu_M + \cdots) \qquad (7\text{-}19)$$

and let ΔG° be the Gibbs energy change when each substance is in its standard state,

$$\Delta G = (q\mu_Q^\circ + r\mu_R^\circ + \cdots) - (l\mu_L^\circ + m\mu_M^\circ + \cdots) \qquad (7\text{-}20)$$

If a_L, a_M, etc., represent the activities in the nonstandard states, then we have

$$l(\mu_L - \mu_L^\circ) = RT \ln a_L^l \qquad (12\text{-}27)$$

and so on.

Combining the several equations, we find

$$\Delta G - \Delta G^\circ = RT \ln \frac{a_Q^q a_R^r \cdots}{a_L^l a_M^m \cdots} \qquad (12\text{-}48)$$

---

† We assume a constant-pressure situation whereupon the molar Gibbs energy and the chemical potential are equal; see Eqs. (10-11), (10-14), and (10-19).

The important quotient appearing on the right side we may call the activity quotient of the reaction. By determining this quotient we may calculate $\Delta G°$ when ΔG is known, and conversely.

Whenever we meet with a case of chemical equilibrium, we immediately acquire important information regarding the Gibbs energy change in the reaction concerned. For in such a reaction, by Eq. (3-41),

$$\Delta G = 0$$

Hence, for equilibrium, by Eq. (12-48),

$$\Delta G° = -RT \ln \frac{a_Q^q a_R^r \cdots}{a_L^l a_M^m \cdots} \qquad (12\text{-}49)$$

At a given temperature $\Delta G°$ is a constant, and therefore the condition of equilibrium is that the activity quotient shall also be constant. The value of this quotient, when the system is in equilibrium, is called the equilibrium constant K and, as noted above in Eq. (7-24),

$$\Delta G° = -RT \ln K \qquad (7\text{-}24)$$

Some of the substances taking part in the reaction, such as pure liquids or solids at approximately 1 bar pressure, have unit activity at a given temperature. It is frequently the custom to retain in the equilibrium constant only the activities of gases and of substances in solution, where the activity varies markedly with the pressure or the composition.

Let us consider a case in which the substances of variable activity are constituents of an ideal solution. Thus, in a homogeneous mixture of methyl bromide, ethyl chloride, ethyl bromide, and methyl chloride we might, from our knowledge of these substances, expect the activity of each substance to be approximately equal to its mole fraction. Hence for the reaction

$$CH_3Cl + C_2H_5Br = CH_3Br + C_2H_5Cl$$

we should expect an equilibrium to be attained that would satisfy the condition

$$K = \frac{x_{CH_3Br} x_{C_2H_5Cl}}{x_{CH_3Cl} x_{C_2H_5Br}} \qquad (12\text{-}49a)*$$

Many equilibrium measurements in early years were made in systems involving gases at moderate pressures and dilute aqueous solutions. In such cases, it was frequently possible, with adequate accuracy, to replace the activities of the gases by their pressures and of the solutes by their molalities. We then have the *mass law* in the form that we owe to Guldberg and Waage[20] and to van't Hoff,[21] a generalization that will always be esteemed as one of the milestones in the progress of chemistry toward an exact science.

After writing the equation for a given reaction, it is the convention to write the equilibrium constant with the activities of the substances produced in the numerator and of the substances consumed in the denominator.

It is sometimes convenient to denote the activity of a gas by its formula in brackets and the activity of any constituent of an aqueous solution by its formula in parentheses. To illustrate these conventions, we write

$$\tfrac{1}{2}N_2(g) + \tfrac{3}{2}H_2(g) = NH_3(g) \qquad K = \frac{[NH_3]}{[N_2]^{1/2}[H_2]^{3/2}}$$

$$N_2(g) + 3H_2(g) = 2NH_3(g) \qquad K = \frac{[NH_3]^2}{[N_2][H_2]^3}$$

$$NH_3(g) = \tfrac{1}{2}N_2(g) + \tfrac{3}{2}H_2(g) \qquad K = \frac{[N_2]^{1/2}[H_2]^{3/2}}{[NH_3]}$$

$$NH_3(aq) = NH_3(g) \qquad K = \frac{[NH_3]}{(NH_3)}$$

If in these equations we substitute for the activities the partial pressures or molalities, the quotients are not strictly constant at finite pressures and concentrations but approach the true equilibrium constant (the quotient of activities) as the gases approach zero pressure and the concentrations of the solutes approach zero.

It is frequently desirable to express the activity of components of solutions in terms of activity coefficients. According to Eq. (12-29),

$$a_i = x_i \gamma_i \Gamma_i \qquad (12\text{-}29)$$

The factor Γ_i is also appropriate for an activity in the molality system or for the activity of a pure solid or liquid at a pressure other than 1 bar. Since the effect of pressure on gases is very large, it is usually best to consider their fugacities directly. If desired, one may use a fugacity coefficient f_i/p_i to express gas imperfection and then $f_i = p_i(f_i/p_i)$. We may illustrate these methods with the reaction

$$2H_2(g) + CH_3COOH(x_1) = C_2H_5OH(x_2) + H_2O(x_3)$$

where the acetic acid, ethanol, and water form a liquid solution of the composition indicated. Combination of the values of Gibbs energy of formation[22] yields $\Delta G° = -19.7$ kJ at 298.15 K, and

$$K = 2.8 \times 10^3 = \frac{a_{C_2H_5OH} a_{H_2O}}{a_{H_2}^2 a_{CH_3COOH}}$$

But $a_{CH_3COOH} = \Gamma_1 \gamma_1 x_1$, and similar expressions hold for ethanol with subscript 2 and water with subscript 3; hence

$$K = \frac{\Gamma_2 \gamma_2 x_2 \Gamma_3 \gamma_3 x_3}{(f/p)_{H_2}^2 p_{H_2}^2 \Gamma_1 \gamma_1 x_1}$$

This expression may be rearranged to

$$K = \frac{x_2 x_3}{p_{H_2}^2 x_1} \frac{\gamma_2 \gamma_3}{(f/p)_{H_2}^2 \gamma_1} \frac{\Gamma_2 \Gamma_3}{\Gamma_1}$$

where the first factor on the right-hand side is the usual approximate equilibrium quotient for the ideal-gas and ideal-solution case and the second quotient contains the correcting factors to the real state at low pressure, i.e., the activity coefficients and f/p for the gas. It is customary to ignore the third factor $\Gamma_2\Gamma_3/\Gamma_1$ unless the pressure is high; let us now check the magnitude of this factor. If we assume that the liquid volumes remain constant, we obtain

$$\ln\left(\frac{\Gamma_2\Gamma_3}{\Gamma_1}\right) = \frac{(V_2^\circ + V_3^\circ - V_1^\circ)(P-1)}{RT}$$

and substitution of numerical values yields

$$\ln\left(\frac{\Gamma_2\Gamma_3}{\Gamma_1}\right) = 0.79 \times 10^{-3}(P-1)$$

if P is in bars. It is evident that a pressure of a few bars has little effect, but the Γ factor becomes very significant for pressures in the 1000-bar range.

CHOICE OF STANDARD STATE WHEN ASSOCIATION OR DISSOCIATION OCCURS

The previous examples have been ones where the choice of components has been rather obvious. In some instances, however, there may be a number of major constituents corresponding to a given component, and the most suitable choice of components may not be immediately obvious. A number of examples will be discussed to illustrate the procedure for choosing components and corresponding standard states.

The Activity of a Solute that Forms Compounds with the Solvent

We frequently have to deal with solutions in which, for one reason or another, it is assumed that the solute forms compounds with the solvent. These are known in general as *solvates*. Thus, many substances dissolved in water are assumed to form hydrates, with one or more molecules of water combined with each molecule of solute. This assumption is sometimes mere hypothesis, but often it rests upon very substantial evidence. It is difficult, however, to determine with any degree of certainty the relative amounts of unhydrated substance and of the various possible hydrates when this interconversion is rapid. How, then, are we to treat such compounds in our thermodynamic work?

The simplest method of disposing of this question would be to ignore the existence of such hydrates, and this would be entirely justifiable, since thermodynamics is not compelled to take cognizance of the various molecular species that may exist in a system, particularly when the existence of such species cannot be absolutely demonstrated. Nevertheless, it will frequently be more convenient, as well as more consistent with chemical usage, to include these hydrates in our consideration, especially since by a simple device we may do this without really complicating our procedure. We may illustrate this device by a concrete case.

When ammonia dissolves in water, it is supposed to form, although in unknown amount, at least one hydrate, the monohydrate, which may be written $NH_3 \cdot H_2O$ or NH_4OH. In other words, we assume the reaction

$$NH_3(aq) + H_2O(l) = NH_4OH(aq)$$

Now, if (NH_3), (H_2O), and (NH_4OH) represent the several activities,

$$\frac{(NH_4OH)}{(NH_3)(H_2O)} = K$$

At high dilution the activity of the water is constant and equal to unity; therefore the ratio of (NH_4OH) to (NH_3) is constant, and the two molalities are also approximately proportional to one another. But we do not know any of these quantities separately; we only know the gross or stoichiometric molality m, as determined, for example, by the number of moles of gaseous ammonia that have been dissolved in 1 kg of water. In very dilute solution this is the same as the number of moles of NH_4OH that would be dissolved in 1 kg of water to produce the same concentration.

In ignorance of the individual concentrations, we may arbitrarily take the standard state of each substance in such manner that at infinite dilution its activity is equal to the gross molality m. Or, in other words we assume such standard states as to make $K = 1$. The activities of the two substances then remain equal as long as $(H_2O) = 1$ (but in concentrated solutions their ratio is the activity of the water).

Such a definition introduces an equal simplicity into the Gibbs energy equation, since it makes $\Delta G° = 0$. The same method may be employed in all similar cases, although it is possible that this method might be abandoned if we ever should succeed in determining quantitatively the actual concentrations of the individual species. If there is no reason to know the amounts of the individual species and as long as equilibrium is maintained among these species, the above procedure is the simplest for treating these solutions.

Dissociating Solutes

One often encounters solutes that dissociate to simple species to an extent that depends upon the temperature, the concentration, and the nature of the solvent. If either the parent molecule or one of the products of dissociation is colored and the other colorless, one can readily determine the degree of dissociation colorimetrically. Even when both are colored, one can use a spectrophotometer and observe each species at the wavelength of its characteristic absorption band. By a simple colorimetric method, Cundall[23] made an exhaustive investigation of the dissociation of N_2O_4 in various solvents, and especially in chloroform. Since chloroform, NO_2, and N_2O_4 might be expected to form nearly perfect solutions, it will be interesting to see how well Cundall's results can be represented by Eq. (12-50)*, which implies $a_i = x_i$.

We have calculated from his data at 0°C the various figures needed for this comparison, and in Table 12-4 we give in the several columns: x_1, the mole fraction of

TABLE 12-4
Dissociation at 0°C of N_2O_4 dissolved in chloroform

x_1	x_2''	x_2'	$\dfrac{(x_2')^2}{x_2''} \times 10^8$
0.00	1.00	0.000 94	88
0.27	0.73	0.000 80	88
0.46	0.54	0.000 67	83
0.70	0.30	0.000 45	68
0.875	0.125	0.000 29	67
0.934	0.066	0.000 19	55
0.950	0.050	0.000 15	45
0.963	0.037	0.000 12	39
0.982	0.018	0.000 10	56

chloroform; x_2'', the mole fraction of N_2O_4; x_2', the mole fraction of NO_2; and the ratio $(x_2')^2/x_2''$, which should be constant according to Eq. (12-49a)*. We see that it changes only twofold in passing from the liquid containing no chloroform to the dilute solution in chloroform; and in the concentrated solution, when the mole fraction of chloroform rises from 0 to 0.5, the ratio changes by only 6 percent. The small variation from constancy that exists is evidently due to the fact that, over this wide range of composition, the activities of NO_2 and N_2O_4 are not quite proportional to their mole fractions.

We have chosen this illustration partly also for the sake of discussing another point that without explanation might prove troublesome. In the table it will be noted that the amounts of NO_2 are so small that x_1 and x_2'' are together practically equal to unity. At higher temperatures this is no longer the case. We might, for example, find a mixture containing 80 moles of $CHCl_3$, 10 moles of NO_2, and 10 moles of N_2O_4, and we might write (1) $x_1 = 0.80$, $x_2' = 0.10$, and $x_2'' = 0.10$. It must be emphasized, however, that this procedure is in some ways arbitrary and due to our foreknowledge of the dissociation of N_2O_4. We would have obtained a mixture of the same composition if we had taken 15 moles of pure N_2O_4 and 80 moles of $CHCl_3$, in which case, if we followed the normal usage and paid no attention to the possible dissociation, we would have written: (2) $x_1 = 80/(80 + 15)$ and $x_2 = 15/(80 + 15)$. Again, we might have reached the same composition had we used 30 moles of NO_2 and 80 moles of $CHCl_3$. In this case, paying no attention to the possible association of the NO_2, we would have written: (3) for $CHCl_3$, $x_1 = 80/(80 + 30)$ and, for NO_2, $x_2 = 30/(80 + 30)$.

Suppose that we were ignorant of this phenomenon of dissociation and were to study the activities of the solvent and the solute in solutions of N_2O_4 in chloroform at 0°C, by any of the methods employed in the preceding chapters. We should of course reckon the mole fractions by method (2) suggested above; and plotting a_2, the activity of N_2O_4, against its mole fraction as in Fig. 12-7, we should undoubtedly find in moderately dilute solutions that a_2 would be roughly proportional to x_2 and might easily be confused with the dashed line "no dissociation." But careful measurements in

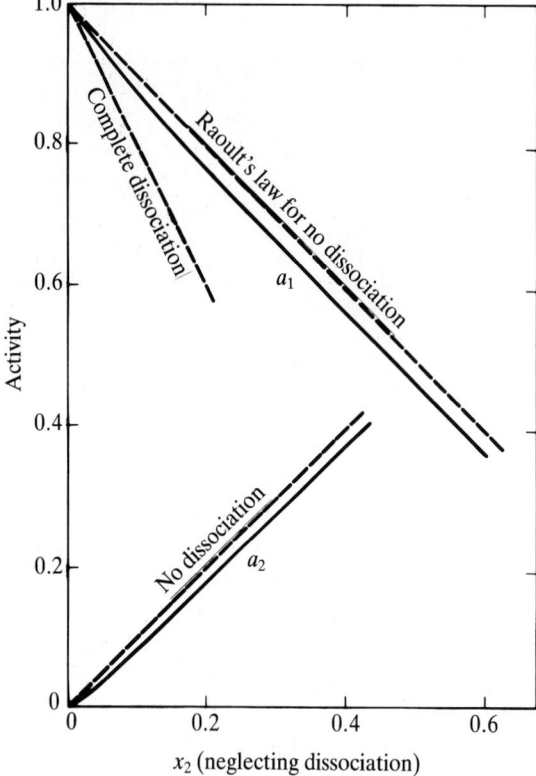

FIGURE 12-7

the very dilute region would yield a curve, such as the solid line in Fig. 12-7, that becomes proportional to x_2^2 at extreme dilution. Thus, at $x_2 = 0$, $da_2/dx_2 = 0$. Likewise a_1 measured at moderate concentration would be found to be near the value calculated from Raoult's law for no dissociation and the difference might be ascribed to experimental error. But at extreme dilution the curve for a_1 approaches the dashed line of slope double the Raoult's-law slope, which is marked "complete dissociation" in Fig. 12-7.

This seems to be, and is in fact, a direct contradiction of our definition above for the activity of an uncharged solute ($x_2/x_2 = 1$ when $x_2 = 0$). However, as we pointed out at that time, there is the exception where the solute dissociates.

Fortunately, in such cases we may eliminate these exceptions to the laws of the dilute solution by considering as solute, not the undissociated substance, but the products of dissociation. Thus, if in place of Fig. 12-7 we had plotted the activity of $CHCl_3$ and of NO_2 against the mole fraction of NO_2, as in the above method (3), we should obtain in extremely dilute solutions such a plot as that of Fig. 12-8, where it is evident that both Henry's law and Raoult's law are obeyed at first, although even in rather dilute solutions marked deviations from these laws would appear, which now would be attributed to association of NO_2.

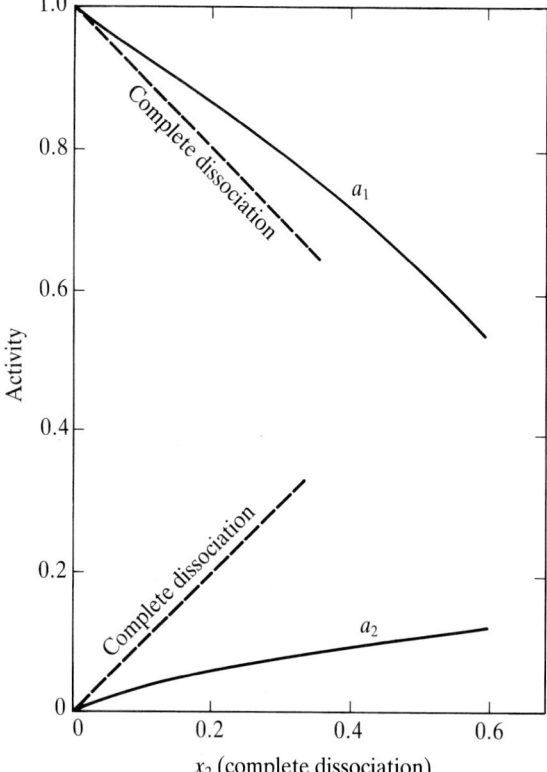

FIGURE 12-8

We have entered at some length into this discussion of the thermodynamic properties of a solution in which dissociation occurs partly because of its intrinsic importance in thermodynamic theory and partly also because this discussion furnishes a simple introduction to the study of solutions of electrolytes.

The above discussion of the NO_2–N_2O_4 solution in chloroform solution is directly applicable to gases that dissociate or polymerize; and, in fact, both NO_2 and N_2O_4 molecules can be observed in gaseous NO_2, and the previous discussion of variation of fugacities in terms of mole fractions can be restated for gaseous NO_2 in terms of the variation of fugacity with pressure or concentration. One need not consider dissociation of N_2O_4 to $2\,NO_2$ but if one does not, then the P–V–T behavior of the gas is most abnormal. Thus it is usually preferable to deal with a dissociating gas as a mixture of two separate molecular species.

Before leaving the problem of dissociation, however, we wish to emphasize that there is an inherent arbitrariness in the dissociation treatment. Thus, one cannot distinguish rigorously between gas imperfection of NO_2 and association to N_2O_4— either can explain a decrease of volume below that calculated from the perfect-gas law for NO_2. Only by assuming a hypothetical behavior for unassociated NO_2 is it possible to calculate the amount of N_2O_4. If the gas density is low, it is reasonable to assume the perfect-gas law and an association treatment is most useful. However, at high densities one would expect substantial gas imperfection even without association. Under these conditions an association treatment becomes quite arbitrary, and it is frequently more useful to apply the virial-coefficient treatment (see Chapter 9) with the realization that abnormal values will be found for the virial coefficients.

THEORIES OF SOLUTION NONIDEALITY

There are many statistical theories for solutions; we consider here only a very few that have proven to be useful for a substantial number of examples.

Equal-size Molecules

There have been a number of studies[24] for the case where the two components have molecules of the same size and are otherwise sufficiently similar to fit into the same spaces in a lattice. They lead to the simple equation presented earlier:

$$G_m^E = RTwx_1x_2 \qquad (12\text{-}12)$$

We now proceed to estimate the parameter w. Let us assume the molecules to be located on a lattice of $N_a + N_b$ sites that has a coordination number of Z. If the components are segregated, then there are $\frac{1}{2}ZN_a$ nearest-neighbor interactions of molecules of component a and $\frac{1}{2}ZN_b$ interactions for component b. Random mixing yields a probability equal to the mole fraction $x_a = N_a/(N_a + N_b)$ that any site contains a molecule of a and a probability x_b for a molecule b. The probabilities of nearest-neighbor interactions of the type a–a, a–b, and b–b are therefore x_a^2, $2x_ax_b$, and x_b^2, respectively, where the factor 2 arises because we are counting both a–b and b–a interactions.

Now let us assume that the interaction energy is $-\varepsilon_{aa}$ when two molecules of type a are adjacent and similarly $-\varepsilon_{ab}$ for an a–b pair and $-\varepsilon_{bb}$ for two molecules of type b. Then the energy of the random mixture is

$$U(\text{mixed}) = -\tfrac{1}{2}Z(N_a + N_b)(\varepsilon_{aa}x_a^2 + 2\varepsilon_{ab}x_ax_b + \varepsilon_{bb}x_b^2)$$

$$= -\tfrac{1}{2}Z(N_a + N_b)^{-1}(\varepsilon_{aa}N_a^2 + 2\varepsilon_{ab}N_aN_b + \varepsilon_{bb}N_b^2) \qquad (12\text{-}50\text{a})*$$

The energy of the unmixed components is

$$U_a + U_b = -\tfrac{1}{2}Z(N_a\varepsilon_{aa} + N_b\varepsilon_{bb}) \qquad (12\text{-}50\text{b})*$$

and the change in energy of mixing is

$$\Delta_{\text{mix}}U = \tfrac{1}{2}Z(N_a + N_b)^{-1}N_aN_b(\varepsilon_{aa} + \varepsilon_{bb} - 2\varepsilon_{ab})$$

$$= (n_a + n_b)x_ax_bwRT \qquad (12\text{-}50\text{c})*$$

where

$$w = \frac{ZN_A(\varepsilon_{aa} + \varepsilon_{bb} - 2\varepsilon_{ab})}{2RT} \qquad (12\text{-}50\text{d})*$$

with N_A the Avogadro number. If the volume of mixing is zero and the entropy of mixing is ideal, then the excess Gibbs energy is equal to $\Delta_{mix} U$ and the w of Eq. (12-50d)* pertains to Eq. (12-12)*.

In general, $2\varepsilon_{ab}$ may be either larger or smaller than the sum $\varepsilon_{aa} + \varepsilon_{bb}$ and w can have either sign. Where there is a specially strong attraction between unlike molecules, w will be negative. Correspondingly, w will be positive if the attractive forces between unlike molecules are weaker than the mean for like molecules, and this can arise for a broad category of cases. The quantum theory for interactions between inert-gas atoms and simple, nonpolar molecules indicates that the attractive potential for unlike molecules is either equal to or less than the geometric mean of the potentials for like-molecule interactions. If we take the case of equality

$$\varepsilon_{ab} = (\varepsilon_{aa}\varepsilon_{bb})^{1/2} \qquad (12\text{-}51)^*$$

Now consider the expression

$$(\varepsilon_{aa}^{1/2} - \varepsilon_{bb}^{1/2})^2 = \varepsilon_{aa} + \varepsilon_{bb} - 2\varepsilon_{aa}^{1/2}\varepsilon_{bb}^{1/2} \qquad (12\text{-}52)^*$$

We note from Eq. (12-51)* that the expression on the left of Eq. (12-52)* is equal to a factor appearing in Eq. (12-50d)*. Thus, for the special case of the geometric mean relationship,

$$w = \frac{ZN_A(\varepsilon_{aa}^{1/2} - \varepsilon_{bb}^{1/2})^2}{2RT} \qquad (12\text{-}53)^*$$

This expression is always positive. Also, we noted above that the geometric mean, Eq. (12-51)* was an upper limit in that theory; hence, w may be more positive than is given by Eq. (12-53)*.

In this simple model, the internal energy increase on vaporization for a pure liquid is

$$\Delta_{vap} U_a = \tfrac{1}{2} Z N_A \varepsilon_{aa} \qquad (12\text{-}54)^*$$

Substitution of this expression yields

$$w = \frac{[(\Delta_{vap} U_a)^{1/2} - (\Delta_{vap} U_b)^{1/2}]^2}{RT} \qquad (12\text{-}55)^*$$

This provides a basis for estimation of w for the simple cases of exactly equal molar volumes and very simple, nonpolar molecules. The analysis of the next section indicates that even small differences in volume should be considered. Thus, for practical purposes it is better in all cases to use the expressions that recognize differences in volume. But this simple model shows the connection between the geometric mean relationship and the form of the expression as the square of the difference in square roots of energies.

Critical Point for Phase Separation

It is interesting to derive the conditions for liquid–liquid (or solid–solid) phase separation for this very simple system with molecules of equal size. The criteria were given in Chapter 10, Eqs. (10-45) and (10-46) in terms of fugacities. On conversion to activities and substitution of Eq. (12-14), one has

$$\frac{\partial \ln a_1}{\partial x_2} = -\frac{1}{1-x_2} + 2x_2 w = 0$$

$$\frac{\partial^2 \ln a_1}{\partial x_2^2} = -\frac{1}{(1-x_2)^2} + 2w = 0$$

$$\frac{\partial \ln a_2}{\partial x_2} = \frac{1}{x_2} - 2(1-x_2)w = 0$$

$$\frac{\partial^2 \ln a_2}{\partial x_2^2} = -\frac{1}{x_2^2} + 2w = 0 \qquad (12\text{-}56)*$$

These equations are satisfied by the conditions $x_1 = x_2 = 0.5$ and $w = 2$. Since w is a function of T, we have $w(T_c) = 2$. At this point the activity coefficient of either component is given by $\ln \gamma = 0.5$ or $\gamma = 1.65$ and $a = 0.82$. One can then predict that the total vapor pressure of a solution at the critical point for phase separation will be $P = 0.82(P_1^\circ + P_2^\circ)$, that is, 82% of the total vapor pressure for the two pure components on a perfect-gas basis. This result may be compared with the two extreme cases. An equimolal ideal solution would have $P = 0.50(P_1^\circ + P_2^\circ)$, whereas two liquids completely insoluble in one another would have a total vapor pressure $P = P_1^\circ + P_2^\circ$.

Regular Solutions of Hildebrand

In 1927 Hildebrand[25] called attention to the very great similarity in the behavior of a class of nonideal solutions that he named *regular solutions*. Such solutions are characterized by the absence of any specific interaction between molecules such as hydrogen bonding, acid–base association, etc. The pure components usually show the properties we associate with normal liquids or fluids (see Appendix 3), although other classes of substances such as metals may yield regular solutions also. These criteria are necessarily somewhat vague because there are no sharp boundaries to this category of solutions. Nevertheless, the classification has proved to be very useful.

Regular solutions differ from ideal solutions in that the intermolecular forces are no longer equal. Also, the molecules may be more unequal in size. However, these differences are sufficiently moderate so that thermal energy still yields very nearly random mixing. Consequently, the ideal entropy of mixing is a good approximation, although differences in volume and intermolecular attraction must have some effect.

Unsymmetrical Systems, Unequal Molal Volumes

The equations derived above for solutions with components having equal molal volumes are found to apply in good approximation to systems where the volumes differ only moderately. Thus the molal volume of cyclohexane is 109.1 cm^3, whereas that of benzene is 89.8 cm^3 at 30°C. In Fig. 12-1 the points for this system fall just slightly to the right of the theoretical curve of Eq. (12-12)* for equal volumes. Where the difference in molal volume of the components is greater, more unsymmetrical curves are obtained.

In 1906 van Laar[13] published a treatment of binary liquid solutions based upon the van der Waals equation. The van Laar equations were considered above on a purely empirical basis. The excess free energy of mixing is given by the equation

$$\frac{G^E}{RT} = a_{12}\left(\frac{b_1 b_2 n_1 n_2}{n_1 b_1 + n_2 b_2}\right) \qquad (12\text{-}41)^*$$

Here b_1 and b_2 are the molecular-volume parameters of the van der Waals equation for components 1 and 2, and a_{12} is a parameter expressing the deviation from ideal-solution behavior, which will be discussed further. The activity coefficients of the two components were given above as Eqs. (12-42a,b)* in terms of the variables z_1 and z_2.

The limitation of van Laar's treatment to the inexact van der Waals equation was unfortunate and unnecessary. In 1931 Scatchard[14] removed this limitation and derived equations that may be obtained from Eqs. (12-41)* and (12-42)* by substitution of the molar volumes $V_1°$ and $V_2°$ for b_1 and b_2. Scatchard's arguments are approximate but are quite plausible for the case of regular solutions. The composition variables, z_1 and z_2, now become the volume fractions

$$\phi_1 = \frac{n_1 V_1°}{n_1 V_1° + n_2 V_2°} \qquad \phi_2 = \frac{n_2 V_2°}{n_1 V_1° + n_2 V_2°} \qquad (12\text{-}40)$$

The excess Gibbs energy of mixing to form 1 mole of solution takes the very simple form

$$\frac{G^E}{RT} = a_{12} V \phi_1 \phi_2 \qquad (12\text{-}57)^*$$

where V is the molar volume of the solution, $x_1 V_1° + x_2 V_2°$. The Scatchard equations fit the experimental data quite well for many binary systems composed of normal liquids.

Prediction of Deviation from Ideality

There is no simple theory at present that will predict accurately the various results in Table 12-1 and those in Figs. 12-5 and 12-6. Nevertheless, it is of interest in some cases to have even a crude estimate of the deviation from Raoult's law for a system that has not been studied experimentally. Thus we wish an estimate of a_{12} or, in the case of systems with components of approximately equal molal volume, of w/V, which is then equal to a_{12}. The van Laar expression in terms of the van der Waals equation parameters is

$$a_{12} = \frac{1}{RT}\left(\frac{a_1^{1/2}}{b_1} - \frac{a_2^{1/2}}{b_2}\right)^2 \qquad (12\text{-}58)^*$$

Here a_1 and a_2 are energies of intermolecular attraction equivalent to the quantities $-\varepsilon_{aa}$ and $-\varepsilon_{bb}$ in Eq. (12-53)*. Thus, for equal-sized molecules with $b_1 = b_2$, the van Laar treatment yielded the same result as the simple calculation of Eqs. (12-47)* to (12-53)*.

Scatchard[14] in his 1931 derivation, which was described in the preceding section, obtained

$$a_{12} = \frac{1}{RT}\left[\left(\frac{\Delta U_1}{V_1}\right)^{1/2} - \left(\frac{\Delta U_2}{V_2}\right)^{1/2}\right]^2 \qquad (12\text{-}59)*$$

where the ΔUs are molar energies of vaporization (approximately $\Delta H_{vap} - RT$) and the Vs are molar volumes of the pure liquid components. Subsequently Hildebrand and Wood[15] derived Eq. (12-59)* by approximate statistical calculations based upon molecular-distribution functions. Equation (12-59)* is to be preferred in practically all cases in comparison to Eq. (12-58)*.

Equation (12-59)* may be simplified for components with equal molal volumes. However, if the volumes differ even slightly, it is best to use the full expression of Eq. (12-59)* for a_{12}. Then the average of V_1 and V_2 may be taken in the calculation of $w = a_{12}V$ for the simple symmetrical equations (12-12) to (12-17).

The comparison of calculated and observed values of a_{12} in Table 12-5 shows that this method yields agreement within a few tenths in $a_{12}V_1$. This is useful for estimates when experimental data are inaccessible. Relatively simple experiments, however, will yield a more reliable value. Also, this treatment implies that G^E and H^E are about equal, with TS^E very small. Actually we see, from a typical case shown in Fig. 12-6, that TS^E is only moderately less than H^E and opposite in sign, so that G^E is relatively much smaller.

The preceding discussions and Eqs. (12-58)* and (12-59)* pertain to molecules with closed electronic shells and with no change in electronic state on vaporization. For metals or other cases with incomplete electronic shells, there is no longer a theoretical basis for the geometric mean estimate for a_{12} (or ε_{ab} in Eq. (12-51)*). Brewer[26-28] has studied many of these more complex systems and suggests alternate expressions. For equal molal volumes, Brewer and Lamoreaux[26] recommend on an empirical basis

$$\varepsilon_{ab} = \tfrac{1}{4}(\varepsilon_{aa}^{1/2} + \varepsilon_{bb}^{1/2})^2 \qquad (12\text{-}60)*$$

TABLE 12-5
Observed and estimated constants for deviation from Raoult's law

System	T/K	$a_{12}V_1$ Obs.	$a_{12}V_1$ Eq. (12-59)*
Benzene–CS$_2$	298	0.47	0.11
SnI$_4$–SiCl$_4$	298	3.8	4.4
SnI$_4$–benzene	298	1.9	1.7

This alternate was extended to unequal molal volumes.[26,27] Where there is a known change in electronic state on vaporization, Brewer[28] recommends the use of an increased value of ε or ΔU corresponding to vaporization to the atomic state present in the liquid or solid state.

POLYMER SOLUTIONS

Let us turn now to a consideration of the properties of solutions of very large nonpolar polymer molecules in ordinary solvents. Rubber in benzene and polystyrene in acetone are examples. The heat of mixing in such cases is approximated by the more general equations of the two preceding sections, but the entropy of mixing deviates grossly from the Raoult's-law value even when the heat of mixing is zero.

The mole fraction is not a satisfactory measure of composition for high-polymer solutions. Frequently the molecular mass of the polymer is not accurately known, but in any case it is so high as to make mole fractions inconvenient. The weight or mass fraction is good for fundamental work, since it is independent of pressure or temperature for a given solution. Theoretical discussions usually yield results in terms of the volume fraction, e.g., Eq. (12-57)*. The volume fraction was defined in Eq. (12-40).

The system rubber–benzene has been studied very carefully and extensively by Gee and collaborators.[29] They measured the osmotic pressure of dilute solutions and the benzene vapor pressure over concentrated solutions. Measurements were made at different temperatures in order to separate heat and entropy effects; also direct calorimetric measurements were made of the heat of mixing. Their results are shown as the solid curves in Figs. 12-9 and 12-10. The molecular mass of the rubber is very large; consequently the slope of the activity and entropy curves should be practically

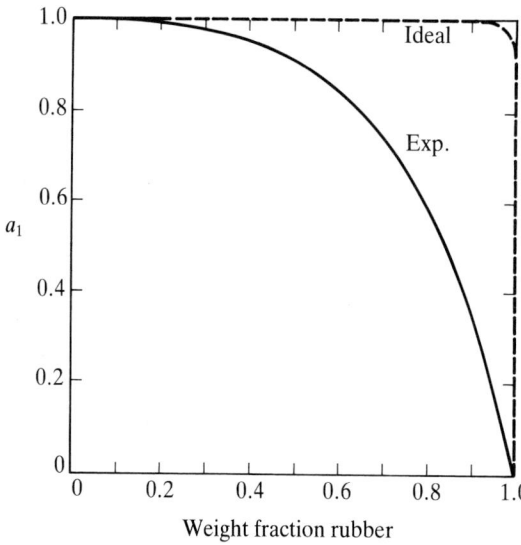

FIGURE 12-9
The activity of benzene in rubber–benzene solutions compared with the prediction for an ideal solution.

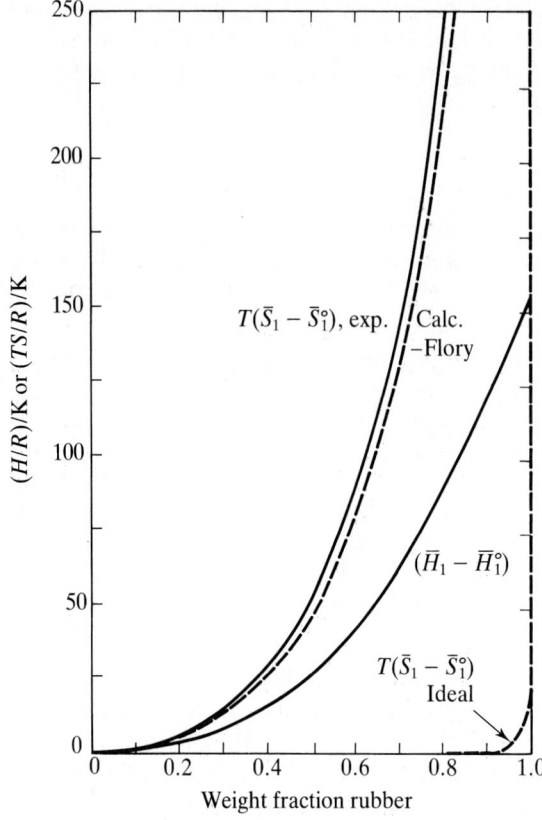

FIGURE 12-10
The partial molar entropy and enthalpy of benzene in rubber–benzene solutions. The predictions for the entropy by ideal solution and Flory–Huggins theories are shown.

zero at the zero-concentration axis. This is observed, but beyond this one point both the entropy and activity curves deviate greatly from those predicted for an ideal solution.

The large discrepancy in concentrated solutions between the observed entropy and the ideal entropy can be explained on the basis that the individual segments of the polymer molecule have considerable freedom of motion. Flory[30] and Huggins[31] assumed a lattice model with a polymer molecule occupying a sequence of solvent-sized sites. They showed that, in the first approximation, the entropy of mixing was given by the replacement of mole fraction by volume fraction in the Raoult's-law expression,

$$\Delta_{mix}S = -R(n_1 \ln \phi_1 + n_2 \ln \phi_2) \qquad (12\text{-}61)^*$$

The partial molar entropy of the solvent is found by differentiation to be

$$\bar{S}_1 - S_1^\circ = -R\left[\ln(1 - \phi_2) + \phi_2\left(1 - \frac{V_1^\circ}{V_2^\circ}\right)\right] \qquad (12\text{-}62)^*$$

We may show that this approaches Raoult's law in very dilute solutions by expanding both expressions in powers of x_2 or ϕ_2. Raoult's law becomes

$$\bar{S}_1 - S_1^\circ = R\left(x_2 + \frac{x_2^2}{2} + \frac{x_2^3}{3} + \cdots\right) \qquad (12\text{-}63)*$$

while Eq. (12-59)* becomes

$$\bar{S}_1 - S_1^\circ = R\left(\phi_2\frac{V_1^\circ}{V_2^\circ} + \frac{\phi_2^2}{2} + \frac{\phi_2^3}{3} + \cdots\right) \qquad (12\text{-}64)*$$

But both $\phi_2 V_1^\circ/V_2^\circ$ and x_2 approach n_2/n_1 as the concentration approaches zero.

In the more concentrated solutions, Eq. (12-62)* may be simplified in a different fashion for very high polymers. Then V_1°/V_2° is negligible as compared with unity, and

$$(\bar{S}_1 - S_1^\circ) = -R[\ln(1 - \phi_2) + \phi_2] \qquad (12\text{-}65)*$$

The curve calculated from this equation is shown in Fig. 12-10 as the dashed line labeled "Flory." The close similarity with the experimental curve is apparent.

Partial molar entropy data for the solvent in several other polymer solution systems are shown in Fig. 12-11. Here it is seen that the agreement with the Flory–Huggins

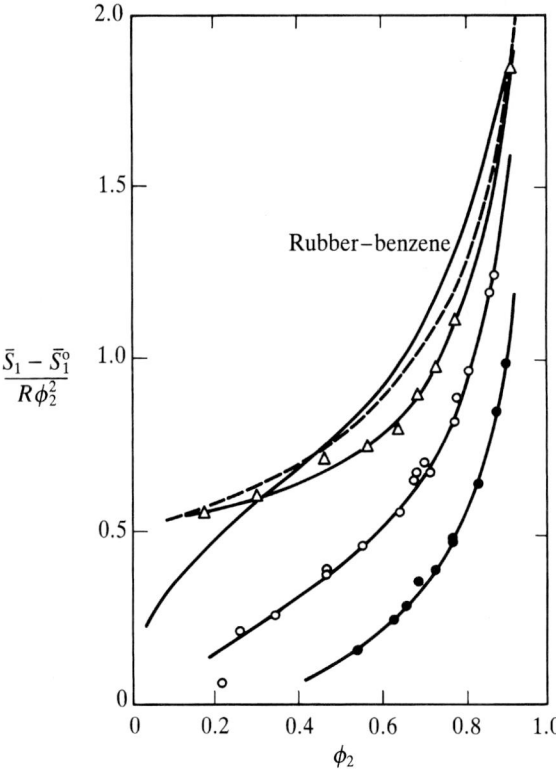

FIGURE 12-11
Comparison of observed partial molar entropies of the solvent (points and solid lines) with the Flory–Huggins curve [Eq. (21-28)*] (broken line). (Paul J. Flory, *Principles of Polymer Chemistry*, Cornell University Press, Ithaca, N.Y., 1953, p. 518, fig. 113). The rubber–benzene curve is from G. Gee and L. R. G. Treloar, *Trans. Faraday Soc.*, **38**, 147 (1942); G. Gee and W. J. C. Orr, *Trans. Faraday Soc.*, **42**, 507 (1946). Others are polydimethylsiloxane in benzene (△) from M. J. Newing, *Trans. Faraday Soc.*, **46**, 613 (1950), and polystyrene in methyl ethyl ketone (●) and in toluene, (○) from C. E. H. Bawn, R. F. J. Freeman, and A. R. Kamaliddin, *Trans. Faraday Soc.*, **46**, 677 (1950).

equation for the rubber–benzene system was apparently accidental and that the true situation is more complex. However, the large deviations from the Flory–Huggins equation are in the direction of the ideal-solution equation. Thus the Flory–Huggins and the Raoult equations for the entropy of mixing appear to give the limits between which will lie the entropy of mixing of solutions of this type.

The excess Gibbs energy of mixing of polymer solutions arising from heat-of-mixing effects is reasonably well represented by Eq. (12-57)*. If we consider a solution with n_1 moles of solvent and write w_1 for the product $a_{12}V_1^\circ$, Eq. (12-57)* becomes

$$\frac{G^{\text{ex}}}{RT} = w_1 n_1 \phi_2 \qquad (12\text{-}66)^*$$

This excess Gibbs energy† is added to the Gibbs energy of mixing that would arise if there were no heat of mixing. For a polymer solution, however, this is given not by Raoult's law but rather by an expression such as the Flory–Huggins equation. For zero heat of mixing $\Delta_{\text{mix}}G = -T\Delta_{\text{mix}}S$; hence from Eqs. (12-61)* and (12-66)*

$$\frac{\Delta_{\text{mix}}G}{RT} = n_1 \ln \phi_1 + n_2 \ln \phi_2 + w_1 n_1 \phi_2 \qquad (12\text{-}67)^*$$

The appropriate temperature derivative gives the corresponding enthalpy of mixing:

$$\Delta_{\text{mix}}H = -RT^2 \frac{dw_1}{dT} n_1 \phi_2 \qquad (12\text{-}68)^*$$

Also, one may derive the activity of the solvent, which is frequently the measurable quantity,

$$\ln a_1 = \ln(1 - \phi_2) + \phi_2\left(1 - \frac{V_1^\circ}{V_2^\circ}\right) + w_1 \phi_2^2 \qquad (12\text{-}69)^*$$

While Eq. (12-69)* reproduces the major deviations of polymer solution properties from ideal behavior, it does not ordinarily yield quantitative agreement with experimental data. Further refinements are discussed in works specially devoted to polymer solution thermodynamics. This is an active research field, and many new developments are being reported.

Solid Solutions

The presence of a regular lattice structure in crystals provides a basis for special theories for solid solutions in which there are different atoms (or ions) in particular types of sites. The general principles are the same as those illustrated above, but

---

† We write this as G^{ex} instead of G^E to remind one that it is an excess from a reference different from Raoult's law.

additional criteria may be involved. Atomic or ionic size is much more important for the energy differences in crystals than for the disordered structures of fluids or gases. Chapter 26 includes several examples involving solid solutions.

MULTICOMPONENT SYSTEMS

The simplest equation for multicomponent liquid or solid solutions is just a generalization of Eq. (12-12) described at the beginning of this chapter. We first illustrate the pattern for three components with the excess Gibbs energy

$$\frac{G^E}{RT} = \frac{w_{12}n_1n_2 + w_{13}n_1n_3 + w_{23}n_2n_3}{n_1 + n_2 + n_3} \qquad (12\text{-}70)*$$

or, per mole,

$$\frac{G^E_m}{RT} = w_{12}x_1x_2 + w_{13}x_1x_3 + w_{23}x_2x_3 \qquad (12\text{-}71)*$$

Then differentiation of Eq. (12-67)* yields for the activity coefficients,

$$\ln \gamma_1 = (1 - x_1)(x_2 w_{12} + x_3 w_{13}) - x_2 x_3 w_{23} \qquad (12\text{-}72a)*$$

$$\ln \gamma_2 = (1 - x_2)(x_1 w_{12} + x_3 w_{23}) - x_1 x_3 w_{13} \qquad (12\text{-}72b)*$$

$$\ln \gamma_3 = (1 - x_3)(x_1 w_{13} + x_2 w_{23}) - x_1 x_2 w_{12} \qquad (12\text{-}72c)*$$

For the accurate representation of many real solutions, a more complex equation is required. A straightforward extension of Eq. (12-71)* yields the multicomponent Margules equation, which is often written as

$$\frac{G^E_m}{RT} = \sum_i \sum_j a_{ij} x_i x_j + \sum_i \sum_j \sum_k a_{ijk} x_i x_j x_k + \cdots \qquad (12\text{-}73)$$

The sums can either be unrestricted—both a_{ij} and a_{ji} included—or restricted. If unrestricted, $a_{ij} = a_{ji} = w_{ij}/2$. Parameters with all indices the same are zero. The higher parameters involving only two components, i.e., $a_{112}, a_{122}, a_{1122}$, etc., are all related to the Redlich–Kister parameters of Eq. (12-33). The parameters involving three or more components are new, however. Thus, one can transfer most parameters from binary solutions, but the equation contains terms to account for real ternary or higher-order effects if present. Also, there is no problem for temperature or pressure derivatives since the mole fractions are clearly independent of those variables. These are important aspects that recommend a Margules-type series for multicomponent systems and favor Margules or Redlich–Kister treatments of binaries. In contrast, unless true volume ratios are used in van Laar treatments, the extension to three or more components is impossible, and even if consistent volume ratios are used there is still the problem of temperature variation of those ratios.

As Eq. (12-73) is written, there are redundancies among the third and higher-order terms. Alternate methods of deleting these redundancies are described in Chapter 26, where expressions for activity coefficients are also given.

A general question of great interest concerns the possibility of obtaining G^E and the other $\ln \gamma_i$ from measurements of the activity coefficient of just one component. Since the mole-fraction dependency of each w_{ij} is different in Eq. (12-72a)* for $\ln \gamma_1$, one concludes that all of the w_{ij} could be determined if the range and accuracy of the measurements were sufficient. One can readily verify that this conclusion remains correct for four or an indefinite number of components, although the actual calculation would become more difficult and the demand for accuracy greater. Similarly, the conclusion can be verified for expressions including higher-order terms after the redundancies are eliminated.

The question of the preceding paragraph can be investigated without use of a Margules or other equation, but rather by means of the multicomponent Gibbs–Duhem equation. Then, at constant T and P, Eq. (10-10) becomes

$$dG = \mu_1 dn_1 + \mu_2 dn_2 + \mu_3 dn_3 + \cdots + \mu_i dn_i \qquad (12\text{-}74)$$

and from Eq. (2-16), one has

$$\left(\frac{\partial \mu_2}{\partial n_1}\right)_{n_2, n_3, \ldots, n_i} = \left(\frac{\partial \mu_1}{\partial n_2}\right)_{n_1, n_3, \ldots, n_i} \qquad \text{etc.} \qquad (12\text{-}75)$$

If μ_1 is known for all compositions, the change in μ_2 can be obtained along any path. The integration can start from the standard state, pure component 2, and then extend to give $\mu_2 - \mu_2^\circ$ at other compositions. This process can be repeated for $\mu_3 - \mu_3^\circ$, and so on.

If the solution phase does not extend over the full composition range, however, there can be limitations. Many aqueous and other liquid solutions extend only to the limits of solid solubility. Then the pure-component standard state cannot be used, and the solute or infinite-dilution standard state is used for one or more solute components. We return to the example of a three-component Margules equation with the solvent as component 1, while components 2 and 3 are solutes. Now, Eq. (12-24) applies for the solutes. Generalization for more than one solute requires that both x_2 and x_3 be zero in the standard state. The effect is to subtract a constant, w_{12}, from Eq. (12-72b), and similarly to subtract w_{13} from Eq. (12-72c)*. Then the activity coefficient of component 2 on the solute state basis is

$$\ln \gamma_2^* = w_{12}(x_1 - x_1 x_2 - 1) + w_{23} x_3 (1 - x_2) - w_{13} x_1 x_3 \qquad (12\text{-}76)*$$

Since we assume that components 2 and 3 have limited solubility, x_2 and x_3 will be small in comparison to x_1. Then it is interesting to expand Eq. (12-76)* by elimination of x_1, and expression in powers of x_2 and x_3. The result is

$$\ln \gamma_2^* = -2x_2 w_{12} + x_3(w_{23} - w_{12} - w_{13}) + \text{higher terms} \qquad (12\text{-}77)*$$

Thus, if γ_2^* is the measured property and only first-order terms are considered, one finds that w_{12} is determined along with $(w_{23} - w_{12} - w_{13})$. Thus, while w_{12} is given unambiguously, w_{13} and w_{23} are not determined individually—only $(w_{23} - w_{13})$. Then γ_1 and γ_3^* remain undetermined.

It is found for this case that the second-order terms for Eq. (12-77)* do determine w_{23} and w_{13} separately. But, if higher-order Margules terms are included, they are of the same order and new redundancies arise. Thus, the situation becomes very complex.

Since much research for limited-solubility solutions makes use of molalities instead of mole fractions, we postpone further discussion of this last topic to Chapters 14 and 17 where simple and unambiguous results are obtained with the molality-based equations.

The treatment of multicomponent nonelectrolyte solutions is continued in Chapter 26.

REFERENCES

1. M. Margules, *Sitzber. Akad. Wiss. Wien., Math-naturw. Kl., Abt. IIa*, **104**, 1243 (1895).
2. K. S. Pitzer and L. Brewer, *Thermodynamics*, revision of 1st edn. by G. N. Lewis and M. Randall, McGraw-Hill, New York, 1961.
3. J. M. Prausnitz, R. N. Lichtenthaler, and E. G. de Azevedo, *Molecular Thermodynamics of Fluid Phase Equilibria*, Prentice-Hall, Englewood Cliffs, N.J., 1986, chapter 6.
4. O. Redlich and A. T. Kister, *Ind. Eng. Chem.*, **40**, 345 (1948).
5. O. Redlich, A. T. Kister, and C. E. Turnquist, *Chem. Eng. Prog. Series*, **48**(2), 49 (1952).
6. E. A. Guggenheim, *Trans. Faraday Soc.*, **33**, 151 (1937).
7. G. Scatchard and C. L. Raymond, *J. Am. Chem. Soc.*, **60**, 1278 (1938).
8. J. A. Barker, *Aust. J. Chem.*, **6**, 207 (1953).
9. M. M. Abbot and H. C. Van Ness, *Fluid Phase Equil.*, **1**, 3 (1977).
10. R. W. Hermsen and J. M. Prausnitz, *Chem. Eng. Sci.*, **18**, 485 (1963).
11. R. V. Orye and J. M. Prausnitz, *Trans. Faraday Soc.*, **61**, 1338 (1965).
12. S. M. Byers, R. E. Gibbs, and H. C. Van Ness, *Am. Inst. Chem. Eng. J.*, **19**, 245 (1973).
13. J. J. van Laar, *Sechs Vortrage uber das Thermodynamische Potential*, Vieweg-Verlag, Brunswick, Germany, 1906; *Z. physik. Chem.*, **72**, 723 (1910).
14. G. Scatchard, *Chem. Rev.*, **8**, 321 (1931).
15. J. H. Hildebrand and S. E. Wood, *J. Chem. Phys.*, **1**, 817 (1933).
16. S. Weissman and S. E. Wood, *J. Chem. Phys.*, **32**, 1153 (1960).
17. G. M. Wilson, *J. Am. Chem. Soc.*, **86**, 127 (1964).
18. R. V. Orye and J. M. Prausnitz, *Ind. Eng. Chem.*, **57**(5), 19 (1965).
19. W. Pfeffer, *Osmotische Untersuchungen*, W. Engelmann, Leipzig, 1877.
20. C. M. Guldberg and P. Waage, *Etudes sur les affinités chimiques*, Brogger and Christie, Christiania, 1867.
21. J. H. van't Hoff, *Z. physik. Chem.*, **1**, 481 (1887).
22. F. D. Rossini et al., "Selected Values of Chemical Thermodynamic Properties," *Natl. Bur. Standards (U.S.) Circ.* 500, 1952; D. D. Wagman et al., "NBS Tables of Thermodynamic Properties," *J. Phys. Chem. Ref. Data*, **11**, supplement no. 2, 1982.
23. J. T. Cundall, *J. Chem. Soc.*, **59**, 1076 (1891); **67**, 7 (1895).
24. K. F. Herzfeld and W. Heitler, *Z. Elektrochem.*, **31**, 536 (1925); W. Heitler, *Ann. Physik*, **80**(4), 630 (1926); see also E. A. Guggenheim, *Mixtures*, Oxford University Press, New York, 1952, chapter IV.
25. J. H. Hildebrand, *Proc. Natl. Acad. Sci., U.S.A.*, **13**, 267 (1927).
26. L. Brewer and R. H. Lamoreaux, "Thermochemical Properties of Molybdenum and its Compounds and Alloys," *Atomic Energy Review Special Issue No. 7*, International Atomic Energy Agency, Vienna, 1980, pp. 156–157.
27. L. Brewer, *Mat. Res. Soc., Symp. Proc.*, **19**, 129 (1983).
28. L. Brewer, *Systematics of the Properties of the Lanthanides*, NATO ASI Series C: Mathematical and Physical Sciences, No. 109, pp. 17–69, S. P. Sinha, ed., D. Reidel Publishing Co., Boston, 1983.

29. G. Gee and L. R. G. Treloar, *Trans. Faraday Soc.*, **38**, 147 (1942); G. Gee and W. J. C. Orr, *Trans. Faraday Soc.*, **42**, 507 (1946).
30. P. J. Flory, *J. Chem. Phys.*, **9**, 660 (1941), **10**, 5 (1942).
31. M. L. Huggins, *Ann. N.Y. Acad. Sci.*, **43**, 1 (1942).

PROBLEMS

12-1. Over the range from $x_2 = 0$ to $x_2 = 0.2$ the activity of the solute is given by
$$\ln(a_2/x_2) = A(x_2 - \tfrac{1}{2}x_2^2)$$
Derive expressions for the activity of the solvent and the excess Gibbs energy of the solution. Also state the range of validity of these results.

12-2. Liquids A and B are miscible in all proportions above a critical temperature. The fugacity of A is given by the equation
$$\log f_A = \text{funct.}(t) + \log x_A + \frac{260}{T} x_B^2 - \frac{3}{T} x_B^3$$
Obtain expressions for (a) the activity coefficient of A based on the solvent standard state; (b) the activity coefficient of B based on the solvent standard state; and (c) the heat of mixing 1 mole of A with 1 mole of B.

12-3. What is the liquid–liquid critical temperature for the solution in Problem 12-2? Solve approximately by neglecting the last term $(3/T)x_B^3$, in the equation for $\log f_A$.

12-4. Derive equations for $\bar{V}_2$ and $\bar{H}_2$ in terms of the Henry's-law constant and its derivatives.

12-5. The change in volume or enthalpy on dilution with additional solvent is zero for an ideal solution. Derive these properties for a nonideal solution that is so dilute that it conforms to Henry's law.

12-6. Derive an expression for the change in solubility (mole fraction of solute) with the temperature for a case in which the solubility is small.

12-7. Calculate the activity of liquid water at 25°C and at 100 bar pressure, assuming it to be incompressible.

12-8. What is the activity of water at 100°C in a solution of which the vapor pressure of water is 700 mmHg?

12-9. At 60°C aniline and water mixtures form two liquids of composition 4.4 and 93.4 percent aniline by weight. Assume that Raoult's law holds for the abundant component and Henry's law holds for the dilute component in each phase. Calculate for each phase the activity coefficient of the dilute component on the basis of a solvent standard state. What is the difference in Gibbs energy between the solute and solvent standard state for each component?

12-10. Predict the total vapor pressure of a solution made from 50 g each of benzene and $C_2H_4Cl_2$ at 323 K. Note the data in Table 12-1.

12-11. Calculate the solubility of solid SnI_4 in liquid $SiCl_4$ at 298 K. Use the observed value of $a_{12}V_1$ given in Table 12-5. The melting point and heat of fusion of SnI_4 are 417.7 K and 18.7 kJ mol^{-1}. Assume ΔC_P° of fusion to be negligible.

12-12. The data for the system benzene–isooctane were fitted to the van Laar equation and to the Redlich–Kister equation with parameters given in the text. Compare the respective values of G^E/RT for 45°C and for x (benzene) = 0.1, 0.5, and 0.9.

12-13. For the system of Problem 12-12, compare the values of γ (isooctane) from the two equations for x (isooctane) = 0.2, 0.5, and 0.8.

12-14. For what values of A and x_2 does the Redlich–Kister equation, Eq. (12-34)*, yield a critical point if $B = 0.1$, and C, D, ... are zero?

12-15. Do the parameters in Table 12-2 predict phase separation for any system listed? If so, what are the compositions of the coexisting phases?

12-16. It has been suggested that molten silver can be used to extract fission product elements from molten uranium. Calculate the solubilities of liquid silver and liquid uranium in one another at 1400 K if the activity coefficients at that temperature are given by $\ln \gamma_{Ag} = 2.83 x_U^2$ and $\ln \gamma_U = 3.50 x_{Ag}^2$. An iterative calculation is appropriate.

12-17. For a solution of nonpolar molecules A and B at 25°C, the measured values of the fugacity of B are as follows:

x_B	f_B (torr)	a_B
0.1	17.5	0.4375
0.2	25.6	0.640
0.4	30.8	0.770
0.6	32.1	0.803
0.8	34.4	0.860
1.0	40.0	1.000

(a) Calculate γ_B relative to pure B for $x_B = 0.1$ and 0.8.

(b) Select the simplest analytical function that you can devise to reasonably represent f_B or γ_B as a function of x_B.

(c) From your analytical function, predict the limiting f_B/x_B ratio as $x_B \to 0$.

CHAPTER 13

MIXED NONIDEAL GASES AND FLUIDS

Having established the properties of ideal solutions and defined various quantities that describe departures from ideality, we next consider mixed nonideal gases and fluids. First, certain basic equations are derived. Then, as was the case for pure fluids, we consider separately mixed gases at the second-virial-coefficient level and then mixed fluids generally, including vapor–liquid phase separation. A primary source of chemical potential information for liquid solutions is the measurement of fugacities for the vapor in equilibrium. While this vapor at low pressure will not depart greatly from the perfect gas law, it is important to determine the corrections from partial pressures to fugacities. For this purpose, the second virial coefficient suffices. In the general presentation, the basic principles and equations are presented for equilibrium between phases when each may be a mixed fluid, but only simple examples are given. Mixed gases at high pressure are considered as well as the case of a mixed nonideal gas in equilibrium with a pure solid phase. The more general case of two mixed fluids is considered for the van der Waals equation. Also, references are given to books devoted to fluid phase equilibria and to articles and reviews describing applications to mixed fluids of equations described in Chapter 9 and Appendix 4.

BASIC EQUATIONS

The equations of Chapter 9 apply to mixed gases of fixed composition as well as to pure gases. Now n and ρ become the total number of moles, $n_T = \sum_i n_i$, and the total density, $\rho_T = \sum_i \rho_i$. The virial coefficients become composition-dependent, but the various equations relating them to the Gibbs energy and other thermodynamic quantities remain unchanged. The equations for the chemical potential or fugacity of each component in a single phase are then derived; these are of primary importance, both for vapor–liquid phase relations and for chemical reactions.

For a mixed gas the generalized virial series becomes

$$\frac{P}{RT} = \sum_i \rho_i + \sum_i \sum_j \rho_i \rho_j B_{ij} + \sum_i \sum_j \sum_k \rho_i \rho_j \rho_k C_{ijk} + \cdots \quad (13\text{-}1)$$

with each ρ_i the density n_i/V; also each sum includes all components. In terms of the total density, the virial series becomes

$$\frac{P}{RT} = \rho_T + \rho_T^2 B_{mix} + \rho_T^3 C_{mix} + \cdots \quad (13\text{-}2)$$

with

$$B_{mix} = \sum_i \sum_j y_i y_j B_{ij} \quad (13\text{-}3a)$$

$$C_{mix} = \sum_i \sum_j \sum_k y_i y_j y_k C_{ijk} \quad (13\text{-}3b)$$

Here y_i, y_j, and y_k are the mole fractions in the gas. It is common practice to use the symbol y_i for mole fraction in the gas or vapor phase and x_i for the liquid phase.

A power series in pressure, instead of density, can be used for mixed gases, but we will use it only at the second virial coefficient level. At that level, the composition dependency is unambiguous and the pressure is given by

$$PV = n_T(RT + PB_{mix}) \quad (13\text{-}4a)^*$$

$$= n_T RT + \frac{P \sum_i \sum_j n_i n_j B_{ij}}{n_T} \quad (13\text{-}4b)^*$$

where B_{mix} was given in Eq. (13-3a). If desired, one can define a B' analogous to that of Eq. (9-4) for pure gases, but this is unnecessary. Then the change in Gibbs energy with pressure at constant temperature is derived

$$\left(\frac{\partial G}{\partial P}\right)_{T,n_i n_j} = V \quad (3\text{-}8)$$

$$G - G° = n_T RT \ln\left(\frac{P}{P°}\right) + P \frac{\sum_i \sum_j n_i n_j B_{ij}}{\sum_i n_i} \quad (13\text{-}5)^*$$

Here $G°$ is the Gibbs energy of the ideal mixed gas at the standard pressure:

$$G° = \sum_i n_i G_i° + RT \sum_i n_i \ln x_i \quad (13\text{-}5a)$$

The chemical potential is then obtained:

$$\mu_i = \left(\frac{\partial G}{\partial n_i}\right)_{P,T,n_{j \neq i}}$$

$$\mu_i - \mu_i^\circ = RT \ln\left(\frac{P}{P^\circ}\right) + P(2\sum_j y_j B_{ij} - B_{\text{mix}}) \quad (13\text{-}6)*$$

whereupon the fugacity and fugacity coefficient are

$$\phi_i = \frac{f_i}{y_i P} = \frac{f_i}{f_i^{\text{id}}} \quad (13\text{-}7a)$$

$$\ln \phi_i = \frac{\mu_i - \mu_i^{\text{id}}}{RT} \quad (13\text{-}7b)$$

$$= \frac{P}{RT}(2\sum_j y_j B_{ij} - B_{\text{mix}}) \quad (13\text{-}8)*$$

and, for two components, Eqs. (13-3) and (13-8)* become

$$B_{\text{mix}} = y_1^2 B_{11} + 2y_1 y_2 B_{12} + y_2^2 B_{22} \quad (13\text{-}9)$$

and

$$\ln \phi_1 = \frac{P}{RT}[(1 - y_2^2)B_{11} + y_2^2(2B_{12} - B_{22})] \quad (13\text{-}10)*$$

The more complex equations, including the third virial coefficient for either the density or the pressure series, are given by Prausnitz et al.[1] and other sources.

We turn now to the basic relationships for equations valid over the full range of density including the liquid. Now we use density as an independent variable instead of pressure. If the composition of a mixed fluid is constant, the relationships for a pure fluid apply. Thus, if the compression factor is $z(T,\rho_T,x)$, where x represents the various mole fractions, then the molar Helmholtz energy is given by integration, Eqs. (9-27b) and (9-28):

$$\frac{A_m^{\text{res}}(T,\rho_T,x)}{RT} = \int_0^{\rho_T} (z - 1)\, d\ln \rho_T' \quad (13\text{-}11a)$$

$$A_m(T,\rho_T,x) = G_m^\circ(T,x) + RT[-1 + \ln(\rho_T RT)] + A_m^{\text{res}} \quad (13\text{-}11b)$$

Appropriate manipulations then give the enthalpy, entropy, etc., as indicated in Eqs. (9-37) through (9-40). To obtain the chemical potential, the Helmholtz energy is converted to the extensive quantity $A(T,V,n_i,n_2, \ldots)$. Then Eq. (10-14) applies.

$$\mu_i = \left(\frac{\partial A}{\partial n_i}\right)_{T,V,n_{j \neq i}} \quad (10\text{-}14)$$

after which the fugacity and fugacity coefficient follow from Eq. (13-7a, b). This process is illustrated in a later section after we have first considered applications at the second virial coefficient level.

SECOND VIRIAL COEFFICIENTS FOR MIXED GASES

From volume measurements on pure and binary mixed gas, one may determine the second virial coefficient as a function of composition and, from Eq. (13-9), obtain B_{12} as well as B_{11} and B_{22}. Then, substitution into Eq. (13-10)* yields the fugacity coefficients needed to correct partial pressure data to fugacities. Also, once the B_{ij} values are determined for various binary mixtures, they suffice through Eq. (13-8)* to determine the fugacities of mixed gases with any number of components.

Before turning to experimental data for the cross coefficients B_{ij}, we note that the form of the basic statistical equation remains unchanged from that of Eq. (9-71) for pure gases. Now,

$$B_{ij} = 2\pi N_A \int_0^\infty \{1 - \exp[-u_{ij}(r_{ij})/kT]\} \, r_{ij}^2 \, dr_{ij} \qquad (13\text{-}12)$$

Thus, the various model equations of Chapter 9 for the interparticle potential can be used without change for the cross coefficient.

When components 1 and 2 are very similar in their molecular characteristics, the intermolecular potential u_{12} will be nearly the same as either of the potentials between like molecules u_{11} or u_{22}, which will also be similar to one another. Then one can expect B_{12} to be approximated by the mean of B_{11} and B_{22}:

$$B_{12} = \tfrac{1}{2}(B_{11} + B_{22}) \qquad (13\text{-}13)*$$

In this case, Eqs. (13-9) and (13-10) simplify to

$$B_{mix} = y_{11} B_{11} + y_{22} B_{22} \qquad (13\text{-}14)*$$

and

$$\ln \phi_1 = \frac{P}{RT} B_{11} \qquad (13\text{-}15)*$$

Substitution of Eq. (13-15)* into Eq. (13-7a) yields for the fugacity

$$f_i = y_i P \exp(B_{ii} P/RT) \qquad (13\text{-}16)*$$

which conforms to the definition of an ideal solution since the fugacity is proportional to the mole fraction at any constant T and P.

It is now interesting to examine the actual properties of gas mixtures with molecules of similar types to determine the range of cases for which Eq. (13-13)* is a good approximation. Figure. 13-1 shows this information for several mixed hydrocarbons. For the butane isomers, B_{12} follows the mean value equation within experimental error, and even for the propane–butane mixture it is a good approximation. But for either ethane–butane or propane–hexane mixtures there are large deviations from Eq. (13-13)*.

It is not sufficient that the intermolecular potential u_{12} be the mean of u_{11} and u_{22}. From Eq. (13-12), one sees that it is the exponential of $-u/kT$ which is linearly related

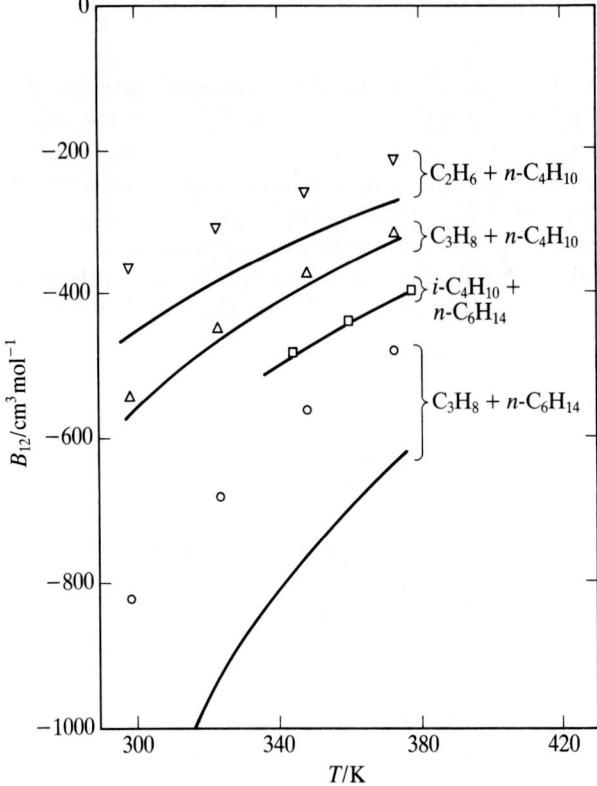

FIGURE 13-1
The cross second virial coefficient B_{12} (symbols) compared with curves showing the mean value of the pure gas coefficients $(B_{11} + B_{22})/2$. This tests the approximation of Eq.(13-13)*

to B. If u/kT is small and a linear expansion of the exponential suffices, then Eq. (13-13)* is valid. But in most cases, u is large compared to kT over a substantial range of r, and Eq. (13-13)* is not expected to be a good approximation.

The data for Fig. 13-1 were taken from the compilation of Dymond and Smith,[2] where many other examples can be found. It is useful also to define an excess quantity for the cross coefficient:

$$\delta_{12} = 2B_{12} - (B_{11} + B_{22}) \qquad (13\text{-}17)$$

If this expression is substituted into Eq. (13-10)*, one obtains for the fugacity coefficient,

$$\ln \phi_1 = \frac{P}{RT}(B_{11} + y_2^2 \delta_{12}) \qquad (13\text{-}18)*$$

Then, at a given T and P the ratio of the fugacity in a mixed gas f_i to that of the pure gas f_i° is

$$\frac{f_1}{f_1^\circ} = y_1 \exp[y_2^2(\delta_{12}P/RT)] \qquad (13\text{-}19)*$$

In Chapter 12 the activity coefficient γ_i was defined as the ratio of the actual fugacity to the ideal fugacity:

$$\gamma_i = \frac{f_i}{y_i f_i^\circ} \tag{13-20}$$

Then

$$\ln \gamma_1 = y_2^2(\delta_{12}P/RT) \tag{13-21}*$$

This relationship with $\ln \gamma$ for one component equal to a constant times the square of the mole fraction of the other component was cited in Chapter 12 as the simplest pattern of nonideal solution behavior, Eqs. (12-12)* through (12-14)*.

We can now examine the question raised at the beginning of this chapter concerning corrections to the vapor-phase fugacities used to determine the nonideality of liquid solutions. In a classic investigation of this type for benzene–cyclohexane, Scatchard et al.[3] measured the vapor compositions and total pressures over the range 303–343 K. They made corrections for gas imperfection on a basis equivalent to the assumption of Eq. (13-13)* or that δ_{12} is zero. From Dymond and Smith,[2] δ_{12} = 66 cm³ mol⁻¹; also P = 0.40 bar at 323 K, whereupon the quantity $\delta_{12}P/RT$ = 0.0010. Thus, the maximum correction needed is 0.1 percent, which is just about their experimental uncertainty. We see from Fig. 13-1 that the values of δ_{12}, the excess in B_{12}, are often much greater than for benzene–cyclohexane, and then this correction will be substantial.

A Useful Approximation

An equation that was found to represent accurately the second virial coefficient for a rather broad class of pure gases of low polarity has been successfully applied to mixed gases[4] in the following manner. This equation (9-66) is in the acentric factor system; it gives B as a function of $T_r = T/T_c$, $V_r = V/V_c$, and ω.

Since the basic statistical equation for the cross coefficient, Eq. (13-12), has the same form as that for the coefficients for like molecules, one expects the pattern of temperature dependency to be the same. Thus, if the equivalent macroscopic parameters for the cross coefficient, the effective critical temperature $T_{c,12}$, the effective critical volume $V_{c,12}$, and the acentric factor can be determined, one expects good estimates of B_{12}.

From Eq. (13-12) it is apparent that the intermolecular distance scale is one important factor. For spherical molecules, each radius is proportional to the cube root of the volume; hence, one adopts

$$V_{c,12} = \tfrac{1}{8}(V_{c,1}^{1/3} + V_{c,2}^{1/3})^3 \tag{13-22}*$$

as the estimate of the effective critical volume. For nonspherical molecules the detailed picture is more complex but Eq. (13-22)* was found to be a satisfactory approximation. For the acentric factor, the simple mean is used:

$$\omega_{12} = \tfrac{1}{2}(\omega_1 + \omega_2) \tag{13-23}*$$

The theory for the energy scale is more complex, but there is good reason to choose the geometric mean as the best simple estimate for the interparticle potential between unlike molecules. Since temperature scales with energy, one can take

$$T_c,12 = (T_{c,1} T_{c,2})^{1/2} \qquad (13\text{-}24)*$$

as a first approximation. This value gives good results in some cases but not in others. While improved estimates of $T_{c,12}$ can be made on the basis of either more sophisticated molecular theory or empirical generalization,[1,4] another approach is possible in many cases.

Examination of the available data for the cross second virial coefficient, as collected by Dymond and Smith,[2] shows a great many cases where B_{12} is known experimentally at a single temperature. In these cases one may evaluate $T_{c,12}$ from this single value, using Eqs. (13-22)* and (13-23)* for V_c and ω, respectively; whereupon Eq. (9-66)* yields B_{12} at other temperatures. Figure 13-2 compares the predictions of this procedure with the experimental data over a range of temperature for several typical examples. $T_{c,12}$ was evaluated at the lowest experimental temperature and B_{12} was calculated for all higher temperatures.

Prausnitz et al.[1] and Tsonopoulos and Heidman[5] describe several somewhat similar methods of estimating B_{12}; some are applicable to different categories of molecules, including cases of greater polarity.

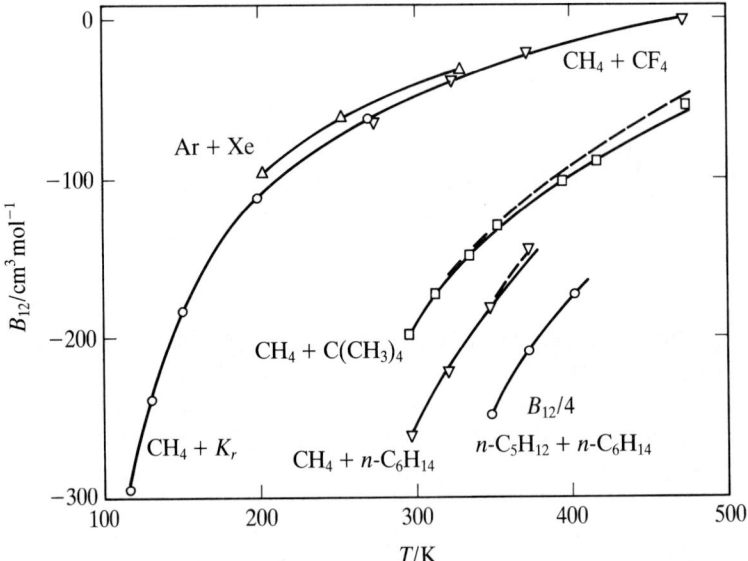

FIGURE 13-2
Curves for B_{12} calculated by the approximate method Eqs. (13-22)*–(13-24)* compared with experimental values. Solid curves were calculated with Eq. (13-22)* for $V_{c,12}$, while dashed curves were calculated for $V_{c,12} = (V_{c,1} + V_{c,2})/2$. For $n\text{-}C_5H_{12} + n\text{-}C_6H_{14}$ the quantity $B_{12}/4$ is shown.

Strong Attractions Between Unlike Molecules

In some cases there are strong attractive forces from hydrogen bonding or similar interactions between unlike species that yield large negative values for the cross coefficient B_{12}. An example is trimethylamine with methanol. A strong O—H--N bond is formed, as is indicated by a large shift in the O—H frequency in the infrared spectrum. There are possibilities of O—H--O bonding in pure methanol, which are discussed in Chapter 9, Eqs. (9-75)* and (9-76)*. But the bonding to the amine is considerably stronger and is not restricted as much by angular constraints. Millen and Mines[6] report for 308 K the values $B_{11} = -1720$ for methanol, $B_{22} = -700$ for trimethylamine, and $B_{12} = -7200 \text{ cm}^3 \text{ mol}^{-1}$. From the temperature coefficient for B_{12}, they calculate $-\Delta H/R = 3480 \text{ K}$ for the energy of the O—H--N hydrogen bond, which is somewhat larger than the 3050 K found for the O—H--O bond in the methanol tetramer and much larger than the enthalpy of association of 1620 K for the methanol dimer.

TREATMENTS VALID FOR THE FULL RANGE OF DENSITY

Phase Equilibria

For the equilibrium in a mixed fluid between the vapor and liquid, the conditions of Eq. (10-47) for any two fluid phases apply. It is convenient to express the equality of the chemical potential for each component by the fugacity; then for phases α and β,

$$T^\alpha = T^\beta \quad P^\alpha = P^\beta \quad f_1^\alpha = f_1^\beta \quad f_2^\alpha = f_2^\beta \quad \text{etc.} \quad (13\text{-}25)$$

If each phase remains distinct throughout the range of T, P, x considered, i.e., there is no critical region, then the properties can conveniently be represented by separate Gibbs energy equations as functions of T, P, x for each phase. The equality of T and P are assured directly, and the fugacities or chemical potentials are readily derived from the Gibbs energy, Eqs. (10-11) and (7-21). Care must be taken that the same reference chemical potential μ_i° is used for each phase. Appropriate equations for a liquid phase are described in Chapter 12 and in certain later chapters. A vapor or gas phase is ordinarily well represented by a virial equation including the second virial coefficient as described above.

If phase compositions have been measured and the vapor equation is known, the parameters in the liquid-phase equation can be optimized by regression methods. If the liquid-phase equation is known independently and compositions are to be predicted, the calculations involve simultaneous equations for f_1 and f_2, but their solution is relatively straightforward even if an iterative method is required. Alternatively, the molar Gibbs energy for the vapor and for the liquid as a function of x_1 (or x_2) can be calculated from the respective equations for a given pressure. Then the common tangent method of Eq. (10-39) and Fig. 10-8 yields the compositions of the two phases at equilibrium.

However, there are many cases, including those with vapor–liquid critical regions, where the same equation of state represents both phases. As noted in Chapter 9, the basic equation is now that for the Helmholtz energy as a function of T, V, and composition.

And only the equality of temperature can be directly imposed, while the equality of P as well as $f_1, f_2, \ldots$ between phases must be obtained by some iterative scheme in which $V, n_1, n_2, \ldots$ are varied. Provided this calculation converges, it is readily accomplished with modern computers, but the matter of convergence is not trivial.

Mixed Fluids and the van der Waals Equation

It is useful to illustrate the treatment of mixed fluids with the simple equation of van der Waals. For pure fluids, it was discussed in Chapter 9 with Eqs. (9-55)* through (9-57)*. If Eq. (9-55)* is generalized for a mixture and converted to the compression factor, it becomes

$$z = \frac{1}{1 - b_{mix}\rho_T} - \frac{a_{mix}\rho_T}{RT} \qquad (13\text{-}26)^*$$

where ρ_T is the total density $\sum_i \rho_i$ and a_{mix} and b_{mix} are functions of composition. This expression may be integrated [see Eqs. (13-11) and (9-56)*] to the molar Helmholtz energy;

$$A_m(T,\rho_m,x) = \sum_i x_i G_i^\circ(T) + RT(\sum_i x_i \ln x_i - 1)$$
$$+ RT \ln\left(\frac{RT\rho_T}{1 - b_{mix}\rho_T}\right) - a_{mix}\rho_T \qquad (13\text{-}27)^*$$

Note the ideal mixing term for Gibbs energy of the perfect gas and the -1 for conversion from Gibbs to Helmholtz energy.

Since we seek the chemical potential, which is given by the derivative of the total (extensive) Helmholtz energy with respect to number of moles n_i, it is useful to rewrite Eq. (13-27)* as follows:

$$A(T,V,n_1,\ldots) = \sum_i n_i G_i^\circ(T) + RT(\sum_i n_i \ln x_i - n_T)$$
$$+ RTn_T \ln\left(\frac{n_T RT}{V - b_{mix}n_T}\right) - \frac{a_{mix}n_T^2}{V} \qquad (13\text{-}28)^*$$

The x-dependency of a_{mix} and b_{mix} can best be considered in terms of the virial expansion of Eq. (13-26)*, which yields

$$B_{mix} = b_{mix} - \frac{a_{mix}}{RT} \qquad (13\text{-}29a)$$

$$C_{mix} = b_{mix}^2 \quad \text{etc.} \qquad (13\text{-}29b)$$

According to Eq. (13-3), the most complex dependency permitted for B_{mix} is quadratic and for C_{mix} is cubic in mole fractions. Thus, a_{mix} can have a quadratic dependency,

$$a_{mix} = \sum_i \sum_j x_i x_j a_{ij} \qquad (13\text{-}30a)$$

For b_{mix} a quadratic dependency would be allowed for B_{mix}, but this would yield a quartic dependency for C_{mix}, which is not allowed. Thus, b_{mix} must be limited to linear dependency:

$$b_{mix} = \sum_i x_i b_i \tag{13-30b}$$

For further discussion of the van der Waals equation, it is convenient to limit the example to a binary fluid, whereupon

$$b_{mix} = x b_1 + x_2 b_2 \tag{13-31a}$$

$$a_{mix} = x_1^2 a_{11} + 2 x_1 x_2 a_{12} + x_2^2 a_{22} \tag{13-31b}$$

and the Helmholtz energy becomes

$$A = n_1 G_1^\circ + n_2 G_2^\circ + RT(n_1 \ln x_1 + n_2 \ln x_2 - n_1 - n_2)$$
$$+ RT(n_1 + n_2) \ln\left(\frac{(n_1 + n_2)RT}{V - n_1 b_1 - n_2 b_2}\right)$$
$$- \frac{n_1^2 a_{11} + 2 n_1 n_2 a_{12} + n_2^2 a_{22}}{V} \tag{13-32}*$$

Then, application of the basic definition of the chemical potential,

$$\mu_i = \left(\frac{\partial A}{\partial n_i}\right)_{T,V,n_{j \neq i}} \tag{10-14}$$

yields, after simplification and conversion to densities,

$$\frac{\mu_1 - G_1^\circ}{RT} = \ln x_1 + \ln\left(\frac{RT\rho_T}{1 - b_{mix}\rho_T}\right) + \frac{b_1 \rho_T}{1 - b_{mix}\rho_T} - \frac{2(a_{11} x_1 + a_{12} x_2)\rho_T}{RT}$$
$$\tag{13-33}*$$

Also, the fugacity and fugacity coefficient are

$$\ln \phi_1 = \ln\left(\frac{f_1}{x_1 P}\right) = \frac{\mu_1 - \mu_1^{id}}{RT} \tag{13-7}$$

$$= -\ln(1 - b_{mix}\rho_T) + \frac{b_1 \rho_T}{1 - b_{mix}\rho_T} - \frac{2(a_{11} x_1 + a_{12} x_2)\rho_T}{RT} - \ln z$$
$$\tag{13-34}*$$

For component 2 the subscripts 1 and 2 are interchanged in Eq. (13-33)* and (13-34)*.

Now the criteria for phase equilibrium are

$$P^\alpha = P^\beta \tag{13-35a}$$

$$x_1^\alpha \phi_1^\alpha = x_1^\beta \phi_1^\beta \tag{13-35b}$$

$$x_2^\alpha \phi_2^\alpha = x_2^\beta \phi_2^\beta \tag{13-35c}$$

with $T^\alpha = T^\beta$ and $x_1 = 1 - x_2$ for each phase, of course; also, P is given by Eq. (13-26)*. With two independent variables ρ_T and x_2 for each phase, one additional quantity can be specified. This can be the pressure or the composition of one phase. Iterative solution of these equations is not difficult with modern computers provided the sequence is convergent. The use of experimental phase composition data to optimize parameters in an equation of state is a more complex calculation, but it too is feasible with appropriate statistical and computational methods.

With increase in temperature along a path maintaining both liquid and vapor phases, the liquid and vapor compositions approach one another and become equal at a point on a critical line in P, T, x space. The mathematical criteria for this critical line are rather complex. Comprehensive discussions of the thermodynamics of critical phenomena in mixed fluids, including those with more than two components, are given by Scott,[7] and by Rowlinson and Swinton,[8] as well as other reviews.

Phase separation into two liquid phases or two supercritical fluid phases has also been observed. The thermodynamic criteria are the same, Eq. (13-25), as for the vapor–liquid case. For certain combinations of parameters, even an equation as simple as that of van der Waals predicts such additional phase separations. This was investigated very thoroughly for the two-component case with the van der Waals equation by van Konynenburg and Scott[9] using the mixing expressions of Eq. (13-31). Diagrams for six types of phase behavior are given by Rowlinson and Swinton,[8] who discuss examples of real systems that display such patterns. Five of these types were found in the calculations with the van der Waals equation.[9]

Improved Equations of State

As was the case for pure fluids, the van der Waals equation is not useful for the quantitative representation of the equilibrium phase compositions or other properties of mixed fluids. But the Redlich–Kwong and other only slightly more complex equations are widely used [see Eq. (9-58)* and Appendix 4]. If fitted to one particular set of properties such as vapor–liquid equilibrium compositions for similar fluids, they give good results, but they are less satisfactory for a global representation of several properties such as supercritical pressures and compressed liquid properties as well as phase compositions. If there are only two composition-dependent parameters, the equations have essentially the same form as that illustrated above for the van der Waals equation and can be solved by the same methods.

There is a very extensive literature on the application of equations of state to problems with mixed fluids. References to books as well as to papers including typical examples are given at the end of this chapter.

SOLUBILITY OF SOLIDS IN GASES

An equilibrium between a pure solid and a mixed gas is an interesting case that is relatively simple to interpret since the chemical potential of the solid is independent of the composition of the gas. Examples that have been studied include CO_2 in air, naphthalene in ethylene and NaCl in steam. Such cases can have practical importance, such as the plugging of flow lines by a solid or the transport of a corrosive material into a turbine.

In these cases one is often interested in quite high pressures; thus, one may require higher virial coefficients or a more comprehensive equation of state. Let us assign the solid as component 2 and the gas as component 1. The fugacity of the solid in its standard state of 1 bar is f_2°, but a higher pressure increases this fugacity to

$$f_2(P) = f_2^\circ \exp\left[\int_1^P \left(\frac{V_{2,m}^s}{RT}\right) dP'\right] \tag{13-36}$$

$$\cong f_2^\circ \exp\left[\frac{PV_{2,m}^s}{RT}\right] \tag{13-37}*$$

where $V_{2,m}^s$ is the molar volume of the solid. Also the approximation on the second line neglects 1 as compared to P and ignores the compression of the solid. This is known as the Poynting correction. The fugacity of a solid of low volatility is, of course, approximately equal to its sublimation pressure, which we assume to be small.

If ϕ_2 is the fugacity coefficient in the gas, one can rearrange its definition, Eq. (13-7), to yield the composition

$$y_2 = \frac{f_2}{\phi_2 P} \tag{13-38}$$

At low pressures where ϕ_2 is near unity, one notes that y_2 is inversely proportional to the pressure. But at high pressure ϕ_2 may become very small and y_2 may reach a minimum and then actually increase with P.

We now substitute Eqs. (13-10) and (13-37)* into (13-38) and assume that the amount of solid dissolved in the gas is very small, so that y_1 is approximately 1.0. Then

$$\ln y_2 = \ln\left(\frac{f_2^\circ}{P}\right) + \frac{P}{RT}(V_{2,m}^s - 2B_{12} + B_{11}) \tag{13-39}*$$

For the system ethylene–naphthalene at 298 K, the volume of the solid, $V_{2,m}^s = 128 \text{ cm}^3 \text{ mol}^{-1}$, while $B_{11} = -140 \pm 1$, and $B_{12} = -681 \pm 10$.[2,10] Thus, the volumetric quantity in the final parentheses of Eq. (13-39)* has the value $1350 \pm 20 \text{ cm}^3 \text{ mol}^{-1}$. The resulting curve for the solubility of naphthalene in ethylene, which is shown in Fig. 13-3, has a minimum near 18 bar. The value of the pressure at minimum solubility is a quantity of practical interest; by differentiation of Eq. (13-39)* it is

$$P(\text{at min. } y_2) = \frac{RT}{V_{2,m}^s - 2B_{12} + B_{11}} \tag{13-40}*$$

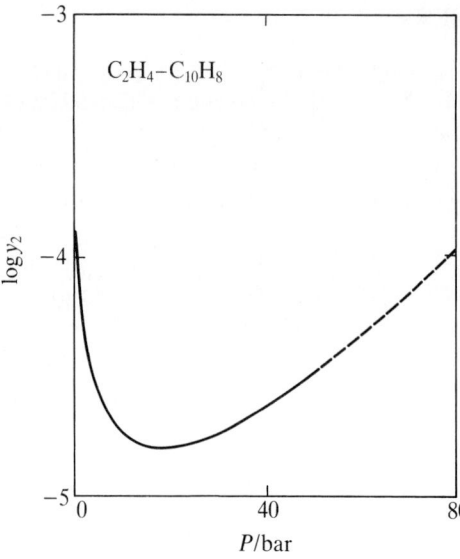

FIGURE 13-3
Solubility of solid naphthalene in ethylene calculated from Eq. (13-37)* including only the second virial coefficient. The third virial coefficient may have a significant effect in the higher pressure range, shown dashed.

Often, it is the concentration of component 2 that is measured instead of the mole fraction. To calculate C_2 we first note that

$$C_2 = \frac{n_2}{V} = y_2 \rho$$

and, with the approximation that y_2 is very small and $y_1 \cong 1$ together with Eqs. (9-3) and (13-9),

$$\ln C_2 = \ln\left(\frac{y_2 P}{RT}\right) - \frac{B_{11}}{RT}$$

Then, the insertion of Eq. (13-39)* for y_2 yields, after simplification,

$$\ln\left(\frac{C_2}{C_2^\circ}\right) = \frac{P}{RT}(V_{2,m}^s - 2B_{12}) \qquad (13\text{-}39a)*$$

where C_2° is the concentration of the pure vapor of component 2. One notes that the shift from mole fraction to concentration has removed the term B_{11} from the final parentheses of Eq. (13-39)*.

In this example, the term $-2B_{12}$ in Eqs. (13-39)* and (13-40)* is considerably larger than either $V_{2,m}^s$ or B_{11} but the other terms are not negligible. There is some confusion in the literature concerning the presence of B_{11} and its coefficient in various expressions related to the solubility of solids in gases; also, it is sometimes implied that $V_{2,m}^s$ may be omitted, whereas it is actually significant but small. These aspects should be considered if full accuracy is desired.

The solubility of sodium chloride in steam is another example of a solid dissolved in a gas, but is one with very different detailed characteristics. That solubility has been measured from 600 to 773 K; it is too small to measure below about 30 bar, and the measurements extend up to the three-phase pressure where solid NaCl, steam containing NaCl, and a liquid solution coexist. The mole fraction of NaCl ranges from 10^{-7} at 600 K and 30 bar to about 0.00023 at 773 K and 324 bar.[11-13] Under these conditions the NaCl in the gas is essentially all in the form of neutral molecules, ion pairs, and at these concentrations there is no need to consider NaCl–NaCl interactions. The steam is far from ideal, however, and one must either use several virial coefficients or a full equation of state for H_2O.

The NaCl–H_2O interaction is best described as a sequence of association reactions of one NaCl with one or more molecules of water. Early treatments[13] of this system fitted the data over a limited range to the equation for a single associated species, but also recognized that no single species would suffice to explain all of the data.

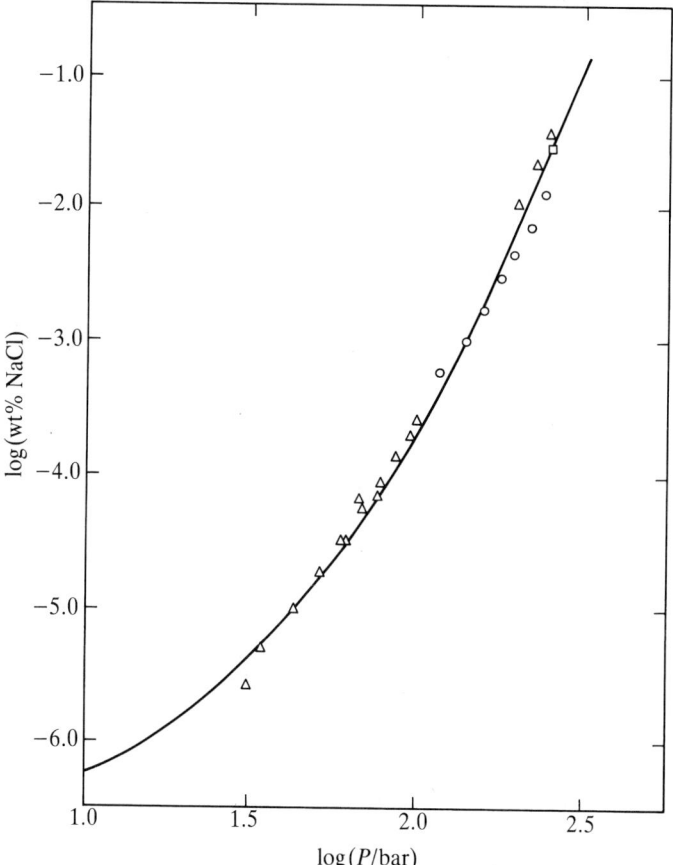

FIGURE 13-4
Solubility of solid NaCl in steam at 450°C. Experimental values are from sources cited in reference 10; the curve is calculated from Eqs. (13-39)–(13-41)*.

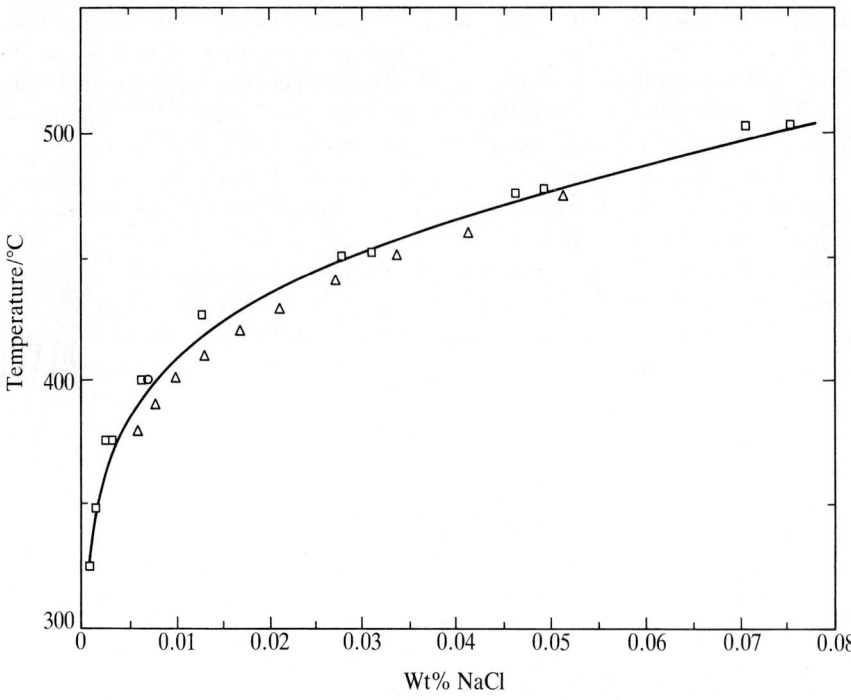

FIGURE 13-5
Solubility of solid NaCl in steam along the three-phase line of equilibrium of solid, liquid, and vapor. Details as in Fig.13-4.

A treatment[14] accounting for the full range of data for NaCl in steam assumes a successive set of hydration reactions in the gas:

$$\text{NaCl(H}_2\text{O)}_{j-1} + \text{H}_2\text{O} = \text{NaCl(H}_2\text{O)}_j \tag{13-41}$$

with the total concentration C of NaCl given by

$$C = C_0(1 + K_1 f_{\text{H}_2\text{O}} + K_1 K_2 f_{\text{H}_2\text{O}}^2 + K_1 K_2 K_3 f_{\text{H}_2\text{O}}^3 + \cdots) \tag{13-42}*$$

and

$$\ln K_j(T) = \frac{\Delta S°(T_r)}{R} - \frac{\Delta H°(T_r)_j}{RT} + \frac{\Delta C_P°}{R}\left[\ln\left(\frac{T}{T_r}\right) - 1 + \frac{T_r}{T}\right] \tag{13-43}*$$

This is, of course, equivalent to a series of virial coefficients. A reference temperature of 500 K was chosen and the values $\Delta S°(T_r)/R = -11.0$ and $\Delta C_P°/R = 3.0$ were found to be satisfactory for all j. The values of $-\Delta H°(T_r)/R$ decrease from 6390 to 3900 K with increasing j; these values could not be determined for each j but were assumed to be the same for groups of steps, 1–3, 4–6, etc. The fugacity of water was taken from the full equation of state of Haar et al.[15] The fugacity of pure solid NaCl is, of course, too small to measure at these temperatures; it is obtained indirectly from a calculation of $\Delta G°$ as a function of temperature. This involves the vapor pressure of the liquid at higher temperatures where it is measurable, the heat capacities, and the heat of fusion.[16] Figures 13-4 and 13-5 show comparisons of calculated mole fractions of NaCl in steam with those experimental values that now appear to be reliable.

MIXED GASES AT HIGH PRESSURES

We conclude this chapter with a general survey of the properties of mixed gases over a wider range of pressure and, in particular, examine whether such mixed gases follow the ideal solution equation even when they depart greatly from the ideal gas law.

Two very simple postulates for mixed gases were made long ago. The first, by Dalton, is that pressure of a gaseous solution is the sum of the pressures that each gas would exert if it alone occupied the entire volume at the same temperature. It is only in the range of nearly ideal gases that this is a useful approximation.

A second and more generally useful postulate is that of Amagat,[17] who proposed that the volume of a gaseous solution was the sum of volumes of the components each at the temperature and total pressure of the solution. In this connection, it is useful to recall the excess volume and define the excess compression factor

$$V^E = V - \sum_i y_i V_i^* = \Delta_{\mathrm{mix}} V \qquad (12\text{-}4)$$

$$z^E = z - \sum_i y_i z_i^* \qquad (13\text{-}44)$$

where V_i^* and z_i^* are the values for the pure gases at the same T and P. Both of the excess quantities are zero if Amagat's law is followed.

It may be shown that, if Amagat's law of additive volumes holds at all pressures up to the pressure of interest, then the ideal-solution law holds for the gaseous solution; i.e., the fugacity of the ith component is

$$f_i = x_i f_i^o \qquad (11\text{-}1)*$$

where x_i is the mole fraction and f_i^o is the fugacity of the pure gas at the temperature and total pressure of the solution. This was demonstrated in Chapter 11, Eq. (11-14).

In 1923 Lewis and Randall[18] argued that gaseous solutions should generally be ideal. They wrote, "It seems reasonable to suppose that the solution of a given pair of substances will be more nearly perfect (ideal) when the density of the solution is less, or, in other words, when the average distance between the molecules is greater." Experimental results have shown that the postulate is incorrect. Indeed, gaseous solutions often follow the Amagat law at high, liquid-like densities as well as at low densities but depart greatly at intermediate, near-critical densities. At the critical point a fluid has infinite compressibility; thus in that region a very small difference in interparticle attraction can cause a large change in average interparticle distance. For a mixed gas, the differences in interparticle potentials for like and unlike interactions can cause large volume changes in this region of intermediate density. In contrast, at very high pressure the molecules are nearly closest packed and difference in attractive forces have little effect on the volume.

This pattern of behavior is illustrated on Fig. 13-6, which shows the compression factor[19] for the mixture CO_2 and n-butane at 411 K. A linear dependency of z on y_2, the mole fraction of butane, implies conformity to Amagat's law with no change in volume on mixing at constant P and T. It is clear that this system follows that principle both at low pressures and at very high pressures, but that it departs substantially at intermediate pressures. This is especially marked at 69 bar where the density is near critical. Indeed, at somewhat smaller pressures near 37 bar, phase separation does occur for pure butane and mixtures very rich in butane, but there is no phase separation at any of the pressures shown in Fig. 13-6.

It is interesting to apply a very simple equation of state to the CO_2–butane system. We use the Redlich–Kwong (RK) equation (9-58)* with parameters a and b

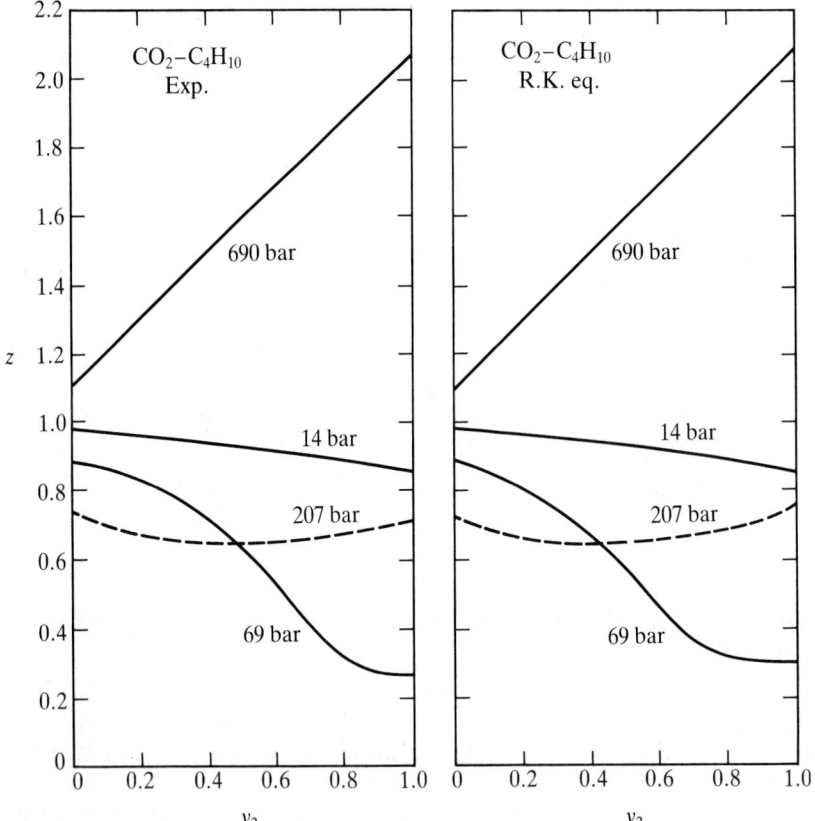

FIGURE 13-6
The compression factor z for the mixed gas CO_2 and n-C_4H_{10} at 411 K. Curves in the left panel are experimental from reference 18; those on the right were calculated from the Redlich–Kwong equation with the very simple mixing rules Eqs. (13-30a,b) and (13-43)*.

evaluated from the observed critical temperature and pressure of CO_2 and n-butane, respectively. For the mixture, the RK parameters are given by the simplest mixing equations that were developed above for the van der Waals equation and for the second virial coefficient:

$$b_{mix} = (1 - y_2)b_1 + y_2 b_2 \qquad (13\text{-}31a)*$$

$$a_{mix} = (1 - y_2)^2 a_{11} + 2y_2(1 - y_2) a_{12} + y_2^2 a_{22} \qquad (13\text{-}31b)*$$

$$a_{12} = (a_{11} a_{22})^{1/2} \qquad (13\text{-}45)*$$

where a_{11} and b_1 are the parameters for CO_2, and a_{22} and b_2 are for n-butane. Equation (13-45)* for a_{12} follows from the same reasoning as that for $T_{c,12}$ in Eq. (13-24)*.

With these values established for the mixed system, the curves on the second section of Fig. 13-6 were calculated. It is apparent that even the very simple RK equation and simplest mixing rules lead to predictions that agree in all general aspects with the experimental properties. While there are differences outside of experimental uncertainties for the intermediate pressures, the agreement is quite good even on a quantitative basis.

While the CO_2-butane system shows negligible volume of mixing at 690 bar, this does not establish ideal solution behavior. The basic relationships, Eqs. (10-9) and (11-8)*, involve the pressure derivative of the Gibbs energy of mixing,

$$\frac{\partial G^E}{\partial P} = V^E \qquad (13\text{-}46)$$

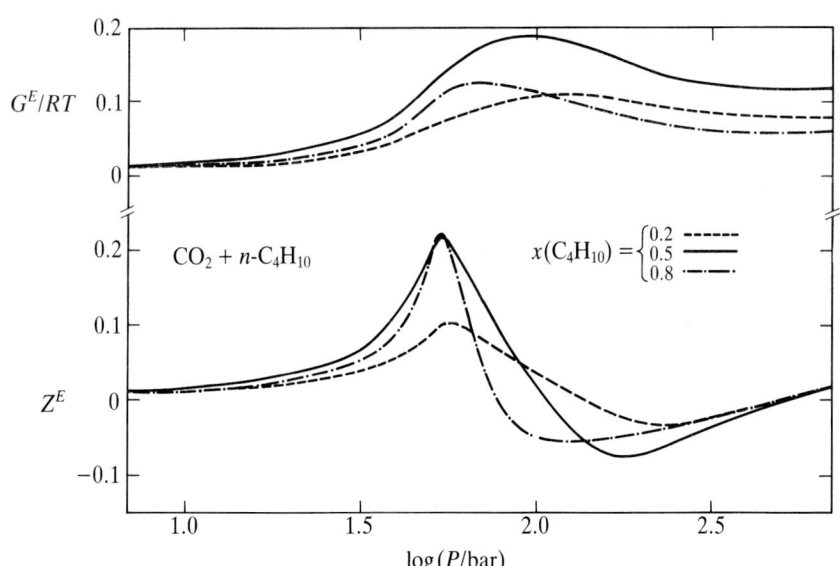

FIGURE 13-7
Excess values of the compression factor (lower section) for $CO_2 + n\text{-}C_4H_{10}$ at 444 K at mole fractions 0.2, 0.5, and 0.8. The upper section shows the excess Gibbs energy derived by integration of Eq. (13-45) for the curves of the lower section.

Thus, at a high pressure the excess Gibbs energy (or the Gibbs energy of mixing) is

$$\frac{G^E}{RT} = \int_0^P \left(\frac{V^E}{RT}\right) dP' = \int_0^P z^E \, d\ln P' \qquad (13\text{-}47)$$

and this has a zero value only if the excess volume remains zero throughout the range of pressure up to the final value or if the nonzero contributions at intermediate pressures cancel. An expression for G^E as a function of composition at the high pressure can be obtained by integration of the volumetric dependence following Eq. (13-47).

Figure 13-7 shows this calculation for the CO_2–butane system at 444 K. At this temperature there is no phase separation at any pressure or composition. We see that the maximum departure from Amagat's law occurs at about 50 bar, while the largest solution nonideality occurs at about 100 bar. At somewhat higher pressures z^E is negative and there is a partial cancellation of the contribution to G^E from the range of near-critical pressure. But this cancellation is only partial and G^E still has a substantial value in the range of high pressure where V^E and z^E have dropped to zero.

REFERENCES

1. Additional equations and examples are given by J. M. Prausnitz, R. N. Lichtenthaler and E. G. de Azevedo, *Molecular Thermodynamics of Fluid Phase Equilibria*, Prentice-Hall, Englewood Cliffs, N.J., 1986, chapter 5.
2. J. C. Dymond and E. B. Smith, *The Second Virial Coefficients of Gases: A Critical Evaluation*, Clarendon Press, Oxford, 1980.
3. G. Scatchard, S. E. Wood, and J. M. Mochel, *J. Phys. Chem.*, **43**, 119 (1939); *J. Am. Chem. Soc.*, **61**, 3206 (1939); **62**, 712 (1940).
4. K. S. Pitzer, *Fluid Phase Equil.*, **59**, 109 (1990).
5. C. Tsonopoulos and J. L. Heidman, *Fluid Phase Equil.*, **57**, 261 (1990).
6. D. J. Millen and G. W. Mines, *J. Chem. Soc. Faraday Trans. II*, **70**, 693 (1974).
7. R. L. Scott, *Ber. Bunsenges. Phys. Chem.*, **76**, 296 (1972).
8. J. S. Rowlinson and F. L. Swinton, *Liquids and Liquid Mixtures*, 3d edn., Butterworth, London, 1982, chapter 6.
9. P. H. van Konynenburg and R. L. Scott, *Phil. Trans. Roy. Soc. London*, **298**, 495 (1980).
10. G. C. Najour and A. D. King, Jr., *J. Chem. Phys.*, **45**, 1915 (1966).
11. J. F. Galobardes, D. R. Van Hare, and L. B. Rogers, *J. Chem. Eng. Data*, **26**, 363 (1981).
12. J. L. Bischoff, R. J. Rosenbauer, and K. S. Pitzer, *Geochim. Cosmochim. Acta*, **50**, 1437 (1986).
13. O. J. Martynova, *Russian J. Inorg. Chem.*, **38**, 587 (1964).
14. K. S. Pitzer and R. T. Pabalan, *Geochim. Cosmochim. Acta*, **50**, 1445 (1986).
15. L. Haar, J. S. Gallagher, and G. S. Kell, *NBS/NRC Steam Tables*, Hemisphere, Washington, D.C., 1984.
16. D. R. Stull and H. Prophet, *JANAF Thermochemical Tables*, 2d edn., National Stand. Ref Data Ser. No. 37, U.S. National Bureau of Standards, 1971.
17. E. H. Amagat, *Ann. Chem. Phys.*, **19**(5), 384 (1880); *Compt. rend.*, **127**, 88 (1898).
18. G. N. Lewis and M. Randall, *Thermodynamics and the Free Energy of Chemical Substances*, McGraw-Hill, New York, 1923, p. 226.
19. R. H. Olds, H. H. Reamer, B. H. Sage, and W. N. Lacey, *Ind. Eng. Chem.*, **41**, 475 (1949); also B. H. Sage and W. N. Lacey, A. P. I. Research Monographs, *Thermodynamic Properties of the Lighter Hydrocarbons and Nitrogen* (1950) and *Properties of the Lighter Hydrocarbons, Hydrogen Sulfide, and Carbon Dioxide* (1955), American Petroleum Institute, New York.

MIXED NONIDEAL GASES AND FLUIDS   243

Additional references for equations of state for mixed fluids

Books on equations of state and phase equilibria

S. Malanowski and A. Anderko, *Modeling Phase Equilibria: Thermodynamic Background and Practical Tools*, Wiley, New York, 1992.
T. G. Squires and M. E. Paulaitis, eds. *Supercritical Fluids*, ACS Symp. Series No. 329, American Chemical Society, Washington, D.C., 1987.
J. M. Prausnitz, R. N. Lichtenthaler, and E. G. de Azevedo, *Molecular Thermodynamics of Fluid Phase Equilibria*, 2d edn. Prentice-Hall, Englewood Cliffs, N.J., 1986.
K.C. Chao and R. L. Robinson, Jr., eds., *Equations of State*, ACS Symposium Series No. 300, American Chemical Society, Washington, D.C., 1986.
S. A. Newman, ed., *Chemical Engineering Thermodynamics*, Ann Arbor Science Pub., Ann Arbor, Mich., 1983.
K. C. Chao and R. L. Robinson, Jr., eds., *Equations of State*, Advances in Chemistry Series No. 182, American Chemical Society, Washington, D.C., 1979.
See also the additional references for Chapter 9.

Articles or reviews treating several mixed systems with equations given in Chapter 9 or Appendix 4

Redlich–Kwong Eq. (9-58)*, Peng–Robinson Eq. (A4-7)*, and five other equations: S. J. Han, H. M. Lin, and K. C. Chao, *Chem. Eng. Sci.*, **43**, 2327 (1988).
Anderko–Pitzer Eq. (9-61)*: A. Anderko and K. S. Pitzer, *Am. Inst. Chem. Eng. J.*, **37**, 1379 (1991).
Benedict–Webb–Rubin Eq. (9-62)*: M. Benedict, G. B. Webb, and L. C. Rubin, *J. Chem. Phys.*, **10**, 747 (1942).

PROBLEMS

13-1. Show from Eq.(11-14) that, if Amagat's law of additive volumes holds for all pressures up to P', then the ideal-solution law holds for the gaseous mixture.

13-2. From Fig. 13-1, it is apparent that for the system $i\text{-}C_4H_{10}(1)\text{-}n\text{-}C_6H_{14}(2)$, $B_{12} = (B_{11} + B_{22})/2$. Then, from Eq.(13-15)* calculate the fugacity coefficients ϕ_1 and ϕ_2 and the fugacities f_1 and f_2 for this system at 500 K and 10 bar using parameters from Table A3-11 and Eq. (9-66).*

13-3. Derive an equation for the enthalpy $H-H°$ for the Redlich–Kwong fluid from Eqs. (9-38) and (9-58)*. Then calculate V and $H-H°$ for 1 mole of an equimolal mixture of CO_2 and propane at 450 K and 140 bar using the parameters of Table A3-11, the relationship of a_i and b_i to critical properties, and the mixing rules of Eq.(13-31). It is necessary to iterate the equation for pressure to find the V that yields 140 bar.

13-4. Derive equations for μ_i and ϕ_i for the Redlich–Kwong equation equivalent to Eqs.(13-33)* and (13-34)* for the van der Waals equation.

13-5. From the results of Problems 13-3 and 13-4, calculate the fugacities of CO_2 and propane for the equimolal mixture at 450 K and 140 bar. Compare this result with that for the ideal-solution equation for this fluid.

CHAPTER 14

SOLUTIONS WITH COMPOSITION GIVEN IN MOLALITY

The basic definitions and many properties of nonideal solutions were presented in Chapter 12 on the basis of compositions given in mole fraction. Much of the literature concerning aqueous solutions, however, uses molality (moles of solute per kilogram solvent) as the measure of composition. Molality is often used for other cases where a nonaqueous or mixed solvent is always the abundant component and the mole fractions of other components are never large. Many of these solutions are electrolytes, and there are additional features associated with the ionic nature of these solutes that are considered in the next chapter. In this chapter we present the definitions and basic relationships when the composition is given in molalities for neutral solutes. These relationships are compared with those on a mole fraction basis.

Standard States

When molality is used for composition, the pure liquid at the given temperature is always the standard state for the solvent, and the infinitely dilute standard state is used for each solute. The latter is a hypothetical state analogous to that defined on a mole fraction basis in Chapter 12.

Activities and Activity Coefficients of Solutes

Each solute is assumed to follow Henry's law in the limit of great dilution. Thus, one defines the activity of solute component i on the molality basis by $a_i/m_i \to 1$ when $m_i \to 0$ and no other solutes are present. The solute fugacity in the standard state is defined as

$$f^\circ_{m,i} = \lim_{m_i \to 0} \frac{f_i}{m_i/m_0} \qquad m_{j \neq i} = 0 \qquad (14\text{-}1)$$

and the chemical potential in the standard state is

$$\mu^\circ_{m,i} = \overline{G}^\circ_{m,i} = \lim_{m_i \to 0} [\overline{G}_i(m_i) - RT \ln(m_i/m_0)] \qquad m_{j \neq i} = 0 \qquad (14\text{-}2)$$

In Eqs. (14-1) and (14-2), we have inserted (m/m_0) where $m_0 = 1 \text{ mol kg}^{-1}$ to cancel the dimensionality of m; this is necessary to retain the correct dimensionality of $f^\circ_{m,i}$ and to provide a dimensionless argument for the logarithm. The condition of zero molality for other components is necessary to avoid cross interactions that would otherwise affect the values of the standard-state properties of the selected solute. In other words, these definitions of the standard state must be applied in binary solutions, even though the resulting f°_i and μ°_i may later be used for multicomponent systems.

If there is any possibility of ambiguity between these and other quantities defined on the molality basis and the corresponding quantities defined on the mole fraction basis, the basis should be designated by a subscript or superscript $f^\circ_{m,i}, \mu^\circ_{m,i}, f^\circ_{x,i}, \mu^\circ_{x,i}$, etc. This designation will be included for each initial definition and whenever quantities in the two systems are compared, but may be omitted when there is no possible ambiguity.

In the limit of infinite dilution, the relationship of the molality to mole fraction is just the number of moles in 1 kg of solvent, which we designate by Ω,

$$\Omega = \frac{1000}{M_s} \qquad (14\text{-}3)$$

with M_s the molar mass (g mol^{-1}) of the solvent. For water $\Omega = 55.51 \text{ mol kg}^{-1}$; then

$$f^\circ_{m,i} = \left(\frac{\Omega}{m_0}\right) f^\circ_{x,i} \qquad (14\text{-}4a)$$

$$\mu^\circ_{m,i} = \mu^\circ_{x,i} + RT \ln(\Omega/m_0) \qquad (14\text{-}4b)$$

The activity coefficient for a solute in the molality system is

$$\gamma_{m,i} = \frac{f_{m,i}}{f^\circ_{m,i}(m_i/m_0)} \qquad (14\text{-}5)$$

Again, m_0 is inserted to make $\gamma_{m,i}$ dimensionless, and from Eq. (14-1) we see that

$$\lim_{m \to 0} \gamma_{m,i} = 1 \qquad (14\text{-}6)$$

One never uses the pure liquid standard state for a solute in the molality system; indeed, that is impossible since the m value becomes infinite. Since the ambiguity between pure liquid and infinitely dilute standard states, which arises for the mole fraction system, is absent for the molality system, there is no need for the γ^* designation.

At finite molality, the solute fugacity and chemical potential are given by the equations

$$f_{m,i} = f^{\circ}_{m,i}\, \gamma_{m,i}(m_i/m_0) \tag{14-7}$$

$$\mu_{m,i} = \mu^{\circ}_{m,i} + RT \ln(f_{m,i}/f^{\circ}_{m,i})$$

$$= \mu^{\circ}_{m,i} + RT \ln[\gamma_{m,i}(m_i/m_0)] \tag{14-8}$$

Again, the m_0 is shown explicitly in these equations. Hereafter, the m_0 will be omitted with the understanding that it is implied whenever required to cancel the dimensionality of molality.

Solvent Activity and the Osmotic Coefficient

For the solvent s the activity is

$$a_s = f_s/f^{\circ}_s \tag{14-9}$$

where f°_s is the fugacity of the pure solvent at the same temperature. Thus the activity of the pure solvent is unity. For dilute solutions the solvent activity remains very close to unity, and its accurate expression would require many significant figures. An accurate expression of an activity coefficient for the solvent would have the same requirement and inconvenience. To overcome this problem and simplify calculations, the *osmotic coefficient* is defined by

$$\phi = -\left(\frac{\Omega}{\sum_i m_i}\right) \ln a_s \tag{14-10}$$

where the sum includes all solute species. This quantity ϕ is sometimes called the "practical" osmotic coefficient to distinguish it from the "rational" osmotic coefficient g defined in terms of mole fraction as

$$\ln a_s = g \ln x_s \tag{14-11}$$

For an ideal solution $g = 1$. One can compare this quantity with ϕ by expressing the mole fraction as $(1 + m/\Omega)^{-1}$ and expanding the logarithm to obtain

$$\phi = g\left(1 - \frac{m}{2\Omega} + \cdots\right) \tag{14-11a}$$

Equation (4-10) can be rearranged to yield $\ln a_s$; then one obtains for the chemical potential

$$\mu_s = \mu^{\circ}_s - \frac{RT\phi}{\Omega} \sum_i m_i \tag{14-12}$$

For a very dilute solution, the activity of the solvent will be given accurately by the ideal solution expression:

$$a_s = x_s = \left(1 + \frac{\sum_i m_i}{\Omega}\right)^{-1} \tag{14-13}$$

$$\ln a_s = -\ln\left(1 + \frac{\sum_i m_i}{\Omega}\right) \cong -\frac{\sum_i m_i}{\Omega} \tag{14-14}$$

Thus, ϕ is unity at infinite dilution and the departure of ϕ from unity is a measure of the departure of the solvent activity from the pattern implied by Eq. (14-14).

If one combines Eq. (12-46)* with Eq. (14-10) one obtains for the osmotic pressure,

$$P - P° = \phi\left(\frac{RT}{\Omega V_s°}\right)\sum_i m_i \tag{14-15}$$

Thus, ϕ is a factor correcting the approximate expression to the true osmotic pressure; this explains the name, osmotic coefficient. In Eqs. (14-10) through (14-15), the ratio m_i/Ω appears, which is dimensionless; hence, the $m_0 = 1\,\text{mol}\,\text{kg}^{-1}$ factor is not needed.

Gibbs Energy

The total Gibbs energy of mixing for the solution is obtained by summation of Eqs. (14-9) and (14-12):

$$\Delta_{\text{mix}}G = \sum_i n_i(\mu_i - \mu_i°) + n_s(\mu_s - \mu_s°)$$

$$= RT\left[\sum_i n_i \ln(m_i \gamma_{m,i}) - \phi\left(\frac{n_s}{\Omega}\right)\sum_i m_i\right]$$

Since $m_i = n_i\Omega/n_s$, this simplifies to

$$\Delta_{\text{mix}}G = RT\sum_i n_i[-\phi + \ln(m_i \gamma_{m,i})] \tag{14-16}$$

It is useful to separate this expression into two parts: one part independent of ϕ or γ_i, which gives the primary effect of the solution composition, and a second part for the "corrective" terms in $(1 - \phi)$ and γ_i. The latter can be called an "excess" quantity; on that basis, one defines

$$G^{Em} = \Delta_{\text{mix}}G + RT\sum_i n_i(1 - \ln m_i) \tag{14-17}$$

The m in the superscript Em is used to distinguish this excess quantity from that in the mole fraction system, which is similar but not exactly the same. Then

$$G^{Em} = RT\sum_i n_i(1 - \phi + \ln \gamma_{m,i}) \tag{14-18}$$

It is often convenient to use expressions based on 1 kg of solvent. If w_s is the mass of solvent in kilograms, $n_i = m_i w_s$, and one obtains

$$\frac{G^{Em}}{w_s RT} = \Delta_{\text{mix}} \frac{G}{w_s RT} + \sum_i m_i (1 - \ln m_i) \qquad (14\text{-}19)$$

$$\frac{G^{Em}}{w_s RT} = \sum_i m_i (1 - \phi + \ln \gamma_{m,i}) \qquad (14\text{-}20)$$

The sums cover all solute species in these and other equations.

A full comparison of Eqs. (14-17) and (14-18) with those for the ideal and excess quantities on a mole fraction basis is complex because the former use the infinitely dilute standard state for all solute species, whereas the usual mole fraction equations are based on the pure liquid for each component. One can show that an ideal solution (on the mole fraction basis) would have the molality activity coefficient

$$\gamma_{m,i}^{\text{id}} = \left(1 + \frac{\sum_i m_i}{\Omega}\right)^{-1} \qquad (14\text{-}21)$$

There is no expectation that most solutions treated on the molality basis would be very close to ideality; hence, Eq. (14-21) has limited value for most systems.

We now reverse the process of generation of the definition for G^{Em} and give the equations yielding ϕ and $\gamma_{m,i}$ by differentiation of G^{Em}:

$$\phi - 1 = -\frac{(\partial G^{Em}/\partial w_s)_{n_i}}{RT \sum_i m_i} \qquad (14\text{-}22)$$

$$\ln \gamma_{m,i} = \left[\frac{\partial (G^{Em}/w_s RT)}{\partial m_i}\right]_{w_s, m_{j, j \neq i}} \qquad (14\text{-}23)$$

Thus, if one has an expression for G^{Em} as a function of w_s and m_i, ϕ and $\ln \gamma_{m,i}$ are readily obtained and consistency with the Gibbs–Duhem equation is assured. If there is only one solute, the sums in various equations reduce to single terms.

A number of other topics could be discussed at this point where there are differences in detail between the equations in the molality system and those for mole fractions. For some of these quantities, however, the effect of change in composition scale is so simple that detailed presentation is unnecessary. Also, the major array of solution data presented in molalities is that for solutions containing ions with or without neutral solutes. Thus, we delay some presentations until the fundamentals of ionic solutions have been introduced; it is easy to delete all terms related to ions and recover an equation for neutral solutes.

TREATMENTS FOR REAL SYSTEMS

As was discussed more fully in Chapter 12, the best and most convenient method of treating a real system is the use of an appropriate analytical equation defined initially for the excess Gibbs energy. By differentiation, equations are derived for the quantities for which measurements are available and the various parameters are determined by appropriate regressions. Then all other quantities can be calculated. In earlier years, however, with limited computational facilities, graphical methods were often used. By use of the Gibbs–Duhem equation or other relationships, conversions can be made from measured quantities to other functions. We describe briefly an example.

Gibbs–Duhem Relationships

The Gibbs–Duhem equation, when converted to osmotic coefficient and activity coefficient in molality, becomes

$$d\ln \gamma_m = d\phi + \frac{\phi - 1}{m} dm \tag{14-24}$$

If ϕ has been measured, Eq. (14-24) is integrated from $m = 0$ to m', yielding

$$\ln \gamma(m') = \phi(m') - 1 + \int_0^{m'} \frac{\phi - 1}{m} dm \tag{14-25}$$

and the integral can be evaluated from a plot of $(\phi - 1)/m$ vs. m. Conversely, the osmotic coefficient can be calculated from activity coefficient data. The integral in Eq. (14-25) was often evaluated graphically in earlier years. Examples of such calculations are available elsewhere.[1]

Expressions of Excess Gibbs Energies

We turn now to the selection of an equation for the excess Gibbs energy. One can assume a simple power series in molality starting with quadratic terms:

$$\frac{G^{Em}}{w_s RT} = \sum_i \sum_j \lambda_{ij} m_i m_j + \sum_i \sum_j \sum_k \mu_{ijk} m_i m_j m_k + \cdots \tag{14-26}$$

This choice has a statistical basis in the theory of McMillan and Mayer,[2] which predicts an equation of this form but in concentrations instead of molalities. But at constant temperature and low concentration, molality is proportional to concentration. And at higher concentration any departure from proportionality in the first term can be carried into higher terms. One can also show that the equations in the mole fraction system, Chapter 12, are consistent with Eq. (14-26) when converted to the infinitely dilute standard state and molalities.

The underlying physical picture is that λ_{ij} represents the binary interaction in the solvent of one solute molecule i with another solute molecule j; the two solute molecules may be of the same species, i.e., $i = j$. It is analogous to the second virial coefficient B_{ij} for an imperfect gas. Then the terms in μ_{ijk} arise from triple interactions of solute molecules again in the solvent at high dilution, so that there is no influence from additional solute molecules. In contrast to Eq. (12-70) for the mole fraction system, terms with all indices the same are not zero.

We now apply the differentiation of Eq. (14-23) to obtain the activity coefficients,

$$\ln \gamma_{m,i} = 2 \sum_j \lambda_{ij} m_j + 3 \sum_j \sum_k \mu_{ijk} m_j m_k + \cdots \qquad (14\text{-}27)$$

where the sums over j and k include the terms $\lambda_{ii} m_i$, $\mu_{iii} m_i^2$, $\mu_{ijj} m_i m_j$, etc. Before applying Eq. (14-22) for the osmotic coefficient, it is convenient to rewrite Eq. (14-26) for an indefinite amount of solution:

$$\frac{G^{Em}}{RT} = w_s^{-1} \sum_i \sum_j \lambda_{ij} n_i n_j + w_s^{-1} \sum_i \sum_j \sum_k \mu_{ijk} n_i n_j n_k + \cdots \qquad (14\text{-}28)$$

Then from Eq. (14-22),

$$\phi - 1 = \left(\sum_i m_i\right)^{-1} \left(\sum_i \sum_j \lambda_{ij} m_i m_j + 2 \sum_i \sum_j \sum_k \mu_{ijk} m_i m_j m_k + \cdots \right) \qquad (14\text{-}29)$$

For a single solute, these equations simplify to

$$\frac{G^{Em}}{w_s RT} = \lambda m^2 + \mu m^3 + \cdots \qquad (14\text{-}30)$$

$$\ln \gamma_m = 2\lambda m + 3\mu m^2 + \cdots \qquad (14\text{-}31)$$

$$\phi - 1 = \lambda m + 2\mu m^2 + \cdots \qquad (14\text{-}32)$$

Expressions for the excess volume, enthalpy, entropy, and heat capacity are readily obtained from the basic relationships involving pressure or temperature derivatives of the excess Gibbs energy. Examples are given in Chapter 18 for solutions containing ionic as well as neutral solutes.

Now we can examine the question whether measurements of $\ln \gamma_i$ for one solute or of ϕ for the solvent allow the calculation of all parameters in Eq. (14-26), which would then determine any other chemical potential. For measurements of ϕ, one finds that all parameters are present in Eq. (14-29) and can be determined if the data are sufficiently wide-ranging and precise. If the experimental information is for one $\ln \gamma_{m,i}$, however, terms in λ_{jj} and μ_{jjj} with $j \neq i$ are absent from Eq. (14-27) and cannot be determined. Thus, $\ln \gamma_j$ cannot be calculated. These conclusions are similar to those reached in the last section of Chapter 12 from the mole-fraction-based equations, but in that case there were complications and qualifications that are absent here.

Osmotic Coefficient and Activity Coefficient from Freezing-Point Data

One of the most sensitive methods of determining solvent activity is by measurement of the lowering of the freezing point of the solvent. This method has been especially useful for aqueous solutions where the temperature is convenient and the ice is pure, with rarely any problem of solid solubility of a solute. It is relatively easy to measure to 10^{-4} degree the temperature difference between the solution in equilibrium with ice and that of pure water with ice, and this precision yields more accurate activity values than can be obtained from vapor pressure measurements. Thus, the freezing-point-depression method is especially useful for dilute solutions where the activity changes are very small.

Since the freezing temperature varies with the composition, a complete treatment involves the effect of temperature on the activity and thereby the heat capacity and heat of dilution of the solution. We present here an analysis that is an excellent approximation for very dilute solutions. Treatments for more concentrated solutions are available.[3]

We take the pure liquid solvent as the standard state of unit activity and consider the activity of the solvent in solution, a_s, which must, of course, equal the activity of the solid at the temperature of equilibrium. Consider difference in chemical potential between solid and pure liquid and its change with temperature:

$$\ln a_s = \frac{\mu_s - \mu^\circ}{RT} = \frac{-\Delta_{fus}G}{RT}$$

$$\left(\frac{\partial \ln a_s}{\partial T}\right)_P = \frac{\Delta_{fus}H}{RT^2} \tag{14-33}$$

Here we note that $\Delta_{fus}H$ is the familiar heat of fusion (or melting). We now designate the freezing point lowering by θ; hence, $T = T_{fus} - \theta$, with T_{fus} the melting temperature of pure solvent. Also, $\Delta_{fus}H$ will change with temperature, but we assume a constant change in heat capacity. Then,

$$\Delta_{fus}H = \Delta_{fus}H^\circ - \Delta C_P \theta \tag{14-34}$$

and on substitution into Eq. (14-33),

$$d\ln a_s = -\frac{(\Delta_{fus}H^\circ - \Delta C_P \theta)}{R(T_{fus} - \theta)^2} d\theta \tag{14-35}$$

After expansion of $(T_{fus} - \theta)$ in powers of θ, and integration, one has

$$\ln a_s = -\left(\frac{\Delta_{fus}H^\circ}{RT_{fus}^2}\right)\left[\theta + \left(\frac{1}{T_{fus}} - \frac{\Delta C_P}{2\Delta_{fus}H^\circ}\right)\theta^2 + \cdots\right] \tag{14-36}$$

We now recall that a_s is the activity of the solvent in solution as well as that of the pure solid, and substitute into the expression for the osmotic coefficient. If there is just one solute of molality m, one has

$$\phi = -\frac{\Omega}{m}\ln a_s$$

$$= \left(\frac{\Omega \Delta_{fus}H^\circ}{RT_{fus}^2}\right)\left[\frac{\theta}{m} + \left(\frac{1}{T_{fus}} - \frac{\Delta C_P}{2\Delta_{fus}H^\circ}\right)\frac{\theta^2}{m} + \cdots\right] \tag{14-37}$$

Next, we substitute the numerical values for the properties of ice and water and obtain

$$\phi = \frac{\theta}{1.860m}(1 + 0.00049\,\theta) \qquad (14\text{-}38)*$$

The second term in the parentheses is usually negligible for dilute solutions. If θ is large, one should consider a more complete expression[3] that includes an additional term in a higher power of θ.

We now recall Eq. (14-32), which may be rearranged to

$$\frac{1-\phi}{m} = \lambda + 2\mu m + \cdots \qquad (14\text{-}39)$$

Now we see that a graph of freezing-point data presented as $(1 - \phi)/m$ vs. m should show a finite intercept λ and a slope of 2μ. Higher terms yield curvature, but this may not be significant for dilute solutions.

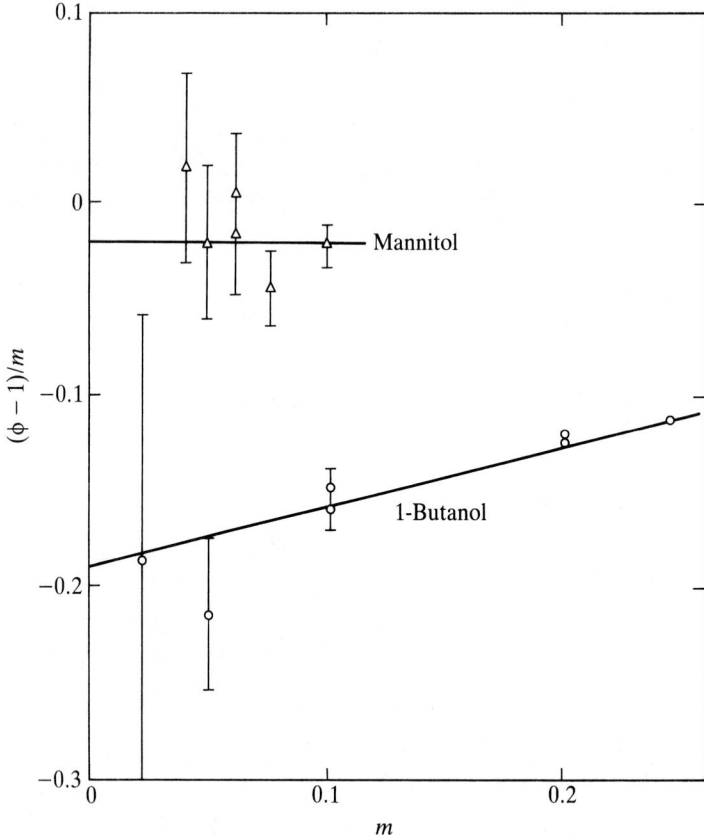

FIGURE 14-1
The freezing-point-lowering expressed as $(\phi - 1)/m$; data of Webb and Lindsley.[4]

Figure 14-1 presents experimental values of $(\phi - 1)/m$ derived from freezing-point measurements of aqueous solutions of mannitol and 1-butanol.[4] The error bars indicate the effect of a change in θ by 0.0002 K. For the parameters in Eq. (14-32), one finds $\lambda = -0.02_0$, $\mu = 0$ for mannitol and $\lambda = -0.19_0$, $\mu = 0.17_0$ for 1-butanol. Then Eq. (14-31) yields for the activity coefficient:

$$\ln \gamma = -0.040m \text{ for mannitol}$$

$$\ln \gamma = -0.38m + 0.51m^2 \text{ for 1-butanol}$$

REFERENCES

1. K. S. Pitzer and L. Brewer, *Thermodynamics*, revision of 1st edn. by G. N. Lewis and M. Randall, McGraw-Hill, New York, 1961, pp. 263–265.
2. W. G. McMillan and J. E. Mayer, *J. Chem. Phys.* **13**, 276 (1950).
3. Reference 1, pp. 406–412.
4. T. J. Webb and C. H. Lindsley, *J. Am. Chem. Soc.*, **56**, 874 (1934).

PROBLEMS

14-1. Show that the volume of a solution containing 1000 g of water is $V = (1000 + mM)/d$, where m is the molality, M is the molal weight of solute, and d is the density of the solution.

14-2. Duke and Iverson [*J. Phys. Chem.*, **62**, 417 (1958)] have determined the solubilities of various metal chromates in fused potassium nitrate–sodium nitrate eutectic mixtures as a function of added sodium chloride. The solubilities of cadmium chromate in the KNO_3–$NaNO_3$ eutectic melt as a function of added sodium chloride are given as follows:

NaCl/m	$CdCrO_4/m, \times 10^2$	
	250°C	300°C
0.0	0.59	0.8
0.1	1.2	1.7
0.2	1.7	2.6
0.3	2.3	3.5
0.4	2.9	4.4
0.5	3.5	5.3

Calculate $\gamma_\pm$ for cadmium chromate in 0.5 molal NaCl, assuming $\gamma_\pm = 1.0$ in the absence of NaCl. Use Eq. (14-27) to calculate values of $\lambda_{CdCrO_4,NaCl}$ between 0.1 and 0.5 molal NaCl, neglecting higher-order terms. Is an equation of the type of Eq. (14-27) adequate with just a linear term in m?

CHAPTER 15

ELECTROLYTE SOLUTIONS: INTRODUCTION AND SIMPLE EXAMPLES

The presence of ions in a solution requires additional definitions to provide for the exact balance of positive and negative electrical charges. Also, it introduces very important changes into the statistical mechanics of solutions, and certain of these features must be included in the equations used to represent the properties of real systems. Indeed, the special type of nonideal behavior arising from the interaction of ions caused confusion and controversy in the interpretation of electrolyte properties in the early years of this century. The solution was found, first empirically by Lewis and others[1-3] and then theoretically by Debye and Hückel[4] in 1923. Following these advances, pure dilute aqueous electrolytes were widely investigated and their properties near room temperature were well established by the middle of the century. In recent years, studies have emphasized mixed electrolytes, more concentrated solutions, and higher temperatures, extending in some cases to fused salts and to near the critical temperature of water.

The older literature and most recent papers use molality as the measure of composition, but for very concentrated solutions one must use mole fraction. The definitions and basic relationships as well as examples in this chapter are given on a molality basis. A brief presentation of definitions and equations on a mole fraction basis is given in Chapter 17, together with examples of extremely concentrated electrolytes.

Strong and Weak Electrolytes, Degree of Dissociation

The electrical conductance is the simplest manifestation of ions in solution. The most useful measure is the conductance per equivalent (per mole of charges) which is called the *equivalent conductance* and given the symbol Λ. Consideration of that quantity for a variety of familiar aqueous electrolytes near room temperature shows two distinctly different types of behavior. One type has a high and nearly constant equivalent conductance that decreases only slowly with increase in concentration. These are called *strong electrolytes* and are assumed to be completely dissociated into ions. The decrease in equivalent conductance with increase in concentration is explained by the long-range electrostatic forces acting between the ions.

A second type of electrolyte has a much smaller equivalent conductance at moderate concentration (near $0.1 \, \text{mol kg}^{-1}$) but this increases rapidly with dilution. Where extrapolation to infinite dilution is feasible, the extrapolated conductance $\Lambda°$ is comparable to that of a strong electrolyte. These are called *weak electrolytes*. They are treated as only partially dissociated into ions, with this dissociation given by a chemical equilibrium treatment. A prototype is acetic acid with the dissociation reaction

$$CH_3COOH = H^+ + CH_3CO_2^-$$

which we abbreviate to $HA = H^+ + A^-$. Then the equilibrium is given by the expression of activities:

$$K = \frac{a_{H^+} \, a_{A^-}}{a_{HA}} \tag{15-1}$$

where K is the dissociation constant. If we ignore, for the present, activity coefficients and take $\Lambda/\Lambda°$ as the fraction ionized, then one may take, approximately,

$$a_{H^+} = a_{A^-} = \frac{m\Lambda}{\Lambda°}$$

$$a_{HA} = m\left(\frac{1 - \Lambda}{\Lambda°}\right)$$

and

$$K_\Lambda = \frac{m(\Lambda/\Lambda°)^2}{(1 - \Lambda/\Lambda°)} \tag{15-2}*$$

It is found that the experimental data for acetic acid at 25°C from $0.1 \, \text{mol kg}^{-1}$ down to $10^{-4} \, \text{mol kg}^{-1}$ or less yield values $K_\Lambda \cong 1.8 \times 10^{-5}$ that are constant within a few percent, which confirms this picture within the expected accuracy of the approximations. Similar results are found for many other weak acids and bases with values of K of 10^{-4} or smaller.

Most aqueous electrolytes are either typical strong electrolytes that can be considered fully dissociated or typical weak electrolytes, but there are intermediate cases. These are relatively infrequent among aqueous systems at room temperature, but are more often found for solvents of lower dielectric constant, including those with water at temperatures increasing above 100°C. While these intermediate cases can, in principle, be treated as weak electrolytes, there are practical difficulties that are discussed in a later section after considering typical strong and weak electrolytes.

ACTIVITY AND ACTIVITY COEFFICIENTS

Symmetrical Strong Electrolytes

Since we assume a strong electrolyte to be completely dissociated into ions, it is not useful to consider an activity of an undissociated species. For an equilibrium with another phase, however, a neutral combination of ions must be transferred. For a symmetrical electrolyte MX comprising the ions M^+ and X^-, the chemical potential for the transfer of MX is

$$\mu_{MX} = \mu_M + \mu_X$$
$$= \mu_M^\circ + \mu_X^\circ + RT(\ln a_M + \ln a_X) \tag{15-3}$$

Thus, it is convenient to define the activity of the salt MX as the product of the ion activities.

$$a_{MX} = a_M a_X \tag{15-4}$$

These definitions are appropriate for any self-consistent system of composition and standard state definitions. Also, they remain satisfactory for a mixed electrolyte with additional ions N^+, Y^-, etc.

In the molality system, one defines the activity coefficient for an ion, as for a neutral species, by the relationship

$$\gamma_{m,i} = \frac{a_{m,i}}{m_i} \tag{15-5}$$

where $a_{m,i}$ now replaces the fugacity ratio f_i/f_i° in Eq. (14-6). The dimensional problem remains the same for an ion as for neutral species, and one implies m_i/m_0 with $m_0 = 1$ mol kg^{-1} wherever m_i appears in this type of expression. Then both $a_{m,i}$ and $\gamma_{m,i}$ are dimensionless; hereafter, we omit the m_0 and the subscript m unless there is ambiguity.

The activity of the salt MX now becomes

$$a_{MX} = \gamma_M \gamma_X m_M m_X \tag{15-6}$$

where the product $\gamma_M \gamma_X$ appears. It is not possible, without special methods or definitions, to measure the absolute value of the activity or the activity coefficient of a single ion. This situation is discussed below; for the present we note that it is customary to define a mean activity coefficient for a symmetrical electrolyte as

$$\gamma_\pm = (\gamma_+ \gamma_-)^{1/2} \tag{15-7}$$

Then the activity of the salt MX becomes

$$a_{MX} = \gamma_{\pm,MX}^2 m_M m_X \tag{15-8}$$

for a pure salt solution $m_M = m_X$, but the two molalities can differ in a mixed electrolyte. Thus, the quantity $a_\pm = (a_+ a_-)^{1/2}$, which is sometimes defined, is useful only for pure salt solutions, and we will not use it.

Unsymmetrical Strong Electrolytes

So far, we have been considering the relationship for symmetrical electrolytes such as KCl or $CuSO_4$. When we treat the more complex types, such as K_2SO_4, $K_4Fe(CN)_6$, and $Al_2(SO_4)_3$, the equations become a little more complicated. If we define an electrolyte Q as one that dissociates into $\nu(=\nu_M + \nu_X)$ ions according to the equation

$$Q = \nu_M M^{z_M+} + \nu_X X^{|z_X|-}$$

electrical neutrality requires

$$\nu_M z_M + \nu_X z_X = 0$$

Now the chemical potential of electrolyte Q is

$$\mu_Q = \mu_Q^\circ + RT(\nu_M \ln a_M + \nu_X \ln a_X) \qquad (15\text{-}9)$$

and it is appropriate to define the activity

$$a_Q = a_M^{\nu_M} a_X^{\nu_X} \qquad (15\text{-}10)$$

Equation (15-5) remains appropriate for the definition of the activity coefficient of a particular ion, but for the salt Q the activity becomes

$$a_Q = \gamma_M^{\nu_M} \gamma_X^{\nu_X} m_M^{\nu_M} m_X^{\nu_X} \qquad (15\text{-}11)$$

and the mean activity coefficient

$$\gamma_\pm = (\gamma_M^{\nu_M} \gamma_X^{\nu_X})^{1/\nu} \qquad (15\text{-}12)$$

whereupon

$$a_Q = \gamma_\pm^\nu m_M^{\nu_M} m_X^{\nu_X} \qquad (15\text{-}13)$$

For a single, pure electrolyte of stoichiometric molality m_Q,

$$m_M = \nu_M m_Q \qquad m_X = \nu_X m_Q$$

and

$$a_q = \nu_M^{\nu_M} \nu_X^{\nu_X} (\gamma_\pm m_Q)^\nu \qquad (15\text{-}14)$$

In mixed electrolytes, these last equations are not valid and one returns to Eq. (15-13).

GIBBS ENERGY AND RELATED QUANTITIES

Solvent Activity and the Osmotic Coefficient

The equations related to solvent activity remain unchanged when some or all of the solute species are ionic; thus, Eqs. (14-10) through (14-15) remain valid. We recall the definition of the osmotic coefficient

$$\phi = -\left(\frac{\Omega}{\sum_i m_i}\right) \ln a_s \qquad (14\text{-}10)$$

where $\Omega = 1000/M_s$ or 55.51 mol kg^{-1} for water. For a pure electrolyte, the substitution $\sum_i m_i = \nu m$ can be made in this equation or any of the others, including Eqs. (14-12) for the chemical potential or (14-15) for the osmotic pressure.

Gibbs Energy

The derivations of Chapter 14 and Eqs. (14-16) through (14-24) remain valid for ionic solutes. For convenience we recall the equations for the excess Gibbs energy per kilogram of solvent in the molality system:

$$\frac{G^{Em}}{w_w RT} = \frac{\Delta_{\text{mix}} G}{w_w RT} + \sum_i m_i (1 - \ln m_i) \qquad (14\text{-}19)$$

$$\frac{G^{Em}}{w_w RT} = \sum_i m_i (1 - \phi + \ln \gamma_i) \qquad (14\text{-}20)$$

where w_w is the amount of solvent in kilograms and the sums cover all solute species. For a single-solute electrolyte one may substitute $\nu_i m$ for m_i and the definition for $\gamma_\pm$, whereupon Eq. (14-20) becomes

$$\frac{G^{Em}}{w_w RT} = \nu m(1 - \phi + \ln \gamma_\pm) \qquad (15\text{-}15)$$

For mixed electrolytes Eq. (14-20) remains the appropriate equation.

If one has an equation for the excess Gibbs energy, appropriate differentiation yields the osmotic and activity coefficients,

$$\phi - 1 = -\frac{(\partial G^{Em}/\partial w_s)_{n_i}}{RT \sum_i m_i} \qquad (14\text{-}22)$$

$$\ln \gamma_i = \left[\frac{\partial (G^{Em}/w_w RT)}{\partial m_i}\right]_{m_j, j \neq i} \qquad (14\text{-}23)$$

These equations are unchanged from those for neutral solutes and there is no complication for Eq. (14-22). For Eq. (14-23), however, there is the problem that a finite change of the molality of a single ion would violate the requirement of electrical neutrality. One can argue that the infinitesimal change implied by the derivative may not cause any significant problem. Alternatively, one may deal with a sum of the terms of Eq. (14-23) for positive and negative ions corresponding to electrical neutrality; this will become an equation for the mean activity coefficient $\gamma_\pm$.

A more detailed analysis indicates that a value of $\ln \gamma_i$ for a single ionic species is uncertain by a quantity $z_i \delta$ related to a possible small net charge on the entire system. For any electrically neutral combination of ions, this term cancels. Thus, it is valid and convenient to use Eq. (14-23) since all rigorous thermodynamic applications will involve transfers of electrically neutral material. The quantity pH is designed to represent the activity of H^+, a single ion. For pH or any similar quantity, this problem related to overall electrical neutrality must be considered.

WEAK ELECTROLYTES

A weak electrolyte comprises a mixed system with ions and a neutral solute species or in certain cases a system with three or more ionic species. Thus, those preceding equations valid for mixed electrolytes may be used for weak electrolytes. The appropriate dissociation reactions must be considered with equilibrium constant equations such as Eq. (15-1).

For the simplest case with a single ionization $HA = H^+ + A^-$, Eqs. (15-3) through (15-8) are valid for the ions with $m_M = m_{H^+}$ and $M_X = m_{A^-}$, the actual dissociated molalities. There will be an activity coefficient γ_{HA} for the undissociated and in this case neutral species. Then Eq. (15-1) becomes

$$K = \frac{\gamma_\pm^2 m_{H^+} m_{A^-}}{\gamma_{HA} m_{HA}} \qquad (15\text{-}16)$$

If the solution is pure HA of molality m and the fraction dissociated is α,

$$m_{H^+} = m_{A^-} = \alpha m$$

$$m_{HA} = (1 - \alpha)m$$

and

$$K = \frac{\gamma_\pm^2 \alpha^2 m}{\gamma_{HA}(1 - \alpha)} \qquad (15\text{-}17)$$

Often investigations include buffer solutions with an added salt MA. Then one returns to Eq. (15-16) with appropriate expressions for m_{A^-}, etc.

The chemical potential of HA is, of course, the same for either the associated or dissociated species at equilibrium, but the standard states are different. Thus,

$$RT \ln K = \mu_{HA}^\circ - \mu_{H^+}^\circ - \mu_{A^-}^\circ \qquad (15\text{-}18)$$

$$\mu_{HA} = \mu_{HA}^\circ + RT \ln(\gamma_{HA} m_{HA})$$

$$= \mu_{H^+}^\circ + \mu_{A^-}^\circ + RT \ln(\gamma_\pm^2 m_{H^+} m_{A^-}) \qquad (15\text{-}19)$$

The activity coefficient that would pertain for a weak electrolyte if it were assumed to be a strong electrolyte is sometimes reported; it is called the *stoichiometric activity coefficient* and for the case HA is

$$\gamma_{\text{stoic}} = \alpha \gamma_\pm \qquad (15\text{-}20a)$$

and the activity related to the ionized standard state is

$$a_{H^+}a_{A^-} = (\gamma_{stoic}m)^2 \qquad (15\text{-}20b)$$

In cases where the dissociation is relatively large and the system could be treated as either a strong or a weak electrolyte, this quantity should be reported, but it is not important when the dissociation is very small.

If the complete formula for a weak acid, base, or salt involves multiple dissociations, there will be an equation analogous to (15-1) for each dissociation and Eqs. (15-16)–(15-20) will be modified.

MEASUREMENTS OF ACTIVITIES FOR STRONG ELECTROLYTES

As we stated in the introduction to this chapter, the experimental data for strong electrolyte solutions do not follow the pattern of nonideality observed for nonionic solutions. During the first two decades of this century there was considerable controversy, first concerning the reality of this difference between ionic and nonionic solutions, and then in establishing what quantitive relationship governed the ionic behavior. While other measurements of activity were made and discussed, freezing-point data gave the most accurate and unambiguous results. Thus, we first examine that method.

Freezing-Point Lowering and the Osmotic Coefficient

The basic relationship between the lowering of the freezing point of a solution θ and the osmotic coefficient was given in Chapter 14. For a strong, single-solute electrolyte we modify Eq. (14-38)* only to replace the single solute molality m by νm to account for the electrolyte dissociation into ν ions. Then, for dilute aqueous solutions, we have

$$\phi = \frac{\theta}{1.860\nu m}(1 + 0.00049\theta) \qquad (15\text{-}21)^*$$

For nonionic solutions we found that $(1 - \phi)/m$ was nearly constant at low molality; thus, $1 - \phi$ varies linearly with m. For electrolytes, however, $1 - \phi$ varies more slowly with m. A possible pattern would be a dependency on m to a fractional power α, i.e., $1 - \phi = \beta m^\alpha$. This is tested in Fig. 15-1, where $\ln(1 - \phi)$ is plotted vs. $\ln m$ for the experimental data of Adams[5] for KNO_3. The slope clearly corresponds to $\alpha = 1/2$, a square-root dependence of $1 - \phi$ on m. For other simple salts such as NaCl, KCl, etc., the data show some curvature at the higher molalities but approach the same linear pattern with slope 1/2 at the lowest molalities.

If the term $\beta m^{1/2}$ is accepted, one still expects some further nonideality and it is reasonable to assume the term linear in m that was found both experimentally and

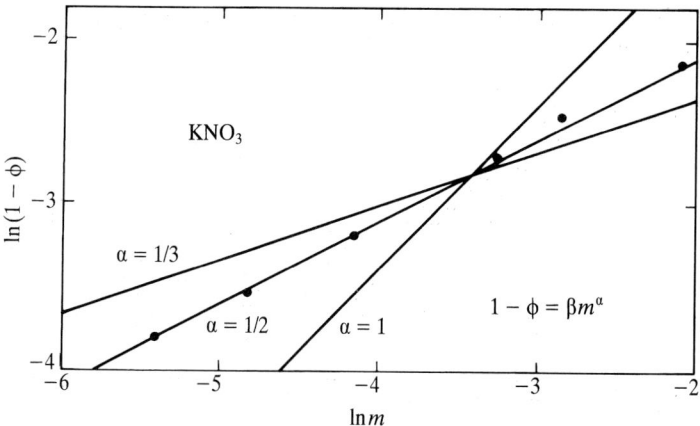

FIGURE 15-1
Test of the expression $1 - \phi = \beta m^\alpha$ and determination of α using freezing-point-depression data for KNO$_3$ from Adams.[5]

theoretically for nonionic solutions. Then, the expression for the osmotic coefficient becomes

$$1 - \phi = \beta m^{1/2} - \lambda m + \cdots \quad (15\text{-}22a)$$

which can be rearranged to

$$\frac{1 - \phi}{m^{1/2}} = \beta - \lambda m^{1/2} + \cdots \quad (15\text{-}22b)$$

Figure 15-2 displays the data of Adams[5] and of Scatchard and Prentiss[6] for KNO$_3$ and KCl on the basis of Eq. (15-22b). The error bars indicate the effect of an

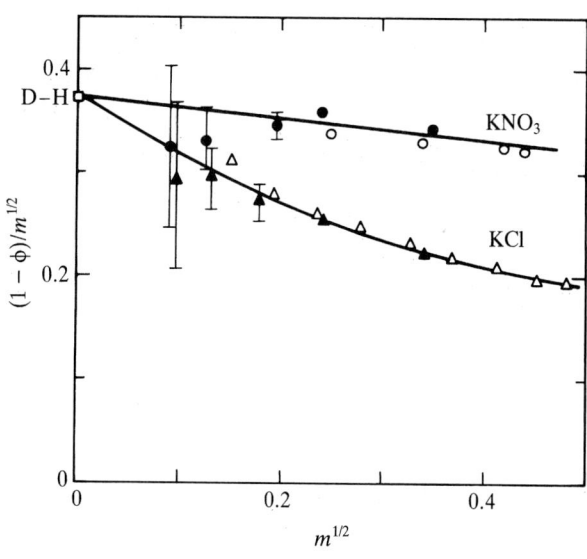

FIGURE 15-2
Osmotic coefficient data for KNO$_3$ and KCl from Adams[5] (solid symbols) and Scatchard and Prentiss[6] (open symbols). The Debye–Hückel theoretical value of the intercept at $m = 0$ is indicated by D-H.

uncertainty of 2×10^{-4} K on the function $(1 - \phi)/m^{1/2}$. These results and those for similar electrolytes are consistent with the pattern shown by the lines in Fig. 15-2. In particular, there is a common intercept (β) at $m = 0$ but different initial slopes (λ in Eq. 15-22B). Over this range to $m^{1/2} = 0.5$, there is significant curvature for KCl but not for KNO_3.

An aspect of primary theoretical interest is the property of a common intercept at $m = 0$. This suggests that the intercept depends only on the ionic charges and their electrical interaction and is independent of other details for the individual solutes. While this result was indicated by earlier work (see Chapter 16), it was the masterful theory of Debye and Hückel[4] in 1923 that clearly demonstrated the principle and gave the numerical value of this intercept in terms of the electrical charges on the ions and the dielectric constant or relative permittivity of the solvent. The Debye–Hückel theory also confirmed the $m^{1/2}$ dependence on molality. The theoretical value of the intercept is indicated by "D-H" in Fig. 15-2.

The Limiting Law and the Ionic Strength

We postpone until Chapter 16 the presentation of the rather complex theory of Debye and Hückel (D-H), but the limiting law, the $\beta m^{1/2}$ term, plays such a central role in interpreting experimental measurements that we state that law here. Furthermore, we generalize from the result for pure, singly charged (1–1) electrolytes to those for cases with multiple charges and for mixed electrolytes. For the latter, one uses the *ionic strength* defined as

$$I = \tfrac{1}{2}\sum_i m_i z_i^2 \qquad (15\text{-}23)$$

where the sum covers all ions. This "effective molality" weighted by the square of the ionic charge was first identified by Lewis and Randall,[1] but was confirmed and clearly established by Debye and Hückel.[4]

In terms of the ionic strength, the initial departure from ideality, the *limiting law*, for the excess Gibbs energy is

$$\frac{G^{Em}}{w_w RT} = -4A_\phi I^{3/2} \qquad (15\text{-}24)*$$

where A_ϕ is a parameter that will be found to be the same as the β defined above for 1–1 electrolytes. The expressions for the activity and the osmotic coefficients are found by differentiation, Eqs. (14-22) and (14-23). The latter derivation is straightforward and yields for a single ion i,

$$\ln \gamma_i = -3z_i^2 A_\phi I^{1/2} \qquad (15\text{-}25)*$$

For a symmetrical electrolyte with ion charges, $z_+ = -z_- = z$, this yields the same expression for the mean activity coefficient,

$$\ln \gamma_\pm = -3z^2 A_\phi I^{1/2} \qquad (15\text{-}26)*$$

The case of an unsymmetrical electrolyte with $z_+ \neq -z_-$ is a little more complex, but Eq. (15-12) yields

$$\ln \gamma_\pm = (v_+ \ln \gamma_+ + v_- \ln \gamma_-)/v \quad (15\text{-}12)$$

$$= -3|z_+ z_-| A_\phi I^{1/2} \quad (15\text{-}27)*$$

These equations are valid for a mixed electrolyte of any complexity. For a single symmetrical solute, Eq. (15-26)* simplifies to

$$\ln \gamma_\pm = -3z^3 A_\phi m^{1/2} \quad (15\text{-}26a)*$$

For the unsymmetrical case, it is easier to first obtain the relationship of I to m ($I = 3m$ for a 2–1 electrolyte, $I = 6m$ for a 3–1 electrolyte, etc.) and then use Eq. (15-27)*.

To derive the osmotic coefficient, one rewrites Eq. (15-24)* to show the complete dependency on the amount of solvent w_w for a fixed amount of solute n_i. Thus,

$$I = \tfrac{1}{2}\sum_i n_i \frac{z_i^2}{w_w}$$

and

$$\frac{G^{Em}}{RT} = -4A_\phi w_w^{-1/2} \left(\frac{\sum_i n_i z_i^2}{2}\right)^{3/2}$$

Then one obtains for ϕ from Eq. (14-22):

$$1 - \phi = \left(\sum_i m_i\right)^{-1} 2 A_\phi I^{3/2} \quad (15\text{-}28)*$$

For a single solute,

$$1 - \phi = |z_+ z_-| A_\phi I^{1/2} \quad (15\text{-}28a)*$$

Thus, we note that the β defined above for a 1–1 electrolyte is just the quantity A_ϕ and that the intercept of $(1 - \phi)/m^{1/2}$ at $m = 0$ for a symmetrical electrolyte will be proportional to z^3.

We conclude this section with the Debye–Hückel expression for A_ϕ. That theory was derived for concentration on a volumetric basis, but for a limiting-law concentration is proportional to molality. After conversion to molality, the result is

$$A_\phi = \frac{1}{3}\left(\frac{2\pi N_A d_w}{1000}\right)^{1/2} \left(\frac{e^2}{4\pi\varepsilon_0 \varepsilon kT}\right)^{3/2} \quad (15\text{-}29)$$

where N_A is the Avogadro number, d_w is the solvent density in g cm^3, e is the electronic charge, ε is the dielectric constant or relative permittivity, k is the Boltzmann constant, and T is the absolute temperature. The factor $4\pi\varepsilon_0$ with ε_0 the permittivity of free space, which is required for SI units, becomes unity and is omitted for esu. Much of the literature on electrolytes uses esu (omitting the $4\pi\varepsilon_0$); also the symbol D is often used for the dielectric constant instead of ε. The value of A_ϕ is 0.391_5 kg$^{1/2}$ mol$^{-1/2}$ at 25°C and increases with temperature to 0.461 at 100°C and to 0.960 at 300°C. A table of values is given in Appendix 7.

If the emphasis is on activity coefficients, the quantity $A_y = 3A_\phi$ is often used. Also, if concentration in molarity is used instead of molality, the forms of all equations of this section remain the same, but the factor d_w drops out of Eq. (15-29).

Activity from the Vapor Pressure of the Solvent

Measurement of the vapor pressure of the solvent, after conversion to fugacity, leads directly to a measure of the activity of the solvent and then to the osmotic coefficient. While vapor pressure measurements are less precise for dilute solutions than those of freezing-point lowering, they can be made at any temperature. Indeed, it becomes easier to measure a small percentage difference in pressure between the solution and pure solvent as the total pressure increases at higher temperatures. Thus, vapor pressure measurements are a very important source of osmotic coefficients, both at room temperature and at higher temperatures up to 350°C in aqueous systems.

At or near 25°C, the vapor pressure of water in equilibrium with an aqueous solution can be determined directly by either static or dynamic methods. The direct static method is perhaps best illustrated by the apparatus of Gibson and Adams[7] as used by Shankman and Gordon.[8] Measurements of the activity of water could be reproduced within 1 part in 2000.

The transpiration method of determining activities of water involves the saturation of an inert gas by the aqueous solution. Measurements by Becktold and Newton[9] indicate that measurements on calcium chloride and barium chloride yield water activities with a probable error of 1 part in 10 000. Brown and Delaney[10] used a method depending upon a temperature difference between solvent and solution to equalize the vapor-pressure difference, using a very sensitive differential manometer to observe the null point. Robinson[11] has analyzed their data and concludes that the activity of water in potassium chloride solutions between 0.1 mol l^{-1} and 2.3 mol l^{-1} as determined by various methods is in agreement within a mean deviation of 0.002 percent.

Very precise measurements up to 573 K were made for NaCl(aq) by Liu and Lindsay.[12] They used a "double thermostat" to maintain equality of temperature to 0.0004 degrees between solution and pure solvent. Also, small differences in pressure were measured precisely even at total pressures up to 86 bar at 300°C. The time required to attain equilibrium varied from 48 hours at 125°C to 6 hours at 300°C. The osmotic coefficient was calculated from the equation

$$\phi = \frac{\Omega}{\nu m}\left[\ln\left(\frac{P°}{P}\right) - \int_P^{P°}\left(\frac{1}{P} - \frac{V_s(g)}{RT}\right)dP - \frac{\overline{V}_s(l)(P° - P)}{RT}\right] \quad (15\text{-}30)$$

The first term in the brackets is, of course, the perfect gas term, with P the pressure over the solution and $P°$ that of pure water, while the integral corrects to fugacities. The final term is a very small correction for the effect of pressure on the chemical potential of the liquid. The resulting osmotic coefficient then pertains to $P°$, the saturation pressure of pure solvent.

The best treatments of the osmotic and activity coefficients and other related properties of aqueous NaCl fit these osmotic coefficients along with heats of dilution, heat capacities, and other data to carefully selected equations. Such treatments are described in Chapter 18. The selection of such an equation, however, is based on theory and experimental results yet to be presented. Hence, at this point we show a sample of the Liu and Lindsay data in a form appropriate for graphical interpolation and for a Gibbs–Duhem integration to yield the activity coefficient of NaCl. Equation (14-25) was the appropriate form of the Gibbs–Duhem equation for a nonionic solution:

$$\ln \gamma(m') = \phi(m') - 1 + \int_0^{m'} \frac{\phi - 1}{m} dm \qquad (14\text{-}25)$$

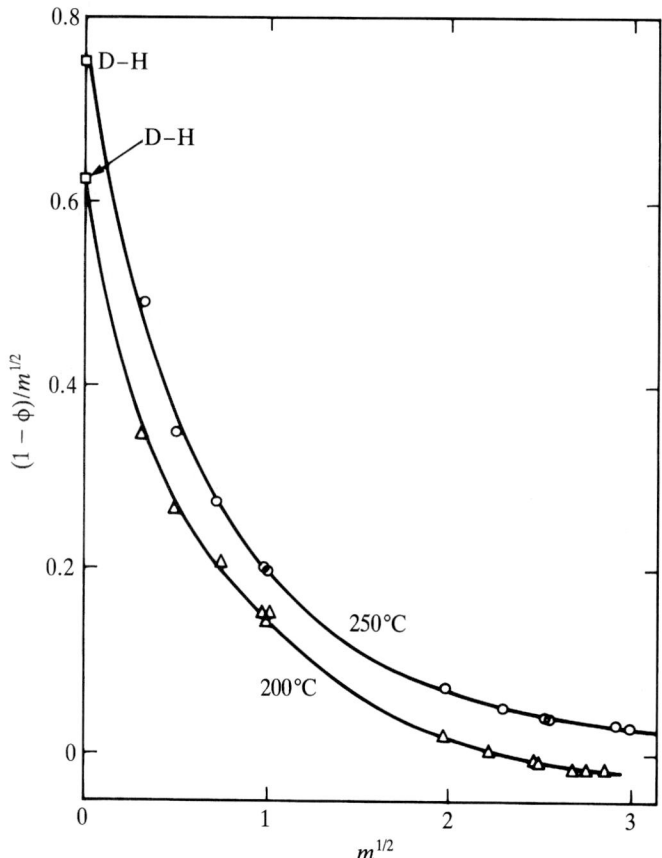

FIGURE 15-3
Osmotic coefficient values from vapor pressure measurements of Liu and Lindsay[12] for NaCl(aq) at 200°C and 250°C. The Debye–Hückel theoretical intercepts are indicated by D-H.

For an electrolyte $(\phi-1)/m$ diverges as $m \to 0$, but this problem is easily solved by a shift to $m^{1/2}$ as the variable:

$$\ln \gamma_\pm(m') = \phi(m') - 1 + 2 \int_0^{m'} \frac{\phi-1}{m^{1/2}} dm^{1/2} \qquad (15\text{-}31)$$

Figure 15-3 shows the experimental values of $(\phi-1)/m^{1/2}$ for 200°C and 250°C as a function of $m^{1/2}$; thus, the area under a curve is the value of the integral in Eq. (15-31).

Isopiestic Method

Once the vapor pressure of water as a function of molality has been determined accurately for one solution, it is not necessary to repeat absolute measurements of the activity of water for other solutions. The isopiestic method as introduced by Bousfield[13] and refined by many investigators,[14] is a simple, convenient method that yields accurate results.

For measurements near room temperature, the experimental arrangement consists of a glass desiccator containing a copper block in good thermal contact with several dishes. The desiccator is initially evacuated to remove the air and then maintained in a thermostat and rocked to agitate the solutions. Depending upon the concentration, at least 1 to 4 days is necessary for the approach to equilibrium within the accuracy of the analysis, which is about 0.1 percent. When equilibrium has been reached, all solutions have the same vapor pressure of water. From analyses of the solutions, including a reference solution for which the vapor pressure of water is accurately known as a function of molality, one then has the partial pressure of water in equilibrium with the solution being studied. Although a more complex apparatus is required, isopiestic measurements are now made at temperatures up to 250°C and many results have been reported by Holmes and others.[15] Isopiestic measurements are often made for mixed electrolytes as well as for solutions with a single solute.

For the measured solution and the reference, the solvent activity is the same. Hence, from the definition of the osmotic coefficient, Eq. (14-11),

$$\phi \sum_i m_i = \phi' \nu' m' \qquad (15\text{-}32)$$

where the primed quantities refer to the reference solution whose properties are known. Then for the unknown,

$$\phi = \frac{\phi' \nu' m'}{\sum_i m_i} \qquad (15\text{-}33a)$$

for a mixed electrolyte or

$$\phi = \frac{\phi' \nu' m'}{\nu m} \qquad (15\text{-}33b)$$

for a single solute.

With ϕ determined, the activity coefficient can be calculated by integration of the Gibbs–Duhem equation in the form Eq. (15-31). Values of $(1 - \phi)/m^{1/2}$ from isopiestic measurements become inaccurate below $0.1\,\text{mol kg}^{-1}$, and there is a considerable contribution to the integral in Eq. (15-31) below this value ($m^{1/2} = 0.316$). The curve for $(1-\phi)/m^{1/2}$ can be drawn smoothly to the Debye–Hückel limit at $m = 0$. But it is important to verify from freezing-point or other measurements valid below $m = 0.1$ that this simple pattern is correct. For typical 1–1, 2–1, and 1–2 strong electrolytes, this simple pattern has been found to be correct, and the uncertainty for the integral is small. But for higher-charge types, this aspect is less certain; also, there is increasing tendency toward hydrolysis of a highly charged cation such as Al^{3+} or Th^{4+}. Isopiestic measurements still yield valuable information for such solutions, but a more complex treatment may be required. An example, $CuSO_4(aq)$, is described below.

Isopiestic measurements are relatively easy, can be made for any nonvolatile solute and at any temperature up to 250°C for aqueous systems. Consequently, a wide array of electrolytes have been investigated by this method. Robinson and Stokes[16] made many of the measurements at 25°C, which are summarized in their excellent book. Isopiestic measurements are also very useful for mixed electrolytes, but the extraction of the various activity coefficients is a more complex procedure to be discussed below.

Activity from Electrochemical Cell Measurements

If the electrodes in a cell react reversibly, one with the cation and the other with the anion, then the cell potential yields the chemical potential of the ionic solute. The equation is

$$\mu_2 - \mu_2^\circ = nF(E^\circ - E)$$

with μ_2 the chemical potential of the solute, n the number of electrons per mole of solute in the cell reaction, F the Faraday constant, and E the cell potential, with μ° and E° the values for the standard state. The definitions with respect to sign of E and the direction of the chemical reaction will be given in Chapter 19, where electrochemical cells are discussed fully. A good example is the cell comprising a hydrogen electrode, a calomel electrode of mercury and mercurous chloride, and an aqueous solution of HCl. For this cell the reaction is $\frac{1}{2}H_2(g) + \frac{1}{2}Hg_2Cl_2(s) = Hg(l) + HCl(aq) = Hg(l) + H^+ + Cl^-$ and we note that this corresponds to one electron transferred in the cell ($n = 1$). Let us assume that the pressure of H_2 is maintained at the standard value of 1 bar; then the activity of HCl, a_2, is related to the cell potential by

$$RT \ln a_2 = \mu_2 - \mu_2^\circ = -F(E - E^\circ) \tag{15-34}$$

where E° is the cell potential when the HCl is at unit activity. Since $a_2 = (\gamma_\pm m)^2$, the expression for the activity coefficient is

$$\ln \gamma_\pm = \frac{F}{2RT}(E^\circ - E) - \ln m \tag{15-35}$$

In order to use cell potentials to calculate activity coefficients, one must determine $E°$ from a limiting process where $\ln \gamma_\pm \to 0$ as $m \to 0$. If we solve Eq. (15-35) for $E°$, we obtain

$$E° = \left[E + \frac{2RT}{F}\ln m\right] + \frac{2RT}{F}\ln \gamma_\pm \qquad (15\text{-}36)$$

Thus, if the quantity in brackets is plotted vs $m^{1/2}$, an extrapolation to $m = 0$ will yield $E°$. Better accuracy can be attained, however, if one uses the D-H limiting law to assist the extrapolation. Then, Eq. (15-26a)* with $z = 1$ yields the limiting behavior as $m \to 0$,

$$\ln \gamma_\pm = -3A_\phi m^{1/2}$$

and we rearrange Eq. (15-36) to

$$E° = \left[E + \frac{2RT}{F}(\ln m - 3A_\phi m^{1/2})\right] + \frac{2RT}{F}(\ln \gamma_\pm + 3A_\phi m^{1/2}) \qquad (15\text{-}37)$$

Now the dependence of the bracketed quantity on $m^{1/2}$ has been removed and one expects an approach linear in m to $E°$ at $m = 0$. Figure 15-4 shows the very precise measurements of Gupta, Hills, and Ives[17] at 25°C on this basis. While there is a little curvature, the extrapolation is unambiguous.

With $E°$ established, the potentials at other molalities yield the activity coefficient and related quantities. Table 15-1 gives these results based on various measurements[17,18] for this cell and for the similar cell with the AgCl, Ag electrode.

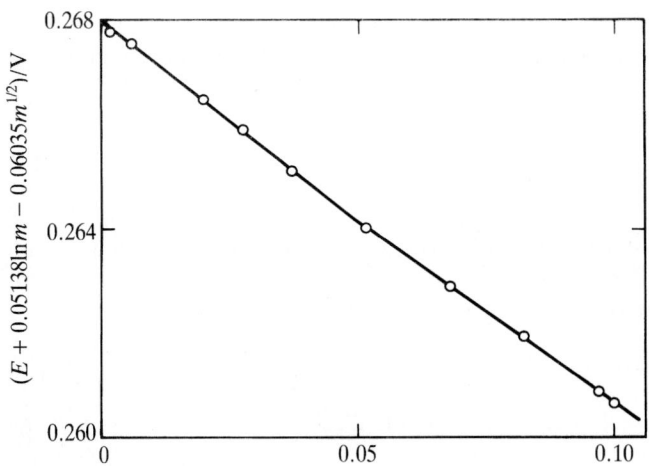

FIGURE 15-4
Electrochemical cell measurements[17] for HCl(aq) presented in accordance with Eq. (15-43) for the determination of $E°$.

TABLE 15-1
Activities in aqueous hydrochloric acid solutions at 25°C

m	$\gamma_\pm$	$a_\pm = m\gamma_\pm$	a_2	$(\mu_2 - \mu_2^\circ)/RT$
0.0005	0.975	0.000 488	0.000 000 24	-15.24
0.001	0.965	0.000 965	0.000 000 93	-13.89
0.002	0.952	0.001 904	0.000 003 63	-12.52
0.005	0.928	0.004 64	0.000 021 54	-10.75
0.01	0.904	0.009 04	0.000 081 7	$- 9.42$
0.02	0.875	0.017 5	0.000 306 0	$- 8.09$
0.05	0.830	0.041 5	0.001 721	$- 6.36_5$
0.1	0.796	0.079 6	0.006 34	$- 5.06_1$
0.2	0.767	0..153 4	0.023 52	$- 3.74_9$
0.3	0.756	0.226 8	0.051 4	$- 2.96_7$
0.4	0.755	0.302 0	0.091 2	$- 2.39_5$
0.5	0.757	0.378 5	0.143 3	$- 1.94_3$
0.6	0.763	0.457 8	0.209 6	$- 1.56_3$
0.7	0.772	0.540	0.291 9	$- 1.23_1$
0.8	0.783	0.634	0.392 8	$- 0.93_5$
0.9	0.795	0.716	0.512	$- 0.66_9$
1	0.809	0.809	0.655	$- 0.42_3$
2	1.009	2.018	4.072	$+ 1.40_4$
3	1.316	3.948	15.59	2.74_6
4	1.762	7.048	49.68	3.90_5
5	2.38	11.90	141.6	4.95_3
6	3.22	19.32	373.3	5.92_2
7	4.37	30.59	935.7	6.84_0
8	5.90	47.20	2 228	7.70_8
9	5.94	71.46	5 106	8.53_8
10	10.44	104.4	10 900	9.29_6
12	17.25	207	42 850	10.66_5
14	27.3	382	146 100	11.89
16	42.4	678	460 000	13.04

Information of comparable range and quality has been obtained for HBr(aq) and HI(aq) from similar cells. But the number of electrolytes is limited for which similar cells exist. Some metals such as copper form good electrodes for measurements on $CuCl_2$(aq), etc. But the number is limited since many metals either react with water or form inert oxide surfaces.

Electrochemical cells have been used successfully for aqueous solutions up to quite high temperatures. The simple system for HCl(aq) with H_2 and Ag,AgCl electrodes was measured up to 523 K by Greeley et al.,[19] but the results become less reliable at the highest temperatures because of increasing solubility of AgCl. Cells[20] with liquid junction have been used satisfactorily to 573 K, but special conditions must be satisfied to limit the junction potential. Various aspects concerning electrochemical cells are considered further in Chapter 19. Butler and Roy recently reviewed various aspects of electrochemical cell measurements.[21]

Activity from Solute Vapor Pressure

Most ionic solutes are nonvolatile at moderate temperatures, but there are exceptions. Concentrated solutions of hydrochloric acid have measurable vapor pressure of HCl(g). If we ignore association in the liquid, the reaction is

$$H^+(aq) + Cl^-(aq) = HCl(g)$$

and

$$P_{HCl} \cong f_{HCl} = Ka_2 = K(m\gamma_\pm)^2 \tag{15-38}$$

Bates and Kirschman[22] measured at 25°C the partial pressure of HCl over the aqueous solutions from 4 to 10 molal. If the value at 8 molal is used to fix K in Eq. (15-38), one obtains the result $K = 5.0_4 \times 10^{-7}$ bar. Then pressures for other molalities can be calculated from the values in Table 15-1. These are compared with the measured pressures in Table 15-2, where agreement is found within experimental uncertainty. More detailed recent studies[23] extend to 350°C and consider ion-association in the liquid as well as the very small ionization of HCl in the vapor.

Before turning to another topic, we remark on the enormous difference between electrolytes such as aqueous HCl and the nonionic solutions considered in previous chapters. There we found departures in activity from ideality by factors of 1.5 or 2 but rarely more than 5, while a_2 for HCl increases by nearly a factor of one million from 1 to 16 mol kg^{-1}. This can be understood in terms of the very strong interaction of ions, and especially H$^+$, with water. Also, we note the success of the strong electrolyte model, rigorously applied, to this remarkable system where the detailed structure is undoubtedly more complex at high molality.

TABLE 15-2
Comparison of measured pressures of HCl with those calculated from cell data

	$10^4 P$/bar	
m	Measured	Ka_2 from Table 15-1
4	0.24$_3$	0.25$_1$
5	0.70$_6$	0.71$_4$
6	1.86$_6$	1.88$_3$
7	4.64	4.72
8	11.2$_4$	(11.24)
9	25.7	25.7$_6$
10	56.0	55.0

Weak Acids and Bases

While measurements of the types just described can be made on weak electrolytes, and they confirm the partial dissociation, only rough estimates of the dissociation constant can be obtained. Accurate values of the dissociation constant for acetic and similar acids or for weak bases can be derived, however, from measurements on buffer solutions using a hydrogen electrode. If there is an electrode responsive to the other ion of the acid, measurements can be made in the ternary buffer solution HA–MA–H_2O, but such an electrode is seldom available. Then one turns to four-component buffer solutions with an added ion for which a measuring electrode exists. A good example is the system HA–NaA–NaCl–H_2O, with HA an acid such as acetic. Here a cell with hydrogen and silver–silver chloride electrodes measures the product of activities of H^+ and Cl^-. A rigorous treatment of such a mixed electrolyte is complex, however, and will be presented in Chapter 19, after the equations and some other topics related to multicomponent ionic systems have been presented in Chapter 17.

Marginally Weak Electrolytes

Some electrolytes that are largely ionized even at high concentrations nevertheless show some association at relatively low concentration. Typical examples are bivalent metal sulfates, $MgSO_4$, $FeSO_4$, etc., which are clearly ionic in the solid state and show high electrical conductance in concentrated aqueous solution, yet indicate some association below 0.1 molal at room temperature. This association is caused primarily by electrical forces, with a double charge on each ion in this case. This effect is also observed for salts with singly charged ions in solvents of low dielectric constant at room temperature or for aqueous solutions at very high temperatures, where the dielectric constant of water is greatly reduced. Thus, NaCl(aq) appears to be a completely dissociated, strong electrolyte even at 573 K as is $CaCl_2$(aq) below about 473 K. But at 573 K and above, $CaCl_2$ shows substantial ion association[24] even at molalities below 0.01 mol kg^{-1}. This information was obtained from heat of dilution measurements that are feasible under extreme conditions where other methods are impractical.

As an example of this type of behavior, we take aqueous $CuSO_4$ with data for 0°C from freezing-point measurements[25], which were converted,[26] by use of heat of dilution data, to 25°C for comparison with isopiestic data.[16] The results are presented in Fig. 15-5 as $(1 - \phi)/m^{1/2}$.

It is interesting to compare Fig. 15-5 with Fig. 15-3, which shows the same function for NaCl(aq) at 200 and 250°C. From the isopiestic data for $CuSO_4$, solid circles at $m = 0.1$ and above, one can draw a smooth curve to the D-H limit in a manner similar to that for NaCl. It is only the open circles from freezing-point-depression measurements near $m = 0.01$ that indicate a departure from the pattern for a typical fully ionized electrolyte. Below the most dilute freezing-point value at $m = 0.005$, the curve, shown dotted, was obtained[26] from an association equilibrium, with its constant evaluated from the freezing-point data in the range 0.005 to 0.05 in m.

The area under the curve in Fig. 15-5 is a term in Eq. (15-31) for the activity coefficient. Thus, the difference in area between the dashed curve and the combined

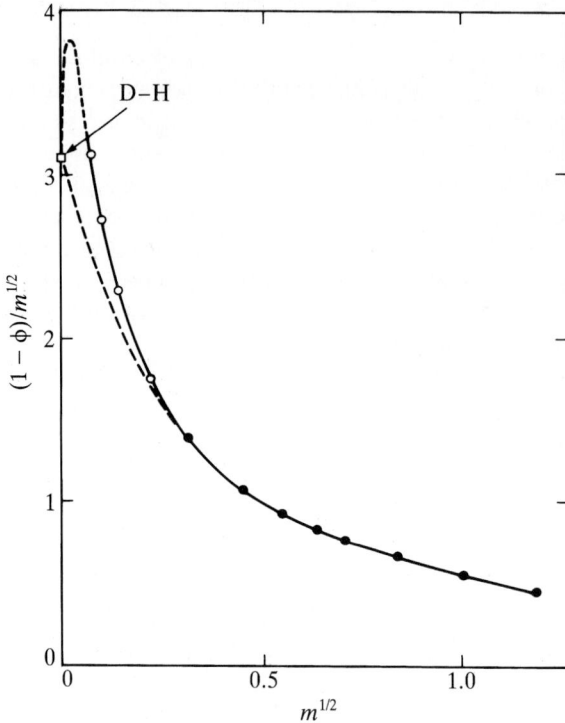

FIGURE 15-5
Osmotic coefficient data for $CuSO_4$(aq) at 25°C from isopiestic measurements (solid circles) and freezing-point-depression measurements converted to 25°C (open circles). See text for discussion of the curves.

solid, dotted curve represents a constant difference, at molalities above 0.1 mol kg^{-1} between the true $\ln \gamma_\pm$ and the value obtained from the simple dashed curve. In this case, it is a significant but not a very large correction.

In the case of $CuSO_4$(aq) there are also electrochemical cell measurements from a cell with copper and Hg, Hg_2SO_4 electrodes. It is not practical to determine the association equilibrium constant, the standard cell potential $E°$, and the ion activity coefficients from cell measurements alone. But the cell data were a useful confirmation in this case for the equation developed from other sources. The data for several other 2–2 electrolytes were treated in a similar manner[26] but with less complete data for other cases.

Representing Properties at High Molalities with Equations

While graphical methods were widely used in the past, the effectiveness of modern computers makes the use of equations more convenient, as well as more precise. For activity and osmotic coefficients one always includes the theoretical term for the Debye–Hückel limiting law, Eqs. (15-25)*, (15-27)*, or (15-28)*. Additional terms can be selected on the basis of more advanced theory or purely for empirical effectiveness. We postpone presentation of theoretically based equations until statistical theory has been developed in Chapter 16. Such equations are especially important for mixed electrolytes, as discussed in Chapter 17. For a single electrolyte in water, a simple series of increasing powers of molality is effective. With the osmotic coefficient as the example,

$$\phi - 1 = -|z_+ z_-| A_\phi I^{1/2} + \sum_i A_i m^{r_i} \qquad (15\text{-}39)$$

The sequences of powers, r_i, can be a regular and complete series of integers from 1 upward as suggested by Eq. (14-27) for neutral solutes. But the use of fractional half-integral or even quarter-integral powers has been found to be advantageous for electrolytes. There is a theoretical basis only for half-integral powers, but no real objection to quarter-integral powers on an empirical basis. In either case, the Gibbs–Duhem transformation yields for the activity coefficient,

$$\ln \gamma_\pm = -3|z_+ z_-| A_\phi I^{1/2} + \sum_i A_i \left(\frac{r_i + 1}{r_i} \right) m^{r_i} \qquad (15\text{-}40)$$

The treatment for $MgCl_2$(aq) by Rard and Miller[27] is interesting in that they give both half-integral and quarter-integral series. In each case they use eight terms with r_i from 1.0 to 4.5 or from 0.75 to 2.50, respectively, to represent the data to saturation at 5.81 molal. Rard and Platford[14] cite many other treatments of this type in their review.

If one is interested in mixed as well as pure electrolytes, then the optimum form of equation should be selected for multicomponent systems. That equation is readily simplified for a single solute. This procedure is presented in Chapter 17 after considering pertinent statistical theory.

REFERENCES

1. G. N. Lewis and M. Randall, *J. Am. Chem. Soc.*, **43**, 1112 (1921).
2. J. N. Bronsted, *J. Am. Chem. Soc.*, **44**, 877, 939 (1922); **45**, 2898 (1923).
3. For a historical review, see K. S. Pitzer, *J. Chem. Ed.*, **61**, 104 (1984).
4. P. Debye and E. Hückel, *Physik. Z.*, **24**, 185, 334 (1923); **25**, 97 (1924).
5. L. H. Adams, *J. Am. Chem. Soc.*, **37**, 481 (1915).
6. G. Scatchard and S. S. Prentiss, *J. Am. Chem. Soc.*, **54**, 2676, 2690 (1932); **55**, 4355 (1933).
7. R. E. Gibson and L. H. Adams, *J. Am. Chem. Soc.*, **55**, 2679 (1933).
8. S. Shankman and A. R. Gordon, *J. Am. Chem. Soc.*, **61**, 2370 (1939).
9. M. F. Becktold and R. F. Newton, *J. Am. Chem. Soc.*, **62**, 1390 (1940).
10. O. L. I. Brown and C. M. Delaney, *J. Phys. Chem.*, **58**, 255 (1954).
11. R. A. Robinson, *J. Phys. Chem.*, **60**, 501 (1956).
12. C.-t. Liu and W. T. Lindsay, Jr., *J. Phys. Chem.*, **74**, 341 (1970); *J. Solution Chem.*, **1**, 45 (1972).
13. W. R. Bousfield, *Trans. Faraday Soc.*, **13**, 401 (1918).
14. For a recent review of the isopiestic method, see J. A. Rard and R. F. Platford, "Experimental Methods: Isopiestic," in *Activity Coefficients in Electrolyte Solutions*, 2d edn., K. S. Pitzer, ed., CRC Press, Boca Raton, Fla, 1991, pp. 209–277.
15. H. F. Holmes and R. E. Mesmer, *J. Phys Chem.*, **87**, 1242 (1983); H. F. Holmes, C. F. Bayes, Jr., and R. E. Mesmer, *J. Chem. Thermo.*, **10**, 983 (1978).
16. R. A. Robinson and R. H. Stokes, *Electrolyte Solutions*, 2d edn. (revised), Butterworths, London, 1959.

17. S. R. Gupta, G. J. Hills, and D. J. G. Ives, *Trans. Faraday Soc.*, **59**, 1874 (1963).
18. G. J. Hills and D. J. G. Ives, *J. Chem. Soc.*, **1951**, 318; H. S. Harned and R. W. Ehlers, *J. Am. Chem. Soc.*, **54**, 1350 (1932); **55**, 2179 (1933); R. G. Bates and V. E. Bower, *J. Res. Natl. Bur. Stand.*, **53**, 283 (1954).
19. R. S. Greeley, W. T. Smith, Jr., M. H. Lietzke, and R. W. Stoughton, *J. Phys. Chem.*, **64**, 1445 (1960).
20. R. E. Mesmer, W. L. Marshall, D. A. Palmer, J. M. Simonson, and H. F. Holmes, *J. Solution Chem.*, **17**, 699 (1988).
21. J. N. Butler and R. N. Roy, "Experimental Measurements: Potentiometric," in *Activity Coefficients in Electrolyte Solutions*, 2d edn., K. S. Pitzer, ed., CRC Press, Boca Raton, Fla, 1991, pp. 155–208.
22. S. J. Bates and H. D. Kirschman, *J. Am. Chem. Soc.*, **41**, 1991 (1919).
23. J. M. Simonson and D. A. Palmer, *Geochim. Cosmochim. Acta*, **57**, 1 (1993); also J. M. Simonson, H. F. Holmes, R. H. Busey, R. E. Mesmer, D. G. Archer, and R. H. Wood, *J. Phys. Chem.*, **94**, 7675 (1990).
24. J. M. Simonson, R. H. Busey, and R. E. Mesmer, *J. Phys. Chem.*, **89**, 557 (1985).
25. P. G. M. Brown and J. E. Prue, *Proc. R. Soc., A*, **232**, 320 (1955).
26. K. S. Pitzer, *J. Chem. Soc. Faraday Trans.*, *II* **68**, 101 (1972).
27. J. A. Rard and D. G. Miller, *J. Chem. Eng. Data*, **26**, 38 (1981).

PROBLEMS

15-1. Prior to the Debye–Hückel theory, the empirical equation $\phi - 1 = -\alpha m^\beta$ was used for electrolytes. Use the Gibbs–Duhem equation to derive the corresponding equation for $\ln \gamma_\pm$. Both α and β are empirical constants.

15-2. The vapor pressure of HX over a 10 molal aqueous solution of HX is 0.142 bar; the activity $a_2 = a_\pm^2 = 1600$ related to the hypothetical ideal molal solute standard state. If the activity coefficient $\gamma_\pm = 0.70$ in 0.5 mol l^{-1} HX, what is the vapor pressure of HX over 0.5 mol l^{-1} HX? Assume HX gas to be perfect.

15-3. For molalities sufficiently dilute that $\gamma_\pm$ may be taken equal to unity, calculate a_2 for solutions of $PbCl_2$, $PbFCl$, $AlCl_3$, and $Al_2(SO_4)_3$ at molality m. For each of these solutions calculate the coefficient of $A_\phi I^{1/2}$ for the Debye–Hückel limiting law. What is a_2 of $Al_2(SO_4)_3$ in a mixture containing $m_{Na_2SO_4}$ and m_{AlCl_3}?

15-4. From Table 15-1, calculate E at 25°C for a HCl concentration cell without liquid junction with $m_1 = 0.05$ and $m_2 = 0.5$ mol kg^{-1}. What is the ratio of HCl pressures for these two solutions?

15-5. Derive the equation

$$\frac{G - G^\circ}{3RT} = -m\phi + m \ln 2^{2/3} m \gamma_\pm$$

for aqueous $CaCl_2$, where $G - G^\circ$ is the total Gibbs energy relative to pure water and $CaCl_2$ in the solute standard state of a solution containing 1 kg water and m moles of $CaCl_2$.

CHAPTER 16

ELECTROLYTE SOLUTIONS: STATISTICAL THEORY

The theory of electrolyte solutions begins, of course, with the brilliant discovery of Arrhenius that certain solutes dissociate into electrically charged species, ions. The next major advance, to which many authors[1-4] contributed during the first two decades of the twentieth century, was the realization that the deviations from ideality in dilute strong electrolytes are fundamentally different from the deviations in nonelectrolyte solutions. In electrolytes the long-range nature of the electrostatic force between ions yields effects that have no counterpart in solutions of neutral molecules. Milner[2] first attempted a theoretical treatment of this phenomenon and obtained an accurate but complicated expression that he was able to evaluate only approximately. Debye and Hückel[5] effected a great simplification by approximations in setting up the problem and obtained simple results of great utility. Their approximations have been the subject of countless further papers, and we shall return presently to a consideration of the Debye and Hückel equations and their region of validity.

Before turning to the quantitative theory, however, let us consider certain general ideas. If the particles of a solute are distributed with reasonable uniformity, then the number of solute molecules or ions within a spherical shell of radius r and thickness

dr about a given particle will be approximately proportional to the volume of the shell, $4\pi r^2\, dr$, and to the concentration of the solute. The effect of the solute within this shell upon the energy of the central solute particle is given by the energy of interaction times the probable number of interacting particles. If the energy of interaction falls off with distance faster than r^{-2}, then it is clear that the dominant effects will come from solute molecules near the central one. This is the situation for all types of nonelectrolyte solutions. But the energy of ionic interaction depends on r^{-1}; hence the net interaction of distant solute ions on a central ion may be very important. Indeed, the total interaction energy in an electrolyte converges to a finite value only because the ratio of positive to negative charge in a shell rapidly approaches unity as the radius increases. It is this long-range nature of the electrostatic interaction that makes it qualitatively different.

The existence of long-range forces between ions by no means excludes the simultaneous existence of short-range forces of the various sorts that operate between neutral molecules. If there is a strong enough short-range attraction, then ions associate just as neutral molecules do. Typical weak electrolytes, such as acetic acid, exemplify this behavior. Such solutions are treated by combining the usual equation for a chemical dissociation equilibrium with that for the special interionic effects of the dissociated ions. It was Bjerrum[6] who emphasized that electrostatic forces alone could cause association. In water, this effect is important only for multiply charged ions, but in solvents of low dielectric constant it is always important. The final example of Chapter 15 considered this association effect for aqueous $CuSO_4$.

Interionic-attraction Theory

Before presenting the theory of Debye and Hückel and the later advances, it is interesting to note which electrically related variables might affect the excess chemical potentials and the activity and osmotic coefficients, and to see how much can be concluded from dimensional reasoning alone. For simplicity, let us consider a binary electrolyte with ionic charges $\pm ze$ in a solvent of dielectric constant or relative permittivity ε. Other factors that might enter are the thermal energy kT and the ion concentration, which may be written N/V, where N is the number of pairs of ions. While the size, shape, and other properties of the ions as well as other properties of the solvent may affect the chemical potential at finite concentrations, we expect that these variables will disappear more rapidly than electrical effects as the concentration is reduced. The only dimensionless combination of the other factors is $(N/V)^{1/3}(z^2e^2/\varepsilon kT)$; consequently the activity coefficient and the osmotic coefficient should be functions of this quantity.

Milner[2] made the first serious effort to apply statistical thermodynamics to the ionic interaction problem. He noted the extreme difficulties of an exact solution and attempted an approximate but direct evaluation of the excess free energy. Milner presented his primary results in the form of a table of values of a quantity very similar to the osmotic coefficient and showed by graphs that his function fitted the freezing-point data for several aqueous strong electrolytes. In a footnote he gave a closed expression, also approximate, for the very dilute solution, which may be rearranged into†

$$1 - \phi = \frac{\pi}{3} \left[\left(\frac{N}{V}\right)^{1/3} \frac{z^2 e^2}{\varepsilon k T} \right]^{3/2} \qquad (16\text{-}1)^*$$

The important feature of this result is the functional dependence on the three-halves power of the bracketed quantity, which we selected from dimensional reasoning. This corresponds to a proportionality of both $1-\phi$ and $\ln \gamma_\pm$ to the half power of the concentration. Had this functional relationship been fully recognized and accepted in 1913, even without the precisely correct coefficient, progress in electrolyte theory would have been more rapid, because the Debye–Hückel result differs from Milner's only by the factor $(2/\pi)^{1/2}$.

In the decade between Milner's work and that of Debye and Hückel,[5] both Bronsted[3] and Lewis and Randall[4] came to adopt on an empirical basis the functional form

$$1 - \phi = \beta m^{1/2}$$

Lewis and Randall found β to be in the range 0.3 to 0.45, while Bronsted assigned β = 0.32 for 1–1 electrolytes in water at 0°C. These values are close to the Debye–Hückel result, which gives $\beta = 0.37_7$.

DEBYE–HÜCKEL THEORY

In 1923 Debye and Hückel[5] (D-H) presented a simple theory of interionic-attraction effects that nevertheless retains all essential features in the limit of infinite dilution. Their result has been of enormous value in the practical treatment of electrolyte solutions by providing a limiting law for the extrapolation to zero concentration of not only activities but also enthalpies, heat capacities, volumes, etc. Although the original presentation involved several approximations, it gives a good physical picture in simple mathematical terms. Other, more mathematically sophisticated theories have shown that the D-H result is exact in the limit of infinite dilution.

For the calculation of the electrical contribution to the chemical potential, the solvent is assumed to constitute only a continuous dielectric in which the ions interact according to Coulomb's law with dielectric constant ε. Furthermore, it is assumed that the average effect on a given ion of all other ions may be obtained from a continuous charge distribution, or "ionic atmosphere," calculated in the following manner.

If the electrical potential at a given point is ψ, the energy of an ion of charge $z_j e$ is $z_j e \psi$. We may expect the concentration of such ions c'_j (particles per cubic

---

† In SI units the factor $4\pi\varepsilon_0$ precedes ε; here ε_0 is the permittivity of free space. In electrostatic units, esu, this factor does not appear. Almost all of the literature, even recent, is presented without the $4\pi\varepsilon_0$ factor and we follow that practice in the following presentation.

centimeter) to be related to the concentration at zero potential c_j by Boltzmann's law [see Eq. 5-11 and Appendix 1]:

$$c'_j = c_j \exp\left(-\frac{z_j e \psi}{kT}\right) \tag{16-2}$$

The local charge density is obtained by summing the ionic charge concentrations over all species of ions,

$$\rho = \sum_j z_j e c_j \exp\left(-\frac{z_j e \psi}{kT}\right) \tag{16-3}$$

At this point we introduce the basic equation of Poisson, relating the electrostatic potential to the charge distribution

$$\nabla^2 \psi = -\frac{4\pi}{\varepsilon}\rho \tag{16-4}$$

∇^2 is a differential operator, which is $\partial^2/\partial x^2 + \partial^2/\partial y^2 + \partial^2/\partial z^2$ in cartesian coordinates. The average charge distribution around a single ion will have spherical symmetry. Consequently, it is convenient to use spherical polar coordinates, whereupon the angles may be omitted and the Poisson equation becomes

$$\frac{1}{r^2}\frac{d}{dr}\left(r^2 \frac{d\psi}{dr}\right) = -\frac{4\pi}{\varepsilon}\rho \tag{16-5}$$

Equation (16-3) for the charge density may now be substituted and a solution sought. The exponential form of Eq. (16-3) is inconvenient, to say the least, for the solution of the differential equation. Since the other approximations can be expected to be valid in the very dilute solutions, where the ions are usually far apart, it is reasonable to expand the exponentials in series:

$$\rho = \sum_j z_j e c_j - \sum_j \frac{c_j z_j^2 e^2 \psi}{kT} + \sum_j \frac{c_j z_j^3 e^3 \psi^2}{2k^2 T^2} + \cdots \tag{16-6}$$

Since the ions come from a neutral solute compound, the first sum must be zero. Also, if the electrolyte is a simple binary type such as NaCl, $CaSO_4$, etc., the third sum is zero. In any case, all terms are dropped except the second, whereupon Eq. (16-5) becomes

$$\frac{1}{r^2}\frac{d}{dr}\left(r^2 \frac{d\psi}{dr}\right) = \kappa^2 \psi \tag{16-7}*$$

with

$$\kappa^2 = \frac{4\pi e^2}{\varepsilon kT}\sum_j c_j z_j^2 \tag{16-8}$$

Equation (16-7)* is a familiar type of differential equation with the general solution

$$\psi = A\frac{e^{\kappa r}}{r} + B\frac{e^{-\kappa r}}{r} \tag{16-9}*$$

where A and B are constants to be determined from the physical conditions of the problem. Since the potential must remain finite at large values of r, A must be zero. It is assumed that there is some distance a of closest approach of other ions to the central ion. For $r < a$ there are no other ions present, and the potential near an ion of charge z_j is given by the familiar electrostatic formula

$$\psi_j = \frac{z_j e}{\varepsilon r} + C_j \qquad (16\text{-}10)*$$

where C_j is a constant that arises because of the presence of other ions outside the region $r < a$. At $r = a$ both the potential ψ and the electric field $-\partial \psi / \partial r$ must be continuous. These two conditions permit the evaluation of both B and C, which are found to be

$$B_j = \frac{z_j e}{\varepsilon(1 + \kappa a)} e^{\kappa a} \qquad (16\text{-}11a)*$$

$$C_j = -\frac{z_j e \kappa}{\varepsilon(1 + \kappa a)} \qquad (16\text{-}11b)*$$

One self-consistency check on this treatment is the calculation of the total charge density associated with the ionic atmosphere of a central ion of charge $z_i e$. Substitution of (16-11)* into (16-6) as simplified yields

$$\rho_i = -\frac{\kappa^2 z_i e}{4\pi r_i} \frac{e^{\kappa a} e^{-\kappa r_i}}{1 + \kappa a} \qquad (16\text{-}12)*$$

and integration over all space with $r > a$ yields

$$\int_a^\infty \rho_i 4\pi r_i^2 \, dr_i = -\frac{z_i e}{1 + \kappa a} e^{\kappa a} \int_a^\infty e^{-\kappa r_i} \kappa^2 r_i \, dr_i = -z_i e$$

Thus the charge of the central ion is exactly compensated by an equal but opposite charge distribution in the solution.

We next turn to the use of this distribution of ions to calculate activity and osmotic coefficients. Also, we shall compare this charge distribution with Monte Carlo calculations for the same model and discuss the effect of improved accuracy at finite concentration. For the calculation of activity and osmotic coefficients we will presently use a method that includes the kinetic effect of the hard spheres in addition to the electrical effects. But we first present a purely electrical method that has simple physical meaning and is of historical interest.

Activity Coefficient from a Charging Process

We calculate the excess chemical potential of ion i, μ^E_i, as the electrical free energy associated with the introduction of one ion into the solution multiplied by Avogadro's number. One assumes that the solution contains the full concentration N_j of each type of ion, which establishes the value of κ. The one additional particle is first introduced in a hypothetical uncharged state and in a second step is imagined to be gradually charged to $z_i e$. The first step requires negligible electrical energy. The energy of adding each increment of charge dq is just that increment times the environmental potential arising from all the other ions. This potential is just the quantity C given in Eq. (16-11b)*; hence

$$\frac{\mu_i^E}{N_A} = \int_0^{z_i e} -\frac{\kappa q \, dq}{\varepsilon(1 + \kappa a)}$$

$$= -\frac{z_i^2 e^2 \kappa}{2\varepsilon(1 + \kappa a)} \tag{16-13}*$$

$$\ln \gamma_i = -\frac{z_i^2 e^2 \kappa}{2\varepsilon kT(1 + \kappa a)} \tag{16-14}*$$

This charging process, which was first proposed by Güntleberg,[7] is simpler than that used by Debye and Hückel, wherein all ions in the solution are charged simultaneously and therefore κ varies during the charging process. Both methods yield the same result.

Limiting Law

Our primary interest in the Debye–Hückel theory is to obtain a limiting law valid in the region of very low concentration for the activity coefficient and other properties. In this region κ becomes very small, and the product κa becomes negligible as compared with unity. Consequently the term $1 + \kappa a$ in the denominator is omitted for limiting-law expressions, although it will be introduced again for some expressions used at higher concentrations. The reason for the introduction of the distance of closest approach a in this initial derivation is to avoid the use of Eq. (16-6) in the region close to an ion where the approximation of expanding the exponential in Eq. (16-3) is clearly incorrect. Since such a finite distance of closest approach is physically reasonable, there is no objection to its introduction, but, as we have seen, the exact value of a drops out of the limiting expression for low concentration.

The mean activity coefficient, which is the thermodynamically defined and measured quantity, was given in Eq. (15-12) for an electrolyte dissociating into ν_+ ions of charge $z_+ e$ and ν_- ions of charge $z_- e$:

$$(\nu_+ + \nu_-) \ln \gamma_\pm = \nu_+ \ln \gamma_+ + \nu_- \ln \gamma_- \tag{15-12}$$

and also

$$\nu_- z_- = -\nu_+ z_+$$

Substitution of Eq. (16-14) with the omission of the $(1 + \kappa a)$ factor yields, after simplification,

$$\ln \gamma_\pm = -\frac{e^2 \kappa |z_+ z_-|}{2\varepsilon kT} \tag{16-15}*$$

Before substituting κ from Eq. (16-8), let us note the definition of ionic strength as proposed originally by Lewis and Randall:[4]

$$I = \tfrac{1}{2}\sum_i m_i z_i^2 \tag{15-23}$$

which is evidently just the molality of a 1–1 electrolyte. The next step is the conversion from the molality to concentration in ions per cubic centimeter. This shift to a volumetric basis should be emphasized because it leads to additional terms in the derived functions for enthalpy, heat capacity, etc.

$$\sum_j c_j z_j^2 = \frac{2N_A d_1}{1000} I = \frac{2N_A M_1}{1000 V_1} I \tag{16-16}$$

Here d_1 is the density of the solvent, M_1 its molar mass, and V_1 its molar volume.

Now we may convert κ to conventional macroscopic units,

$$\kappa^2 = \frac{8\pi e^2 N_A d_1}{1000 \varepsilon k T} I \tag{16-17}$$

and the final equation for the activity coefficient is

$$\log \gamma_\pm = -A_\gamma |z_+ z_-| I^{1/2} \tag{16-18}*$$

$$A_\gamma = \left(\frac{2\pi N_A d_1}{1000}\right)^{1/2} \left(\frac{e^2}{\varepsilon k T}\right)^{3/2} \tag{16-19}$$

In the brief presentation of the limiting law in Chapter 15, the equations for $\ln \gamma_\pm$, $1 - \phi$, and the excess Gibbs energy G^{Em} were written in terms of the parameter for the osmotic coefficient A_ϕ. On that basis

$$A_\gamma = 3A_\phi \tag{16-20}$$

and we recall that

$$\frac{G^{Em}}{w_w RT} = -4A_\phi I^{3/2} \tag{15-24}*$$

and for a single solute

$$1 - \phi = |z_+ z_-| A_\phi I^{1/2} \tag{15-28a}*$$

The excess enthalpy is also called the relative enthalpy $L = H^E = H - H°$. It is obtained from the temperature derivative of the excess Gibbs energy,

$$\begin{aligned}
\frac{L}{w_w R} &= -T^2 \left[\frac{\partial (G^E/w_w RT)}{\partial T}\right]_P \\
&= 4T^2 \left(\frac{\partial A_\phi}{\partial T}\right)_P I^{3/2} \\
&= \frac{A_L}{R} I^{3/2}
\end{aligned} \tag{16-21}*$$

with

$$\frac{A_L}{RT} = -6A_\phi \left[1 + T\left(\frac{\partial \ln \varepsilon}{\partial T}\right)_P + \frac{T\alpha_w}{3} \right] \quad (16\text{-}22)$$

where $\alpha_w = (\partial \ln V/\partial T)_P$ is the coefficient of thermal expansion of water.

Experimental measurements of enthalpy for a single solute usually yield the apparent molar enthalpy, ϕL, defined as

$$\phi L = \frac{L}{n_2} \quad (16\text{-}23)$$

where n_2 is the number of moles of solute. Then the limiting law yields

$$\phi L = \frac{\nu|z_+ z_-|}{2} A_L I^{1/2} \quad (16\text{-}24)*$$

Similar derivations for the heat capacity and the volume yield, with $J = C_P - C_P^\circ$,

$$\frac{C_P}{w_w R} = \frac{A_J}{R} I^{3/2} \quad (16\text{-}25)*$$

$$\phi_J = \frac{C_P - C_P^\circ}{n_2} = \frac{\nu|z_+ z_-|}{2} A_J I^{1/2} \quad (16\text{-}26)*$$

$$A_J = \left(\frac{\partial A_L}{\partial T}\right)_P \quad (16\text{-}27)$$

$$V - V^\circ = \left(\frac{\partial G^{Em}}{\partial P}\right)_T \quad (16\text{-}28)$$

$$\phi_V = \frac{V - V^\circ}{n_2} = \frac{\nu|z_+ z_-|}{2} A_V I^{1/2} \quad (16\text{-}29)*$$

$$\frac{A_v}{RT} = 2A_\phi \left[3\left(\frac{\partial \ln \varepsilon}{\partial P}\right)_T + \left(\frac{\partial \ln V_w}{\partial P}\right)_T \right] \quad (16\text{-}30)$$

THEORY FOR PROPERTIES AT FINITE MOLALITY

The theory for the D-H limiting law is extremely valuable since it gives more accurate values of the coefficient of the $I^{1/2}$ term than can be obtained from measurements of osmotic coefficients, apparent molar enthalpies, etc. We turn now to theory for thermodynamic properties at higher concentrations. There are many equations from statistical mechanics that assume potential models for ion–ion and ion–solvent potentials and yield, after some approximations, thermodynamic properties. It is beyond the scope of this chapter to derive or even to describe comprehensively these results. Even if parameters in the potential models are evaluated empirically, few of these theories yield agreement with thermodynamic properties within experimental uncertainty. The more fruitful approach has been the use of these theories to suggest a form of equation for the excess Gibbs energy and then to evaluate the coefficients in this equation (other than the limiting law) from the experimental thermodynamic properties.

Since the final justification of these equations is their empirical success in fitting data, we shall not undertake the complete statistical-mechanical derivations, which are complex. Rather, we describe the primary features, with references to literature where rigorous derivations can be found.

Radial Distribution Function

We now consider the function $g_{ij}(r)$, which gives the relative probability of finding one ion of charge z_i at a distance r from an ion of charge z_j as a ratio to the probability of a random distribution. At large r, the distribution will become random and $g_{ij}(r)$ becomes unity. We wish to compare the D-H result with that calculated by more exact methods and then make an optimum selection of equations for use in representing experimental data. For this comparison, we adopt the "restricted primitive model" (RPM) with hard spheres of diameter a and charge $\pm ze$.

The concentration of ionic species i near an ion of species j is given by the Boltzmann expression, Eq. (16-2) above,

$$c'_i = c_i \exp(-z_i e \psi_j / kT)$$

where c_i is the average concentration and $\psi_j(r)$ is an electrostatic potential associated with the ion j. This implies a radial distribution function

$$g_{ij(r)} = \exp(-z_i e \psi_j(r)/kT) \qquad (16\text{-}31)$$

After substitution of Eqs. (16-10)* and (16-11)*, this becomes

$$g_{ij} = \exp(-q_{ij}) \qquad (16\text{-}32)^*$$

with

$$q_{ij} = \frac{z_i z_j e^2 \exp(\kappa a)}{\varepsilon kT (1 + \kappa a)} \frac{\exp(-\kappa r)}{r} \qquad (16\text{-}33)^*$$

Clearly the required symmetry with exchange of i and j is present. However, for the Poisson equation the exponentials in ρ were expanded and only the second term was used. Thus it can be argued that it is inconsistent to retain the exponential form for g_{ij}. Since this entire treatment is approximate, this objection has limited significance. Alternatively, one may note that the third term in the expanded form for the charge density is zero for a symmetrical electrolyte. Thus, without any inconsistency for symmetrical electrolytes, one may retain the third term in the expansion and the corresponding expression for the radial distribution function is[8]

$$g_{ij}(r) = 1 - q_{ij} + \frac{q_{ij}^2}{2} \qquad (16\text{-}34)*$$

The original Debye–Hückel approximation is given by the first two terms:

$$g_{ij}(r) = 1 - q_{ij} \qquad (16\text{-}35)*$$

For comparison we have the results of Monte Carlo simulations for the same hard-sphere ionic model. These results are exact insofar as the interionic forces are concerned but are subject to approximations concerning the limited number of particles considered, together with statistical uncertainties inherent in the Monte Carlo method. These uncertainties are considered in detail by Sørensen,[9] who reports numerous examples.

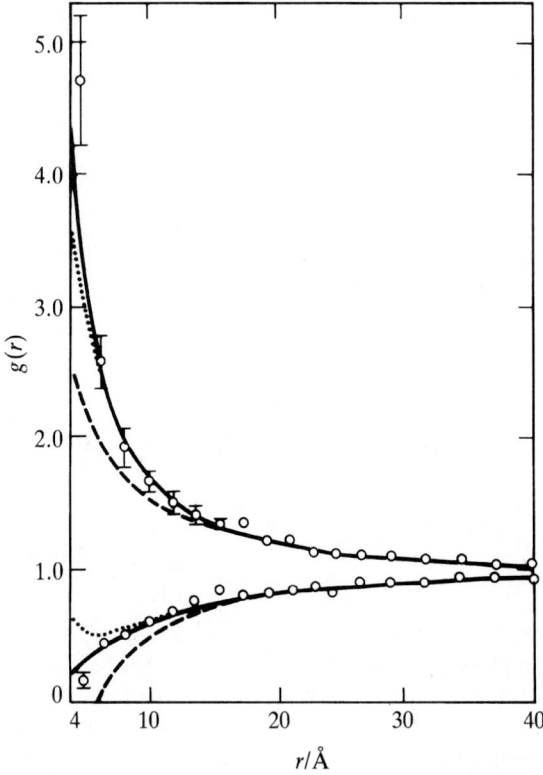

FIGURE 16-1

The radial distribution functions (g_{+-} above, $g_{++} = g_{--}$ below) for 0.00911 mol l^{-1} aqueous solution (1–1 type, $a = 4.25$ Å). The points are Monte Carlo calculations, the solid curve is the exponential D-H expression, the dotted and dashed curves are, respectively, the three-term and two-term D-H expressions.

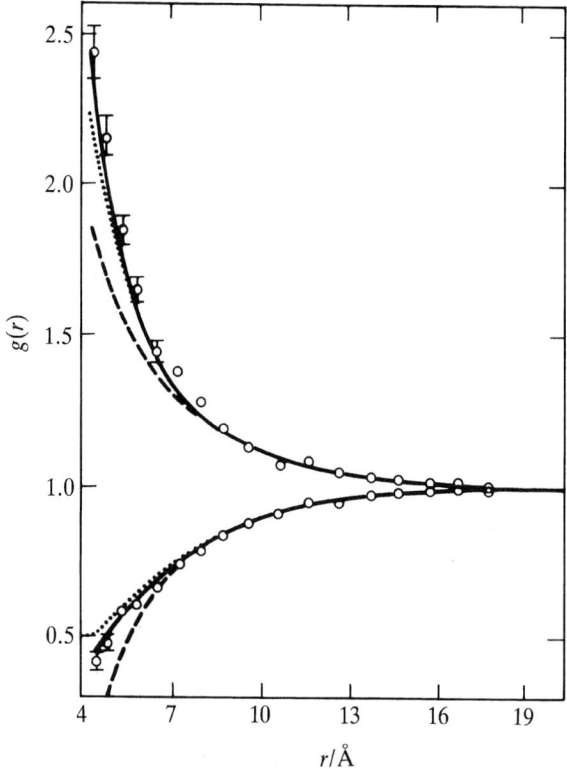

FIGURE 16.2
The radial distribution functions for 0.425 mol l^{-1} solution; other aspects the same as in Fig. 16-1.

In Figs. 16-1 and 16-2 we compare the D-H curves (dashed) with the Monte Carlo results (circles) from Card and Valleau[10] for the case $a = 4.25$ Å, the solvent properties of water at 25°C, and singly charged ions at molalities of 0.00911 and 0.425 mol kg^{-1}. At each molality the two-term D-H curves for g_{+-} and for g_{++} lie well below the Monte Carlo results at the smaller values of r. Also, for the dilute solution, the g_{++} curve goes below zero, which is impossible for the true curve.

The dotted curves for the three-term expression[8] for g_{ij}, Eq. (16-34)*, agree very well with the Monte Carlo results except just outside of the hard core diameter at 4.25 Å. Thus, thermodynamic properties based on this three-term expression should constitute a great improvement from the D-H results and provide guidance in the selection of a semi-empirical equation to represent experimental data.

The solid curves in Figs. 16-1 and 16-2 are for the exponential expression, Eq. (16-32)*. The agreement of these curves with the Monte Carlo data is excellent, but there are inconsistencies in using the exponential expression for $g_{ij}(r)$ after having dropped the higher terms in the earlier expression, Eq. (16-6), when solving the Poisson equation. By complex methods, the exponential expression can be used at each step and, as Sørensen[9] shows, the results are excellent. But this treatment does not yield simple equations that are useful for empirical purposes.

Thermodynamic Properties from the Radial Distribution Function.

Statistical mechanics provides several methods for obtaining the Gibbs energy or an equivalent quantity from the interparticle potential and the radial distribution function. For our purpose, the "pressure equation" is convenient in that it yields the osmotic coefficient. For an imperfect gas, the equation yields the pressure, while in the McMillan–Mayer solution theory it yields the osmotic pressure. Several books on statistical mechanics give derivations of the pressure equation, together with applications to nonionic fluids.[11–15]

Rasaiah and Friedman[16] were the first to take full advantage of this theory for application to electrolytes. The pressure equation now yields the osmotic coefficient for the solution,

$$\phi - 1 = (\Pi/ckT) - 1$$
$$= -(6ckT)^{-1} \sum_i \sum_j c_i c_j \int_0^\infty r \frac{\partial u_{ij}}{\partial r} g_{ij}(r)(4\pi r^2)\, dr \qquad (16\text{-}36)$$

Here Π is the osmotic pressure, u_{ij} is the potential of mean force in the solvent, c_i is the concentration of the ith ion and c is the total concentration of solute species, $c = \Sigma_i c_i$. Once ϕ is known as a function of concentration, the activity coefficient can be calculated from the Gibbs–Duhem equation.

For the RPM the potential of mean force is just the Coulomb law in the dielectric solvent together with the hard core

$$u_{ij}(r) = \infty \qquad r < a$$
$$= \frac{z_i z_j e^2}{\varepsilon r} \qquad r \geq a$$

In applying this potential and any of the accompanying radial distribution functions, one faces an anomaly in the integral at $r = a$ where $\partial u/\partial r$ is infinite. This has been resolved mathematically and yields[17] for the osmotic coefficient

$$\phi - 1 = (\Pi/ckT) - 1$$
$$= -(6ckT)^{-1} \sum_i \sum_j c_i c_j \int_a^\infty r \frac{\partial u_{ij}}{\partial r} g_{ij}(4\pi r^2)\, dr$$
$$+ \frac{2\pi a^3}{3c} \sum_i \sum_j c_i c_j g_{ij}(a) \qquad (16\text{-}37)$$

where $g_{ij}(a)$ is the value of $g_{ij}(r)$ for r infinitesimally greater than a.

When either the two-term (the D-H) or the three-term expression, Eq. (16-35)* or Eq. (16-34)*, is used for g_{ij}, the value of the integral in Eq. (16-37) is

$$\phi - 1 = -\frac{\kappa^3}{24\pi c(1 + \kappa a)} \qquad (16\text{-}38)*$$

It is interesting to compare this result, when simplified for the RPM, with that obtained from the charging process. It is convenient to use the "length" that was first defined by Bjerrum[6]

$$b = \frac{z^2 e^2}{2\varepsilon kT} \tag{16-39}$$

whereupon the D-H expression for κ^2 becomes

$$\kappa^2 = 8\pi bc \tag{16-40}$$

Then Eq. (16-38)* simplifies to

$$\phi - 1 = -\frac{b\kappa}{3}\frac{1}{1 + \kappa a} \tag{16-41}*$$

and a Gibbs–Duhem transformation, Eq. (15-37), yields for the activity coefficient

$$\ln \gamma_\pm = -b\kappa \left[\frac{1}{3(1 + \kappa a)} + \frac{2}{3\kappa a}\ln(1 + \kappa a) \right] \tag{16-42a}*$$

$$= -b\kappa(1 - \tfrac{2}{3}\kappa a + \tfrac{5}{9}\kappa^2 a^2 \cdots) \tag{16-42b}*$$

The equivalent expression from the charging process, Eq. (16-14), is

$$\ln \gamma_\pm = -b\kappa \left(\frac{1}{1 + \kappa a} \right) \tag{16-43a}*$$

$$= -b\kappa(1 - \kappa a + \kappa^2 a^2 \cdots) \tag{16-43b}*$$

It is clear that, in the limit $\kappa \to 0$, the two expressions become identical for the limiting law. This is consistent with the widely accepted conclusion that the limiting law is exact. For the behavior at finite concentration, finite κa, the two expressions have similar character but differ quantitatively. However, for our primary purpose in selecting an appropriate equation for empirical use, the difference between the two functions, Eqs. (16–42)* and (16–43)*, at finite κa is unimportant.

The important advantage in the use of the pressure equation lies in the second term, which gives the contribution from the repulsive interactions of the hard cores. This term would remain if the charges were removed from the solute particles. The complete result for the three-term expression for $g_{ij}(r)$ is[8]

$$\phi - 1 = \frac{-\kappa^3}{24\pi c(1 + \kappa a)} + c\left[\frac{2\pi a^3}{3} + \frac{\kappa^4 a}{48\pi c^2(1 + \kappa a)^2} \right] \tag{16-44}*$$

Here the first term in brackets, $2\pi a^3/3$, comes from the last term in Eq. (16-37)*, while the final term comes from the third term in Eq. (16-34)*.

Equation (16-44)* is valid for multicomponent or unsymmetrical electrolytes with ions of different charges provided $g_{ij}(r)$ is given by Eq. (16-34)*. For further discussion, however, we consider the form simplified for the RPM.

$$\phi - 1 = \frac{-b\kappa}{3(1 + \kappa a)} + c\left[\frac{2\pi a^3}{3} + \frac{4\pi b^2 a}{3(1 + \kappa a)^2}\right] \qquad (16\text{-}45)^*$$

The restricted primitive model is a rough approximation, and there are several more approximations in the derivation of Eq. (16-45)*; nevertheless, that equation fits experimental data for simple aqueous electrolytes quite well. This is apparent from Fig. 16-3, which shows the osmotic coefficient from experiment for HBr(aq) and values for $a = 4.25$ Å from Monte Carlo calculations, from Eq. (16-45)* and from the original D-H equation derived via the charging process.† The excellent agreement shown in Fig. 16-3 indicates that Eq. (16-45)* contains all of the features that are important for the concentration range up to one molal. When it is applied empirically with adjustment of a to an array of accurate data for several aqueous electrolytes, however, one finds discrepancies that are clearly outside of experimental uncertainty. Thus, it is best to regard Eq. (16-45)* as a guide to the selection of a semi-empirical equation that does represent experimental data within experimental uncertainty.

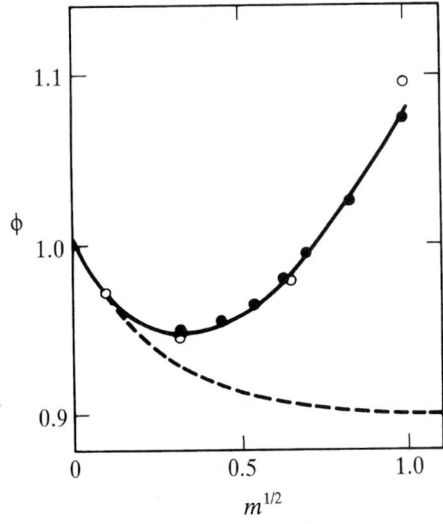

FIGURE 16.3

The osmotic coefficient for a 1–1 type aqueous electrolyte at 25°C. Solid circles are experimental values for HBr. Open circles are calculated by the Monte Carlo method ($a = 4.25$ Å). The solid curve is based on Eq. (16-45)*, while the dashed curve is the traditional D-H expression Eq. (16-46a)*.

---

† The charging process yields $\ln \gamma_\pm$, Eq. (16-43a)*, which is transformed by the Gibbs–Duhem equation to

$$\phi - 1 = -(b\kappa/3)\,\sigma(\kappa a) \qquad (16\text{-}46a)^*$$

$$\sigma(x) = (3/x^3)[1 + x - (1 + x)^{-1} - 2\ln(1 + x)] \qquad (16\text{-}46b)$$

While one could proceed from Eq. (16-45)* and the associated concepts to a semi-empirical equation for a pure electrolyte, it is for mixed electrolytes that these theoretical concepts are particularly important. In the next chapter, we consider semi-empirical equations for both pure and mixed electrolytes.

REFERENCES

1. G. N. Lewis, *Z. physik chem.*, **70**, 212 (1909).
2. S. R. Milner, *Phil. Mag.*, (6) **23**, 551 (1912); (6) **25**, 742 (1913).
3. J. N. Bronsted, *J. Am. Chem. Soc.*, **44**, 983 (1922).
4. G. N. Lewis and M. Randall, *J. Am. Chem. Soc.*, **43**, 1112 (1921); see also *Thermodynamics and the Free Energy of Chemical Substances*, 1st edn., McGraw-Hill, New York, 1923, pp. 342–346.
5. P. Debye and E Hückel, *Physik. Z.*, **24**, 185, 334 (1923); **25**, 97 (1924).
6. N. Bjerrum, *Kgl. Danske Videnskab. Selskab, Mat.-fys. Medd.*, **7** (9) (1926).
7. E. Güntleberg, page 155 in N. Bjerrum, *Z. physik. Chem.*, **119**, 145 (1926).
8. K. S. Pitzer, *J. Phys. Chem.*, **77**, 268 (1973).
9. T. S. Sørensen, *J. Chem. Soc., Faraday Trans.*, **86**, 1815 (1990); **87**, 479 (1991); *Mol. Simulation*, **11**, 1, 267 (1993).
10. D. N. Card and J. P. Valleau, *J. Chem. Phys.*, **52**, 6232 (1970).
11. D. Henderson and S. G. Davison, in *Physical Chemistry, an Advanced Treatise*, vol. II, H. Eyring, Ed., Academic Press, New York, 1967, p. 359.
12. N. Davidson, *Statistical Mechanics*, McGraw-Hill, New York, 1962, p. 473.
13. T. L. Hill, *Statistical Mechanics*, McGraw-Hill, New York, 1956, p. 190.
14. T. L. Hill, *An Introduction to Statistical Thermodynamics*, Addison-Wesley, Reading, Mass, 1960, p. 304.
15. D. A. McQuarrie, *Statistical Mechanics*, Harper and Row, New York, 1976, p. 261.
16. J. C. Rasaiah and H. L. Friedman, *J. Chem. Phys.*, **48**, 2742 (1968); **50**, 3965 (1969).
17. J. A. Barker and D. Henderson, *Rev. Mod. Phys.*, **48**, 591 (1976).

PROBLEMS

16-1. Verify the derivation of Eq. (16-15)* from (16-14)* for the limit of low concentration with the general electrolyte $A_{\nu_+} B_{\nu_-}$.

16-2. Evaluate Eq.(16-33)* for q_{ij} as a function of interionic distance r for an aqueous solution at 25°C. At what distance r is the term $q_{ij}^2/2$ reduced to 0.1, i.e., a 10 percent correction to $g_{ij}(r)$? Relate this distance to the size of a water molecule and to the probable distance between ions as a function of molar concentration.

CHAPTER 17

SEMI-EMPIRICAL EQUATIONS FOR PURE AND MIXED ELECTROLYTES

In this chapter we consider equations to represent the results of measurements on both pure and mixed electrolytes and methods for the prediction of properties of other mixtures. The fundamental relationships for activity, activity coefficient, excess Gibbs energy, osmotic coefficient, and ionic strength, as given in Chapters 14 and 15, are valid for multisolute systems, as is the derivation in Chapter 16 of the Debye–Hückel limiting law. We seek an equation that also accommodates reasonably accurately the concepts of the last section of Chapter 16, yet is not too complex for practical application to mixed electrolytes with many components, such as seawater. Most, if not all, of the parameters should be related to single salts so that they can be evaluated from data for pure electrolytes. But mixing parameters should be included, if really needed, to represent mixed electrolyte properties. We first note the system that Guggenheim[1] proposed in 1935 with these objectives in mind.

GUGGENHEIM EQUATIONS

The assumption underlying these equations is a principle first advanced by Bronsted[2] that states that short-range interactions need be considered only between unlike charged ions. The reasoning is that the repulsion of like charged ions keeps them so well separated that only their electrical interaction is significant. Guggenheim[1] adopted the D-H electrical term in the original form, Eqs. (16-43a) and (16-46a), as generalized for mixed electrolytes with a standard value of a so that κa becomes $I^{1/2}$. He added a simple short-range term proportional to molality with a parameter specific to the two ions interacting. Then, in the general form for a mixed electrolyte, one has

$$\ln \gamma_{MX} = -\frac{A_\gamma |z_M z_X| I^{1/2}}{1 + I^{1/2}} + \frac{2\nu_M}{\nu_M + \nu_X} \sum_a m_a \beta_{Ma} + \frac{2\nu_X}{\nu_M + \nu_X} \sum_c m_c \beta_{cX} \quad (17\text{-}1)^*$$

$$\phi - 1 = \left(\sum_c m_c + \sum_a m_a\right)^{-1} [-2A_\phi I^{3/2} \sigma(I^{1/2})] + \sum_c \sum_a m_c m_a \beta_{ca} \quad (17\text{-}2)^*$$

where the sums cover all cations c and all anions a and the function $\sigma(x)$ and the D-H parameters A_γ and A_ϕ have been given above, Eqs. (16-46b), (16-19), (15-35). The parameters β_{ca} are constants at a given T.

For a single 1–1 electrolyte, Eqs. (17-1)* and (17-2)* reduced to

$$\ln \gamma = -\left[\frac{A_\gamma m^{1/2}}{1 + m^{1/2}}\right] + 2\beta_{MX} m \quad (17\text{-}3)^*$$

$$\phi - 1 = -A_\phi m^{1/2} \sigma(m^{1/2}) + \beta_{MX} m \quad (17\text{-}4)^*$$

Guggenheim and Turgeon[3] showed that Eqs. (17-3)* and (17-4)* fitted, essentially within experimental error, the data for 1–1 electrolytes in water at 0°C and at room temperature at concentrations up to 0.1 mol kg^{-1}. They acknowledge, however, and others have shown that substantial discrepancies arise at higher concentrations. In seeking equations that can be used at higher concentration, one may examine separately the first and second terms on the right side. The first terms in Eq. (17-1)* and Eq. (17-2)* must remain general functions of ionic strength if these equations are to maintain their simplicity and utility for mixed electrolyte solutions. We shall return later to the question of an improved mathematical form of this first term. Next we note that, by taking differences between the properties of different electrolytes of the same type and at the same concentration, the first term cancels and we have (for 1–1 electrolytes)

$$\ln \gamma_{M''X''} - \ln \gamma_{M'X'} = 2m(\beta_{M''X''} - \beta_{M'X'})$$

$$\phi_{M''X''} - \phi_{M'X'} = m(\beta_{M''X''} - \beta_{M'X'})$$

If the β's are constants, the differences in $\ln \gamma$ and in ϕ must be proportional to molality. That approximation is tested in Fig. 17-1. Clearly, the assumption of constant β's is not correct; rather one needs for the osmotic coefficient an expression that decreases rapidly and linearly in $m^{1/2}$ at low molality and then becomes approximately constant at higher molality. It is interesting to note that the two terms in brackets on the right of Eq. (16-45)* have exactly that character. The term $2\pi a^3/3$ is constant, while the other term decreases with κ, i.e., with $m^{1/2}$.

With these ideas in mind, an improved set of equations for electrolytes was proposed in 1973 by Pitzer.[4] These equations, which are now widely used and are commonly called ion-interaction or Pitzer equations, will be described next.

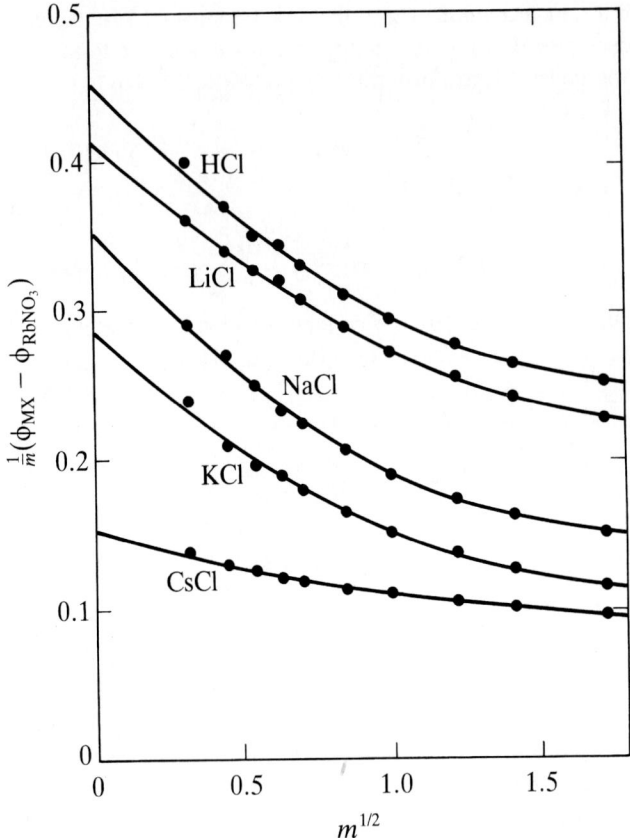

FIGURE 17-1
Difference in osmotic coefficient of several electrolytes from that of RbNO$_3$.

ION INTERACTION (PITZER) EQUATIONS

First a general structure was selected for the equation; then the exact forms for certain functions were chosen by comparison with experimental data. The total excess Gibbs energy for a solution containing w_w kg of solvent and $n_i, n_j, \ldots$ moles of solute species $i, j, \ldots$ was taken to be

$$\frac{G^{Em}}{RT} = w_w f(I) + \frac{1}{w_w}\sum_{ij}\lambda_{ij}(I)n_i n_j + \frac{1}{w_w^2}\sum_{i,j,k}\mu_{ijk}n_i n_j n_k \qquad (17\text{-}5)*$$

Here $f(I)$ is a function of ionic strength (also temperature and solvent properties) expressing the effect of the long-range electrostatic forces; $f(I)$ may have a form corresponding either to the D-H term in Eq. (16-44)* from the pressure equation or to Eq. (16-14)* from the charging process; $\lambda_{ij}(I)$ is a function of ionic strength with the qualitative behavior indicated by Fig. 17-1 or the term in brackets in Eq. (16-45)*. The effect of short-range forces between species i and j is, of course, the primary basis for $\lambda_{ij}(I)$. Also included is a term μ_{ijk} for triple particle interaction, which may become significant at high concentration. While μ_{ijk} may, in principle, be a function of ionic strength, there is no indication from simple theory or simulations for μ_{ijk}, while there was clear guidance of that type for λ_{ij} from Eqs. (16-44)* and (16-45)* and from Figs. 16-1 and 16-2. Thus, μ_{ijk} is usually taken as a constant at a given temperature. Also, μ_{ijk} is ordinarily neglected when all ions have the same sign because the mutual repulsion will make their short-range interaction negligible. The λ_{ij} and μ_{ijk} matrices are symmetric with $\lambda_{ij} = \lambda_{ji}$, etc.

The λ_{ij} and μ_{ijk} terms are also appropriate for ion–neutral and neutral–neutral interactions. For the ion–neutral case, an ionic strength dependence is allowed in theory at least when the neutral species has a significant dipole moment, but these parameters are ordinarily taken as constants at a given temperature.

For electrolytes with cations $c, c', \ldots$ and anions $a, a', \ldots$, only certain combinations of λ's and μ's are measurable. The following definitions are appropriate:

$$B_{ca} = \lambda_{ca} + \frac{\nu_c}{2\nu_a}\lambda_{cc} + \frac{\nu_a}{2\nu_c}\lambda_{aa} \tag{17-6a}$$

$$\Phi_{cc'} = \lambda_{cc'} - \frac{z_{c'}}{2z_c}\lambda_{cc} - \frac{z_c}{2z_{c'}}\lambda_{c'c'} \tag{17-6b}$$

$$C_{ca} = \frac{3}{2}\left(\frac{\mu_{cca}}{z_c} + \frac{\mu_{caa}}{z_a}\right) \tag{17-6c}$$

$$\Psi_{cc'a} = 6\mu_{cc'a} - \frac{3z_{c'}}{z_c}\mu_{cca} - \frac{3z_c}{z_{c'}}\mu_{c'c'a} \tag{17-6d}$$

with corresponding expressions for $\Phi_{aa'}$ and $\Psi_{aa'c}$. Since the λ_{ij} are functions of I, the $B_{ca'}$, $\Phi_{cc'}$, and $\Phi_{aa'}$ are also functions of I. We omit the explicit notation $B(I)$, etc., for brevity, however, retaining it only for $f(I)$.

In terms of these quantities the excess Gibbs energy becomes

$$\frac{G^{Em}}{RT} = w_w f(I) + \frac{2}{w_w}[\sum_c\sum_a n_c n_a B_{ca} + \sum_{c>c'}\sum n_c n_{c'}\Phi_{cc'} + \sum_{a>a'}\sum n_a n_{a'}\Phi_{aa'}]$$

$$+ \frac{1}{w_w^2}[2\sum_c\sum_a n_c n_a(\sum_c n_c z_c)C_{ca} + \sum_{c>c'}\sum_a n_c n_{c'} n_a \Psi_{cc'a}$$

$$+ \sum_{a>a'}\sum_c n_a n_{a'} n_c \Psi_{aa'c}] \tag{17-7)*}$$

Next the composition is converted to molalities ($m_i = n_i/W_w$)

$$\frac{G^{Em}}{w_w RT} = f(I) + 2 \sum_c \sum_a m_c m_a [B_{ca} + (\sum_c m_c z_c) C_{ca}]$$

$$+ \sum_{c>c'} \sum m_c m_{c'} [2\Phi_{cc'} + \sum_a m_a \Psi_{cc'a}]$$

$$+ \sum_a \sum_{a'} m_a m_{a'} [2\Phi_{aa'} + \sum_c m_c \Psi_{aa'c}] \qquad (17\text{-}8)*$$

If neutral solute species are present, the λ and μ terms in Eq. (17-5)* for these species are added to Eqs. (17-7)* and (17-8)*.

Equations for the activity and osmotic coefficients are obtained by taking the appropriate derivative of Eq. (17-7)* and then converting to a molality basis. Before taking these steps for the general case, however, it is convenient to consider the case of a pure electrolyte.

Equations for a Single Solute

For a single ionic solute, $M_{\nu_M}^{z_{M^+}} X_{\nu_X}^{z_{X^-}}$ Eq. (17-7) simplifies to

$$\frac{G^{Em}}{RT} = w_w f(I) + \frac{2}{w_w} n_M n_X \left(B_{MX} + \frac{1}{w_w} n_M z_M C_{MX} \right) \qquad (17\text{-}9)*$$

Then the appropriate differentiation followed by conversion to molalities yields

$$\phi - 1 = -(\sum_i m_i)^{-1} \left(\frac{\partial G^{Em}/RT}{\partial w_w} \right)_{T,P,n_i}$$

$$= |z_+ z_-| f^\phi + m \left(\frac{2\nu_M \nu_X}{\nu} \right) B_{MX}^\phi + m^2 \left[\frac{2(\nu_M \nu_X)^{3/2}}{\nu} \right] C_{MX}^\phi$$

$$(17\text{-}10)*$$

with

$$f^\phi = \frac{1}{2}\left(f' - \frac{f}{I} \right) \qquad (17\text{-}11)$$

$$B_{MX}^\phi = B_{MX} + I B_{MX}' \qquad (17\text{-}12)$$

$$C_{MX}^\phi = 2|z_M z_X|^{1/2} C_{MX} \qquad (17\text{-}13)$$

and

$$\nu = \nu_M + \nu_X \qquad f' = \frac{\partial f}{\partial I} \qquad B' = \frac{\partial B}{\partial I}$$

The function f^ϕ must contain the Debye–Hückel limiting law but an extended form, such as that adopted by Guggenheim. Eq. (17-2)*, or one derived from the first term of Eq. (16-45), can be selected. Similarly, the ionic-strength dependence of B^ϕ should have the general character of the second term of Eq. (16-45)* or of the curves of Fig. 17-1, but alternate mathematical forms can be considered. As explained in reference 4, various forms for f^ϕ and B^ϕ were tested for an array of accurate osmotic coefficient data for 13 salts, eight of the 1–1 type and five of the 2–1 type. A clear preference was found for the expressions

$$f^\phi = \frac{A_\phi I^{1/2}}{1 + bI^{1/2}} \tag{17-14}$$

$$B^\phi_{MX} = B^{(0)}_{MX} + B^{(1)}_{MX} \exp(-\alpha I^{1/2}) \tag{17-14a}$$

with a universal value of $1.2\,\text{kg}^{1/2}\text{mol}^{-1/2}$ for b. A_ϕ is the Debye–Hückel parameter given by Eq. (15-35), while $\beta^{(0)}_{MX}$ and $\beta^{(1)}_{MX}$ are solute-specific parameters. Although a solute-specific value could also be used for α, the uniform value $2.0\,\text{kg}^{1/2}\,\text{mol}^{-1/2}$ was very satisfactory for all of the test solutions and for a very extensive array of other electrolytes.[5] This value is assumed unless an explicit statement is made about a different value. C^ϕ_{MX}, is, of course, a numerical constant specific to the solute MX.

When these equations were applied to 2–2 electrolytes, MgSO$_4$, etc., there were special problems related to the weak but significant ion-association in very dilute solutions (see Chapter 15, Fig. 15-5 and associated text). It was found[6] that the addition of a third term to B^ϕ of the same form as the second term sufficed to represent this behavior. Thus the final expression for B^ϕ is

$$B^\phi_{MX} = B^{(0)}_{MX} + B^{(1)}_{MX} \exp(-\alpha_1 I^{1/2}) + B^{(2)}_{MX} \exp(-\alpha_2 I^{1/2}) \tag{17-15}$$

For 2–2 electrolytes, α_1 is shifted to 1.4 and $\alpha_2 = 12\,\text{kg}^{1/2}\,\text{mol}^{-1/2}$; also $\beta^{(2)}_{MX}$ is large and negative. It was further shown[6] that $\beta^{(2)}$ is related to the association constant with $\beta^{(2)} = -K_{assoc}/2$.

From the relationship of the osmotic coefficient to the excess Gibbs energy, one obtains the following expressions:

$$f = -\frac{4IA_\phi}{b} \ln(1 + bI^{1/2}) \tag{17-16}$$

$$B_{MX} = \beta^{(0)}_{MX} + \beta^{(1)}_{MX} g(\alpha_1 I^{1/2}) + \beta^{(2)}_{MX} g(\alpha_1 I^{1/2}) \tag{17-17}$$

$$g(x) = \frac{2[1 - (1 + x)\exp(-x)]}{x^2} \tag{17-18}$$

Then the mean activity coefficient for the pure electrolyte is obtained by differentiation of Eq. (17-9)* followed by an appropriate combination of γ_M with γ_X to yield $\gamma_\pm$:

$$\ln \gamma_i = \left(\frac{\partial (G^{Em}/RT)}{\partial n_i}\right)_{T,P,w_w,n_{j \ne i}} \tag{17-19}$$

$$\ln \gamma_\pm = |z_M z_X| f^\gamma + m\left(\frac{2\nu_M \nu_X}{\nu}\right) B^\gamma_{MX}$$

$$+ m^2 \left[\frac{2(\nu_m \nu_X)^{3/2}}{\nu}\right] C^\gamma_{MX} \tag{17-20}*$$

with

$$f^\gamma = -A_\phi\left(\frac{I^{1/2}}{1 + bI^{1/2}} + \frac{2}{b}\ln(1 + bI^{1/2})\right) \quad (17\text{-}21)$$

$$B^\gamma_{MX} = B_{MX} + B^\phi_{MX} \quad (17\text{-}22a)$$

$$C^\gamma_{MX} = \frac{3C^\phi_{MX}}{2} = 3|z_M z_X|^{1/2} C_{MX} \quad (17\text{-}22b)$$

The parameters for several 1–1, 2–1, 1–2, and 2–2 electrolytes for 25°C are listed in Table 17-1. A more extensive table is given in Appendix 8 and an even more extensive table in reference 7. Also, a table of Debye–Hückel parameters for various temperatures is given in Appendix 7. Various experimental methods of measurement of the osmotic or activity coefficients were discussed in Chapter 15. Isopiestic measurements are the basis of the largest number of parameters, but these are relative to a reference salt, usually NaCl. Hence, other methods are involved for NaCl and for solutes such as HCl for which an excellent electrochemical cell is available.

TABLE 17-1
Ion-interaction parameters for several electrolytes for 25°C[a]

	\multicolumn{5}{c}{1–1 type}				
	$\beta^{(0)}$	$\beta^{(1)}$	$C^\phi = 2C$	Max m	σ
HCl	0.177 5	0.294 5	0.000 80	6	0.001
LiCl	0.149 4	0.307 4	0.003 59	6	0.001
NaCl	0.076 5	0.266 4	0.001 27	6	0.001
NaBr	0.097 3	0.279 1	0.001 16	4	0.001
NaI	0.119 5	0.343 9	0.001 8	3.5	0.001
NaOH	0.086 4	0.253	0.004 4	6	0.002
NaClO$_4$	0.055 4	0.275 5	−0.001 18	6	0.001
NaNO$_3$	0.006 8	0.178 3	−0.000 72	6	0.001
NaH$_2$PO$_4$	−0.053 3	0.039 6	0.007 95	6	0.003
KCl	0.048 35	0.212 2	−0.000 84	4.8	0.000 5
KNO$_3$	−0.081 6	0.049 4	0.006 60	3.8	0.001
	\multicolumn{5}{c}{2–1 or 1–2 type}				
	$\beta^{(0)}$	$\beta^{(1)}$	$C^\phi = 2^{3/2}C$	Max m	σ
MgCl$_2$	0.350 93	1.650 8	0.006 507	4.0	0.003
Mg(NO$_3$)$_2$	0.367 1	1.584 8	−0.020 624	2.0	0.003
CaCl$_2$	0.305 3	1.708 5	0.002 153	4.3	0.003
Na$_2$SO$_4$	0.018 69	1.099 4	0.005 549	3.8	0.002
K$_2$SO$_4$	0.049 95	0.779 3	–	0.7	0.002

[a] All values are from Pitzer and Mayorga[5] except those for MgCl$_2$ and Na$_2$SO$_4$, which are from J. A. Rard and D. G. Miller [*J. Chem. Eng. Data*, **26**, 33, 38 (1981)], and those for CaCl$_2$ from R. C. Phutela and K. S. Pitzer [*J. Solution Chem.*, **12**, 201 (1983)].

From the discussion above concerning the ionic strength dependency of B^ϕ_{MX}, together with that in Chapter 16 related to the second term of Eq. (16-45)*, we expect that the two parameters $\beta^{(0)}$ and $\beta^{(1)}$ in Eqs. (17-14a) and (17-15) will be interrelated. Indeed, examination of the values in Table 17-1 indicates that there is a general correlation, but it is far from exact. This relationship is discussed more fully in reference 4.

In addition to the values in Table 17-1 and Appendix 8, which are for 25°C, there is a growing database for other, primarily higher, temperatures. This is considered in Chapter 18 along with heat capacities, heats of dilution, and other thermal properties.

We note also that in recent, very extensive evaluations of various interrelated data for NaCl and NaBr, Archer[8,9] has found the inclusion of an ionic-strength dependence for the C_{MX} parameter to be significant. He assumes for the ionic strength the same form as was adopted for B^ϕ_{MX},

$$C^\phi_{MX} = C^{(0)}_{MX} + C^{(1)}_{MX}\exp(-\alpha_c I^{1/2})$$

with $\alpha_c = 2.5$ kg$^{1/2}$ mol$^{-1/2}$ the optimum for NaCl[8] and acceptable for NaBr, where 2.0 was initially used.[9] Archer[8] also states that this pattern for C^ϕ_{MX} removes a small but systematic discrepancy that had been noted for Na_2SO_4.

From the maximum m values in Table 17-1, it is apparent that the equations are valid to molalities approaching saturation with the solid salt. The only exceptions among the solutes listed in Table 17-1 are LiCl and $CaCl_2$, which have very high saturation molalities, and HCl where no solid saturation occurs. If the chemical potential of the solid were known independently with sufficient precision, one could calculate the activity coefficient at saturation molality, but this is rarely if ever the case. Rather the reverse calculation is made; the chemical potential of the solid salt is calculated from the activity coefficient and molality at saturation together with the chemical potentials of the ions involved. Once the chemical potentials of the solids are determined in this manner, however, the information is very useful for mixed electrolyte solutions at saturating molality. This will be discussed after the basic relations for mixed electrolytes are derived.

Mixed Electrolytes

For future reference in this and later chapters, it is useful to restate the excess Gibbs energy for a mixed electrolyte incorporating the choices of the preceding section for various terms. Then Eq. (17-8)* becomes

$$\begin{aligned}\frac{G^{Em}}{w_w RT} = &-\frac{4IA_\phi}{b}\ln(1+bI^{1/2}) + \sum_c\sum_a m_c m_a [B_{ca} + (\sum_c m_c z_c)C_{ca}] \\ &+ \sum_{c<c'}\sum m_c m_{c'}[2\Phi_{cc'} + \sum_a m_a \Psi_{cc'a}] \\ &+ \sum_{a<a'}\sum m_a m_{a'}[2\Phi_{aa'} + \sum_c m_c \Psi_{caa'}] + 2\sum_n\sum_c m_n m_c \lambda_{nc} \\ &+ 2\sum_n\sum_a m_n m_a \lambda_{na} + 2\sum_{n<n'}\sum m_n m_{n'}\lambda_{nn'} + \sum_n m_n^2 \lambda_{nn} + \cdots \quad (17\text{-}23)^*\end{aligned}$$

The second-order terms for neutral species are included but not the third-order terms; also the expression for $f(I)$ is shown in full while B_{ca} refers back to Eqs. (17-17) and (17-18) for details.

Equations for the activity and osmotic coefficients for mixed electrolytes are obtained[7,10] from Eq. (17-7)* by the same differentiation and substitution processes that were used for the single solute systems in Eqs. (17-10)* and (17-16)*. In this case it is preferable to obtain initially the expressions for the activity of individual ions M^{z_M+} and X^{z_X-} and to obtain the expression for the mean activity coefficient in a subsequent step. Indeed, for many practical applications to complex mixtures, it is simpler to use the single-ion expressions. The results are:

$$(\phi - 1) = -(\sum_i m_i)^{-1} \left(\frac{\partial (G^{Em}/RT)}{\partial w_w} \right)_{T,P,n_i}$$

$$= \frac{2}{\sum_i m_i} \left(\frac{-A_\phi I^{3/2}}{1 + bI^{1/2}} + \sum_c \sum_a m_c m_a (B^\phi_{ca} + ZC_{ca}) \right.$$

$$+ \sum_{c<c'} \sum m_c m_{c'} (\Phi^\phi_{cc'} + \sum_a m_a \Psi_{cc'a})$$

$$+ \sum_{a<a'} \sum m_a m_{a'} (\Phi^\phi_{aa'} + \sum_c m_c \Psi_{caa'})$$

$$+ \sum_n \sum_c m_n m_c \lambda_{nc} + \sum_n \sum_a m_n m_a \lambda_{na} + \sum_{n<n'} \sum m_n m_{n'} \lambda_{nn'}$$

$$\left. + \tfrac{1}{2} \sum_n m_n^2 \lambda_{nn} \cdots \right) \qquad (17\text{-}24)^*$$

$$\ln \gamma_M = \left(\frac{\partial (G^{Em}/RT)}{\partial n_M} \right)_{T,P,w_w,n_i \neq M}$$

$$= z_M^2 F + \sum_a m_a (2B_{Ma} + ZC_{Ma}) + \sum_c m_c (2\Phi_{Mc} + \sum_a m_a \Psi_{Mca})$$

$$+ \sum_{a<a'} \sum m_a m_{a'} \Psi_{Maa'} + z_M \sum_c \sum_a m_c m_a C_{ca} + 2 \sum_n m_n \lambda_{nM} + \cdots \qquad (17\text{-}25)^*$$

$$\ln \gamma_X = \left(\frac{\partial (G^{Em}/RT)}{\partial n_X} \right)_{T,P,w_w,n_i \neq X}$$

$$= z_X^2 F + \sum_c m_c (2B_{cX} + ZC_{cX}) + \sum_a m_a (2\Phi_{Xa} + \sum_c m_c \Psi_{cXa})$$

$$+ \sum_{c<c'} \sum m_c m_{c'} \Psi_{cc'X} + |z_X| \sum_c \sum_a m_c m_a C_{ca} + 2 \sum_n m_n \lambda_{nX} + \cdots \qquad (17\text{-}26)^*$$

where the second-virial terms for neutral species have been added but the third-virial terms for neutrals are omitted.

The quantity F includes the Debye–Hückel term and other terms as follows:

$$F = f^y + \sum_c\sum_a m_c m_a B'_{ca} + \sum_{c<c'}\sum m_c m_{c'} \Phi'_{cc'} + \sum_{a<a'}\sum m_a m_{a'} \Phi'_{aa'} \qquad (17\text{-}27)$$

Also, Φ' is the ionic strength derivative of Φ, and

$$Z = \sum_i m_i |z_i| \qquad (17\text{-}28)$$

$$\Phi^\phi_{cc'} = \Phi_{cc'} + I\Phi'_{cc'} \qquad (17\text{-}29)$$

The mean activity coefficient for the electrolyte $M_{\nu^+} X_{\nu^-}$ in a mixture is readily obtained from Eqs. (17-25) and (17-26).

$$\begin{aligned}
\ln \gamma_{MX} = \; & |z_M z_X| F + \frac{\nu_M}{\nu} \sum_a m_a \left(2B_{Ma} + ZC_{Ma} + 2\frac{\nu_X}{\nu_M} \Phi_{Xa} \right) \\
& + \frac{\nu_X}{\nu} \sum_c m_c \left(2B_{cX} + ZC_{cX} + 2\frac{\nu_M}{\nu_X} \Phi_{Mc} \right) \\
& + \sum_c\sum_a m_c m_a \nu^{-1} (2\nu_M z_M C_{ca} + \nu_M \psi_{Mca} + \nu_X \psi_{caX}) \\
& + \sum_{c<c'}\sum m_c m_{c'} \frac{\nu_X}{\nu} \psi_{cc'X} + \sum_{a<a'}\sum m_a m_{a'} \frac{\nu_M}{\nu} \psi_{Maa'} \\
& + \frac{2}{\nu} \sum_n m_n (\nu_M \lambda_{nM} + \nu_X \lambda_{nX})
\end{aligned} \qquad (17\text{-}30)^*$$

It is interesting to note that in the equations for mixtures the original third-virial quantity C_{ca} appears rather than the secondary quantities of C^ϕ_{ca} of Eq. (17-13) and C^y_{ca} of Eq. (17-22b). Since C^ϕ_{ca} is the parameter often listed in published tables, this difference between C_{ca} and C^ϕ_{ca} should be kept in mind.

There are many somewhat simplified forms of Eq. (17-30) for cases where all solutes are of the same valence type, and further where there is a simple type or where there is a common cation or common anion or where there are only two solutes. Since these transformations are quite straightforward and in many cases do not shorten the expression very much, they will be omitted.

Another observable combination of activity coefficients is that for the exchange of an amount of one ion by an equal electrical charge of a different ion of the same sign. This occurs, for example, with exchange between two liquid phases when positive ions are complexed to form neutral molecular species in the nonaqueous phase and in certain electrical cells. The pertinent combination of activity coefficients may be written, for M^{z_M+} and N^{z_N+},

$$\begin{aligned}
z_N \ln \gamma_M - z_M \ln \gamma_N = \; & z_N z_M (z_M - z_N) F \\
& + \sum_a m_a [2z_N B_{Ma} - 2z_M B_{Na} + Z(z_N C_{Ma} - z_M C_{Na})] \\
& + 2\sum_c m_c (z_N \Phi_{Mc} - z_M \Phi_{Nc}) + \sum_c\sum_a m_c m_a (z_N \psi_{Mca} - z_M \psi_{Nca}) \\
& + \sum_{a<a'}\sum m_a m_{a'} (z_N \psi_{Maa'} - z_M \psi_{Naa'})
\end{aligned} \qquad (17\text{-}31)^*$$

The corresponding equation for the difference in activity coefficients of anions is readily obtained by transposing symbols in Eq. (17.31)*.

It is interesting to ask whether it is possible to determine the chemical potentials of all other components of a solution from measurements of the chemical potential of one component over the entire range of solution compositions. If the measurements are of the solvent, one obtains the osmotic coefficient, and one notes from Eq. (17-24)* that all parameters are included. Thus with sufficient precision of measurement and range of composition, all parameters can be determined, and by Eq. (17-30)*, the chemical potential of any solute component can be calculated.

If the measurements are of a solute component MX, however, an examination of Eq. (17-30)* shows that terms in $\beta_{ca}^{(0)}$ do not appear unless either c is M or a is X. Thus, it is not possible to evaluate $\beta_{ca}^{(0)}$, a very important parameter, for any component ca not involving either M or X, and consequently the chemical potential of that component. These conclusions are the same as those reached for nonelectrolytes in Chapter 14.

In the case where there is a common ion and $\nu_M = \nu_X$, one finds that the quantities $\beta_{Ma}^{(0)}$ and Φ_{Xa} (and $\beta_{cX}^{(0)}$ and Φ_{Xc}) appear in Eq.(17-30)* with exactly the same composition dependence. Thus, only the sum can be determined from measurements of γ_{MX}. Again, the activity coefficient or chemical potential of another solute component Ma (or cX) cannot be determined without some additional measurement. This topic is discussed further in reference 7.

Experimental measurements for mixed electrolytes have been made by the isopiestic method, which yields the osmotic coefficient, and in appropriate cases by electrochemical cells, which yield the activity coefficient of one component. Another source is solid salt solubility in the mixed electrolyte solution. There is an extensive array of solubility information that Harvie and Weare[11] considered in 1980 along with the mixing parameters already available from isopiestic and EMF measurements of solutions at lower molality; the solubility data are considered in a separate section below.

The first question is whether the mixing parameters Φ and Ψ are needed at all; the Bronsted principle and the Guggenheim treatment assumes that they can be neglected. This question was addressed by Pitzer and Kim[10] who examined an extensive array of data on the basis $\Phi = \psi = 0$ and with Φ and possibly also ψ evaluated. First considered were simple MX–NX mixtures, where MX is a 1–1 electrolyte but NX can be of any charge type. First one calculates $\Delta\phi$, and the difference between the measured osmotic coefficient for the mixture and that calculated with $\Phi_{MN} = \psi_{MNX} = 0$. Then Eq. (17-24)* is simplified and rearranged to yield

$$\Delta\phi \left[\frac{\sum_i m_i}{2m_M m_N} \right] = \Phi_{MN} + m_X \psi_{MNX} \qquad (17\text{-}32)^*$$

Thus a plot of the left side of Eq. (17-32)* vs. m_X will yield Φ_{MN} as the intercept at $m_X = 0$ and ψ_{MNX} as the slope. If neither parameter is significant, all values of $\Delta\phi$ will be zero within the experimental uncertainty.

A similar calculation is possible where emf measurements yield the activity coefficient of MX. Now $\Delta \ln \gamma_{MX}$ is the corresponding difference of the experimental value from that calculated without Φ and ψ, and one obtains

$$\frac{\Delta \ln \gamma_{MX}}{m_N} = \Phi_{MN} + \tfrac{1}{2}(m_M + m_X)\psi_{MNX} \quad (17\text{-}33)^*$$

Then the corresponding graph gives the desired information. For a somewhat more complex charge type of MX in mixtures, Eq. (17-32)* remains unchanged but Eq. (17-33)* must be modified.

Table 17-2 gives some typical results for common solutes from reference 10. It is evident that for many cases Φ and ψ are either negligible or so small that their omission in practical calculations would be acceptable in many cases. But there are exceptions; HCl–CsCl and LiCl–CsCl are cases where there are large deviations that are eliminated when the mixing terms are included.

The initial results for the osmotic coefficient in a few cases of 1–2 mixing gave results similar to those for 1–1 mixing in Table 17-2. But for 1–3 mixing, HCl–AlCl$_3$,

TABLE 17-2
Mixing parameters for binary symmetrical mixtures with a common ion at 25°C

System	Experimental	MAX I	σ with $\Phi = \psi = 0$	Φ	ψ	σ with Φ and ψ
HCl–LiCl	ln γ	5	0.023	0.015	0.000	0.007
HBr–LiBr	ln γ	2.5	0.027	0.015	0.000	0.011
HClO$_4$–LiClO$_4$	ϕ	4.5	0.006	0.015	-0.001_7	0.001
HCl–NaCl	ln γ	3	0.040	0.036	-0.004	0.002
HBr–NaBr	ln γ	3	0.028	0.036	-0.012	0.002
HClO$_4$–NaClO$_4$	ϕ	5	0.025	0.036	-0.016	0.002
HCl–KCl	ln γ	3.5	0.014	0.005	-0.007	0.010
HBr–KBr	ln γ	3	0.030	0.005	-0.021	0.008
HCl–CsCl	ln γ	3	0.082	-0.044	-0.019	0.005
LiCl–NaCl	ϕ	6	0.002	0.012	-0.003	0.001
LiNO$_3$–NaNO$_3$	ϕ	6	0.014	0.012	-0.007_2	0.002
LiClO$_4$–NaClO$_4$	ϕ	2.6	0.003	0.012	-0.008_0	0.001
LiOAc–NaOAc	ϕ	3.5	0.004	0.012	-0.004_3	0.002
LiCl–KCl	ϕ	4.8	0.045	-0.022	-0.010	0.003
LiCl–CsCl	ϕ	5	0.100	-0.095	0.009_4	0.004
NaCl–KCl	ϕ	4.8	0.014	-0.012	-0.001_8	0.001
NaBr–KBr	ϕ	4	0.009	-0.012	-0.002_2	0.003
NaNO$_3$–KNO$_3$	ϕ	3.3	0.008	-0.012	-0.001_2	0.001
KCl–CsCl	ϕ	5	0.003	0.000	-0.001_3	0.001
NaCl–NaBr	ϕ	4.4	0.001	0.000	0.000	0.001
KCl–KBr	ϕ	4.4	0.002	0.000	0.000	0.002
LiCl–LiNO$_3$	ϕ	6	0.008	0.016	-0.003	0.004
NaCl–NaNO$_3$	ϕ	5	0.007	0.016	-0.006	0.001
KCl–KNO$_3$	ϕ	4	0.003	0.016	-0.006	0.001

with $\ln \gamma_{HCl}$ measured by cell emf, the results did not at all conform to the linear expectation of Eq. (17-33).

A complete statistical theory[12] of the electrical interactions of ions yields terms in addition to that of Debye and Hückel. All are higher-order in that they depend on ionic strength to a power higher than 1/2. Thus, they become negligible in comparison with the Debye–Hückel term in the limit of infinite dilution. One of these higher-order terms[12,13] arises for unsymmetrical mixing of ions of the same sign. The theory for this particular term is given in Appendix 9, where it is shown that the result is a theoretical contribution to the parameter Φ_{ij}. This contribution is dependent on the ionic strength and is given the symbol $^E\theta$. Then

$$\Phi_{ij} = \theta_{ij} + {}^E\theta_{ij} \qquad (17\text{-}34)$$

with θ_{ij} a contribution independent of I but specific to the ions i and j while $^E\theta_{ij}$ depends only on the charges z_i, z_j of the ions but depends also on I. For symmetrical mixing $^E\theta$ disappears. Indeed, tables often list values for symmetrical mixing under θ_{ij} instead of Φ_{ij}; the two are equal in that case. Since $^E\theta$ has a nonzero derivative with respect to I, one also has

$$\Phi'_{ij} = {}^E\theta'_{ij} = \frac{d({}^E\theta_{ij})}{dI} \qquad (17\text{-}35)$$

$$\Phi^\phi_{ij} = \theta_{ij} + {}^E\theta_{ij} + I {}^E\theta'_{ij}$$

Now Φ_{MN} should be replaced with θ_{MN} in Eqs. (17-32) and (17-33) with the $^E\theta$ term included in the calculation of $\Delta\phi$ or $\Delta\ln\gamma$.

The calculation of the numerical value of $^E\theta$ is rather complex, but Appendix 9 gives a simple, approximate expression that is adequate for most purposes as well as references to other sources. Figure 17-2 shows that the inclusion of the $^E\theta$ term in the calculation of $\Delta\ln\gamma_{MX}$ removes the discrepancy for the system $AlCl_3$–HCl and yields a pattern from which θ_{MN} and ψ_{MN} can be determined accurately. From Fig. 17-2 the parameters[13] for HCl–$AlCl_3$ were $\theta = 0.185$, $\psi = 0.013$ with $\sigma = 0.006$; while for HCl–$SrCl_2$ they were $\theta = 0.065$, $\psi = 0.003$, $\sigma = 0.002$. Recently the system HCl–$ThCl_4$ has been investigated[14] in which the $^E\theta$ term for 4–1 mixing is even several times larger than that for 3–1 mixing.

For mixtures without a common ion, the equations become much more complex. When the mixing parameters are already known from information on common-ion mixtures, the predictions for non-common-ion systems are excellent. Even without mixing parameters, the results are often quite useful. Comparisons of both bases are given in Table 17-3.

The value of σ with $\theta = \psi = 0$ gives an estimate of the accuracy to be obtained without difference parameters. However, these deviations are usually proportional to molality. Hence, a better estimate is obtained by noting the effect of $\theta_{cc'}$ or $\theta_{aa'}$, in Eq. (17-24)* or (17-30)*. Thus, in the former one finds the result $(m_c m_{c'}/\Sigma_i m_i)\,\theta_{cc'}$; and if one has an equimolal mixture of $HClO_4$ and $LiClO_4$, for example, $(m_c m_{c'}/\Sigma_i m_i)$ is $m/8$ where m is the total molality of ClO_4^-. With $\theta_{H,Li} = 0.015$ one calculates the effect

SEMI-EMPIRICAL EQUATIONS FOR PURE AND MIXED ELECTROLYTES 303

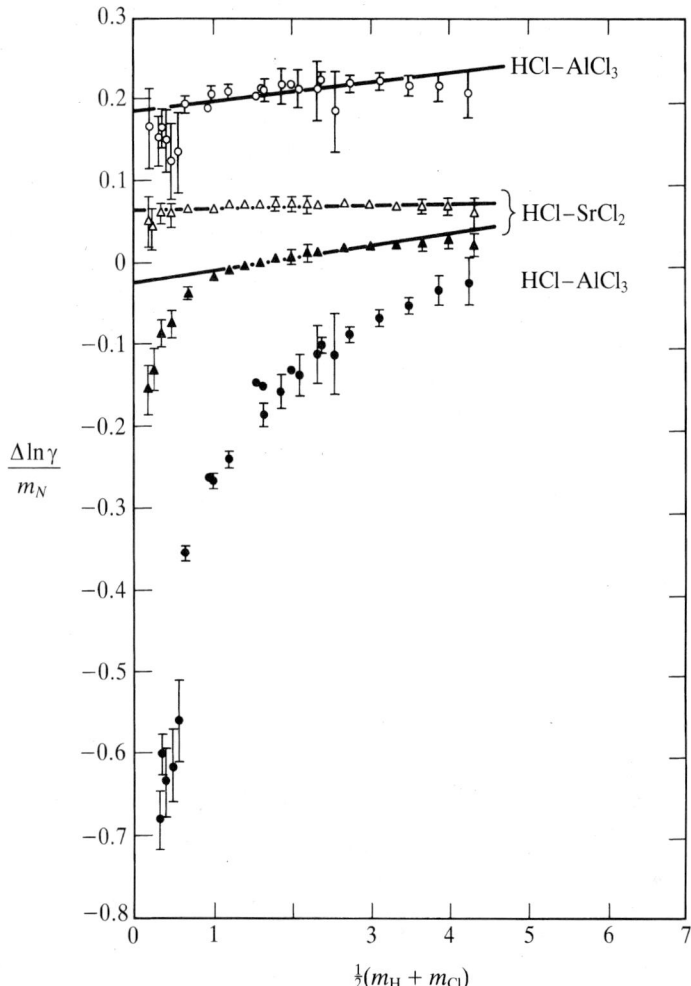

FIGURE 17-2
Evaluation of mixing terms for the systems HCl–SrCl$_2$ and HCl–AlCl$_3$. The solid symbols represent values of $\Delta \ln \gamma_{HCl}/m_N$ calculated without $^E\theta_{HN}$, with open symbols for values after inclusion of $^E\theta_{HN}$.

TABLE 17-3
Binary mixtures without common ion at 25°C

System	Experimental	MAX I	σ with $\Phi = \Psi = 0$	σ with Φ and Ψ included
NaCl–KBr	ϕ	4	0.012	0.002
KCl–NaBr	ϕ	4	0.012	0.001
NaCl–KNO$_3$	ϕ	4	0.007	0.001
NaNO$_3$–KCl	ϕ	4	0.007	0.002

on the osmotic coefficient to be $0.0019m$. This is negligible for most purposes unless m is large (considerably above 1 mol kg^{-1}). The example chosen has a typical θ; in some cases the effect of θ and ψ is even smaller, while in a few cases it is somewhat larger.

Solid Solubilities in Mixed Electrolytes

With the mixing parameters determined independently and the chemical potentials of saturating solids determined from the solubility in pure solute solution, the equations predict solubilities in mixed solutions. Alternatively, the solubility data allow refinements in the ψ_{ijk} parameters that become important only for high molalities of various ions. Harvie and Weare[11] initiated studies of this type in a landmark paper of

TABLE 17-4
The chemical potentials, enthalpies and entropies of the ions and certain minerals of the Na–K–Mg–Ca–Cl–SO$_4$–H$_2$O system at 25°C^a

Species or mineral	Formula	$-\mu°/RT$	$-\Delta_f H°/RT$	$S°/R$
Water	H$_2$O	95.6635	115.304	8.409
Hydrogen ion	H$^+$	0.0	0.0	0.0
Sodium ion	Na$^+$	105.651	96.865	7.096
Potassium ion	K$^+$	113.957	101.81	12.33
Magnesium ion	Mg^{+2}	183.468	188.329	−16.64
Calcium ion	Ca^{+2}	223.30	219.0	−6.4
Hydroxide ion	OH$^-$	63.435	95.666	−1.29
Chloride ion	Cl$^-$	52.955	67.432	6.778
Sulfate ion	SO$_4^{2-}$	300.386	366.800	2.42
Arcanite	K$_2$SO$_4$	532.39	580.01	21.12
Bischofite	MgCl$_2$·6H$_2$O	853.1	1008.11	44.03
Bloedite	Na$_2$Mg(SO$_4$)$_2$·4H$_2$O	1383.6	–	–
Carnallite	KMgCl$_3$·6H$_2$O	1020.3	1184.85	55.53
Epsomite	MgSO$_4$·7H$_2$O	1157.83	(1366.3)	44.79
Glaserite	NaK$_3$(SO$_4$)$_2$	1057.05	(1151.5)	–
Gypsum	CaSO$_4$·2H$_2$O	725.67	815.9	23.35
Halite	NaCl	154.99	165.88	8.676
Hexahydrite	MgSO$_4$·6H$_2$O	1061.60	(1244.8)	41.87
Kainite	KMgClSO$_4$·3H$_2$O	938.2	–	–
Kieserite	MgSO$_4$·H$_2$O	579.80	649.34	(14.99)
Leonite	K$_2$Mg(SO$_4$)$_2$·4H$_2$O	1403.97	(1592.4)	–
Mirabilite	Na$_2$SO$_4$·10H$_2$O	1471.15	1475.75	71.21
Schoenite	K$_2$Mg(SO$_4$)$_2$·6H$_2$O	1596.1	(1831.2)	–
Sylvite	KCl	164.84	176.034	9.934
Thenardite	Na$_2$SO$_4$	512.35	559.55	17.99

^{a} Values for $\mu° = \Delta_f G°$ are from Harvie et al.[11]; most values for $\Delta_f H°$ and $S°$ are from R. T. Pabalan and K. S. Pitzer, *Geochim. Cosmochim. Acta*, **51**, 2429 (1987), with the remainder from D. D. Wagman et al., *J. Phys. Chem. Ref. Data*, **11**, Supplement No. 2 (1989); values in parentheses have large uncertainties; the values above for NaCl and KCl differ slightly from those in Table A6-3 of Appendix 6 because the latter were taken from a different source.

1980 concerning the seawater-related system Na–K–Mg–Ca–Cl–SO$_4$–H$_2$O. We present a few examples from this important study.

Consider the reaction

$$M_{\nu^+}X_{\nu^-}\cdot nH_2O(cr) = \nu_+ M^{z_+} + \nu_- X^{z_-} + nH_2O(\ell)$$

Then

$$\Delta\mu° = \nu_+\mu°_M + \nu_-\mu°_X + n\mu°_{H_2O} - \mu°_{cr} \qquad (17\text{-}36)$$

where each $\mu°$ is the change in chemical potential for formation from the elements. Then

$$-\frac{\Delta\mu°}{RT} = \ln(m_M^{\nu_M} m_X^{\nu_X}) + (\nu_M + \nu_X)\ln\gamma_{MX} + n\ln a_{H_2O} \qquad (17\text{-}37)$$

Next the expression for $\ln\gamma_{MX}$ is obtained from Eq. (17-30)*, while $\ln a_{H_2O}$ is related to the osmotic coefficient by Eq. (14-11) and Φ is given by Eq. (17-24)*.

The chemical potentials, enthalpies, and entropies for certain ions and solid minerals in their standard states are given in Table 17-4. The chemical potential values are from Harvie et al.,[11] whose choices of mixing parameters are listed in Table 17-5. In all cases their θ_{ij} values are the same as those in Table 17-2, which had already been determined from measurements on unsaturated solutions, and in

TABLE 17-5
Electrolyte mixing parameters from Harvie and Weare[11] for the Na–K–Mg–Ca–Cl–SO$_4$–H$_2$O system at 25°C

i	j	k	θ_{ij}	Ψ_{ijk}
Na	K	Cl	−0.012	−0.0018
		SO$_4$		−0.010
Na	Mg	Cl	0.07	−0.012
		SO$_4$		−0.015
Na	Ca	Cl	0.07	−0.014
		SO$_4$		−0.023
K	Mg	Cl	0.0	−0.022
		SO$_4$		−0.048
K	Ca	Cl	0.032	−0.025
		SO$_4$		0
Mg	Ca	Cl	0.007	−0.012
		SO$_4$		0.05
Cl	SO$_4$	Na	0.02	0.0014
		K		0
		Mg		−0.004
		Ca		0

most cases they left Ψ_{ijk} unchanged; but in a few cases Ψ_{ijk} was adjusted within the range of uncertainty of the more dilute solution treatment. For all cases of unsymmetrical mixing the $^E\theta$ and $^E\theta'$ terms were included; indeed, Harvie et al.[11] found their inclusion to be essential for an even qualitatively correct result for solubility of $CaSO_4$ in NaCl(aq).

Figures 17-3 a–d give four typical examples; the various solubility measurements are cited by Harvie et al.[11] The system KCl–NaCl in Fig. 17-3a is simple with no intermediate mineral phase. In this case both the parameters and the measured solubilities are of high accuracy and the agreement is nearly perfect. No adjustment was needed for Ψ_{ijk} in this or in any other case shown in these figures. The KCl–$MgCl_2$

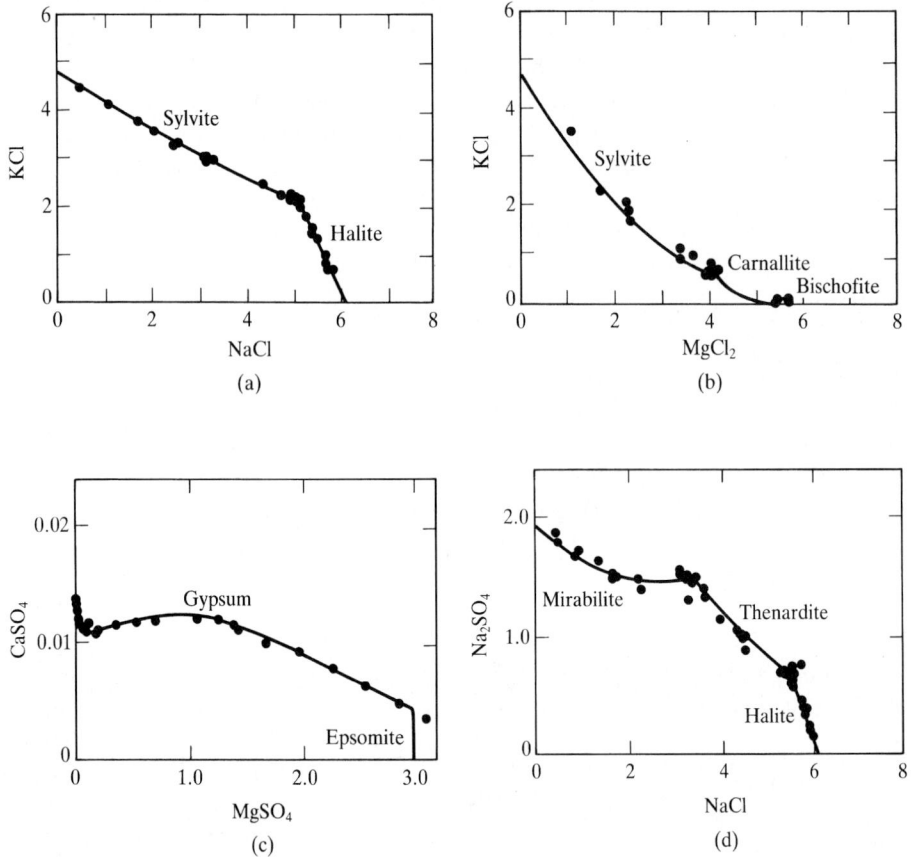

FIGURE 17-3
Mineral solubility predictions of Harvie and Weare[11]: (a) NaCl (halite) and KCl (sylvite); (b) KCl (sylvite) and $MgCl_2 \cdot 6H_2O$ (bischofite); (c) $CaSO_4 \cdot 2H_2O$ (gypsum) and $MgSO_4 \cdot 7H_2O$ (epsomite); and (d) $Na_2SO_4 \cdot 10H_2O$ (mirabilite) and NaCl (halite). The intermediate phases are carnallite ($KMgCl_3 \cdot 6H_2O$) and thenardite (Na_2SO_4).

system (b) has the intermediate mineral carnallite, $KMgCl_3 \cdot 6H_2O$; also the pure $MgCl_2 \cdot 6H_2O$ becomes the stable solid only at very low KCl molality. In (c) the complex curve of $CaSO_4$ solubility is surprising but is predicted accurately by the equations. An interesting aspect of (d) is the shift of the saturating solid from $Na_2SO_4 \cdot 10H_2O$ to anhydrous Na_2SO_4 with the increase in ionic strength and resulting reduction of water activity.

In each of the cases in Fig. 17-3 or in any case of two solid salts with a common ion, the calculations are quite simple. Equations (17-36) and (17-37) are applied for the saturating salt. The molality of one of the non-common ions is fixed and Eq. (17-37) is solved for the other molality. An iterative or similar method is needed since all three terms on the right side depend on both of the molalities. Even if the solid is not hydrated and the osmotic coefficient is not needed, it is still not possible to solve directly for the molality.

Repetition of this process for various assumed values of one molality gives the curves on Fig. 17-3; also, the intersections of two curves give the fixed-point compositions where two solids are in equilibrium with the solution phase. A curve extended beyond an intersection represents a metastable equilibrium of the particular solid with a solution supersaturated with respect to the other solid. In some cases this solubility can be measured; indeed, in a few cases the precipitation of one salt under conditions supersaturated with respect to another is an industrially successful process (an example is the precipitation of KCl in a solution supersaturated with respect to borax, $Na_2B_4O_7 \cdot 10H_2O$).

If the two salts do not have a common ion or if there are more than two salts, the calculations become more complex and often involve the simultaneous solution of two or more equations. Also, the graphical representation of the results is more complex, and the presentations that are possible give less complete information than that of Fig. 17-3. With a large number of components and ions it becomes preferable to work with individual ion molalities as the primary variables. Then an equation for electrical neutrality must be added to the set to be solved simultaneously.

For multicomponent cases Harvie and Weare[11] found it preferable to minimize the total Gibbs energy subject to the required series of conditions by a Lagrangian method that they describe in some detail. The best method for a particular investigation depends to some degree on the computer facilities and optimization codes available.

It is an important characteristic of the ion-interaction (Pitzer) equations as described above that all parameters can be evaluated from measurements on pure electrolytes and common-ion binary mixtures. No new parameters appear for more complex mixtures. Consequently, calculations for these more complex systems are strictly predictions and rigorous tests of the equations.

For the full array of major seawater ions, the system Na–K–Mg–Cl–SO_4–H_2O with five components, there are 13 fixed points with different sets of four solids. Harvie and Weare[11] presented predictions for all 13 sets and compared them with the experimental values recommended by Braitsch.[15] This comparison is shown in Table 17-6; the agreement is within experimental uncertainty in most if not all cases.

TABLE 17-6
Invariant points for the quinary system Na–K–Mg–Cl–SO$_4$–H$_2$O at 25°C. Experimental values[15] are in parentheses; chemical formulas for the minerals are in Table 17-4

m_{Na}	m_K	m_{Mg}	m_{SO_4}	Solid phases
2.62 (2.69)	1.63 (1.58)	2.08 (1.97)	0.84 (0.78)	Halite + sylvite + glaserite + schoenite
2.52 (2.41)	1.59 (1.51)	2.16 (2.19)	0.84 (0.78)	Halite + leonite + sylvite + schoenite
1.16 (1.31)	1.01 (1.10)	3.40 (3.17)	0.86 (0.79)	Halite + sylvite + leonite + kainite
0.48 (0.49)	0.57 (0.64)	4.21 (4.05)	0.32 (0.29)	Halite + sylvite + carnallite + kainite
0.30 (0.26)	0.22 (0.20)	4.75 (4.81)	0.40 (0.35)	Halite + kieserite + carnallite + kainite
5.20 (5.32)	1.04 (0.90)	0.95 (0.89)	1.31 (1.24)	Halite + thenardite + glaserite + bloedite
3.08 (3.21)	1.31 (1.30)	2.00 (1.85)	1.14 (1.06)	Halite + glaserite + bloedite + schoenite
2.49 (2.96)	1.18 (1.22)	2.40 (2.01)	1.11 (1.05)	Halite + bloedite + schoenite + leonite
1.43 (1.25)	0.85 (0.72)	3.31 (3.46)	1.14 (1.10)	Halite + epsomite + bloedite + leonite
1.16 (1.18)	0.85 (0.75)	3.51 (3.54)	1.03 (1.04)	Halite + epsomite + leonite + kainite
0.76 (0.66)	0.50 (0.68)	3.95 (4.01)	0.81 (0.79)	Halite + kainite + hexahydrite + epsomite
0.40 (0.33)	0.23 (0.32)	4.57 (4.63)	0.59 (0.45)	Halite + kainite + hexahydrite + kieserite
0.09 (0.07)	0.02 (0.02)	5.74 (5.83)	0.06 (0.05)	Halite + bischofite + kiserite + carnallite

Extremely Concentrated Electrolytes

The equations described above are adequate to solid saturation for many salts, but there are exceptions. Calcium chloride is one, LiCl and ZnCl$_2$ are others, and in extreme cases, especially at high temperatures, there are aqueous systems continuously miscible to the pure fused salt. In the extreme case mentioned last, one must abandon molality as the composition variable because it would become infinite. Equations for electrolytes in terms of mole fraction are considered below. First we consider extensions within the molality system.

Equation (17-5)* as written was just the first few terms of a series in increasing powers of the number of moles of solute (or in molality for Eq. (17.8)*, etc.). One can just add higher-order terms. Ananthaswamy and Atkinson[16] did exactly that for $CaCl_2$(aq) using three terms in addition to those in Eq. (17.10)*. New possibilities arise concerning the mixing terms that appear for mixed systems. Filippov et al.[17] pioneered one approach for $CaCl_2$ and its mixtures with $CdSO_4$, $NaCl$, and Na_2SO_4. Anstiss and Pitzer[18] used the first method for LiCl and the second for $ZnCl_2$ and its mixture with NaCl. Since these extensions are relatively straightforward and the number of probable applications seems to be limited, they will not be presented in detail.

EQUATIONS FOR ELECTROLYTES BASED ON MOLE FRACTIONS

There is no particular difficulty in translating an equation for the excess Gibbs energy for an electrolyte from molalities to mole fractions. And the derivation of equations for $\ln \gamma_{MX}$ and $\ln \gamma_{H_2O}$ (replacing the osmotic coefficient) is straightforward for pure electrolytes ($MX-H_2O$ binaries). But for mixed electrolytes with many components the mole-fraction-based equations for activity coefficients are found to be much more complicated than their molality-based counterparts. To understand this we recall Eq. (17-5)* and note that in differentiation of G^{Em} with respect to a number of moles n_M, only terms that involved n_M are present in the result. But, because a mole fraction includes all mole numbers in the denominator ($x_M = n_M/\Sigma_i n_i$), every term in an equivalent G^E yields a term in the result for an activity coefficient. This formulation and set of derivations has been carried out by Clegg et al.[19] for the general case and has been used successfully for a few examples.

For our presentation[20] here we consider only the simplified case of two symmetrically ionized common anion components and one neutral component (solvent). Thus the species are M^+, N^+, X^-, and 1. Conversion to the case M^+, X^-, Y^-, 1 is, of course, trivial. With the definition of the mole fraction of any species,

$$x_j = \frac{n_j}{\sum_i n_i} \qquad (17\text{-}38)$$

where n_i is a number of moles and the sum covers all species. We next note that x_1 for the pure solvent is $x_1^\circ = 1.0$, but for a pure liquid salt the mole fraction of either ion is 1/2 and not 1.0. Thus, in a pure salt reference state we have $x_i^\circ = 1/2$ and the ideal entropies and Gibbs energies of mixing are

$$\frac{\Delta_m S^{id}}{R} = -\sum_i n_i \ln\left(\frac{x_i}{x_i^\circ}\right) \qquad (17\text{-}39)$$

$$\frac{\Delta_m G^{id}}{RT} = \sum_i n_i \ln\left(\frac{x_i}{x_i^\circ}\right) \qquad (17\text{-}40)$$

This simple result is readily derived for the case that all particles mix randomly without respect to charge but that charge neutrality must be maintained for each pure component and for the mixture. Such completely random mixing is, of course, a poor structural model for fused salts where there is a strong pattern of alternation of charge.

But for the mixing of salts where the ions have the same magnitude of charge, the assumption that there is no cation–anion mixing while cations mix randomly and anions mix randomly (Temkin model) yields the same entropy of mixing as is given by Eq. (17-39). This was demonstrated by Laity[21] and by Blander.[22] Thus, we take Eqs. (17-39) and (17-40) as definitions of ideal mixing for the present treatment, which is limited to singly charged ions and neutral molecules.

The excess Gibbs energy for any amount of material is G^E and, per mole of particles, is g^E. Here and in some later sections, we follow the literature,[19,20] with g for the molar quantity instead of G_m. Thus for the Gibbs energy of mixing,

$$\Delta_m G = \Delta_m G^{id} + G^E \quad (17\text{-}41)$$

$$g^E = \frac{G^E}{\sum_i n_i} \quad (17\text{-}42)$$

The activities a_j and activity coefficients γ_j are related by

$$\ln a_j = \ln\left(\frac{x_j \gamma_j}{x_j^0}\right) \quad (17\text{-}43)$$

and for a salt MX

$$\ln a_{MX} = \ln(a_M a_X) \quad (17\text{-}44)$$

Differentiation of G^E with respect to n_j at constant T, P, and other n_i yields the excess chemical potential μ_j^E; also

$$\mu_j^E = RT \ln \gamma_j \quad (17\text{-}45)$$

$$g^E = \sum_i x_i \mu_i^E = RT \sum_i x_i \ln \gamma_i \quad (17\text{-}46)$$

For electrolytes the activities and activity coefficients can ordinarily be measured for neutral combinations of ions. Thus for a salt mixture containing M^+, N^+, X^-, one can determine $(a_M a_X)$ and $(a_N a_X)$, but not any of the individual a's.

For systems with all pure components liquid at the conditions of interest, the normal standard states are the pure liquid components. This yields $\gamma_j = 1$ and $a_j = 1$ for the pure liquid in Eqs. (17-43) and (17-44). One may wish to use an infinitely dilute reference state for an ionic component and to relate that state to the usual molality-based reference state. These relationships are given in references 19 and 20.

We turn now to the excess Gibbs energy, which arises from inequalities in interparticle forces. If there is a substantial ionic concentration, interionic forces are effectively screened from the R^{-2} long-range to short-range. Then interionic forces can be combined with all other interparticle forces on the same basis and one expects the same type of expression for excess Gibbs energy as was found for nonelectrolytes. At very low ionic concentration, however, the alternating charge pattern and its accompanying screening effect is lost and the long-range nature of ionic forces must be considered. This is the effect described by the Debye–Hückel treatment. For the full range of composition one expects the excess Gibbs energy to comprise two terms: a short-range-force term G^S and a Debye–Hückel term G^{DH}; thus

$$G^E = G^S + G^{DH} \qquad (17\text{-}47\text{a})$$

$$g^E = g^S + g^{DH} \qquad (17\text{-}47\text{b})$$

where the last equation refers to one mole of particles. The logarithms of the activity coefficients are similarly sums of terms for short-range forces and for the Debye–Hückel effect.

For the short-range force term we take the three-suffix Margules expression given in Chapter 12:

$$\frac{g^S}{RT} = \sum_i \sum_j a_{ij} x_i x_j + \sum_i \sum_j \sum_k a_{ijk} x_i x_j x_k + \cdots \qquad (17\text{-}48)^*$$

where a's with all suffixes equal are zero. If there are only two components, one has

$$\frac{g^S}{RT} = x_1 x_2 (2 a_{12} + 3 x_1 a_{112} + 3 x_2 a_{122}) = x_1 x_2 [w_{12} + u_{12}(x_1 - x_2)]$$

$$(17\text{-}49)^*$$

with

$$w_{12} = 2 a_{12} + \tfrac{3}{2}(a_{112} + a_{122}) \qquad (17\text{-}50\text{a})$$

$$u_{12} = \tfrac{3}{2}(a_{112} - a_{122}) \qquad (17\text{-}50\text{b})$$

where the second equality for Eq. (17-49)* shows that only two of the three parameters in the first line are independent.

If one generalizes the definitions of Eq. (17-49)* and defines a third quantity related to a_{123}, one can write for the general case

$$\frac{g^S}{RT} = \sum \sum_{j>1} x_i x_j [w_{ij} + u_{ij}(x_i - x_j)] + \sum \sum \sum_{k>j>1} x_i x_j x_k C_{ijk} - \frac{g^{S^\circ}}{RT} \qquad (17\text{-}51)^*$$

$$w_{ij} = 2 a_{ij} + \tfrac{3}{2}(a_{iij} + a_{ijj}) \qquad (17\text{-}51\text{a})$$

$$u_{ij} = \tfrac{3}{2}(a_{iij} - a_{ijj}) \qquad (17\text{-}51\text{b})$$

$$C_{ijk} = 6 a_{ijk} - \tfrac{3}{2}(a_{iij} + a_{ijj} + a_{iik} + a_{ikk} + a_{jjk} + a_{jkk}) \qquad (17\text{-}51\text{c})$$

Note that interchange of subscripts leaves w_{ij} or C_{ijk} unchanged but changes the sign of u_{ij}. Also g^{S° is the value for the same material in the reference states of the various components. The values are zero for neutral species but not for electrolytes, where terms involving $w_{M,X}$ remain for a pure fused salt MX.

A notation used by others is related by $A'_{ij} = w_{ij} - u_{ij}$ and $A'_{ji} = w_{ij} + u_{ij}$. We prefer a form in which one parameter (u_{ij}) will become zero in a symmetrical case rather than two parameters becoming equal. Our third parameter, C_{ijk}, is equivalent, except for sign, to the C^* or Q' used by others. Here we prefer a positive sign in Eq. (17-51)*.

The expression for the short-range force contribution to an activity coefficient is obtained by differentiation of Eq. (17-51)*.

$$\ln \gamma_i^S = \sum_j{}' x_j[(1 - x_i)w_{ij} + (2x_i - 2x_i^2 + 2x_ix_j - x_j)u_{ij}]$$

$$- \sum_{k>j}{}'\sum{}' x_jx_k[w_{jk} + 2(x_j - x_k)u_{jk} - (1 - 2x_i)C_{ijk}]$$

$$- \sum_{l>k>j}{}'\sum{}'\sum{}' 2x_jx_kx_lC_{jkl} - \ln \gamma_i^{S\circ} \qquad (17\text{-}52)^*$$

The prime on the summations is a reminder that terms with any two indices equal are omitted and that neither j nor k nor l may equal i in the multiple sums. For ions, the last term is a reference-state correction that is best considered for complete salts; it is zero for neutral species.

For a salt MX one wishes the product $\gamma_M\gamma_X$,

$$\ln(\gamma_M\gamma_X)^S = \ln \gamma_M^S + \ln \gamma_X^S - \ln(\gamma_M\gamma_X)^0 \qquad (17\text{-}53)$$

where the last term is the value for the pure fused salt, which may not be zero, and must be subtracted. An appropriate expression for $g^{S\circ}/RT$ would eliminate the need for the correction in Eq. (17-53), but it may be easier to handle this problem for the activity coefficient. A shift to the infinitely dilute reference state merely changes $\ln \gamma_i^{S\circ}$.

At this point it is interesting to consider a few examples that illustrate the extent to which the parameters measurable from simple systems are able to determine the properties of more complex systems. Consider first the common ion mixture of two salts MX and NX with cation fraction F of ion M. Then the mole fractions are $x_M = F/2$, $x_N = (1 - F)/2$, and $x_X = 1/2$ and

$$\ln(\gamma_M\gamma_X)^S = \tfrac{1}{4}\{2w_{M,X} + (1 - F)^2[2w_{M,N} + C_{M,N,X} - u_{M,X} - u_{N,X}$$

$$- (1 - 4F)u_{M,N}]\} - \ln \gamma_{MX}^{S\circ} \qquad (17\text{-}54)^*$$

For pure liquid MX, $F = 1$, $\gamma_{MX}^S = 1$, and $\ln \gamma_{MX}^{S\circ} = w_{M,X}/2$. It is convenient to define

$$W_{MX,NX} = \tfrac{1}{4}(2w_{M,N} + C_{M,N,X} - u_{M,X} - u_{N,X}) \qquad (17\text{-}55a)$$

$$U_{M,N} = \tfrac{1}{4}u_{M,N} \qquad (17\text{-}55b)$$

whereupon

$$\frac{g^S}{RT} = \tfrac{1}{2}F(1 - F)[W_{MX,NX} - (1\text{-}2F)U_{M,N}] \qquad (17\text{-}56a)^*$$

$$\ln \gamma_{MX}^S = (1 - F)^2[W_{MX,NX} - (1 - 4F)U_{M,N}] \qquad (17\text{-}56b)^*$$

Measurements at a variety of values of F allow the two parameters $W_{MX,NX}$ and $U_{M,N}$ to be determined.

Consider next the binary mixture of a neutral species, 1, with an ionic component MX. Now $x_M = x_X = (1 - x_1)/2$ and

$$\frac{g^S}{RT} = \tfrac{1}{4}x_1(1 - x_1)[2w_{1,M} + 2w_{1,X} - w_{M,X}$$
$$+ (3x_1 - 1)(u_{1,M} + u_{1,X}) + (1 - x_1)C_{1,M,X}] \quad (17\text{-}57a)*$$

$$\ln \gamma_1^S = \tfrac{1}{4}(1 - x_1)^2[2w_{1,M} + 2w_{1,X} - w_{M,X}$$
$$+ (6x_1 - 1)(u_{1,M} + u_{1,X}) + (1 - 2x_1)C_{1,M,X}] \quad (17\text{-}57b)*$$

$$\ln(\gamma_M \gamma_X)^S = x_1^2[w_{1,M} + w_{1,X} - \tfrac{1}{2}w_{M,X} + (3x_1 - 2)(u_{1,M} + u_{1,X})$$
$$+ (1 - x_1)C_{1,M,X}] \quad (17\text{-}57c)*$$

Now define

$$W_{1,MX} = \tfrac{1}{4}(2w_{1,M} + 2w_{1,X} - w_{M,X} + 2u_{1,M} + 2u_{1,X}) \quad (17\text{-}58a)$$

$$U_{1,MX} = -\tfrac{3}{4}(u_{1,M} + u_{1,X}) + \tfrac{1}{4}C_{1,M,X} \quad (17\text{-}58b)$$

whereupon

$$\frac{g^S}{RT} = x_1(1 - x_1)[W_{1,MX} + (1 - x_1)U_{1,MX}] \quad (17\text{-}59a)*$$

$$\ln \gamma_1^S = (1 - x_1)^2[W_{1,MX} + (1 - 2x_1)U_{1,MX}] \quad (17\text{-}59b)*$$

$$\ln(\gamma_M \gamma_X)^S = x_1^2[2W_{1,MX} + 4(1 - x_1)U_{1,MX}] \quad (17\text{-}59c)*$$

Again we note in Eq. (17-58ab) the combinations of the original parameters that can be measured with this system.

Next, consider the three-component system with a neutral species (1) and two salts with a common ion MX, NX and with cation fraction F of M. Also define a total mole fraction of ions $x_I = (1 - x_1)$, whereupon $x_M = Fx_I/2$, $x_N = (1 - F)x_I/2$, and $x_X = x_I/2$ and use the definitions of Eqs. (17-55ab) and (17-58ab):

$$\frac{g^S}{RT} = x_1 x_I \{FW_{1,MX} + (1 - F)W_{1,NX}$$
$$+ x_I[FU_{1,MX} + (1 - F)U_{1,NX} + F(1 - F)Q_{1,MX,NX}]\}$$
$$+ x_I^2 F(1 - F)[W_{MX,NX} + x_I(2F - 1)U_{M,N}]/2 \quad (17\text{-}60a)*$$

$$\ln \gamma_1^S = x_I^2 \{FW_{1,MX} + (1 - F)W_{1,NX} - F(1 - F)W_{MX,NX}/2$$
$$+ (x_I - x_1)[FU_{1,MX} + (1 - F)U_{1,NX} + F(1 - F)Q_{1,MX,NX}]$$
$$+ x_I F(1 - F)(1 - 2F)U_{M,N}\} \quad (17\text{-}60b)*$$

$$\ln(\gamma_M\gamma_X)^S = 2x_1\{(1-x_IF)W_{1,MX}$$
$$+ x_I(1 - F + 2Fx_1)U_{1,MX} - x_I(1-F)[W_{1,NX} + (x_I - x_1)U_{1,NX}$$
$$- (1 - 2x_IF)Q_{1,MX,NX}]\} + x_I(1-F)\{(1-x_IF)W_{MX,NX}$$
$$+ x_I[3F - 1 + 2x_IF(1 - 2F)]U_{M,N}\} \qquad (17\text{-}60c)*$$

$$Q_{1,MX,NX} = \tfrac{1}{4}(u_{1,M} + u_{1,N} + C_{1,M,N}) + \tfrac{1}{8}(u_{M,X} + u_{N,X} - C_{M,N,X}) \qquad (17\text{-}60d)$$

Thus, we note that, for this ternary system, all but one of the parameters of Eq. (17-60)* can be determined from the binary systems. The additional parameter $Q_{1,MX,NX}$ enters only with rather small coefficients and its effect may be rather small, but that remains to be determined from measurements on real systems.

We turn next to the Debye–Hückel term in Eq. (17-47). In dilute ionic solutions the distribution changes from a random pattern at extremely low concentration to one of alternating positive and negative charges at moderately higher concentration. Debye and Hückel gave a simple treatment that describes this effect. The limiting term at low concentration is given rigorously; the effect at higher concentration is given approximately. There are various alternate forms with respect to the higher concentration range. We choose a form analogous to Eq. (17-14), which has its origin in the pressure equation of statistical mechanics and was selected in tests described above for the molality-based equations.

The basic variable is the ionic concentration weighted by the square of the charge. Concentration is replaced by molality for most work at moderate dilution but is replaced by mole fraction for our present problem of systems continuously miscible to the fused salt. Thus, ionic strength on a mole fraction basis is defined as

$$I_x = \tfrac{1}{2}\sum_i x_i z_i^2 \qquad (17\text{-}61)$$

where z_i is the charge on the ith species of ions. Thus for singly charged ions $I_x = x_1/2$. Then the activity coefficient of a single uncharged solvent is[23]

$$(\ln \gamma_1)^{DH} = \frac{2A_x I_x^{3/2}}{1 + \rho I_x^{1/2}} \qquad (17\text{-}62)$$

The excess Gibbs energy is

$$\frac{g^{DH}}{RT} = -\frac{4A_x I_x}{\rho}\ln\left(\frac{1 + \rho I_x^{1/2}}{1 + \rho (I_x^\circ)^{1/2}}\right) \qquad (17\text{-}63)$$

where I_x° represents I_x for pure fused salt, which is 1/2 for singly charged ions. The Debye–Hückel parameter A_x is

$$A_x = \frac{1}{3}\left(\frac{2\pi N_A d_1}{M_1}\right)^{1/2}\left(\frac{e^2}{4\pi\varepsilon_0\varepsilon kT}\right)^{3/2} \qquad (17\text{-}64)$$

where d_1, M_1, and ε are the density, molecular mass, and dielectric constant (relative permittivity) of the solvent, while N_A is Avogadro's number, e is the electronic charge,

k is Boltzmann's constant, and ε_0 is the permittivity of free space. Most of the literature on electrolytes is still written for esu; in that case $\varepsilon_0 = (4\pi)^{-1}$.

The parameter ρ in Eqs. (17-62) and (17-63) is related to a hard-core collision diameter a in the approximate theory:

$$\rho = a\left(\frac{2e^2 N_A d_1}{M_1 \varepsilon_0 \varepsilon kT}\right)^{1/2} \quad (17\text{-}65)$$

In practice, however, the hard core is not an accurate model and is not independently known; hence, ρ is treated empirically. In multicomponent systems it complicates the equations greatly to make ρ dependent on the composition of ionic components. It has been found satisfactory to take a standard value for ρ and let the short-range force terms accommodate any composition dependency of ρ. It is less clear whether ρ should be made dependent on the variables d_1, ε, and T that also appear in Eq. (17-65) and are known. For the NaCl–H$_2$O system over an extremely wide range of temperature and density, ρ was defined[24] to recognize these variables as follows:

$$\rho = 2150\left(\frac{d_1}{\varepsilon T}\right)^{1/2} \quad (17\text{-}66)$$

with d_1 in g cm^{-3} and T in K. In calculations for several metal nitrates[23] in water near 373 K, ρ was given a fixed value of 14.9. While the numerical parameter 2150 in Eq. (17-66) was selected to best fit data for NaCl–H$_2$O over a wide range of 373–573 K, it yields at 373 K a value (14.6) very close to that chosen in the other research.

A Gibbs–Duhem transformation of Eq. (17-62) or differentiation of Eq. (17-63) yields the mean activity coefficient of an ionic component of charge z_j for the infinitely dilute reference state,

$$(\ln \gamma_j^*)^{DH} = -z_j^2 A_x \left\{\frac{2}{\rho}\ln(1 + \rho I_x^{1/2}) + I_x^{1/2}\left(\frac{1 - 2I_x/z_j^2}{1 + \rho I_x^{1/2}}\right)\right\} \quad (17\text{-}67)$$

or for the pure fused salt reference state

$$(\ln \gamma_j)^{DH} = -z_j^2 A_x \left[\frac{2}{\rho}\ln\left(\frac{1 + \rho I_x^{1/2}}{1 + \rho(I_x^\circ)^{1/2}}\right) + \frac{I_x^{1/2}(1 - 2I_x/z_j^2)}{1 + \rho I_x^{1/2}}\right] \quad (17\text{-}68)$$

Here I_x° is the ionic strength of the pure fused salt, which is $z_j^2/2$ for the MX type. One may note that when z_j^2 reduces to zero, Eqs. (17-67) and (17-68) reduce to Eq. (17-62). We have written these equations for ions of charge z_j for general interest, even though our applications will involve only singly charged ions.

For a single solvent of known properties and a defined value for ρ, the various Debye–Hückel terms are fully determined and the only disposable parameters are those in the short-range force function. Also, as noted above, there is a relationship between ρ and these last parameters. Consequently, the value of ρ and the form of the extended Debye–Hückel expression should be stated clearly in all cases.

It should be mentioned that this entire treatment is designed for liquid systems well removed from the critical region. Near the critical point or critical curve the

TABLE 17-7
Parameters for Eqs. (17-60)* and (17-69) for the system $LiNO_3$–KNO_3–H_2O

Parameter[a]	a	$b \times 10^2/K$
$W_{1,MX}$	−3.582 (±0.017)	1.156 (±0.041)
$U_{1,MX}$	0.759 (±0.050)	0.55 (±0.10)
$W_{1,NX}$	0.688 (±0.019)	−1.007 (±0.044)
$U_{1,NX}$		0.899 (±0.070)
$W_{MX,NX}$	−2.865 (±0.073)	0.253
$Q_{1,MX,NX}$	−1.88 (±0.11)	

[a] $1 = H_2O$, $M = Li^+$, $N = K^+$, $X = NO_3^-$.

compressibility becomes very large and special effects arise. Also the Debye–Hückel derivation basically involves the Helmholtz energy and the ionic concentrations. The simple conversion to Gibbs energy and mole fraction is a satisfactory approximation only for relatively incompressible fluids.

These equations have now been applied to a number of systems, including several involving nitric acid. As an example we present here the system $LiNO_3$–KNO_3–H_2O, for which there were measurements[20] of the vapor pressure of water at temperatures from 373 K to 436 K. For ρ in the Debye–Hückel term the fixed value 14.9 was selected on the basis of some earlier experiments,[23] and it was found to be satisfactory. The other parameters were assumed to have a linear temperature dependence

$$P = a + b(T - 373.15) \qquad (17\text{-}69)$$

Initial regressions of the full data set for all of the parameters of Eq. (17-60)* except the independently known A_x and ρ yielded well-defined values of all but the temperature dependence of $W_{MX,NX}$. This is a property of the anhydrous fused salt, for which the enthalpy of mixing was known from Kleppa and Hersh.[25] That information provided a value for b in $W_{MX,NX}$, which is included in Table 17-7, along with the results of a regression of other parameters with this value fixed. Figure 17-4 compares the resulting curves with the experimental data for the highest temperatures. The results at lower temperatures are similar but cover a somewhat reduced range of composition.

SINGLE-ION ACTIVITIES; pH

Since the transfer of a single charged species generates a very large space-charge potential, its activity cannot be measured by ordinary thermodynamic methods. Nevertheless, the quantity pH = $-\log a_{H^+}$ is commonly used. While other types of electrochemical cells are used more frequently, the essential features are most clearly illustrated with the cell comprising hydrogen and silver–silver chloride electrodes. As

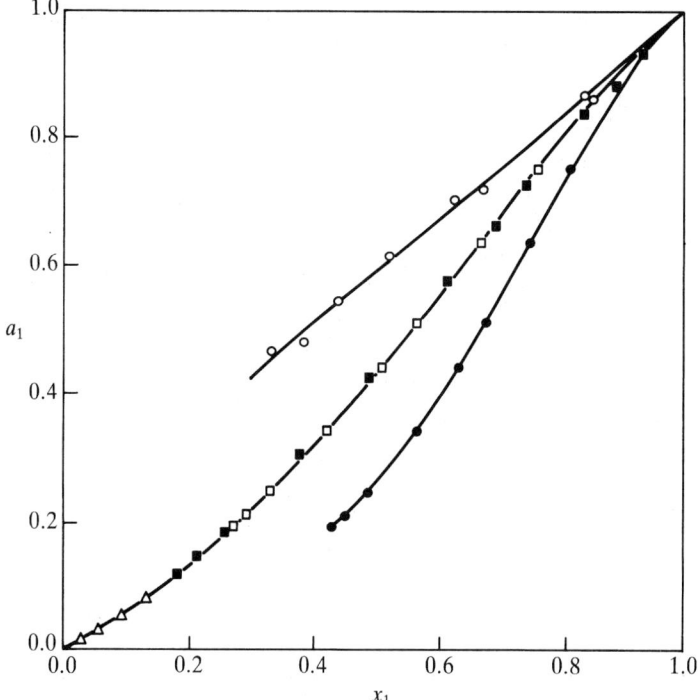

FIGURE 17-4
Water activity at 423–436 K for the system $LiNO_3$–KNO_3–H_2O. The calculated curves and experimental data are for KNO_3–H_2O, $F = 0$ (top curve); $F = 0.5034$ (center); $F = 1$, $LiNO_3$–H_2O (bottom); ■, ○, reference 20 (436 K); □, ●, H. Braunstein and J. Braunstein, *J. Chem. Thermodyn.*, **3**, 417 (1971) (423 K); △, T. B. Tripp and J. Bernstein, *J. Phys. Chem.*, **73**, 1984 (1969) (423 K, $F = 0.5000$). (From Simonson and Pitzer[20].)

noted in Chapter 15, that cell yields the activity of HCl and a simple rearrangement gives (with $f_{H_2} = 1$ bar)

$$\ln a_{H^+} = \frac{(E° - E)F}{RT} - \ln(m_{Cl^-}\gamma_{Cl^-}) \qquad (17\text{-}70)$$

Everything in Eq. (17-70) is measurable except the activity coefficient of chloride ion. For a dilute solution it is a good approximation to assume an extended Debye–Hückel expression such as

$$\ln \gamma_{Cl^-} = -\frac{3A_\phi I^{1/2}}{1 + 1.5 I^{1/2}} \qquad (17\text{-}71)$$

This is known as the Bates–Guggenheim[26,27] convention and was widely endorsed for use up to ionic strength of 0.1 mol kg^{-1}.

Recently, there has been increased interest in pH for more concentrated solutions. No expression in terms of ionic strength as the only variable suffices, and expressions using the ion-interaction (Pitzer) equations have been discussed by Knauss et al.,[28] by Mesmer,[29] and by Covington and Ferra.[30] Equation (17-26)* can be used for $\ln \gamma_{Cl^-}$ with the molalities of the various solute species substituted. This gives a definite result that Knauss et al.[28] found to be useful for pH applications. But the basic uncertainty of a single ion activity remains.

Mesmer[29] suggests an alternative that merits serious consideration. He proposes that the molality of H^+ rather than the activity of H^+ be the basis, i.e., pH $= -\log m_{H^+}$. Then for the H_2–AgCl/Ag cell,

$$\ln m_{H^+} = \frac{(E° - E)F}{RT} - \ln m_{Cl^-} - 2 \ln \gamma_{HCl} \qquad (17\text{-}70a)$$

Now it is the activity coefficient of HCl which is involved, and this is a rigorously based quantity. For a mixed electrolyte of moderate or even high molality, it is given by Eq. (17-30)* in the ion-interaction system. For most, if not all, practical purposes the Mesmer alternative should be satisfactory, although some adjustment from current practice would be required. Possibly a modified symbol such as pHm should be used for the Mesmer alternative.

NEUTRAL SOLUTES

All of the examples of this chapter have been for strong electrolytes, but the equations are not restricted to a single neutral (solvent) species. Equations (17-5), (17-24)*, (17-25)* and (17-26)* explicitly include λ_{ij} terms for neutral solute species, and the introduction of such terms as well as μ_{ijk} terms is straightforward for other equations. Also, the mole-fraction-based equations are readily extended for a neutral solute species. A neutral solute makes no contribution to the ionic strength or to the Debye–Hückel term.

If the neutral solute species has no chemical interaction with any of the ions, e.g. nitrogen, and its concentration is measured, then the additional term is directly evaluated and no complications arise. Long and McDevit[31] have presented an excellent review of the literature for this type of system.

The cases of weak acids or bases and others that are similar in having a chemical equilibrium between certain ions and the neutral species are interesting and important. Typical are systems involving a weak acid HA, its salt MA, and another common cation salt MCl, selected in order to use an electrochemical cell active to H^+ and Cl^- to monitor the dissociation of HA via the activity of H^+. Clearly, this is a moderately complex mixed electrolyte (H^+, M^+, A^-, Cl^-) with the neutral species HA. Since electrochemical cells play a major role, we have decided to place the detailed treatment of this type of system in Chapter 19, which concerns such cells.

Differentiation of the excess Gibbs energy, Eq. (17-5) yields for the activity coefficient of a particular neutral species N the expression

$$\ln \gamma_N = 2(\sum_c m_c \lambda_{Nc} + \sum_a m_a \lambda_{Na} + \sum_n m_n \lambda_{Nn}) \qquad (17\text{-}72)*$$

where only the second-order terms are included. Addition of the third order μ_{ijk} terms is straightforward. For a single symmetrical salt MX of molality m, the effect of the salt on a very dilute solution of the neutral can be expressed by

$$\ln \gamma_N = 2m(\lambda_{NM} + \lambda_{NX}) \qquad (17\text{-}73)^*$$

Here the effect of neutral–neutral interaction is neglected.

Before proceeding further, we consider the possibility of an ionic-strength dependence of λ_{NM} and λ_{NX}. There is a theoretical basis if the neutral has a dipole moment (see reference 7, Appendix D), but the magnitude is very small, and it is usually negligible. It is impossible to determine the $\lambda_{N,c}$ and $\lambda_{N,a}$ values individually; only sums for electrically neutral combinations of ions can be measured, e.g. $\lambda_{N,H} + \lambda_{N,Cl}$ for HCl(aq).

In much of the literature the equation equivalent to Eq. (17-73)* for a single salt is written

$$\log \gamma_N = k_{MX} m_{MX} \qquad (17\text{-}74)^*$$

where the use of base-10 log should be noted. Then, for a symmetrical electrolyte $k_{MX} = 2(\lambda_{N,M} + \lambda_{N,X})/2.303$. This k_{MX} symbolism conceals the fact that these quantities should be ion additive; thus for any given N,

$$k_{HCl} - k_{HNO_3} = k_{NaCl} - k_{NaNO_3} = \frac{2(\lambda_{N,Cl} - \lambda_{N,NO_3})}{2.303}$$

Also, one notes from Eq. (17-30)* the effect of $\lambda_{N,M}$ and $\lambda_{N,X}$ on the activity coefficient of the electrolyte MX. Haugen and Friedman[32] verified both of these last relationships for systems with nitromethane as the neutral solute. They take $\lambda_{N,H^+} = 0$ and give individual values for other ions.

Table 17-8 gives a few typical values on the $\lambda_{N,i}$ basis. Corti et al.[33] have recently examined the CO_2–NaCl–H_2O system on the basis of the treatment given in this chapter. They consider a very wide range of pressure, of molality, and of temperature, 25–250°C, and find a detectable but very small contribution from third-order μ_{ijk} interactions.

TABLE 17-8
Parameters for $\lambda_{N,i}$ for ion–neutral interactions

Ion	CH_3NO_2[a]	Benzene[b]	Benzoic acid[b]
H^+	(0)	(0)	(0)
Li^+	0.18	0.107	0.123
Na^+	0.16	0.169	0.115
K^+	0.06	0.136	0.071
Cl^-	−0.04	0.055	0.094

[a] Haugen and Friedman[32]
[b] Long and McDevit[31]

REFERENCES

1. E. A. Guggenheim, *Phil. Mag.* **19**(7), 558 (1935).
2. J. N. Bronsted, *J. Am. Chem. Soc.*, **44**, 877 (1922); **45**, 2898 (1923).
3. E. A. Guggenheim and J. C. Turgeon, *Trans. Faraday Soc.*, **51**, 747 (1955).
4. K. S. Pitzer, *J. Phys. Chem.*, **77**, 268 (1973).
5. K. S. Pitzer and G. Mayorga, *J. Phys. Chem.*, **77**, 2300 (1973); **78**, 2698 (1974)
6. K. S. Pitzer and G Mayorga, *J. Solution Chem.*, **3**, 539 (1974).
7. K. S. Pitzer, "Ion-interaction Approach: Theory and Data Correlation," in *Activity Coefficients in Electrolyte Solutions*, 2d edn., K. S. Pitzer, ed., CRC Press, Boca Raton, Florida, 1991, Chapter 3.
8. D. G. Archer, *J. Phys. Chem. Ref. Data*, **21**, 793 (1992).
9. D. G. Archer, *J. Phys. Chem. Ref. Data*, **20**, 509 (1991).
10. K. S. Pitzer and J. J. Kim, *J. Am. Chem. Soc.*, **96**, 5701 (1974).
11. C. E. Harvie, N. Møller, and J. H. Weare, *Geochim. Cosmochim. Acta*, **48**, 723 (1984); also, C. E. Harvie and J. H. Weare, *ibid*, **44**, 981 (1980).
12. H. L. Friedman, *Ionic Solution Theory*, Wiley-Interscience, New York, 1962.
13. K. S. Pitzer, *J. Solution Chem.*, **4**, 249 (1975).
14. R. N. Roy, K. M. Vogel, C. E. Good, W. B. Davis, L. N. Roy, D. A. Johnson, A. R. Felmy, and K. S. Pitzer, *J. Phys. Chem.*, **96**, 11065 (1992).
15. O. Braitsch, *Salt Deposits: Their Origin and Composition*, Springer-Verlag, New York, 1971.
16. A. Ananthaswamy and G. Atkinson, *J. Chem. Eng. Data*, **30**, 120 (1985).
17. V. K. Filippov, N. A. Charykov, and V. Rumyantsev, *Dokl. Akad. Nauk SSSR, Fiz. Khim.*, **296**, 665 (1987); Engl. Transl. **296**, 936 (1987).
18. R. G. Anstiss and K. S. Pitzer, *J. Solution Chem.*, **20**, 849 (1991).
19. S. L. Clegg, K. S. Pitzer, and P. Brimblecombe, *J. Phys. Chem.*, **96**, 9470 (1992).
20. K. S. Pitzer and J. M. Simonson, *J. Phys. Chem.*, **90**, 3005 (1986); J. M. Simonson and K. S. Pitzer, *J. Phys. Chem.*, **90**, 3009 (1986); and earlier references there cited.
21. R. W. Laity, in *Reference Electrodes*, D. J. G. Ives and G. J. Janz, eds., Academic Press, New York, 1961, pp. 544–546.
22. M. Blander, *Molten Salt Chemistry*, Interscience, New York, 1964, pp. 130–135.
23. K. S. Pitzer, *J. Am. Chem. Soc.*, **102**, 2902 (1980); *Ber. Bunsenges, Phys. Chem.*, **85**, 952 (1981).
24. K. S. Pitzer and Y.-g. Li, *Proc. Natl Acad. Sci., U.S.A.*, **80**, 7689 (1983).
25. O. J. Kleppa and L. S. Hersh, *J. Chem. Phys.*, **34**, 351 (1961).
26. R. G. Bates and E. A. Guggenheim, *Pure and Appl. Chem.*, **1**, 163 (1960).
27. R. G. Bates, *Pure and Appl. Chem.*, **36**, 407 (1973).
28. K. G. Knauss, T. J. Wolery, and K. J. Jackson, *Geochim. Cosmochim. Acta*, **54**, 1519 (1990); **55**, 1177 (1991).
29. R. E. Mesmer, *Geochim. Cosmochim. Acta*, **55**, 1175 (1991).
30. A. K. Covington and M. I. Ferra, Abstract for 12th IUPAC Conference, Snowbird, Utah, 1992.
31. F. A. Long and W. F. McDevit, *Chem. Rev.*, **51**, 119 (1952).
32. G. R. Haugen and H. L. Friedman, *J. Phys. Chem.*, **60**, 1363 (1956).
33. H. R. Corti, J. J. de Pablo, and J. M. Prausnitz, *J. Phys. Chem.*, **94**, 7876 (1990).

PROBLEMS

17-1. Use the data of Table 17-1 to calculate the ratio of the vapor pressure above a 0.1 mol kg^{-1} solution of NaClO$_4$ to the vapor pressure of pure water at 25°C.

17-2. Properties for KCl(aq) (K$^+$ + Cl$^-$) are given in Table 17-4, while those for KCl(s) are listed in Appendix 6. (a) Calculate $\Delta H°$, $\Delta S°$, and $\Delta G°$ for KCl(s) = KCl(aq) at 298 K. (b) Use the activity coefficient equation with parameters from Table 17-1 to calculate the solubility of KCl in water at 298 K.

17-3. The solubility of $KBrO_3$ is $0.48 \, \text{mol kg}^{-1}$ at 25°C. Use the data of Appendix 8 to calculate ΔG°_{298} for $KBrO_3(s) = KBrO_3(aq)$.

17-4. For $Ba(NO_3)_2(s) = Ba(NO_3)_2(aq)$, $\Delta G^\circ_{298} = 3200 \, \text{cal}$. Calculate the molality of saturated $Ba(NO_3)_2$ solution, using the data of Table A8-2.

17-5. Make a table of the particular forms of Eqs. (17-1)* and (17-2)* for 1–1, 1–2, 1–3, and 2–2 pure electrolytes in terms of the molality of the complete salt.

17-6. Show that the Bronsted specific-interaction terms in Eqs. (23-36)* and (23-37)* are consistent with Eq. (23-33).

17-7. Estimate the activity coefficient of dilute $TlCl$ in an aqueous solution at 25°C as a function of the molality of added $NaNO_3$. First consider just the extended Debye–Hückel term; then consider also the ion-interaction terms with parameters from Appendix 8. How would the solubility of $TlCl$ in aqueous $NaNO_3$ differ from that in pure water?

CHAPTER 18

ELECTROLYTES AT DIFFERENT TEMPERATURES AND PRESSURES; THERMAL AND VOLUMETRIC PROPERTIES

For nonelectrolytes the relationship between thermal properties and the temperature change in the excess Gibbs energy has been considered at several points. Since the situation is a little more complex for electrolytes, it has been postponed until this point where it is the primary focus. The situation for volumetric properties and pressure effects is, of course, similar.

As we have noted above, the molality basis is used for most work with electrolytes together with the infinitely dilute standard state for all solutes. Many measurements yield the difference in the measured quantity between the solution and the same amount of pure water. This is termed a "relative" quantity and the symbols are L for enthalpy and J for heat capacity. Thus

$$L = H - n_1 H_1^\circ = -T^2 \left(\frac{\partial (G^{Em}/T)}{\partial T} \right)_{P,m} \tag{18-1}$$

$$J = C_P - n_1 C_{P_1}^\circ = \left(\frac{\partial L}{\partial T} \right)_{P,m} \tag{18-2}$$

Here the familiar relationships between G, H, and C_P are applied to the excess Gibbs energy in the molality system as defined in Chapter 14, and the subscript m indicates constancy of composition.

For the relative volume there is no special symbol in general use, but the volume difference between the solution and its pure water per mole of solute is called the "apparent molar volume." The molar values of L and J are similarly defined and used; thus

$$\phi_L = \frac{L}{n_2} \tag{18-3}$$

$$\phi_J = \frac{J}{n_2} \tag{18-4}$$

$$\phi_V = \frac{V - n_1 V_1^\circ}{n_2} = \overline{V}_2^\circ + \frac{1}{n_2}\left(\frac{\partial G^{Em}}{\partial P}\right)_{T,m} \tag{18-5}$$

In the several equations above, the n_1 and n_2 are the numbers of moles of solvent and solute, respectively, and the molar pure water property is indicated by the usual degree symbol. The standard symbol for an apparent molar property is ϕ as either a subscript or superscript. We prefer the preceding superscript location because it avoids confusion with subscripts indicating temperature, substance, or other information. But the subscript location is now in common use.

One could first apply the equations above to the Debye–Hückel limiting law, but instead we proceed at once to use the full ion-interaction equation for a pure electrolyte that includes the Debye–Hückel term. Then one starts with Eq. (17-23)* simplified for a pure electrolyte,

$$\frac{G^{Em}}{w_w RT} = -\frac{4IA_\phi}{b}\ln(1 + bI^{1/2}) + 2m_M m_X(B_{MX} + m_M z_M C_{MX}) \tag{18-6)*}$$

and with substitution for a salt with ν_M and ν_X ions in the formula, with charges z_M and z_X, with $\nu = \nu_M + \nu_X$, and with m the overall molality

$$\frac{G^{Em}}{n_2 RT} = \nu|z_M z_X|\frac{4A_\phi}{b}\ln(1 + bI^{1/2}) + 2\nu_M \nu_X m(B_{MX} + \nu_M z_M m C_{MX})$$

$$\tag{18-7)*}$$

We recall the universal value $1.2\,\text{kg}^{1/2}\,\text{mol}^{-1/2}$ for b.

Then differentiation as indicated in Eq. (18-5) yields for the apparent molar volume,

$$\phi V_{MX} = \overline{V}_{MX}^\circ + \nu|z_M z_X|\frac{A_V}{2b}\ln(1 + bI^{1/2})$$

$$+ 2\nu_M \nu_X RT[mB_{MX}^V + m^2(\nu_M z_M)C_{MX}^V] \tag{18-8)*}$$

where

$$B^V_{MX} = \beta^{(0)V} + \beta^{(1)V}g(\alpha_1 I^{1/2}) + \beta^{(2)V}g(\alpha_2 I^{1/2}) \qquad (18\text{-}9a)$$

$$\beta^{(i)V} = \left(\frac{\partial \beta^{(i)}}{\partial P}\right)_T \qquad i = 0,1,2 \qquad (18\text{-}9b)$$

$$C^V_{MX} = \left(\frac{\partial C_{MX}}{\partial P}\right)_T = \frac{(\partial C^\phi_{MX}/\partial P)_T}{2|z_M z_X|^{1/2}} \qquad (18\text{-}10)$$

$$g(x) = \frac{2[1 - (1+x)\exp(-x)]}{x^2} \qquad (17\text{-}18)$$

Equation (15-35) for A_ϕ, when differentiated, yields the Debye–Hückel parameter for volume

$$A_V = 2A_\phi RT\left[3\left(\frac{\partial \ln \varepsilon}{\partial P}\right)_T + \left(\frac{\partial \ln V_w}{\partial P}\right)_T\right] \qquad (18\text{-}11)$$

In an application one has experimental values of $^\phi V_{MX}$ for a series of values of m from which values of $\beta^{(0)V}_{MX}$, $\beta^{(1)V}_{MX}$, $\beta^{(2)V}_{MX}$, if present; C^V_{MX}; and $\overline{V}^\circ_{MX}$ are derived by regression or other appropriate means. One then has values of the pressure derivatives of the parent quantities $\beta^{(0)}_{MX}$, $\beta^{(1)}_{MX}$, etc., which are important components of a comprehensive treatment, as we shall see below.

The procedure for enthalpy is similar; from Eqs. (18-1), (18-3), and (18-7)* one finds

$$^\phi L_{MX} = \nu|z_M z_X|\frac{A_L}{2b}\ln(1 + bI^{1/2}) - 2\nu_M \nu_X RT^2[mB^L_{MX} + m^2(\nu_M z_M)C^L_{MX}] \qquad (18\text{-}12)^*$$

where

$$B^L_{MX} = \left(\frac{\partial B_{MX}}{\partial T}\right)_{P,I}$$

$$= \beta^{(0)L} + \beta^{(1)L}g(\alpha_1 I^{1/2}) + \beta^{(2)L}g(\alpha_2 I^{1/2}) \qquad (18\text{-}13a)$$

$$\beta^{(i)L} = \left(\frac{\partial \beta^{(i)}}{\partial T}\right)_P \qquad i = 0,1,2 \qquad (18\text{-}13b)$$

and

$$C^L_{MX} = \left(\frac{\partial C_{MX}}{\partial T}\right)_P = \frac{(\partial C^\phi_{MX}/\partial T)_P}{2|z_M z_X|^{1/2}} \qquad (18\text{-}14)$$

Also, the Debye–Hückel parameter for enthalpy is

$$\frac{A_L}{RT} = 4T\left(\frac{\partial A_\phi}{\partial T}\right)_P = -6A_\phi\left[1 + T\left(\frac{\partial \ln \varepsilon}{\partial T}\right)_P + \frac{T\alpha_w}{3}\right] \qquad (18\text{-}15)$$

where $\alpha_w = (\partial \ln V/\partial T)_P$ is the coefficient of thermal expansion of water.

A common type of enthalpy measurement is the heat of dilution. If this is reported for the dilution of solution containing 1 mole of solute from m_1 to m_2, it is related as follows to the apparent molal enthalpy:

$$\Delta_{dil} H(m_1 \to m_2) = {}^\phi L_2 - {}^\phi L_1 \tag{18-16}$$

The integral heat of solution of a solid salt MX is taken as the heat effect for the reaction,

$$n_2 MX(cr) + n_1 H_2O(l) = n_2 MX(aq, m)$$

The enthalpy change for this reaction is given as

$$\Delta_{sol} H = n_1 \overline{H}_1 + n_2 \overline{H}_2 - n_1 H_1^\circ - n_2 H_2^\circ(cr) \tag{18-17}$$

which may be rewritten as

$$\Delta_{sol} H = L + n_2[\overline{H}_2^\circ - H_2^\circ(cr)]$$

As the concentration m approaches zero, we have

$$\lim_{m \to 0} (\Delta_{sol} H / n_2) = \Delta_{sol} \overline{H}_2^\circ - H_2^\circ(cr) \tag{18-18}$$

where $\Delta_{sol} \overline{H}^\circ$ is the heat of solution per mole of salt at infinite dilution. At finite concentrations we therefore have

$$\Delta_{sol} H = \Delta_{sol} \overline{H}^\circ + {}^\phi L \tag{18-19}$$

The value of $\Delta_{sol} \overline{H}^\circ$ at a given temperature may be found by fitting the experimental values of $\Delta_{sol} \overline{H}$ to Eqs. (18-12)* and (18-19), treating $\Delta_{sol} \overline{H}^\circ$ as an adjustable parameter.

The apparent molal heat capacity, ${}^\phi C_P$, is defined to be

$$^\phi C_P = \frac{C_P - n_1 C_{P_1}^\circ}{n_2} \tag{18-20}$$

From Eqs. (18-2) and (18-3), we find that

$$^\phi C_P - \overline{C}_{P_2}^\circ = \left(\frac{\partial {}^\phi L}{\partial T}\right)_{P,m} \tag{18-21}$$

From Eq. (18-12)* one then obtains

$$^\phi C_{P,MX} = \overline{C}_{P,MX}^\circ + \nu|z_M z_X| \frac{A_J}{2b} \ln(1 + bI^{1/2})$$
$$- 2\nu_M \nu_X RT^2 [mB_{MX}^J + m^2(\nu_M z_M) C_{MX}^J] \tag{18-22}*$$

where

$$B^J_{MX} = \left(\frac{\partial^2 B_{MX}}{\partial T^2}\right)_{P,I} + \frac{2}{T}\left(\frac{\partial B_{MX}}{\partial T}\right)_{P,I} \quad (18\text{-}23)$$

$$C^J_{MX} = \left(\frac{\partial^2 C_{MX}}{\partial T^2}\right)_P + \frac{2}{T}\left(\frac{\partial C_{MX}}{\partial T}\right)_P \quad (18\text{-}24)$$

$$A_J = \left(\frac{\partial A_L}{\partial T}\right)_P \quad (18\text{-}25)$$

The superscript J is used in view of its definition as the relative heat capacity, $J = (\partial L/\partial T)_P$, and to minimize confusion between the use of the letter C with different subscripts for the third virial coefficient and for the heat capacity.

Heat capacity data on electrolytes are obtained by direct measurements on the solution using a calorimeter.

As described in Chapter 4, a twin flow calorimeter is often used, which measures the heat capacity of the solution relative to that of an equal volume of pure water. Then a simple calculation yields the apparent molar heat capacity without having to subtract two nearly equal quantities. The resulting values of $^\phi C_P$ may then be fitted to Eq. (18-22)*, treating $\overline{C}^\circ_{P_2}$ as an adjustable parameter. Values of $^\phi C_P$ may also be obtained from heat of solution data according to

$$\left(\frac{\partial \Delta_{sol} H}{\partial T}\right)_P = {}^\phi C_P - C^\circ_{P_2}(cr) \quad (18\text{-}26)$$

provided $C^\circ_{P_2}(cr)$, the heat capacity of the pure salt in the crystalline solid phase, is known.

Equations for the partial molar enthalpies, heat capacities, and volumes are readily derived from the appropriate derivatives of equations given above, while entropies are readily obtained from enthalpies and Gibbs energies. In past years, partial molal enthalpies and volumes were used in calculating the effects of temperature and pressure on activity and osmotic coefficients. Now that task is best accomplished by the use of the temperature and pressure derivatives of the virial coefficients, which are more directly determined from the apparent molal quantities as described above. Consequently, equations for partial molal enthalpies and volumes will be omitted here, but some of these expressions are given in References 1 and 2.

Values of the volume and enthalpy parameters for several common electrolytes at 25°C are given in Tables 18-1 and 18-2. They were calculated by Monnin[3] and Silvester and Pitzer[4] from original measurements cited there by the methods described above. While this type of information for 25°C is useful, it is more interesting and more important to consider a wide range of temperature, and we turn now to that type of application of these important thermodynamic relationships.

TABLE 18-1
Enthalpy parameters for several electrolytes at 25°C[a]

	1–1 type[b]			
	$10^4 \beta^{(0)L}$	$10^4 \beta^{(1)L}$	$10^5 C^{\phi L}$	Max m
HCl	-3.08_1	1.41_9	-6.21_3	4.5
LiCl	-1.68_5	5.36_6	-4.52_0	6.4
NaCl	7.15_9	7.00_5	-10.5_4	6.0
NaBr	7.69_2	10.7_9	-9.30	9.0
NaI	8.35_5	8.28	-8.35	6.0
NaOH	7.00	1.34	-18.9_4	4.2
NaClO$_4$	12.96	22.9_7	-16.2_3	6.0
NaNO$_3$	12.66	20.6_0	-23.1_6	2.2
KCl	5.79_4	10.71	-5.09_5	4.5
KNO$_3$	2.06	64.5	39.7	2.4

	2–1 or 1–2 type[b]			
	$10^3 \beta^{(0)L}$	$10^3 \beta^{(1)L}$	$10^4 C^{\phi L}$	Max m
MgCl$_2$	-0.19_4	2.77	-1.65	2.0
Mg(NO$_3$)$_2$	0.51_5	4.49	–	0.1
CaCl$_2$	-0.17_3	3.9_0	–	0.1
Na$_2$SO$_4$	2.36_7	5.63	-4.88	3.0
K$_2$SO$_4$	1.44	6.70	–	0.1

[a] All values are from L. S. Silvester and K. S. Pitzer, *J. Solution Chem.*, **7**, 327 (1978); dimensions are kg mol^{-1} K^{-1} for $\beta^{(0)L}$ and $\beta^{(1)L}$ and kg^2 mol^{-2} K^{-1} for C^L.
[b] For 1–1 type $C^{\phi L} = 2C^L$; for 2–1 or 1–2 type $C^{\phi L} = 2^{3/2} C^L$.

TABLE 18-2
Volume parameters for some electrolytes at 25°C[a]

	$\overline{V}°$	$10^5 \beta^{(0),V}$	$10^4 \beta^{(1),V}$	$10^7 C^V$	
	cm^3 mol^{-1}	kg mol^{-1} bar^{-1}	kg mol^{-1} bar^{-1}	kg^2 mol^{-2} bar^{-1}	m max
NaCl	16.68	1.234	–	-6.45	–
Na$_2$SO$_4$	11.787	5.308	1.234	-27.94	2.2
NaHCO$_3$	23.181	-1.162	1.78	–	1.0
Na$_2$CO$_3$	-6.48	5.98	0.816	-32.5	1.7
KCl	26.87	1.3949	0.0235	-8.70	4.7
K$_2$SO$_4$	32.167	3.348	2.38	–	0.65
KHCO$_3$	33.371	-0.2705	1.695	–	1.0
K$_2$CO$_3$	13.90	3.2889	2.07	-7.47	7.6
CaCl$_2$	17.419	1.3287	–	-2.17	–
MgCl$_2$	13.734	1.3833	–	-3.57	–

[a] From C. Monnin, *Geochim. Cosmochim. Acta*, **53**, 1177 (1989); for 1–1 type $C^{\phi V} = 2C^V$; for 2–1 or 1–2 type $C^{\phi V} = 2^{3/2} C^V$.

As one proceeds to consider data at other temperatures it is often desirable to convert measurements made at different pressures to a single pressure for comparison and for further calculations. The basic relationships are

$$\left(\frac{\partial G}{\partial P}\right)_T = V \tag{3-8}$$

$$\left(\frac{\partial H}{\partial P}\right)_T = V - T\left(\frac{\partial V}{\partial T}\right)_P \tag{3-15}$$

$$\left(\frac{\partial C_P}{\partial P}\right)_T = -T\left(\frac{\partial^2 V}{\partial T^2}\right)_P \tag{3-35}$$

With an equation for the volume as a function of T and P, it is straightforward to convert enthalpy and heat capacity data, as well as osmotic coefficients, to a different pressure. Rogers and Pitzer[5] give the equations needed for this conversion. Further details for this conversion are given for NaCl(aq) and KCl(aq) in references 1 and 6, respectively.

COMPREHENSIVE EQUATIONS OF STATE FOR ELECTROLYTES

With recent calorimetric developments allowing accurate heat capacity[7,8] and heat of dilution[9] measurements under pressure and at high temperatures, it has become possible to develop comprehensive equations that interrelate the activity and osmotic coefficients as well as thermal and volumetric properties over wide ranges of conditions. Typically, the heat capacity is measured at at least one uniform pressure from near 25°C to 300°C and sometimes higher temperatures for a wide composition range. The volume (or density) is measured over this same temperature and composition range and preferably at more than one pressure. In principle, with Gibbs energy and enthalpy information at 25°C, the heat capacity allows integration of these properties upward in temperature. And from the Gibbs energy as a function of molality, the osmotic and activity coefficients can be calculated.

For the liquid under conditions not too close to the critical point of water, the Gibbs energy is the preferred basis function with T, P, and m as independent variables. Very close to the critical point or where the vapor–liquid equilibrium is to be considered, the Helmholtz energy would be the better basis, as discussed in Chapters 9 and 13, and it is used for NaCl–H$_2$O under those conditions.[10] But we will not consider the near-critical or supercritical range at this point.

In actual applications one often has high-temperature measurements of the osmotic coefficient and/or the heat of dilution, and then one can regress the entire data set with appropriate weights to generate the comprehensive equation. Table 18-3 lists several systems[11–21] for which there are comprehensive equations of this type. The temperature dependency of certain quantities is so complex and requires so many parameters for accurate representation that complete expressions will not be given here.

TABLE 18-3
Systems for which comprehensive, high-temperature equations are available

System	Max $T/°C$	Max P/bar	Max $m/\text{mol kg}^{-1}$	Measured properties	Reference
HCl	a	400	a	C_P, ϕL, γ	11
NaCl	300	1000	6.0	V, C_P, ϕL, ϕ	1, 12
NaCl	325	1000	m_{sat}	V, C_P, ϕL, ϕ	13
NaBr	325	1500	b	V, C_P, ϕL, ϕ	14
NaOH	250	400	6.3	V, C_P, ϕL, ϕ	15
KCl	325	500	c	V, C_P, ϕL, ϕ	16
Na$_2$SO$_4$	300	200	m_{sat}	V, C_P, ϕL, ϕ	17
CaCl$_2$	250	400	4.6^d	C_P, ϕL, ϕ	18
MgSO$_4$	e	e	m_{sat}	C_P, ϕL, ϕ	21

[a] The basic equation for HCl is valid to 250°C and 7 mol kg^{-1}, but equations with extra terms have extended validity to 375°C and 15.6 mol kg^{-1}. $\bar{V}°_{HCl}$ is not included.

[b] At the lower temperatures the equation for NaBr is valid to saturation m_s but the C_P and ϕL basis extends only to 8 mol kg^{-1} below 200°C and to only 3 mol kg^{-1} above 200°C.

[c] To saturation below 220°C, but the C_P basis above 220°C is only to 3.0 mol kg^{-1}.

[d] A less comprehensive treatment[19,20] for CaCl$_2$ at saturation pressure extends to solid-saturation molality and to the equilibria with the saturating solids.

[e] The solubility of MgSO$_4$ becomes very small above 200°C, but some extrapolation is probably valid; this equation is for saturation pressure only.

For example, the Archer[13] equations for NaCl(aq) have 54 parameters with 810 digits in all. Additional information about some of these equations is given in the footnote to Table 18-3. For many purposes, less comprehensive equations suffice and they will be considered next.

Gibbs Energies at Various Temperatures.

If one is interested only in the activity and osmotic coefficients at saturation pressure and for temperatures below about 250°C, then the pressure effect is small and relatively simple equations suffice. Isopiestic measurements have been made for a substantial number of aqueous electrolytes for temperatures up to 250°C with NaCl–H$_2$O, the osmotic coefficients of other systems can be calculated and from them the ion-interaction parameters evaluated.

For example, Holmes and Mesmer[22] reported isopiestic measurements for KCl–H$_2$O and fitted Eq. (17-10)* with the following temperature-dependent expression for each Pitzer parameter $\beta^{(0)}$, $\beta^{(1)}$, C^ϕ:

$$f(T) = Q1 + Q2\left(\frac{1}{T} - \frac{1}{T_r}\right) + Q3\ln\left(\frac{T}{T_r}\right) + Q4(T - T_r)$$

$$+ Q5(T^2 - T_r^2) + Q6\ln(T - 260) \qquad (18\text{-}27)*$$

with $T_r = 298.15$ K. The range of validity is 237–523 K and to 7 mol kg^{-1}. The resulting Q parameters are given in Table 18-4.

TABLE 18-4
Parameters for ion-interaction terms for KCl–H$_2$O at temperatures from 273 to 523 K[a]

$$f(T) = Q1 + Q2\left(\frac{1}{T} - \frac{1}{T_r}\right) + Q3\ln\left(\frac{T}{T_r}\right) + Q4(T - T_r) + Q5(T^2 - T_r^2) + Q6\ln(T - 260)$$

	$\beta^{(0)}$	$\beta^{(1)}$	$C^\phi = 2C$
Q1	0.048 08	0.0476	-7.88×10^{-4}
Q2	-758.48	303.9	91.270
Q3	$-4.706\ 2$	1.066	0.586 43
Q4	0.010 072	0	$-0.001\ 2980$
Q5	-3.7599×10^{-6}	0	4.9567×10^{-7}
Q6	0	0.0470	0

[a] From H. F. Holmes and R. E. Mesmer, *J. Phys. Chem.*, **87**, 1242 (1983).

Holmes and Mesmer[22] also evaluated various published data and developed the following equation for the standard-state heat capacity:

$$\frac{C_{P,2}^\circ}{R} = -119.251 + 0.669\ 25T - 0.001\ 0259T^2 - \frac{82.5}{(T-270)} \quad (18\text{-}28)*$$

Appendix 10 presents equations of this type for ten common and important electrolytes and gives references to similar information for other systems both pure and mixed.

Solid Solubility in Electrolytes at High Temperatures.

With equations for the activity and osmotic coefficient at high temperatures, it is possible to calculate the solubility of solid salts (minerals) provided the chemical potential of the solid is also known at the same temperature. In many cases the heat capacity of the solid is known and thereby the entropy and enthalpy over a range of temperature. Then the chemical potential can be converted from 298.15 K to the temperature of interest. Thus, at any temperature

$$\frac{\mu_i^\circ}{RT} = \frac{H_i^\circ}{RT} - \frac{S_i^\circ}{R} \quad (18\text{-}29)$$

Then Eqs. (3-9) and (3-18) give the temperature derivatives of S_i° and H_i°, which can be integrated to yield

$$\frac{\mu_{i,T_f}^\circ}{RT_f} = \frac{\mu_{i,T_r}^\circ}{RT_f} - \frac{S_{i,T_r}^\circ}{R}\left(1 - \frac{T_r}{T_f}\right) + \int_{T_r}^{T_f} \frac{C_{P,i}^\circ}{R}\left(\frac{1}{T_f} - \frac{1}{T}\right)dT \quad (18\text{-}30)$$

with T_f the final temperature and $T_r = 298.15$ K at which μ_i° and S_i° are assumed to be known. Values of μ_i° and S_i° at T_r for several ions and minerals are given in Table 17-4.

The sources listed in Chapter 8 include many more solids and some give values for additional aqueous ions.

For anhydrous solids the heat capacity and entropy are almost always known, but sometimes estimates are needed for a hydrated salt. Pabalan and Pitzer[23] present estimates of the contribution per H_2O of hydration based on data for the hydrates of $MgCl_2$ and $MgSO_4$, as well as heat capacity equations, entropies, enthalpies, and chemical potentials for several minerals as well as their aqueous ions.

Once the chemical potentials are known for the temperature of interest, the equilibrium calculation is just that discussed in Chapter 17 and Eqs. (17-36) and (17-37) apply. We next consider, as an example, the solubility of Na_2SO_4 in pure aqueous solution using experimental values from Linke and Seidell.[25] The calculation is a prediction for higher temperatures since the 298 K value of the solubility determined the difference of the chemical potential between solid and solution. Figure 18-1 shows the results for both the comprehensive equation of state of

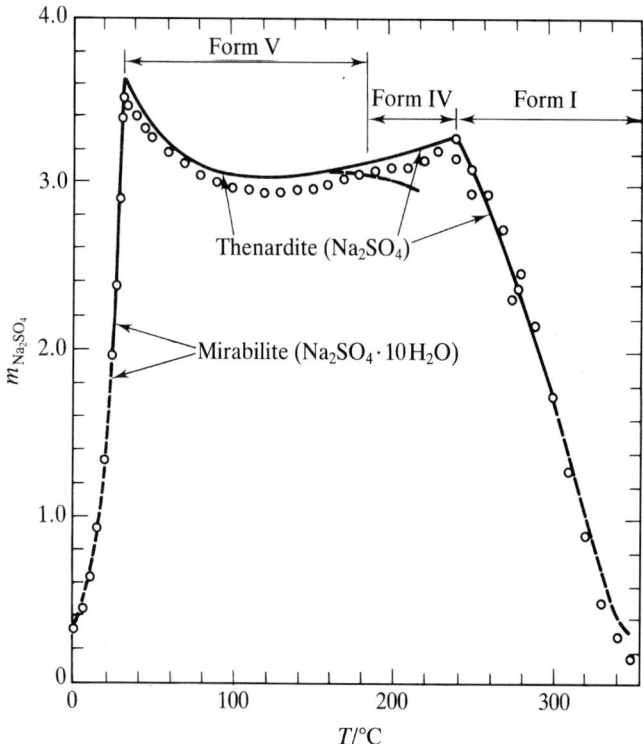

FIGURE 18-1
Solubilities of mirabilite ($Na_2SO_4 \cdot 10H_2O$) and thenardite (Na_2SO_4) in water as a function of temperature. The symbols are experimental data tabulated by Linke and Seidell,[25] and the curves are predicted values. The solid curve is for the comprehensive equation of Pabalan and Pitzer,[17] while the dashed curve near 200°C is for the earlier equation of Holmes and Mesmer.[24] The dashed lines below 25°C and above 300°C are for extrapolations.

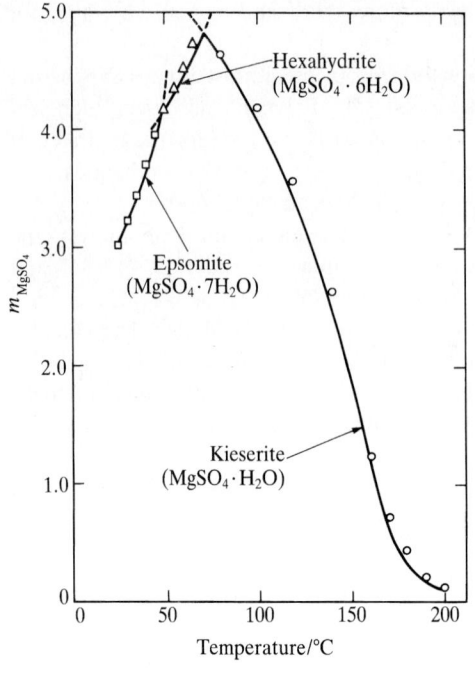

FIGURE 18-2
Calculated solubilities of Pabalan and Pitzer[21] for the system $MgSO_4-H_2O$ compared to experimental data from Linke and Seidell.[25]

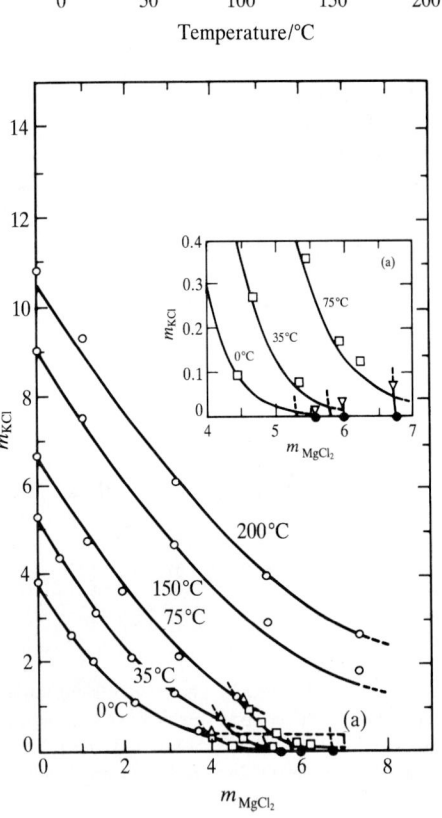

FIGURE 18-3
Calculated solubilities of Pabalan and Pitzer[23] for the system $KCl-MgCl_2-H_2O$ compared with experimental data from Linke and Seidell.[25] The inset shows an expanded view of the area delineated by box (a). The values of $\psi_{K,Mg,Cl}$ above 25°C were adjusted on the basis of the experimental data. Solid phases: ○, sylvite (KCl); □, carnallite ($KCl \cdot MgCl_2 \cdot 6H_2O$); ●, bischofite ($MgCl_2 \cdot 6H_2O$); △, sylvite + carnallite; ▷, carnallite and bischofite.

Pabalan and Pitzer[17] and for the earlier and simpler equation of Holmes and Mesmer,[24] which is included in Appendix 10. The agreement is excellent over the full range to 300°C for the comprehensive equation; indeed, the extrapolation to somewhat higher temperature shows surprising agreement. Below about 180°C the agreement is equally good for the simpler equation, but there is increasing deviation above 200°C.

Pabalan and Pitzer[23] present similar predictive calculations for NaCl, KCl, $MgCl_2$, K_2SO_4, and $MgSO_4$ and find comparably good agreement for all cases to 200°C and to 300°C for NaCl and KCl. The results for $MgSO_4$ are shown in Fig. 18-2.

We turn now to mineral solubility in mixed electrolytes. For these cases the mixing terms Φ_{ij} and ψ_{ijk} must be considered. Initial calculations for higher temperatures were made[23] using 298 K values for these parameters and then small adjustments were made in ψ if needed. In no case was a large change needed from the values at 298 K (see Tables 17-2 and 17-5). The results for the systems $KCl-MgCl_2-H_2O$ and $KCl-K_2SO_4-H_2O$ are shown in Fig. 18-3 and 18-4, respectively.

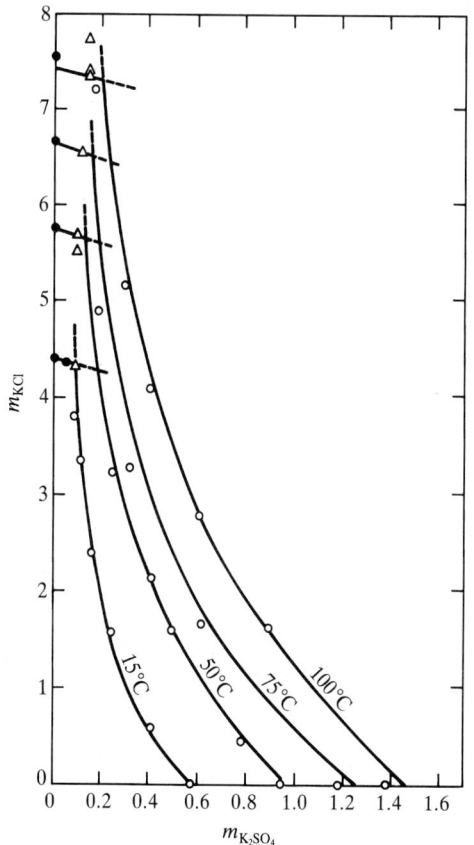

FIGURE 18-4
Calculated curves of Pabalan and Pitzer[23] compared with experimental solubilities from Linke and Seidell[25] for the $KCl-K_2SO_4-H_2O$ system. Solid phases: ○, arcanite (K_2SO_4); ●, sylvite (KCl); △, arcanite + sylvite.

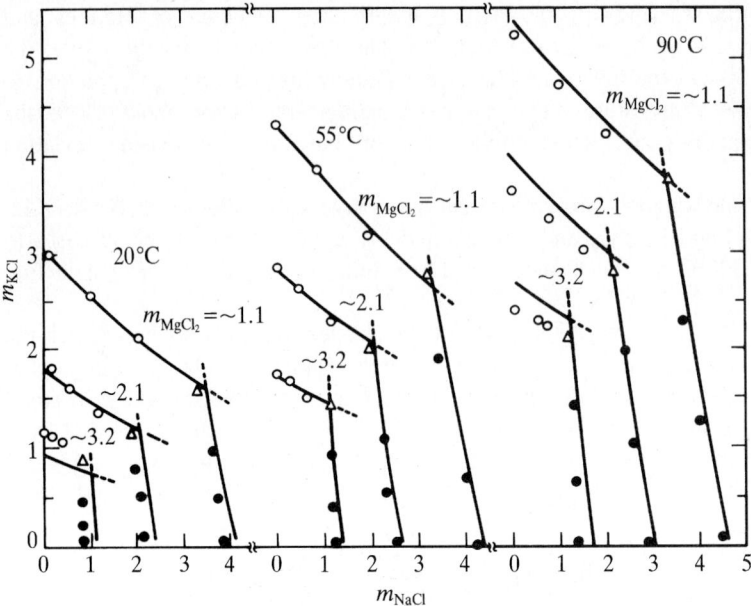

FIGURE 18-5
Predicted and experimental solubilities of halite and/or sylvite in the quaternary system NaCl–KCl–MgCl$_2$–H$_2$O at 20, 55, and 90°C and at MgCl$_2$ molalities of approximately 1.1, 2.1, and 3.2. Experimental data are from [Kayser, Kali, **17**, 1 (1923)]; calculations of Pabalan and Pitzer.[23] Solid phases: ○, sylvite (KCl); ●, halite (NaCl); △, sylvite + halite.

Once the Φ_{ij} and Ψ_{ijk} values are determined from common-ion, two-salt systems, no parameters remain to be adjusted, and the results for more complex mixtures are strictly predictions. Figure 18-5 shows the results for NaCl–KCl–MgCl$_2$–H$_2$O at several temperatures. The numerical values on the graph are the molalities of MgCl$_2$, which are constant for a series of values of the molalities of NaCl and KCl at saturation of one or both of the latter salts. Greenberg and Møller[26] and Møller[27] give several other examples of mineral solubility calculations for multicomponent electrolyte solutions.

Aqueous Electrolytes at Supercritical Temperatures

At still higher temperatures than the maxima for the examples in Table 18-3, the decrease in dielectric constant of water leads to strong association of ions. Thus NaCl(aq) becomes dominated by neutral molecules that have a very large dipole moment. The equations appropriate for these conditions are related to those described in Chapter 13, but special modifications and features are needed for good representation of various experimental properties. This is a relatively unexplored field. A recent study[28] concerns NaCl–H$_2$O, KCl–H$_2$O, and the ternary NaCl–KCl–H$_2$O.

Solubilities of Salts and of Ice at Low Temperatures

While no new principles are involved for temperatures below 0°C, there is the additional experimental opportunity of measurement of the equilibrium with ice. This is usually called the *freezing-point lowering*. It yields the osmotic coefficient at the temperature of equilibrium. Because high precision is readily obtained, freezing-point measurements were, for many years, an active research area.

Both enthalpy and heat capacity data are required, however, for conversion of these osmotic coefficients from the freezing temperature to 25°C or another temperature of interest. Since these thermal data for the range below 25°C are often missing or inaccurate, most recent research has emphasized precise measurements at 25°C rather than the freezing-point lowering as an indirect source of information at 25°C.

A treatment of freezing-point lowering of nonelectrolytes in the molality system was given in Chapter 14. This treatment was converted for a dilute electrolyte in Eq. (15-27)*. A treatment including higher-order terms is available.[29]

We prefer to present here a more general treatment considering the activity as well as the osmotic coefficient; this allows treatment of solid salt solubility as well as freezing-point lowering. The basic equations are exactly the same as those given above for high temperatures. One needs only to obtain temperature-dependency expressions valid to the low temperatures of interest. A comprehensive treatment for the Na–K–Mg–Ca–Cl–SO$_4$–H$_2$O system at temperatures below 25°C was presented by Spencer et al.[30] We give here only the equations and parameters for Na–K–Cl–H$_2$O as an illustration of this type of treatment.

For any parameter, $\beta_{ij}^{(0)}$, $\beta_{ij}^{(1)}$, C_{ij}^{ϕ}, θ_{ij}, ψ_{ijk}, and $\Delta_r\mu°$, Spencer et al. use the equation

$$P(T) = a_1 + a_2T + a_6T^2 + a_9T^3 + \frac{a_3}{T} + a_4 \ln T \qquad (18\text{-}31)*$$

Note that $\Delta_r\mu°$ is for the solubility reaction (also that $\Phi_{ij} = \theta_{ij}$ for this case),

$$H_2O(cr) = H_2O(l)$$

$$MCl(cr) = M^+(aq) + Cl^-(aq)$$

The numerical values of these parameters are given in Table 18-5. For the range 0–25°C the available knowledge from comprehensive treatments was used; also, the ice–water standard chemical potential difference is from Speedy[31] and is based on enthalpy and heat capacity data. But below 0°C, most of the other parameters were determined to fit the mineral solubility and the freezing-point lowering data. Thus, this is a more "empirical" treatment with respect to temperature dependency below 0°C than those discussed above for high temperatures, but it is still constrained by the molality dependence of the basic equations. The results are shown in Figs. 18-6, 18-7, and 18-8. Here the solid solubility data are from a compilation of Bukshtein,[32] while freezing-point lowering data are from Hall.[33]

TABLE 18-5
Values of the fitting constants, Eq. (18-31)*, for the Debye–Hückel model parameter, A_ϕ, and for other parameters as a function of temperature (K)[29]

Parameter	a_1	a_2	a_6	a_9	a_3	a_4
A_ϕ	8.66836498×10^1	$8.48795942 \times 10^{-2}$	$-8.88785150 \times 10^{-5}$	$4.88096393 \times 10^{-8}$	$-1.32731477 \times 10^{-3}$	-1.76460172×10
$\beta^{(0)}_{Na,Cl}$	7.87239712	$-8.3864096 \times 10^{-3}$	$1.44137774 \times 10^{-5}$	$-8.7820301 \times 10^{-9}$	-4.96920671×10^2	$-8.20972560 \times 10^{-1}$
$\beta^{(1)}_{Na,Cl}$	8.66915291×10^2	$6.06166931 \times 10^{-1}$	$-4.80489210 \times 10^{-4}$	$1.88503857 \times 10^{-7}$	-1.70460145×10^4	-1.67171296×10^2
$C\phi_{Na,Cl}$	1.70761824	$2.32970177 \times 10^{-3}$	$-2.46665619 \times 10^{-6}$	$1.21543380 \times 10^{-9}$	-1.35583596	$-3.87767714 \times 10^{-1}$
$\beta^{(0)}_{K,Cl}$	2.65718766×10^1	$9.92715099 \times 10^{-3}$	$-3.62323330 \times 10^{-6}$	$-6.28427180 \times 10^{-11}$	-7.55707220×10^2	-4.67300770
$\beta^{(1)}_{K,Cl}$	1.69742977×10^3	1.22270943	$-9.99044490 \times 10^{-4}$	$4.04786721 \times 10^{-7}$	-3.28684422×10^4	-3.28813848×10^2
$C\phi_{K,Cl}$	-3.27571680	$-1.27222054 \times 10^{-3}$	$4.71374283 \times 10^{-7}$	$1.1162507 \times 10^{-11}$	9.07747666×10	$5.80513562 \times 10^{-1}$
$\theta_{Na,K}$	-1.82266741×10^1	$-3.69038470 \times 10^{-3}$	0	0	6.12415011×10^2	3.02994981
$\psi_{Na,K,Cl}$	6.48108127	$1.46803468 \times 10^{-3}$	0	0	-2.04354019×10^2	-1.09448043
$\Delta_r\mu°$, ice	7.875060393×10^3	1.169118490×10	$-1.7183789 \times 10^{-2}$	$1.24395543 \times 10^{-5}$	-9.3314790×10^4	-1.7287461×10^3
$\Delta_r\mu°$, halite	9.14839001×10^3	8.22348745	$-8.1288759 \times 10^{-3}$	$3.95552403 \times 10^{-6}$	-1.54040868×10^5	-1.83624247×10^3
$\Delta_r\mu°$, hydrohalite	-1.2222551×10^4	-9.8806459	$8.46685083 \times 10^{-3}$	$-3.4459117 \times 10^{-6}$	2.09823965×10^5	2.42328528×10^3
$\Delta_r\mu°$, sylvite	-1.62917341×10^3	-1.51940390	$1.45249679 \times 10^{-3}$	$-6.9427505 \times 10^{-7}$	2.26012743×10^4	3.33075506×10^2

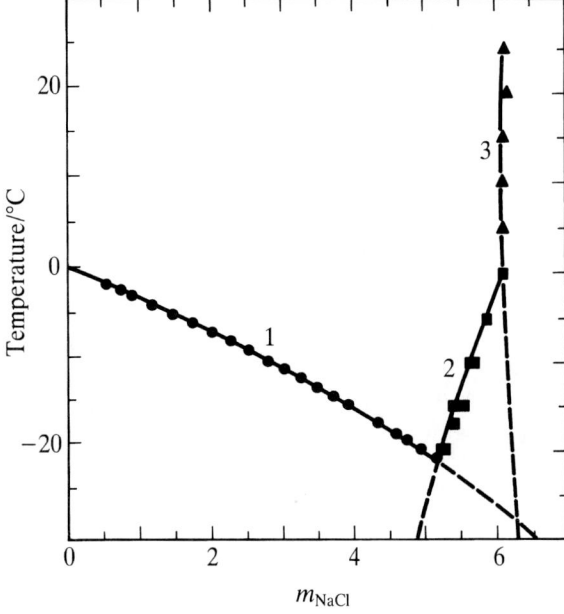

FIGURE 18-6
Calculated curves and experimental data in the system NaCl–H_2O are shown for solutions in equilibrium with ice (1, circles), hydrohalite (NaCl–H_2O) (2, squares), and halite (3, triangles); from Spencer et al.[30]

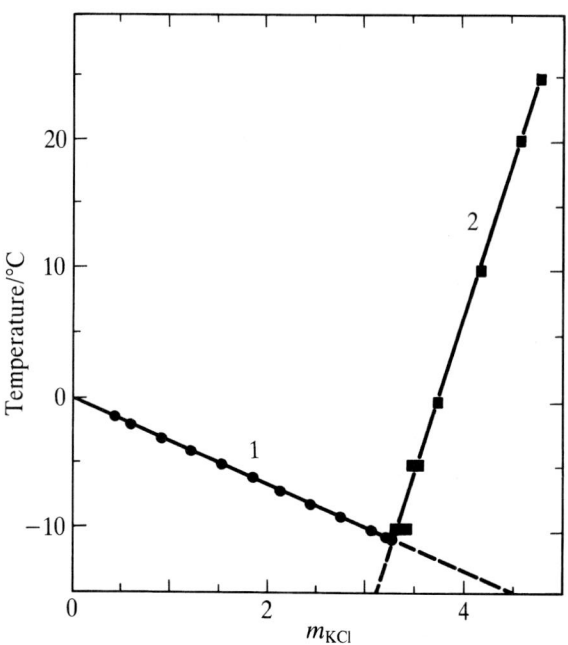

FIGURE 18-7
Calculated curves and experimental data in the system KCl–H_2O are shown for solutions in equilibrium with ice (1, circles) and sylvite (2, squares); from Spencer et al.[30]

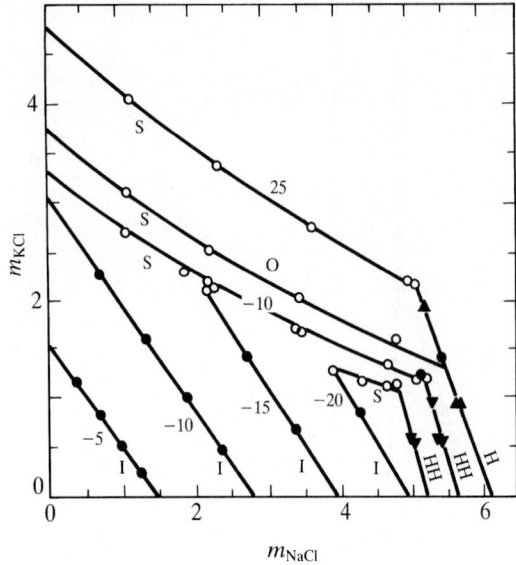

FIGURE 18-8
Isothermal sections for the system NaCl–KCl–H$_2$O showing calculated curves and experimental data for sylvite solubility (S, open circles) at –20, –10, 0, and 25°C; hydrohalite solubility (HH, inverted triangles) at –20 and –10°C; and halite solubility (H, upright triangles) at 0 and 25°C. Ice melting curves (I) and data (filled circles) are shown at –20, –15, –10 and –5°C (from Spencer et al.[30]).

For some of the other systems treated by Spencer et al.,[30] the results are generally similar, although some of the data are less accurate. Indeed, good agreement was obtained for solutions containing CaCl$_2$ with or without other salts to temperatures below –50°C.

REFERENCES

1. K. S. Pitzer, J. C. Peiper, and R. H. Busey, *J. Phys. Chem. Ref. Data*, **13**, 1 (1984).
2. K. S. Pitzer and L. Brewer, *Thermodynamics*, 2nd edn., revision of 1st edn. by G. N. Lewis and M. Randall, McGraw-Hill, New York, 1961.
3. C. Monnin, *Geochim. Cosmochim. Acta*, **53**, 1177 (1989).
4. L. F. Silvester and K. S. Pitzer, *J. Solution Chem.*, **7**, 327 (1978).
5. P. S. Z. Rogers and K. S. Pitzer, *J. Phys. Chem. Ref. Data*, **11**, 15 (1982).
6. R. T. Pabalan and K. S. Pitzer, *J. Chem. Eng. Data*, **33**, 354 (1988).
7. D. Smith-Magowan and R. H. Wood, *J. Chem. Thermodyn.*, **13**, 1047 (1981).
8. P. S. Z. Rogers and K. S. Pitzer, *J. Phys. Chem.*, **85**, 2886 (1981).
9. R. H. Busey, H. F. Holmes, and R. E. Mesmer, *J. Chem. Thermodyn.*, **16**, 343 (1984).
10. K. S. Pitzer and J. C. Tanger, IV, *Int. J. Thermophys.*, **9**, 635 (1988); A. Anderko and K. S. Pitzer, *Geochim. Cosmochim. Acta*, **57**, 1657 (1993).
11. H. F. Holmes, R. H. Busey, J. M. Simonson, R. E. Mesmer, D. G. Archer, and R. H. Wood, *J. Chem. Thermodyn.*, **19**, 863 (1987).
12. P. S. Z. Rogers and K. S. Pitzer, *J. Phys. Chem. Ref. Data*, **11**, 15 (1982).
13. D. G. Archer, *J. Phys. Chem. Ref. Data*, **21**, 793 (1992).
14. D. G. Archer, *J. Phys. Chem. Ref. Data*, **20**, 509 (1991).
15. J. M. Simonson, R. E. Mesmer, and P. S. Z. Rogers, *J. Chem. Thermodyn.*, **21**, 561 (1989).
16. R. T. Pabalan and K. S. Pitzer, *J. Chem. Eng. Data*, **33**, 354 (1988).
17. R. T. Pabalan and K. S. Pitzer, *Geochim. Cosmochim. Acta*, **52**, 2393 (1988).

18. H. F. Holmes, R. H. Busey, J. M. Simonson, and R. E. Mesmer, *J. Chem. Thermodyn.*, **26**, 271 (1994).
19. K. S. Pitzer and Y. Shi, *J. Solution Chem.*, **22**, 99 (1993).
20. K. S. Pitzer and C. S. Oakes, *J. Chem. Eng. Data*, in press.
21. R. C. Phutela and K. S. Pitzer, *J. Phys. Chem.*, **90**, 895 (1986).
22. H. F. Holmes and R. E. Mesmer, *J. Phys. Chem.*, **87**, 1242 (1983).
23. R. T. Pabalan and K. S. Pitzer, *Geochim. Cosmochim. Acta*, **51**, 2429 (1987).
24. H. F. Holmes and R. E. Mesmer, *J. Solution Chem.*, **15**, 495 (1986).
25. W. F. Linke and A. Seidell, *Solubilities of Inorganic and Metal Organic Compounds*, 4th edn., American Chemical Society, 1965.
26. J. P. Greenberg and N. Møller, *Geochim. Cosmochim. Acta*, **53**, 2503 (1989).
27. N. Møller, *Geochim. Cosmochim. Acta*, **52**, 821 (1988).
28. A. Anderko and K. S. Pitzer, *Geochim. Cosmochim. Acta*, **57**, 1657, 4885 (1993).
29. Reference 2, pp. 406–412.
30. R. J. Spencer, N. Møller, and J. H. Weare, *Geochim. Cosmochim. Acta*, **54**, 575 (1990).
31. R. J. Speedy, *J. Phys. Chem.*, **91**, 3354 (1987).
32. V. M. Bukshtein, M. G. Valyashko, and A. D. Pel'sh, eds., *Spravochnik po rastvorimosti solevykh sistem*, vols. I and II, Izd. Vses. Nauch.-Issled. Inst. Goz., Goskhimizdat., Moscow-Leningrad, 1953, p. 1270.
33. D. L. Hall, S. M. Sterner, and R. J. Bodnar, *Econ. Geol.*, **83**, 197 (1988).

PROBLEMS

18-1.
  (a) From Table 17-4 find $\Delta H°$ for the reaction $Na_2SO_4(s) = Na_2SO_4(aq)$.
  (b) From Table 18-1 and the result of part (a) find ΔH for the reaction $Na_2SO_4(s) + 55.5\ H_2O(l)$ = solution of Na_2SO_4 in $55.5\ H_2O$.

18-2. Calculate the apparent molar enthalpy $^\phi L_{MX}$ at 25°C for aqueous $NaClO_4$ at $6\ mol\ kg^{-1}$. Relate this to the heat of solution of solid $NaClO_4$ and to the heat of dilution of this solution.

18-3. Calculate the apparent molar volume and the density of 4 molal KCl (aq) at 25°C.

CHAPTER 19

ELECTROCHEMICAL CELLS AND ELECTRODE POTENTIALS

We have already considered the relationship between the potential of an electrochemical or galvanic cell and the Gibbs-energy change of the chemical reaction which accompanies passage of current through the cell. In this chapter a number of additional topics related to cells will be discussed. These include a detailed analysis of the various chemical potentials in a cell, the definition of half-cell potentials, and a number of examples of the use of cells to determine important thermodynamic quantities.

Half-cell potentials are useful in the discussion of oxidation–reduction chemistry since they are proportional to Gibbs-energy changes per equivalent and therefore indicate immediately the possible direction of a given redox reaction. Thus many authors appropriately include in tables of half-cell potentials many values obtained by indirect methods. Since our interest is primarily the use of cells to obtain quantitative Gibbs-energy values, we merely recognize these indirect methods here but will discuss them elsewhere as methods of obtaining Gibbs energies, entropies, etc.

To be useful for accurate thermodynamic purposes, a galvanic cell must be reversible, and the chemical change that occurs on passage of current must be known and precisely defined. Not all cells meet these criteria even though they may yield a measurable voltage. Other cells, which would seem to be satisfactory in theory, are unsatisfactory because one or both electrodes polarize, i.e., fail to react reversibly when a minute current is passing. Electrode polarization is a complex subject that falls into the general domain of chemical kinetics.[1] For thermodynamic purposes we confine our attention to electrodes that are reversible, at least with the small current necessary for precise potentiometric measurement.

One more possible difficulty that must be mentioned is the physical state of the electrode materials. Solids, and in particular solid metals, can differ in physical state because of strains or crystal imperfections to a degree that appreciably affects the measured voltage of the cell. The standard state of the metal is the most stable state; hence these effects can be in only one direction. We cannot hope to give a full discussion of the means by which unstrained solids may be produced, but in general spongy metal produced by electrolysis is less likely to be strained than rolled or polished bar or sheet material. The difficulty of reproducing the physical state of solid Ag and AgCl is such that Bates et al.[2] suggest the use of y_{HCl} as a means of standardization of individual electrodes. They recommend 0.904 for y_{HCl} at 25°C and 0.01 mol kg^{-1}.

CHEMICAL POTENTIAL IN AN ELECTROCHEMICAL CELL

It is assumed that the reader is familiar with the general nature of electrochemical (galvanic) cells. Nevertheless, the exact relationship of the various thermodynamic quantities in a simple example will be carefully considered because there are novel aspects that have not arisen in other systems.

The measured emf of a galvanic cell is the electrical potential difference between two wires of the same metal attached to the two electrodes. This will be written $V_2 - V_1$. Our task is to relate this potential difference to the properties of other components of the cell. This is best done by considering the escaping tendency of electrons, which may be measured by μ_{e^-}, their partial molar Gibbs energy, or chemical potential.

When electrical connection is made between two pieces of different metals, electrons flow instantly until the escaping tendency of electrons is the same in each metal. While this requires that initially neutral metals become charged, the magnitude of the electron excess or deficiency is extremely small. Likewise, when a zinc electrode is placed in an electrolyte solution, zinc ions dissolve from or deposit onto the electrode until the escaping tendency of the zinc ions is the same in the electrode and the solution. Again the charge transferred is so small that the ratio of valence electrons to zinc ions in the zinc metal is still almost exactly 2. Consequently, one may write

$$\mu_{Zn^{2+}} + 2\mu_{e^-} = \mu_{Zn}$$

where μ_{Zn} is a constant at a given temperature and pressure. Then, in the solution

$$\mu_{Zn^{2+}} \text{ (in soln)} = \mu_{Zn^{2+}} \text{ (in metal)}$$

$$\mu_{e^-} = \tfrac{1}{2}[\mu_{Zn} - \mu_{Zn^{2+}} \text{ (in soln)}]$$

This result shows the relationship between the escaping tendency of the zinc ion in solution and that of electrons in the electrode or a wire attached to the electrode.

Consider next a complete cell A such as is indicated by the following succession of symbols for the various parts:

$$\text{Pt}, \text{H}_2(\text{g}) | \text{HCl(aq, sat with H}_2) | \text{HCl(aq, sat with AgCl)} | \text{AgCl}, \text{Ag} \qquad \text{(A)}$$

The concentration of HCl in the two aqueous phases is the same. Ordinarily the solubilities of both AgCl and H_2 in the aqueous HCl are assumed to be so small that the two aqueous solutions may be regarded as identical. Thus the cell may be described as 'without liquid junction.' Nevertheless, it is important to realize that the solution near the silver chloride electrode is saturated with AgCl and substantially free from H_2, whereas the solution near the hydrogen electrode is saturated with hydrogen and free from AgCl. If the same portion of the solution contained both H_2 and AgCl, these might spontaneously react, although the rate would probably be too slow to affect the electrical-potential measurements in this case.

At the silver chloride electrode the following equations apply:

$$\mu_{\text{Cl}^-} \text{ (in AgCl)} = \mu_{\text{Cl}^-} \text{ (in aq HCl)}$$

$$\mu_{\text{Ag}^+} \text{ (in Ag metal)} = \mu_{\text{Ag}^+} \text{ (in AgCl)}$$

$$\mu_{\text{Ag}^+} + \mu_{\text{Cl}^-} = \mu_{\text{AgCl}}$$

$$\mu_{\text{Ag}^+} + 2\mu_{\text{e}^-} = \mu_{\text{Ag}}$$

These yield

$$\mu_{\text{e}^-} \text{ (in Ag)} = \mu_{\text{Ag}} - \mu_{\text{AgCl}} + \mu_{\text{Cl}^-}$$

At the hydrogen electrode one has

$$\mu_{\text{H}^+} \text{ (in aq HCl)} + \mu_{\text{e}^-} \text{ (in Pt)} = \tfrac{1}{2}\mu_{\text{H}_2}$$

These may be combined to yield

$$\mu_{\text{e}^-} \text{ (in Ag)} - \mu_{\text{e}^-} \text{ (in Pt)} = \mu_{\text{Ag}} + (\mu_{\text{H}^+} + \mu_{\text{Cl}^-}) - \mu_{\text{AgCl}} - \tfrac{1}{2}\mu_{\text{H}_2} \qquad (19\text{-}1)$$

The possibility of electrical charging of substances, provided that the charges are small, does not affect the escaping tendency of neutral species; hence μ_{H_2}, μ_{AgCl}, and μ_{Ag} may be identified with the ordinary chemical potentials of these substances. The escaping tendency of a charged species is affected by the net electrical charge of the phase in which it is located.† However, whatever the effect of such a net charge in the aqueous solution may be upon the escaping tendency of the H^+ ion, it will have exactly the opposite effect on the escaping tendency of the Cl^- ion. Thus the sum $\mu_{\text{H}^+} + \mu_{\text{Cl}^-}$

---

† In order to emphasize this additional effect upon the escaping tendency of ions, Guggenheim uses the term *electrochemical potential* [E. A. Guggenheim, *J. Phys. Chem.*, **33**, 842 (1929)].

is not affected by such a net charge and may be identified with the ordinary chemical potential, or partial molar Gibbs energy, of HCl for the aqueous solution present in the cell.

There is a second complication in the interpretation of the escaping tendency of H^+ and Cl^-. The solution near the Ag,AgCl electrode is saturated with AgCl but presumably free from H_2, whereas the solution near the hydrogen electrode is saturated with H_2 but free from Ag^+. Thus there will be diffusion of Ag^+ toward the hydrogen electrode (where at equilibrium it will be reduced) and a diffusion of H_2 in the opposite direction. Also, the dissolving of AgCl increases the concentration of Cl^- above that of H^+ in the solution near the Ag,AgCl electrode. All these effects approach zero as the solubilities of AgCl and H_2 approach zero and are probably negligible in this particular cell at 25°C. At higher temperature or for other cells of similar type, but in which the substances are more soluble, the errors from these complications become more significant.

Let us now complete the analysis of the cell indicated by the diagram, now abbreviated in a conventional manner to Pt,H_2|HCl(aq)|AgCl,Ag. If positive electricity passes from left to right (or negative electricity from right to left) through the cell in the amount of 1 equiv, i.e., 1 mole of electrons, the net reaction is

$$\tfrac{1}{2}H_2(g) + AgCl(s) = Ag(s) + H^+(aq) + Cl^-(aq)$$

The Gibbs-energy change for this chemical reaction is clearly given by the right side of Eq. (19-1) and is thus measured by the difference in escaping tendency of electrons between the two electrodes. But the measured emf of the cell E is the difference in electrical potential between wires attached to the two electrodes and is conventionally taken as

$$E = V_2 - V_1 \qquad (19\text{-}2)$$

where subscript 2 refers to the right electrode and 1 to the left electrode. Also the electrical potential is defined for a unit positive test charge, whereas electrons are negative and have a charge F (the Faraday constant) per mole. Hence

$$E = V_2 - V_1 = -\frac{\mu_{e^-}(\text{in Ag}) - \mu_{e^-}(\text{in Pt})}{F} \qquad (19\text{-}3)$$

or in general

$$E = V_2 - V_1 = -\frac{\mu_{e^-}(2) - \mu_{e^-}(1)}{F} = -\frac{\Delta G}{nF} \qquad (19\text{-}4)$$

The last equation introduces the Gibbs-energy change for the cell reaction, ΔG, divided by the number of equivalents of electricity n flowing through the external circuit when that amount of reaction occurs in the cell.

The final result of Eq. (19-4) conforms to the basic thermodynamic principles, since nFE is the electrical work done in the external circuit when the cell changes Gibbs-energy content by $-\Delta G$.

CONVENTIONS REGARDING THE SIGN OF ELECTROMOTIVE FORCE

Let us now direct our attention specifically to the name and sign to be ascribed to the measured emf and to the related definitions and conventions.[3] We continue to consider as an example the cell A:

$$Pt, H_2 | HCl(aq) | AgCl, Ag \tag{A}$$

together with the definitions associated with Eq. (19-2). For cell A in particular this yields

$$E_A = V_{Ag, AgCl} - V_{H_2} \tag{19-2a}$$

If all substances involved have their standard partial molal free energies, then these potentials are *standard* potentials and are so indicated.

$$E_A^\circ = V_{Ag, AgCl}^\circ - V_{H_2}^\circ = 0.222 \text{ V} \tag{19-2b}$$

The chemical reaction of a cell is, by convention, written in the direction such that it corresponds to positive electricity flowing through the cell from left to right (or negative electricity flowing through the cell from right to left). Thus the reaction for cell diagram A is

$$\tfrac{1}{2}H_2(g) + AgCl(s) = H^+(aq) + Cl^-(aq) + Ag(s) \tag{A}$$

If the cell emf E is positive (as it will be for this cell at 1 bar of hydrogen pressure and 1 molal hydrochloric acid), then positive electricity spontaneously tends to flow from the right electrode to the left electrode in the external circuit and from left to right within the cell. Consequently a positive E corresponds to a chemical reaction that tends to take place spontaneously and to a negative ΔG for the reaction. As noted in the preceding section, electrical work received by the cell at constant temperature and pressure is ΔG, and

$$\Delta \mu = \Delta G = -nFE \tag{19-4a}$$

where F is the Faraday constant and n is the number of equivalents of electricity flowing per mole of reaction as written.

The conventions and equations given above for the *cell* emf E are entirely consistent with those of the earlier editions of this book. The electrical potentials V were first introduced by Gibbs[4] in 1877 with the words 'the electrical potentials in pieces of the same kind of metal connected with the two electrodes.' Since each quantity has certain advantages in further applications, we shall continue to use both and wish to emphasize certain differences at this point.

If we write cell A in the reverse direction, securing cell B,

$$Ag, AgCl | HCl(aq) | H_2, Pt \tag{B}$$

the emf E has the opposite sign, that is, $E_B = -E_A$, but the expression $V_{Ag, AgCl} - V_{H_2}$ has a definite and unchanged value regardless of the diagram used or any other adequate description of the cell. Also, the relationship of E to ΔG is fully specified in

Eq. (19-4) and the associated definitions. It remains unchanged when the cell is reversed, but the relationship of $V_{Ag,AgCl} - V_{H_2}$ to ΔG becomes definite only after the chemical reaction has been written in one direction or the other. If one accepts the cell diagram and the convention that the chemical reaction corresponds to positive electricity flowing through the cell from left to right, then one may write

$$\Delta G = nF(V_{left} - V_{right}) \quad (19\text{-}5)$$

which follows at once from the rearrangement of Eq. (19-4). Thus we see that the Gibbs electrical potentials have certain advantages in dealing directly with galvanic cells, but that the definitions associated with the cell emf are necessary in relating cell information unambiguously to chemical reactions.

HALF CELLS AND ELECTRODE POTENTIALS

Galvanic cells can always be divided into two parts, one associated with each electrode, and these respective parts or half cells may then be recombined in different pairs to yield additional cells. There are, of course, problems associated with the joining of half cells. Sometimes their electrolytes are incompatible and junction is possible, if at all, only by interposing a third electrolyte. We shall lay aside for the present these problems and those of the electrical potentials associated with complex junctions and assume that the half cells to be considered can be combined without appreciable junction potential into complete cells.

It is obvious that there is a great simplification if a table of half-cell potentials will suffice to allow the calculation of the emf of all cells. It is well known that this is possible provided that a particular half cell is selected as a reference point. The hydrogen electrode is generally accepted as this reference electrode, whose potential is assigned the value zero when the fugacity of hydrogen gas is 1 bar and the activity of hydrogen ion is 1 mol kg^{-1}. Actually one does not determine the activity of a single ion; when a whole cell is considered, we have seen that its emf depends on a combination of activities corresponding to electroneutrality.

Thus we may write

$$Pt, H_2 | H^+: \quad E^\circ = V^\circ = \Delta\mu^\circ = 0.000$$

with the chemical half reaction

$$\tfrac{1}{2}H_2 = H^+ + e^-$$

The electron indicated by e$^-$ is, of course, delivered through the connecting wire into the external circuit. Since the electrons always cancel when half reactions are combined into whole-cell reactions, it is necessary only that their state be the same. However, in view of the zero value of $\Delta\mu^\circ$ for the hydrogen half reaction, it is possible in a formal way to say that the chemical potential of the electron in its standard state is that which exists in equilibrium in any wire connected to a standard hydrogen half cell.

If we subtract the reaction above for the hydrogen half cell from the complete reaction (A) above, one has for the AgCl,Ag half cell the reaction

$$AgCl(s) + e^- = Ag(s) + Cl^-(aq)$$

and we attribute to this half cell the standard potential $E^\circ = 0.222$ V and the $\Delta\mu^\circ = 21.4$ kJ mol^{-1} of cell A. Since this reaction is a reduction of AgCl(s) to Ag(s), the term "reduction potential" is an unambiguous description. The term "electrode potential" is, however, in more common use.

It is possible to reverse the direction of reaction (A) and of each half reaction. This changes the sign of the cell potential to -0.222 V and is associated with the reaction

$$Ag(s) + Cl^-(aq) = AgCl(s) + e^-$$

Since this is an oxidation of Ag(s) to AgCl(s), the appropriate term is "oxidation potential", and the value -0.222 is associated with the AgCl,Ag electrode.

In his extensive studies of cell potentials and his important book *Oxidation Potentials*, Latimer[5] chose this second convention. The information in Latimer's book continues to be of great value, and should be consulted widely but with recognition of his sign convention.

The Gibbs electrical potentials can also be considered in this connection. If we define the standard hydrogen half-cell potential to be zero ($V^\circ_{H_2} = 0$), then there is no ambiguity in the sign of the remaining half-cell potentials. Thus from Eq. (19-2b) $V^\circ_{Ag,AgCl} = +0.222$ V. Since this potential value is associated with the electrode rather than the direction of the assumed chemical half reaction, the term *electrode potential* seems appropriate. We follow the definitions $E = V^*_{right} - V^*_{left} = V^*_{cathode} - V^*_{anode}$ and then use the quantity E in the thermodynamic treatment of whole cells. Note that the anode is an electrode where oxidation is occurring or is assumed to occur by the chemical reaction written, and a cathode an electrode where reduction is occurring or is assumed to occur, irrespective of whether the reaction is spontaneous or is forced by an external emf. Let us illustrate this procedure with an example. Table 19-3 near the end of this chapter contains the entries

$$\tfrac{1}{2}Cl_2 + e^- = Cl^- \qquad E^\circ = 1.3583 \text{ V}$$

$$AgCl + e^- = Ag + Cl^- \qquad E^\circ = 0.2223 \text{ V}$$

$$H^+ + e^- = \tfrac{1}{2}H_2 \qquad E^\circ = 0.0000 \text{ V}$$

Consider a cell made up of a chlorine electrode, that is, Cl_2, at 1 bar on an inert electrode M, an aqueous chloride solution (e.g., sodium chloride), and a silver–silver chloride electrode. The diagram for the cell is

$$M,Cl_2 | Cl^-(aq) | AgCl,Ag \qquad\qquad (C)$$

and the chemical equation that corresponds to it is

$$AgCl(s) = Ag(s) + \tfrac{1}{2}Cl_2(g)$$

The chlorine electrode may be seen to be the positive electrode of the cell directly from the tabulated $E°$ values since the value 1.3583 is more positive than 0.2223. But the cell emf is

$$E° = V°_{right} - V°_{left} = 0.2223 - 1.3583 = -1.1360 \text{ V}$$

where the negative value indicates at once that the chemical reaction tends to go in the reverse direction; i.e., silver tends to react spontaneously with chlorine to yield silver chloride.

We note further that the convention of positive electricity flowing through the cell from left to right implies that the left electrode is the anode and the right electrode the cathode. Thus one could instead have described the cell as a chlorine anode and a silver–silver chloride cathode with an aqueous chloride electrolyte. Then the same calculation follows,

$$E° = V°_{cathode} - V°_{anode} = 0.2223 - 1.3583 = -1.1360 \text{ V}$$

and one concludes from the negative result that spontaneous reaction in this direction is impossible but that electrolysis is possible if the applied potential exceeds 1.136 V.

EFFECT OF TEMPERATURE AND PRESSURE ON THE ELECTROMOTIVE FORCE OF A CELL

The well-known temperature derivative of the Gibbs energy [Eq. (3-9)] when applied to the emf of a cell [Eq. (19-4) or (19-4a)] yields an expression for the entropy change of the cell reaction:

$$\Delta S = nF \left(\frac{\partial E}{\partial T} \right)_P \quad (19\text{-}6)$$

and by Eq. (3-16b)

$$\Delta H = -nF \left[E - T \left(\frac{\partial E}{\partial T} \right)_P \right] \quad (19\text{-}7)$$

Cell potentials are frequently measured to sufficiently high precision and over a sufficient range to make the temperature derivative of useful accuracy. Indeed, a further derivative of Eq. (19-7) yields heat-capacity values that sometimes have useful accuracy.

$$\Delta C_P = nFT \left(\frac{\partial^2 E}{\partial T^2} \right)_P \quad (19\text{-}8)$$

Again the cell $Pt, H_2 | HCl(aq) | AgCl, Ag$ is taken as an example, with the chemical reaction

$$\tfrac{1}{2}H_2(g) + AgCl(s) = Ag(s) + H^+(aq) + Cl^-(aq)$$

Bates and Bower[6] measured these cells with particularly great care at 5° intervals over the range 0 to 90°C, and their results at 0.1 mol kg^{-1} HCl may be expressed by the equation

$$E = 0.35510 - 0.3422 \times 10^{-4}t - 3.2347 \times 10^{-6}t^2 + 6.314 \times 10^{-9}t^3 \text{ V}$$

where t is the temperature in degrees Celsius. From Eq. (19-7) one obtains

$$\Delta H = -35.17 - 0.1705t + 1.870 \times 10^{-4}t^2 + 1.217 \times 10^{-6}t^3 \text{ kJ mol}^{-1}$$

$$\Delta H = -39.29 \text{ kJ mol}^{-1} \text{ at } 25°C$$

This result is probably about as accurate as that obtained by indirect combination of calorimetric measurements. Application of Eq. (19-8) yields

$$\Delta C_P = -170.5 + 0.374t + 3.65 \times 10^{-3}t^2 \text{ J K}^{-1}\text{mol}^{-1}$$

$$\Delta C_P = -158.9 \text{ J K}^{-1}\text{mol}^{-1} = -19.1R$$

The partial molar heat capacity of aqueous hydrochloric acid at 25°C has been measured calorimetrically by Gucker and Schminke,[7] and the value for 0.1 molal solution is $-3.49R$. The heat capacities of hydrogen, silver, and silver chloride have been measured accurately,[8] and, combining all these values, one obtains, at 25°C, $\Delta C_P = -19.37R$. In this case the calorimetric value is probably more accurate. But it is impressive that a value obtained from the second temperature derivative of a cell potential comes as close as this.

It should be remembered that the example just cited represents unusually precise results. Only a few cell measurements yield comparably reliable second derivatives and heat capacities.

The pressure derivative of the Gibbs energy yields the volume change for the reaction. In terms of the cell emf this becomes

$$\Delta V = -nF \left(\frac{\partial E}{\partial P} \right)_T \tag{19-9}$$

Here V is volume, not electrical potential. Large volume changes are always associated with the consumption or evolution of gas, and these effects are normally treated by the method of the next section. Nevertheless Eq. (19-9) remains available for the calculation of ΔV whenever desired.

ACTIVITY, ELECTROMOTIVE FORCE AND EQUILIBRIUM CONSTANT

The definitions and discussion of activity in Chapters 12 and 15 may be applied directly to cell emf through Gibbs-energy relationships. Let us consider a cell such that, when n equivalents of electricity pass in the defined direction, the reaction is

$$b\text{B} + c\text{C} + \cdots = q\text{Q} + r\text{R} + \cdots$$

When each of these substances has unit activity, we write the change of Gibbs energy as $\Delta G°$ and the electromotive force as $E°$. In the more general case, where the activities are not unity, the corresponding values of ΔG and of E are given immediately by Eq. (19-10), which reads

$$\Delta G = \Delta G° + RT \ln \frac{a_Q^q a_R^r \cdots}{a_B^b a_C^c \cdots} \qquad (19\text{-}10)$$

and since $\Delta G = -nFE$,

$$E = E° - \frac{RT}{nF} \ln \frac{a_Q^q a_R^r \cdots}{a_B^b a_C^c \cdots} \qquad (19\text{-}11)$$

If we avoid those irregularities in the surface conditions of metal electrodes that we have discussed in a preceding section, the activities of the various cell constituents will ordinarily depend, at a given temperature, only upon pressure and concentration. The pressure effect is usually negligible except when one of the substances concerned is a gas. In such a case the activity, which is equal to the fugacity, may be found by the methods of Chapter 9. It frequently suffices to assume a perfect gas and to set the activity equal to the partial pressure.

Since the standard Gibbs-energy change of a reaction may be calculated at once from an equilibrium constant by Eq. (7-26) and since, moreover, it is related to the standard emf of a galvanic cell by the equation $\Delta G° = -nFE°$, we may determine $E°$ for a cell from the equilibrium constant of the reaction occurring within the cell. Thus,

$$-RT \ln K = \Delta G° = -nFE° \qquad (19\text{-}12)$$

and

$$E° = \frac{RT}{nF} \ln K \qquad (19\text{-}13)$$

or, at $298.15\,K = 25°C$,

$$E°_{298} = \frac{0.05916}{n} \log K_{298} \qquad (19\text{-}13\text{a})$$

EXAMPLES AND APPLICATIONS OF CELLS

Liquid Junctions

In general, cells that involve a junction between two significantly different solutions give an emf that varies with the physical nature of the liquid junction, as well as the properties of the electrodes and solutions in each half cell. In the junction region a diffusive, nonequilibrium process is occurring. There is an extensive theory of liquid-junction effects; Bayes and Mesmer[9] and Bard and Faulkner[10] give useful summaries. If the ionic compositions of the two half-cell solutions differ only in minor components, with a substantial and uniform concentration of a major electrolyte component, then the junction potential is small and a simple approximation suffices.[9] Another special case where the theory is simple and accurate is that of a single electrolyte of varying concentration—a concentration cell.

In contrast to HCl(aq), where the simple cell A above yields the properties in a straightforward manner, there is no simple electrode for Na^+, K^+, and similar ions. Brown and MacInnes[11] and Janz and Gordon[12] measured very carefully the cell

$$\text{Ag,AgCl} | \text{MCl}(m_1) \vdots \text{MCl}(m_2) | \text{AgCl,Ag} \tag{D}$$

Here the dashed vertical line designates the junction between miscible solutions of different compositions where diffusion is taking place and a potential difference E_j is generated. In this case, E_j is substantial but can be calculated accurately if the transference number is also known as a function of molality. While this method yielded accurate results, the calculations are cumbersome. Thus, it has not been widely used and will not be described in detail.[13]

It is possible to formulate a cell in which the principal ionic components have the same concentration in each half cell and the components that differ have much smaller concentrations. When this criterion is fulfilled, the junction potential is small and the following expression[9] gives a good approximation:

$$E_j = \sum_i D_i (m_{i,A} - m_{i,B}) \tag{19-14a}$$

with

$$D_i = \frac{RT}{F} \frac{z_i \lambda_i}{|z_i| \sum_j |z_j| \lambda_j m_j} \tag{19-14b}$$

Here $m_{i,A}$ and $m_{i,B}$ are the molalities in the half cells of ionic species i with charge z_i and equivalent conductance λ_i. The sum in the denominator of Eq. (19-14b) includes all species and must have approximately the same value for half-cell solutions A and B for this method to be valid. An application of this method is described below, where it was used to interpret the measurements of the dissociation constant for acetic acid at high temperatures.

Dissociation Constant of a Weak Acid

Harned and Ehlers[14] measured the potential of the cell without liquid junction

$$\text{Pt,H}_2 | \text{HOAc}(m_1), \text{NaOAc}(m_2), \text{NaCl}(m_3) | \text{AgCl,Ag} \tag{E}$$

where OAc^- is the acetate ion. The ratios of the molalities $m_1:m_2:m_3$ were held constant in a series of solutions. The potential of this cell, like that of the cell discussed at the beginning of this chapter, depends on the activity product $a_{H^+} a_{Cl^-}$. Substitution of the dissociation constant of acetic acid,

$$K_a = \frac{a_{H^+} a_{OAc^-}}{a_{HOAc}} = \frac{m_{H^+} m_{OAc^-}}{m_{HOAc}} \frac{\gamma_{H^+} \gamma_{OAc^-}}{\gamma_{HOAc}} \tag{19-15}$$

into the usual equation for the cell emf yields

$$E = E° - \frac{RT}{F}\ln\left(\frac{m_{Cl^-}m_{HOAc}}{m_{OAc^-}}\right) - \frac{RT}{F}\ln\left(\frac{\gamma_{Cl^-}\gamma_{HOAc}}{\gamma_{OAc^-}}\right) - \frac{RT}{F}\ln K_a \quad (19\text{-}16)$$

where $E°$ is the standard potential of the cell

$$\text{Pt},\text{H}_2|\text{HCl}(\text{aq})|\text{AgCl},\text{Ag}$$

considered earlier.

The function of molalities in the second term in Eq. (19-16) may be written $m_3(m_1 - m_{H^+})/(m_2 + m_{H^+})$. The molality of the hydrogen ion constitutes a very small correction. It suffices in this case to take

$$m_{H^+} \cong K_a \frac{m_{HOAc}}{m_{OAc^-}} \cong K_a \frac{m_1}{m_2} \quad (19\text{-}17)^*$$

with an approximate value of K_a.

The third term in Eq. (19-16) involves the ratio of γ_{Cl^-} to γ_{OAc^-}; consequently the D-H (Debye–Hückel) limiting-law terms cancel. This cancellation remains valid for an extended but universal D-H expression such as that contained in Eq. (17-26)*. The remaining effects from the terms in $B_{Na,Cl}$, $B_{Na,OAc}$, and the various $\lambda_{HOAc,i}$ in Eq. (17-26)* all yield an initial dependence on molality to the first power. Hence we define

$$E' = E + \frac{RT}{F}\ln\left(\frac{m_3(m_1 - m_{H^+})}{m_2 + m_{H^+}}\right) \quad (19\text{-}18)$$

and plot E' versus m_1 or any other of the molalities. The result is a linear plot in the dilute region whose intercept at $m = 0$ is $E° - (RT/F)\ln K$. Substitution of the value of $E°$ then gives K_a, which is found to be 1.75×10^{-5} at 25°C. Harned and Ehlers[14] report results over the range 5–55°C.

If there is sufficient independent knowledge of the $\beta^{(0)}$, $\beta^{(1)}$, and λ terms available, the slope of this plot can be predicted. Otherwise the observed slope yields one equation related to these quantities.

In the case of acetic acid the approximate procedure of Eq. (19-17)* for m_{H^+} suffices, but for acids with larger values of K_a this is no longer the case. Hamer[15] discussed this problem in some detail. Phosphoric acid, with K_a about 10^{-2} requires quite a different treatment, which is described below.

While the procedure of Harned and Ehlers[14] with cell E was satisfactory at 55°C, at higher temperatures the solubility of AgCl becomes too large. Thus, for measurements to 295°C, Mesmer et al.[16] used the following cell with liquid junction:

$$\text{Pt},\text{H}_2|\text{HCl}(m_1),\text{NaCl}(m_2) \; \| \; \text{HOAc}(m_1),\text{NaOAc}(m_1),\text{NaCl}(m_2)|\text{H}_2,\text{Pt} \quad (F)$$

Here the double lines ($\|$) indicate a liquid junction with zero or very small potential. In this case the ratio m_1/m_2 was 0.01 or less; thus, the dominant terms in Eqs. (19-14a,b) were those for Na$^+$ and Cl$^-$. Then E_j is in the range 0.2 to 0.5 mV and the remaining uncertainty in this term is no larger than other uncertainties. The final equation is

$$\ln Q = -\frac{(E - E_j)F}{RT} + \ln(m_{H^+}) + \ln\left(\frac{m_{Ac^-}}{m_{HAc}}\right) \qquad (19\text{-}19)$$

Note that this $m_{H^+} = m_1$, that of the reference cell. Correction was made for the excess of m_{Ac^-} and the deficiency of m_{HAc} from m_1. Measurements extend to 568 K and were made at the saturation pressure and at a higher pressure at each temperature. The equilibrium quotient is defined as $Q = m_{H^+} m_{Ac^-}/m_{HAc}$ and the ionization constant is related by

$$\ln K = \ln Q + \ln\left(\frac{\gamma_{H^+} \gamma_{Ac^-}}{\gamma_{HAc}}\right) \qquad (19\text{-}20)$$

With the relatively high molality of NaCl and high ionic strength, the activity coefficient term can be quite large. Also, the solutions are mixed electrolytes, so a complete treatment is complex. Two comprehensive treatments are presented by Mesmer et al.[16] in which literature data for densities and other types of measurements were included along with the cell data of Harned and Ehlers[14] and the new results. One of these treatments was based on the ion-interaction equations given in Chapters 17 and 18 and in the section below on phosphoric acid. The final results for $\ln Q$ at saturation pressure are given in Table 19-1. In the limit $I = 0$, $Q = K$, and the values in the column so designated pertain to $\ln K$ for pure acetic acid. Values in the columns for $m \geq 0.5 \,\text{mol kg}^{-1}$ and for 50°C and higher temperatures are essentially equal to those

TABLE 19-1
Values of ln Q for the ionization of acetic acid based primarily on cell measurements[14,16]

	Ionic strength, $I/\text{mol kg}^{-1}$						
$t/°C$	0	0.1	0.5	1.0	3.0	5.0	$\sigma\,(I=0)$
0	−4.780	−4.581	−4.518	−4.548	−4.858	−5.275	0.004
25	−4.757	−4.547	−4.469	−4.484	−4.737	−5.086	0.002
50	−4.786	−4.564	−4.472	−4.474	−4.675	−4.965	0.002
75	−4.848	−4.613	−4.506	−4.494	−4.643	−4.877	0.004
100	−4.937	−4.684	−4.560	−4.533	−4.630	−4.811	0.005
125	−5.047	−4.774	−4.629	−4.586	−4.628	−4.757	0.007
150	−5.180	−4.881	−4.710	−4.647	−4.633	−4.712	0.008
175	−5.334	−5.005	−4.802	−4.716	−4.643	−4.673	0.011
200	−5.515	−5.147	−4.904	−4.791	−4.653	−4.639	0.015
225	−5.727	−5.311	−5.015	−4.868	−4.662	−4.605	0.020
250	−5.978	−5.501	−5.136	−4.947	−4.665	−4.571	0.028
275	−6.282	−5.725	−5.266	−5.021	−4.653	−4.528	0.038
300	−6.664	−5.995	−5.403	−5.081	−4.609	−4.465	0.052

originally measured with cell F, but have been smoothed by the comprehensive treatment, which also yields volumetric and thermal properties.

Earlier investigations of this type using cells with liquid junction are listed and reviewed by Mesmer et al.[17,18]

Dissociation Constant of Water

Measurements of cells without liquid junction of the type

$$\text{Pt},\text{H}_2 | \text{MOH}(m_1), \text{MCl}(m_2) | \text{AgCl}, \text{Ag} \quad \text{(G)}$$

with M an alkali metal, may be used in connection with $E°$ for cell A with the same electrodes and HCl electrolyte to obtain the ionic dissociation constant of water:

$$K_w = a_{H^+} a_{OH^-} = m_{H^+} m_{OH^-} \gamma_{H^+} \gamma_{OH^-} \quad (19\text{-}21)$$

In this case the emf is

$$E = E° - \frac{RT}{F} \ln\left(\frac{m_2}{m_1}\right) - \frac{RT}{F} \ln\left(\frac{\gamma_{Cl^-} a_{H_2O}}{\gamma_{OH^-}}\right) - \frac{RT}{F} \ln K_w \quad (19\text{-}22)$$

and the effect of water ionization on m_{OH^-} is negligible. Again the D-H limiting-law terms cancel in the activity-coefficient factor.

We define K' by

$$-\log K' = \frac{F(E - E°)}{2.303 RT} + \log\left(\frac{m_2}{m_1}\right) \quad (19\text{-}23)$$

Hence

$$-\log K' = -\log K_w - \log\left(\frac{\gamma_{Cl^-} a_{H_2O}}{\gamma_{OH^-}}\right) \quad (19\text{-}24)$$

The Guggenheim approximation [Eq. (17-1)*] suffices for the ion activity coefficients and the Raoult's-law approximation for the water activity.

$$\log a_{H_2O} \cong -\frac{2(m_1 + m_2)}{2.303 \times 55.5} = -0.016(m_1 + m_2)$$

The second term in Eq. (19-24) is then

$$-\log\left(\frac{\gamma_{Cl^-} a_{H_2O}}{\gamma_{OH^-}}\right) = (B_{MOH} - B_{MCl} + 0.016)(m_1 + m_2) \quad (19\text{-}25)$$

with the B_{ij} used here related to the β_{ij} defined earlier by the factor (2/2.303). Then a plot of $-\log K'$ versus $m_1 + m_2$ should have the slope $B_{MOH} - B_{MCl} + 0.016$.

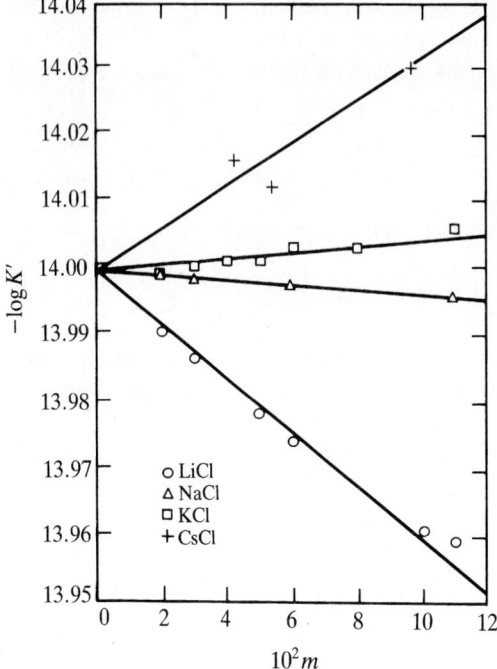

FIGURE 19-1
The calculation of the dissociation constant of water using Eq. (19-24).

Guggenheim and Turgeon[19] treated the data of Harned and his collaborators[20] in this manner with the results shown in Fig. 19-1. The resulting K_w is 1.002×10^{-14} at 25°C. The slopes of the various curves agree, within experimental error, with the available knowledge of the Bs for the various chlorides and hydroxides. In this case the use of the more detailed ion-interaction equations and parameters would yield the same result.

Phosphoric Acid[21]

There are experimental cell data for certain buffer solutions as well as for pure phosphoric acid. For the pure acid the osmotic-coefficient data based on solvent vapor pressure[22,23] are available. Among various cell measurements on buffer solutions, the data of Bates[24] on HCl–KH_2PO_4 mixtures are particularly extensive and of the highest accuracy. For the buffer-solution calculations, the parameters for the pure electrolytes HCl, KCl, and KH_2PO_4 are known, see Appendix 8. Values are also needed for the parameters $\Phi_{H,K}$ and $\Phi_{Cl,A}$ (A^- is $H_2PO_4^-$) from appropriate mixtures. The former is given in Table 17-2; the latter was determined to be $\Phi_{Cl,A} = 0.10$ from the isopiestic measurements of Childs, Downes, and Platford[25], which also yield $\psi_{K,Cl,A} = -0.010_5$.

While Chapter 17 provides a complete basis for the equations for the present system, some discussion is needed. In view of the high molality of the pure H_3PO_4, the third-order term $2m_{HA}^3 \mu_{HA,HA,HA}$ is added to Eq. (17-24)* for the osmotic coefficient and the corresponding term $3m_{HA}^2 \mu_{HA,HA,HA}$ to Eq. (17-72)* for the activity coefficient of the neutral species y_{HA}. These terms are derived from the basic equation for the excess Gibbs energy Eq. (17-5)* by the appropriate differentiation. Also, we note that the inclusion of the association reaction of $H^+ + A^-$ supersedes the binary ion-interaction term B_{HA} and that for the pure H_3PO_3 solution the ion molality is never large and the third-order term in C_{HA} can be neglected.

For pure H_3PO_4 the pertinent equations are given below, where ϕ' is an osmotic coefficient calculated from the water activity on the basis of dissociation into two ions, $\nu = 2$ (note that Platford[23] reports his data on the basis of $\nu = 4$). If the actual dissociation of acid of molality m yields molality m_H of H^+ and A^-, the osmotic coefficient ϕ on that basis is

$$(\phi - 1)(m + m_H) = 2m_H f^\phi + 2m_H(m - m_H)(\lambda_{H,HA} + \lambda_{A,HA})$$
$$+ (m - m_H)^2 \lambda_{HA,HA} + 2(m - m_H)^3 \mu_{HA,HA,HA} \tag{19-26)*}$$

$$\phi(m + m_H) = 2m\phi' \tag{19-27}$$

In addition, we have the dissociation equilibrium

$$\ln K = \ln\left(\frac{m_H^2}{m - m_H}\right) + \ln\left(\frac{y_H y_A}{y_{HA}}\right) \tag{19-28}$$

and the expressions for $\frac{1}{2}\ln(y_H y_A) = \ln y_\pm$ and $\ln y_{HA}$ which are obtained as described above:

$$\tfrac{1}{2}\ln(y_H y_A) = f^\gamma + (m - m_H)(\lambda_{H,HA} + \lambda_{A,HA}) \tag{19-29)*}$$

$$\ln y_{HA} = 2m_H(\lambda_{H,HA} + \lambda_{A,HA}) + 2(m - m_H)\lambda_{HA,HA}$$
$$+ 3(m - m_H)^2 \mu_{HA,HA,HA} \tag{19-30)*}$$

Equation (19-29)* is a simplification of Eq. (17-30)* for this situation. Note that f^γ is the only term in F which remains.

For the buffer solutions with $HCl(m_1)$ and $KH_2PO_4(m_2)$ the measured quantity is the potential E of a cell with hydrogen and silver–silver chloride electrodes. If the hydrogen ion molality is m_H and the standard potential is $E°$, we obtain

$$E - E° = -\frac{RT}{F}\ln(m_H m_1 y_H y_{Cl}) \tag{19-31}$$

Other symbols have their usual meaning with $E° = 0.2223$ V. Since the buffer-solution measurements extend only to a maximum concentration of 0.4 molal, third virial coefficients are omitted for those calculations.

Then one has the ionic strength, $I = m_2 + m_H$, and also $m_{HA} = m_1 - m_H$, $m_{Cl} = m_1$, $m_K = m_2$, and $m_A = m_2 - m_1 + m_H$. For these buffer solutions,

$$\ln \gamma_{HCl} = \tfrac{1}{2}\ln(\gamma_H \gamma_{Cl})$$
$$= f^\gamma + (m_1 + m_H)B_{HCl} + m_2(B_{KCl} + \Phi_{HK}) + m_H m_1 B'_{HCl}$$
$$+ m_1 m_2 B'_{KCl} + m_2(m_2 - m_1 + m_H)B'_{KA}$$
$$+ (m_2 - m_1 + m_H)\Phi_{Cl,A} + (m_1 - m_H)(\lambda_{H,HA} + \lambda_{Cl,HA})$$
$$(19\text{-}32)*$$

$$\tfrac{1}{2}\ln(\gamma_H \gamma_A) = f^\gamma + m_1(B_{HCl} + \Phi_{Cl,A}) + m_2 B_{KA} + m_H m_1 B'_{HCl}$$
$$+ m_1 m_2 B'_{KCl} + m_2(m_2 - m_1 + m_H)B'_{KA} + m_2 \Phi_{H,K}$$
$$+ (m_1 - m_H)(\lambda_{H,HA} + \lambda_{A,HA}) \quad (19\text{-}33)*$$

$$\ln \gamma_{HA} = 2[m_H \lambda_{H,HA} + m_1 \lambda_{Cl,HA} + m_2 \lambda_{K,HA}$$
$$+ (m_2 - m_1 + m_H)\lambda_{A,HA} + (m_1 - m_H)\lambda_{HA,HA}] \quad (19\text{-}34)*$$

For either the pure acid or the buffer solutions, it was necessary to eliminate the hydrogen ion molality m_H by an iterative solution of the appropriate set of equations.

In Eqs. (19-32)* and (19-33)* all the B and Φ parameters are independently known. Thus, in the full array of equations (19-26)* to (19-34)* there are six parameters to be evaluated. While all might be simultaneously evaluated from the full array of data for both pure acid and buffer solutions, that calculation would be very complex and the assignment of appropriate weights would be difficult. Instead three parameters were evaluated from each set of data. From the pure-acid data, the sum $(\lambda_{H,HA} + \lambda_{A,HA})$ and the second and third virial coefficients for neutral–neutral interaction were determined by appropriate least-squares methods, while K and the second virial coefficients for ion-neutral interactions were determined from data on buffer solutions on the basis $\lambda_{Cl,HA} = 0$. This process was iterated until all results were consistent. The resulting parameters are given in Table 19-2. The standard deviation of

TABLE 19-2
Parameters for phosphoric acid solutions at 25°C (A^- is $H_2PO_4^-$)

$K = 7.142_5 \times 10^{-3}$	($pK = 2.1461_5$)
$\lambda_{HA,HA} = 0.0503_1$	$\mu_{HA,HA,HA} = 0.0109_5$
$\lambda_{A,HA} = -0.400$	$\lambda_{H,HA} = 0.290$
$\lambda_{K,HA} = -0.070$	($\lambda_{Cl,HA} = 0$)

Also determined from KCl-KH_2PO_4 mixtures:
$\Phi_{Cl,A} = 0.10$ $\psi_{K,Cl,A} = -0.010_5$

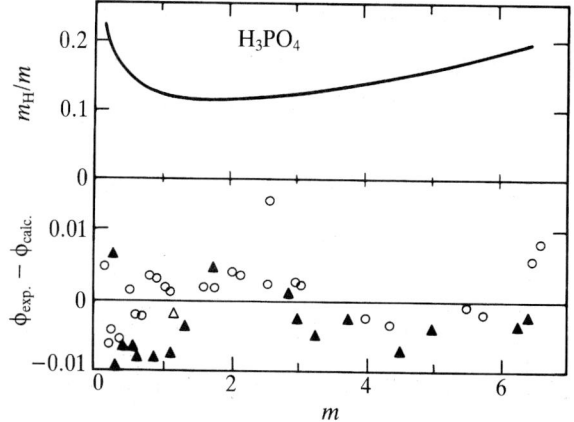

FIGURE 19-2
Fractional dissociation of H_3PO_4 in upper section; deviation of calculated from experimental osmotic coefficients in lower section (circles, Elmore et al.[22]; triangles, Platford[23]); from Pitzer and Silvester.[21]

fit for the buffer solutions was 0.0002 V in the cell potential, while for the pure acid to 6 molal it was 0.005 in the osmotic coefficient. In each case, the fit is essentially within the experimental uncertainty. Figure 19-2 shows in the lower section the deviations in osmotic coefficient as a function of total molality for each set of experimental measurements. In the upper portion of Fig. 19-2 the fraction dissociated, m_H/m, is shown. The increase in dissociation at very low concentration is expected; the minimum near $2\,\mathrm{mol\,kg^{-1}}$ and the increase at higher concentration are surprising and are caused primarily by the large, negative value of the sum of ion–neutral second virial coefficients ($\lambda_{H,HA} + \lambda_{A,HA}$).

The second virial coefficient $B_{H,A}$ for H^+ interaction with A^- was omitted on the basis that it was redundant once the association equilibrium was introduced. This redundancy was verified by test calculations. In the original paper on this calculation, Pitzer and Silvester[21] also discuss the difficulties that Bates[24] and others encountered in treating the dissociation of H_3PO_4 by the buffer method and their avoidance by the simultaneous treatment of other types of measurements as described above.

Other Types of Electrodes

There are special electrodes that are sensitive primarily or exclusively to a particular ionic species. The "glass electrode" sensitive to H^+ is best known, and it is widely used instead of the hydrogen electrode when less than the highest precision suffices. Other ion-sensitive electrodes exist and more are being developed. Butler and Roy[26] describe the status of this field as of 1991 with many examples. Many other reviews are also available.[27]

Cells with Solid Electrolyte

Some crystalline solids show ionic conductance at high temperatures, and such materials may serve as electrolytes in galvanic cells.[27] A good example is silver iodide, which becomes an electrolytic conductor above 146°C. The cell

$$\text{Ag(s)} | \text{AgI(s)} | \text{Ag}_2\text{S(s)}, \text{S(l)}, \text{graphite}$$

has been found[28] to give reversible potentials in the range 150 to 425°C. The cell reaction is

$$2\text{Ag(s)} + \text{S(l)} = \text{Ag}_2\text{S(s)}$$

Hence, the cell potential gives immediately the Gibbs energy of formation of Ag_2S. Such cells have also been employed[29] to study solid compounds and solutions such as Ag in Ag–Sb alloy and Ag_2S in Ag_2S–Sb_2S_3. In the former case the cell is Ag|AgI|Ag–Sb alloy and in the latter Pt,S(l),Ag_2S|AgI|Ag_2S–Sb_2S_3, S(l), Pt. In either case 1 mole of Ag^+ is transferred per equivalent through the AgI from the left to the right electrode. The cells with the Ag_2S–Sb_2S_3 electrode gave a constant potential of 43 mV in the composition range Ag_3SbS_3 to AgSbS_2 and 98 mV in the range AgSbS_2 to Sb_2S_3 at 275°C. These results show that no significant solid solubility arises in this system and give the Gibbs energies of formation of the compounds. For example, the reaction

$$\text{Ag}_2\text{S} + \text{Sb}_2\text{S}_3 = 2\text{AgSbS}_2$$

involves two equivalents of Ag^+ transferred through the cell, and

$$\Delta G = -2FE = -18.9 \text{ kJ mol}^{-1} \text{ at } 548 \text{ K}$$

Cells with solid electrolyte frequently give more satisfactory results at high temperature than those with liquid electrolyte. All diffusion processes become more rapid at high temperature, and the unwanted diffusion of material from one electrode to the other may become troublesome in the liquid. This diffusion is, of course, greatly reduced in a solid electrolyte.

Electronic conductance is commonly more important than ionic conductance in solids, and electronic conductance effectively short-circuits the cell; consequently only a limited number of solids are suitable for solid electrolytes. If the electronic conductance is a small but not negligible fraction of the ionic conductance, then a correction may be made. But the electronic-transference number is usually a sensitive function of impurity concentration or deviation from simple salt composition; consequently the corrections for electronic conductance are complex and beyond the scope of the present discussion.[30]

STANDARD ELECTRODE POTENTIALS

Many types of cells have been employed to obtain the standard electrode potentials of the elements and the potentials of other important electrodes such as Hg_2Cl_2, Hg, etc. Liquid junctions are to be avoided if possible; otherwise some limiting extrapolation is made that should eliminate the junction potential. It is impractical to survey this field, but a few examples may suggest the types of cells employed. The treatment of the results should be apparent in most cases from the discussions already presented.

Electrode Potentials of Metals

If the metal is stable in contact with the aqueous solution and the metal chloride is soluble, then a cell of the type

$$M \mid MCl_z(aq) \mid Hg_2Cl_2, Hg$$

yields $E° = V°_{Hg_2Cl_2,Hg} - V°_M$, and $E°_M = V°_M$ can readily be calculated. Also, from the molality dependence of the emf the activity coefficient of MCl_z can be obtained.

Sometimes it is preferable to use a saturated amalgam of M in place of the solid metal. If mercury is insoluble in the solid metal, the result is unchanged; otherwise a correction must be made for the reduction of activity of M by solution of Hg.

These methods are successful for such metals as Tl, Cd, Zn, Fe, etc. In case the metal chloride is insoluble but the sulfate is soluble, an analogous method may be used with the $PbSO_4$, Pb electrode. Thus one may measure $E°$ for the half cell Cu, Cu^{2+} in the cell

$$Cu \mid CuSO_4(aq) \mid PbSO_4, Pb$$

Extrapolation to the solute standard state is more difficult with doubly or more highly charged ions present; hence the chloride system is to be preferred wherever it is applicable.

Some metals react spontaneously with water. But in some such cases a dilute amalgam of the metal reacts reversibly with the aqueous metal ion without any appreciable spontaneous reaction with water. This is the case for sodium. Hence one may measure the cell

$$Na \text{ in } Hg(x_2) \mid NaCl(aq) \mid Hg_2Cl_2, Hg$$

and obtain thereby the potential of the sodium amalgam electrode. To relate this to sodium metal one must obtain the activity of Na in the amalgam. If another solvent[31] is available, which does not react with the metal but which dissolves a salt of the metal with some ionization, then one may measure a second cell that relates the metal to the amalgam. In the case of sodium a suitable solvent is ethylamine with NaI the salt, and the cell is

$$Na \mid NaI \text{ in ethylamine} \mid Na \text{ in } Hg(x_2)$$

In this case the potential is independent of the NaI concentration.

By this two-stage method the standard potentials have been measured for Na, K, and some other active metals. There is always a risk that there is still some spontaneous reaction with water, which causes error in this type of cell; hence some caution is appropriate in accepting the results.

Electrode Potentials of the Halogens

An inert electrode, such as platinum, in the presence of chlorine and chloride ion yields a reversible emf. Possible examples of complete cells are Pt,$H_2 \mid HCl(aq) \mid Cl_2$,Pt and Ag,AgCl $\mid HCl(aq) \mid Cl_2$,Pt. In the former the $E°$ gives $E°_{Cl_2,Cl^-}$ directly, but the actual cell potentials at finite concentrations of HCl must be extrapolated to the solute standard state at infinite dilution. The second cell potential is independent of the activity of Cl⁻ and yields $E^\circ_{Cl_2,Cl^-} - E^\circ_{AgCl,Ag}$. By the latter method Randall and Young[32] found $E^\circ_{Cl_2,Cl^-} = 1.358$ V.

The principal complication in measurements of the chlorine electrode relates to the very substantial solubility of chlorine in water and the subsequent reactions $Cl_2 + Cl^- = Cl_3^-$ and

$$Cl_2 + H_2O = HOCl + H^+ + Cl^-$$

It is possible to avoid these difficulties by keeping the partial pressure of Cl_2 very low and then correcting the potential to the standard value for 1 bar fugacity of Cl_2.

The potentials of the bromine and iodine electrodes may be measured by methods analogous to those for the chlorine system but modified in view of the liquid and solid states, respectively.

Electrode Potentials by Indirect Methods

Many active metals as well as the halogen fluorine react spontaneously with water, and their potentials in aqueous-solution systems have not been measured. Indirect methods may be used, however, to obtain the Gibbs energy of the equivalent reaction. In the case of fluorine this is

$$\tfrac{1}{2}H_2(g) + \tfrac{1}{2}F_2(g) = H^+(aq) + F^-(aq)$$

and

$$\Delta G^\circ = -FE^\circ_{F_2,F^-}$$

For a metal yielding the ion M^{z+} the reaction is

$$M^{z+} + \frac{z}{2}H_2 = M + zH^+$$

with

$$\Delta G^\circ = -FE^\circ_{M,M^{z+}}$$

In these cases the heat of the reaction may be obtained from thermal measurements and the entropies of all substances from the third law by methods described in Chapters 5, 6, 18 and 20.†

---

† In this and earlier sections, most reactions are written in terms of H⁺(aq) rather than e⁻. This procedure is strongly recommended since all major tables of ionic properties are based on the definition of zero values for $\Delta_f H^\circ$, S°, C_P°, as well as $\Delta_f G^\circ$ for H⁺(aq). Consistency with this basis yields for e⁻, $\Delta_f G^\circ = \Delta_f H^\circ = 0$ but $S^\circ = \tfrac{1}{2}S^\circ[H_2(g)]$ and $C_P^\circ = \tfrac{1}{2}C_P^\circ[H_2(g)]$.

Summary of Standard Electrode Potentials

Table 19-3 includes values for the standard potentials for a number of half reactions. Sources include Latimer,[5] Vanysek,[33] and Bard et al.[34]; also Cox et al.[35] give the internationally recommended CODATA values for standard Gibbs energies, from which potentials can be calculated. The agreement is generally good among these sources; hence individual references are not given. Vanysek[33] and Bard et al.[34] give especially long and complete lists.

TABLE 19-3
Potentials of half cells at 25°C

Reaction	$E°/V$
$Li^1 + e^- = Li$	−3.04
$K^+ + e^- = K$	−2.93
$Ba^{2+} + 2e^- = Ba$	−2.92
$Ca^{2+} + 2e^- = Ca$	−2.85
$Na^+ + e^- = Na$	−2.71
$Mg^{2+} + 2e^- = Mg$	−2.36
$Al^3 + 3e^- = Al$	−1.66
$2H_2O + 2e^- = H_2 + 2OH^-$	−0.828
$Zn^{2+} + 2e^- = Zn$	−0.762
$Cr^{3+} + 3e^- = Cr$	−0.744
$Ga^{3+} + 3e^- = Ga$	−0.56
$Fe^{2+} + 2e^- = Fe$	−0.44
$Cd^{2+} + 2e^- = Cd$	−0.403
$PbSO_4 + 2e^- = Pb + SO_4^{2-}$	−0.356
$In^{3+} + 3e^- = In$	−0.338
$Tl^+ + e^- = Tl$	−0.336
$AgI + e^- = Ag + I^-$	−0.152 2
$Sn^{2+} + 2e^- = Sn$	−0.137
$Pb^{2+} + 2e^- = Pb$	−0.126
$2H^+ + 2e^- = H_2$	0.000 00
$AgBr + e^- = Ag + Br^-$	0.71
$Sn^{4+} + 2e^- = Sn^{2+}$	0.15
$AgCl + e^- = Ag + Cl^-$	0.2223
$Hg_2Cl_2 + 2e^- = 2Hg + 2Cl^-$	0.2681
$Cu^{2+} + 2e^- = Cu$	0.34
$I_2 + 2e^- = 2I^-$	0.5355
$Fe^{3+} + e^- = Fe^{2+}$	0.771
$Hg_2^{2+} + 2e^- = 2Hg$	0.797
$Ag^+ + e^- = Ag$	0.799
$2Hg^{2+} + 2e^- = Hg_2^{2+}$	0.92
$AuCl_4^- + 3e^- = Au + 4Cl^-$	1.002
$Br_2(l) + 2e^- = 2Br^-$	1.066
$Tl^{3+} + 2e^- = Tl^+$	1.25
$Cl_2(g) + 2e^- = Cl^-$	1.3583
$PbO_2 + SO_4^{2-} + 4H^+ + 2e^- = PbSO_4 + 2H_2O$	1.69
$Ag^{2+} + e^- = Ag^+$	1.98
$O_3 + 2H^+ + 2e^- = O_2 + H_2O$	2.076
$F_2 + 2e^- = 2F^-$	2.87

REFERENCES

1. For a recent review of electrode kinetics, see B. E. Conway, *Modern Aspects of Electrochemistry*, No. 16, Plenum Press, New York, 1985, pp. 103–188.
2. R. G. Bates, E. A. Guggenheim, H. S. Harned, D. J. G. Ives, G. J. Janz, C. B. Monk, R. A. Robinson, R. H. Stokes, and W. F. K. Wynne-Jones, *J. Chem. Phys.*, **25**, 361 (1956).
3. These definitions are essentially those recommended by the International Union of Pure and Applied Chemistry in 1988 (*Quantities, Units and Symbols in Physical Chemistry*, I. Mills, ed., Blackwell Scientific, Oxford, 1988).
4. *Collected Works of J. Willard Gibbs*, Yale University Press, New Haven, Conn., reprinted 1948, pp. 332–349; also p. 429.
5. W. M. Latimer, *The Oxidation States of the Elements and Their Potentials in Aqueous Solutions*, 2d edn., Prentice-Hall, New York, 1952.
6. R. G. Bates and V. E. Bower, *J. Res. Natl. Bur. Stand.*, **53**, 283 (1954).
7. F. T. Gucker and K. H. Schminke, *J. Am. Chem. Soc.*, **54**, 1358 (1932).
8. See F. D. Rossini et al., "Selected Values of Chemical Thermodynamic Properties," *National Bureau of Standards (U.S.) Circ. 500*, 1952.
9. C. F. Bayes, Jr., and R. E. Mesmer, *The Hydrolysis of Cations*, Wiley, New York, 1976, section 2.1.4 and appendix I.
10. A. J. Bard and L. R. Faulkner, *Electrochemical Methods*, Wiley, New York, 1980, section 2.3.
11. A. S. Brown and D. A. MacInnes, *J. Am. Chem. Soc.*, **57**, 1356 (1935).
12. G. J. Janz and A. R. Gordon, *J. Am. Chem. Soc.*, **65**, 218 (1943), and earlier paper there cited.
13. The treatment for cell D is given in the second edition of this book: K. S. Pitzer and L. Brewer, *Thermodynamics*, revision of 1st edn. by G. N. Lewis and M. Randall, McGraw-Hill, New York, 1961, pp. 362–364.
14. H. S. Harned and R. W. Ehlers, *J. Am. Chem. Soc.*, **54**, 1350 (1932).
15. W. J. Hamer, *The Structure of Electrolytic Solutions*, Wiley, New York, 1959, p. 236.
16. R. E. Mesmer, C. S. Patterson, R. H. Busey, and H. F. Holmes, *J. Phys. Chem.*, **93**, 7483 (1989).
17. R. E. Mesmer, W. L. Marshall, D. A. Palmer, J. M. Simonson, and H. F. Holmes, *J. Solution Chem.*, **17**, 699 (1988).
18. R. E. Mesmer, D. A. Palmer, and J. M. Simonson, "Ion Association at High Temperatures and Pressures," in *Activity Coefficients in Electrolyte Solutions*, 2d edn., K. S. Pitzer, ed., CRC Press, Boca Raton, Fla, 1991, Chapter 8.
19. E. A. Guggenheim and J. C. Turgeon, *Trans. Faraday Soc.*, **51**, 747 (1955).
20. H. S. Harned and G. E. Mannweiler, *J. Am. Chem. Soc.*, **57**, 1873 (1935); and earlier papers there cited.
21. K. S. Pitzer and L. F. Silvester, *J. Solution Chem.*, **5**, 269 (1976).
22. K. L. Elmore, C. M. Mason, and J. H. Christensen, *J. Am. Chem. Soc.*, **68**, 2528 (1946).
23. R. F. Platford, *J. Solution Chem.*, **4**, 591 (1975).
24. R. G. Bates, *J. Res. Natl. Bur. Stand.*, **47**, 127 (1951).
25. C. W. Childs, C. J. Downes, and R. F. Platford, *J. Solution Chem.*, **3**, 139 (1974).
26. J. N. Butler and R. N. Roy, "Experimental Methods: Potentiometric," in *Activity Coefficients in Electrolyte Solutions*, 2d edn., K. S. Pitzer, ed., CRC Press, Boca Raton, Fla, 1991, Chapter 4.
27. P. L. Bailey, *Analysis with Ion-Selective Electrodes*, 2d edn., Heyden, London, 1980.
28. K. Kiukkola and C. Wagner, *J. Electrochem. Soc.*, **104**, 308, 379 (1957).
29. A. G. Verduch and C. Wagner, *J. Phys. Chem.*, **61**, 558 (1957).
30. For a discussion of these problems, see K. Kiukkola and C. Wagner, *J. Electrochem. Soc.*, **104**, 308 (1957).
31. There is an extensive literature on cells with nonaqueous electrolyte solutions; for a summary see H. Strehlow, *Z. Elektrochem.*, **56**, 827 (1952).
32. M. Randall and L. E. Young, *J. Am. Chem. Soc.*, **50**, 989 (1928).
33. P. Vanysek, "Electrochemical Series," in *CRC Handbook of Chemistry and Physics*, 71st edn., D. R. Lide, ed., CRC Press, Boca Raton, Fla, 1990.

34. A. J. Bard, R. Parsons, and J. Jordan, *Standard Potentials in Aqueous Solutions*, Marcel Dekker, New York, 1985.
35. J. D. Cox, D. D. Wagman, and V. A. Medvedev, *CODATA Key Values for Thermodynamics*, Hemisphere, New York, 1989.

PROBLEMS

19-1. Calculate from the data in Table 19-3 the equilibrium constants for the reactions

$$O_3 + 2Ag^+ + 2H^+ = O_2 + H_2O + 2Ag^{2+} \quad \text{and} \quad 3Fe^{2+} = Fe + 2Fe^{3+}.$$

19-2. The quantity pH may be defined by pH $= -\log a_{H^+}$, but its precise measurement is beset by the difficulties of determination of single-ion activities. Consider the cell $H_2,Pt|$ aq soln $HA(m_1),NaA(m_2),NaCl(m_3)|AgCl,Ag$, where HA is any weak acid. Relate the pH as defined above to the potential of this cell and any other quantities required. Discuss the possible assumptions with respect to any of these additional quantities in relation to the section on pH in Chapter 17.

19-3. In Table 17-4 the value 0.0 is shown for $\mu°$, $\Delta_f H°$, and $S°$ for H^+, but no value is given there or in this chapter for these quantities for the e^- in a half-cell reaction. From the reaction $\frac{1}{2}H_2(g) = H^+(aq) + e^-$, derive expressions for $\mu°$, $\Delta_f H°$, $S°$, and $C_P°$ for e^-.

19-4. What is the reaction for the cell $Pt,S(l),Ag_2S|AgI|(Ag_3SbS_3 + AgSbS_2),S(l),Pt$? Given the potential $E = 43\,mV$ of this cell at 548 K and the data in the text for related cells, calculate ΔG for the reaction $3Ag_2S + Sb_2S_3 = 2Ag_3SbS_3$

19-5. Harned and Nims† measured the following type of cell:

$$Ag|AgCl|NaCl(m)|Na \text{ in } Hg|NaCl \text{ (0.1 mol kg}^{-1})|AgCl|Ag$$

The $|Na \text{ in } Hg|$ is a sodium amalgam that is sufficiently dilute that the spontaneous reaction with water is negligible; thus it serves to transfer the Na^+ ion from one aqueous solution to the other. The cell reaction is $NaCl(aq,m) = NaCl(aq, 0.1 \text{ mol kg}^{-1})$. Take the cell potentials listed below, extrapolate as required to $m = 0$, and calculate $\gamma_\pm$ at 0.1 and 0.5 molal and 25°C. Compare the results with those given by Eq. (17-20)* and Table 17-1.

m	E/V
0.05	−0.032 50
0.10	(0.0)
0.20	+0.032 51
0.50	0.075 71
1.00	0.109 55

---

† H. S. Harned and L. F. Nims, *J. Am. Chem. Soc.*, **54**, 423 (1932).

CHAPTER 20

STATISTICAL CALCULATIONS OF THERMODYNAMIC PROPERTIES: ADVANCED TREATMENT

In Chapter 5 the general principles were developed for statistical calculations of thermodynamic properties from atomic or molecular data. Also, a few simple examples illustrated these principles. While the treatment for atoms was complete, that for linear molecules involved approximations, and nonlinear molecules were not considered. We continue that presentation here; in addition, there are sections on radiation and plasmas.

The reader will need to refer frequently to Chapter 5 for background; also, assembled below are a few basic relationships from Chapter 5. The perfect gas law is assumed as before. The partition function is defined by

$$Q = \sum_i g_i \exp(-\varepsilon_i/kT) \qquad (5\text{-}14)$$

with ε_i the energy above that of the lowest quantum state. Then the various contributions of internal energy levels to the molar thermodynamic functions are related to Q by

$$-\left[\frac{G_T - H_0}{RT}\right]_{\text{int}} = \ln Q \qquad (5\text{-}25)$$

$$\left[\frac{H_T - H_0}{RT}\right]_{int} = T\, d\ln Q\, dT \qquad (5\text{-}16)$$

$$\left[\frac{S_T - S_0}{R}\right]_{int} = \ln Q + T\frac{d\ln Q}{dT} \qquad (5\text{-}18)$$

$$\left(\frac{C}{R}\right)_{int} = T^2\frac{d^2\ln Q}{dT^2} + 2T\frac{d\ln Q}{dT} \qquad (5\text{-}17)$$

The contributions of translation to each of the thermodynamic functions as given in Chapter 5 were exact and are used here without change [Eqs. (5-7) through (5-10)].

NONLINEAR MOLECULES

For nonlinear molecules the rotational energy-level pattern is usually much more complicated than that for linear molecules. Consequently, the necessary calculations to obtain the partition function are much longer, even though they involve only the same principles as were employed for linear molecules. The result is best expressed in terms of the principal moments of inertia I_A, I_B, I_C or the corresponding rotational constants A, B, C.

Following the pattern for linear molecules, we define a set of three quantities with the dimension temperature:

$$\theta_A = \frac{hcA}{k} = \frac{h^2}{8\pi^2 I_A k} \qquad (20\text{-}1a)$$

$$\theta_B = \frac{hcB}{k} = \frac{h^2}{8\pi^2 I_B k} \qquad (20\text{-}1b)$$

$$\theta_C = \frac{hcC}{k} = \frac{h^2}{8\pi^2 I_C k} \qquad (20\text{-}1c)$$

Then the high-temperature approximation for the rotational partition function is found to be[1]

$$Q_r = \frac{\pi^{1/2}}{\sigma}\left(\frac{T^3}{\theta_A \theta_B \theta_C}\right)^{1/2} \qquad (20\text{-}2)$$

where σ is again the symmetry number. In this general case it is the number of rotational orientations that differ only in the exchange of identical particles. Thus, a symmetrical, bent triatomic molecule such as H_2O has $\sigma = 2$ and a pyramidal molecule such as NH_3 has $\sigma = 3$. But the planar molecule SO_3 has $\sigma = 6$ because it may be turned over by rotation about an S–O axis in addition to three rotational orientations about the threefold axis through the S. A tetrahedral molecule such as CH_4 has $\sigma = 12$, while an octahedral molecule such as SF_6 has $\sigma = 24$.

On insertion of the partition function [Eq. (20-2)] into the equations recalled above, we obtain the final equations for the rotation of nonlinear molecules:

$$-\left(\frac{G - H_0}{RT}\right)_r = \tfrac{3}{2}\ln T - \ln \sigma - \tfrac{1}{2}\ln(\theta_A \theta_B \theta_C) + \tfrac{1}{2}\ln \pi \qquad (20\text{-}3)$$

$$\left(\frac{H - H_0}{RT}\right)_r = \frac{3}{2} \qquad (20\text{-}4)$$

$$\left(\frac{S}{R}\right)_r = \tfrac{3}{2}\ln T - \ln \sigma - \tfrac{1}{2}\ln(\theta_A \theta_B \theta_C) + \tfrac{3}{2} + \tfrac{1}{2}\ln \pi \qquad (20\text{-}5)$$

$$\left(\frac{C}{R}\right)_r = \frac{3}{2} \qquad (20\text{-}6)$$

These equations are valid at sufficiently high temperatures for all nonlinear molecules. Because of the great variety of types of polyatomic molecules, it is difficult to give a precise statement of the low-temperature limit of validity of these formulas. There is certainly no significant error for any polyatomic molecule at room temperature (300 K), and calculations for various individual cases indicate that these equations are usually reliable down to 100 K or lower. Higher-order terms have been derived[2] for the symmetrical case with two equal moments of inertia, $A = B$.

In some cases, the geometry of a molecule may be known but the principal axes are not obvious from symmetry. In that case, the product of the principal moments of inertia, which suffices to calculate the product $\theta_A \theta_B \theta_C$, can be obtained from the relationship[3]

$$I_A I_B I_C = \begin{vmatrix} I_{xx} & -I_{xy} & -I_{xz} \\ -I_{xy} & I_{yy} & -I_{yz} \\ -I_{xz} & -I_{yz} & I_{zz} \end{vmatrix} \qquad (20\text{-}7)$$

Here I_{xx}, I_{xy}, etc., are moments and products of inertia with respect to any Cartesian coordinate system having the center of mass at the origin; that is,

$$I_{xx} = \sum_i m_i(y_i^2 + z_i^2) \quad \text{etc.}$$

$$I_{xy} = \sum_i m_i x_i y_i \quad \text{etc.}$$

where m_i is the mass of the atom i whose coordinates are x_i, y_i, z_i.

DIATOMIC MOLECULES: HIGHER APPROXIMATIONS

The first approximation to the thermodynamic functions for a diatomic molecule is given by the rotational and vibrational terms developed in Chapter 5 provided that the molecule has but a single electronic state within the thermal-energy range. This is commonly called the rigid rotator harmonic oscillator approximation. This approximation is excellent at low temperatures, but at higher energies the spectroscopic energy levels deviate from the pattern assumed in the earlier treatment in three significant respects. Consequently, higher approximations are needed for calculations at high temperatures. An improved expression for the energy levels, which includes these three additional terms, is usually an adequate approximation, although still more terms can be included.[4] The symbols for the vibrational and rotational constants, ω and B, are now given the subscript e to show that they pertain to the equilibrium interatomic distance of minimum potential energy:

$$\frac{\varepsilon}{hc} = \omega_e(v + \tfrac{1}{2}) - x_e\omega_e(v + \tfrac{1}{2})^2 + B_e J(J + 1)$$
$$- DJ^2(J + 1)^2 - \alpha(v + \tfrac{1}{2})J(J + 1) \tag{20-8}$$

Here the first and third terms are the familiar vibration and rotation terms, Eqs. (5-26) and (5-36), which are strictly correct only for harmonic forces in vibration and for a rigid body in rotation. The first correction term, which involves x_e, arises from deviation of the actual forces in the molecule from the harmonic-force law. The fourth term, $DJ^2(J + 1)^2$, arises because the centrifugal force of rotation stretches the molecule and increases its moment of inertia. From Eq. (5-27) we see that an increase in moment of inertia has the effect of decreasing B. Since the stretching may be related to the centrifugal force and to the force holding the atoms together, one may relate D to B_e and ω_e. The relation obtained is

$$D = \frac{4B_e^3}{\omega_e^2} \tag{20-9}$$

If the interatomic force does not follow the harmonic law, then the mean distance between atoms may vary with the vibrational energy. This effect leads to the last term.

Table 20-1 lists the various constants for a few diatomic molecules. Terms of still higher order are known for several of these molecules.[4] One may now proceed in either of two ways. With fast computers, one may carry out a direct summation of the basic equations,

$$Q = \sum_J \sum_v (2J + 1) \exp(-\varepsilon_{Jv}/kT)$$

$$Q' = \sum_J \sum_v (2J + 1)(\varepsilon_{Jv}/kT) \exp(-\varepsilon_{Jv}/kT)$$

$$Q'' = \sum_J \sum_v (2J + 1)(\varepsilon_{Jv}/kT)^2 \exp(-\varepsilon_{Jv}/kT)$$

with the energy scale adjusted to make $\varepsilon_{00} = 0$. The upper limits of the sums must correspond to energies not exceeding the dissociation limit, but in many cases fewer terms will suffice. In the direct summation method, it is straightforward to use the entire expression for the energy including terms beyond those in Eq. (20-8) or Table 20-1; see comments on calculations for H_2, N_2, etc., by Chase et al.[5] The equations relating Q, Q', Q'' to the thermodynamic functions were given in Chapter 5 [Eqs. (5-22) to (5-24)].

TABLE 20-1
Constants for a few diatomic molecules[4] in reciprocal centimeters (cm^{-1})

	ω_e	$x_e\omega_e$	B_e	α	ν_{00}[a]	Electronic state
H$_2$	4401.21	121.34	60.853	3.062	0	$^1\Sigma_g^+$
HCl	2990.95	52.819	10.593	0.3072	0	$^1\Sigma^+$
N$_2$	2358.57	14.324	1.99824	0.01732	0	$^1\Sigma_g^+$
NO	1904.20	14.075	1.67195	0.0171	0	$^2\Pi_{1/2}$
	1904.04	14.100	1.72016	0.0182	119.82	$^2\Pi_{3/2}$
O$_2$	1580.19	11.98	1.44563	0.01593	0	$^3\Sigma_g^-$
	1483.5	12.9	1.4264	0.0171	7918.1	$^1\Delta_g$
	1432.77	14.00	1.40037	0.01820	13195.1	$^1\Sigma_g^+$
Cl$_2$	559.72	2.675	0.2440	0.0015	0	$^1\Sigma_g^+$
I$_2$	214.50	0.615	0.03737	0.00011	0	$^1\Sigma_g^+$

[a] ν_{00} gives the energy of this electronic state above the lowest electronic state as measured between the ground vibration–rotation levels.

An alternate procedure is appropriate for use with simple calculators; it is based on appropriate expansions in series of the smaller terms in Eq. (20-8).

Before proceeding, let us rearrange the energy-level expression into a form more convenient for our purpose:

$$\frac{\varepsilon - \varepsilon_0}{hc} = \omega_0 v - x\omega_0 v(v-1) + B_0 J(J+1) - 4\frac{B_e^3}{\omega_e^2} J^2(J+1)^2 - \alpha v J(J+1)$$

$$(20\text{-}10)$$

where

$$\omega_0 = \omega_e - 2x_e\omega_e$$

$$B_0 = B_e - \frac{\alpha}{2}$$

$$x = \frac{x_e}{1 - 2x_e} \cong x_e$$

This form has the practical advantage that ω_0 gives the actual energy of the first excited vibrational state above the ground state and similarly B_0 gives the actual rotational-level spacing in the ground vibrational state. Also we have included the theoretical value for D.

Now let us introduce the temperature and define additional quantities as follows:

$$\frac{\varepsilon - \varepsilon_0}{kT} = uv - xuv(v-1) + \frac{\theta}{T}J(J+1) - 4y^2\frac{\theta}{T}J^2(J+1)^2 - \delta\frac{\theta}{T}vJ(J+1)$$

$$(20\text{-}11)$$

where

$$u = \frac{hc\omega_0}{kT} \tag{20-11a}$$

$$\theta = \frac{hcB_0}{k} \tag{20-11b}$$

$$y = B_e/\omega_e = \left(\frac{D}{4B_e}\right)^{1/2} \tag{20-11c}$$

$$\delta = \frac{\alpha}{B_0} \tag{20-11d}$$

The partition function for vibration and rotation is

$$Q = \sum_v \sum_J (2J + 1)\exp(-\varepsilon/kT) \tag{20-12}$$

where the exponent is given in Eq. (20-11). We deal first with the sum over J. Let us examine further the exponential factor, which we must sum,

$$\exp\left(-\frac{\theta}{T}J(J+1) + 4y^2\frac{\theta}{T}J^2(J+1)^2 + \delta\frac{\theta}{T}vJ(J+1)\right)$$

$$= \exp\left(-\frac{\theta}{T}J(J+1)\right)\exp\left(4y^2\frac{\theta}{T}J^2(J+1)^2 + \delta\frac{\theta}{T}vJ(J+1)\right)$$

The exponent in the second exponential factor is always small at values of J where the first factor is substantial. Consequently, we may expand the second exponential in series and retain only the first two terms. Also, since θ/T is small, we may replace the sum over J by an integral. We further substitute $z = J(J+1)$, $dz = (2J+1)dJ$.

$$Q = \sum_v e^{-uv + xuv(v-1)} \int_0^\infty e^{-z\theta/T}\left(1 + 4\frac{y^2\theta^2 z^2}{T} + \frac{\delta\theta vz}{T}\right)dz$$

The integration is straightforward. We may also add the small terms given in Eq. (5-31) that correct for the difference between the true sum over J and the integral provided that $\theta/T \leq 0.3$ and introduce the symmetry number. We then have

$$Q = \sum_v e^{-uv + xuv(v-1)}\left(\frac{T}{\sigma\theta}\right)\left(1 + \frac{8y^2 T}{\theta} + \delta v + \frac{\theta}{3T} + \frac{\theta^2}{15T^2}\right) \tag{20-13}$$

Again we expand the exponential of the small term which involves x and neglect all cross terms involving products of small quantities.

$$Q = \left(\frac{T}{\sigma\theta}\right)\sum_v e^{-uv}\left[1 + xuv(v-1) + \delta v + \frac{8y^2 T}{\theta} + \frac{\theta}{3T} + \frac{\theta^2}{15T^2}\right]$$

$$\tag{20-14}$$

Since u is not necessarily a small quantity, the sum over v must be taken exactly. In order to do so we note that

$$\sum_{v=0}^{\infty} e^{-uv} = (1 - e^{-u})^{-1}$$

$$-\frac{d}{du}\sum_v e^{-uv} = \sum_v v e^{-uv} = e^{-u}(1 - e^{-u})^{-2}$$

$$\frac{d^2}{du^2}\sum_v e^{-uv} = \sum_v v^2 e^{-uv} = e^{-u}(1 - e^{-u})^{-2} + 2e^{-2u}(1 - e^{-u})^{-3}$$

By use of these expressions, we obtain the exact sum over the vibrational quantum number v:

$$Q = \frac{T}{\sigma\theta(1 - e^{-u})}\left(1 + \frac{2xu}{(e^u - 1)^2} + \frac{\delta}{e^u - 1} + \frac{8y^2 T}{\theta} + \frac{\theta}{3T} + \frac{\theta^2}{15T}\right) \tag{20-15}$$

Finally, let us obtain the logarithm of Q, which we shall separate into three terms: the rigid rotator term, the harmonic oscillator term, and a term including all the higher terms introduced in this section. We recall that

$$\ln(1 + x) \cong x + \frac{x^2}{2} + \cdots \qquad \text{for } x \ll 1$$

$$\ln Q = \ln T - \ln \sigma - \ln \theta + \frac{\theta}{3T} + \frac{\theta^2}{90T^2} - \ln(1 - e^{-u})$$

$$+ \left(\frac{2xu}{(e^u - 1)^2} + \frac{\delta}{e^u - 1} + \frac{8y^2 T}{\theta}\right) \tag{20-16}$$

$$= \ln Q_r + \ln Q_{\text{vib}} + \ln Q_{\text{cor}}$$

$$\ln Q_{\text{cor}} = \frac{2xu}{(e^u - 1)^2} + \frac{\delta}{e^u - 1} + \frac{8y^2 T}{\theta} \tag{20-17}$$

We may simplify the expression for the correction terms further by the substitution $\theta/T = yu/T$, which is exact except for the small differences between B_e and B_0, ω_e and ω_0. Then we proceed to obtain the corrections to the various thermodynamic functions:

$$-\left(\frac{G - H_0}{RT}\right)_{\text{cor}} = \ln Q_{\text{cor}} = \frac{2xu}{(e^u - 1)^2} + \frac{\delta}{e^u - 1} + \frac{8y}{u} \tag{20-18}$$

The remaining functions can be obtained by appropriate differentiation. Table 20-2 gives an illustration of the complete calculation for N_2 at 2000 K.

TABLE 20-2
Sample calculation of the thermodynamic properties of a diatomic molecule including anharmonicity effects
N_2 at 2000 K. See Table 20-1 for molecular data

	$-(G° - H_0°)$	$C_P°$
	RT	R
Translation, M = 28.016	20.350	2.500
Rotation (rigid), θ/T = 0.001 4375	5.852	1.000
Vibration (harmonic), u = 1.6767	0.207	0.795
Corrections		
Vibration anharmonicity, x = 0.006 07	0.001	0.011
Vibration–rotation, δ = 0.008 67	0.002	0.010
Rotation stretching, y = 0.000 847	0.004	0.008
Total	26.416	4.324

POLYATOMIC MOLECULES: HIGHER APPROXIMATIONS

The energy levels of a polyatomic molecule deviate from the pattern calculated on the rigid rotator harmonic oscillator model for exactly the same reasons that were mentioned for diatomic molecules in the preceding section. However, the number and variety of terms are very great for the general case. Consequently, it is preferable to develop the appropriate formulas for various classes of molecules. Also, there is much less complete spectroscopic information with respect to these higher terms for polyatomic than for diatomic molecules. Pennington and Kobe[6] and Woolley[7] have presented formulas and tables for the thermodynamic functions that correspond to a sufficiently complete energy-level formula for molecules such as carbon dioxide, nitrous oxide, acetylene, or water. These higher approximations have been included for the molecules just mentioned and a few others, but the thermodynamic data for most polyatomic molecules are limited to the rigid rotator harmonic oscillator approximation. There is one additional type of motion, however, which arises very commonly in moderately large molecules and therefore deserves attention at this point.

INTERNAL ROTATION

If a molecule contains groups of atoms connected by single electron-pair bonds, there arise internal rotational motions. Simple examples are H_3C—CH_3 and H_2N—OH, where the rotations are CH_3 versus CH_3 and NH_2 versus OH, respectively. Since the rotating groups are relatively close to one another, it is reasonable that there should be a significant change of potential energy (and therefore a force) associated with the internal rotational coordinate. We may divide internal-rotation problems into three categories depending on the magnitude of this potential.

If the potential is very large in comparison with kT, then we have a torsional vibration that may be treated like any other vibration. Twisting about a double bond, as in $H_2C = CH_2$, normally falls in this category.

If the potential change with angle of rotation is very small as compared with kT, then it may be ignored and we have a free rotation. The energy levels, from quantum mechanics, are

$$\varepsilon = \frac{h^2 K^2}{8\pi^2 I_r} \tag{20-19}$$

where K is a quantum number that can have all integral values, positive and negative including zero, and I_r is the reduced moment of inertia for the rotation. If the molecule consists of a pair of symmetrical coaxial tops, such as H_3C—CH_3 or H_3C—CCl_3, then

$$I_r = \frac{AB}{A + B} \tag{20-20}$$

where A and B are the moments of inertia of the respective tops about the axis of internal rotation.

If one or both of the rotating groups has symmetrically placed atoms, then only certain values of K are allowed. It is always found that $1/n$ of the K values are allowed if there are n equivalent orientations. Thus the symmetry number for internal rotation, n, plays an entirely equivalent role to that of the symmetry number for overall rotation. The partition function for free rotation is then

$$Q_f = \frac{1}{n} \sum_{K=-\infty}^{+\infty} \exp\left(\frac{-K^2 h^2}{8\pi^2 I_r kT}\right) \tag{20-21}$$

We again find that it is an appropriate approximation to replace the sum by an integral and obtain

$$Q_f = \frac{1}{n} \left(\frac{8\pi^3 I_r kT}{h^2}\right)^{1/2} \tag{20-22}$$

The thermodynamic functions can be readily obtained from this partition function. As expected, we find $C/R = 1/2$. One of the few examples where free internal rotation has been observed is dimethyl cadmium, H_3C—Cd—CH_3. Here the rotating groups are separated by two bond lengths instead of one. Table 20-3 gives the numerical results for this substance.[8]

The more common and more complex case of internal rotation involves a potential barrier that is neither very large nor very small compared with kT. In this case one must seek the energy levels for an appropriate molecular model. Since we have a rotational coordinate, the potential energy must be a periodic function with period $2\pi/n$, where n is the number of times per revolution that the molecule returns to an

TABLE 20-3
Entropy of cadmium dimethyl, an example of free internal rotation

	$S°/R$
Translation and overall rotation	30.50
Vibration	4.41
Free internal rotation	1.47
Total, calculated	36.38
Experimental from third law[a]	36.43 ± 0.10

[a] J. C. M. Li, *J. Am. Chem. Soc.*, **78**, 1081 (1956).

equivalent position. Thus, for ethane $n = 3$. We can expand the potential in a Fourier series, and it is found[9,10] that the first two terms suffice to give an excellent approximation:

$$V = C_0 + C_1 \cos n\phi + \cdots$$
$$= \tfrac{1}{2}V_0(1 - \cos n\phi) \qquad (20\text{-}23)$$

In the second line the expression is arranged so that the potential is zero when the angle ϕ is zero and the height of the potential peaks is V_0.

The quantum-mechanical problem of a rotation subject to this potential barrier has been solved, but no simple expression can be given for the energy levels. Rather the levels must be obtained from infinite but convergent continued fractions or some equivalent expressions. Also, there are complexities arising from the relationships between the rotation of the entire molecule and the internal rotation. It is beyond the scope of this presentation to discuss all these problems, although any particular case is readily solved with a modern computer.

TABLE 20-4
Thermodynamic functions for methyl chloroform, an example of restricted internal rotation
$T = 286.53$ K

	$S°/R$
Translation + rotation	33.20
Vibration	4.06
Internal rotation	1.09
Total, calculated	38.35
Experimental[a]	38.36 ± 0.08

[a] T. R. Rubin, B. H. Levedahl, and D. M. Yost, *J. Am. Chem. Soc.*, **66**, 279 (1944).

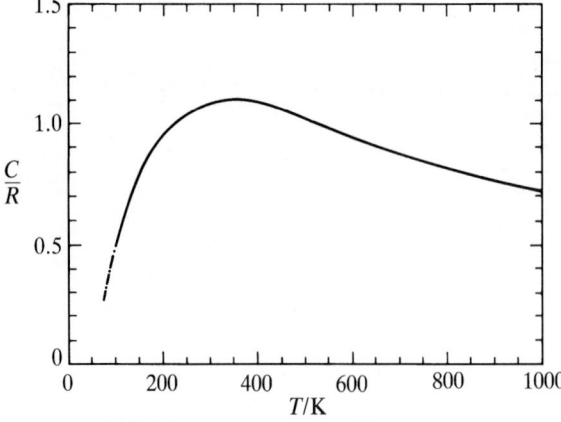

FIGURE 20-1
The heat capacity of internal rotation in CH_3CCl_3 as a function of temperature.

Fortunately it is found,[11] for most real examples, that the final results for the contribution of an internal rotation to the thermodynamic properties can be presented in general tables as the functions of two variables. One variable is the ratio of the potential barrier to thermal energy, V_0/RT. While there are alternatives for the second variable, the favored choice is the reciprocal of the partition function that would apply if the rotation were free, $1/Q_f$.

Appendix 12 includes tables of numerical values of the contributions of restricted internal rotation to thermodynamic functions. Also included there is a more general definition for the reduced moment of inertia.

Table 20-4 gives an example of the calculations for a restricted internal rotation. The moments of inertia are calculated for rotation of the entire rigid molecule, i.e., as if the internal rotation were frozen. Other parameters are obtained as described in this and previous sections. In this example,[10] CH_3CCl_3, the product of the principal moments of inertia for the rigid molecule is $I_A I_B I_C = 6.14 \times 10^{-113}$ g cm^2 while the reduced moment for internal rotation is 5.25×10^{-40} g cm^2. The potential barrier to internal rotation was found from the infrared spectrum to be well fitted by Eq. (20-23) with $V_0/R = 1493$ K. The results are given in Table 20-4, which shows excellent agreement with the experimental entropy.

The heat capacity of internal rotation in CH_3CCl_3 is shown in Fig. 20-1. Note that the heat capacity rises initially along the pattern of a somewhat anharmonic oscillator. But instead of remaining somewhat above 1.0 in C/R, it decreases at still higher temperatures and eventually approaches the free-rotation value of 0.5 as an asymptote. The other functions are, of course, integrals of the heat capacity.

ELECTRONIC STATES FOR MOLECULES

A large majority of molecules of importance at room temperature have electronic states in which the electrons are paired in a fashion that yields zero electron spin and zero orbital angular momentum. Thus the electronic state is a single quantum state, and there is no electronic contribution to the thermodynamic properties at moderate temperatures. Exceptions among common substances are O_2, NO, and NO_2. Each of these three presents a slightly different situation.

Oxygen in its lowest energy form is said to have a $^3\Sigma_g^-$ electronic state. For the complete meaning of this symbol the reader should consult a text on molecular spectra. For our purpose it is sufficient to know that the symbol Σ means that there is no orbital angular momentum, while the preceding superscript 3 means that it is a triplet state with respect to electron spin; i.e., there are two spins coupled parallel, yielding a spin vector $S = 1$ and multiplicity $2S + 1 = 3$. While the electron spin introduces many complexities into the detailed spectrum, the net effect is to yield just an additional multiplicity factor of 3 for all rotational energy levels. Thus, the effect of the electron spin is to add $\ln 3$ to $S°/R$ and to $-(G° - H_0°)/RT$ but to make no change in either enthalpy or heat capacity. These results apply in the absence of a magnetic field; the net electron spin has a major effect on the magnetic susceptibility.

At very high temperatures one must consider the excited electronic states of oxygen, which are designated as $^1\Delta_g$ and $^1\Sigma_g^+$ and which lie 7918 and 13 195 cm^{-1}, respectively, above the ground state. The symbol $^1\Sigma_g^+$ implies the normal type of singlet state, but the symbol $^1\Delta_g$ indicates 2 units of orbital angular momentum ($\Lambda = 2$) and no net spin angular momentum. Again there are various detailed effects, but in the absence of a magnetic field the net result is just a multiplicity factor of 2 for all of the vibration–rotation states of the $^1\Delta_g$ electronic state.

The distribution of the oxygen among the three electronic states is governed by the thermodynamic properties of the three states. If the vibrational and rotational properties were the same, one could use the multiplicities of 3, 2, and 1 together with the energy differences to calculate the distribution. This electronic partition function would be

$$Q_{el} = 3 + 2e^{-7918hc/kT} + e^{-13195hc/kT} \qquad (20\text{-}24)$$

In the case of oxygen, however, this calculation would not be very accurate, for we see from Table 20-1 that the molecular constants are appreciably different in the three states. An accurate calculation would require the treatment of each electronic state as a distinct species, with a sum of the complete partition functions for each of the states yielding the total population.

The case of NO_2 is much simpler. Here we have just double multiplicity because of the one unpaired electron spin. This adds $\ln 2$ to $S°/R$ and $-(G° - H_0°)/RT$ functions but has no effect on heat capacity or enthalpy.

The NO molecule represents a very interesting case. It has both orbital and spin angular momentum—1 unit of the former and 1/2 unit of the latter. This yields the pair of states $^2\Pi_{1/2}$ and $^2\Pi_{3/2}$, which are separated by only 119.8 cm^{-1} in energy. The complete theory for this case is somewhat more complex, but the basic ideas are just those already illustrated.

One must always base a calculation of this type on the complete internal partition function for the substance. The differences arise in the energy-level patterns for various molecules and the possibilities of factoring the partition function. The equations given in this chapter suffice for most gases at temperatures up to about 2000 K, but there are further complexities that require extension of these methods in certain cases.

Free radicals and molecules of importance at high temperatures normally have much more loosely bound electrons, and low-lying electronic levels are much more common than for molecules stable at room temperature. Important electronic contributions to the thermodynamic properties can be expected for such molecules.

RADIATION; BOSE–EINSTEIN AND FERMI–DIRAC STATISTICS

While for all ordinary gases the Boltzmann distribution law, Eq. (A1-5) or (5-11), is a good approximation, this is not true for radiation, a photon gas. It was the special properties of radiation that led Planck to first propose quantum theory. As now developed, quantum statistical mechanics considers the quantum states for multi-particle systems of truly identical, "indistinguishable" particles. It is found that their wave functions must be either symmetric or antisymmetric to an interchange of any two particles. Particles of zero or integral nuclear spin, photons or ^{4}He nuclei, etc., fall in the first group, while electrons, protons and other particles of half-integral spin are in the second group. Functions are then derived for the distributions of single-particle energy states in each system. These derivations are similar to that in Appendix 1 for the Boltzmann case, and are available in most if not all books on statistical mechanics. Thus, we omit the derivations and give just the final results, which are

$$N_i = \frac{g_i}{e^{\alpha + \varepsilon_i/kT} \pm 1} \qquad (20\text{-}25)$$

where the + sign applies to the half-integral-spin systems and the − sign to the integral-spin systems. These and the subsequent equations are known as Fermi–Dirac and Bose–Einstein statistics, respectively. In Eq. (20-25) the quantity α is adjusted to yield the correct total number of particles, as was the case for Eq. (A1-5). At high temperature the ±1 becomes negligible and the Boltzmann distribution results.

Although the properties of Fermi–Dirac statistics are very important for the conduction electrons in metals, the strong coulombic forces are also important; hence, this is not a simple gas. The full treatment of the properties of conduction electrons is complex and will not be presented here. For the photon gas—radiation—however, there are no complicating forces, and the properties follow directly after the special characteristics of a photon are inserted.[12]

First, we note that the number of photons is not fixed. Thus, the quantity α drops out of Eq. (20-25) and one has the Planck distribution law

$$N_i = \frac{g_i}{e^{\varepsilon_i/kT} - 1} \qquad (20\text{-}26)$$

Also, one notes that the photon energy is $\varepsilon_i = h\nu_i$.

The expression for g_i in Eq. (20-26) is less easily obtained. The rigorous derivation involves a rather complex geometrical consideration,[12,13] but Hill[14] gives a simpler argument via the pressure. In either case the result is

$$g(\nu)\,d\nu = \frac{8\pi V}{c^3}\nu^2\,d\nu \qquad (20\text{-}27)$$

where $g(\nu)\,d\nu$ is the number of photon states between ν and $\nu + d\nu$ in frequency, and c is the velocity of light. Then after substitution into Eq. (20-26), one obtains an expression for the energy density of photons as a function of frequency:

$$\rho(\nu,T)\,d\nu = \frac{8\pi h\nu^3}{c^3}\frac{1}{e^{h\nu/kT}-1}\,d\nu \qquad (20\text{-}28)$$

This can be integrated to yield the total energy density

$$\frac{U}{V} = \frac{8\pi^5(kT)^4}{15(hc)^3} \qquad (20\text{-}29)$$

After consideration of geometrical factors,[12] the energy flux per unit area and time for a black body is

$$e(\nu,T)\,d\nu = \frac{2\pi h\nu^3}{c^2}\frac{1}{e^{h\nu/kT}-1}\,d\nu \qquad (20\text{-}30)$$

or, in terms of wavelength, since $\nu = c/\lambda$, $d\nu = -(c/\lambda^2)\,d\lambda$, the energy flux between λ and $\lambda + d\lambda$ is

$$e(\lambda,T)\,d\lambda = \frac{2\pi hc^2}{\lambda^5}\frac{1}{e^{hc/\lambda kT}-1}\,d\lambda \qquad (20\text{-}30\text{a})$$

Equation (20-30) or (20-30a) provides the basis for temperature measurement at very high temperatures. Both the absolute intensity and the relative intensity between substantially different frequencies vary strongly with temperature. The optical measuring instrument can be calibrated at a fixed point such as the melting point of gold.

It is also interesting to note that the total energy can be expressed as aVT^4 over the full range of temperature, with $a = 8\pi^5 k^4/15(hc)^3$. Then from Eq. (3-10), $(\partial S/\partial U)_V = 1/T$, and the knowledge that the entropy is zero at $T = 0$,

$$dS \text{ (at constant } V) = 4aT^2 V\,dT \qquad (20\text{-}31)$$

$$S = \frac{4}{3}aT^3 V = \frac{4U}{3T} \qquad (20\text{-}32)$$

Other thermodynamic functions then follow from familiar relationships:

$$A = U - TS = -\frac{1}{3}aVT^4 = -\frac{U}{3} \qquad (20\text{-}33)$$

$$P = -\left(\frac{\partial A}{\partial V}\right)_T = \frac{1}{3}aT^4 = \frac{U}{3V} \qquad (20\text{-}34)$$

$$G = A + PV = 0 \qquad (20\text{-}35)$$

We see that these various properties are easily derived by thermodynamics once the internal energy of the radiation has been obtained from quantum statistics. The last result that G remains zero is connected to the absence of a fixed number of particles in the system and the absence of the parameter α in the Planck distribution function.

IONIZATION IN GASES; PLASMAS

At extremely high temperatures a special situation arises. For atomic hydrogen as an example, the electronic energy states and their multiplicities are accurately known from theory. There are an infinite number of states of finite energy; hence, the sum for the partition function, Eq. (5-14), fails to converge. At ordinary temperatures the individual terms become very small and the sum can be terminated to yield a finite value without appreciable uncertainty. At temperatures where there is significant ionization, i.e., in a plasma, however, there is a serious difficulty. This situation was discussed by Pitzer;[15] other references are cited and recommendations are made for a self-consistent and practical solution.

REFERENCES

1. J. E. Mayer and M. G. Mayer, *Statistical Mechanics*, 2d edn., Wiley, New York, 1977, p. 184.
2. R. Kayser, Jr. and J. E. Kilpatrick, *J. Chem. Phys.*, **68**, 1511 (1978).
3. J. O. Hirschfelder, *J. Chem. Phys.*, **8**, 431 (1940).
4. K. P. Huber and G. Herzberg, *Molecular Spectra and Molecular Structure IV. Constants of Diatomic Molecules*, Van Nostrand Reinhold, New York, 1979.
5. M. W. Chase, Jr., C. A. Davies, J. R. Downey, Jr., D. J. Frurip, R. A. McDonald, and A. N. Syverud, *JANAF Thermochemical Tables*, 3d edn., *J. Phys. Chem. Ref. Data*, **14** (suppl. no. 1) (1985).
6. R. E. Pennington and K. A. Kobe, *J. Chem. Phys.*, **22**, 1442 (1954).
7. H. W. Woolley, *J. Res. Natl. Bur. Stand.*, **56**, 105 (1956).
8. J. C. M. Li, *J. Am. Chem. Soc.*, **78**, 1081 (1956).
9. K. S. Pitzer, *Discuss. Faraday Soc.*, **1951** (10), p. 66; D. R. Herschbach, *J. Chem. Phys.*, **27**, 1420 (1957).
10. K. S. Pitzer and J. L. Hollenberg, *J. Am. Chem. Soc.*, **75**, 2219 (1953).
11. K. S. Pitzer and W. D. Gwinn, *J. Chem. Phys.*, **10**, 428 (1942); K. S. Pitzer, *J. Chem. Phys.*, **5**, 469 (1937); J. C. M. Li and K. S. Pitzer, *J. Phys. Chem.*, **60**, 466 (1956).
12. For an especially clear and complete treatment of radiation, see N. Davidson, *Statistical Mechanics*, McGraw-Hill, New York, 1962.
13. C. Kittel and H. Kroemer, *Thermal Physics*, W. H. Freeman, San Francisco, 1980.
14. T. L. Hill, *An Introduction to Statistical Thermodynamics*, Addison-Wesley, Reading, Mass., 1960.
15. K. S. Pitzer, *J. Chem. Phys.*, **70**, 393 (1979).

PROBLEMS

20-1. Estimate the entropy of the radical CH_2Cl at 350 K. The odd electron will contribute $R \ln 2$ of electronic entropy. Assume 120° angles at carbon with the distances 1.1 Å for C—H and 1.7 Å for C—Cl and that the C—Cl stretching vibration frequency is 730 cm^{-1} as in CH_3Cl. How many other vibrations has this radical, and what is the possible range in entropy if their frequencies are all above 1000 cm^{-1}?

20-2. Estimate the importance of the $^1\Delta$ state of O_2 for the function $-(G° - H_0°)RT$ at 3000 K. Do not make a detailed calculation; accuracy to a factor of 2 in the contribution of the $^1\Delta$ state is sufficient.

20-3. Sodium vapor at the boiling point at 1163 K is a mixture of atoms and diatomic molecules. The molecules have single electronic states and the molecular constants $\omega = 159$, $\omega x = 0.726$, $B = 0.1547$, $\alpha = 0.00079$, $D = 0.584 \times 10^{-6}$, all in cm^{-1}. The dissociation energy is $\Delta H_0°/R = 8810$ K. The lowest atomic energy level is $^2S_{1/2}$; the next in importance are $^2P_{1/2}$ and $^2P_{3/2}$ at 16956 and 16973 cm^{-1}, respectively. Calculate the Gibbs-energy function for both Na and Na_2 at 1163 K, and find the proportion of Na_2 in the saturated vapor.

20-4. The Sackur–Tetrode formula [Eq. (5-7)] may be used to calculate the entropy and other properties of a gas of electrons, provided that the density is low enough and the temperature high enough and provided that the space charge is canceled by appropriate positive charges. Calculate the value of $S°/R$ for electrons at 5000 K.

20-5. From the spectrum of Cs gas, $\Delta U_0°$ for $Cs(g, {}^2S_{1/2}) = Cs^+(g, {}^1S_0) + e^-(g)$ is 31406.71 cm^{-1} or 3.893 eV. In Problem 20-4, we note that the entropy and heat capacity of electrons at low pressures can be calculated in the same manner as for other monatomic gases. Thus, for the ionization of Cs(g), $\Delta C_P = \frac{5}{2}R$ at temperatures below that where the first excited electronic level of Cs(g) has an appreciable population. From Eq. (5-7), calculate $\Delta S°$ and $\Delta G°$ for the ionization of Cs at 2000 K and the percent ionization of Cs vapor at 2000 K and 10^{-4} bar. Note the spin multiplicity of Cs(g) and $e^-(g)$.

CHAPTER 21

RESIDUAL ENTROPIES AT ZERO KELVIN

It was noted in Chapter 6 that there were examples where disorder present at higher temperatures is not removed as the temperature decreases but rather becomes "frozen" and remains as residual entropy as $T \to 0$. In the case of a glassy material the situation is obvious: it fails to crystallize. Likewise, a solid solution may or may not transform on cooling to one or two pure substances of exact composition and regular structure. If it does not, then the randomness of mixing is retained as $T \to 0$. In any particular case, the measurements cease at some low but finite temperature. Thus, one must also consider the possibility that the disorder might be removed at an even lower temperature.

We consider first molecular crystals with a regular array of molecules and consider possible randomness of orientation or of exact position. As pointed out in Chapter 6, the overall thermodynamic properties, entropies, Gibbs-energy functions, etc., are best obtained by statistical calculation from molecular parameters whenever that information is adequate, as it often is. Thus, our primary interest at this point is the exploration of structural behavior using thermal properties as the source of information. The pattern of results is of interest, however, in making a judgment about the probability of residual entropy in the cases of molecular crystals where the molecular parameters are unknown or incomplete.

The simplest case of possible rotational disorder is that of a linear or quasilinear molecule that is not symmetric to end-for-end rotation. Linear or quasilinear molecules that are symmetric, such as the diatomic halogens, N_2, O—C—O, the normal paraffins

C_2H_6, C_3H_8, ..., CH_3—C≡C—CH_3, etc., are all known to have entropies in the solid state approaching the practical zero as $T \to 0$. Thus, any residual entropy for nonsymmetric, linear, or quasilinear molecules will be just that of end-for-end randomness, as indicated below for CO:

```
CO      OC      OC      CO      OC      CO      CO
   CO      OC      CO      CO      OC      OC
OC      OC      CO      OC      CO      CO
```

The multiplicity is 2 for each molecule or 2^{N_A} for N_A molecules in a crystal. Thus,

$$S = \frac{R}{N_A} \ln W = \frac{R}{N_A} \ln 2^{N_A} = R \ln 2$$

$$\frac{S}{R} = \ln 2 = 0.693$$

The actual residual entropy found experimentally is usually a little smaller, which indicates that the molecular orientation is not quite fully random. For CO,[1,2] $S/R = 0.5$ instead of 0.69.

For CO, which might rotate rather easily in the crystal and has a very small dipole moment, one can ask whether this residual entropy is really frozen or whether there might be a transition or anomaly at a very low temperature. This was first investigated by Melhuish and Scott,[3] who estimated that ordering would occur near 5 K on the basis of the dipole moment of CO. This lies below the lowest temperatures of measurement of Clayton and Giauque[1] of 14 K. Subsequently, Gill and Morrison[2] extended the measurements down to 2.5 K, which is one-half of the estimated transition temperature, without finding any anomaly. Then calculations were made of the rate of ordering by Curl et al.[4] who found, for the best estimate of the potential restricting rotation (600 cm^{-1}), a rate of conversion of 5×10^{-7} per year. This rate is very sensitive to the barrier height and becomes rapid by quantum tunneling for barriers of 200 cm^{-1} or less. Unless the estimates of rotation potential and of ordering temperature are both in error by factors of more than 2, however, the conclusion is that the randomness is frozen in CO below 5 K.

The situation for longer molecules such as N_2O (NNO) is entirely analogous to that of CO except that the rate of rotation in the crystal would be much smaller; hence, disorder, if present, is certainly frozen. The possibility of disorder in 1-olefins was studied by McCullough et al.,[5] who found that the shorter 1-olefin molecules attain perfect order in their crystals, but those longer than 1-decene have end-for-end disorder. The case of NO is similar since we know that it exists as a dimeric N_2O_2 unit in the crystal.[6] The N_2O_2 molecule is rectangular and presumably has the pairs of similar atoms in the diagonally opposite corners. Thus there are two possible orientations in the crystal for the dimeric unit. The observed[7] residual entropy is $S/R = 0.7$ for N_2O_2, which is close to the value of $\ln 2$ expected for this model.

There are cases of disorder involving the location of the protons of water molecules that can be described in terms of rotation of those molecules but are more easily visualized as shifts of protons along O—H----O hydrogen bonds. In these cases the proton is always closer to one oxygen than the other and there are two protons close to each oxygen so that H_2O molecules still exist. But the possible ordering process would probably be shifts of protons along the O—H---O axis to the O---H—O position and vice versa.

This randomness of proton location was first observed for ice, where Giauque and Stout[8] found a residual entropy $S/R = 0.4$. Long and Kemp[9] found the same value for D_2O. The structural model was proposed by Pauling.[10] Each oxygen is known to have four nearest-neighbor oxygens, and the protons are close to the central oxygen along two links and farther away in the other two. If no further restriction is placed on the proton locations, Pauling calculates that there are $(3/2)^{N_A}$ arrangements for N_A water molecules. Hence $S/R = \ln(3/2) = 0.405$.

Most hydrated salt crystals show no residual entropy. An exception is $Na_2SO_4 \cdot 10\ H_2O$, which shows a residual entropy[11,12] near $S/R = \ln 2$ per mole of crystal containing 10 H_2O. This effect is clearly explained by the crystal structure.[13] One four-membered hydrogen-bonded ring is formed per 10 water molecules. If the protons are unsymmetrically located in the hydrogen bonds, there are two sets of proton positions, shown by diagrams A and B, which retain water-molecule units with two hydrogens close to each oxygen.

If the various rings in the crystal have structures A and B at random, then the calculated entropy of disorder is $R \ln 2$. The isomorphous crystal $Na_2SeO_4 \cdot 10\ H_2O$ has the same ring structure and presumably would have the same residual entropy.

Again, it is interesting to ask about the rate of ordering and the temperature at which it would occur. Pitzer and Polissar[14] estimated that the transition in ice to an ordered state would be below 100 K and probably near 60 K. Subsequently, it was shown by experiments with ice doped with HF or KOH[15,16] that H_3O^+ ions or OH^- ions catalyzed the ordering of the protons in ice. KOH was the more effective dopant, and with it a substantial portion of the ice converted to an ordered structure on cooling and then showed a sharp transition at 72 K on heating.[15] It was not possible to obtain complete conversion, however. Various theoretical studies have been made of the structure of ice in this transition region; a recent example is that of Barkema and de Boer.[17]

In clathrate crystals the cage structures are comprised of hydrogen-bonded H_2O and there are possibilities of icelike randomness of proton location. Suga and associates investigated several examples[16] and found evidence for this disorder. If the molecules in the cages have appropriate properties, there are also possibilities of disorder in their angular orientation, which was also considered.[16]

Ionic cyanide crystals have the possibility of end-for-end disorder of the CN^- ion analogous to the situation for CO, NNO, etc. NaCN, KCN, RbCN and several other cyanides have also been studied by Suga and Matsuo and their associates.[18,19] The disorder is removed in a transition at 172 K in NaCN and at 83 K in KCN. This trend suggests that the transition in RbCN would tend to occur at a much lower temperature; however, at that temperature the rate might be very slow. Indeed, no transition is observed,[19] but on cooling, slow heat evolution is observed in RbCN just above 30 K. Also, a sample held at very low temperature and then heated shows slow heat absorption just below 30 K. Evidently, the disorder is essentially frozen in RbCN. Suga[20] has recently reviewed this and other types of disorder and slow relaxation in crystals.

We turn now to crystals without discrete molecules but with covalent or ionic bonding extending throughout the crystal. Several types of disorder are known at higher temperatures, and in some cases this disorder may or may not be removed as the temperature is reduced on cooling toward zero kelvin. In these systems, chemical equilibria are often attained only at very high temperatures, and the accuracy of measurements is not sufficient to detect an entropy difference of the size of the possible disorder as $T \rightarrow 0\,K$. This was the situation for Fe_3O_4, magnetite, until very recently. Fe_3O_4 has a cubic structure at room temperature, with iron atoms on two types of sites. Half of the Fe^{3+} ions are on one type of site, while the other set of sites contains equal numbers of Fe^{2+} and Fe^{3+} ions. Thus, at complete randomness the ion-mixing entropy would be $2R \ln 2$ or $1.39R$. The electrostatic repulsive effects are so strong, however, that short-range ordering is expected, and Anderson[21] showed that the plausible residual disorder would be the same as that in ice, $S/R = \ln(3/2) = 0.405$. In this case, ordering requires only the movement of electrons, not protons or iron atoms; hence, it should be more rapid than in ice. There is a transition at 121 K, below which the structure becomes one of lower symmetry. From the structural information, it is not certain whether the ice-like entropy is removed or not, nor, until recently, was this question resolved by the thermodynamics of reactions involving Fe_3O_4. In 1987 and 1988, O'Neill[22] reported electrochemical cell measurements for the reactions $3Fe + 2O_2 = Fe_3O_4$ and $3Fe_2SiO_4 + O_2 = 2Fe_3O_4 + 2SiO_2$ which, in each case, agree with treatments using the measured third-law entropy of Fe_3O_4 and are accurate enough to preclude a residual entropy of $R\ln(3/2)$.

Another iron oxide, wustite, departs significantly from the nominal composition FeO and has also been studied by O'Neill[22] and by Holmes et al.[23] with the cell

$$Pt, Fe + Fe_xO \,|\, CSZ \,|\, O_2, Pt$$

where CSZ is calcia-stabilized zirconia, a solid electrolyte transporting oxide ions. The iron oxide in contact with iron has a composition varying only slightly with temperature and with $x \cong 0.955$, but the cell measurements were converted to $x = 0.947$ for comparison with the third-law entropy. The resulting $\Delta_r G°$ values are in good agreement with the third-law entropy measured by Todd and Bonnickson[24] without any addition for a residual entropy at 0 K. Indeed, Stull and Prophet[25] also concluded, on a tentative basis, that the residual entropy in $Fe_{0.947}O$ was zero from a consideration of earlier measurements of equilibria of Fe_xO with Fe including the reaction

$$Fe_xO + H_2(g) = xFe + H_2O(g)$$

This absence of significant residual entropy in a significantly nonstoichiometric solid initially seems surprising and suggests that the material departing from the simple FeO structure agglomerates into regions containing several Fe vacancies. Indeed, recent neutron diffraction measurements[26] show that the vacancies are strongly clustered even at 1373 K and that the clusters grow still larger in size and fewer in number with decrease in temperature. Thus, it is reasonable that the residual entropy in wustite is small in comparison with the uncertainties of thermodynamic measurements. There is an extensive literature on the oxides of iron, which was reviewed in 1978 by Spencer and Kubaschewski.[27] A recent paper of Gronvold et al.[28] reports heat-capacity measurements of wustite extending to 5 K that confirm the entropy value of Todd and Bonnickson[24]; recent references on iron oxides are also included.

While the absence of significant residual entropies in these two iron oxides may or may not be typical, it strongly suggests that it is better to neglect residual entropies for solids of this general type than to add the values corresponding to the random distribution of vacancies or defects. The possible residual entropy should, however, be considered as an uncertainty in the experimental entropy based on the third law until its absence is established by separate thermodynamic measurements or structural studies such as those discussed.

REFERENCES

1. J. O. Clayton and W. F. Giauque, *J. Am. Chem. Soc.*, **54**, 2610 (1932).
2. E. K. Gill and J. A. Morrison, *J. Chem. Phys.*, **45**, 1585 (1966).
3. M. W. Melhuish and R. L. Scott, *J. Phys. Chem.*, **68**, 2301 (1964).
4. R. F. Curl, Jr., H. P. Hopkins, Jr., and K. S. Pitzer, *J. Chem. Phys.*, **48**, 4064 (1968).
5. J. P. McCullough, H. L. Finke, M. E. Gross, J. F. Messerly, and G. Waddington, *J. Phys. Chem.*, **61**, 289 (1957).
6. W. J. Dulmage, E. A. Myers, and W. N. Lipscomb, *J. Chem. Phys.* **19**, 1432 (1951); *Acta Crystallogr.*, **6**, 760 (1953).
7. H. L. Johnston and W. F. Giauque, *J. Am. Chem Soc.*, **51**, 3194 (1929).
8. W. F. Giauque and W. Stout, *J. Am. Chem. Soc.*, **58**, 1144 (1936).
9. E. A. Long and J. D. Kemp, *J. Am. Chem. Soc.*, **58**, 1829 (1936).
10. L. Pauling, *J. Am. Chem. Soc.*, **57**, 2680 (1935).
11. K. S. Pitzer and L. V. Coulter, *J. Am. Chem. Soc.*, **60**, 1310 (1938).
12. G. E. Brodale and W. F. Giauque, *J. Am. Chem. Soc.*, **80**, 2042 (1958).
13. H. W. Ruben, D. H. Templeton, R. D. Rosenstein, and I. Olovsson, *J. Am. Chem. Soc.*, **83**, 820 (1961).

14. K. S. Pitzer and J. Polissar, *J. Phys. Chem.*, **60**, 1140 (1956).
15. Y. Tajima, T. Matsuo, and H. Suga, *J. Phys. Chem. Solids*, **45**, 1135 (1984).
16. H. Suga, T. Matsuo, and O. Yasamuro, *Pure and Appl. Chem.*, **64**, 17 (1992).
17. G. T. Barkema and J. de Boer, *J. Chem. Phys.*, **99**, 2059 (1993).
18. H. Suga, T. Matsuo, and S. Seki, *Bull. Chem. Soc. Japan*, **38**, 1115 (1965).
19. T. Shimada, T. Matsuo, H. Suga, and F. Lutz, *J. Chem. Phys.*, **85**, 3530 (1986).
20. H. Suga, *J. Chem. Thermodyn.*, **25**, 463 (1993).
21. P. W. Anderson, *Phys. Rev.*, **102**, 1008 (1956).
22. H. St. C. O'Neill, *Am. Mineral.*, **72**, 67 (1987); **73**, 470 (1988).
23. R. D. Holmes, H. St. C. O'Neill, and R. J. Arculus, *Geochim. Cosmochim. Acta*, **50**, 2439 (1986).
24. S. S. Todd and K. R. Bonnickson, *J. Am. Chem. Soc.*, **73**, 3894 (1951).
25. D. R. Stull and H. Prophet, *JANAF Thermochemical Tables*, 2d edn., NSRDS-NBS-37 (1971) and subsequent editions of the JANAF Tables.
26. M. J. Radler, J. B. Cohen, and J. Faber, Jr., *J. Phys. Chem. Solids*, **51**, 217 (1990).
27. P. J. Spencer and O. Kubaschewski, *CALPHAD*, **2**, 147 (1978).
28. F. Gronvold, S. Stolen, P. Tolmach, and E. F. Westrum, Jr., *J. Chem. Thermodyn.*, **25**, 1089 (1993).

CHAPTER 22

HYDROGEN, HELIUM, AND METHANE AT LOW TEMPERATURES

Both hydrogen and helium show special effects at low temperatures that are, for the most part, unique to the various isotopes of these two elements. Methane in its various isotopic species also shows very interesting effects in the solid state below 30 K. The underlying cause in all cases is to be found in quantum mechanics, but we shall not pretend to discuss that aspect in any considerable detail.[1] However, it is of interest to note that none of these peculiar phenomena shows any disagreement with thermodynamic principles. Here again we are impressed by the general validity of thermodynamics with respect to any system containing many particles.

For any of these substances, the best values for the entropy and other properties in the gas are those calculated from molecular properties as described in Chapter 5 for H_2 and He and in Chapter 20 for CH_4. It is interesting, nevertheless, to see whether a particular species approaches the practical zero of entropy (i.e., with nuclear spin entropy still present) at some low but finite temperature. We shall also see that there are special cases and circumstances where nuclear-spin entropy is lost at relatively high temperatures.

HYDROGEN

The unique properties of hydrogen arise from the fact that the moment of inertia of the molecule is smaller than that of the other familiar diatomic molecules by a factor of about 10 or more. The rotational energy levels are given by the combination of Eqs. (5-26) and (5-27).

$$\varepsilon_J = \frac{h^2 J(J+1)}{8\pi^2 I} \qquad (22\text{-}1)$$

where J is the quantum number and I the moment of inertia. The successive ε_J values are spaced so far apart for hydrogen that this spacing is much larger than kT at low temperatures of practical interest. In addition, there are the three isotopes, stable ^{1}H (or H) and ^{2}H (or D) and radioactive ^{3}H (or T), each of which has nuclear spin. Two or more nuclei with spin in the same molecule will have their angular momenta coupled together in a quantized fashion. The $\frac{1}{2}$-unit spins of ^{1}H couple parallel or antiparallel. It is found that only rarely will this coupling be changed unless a magnetic catalyst is present. Consequently H_2 comprises two species, depending on the relative coupling, and the amount of each species will be relatively permanent and subject only to slow change in the absence of an appropriate catalyst.

In Chapter 5 we noted that a particular symmetrical diatomic molecule shows only the rotational energy levels corresponding to even values or to odd values of J, but not both. It is found that the nuclear-spin coupling determines which set of J values is permitted. For most substances the macroscopic properties of the species with even and with odd J values are indistinguishable, and these characteristics are detectable only in the spectra and in other microscopic properties. However, for hydrogen we noted that the rotational energy spacing was large, and therefore we expect a substantial difference in various thermodynamic properties for the two species.

Ortho and Para Hydrogen

With the development of quantum mechanics and the emergence of the ideas just discussed, several investigators[2-4] sought to show that these two species of hydrogen existed and that their relative amounts could be changed. Bonhoeffer and Harteck,[3] and independently Eucken and Hiller,[4] in 1929 successfully catalyzed the conversion and measured the difference in thermal conductivity, heat capacity, etc., of the two species of H_2. The species with even J values, which has antiparallel nuclear-spin orientation, is called *para* hydrogen, and the other species with parallel spins and odd J values is called *ortho* hydrogen. Since the lowest rotational energy level has $J = 0$, the para species becomes the stable form at low temperature. It is found, however, that the parallel spin coupling is three times as probable as the antiparallel coupling†; hence ortho is three times as abundant as para at equilibrium at high temperatures. This mixture of 1/4 para and 3/4 ortho is known as normal hydrogen (n-H_2).

---

† The 3:1 probability ratio arises from the number of nuclear-spin states. The parallel coupling of two spins, each of $\frac{1}{2}$ unit, yields a net spin of 1, which may have a component along a given axis of $+1$, 0, or -1 and hence three states, whereas a net spin of zero can only have a component of zero.

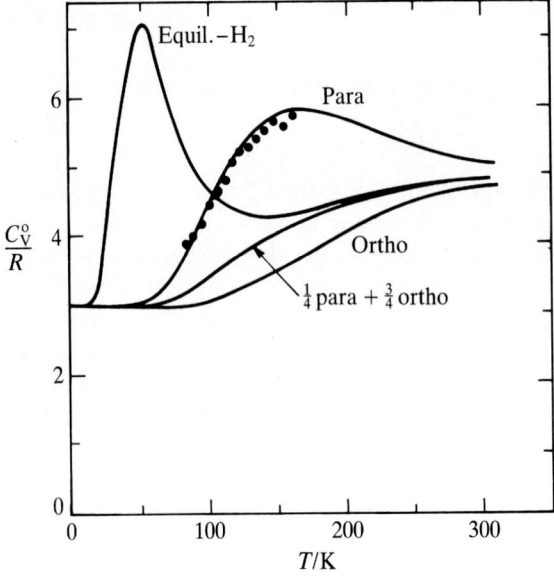

FIGURE 22-1
The heat capacity of hydrogen gas. The experimental points shown are for composition approximately 95 percent para-H_2.

The thermodynamic functions C_P°, S°, $H^\circ - H_0^\circ$, and $(G^\circ - H_0^\circ)/T$ may be calculated for each species of hydrogen by substituting the appropriate spectroscopic data into the formulas of Chapter 5. At low temperatures it is easier to sum Eq. (5-29) over the rotational states directly than it is to correct adequately the integral form in Eq. (5-31). The resulting heat capacities are shown in Fig. 22-1, which shows also the experimental values of Clusius and Hiller[5] for a sample of about 95 percent para-H_2. The curve for n-H_2 is just $\tfrac{3}{4}C_P^\circ$ (ortho) + $\tfrac{1}{4}C_P^\circ$ (para), and it is well confirmed by experiment.

One may also consider the case of H_2 in the presence of an ortho–para conversion catalyst and calculate the properties of an equilibrium mixture. Now one sums over both even and odd values of J for the rotational partition function, Eq. (5-29), but with $g_J = 1$ for even J and $g_J = 3$ for odd J. The resulting heat capacity is shown in Fig. 22-1.

Alternatively, one can treat ortho and para hydrogen as separate species and consider a chemical equilibrium between them. In this simple case of two species of the same formula, that is para-H_2 = ortho-H_2, the equilibrium constant is just

$$K = e^{-\Delta\varepsilon_0/kT}\,\frac{Q(\text{ortho})}{Q(\text{para})} \tag{22-2}$$

where $\Delta\varepsilon_0/k = (\varepsilon_0/k)(\text{ortho}) - (\varepsilon_0/k)(\text{para}) = h^2/4\pi^2 Ik = 170\,\text{K}$ and the Qs are the partition functions defined by Eq. (5-29). The equilibrium composition is shown in Fig. 22-2. The heat capacity of the equilibrium mixture includes a term for the heat of

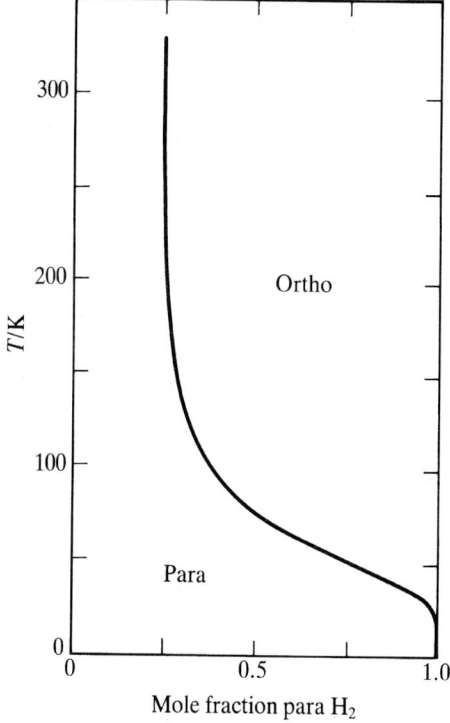

FIGURE 22-2 The equilibrium ortho–para composition of hydrogen.

conversion multiplied by the change of composition. Likewise, the entropy of ortho–para mixtures contains an entropy of mixing term. The final values for various properties of equilibrium H_2 are the same, of course, for either method of calculation.

Various properties of para hydrogen, normal H_2 (75 percent ortho) and of other ortho–para mixtures have been measured for the gas, liquid, and solid states. The differences for the gas have been considered above. For the liquid or the solid above 12 K, there are significant quantitative differences, but the general pattern is the same provided the ortho–para conversion rate is negligible. We shall not discuss this region further but rather turn our attention to the very interesting properties of the solid below 12 K.

The heat capacity of H_2 with various proportions of ortho and para was measured by Hill and Ricketson[6] and by Ahlers[7] and Ahlers and Orttung.[8] From Fig. 22-3 it is apparent that the heat capacity of para-H_2 decreases smoothly with decrease in temperature in a normal manner but that some striking effect occurs in ortho-H_2.

Para-H_2 has fallen almost entirely into the $J = 0$ rotational state while still in the gas above 20 K; hence, it behaves in the liquid and solid as if it had spherical molecules. No thermal effects related to rotation occur, and the heat capacity arises only from lattice vibrations such as occur for neon.

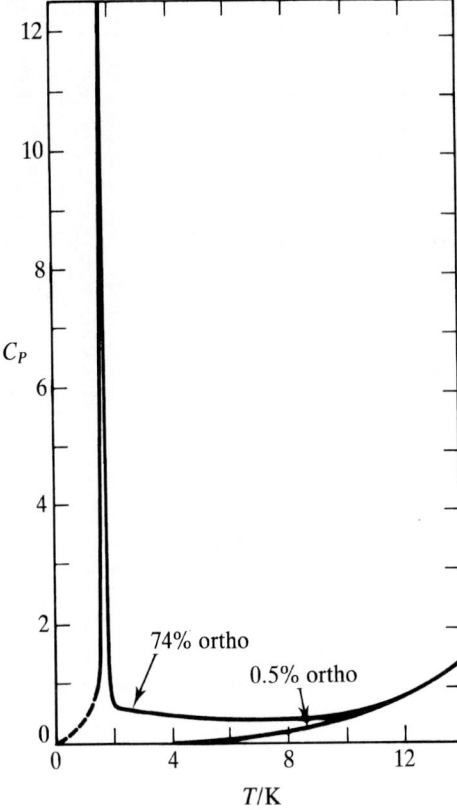

FIGURE 22-3
The heat capacity of solid hydrogen.

Ortho hydrogen, however, is entirely in the $J = 1$ state when it condenses and freezes. Since these effects above 12 K show no marked differences from para-H_2, one presumes that ortho-H_2 remains equally distributed among the three quantum states with $J = 1$. Detailed investigations of nuclear magnetic resonance, neutron diffraction, and other properties[9] confirm this picture. In asymmetric surroundings, the threefold degeneracy of the $J = 1$ states can be broken. Evidently, this is the origin of the transition near 2 K and the high heat capacity both below and above that peak.

One finds on detailed analysis[9] that this is a very complex situation. The lattice has a "closest packed" pattern with 12 neighbors symmetrically located around a given molecule. For an isolated ortho molecule surrounded by para molecules, the threefold degeneracy is broken by only a very small energy difference. For two adjacent ortho molecules surrounded by the para species, the nine states are distributed over an energy range of several cm^{-1} but the lowest two states remain very close in energy. A few larger clusters have been studied theoretically as well as the lattice with 100 percent ortho molecules.[9] It is clear that the loss of the $J = 1$ rotational entropy will be distributed over a wide range of temperature. A substantial portion is lost in a rather sharp transition near 2 K for compositions near 75 percent ortho. For compositions below 60 percent ortho, however, there is only a broad region of high heat capacity.

Deuterium

There are also ortho and para species for D_2 with properties similar to those of H_2, but there are quantitative differences. The deuteron has unit spin and D_2 can have spins 0, 1, or 2 with a total multiplicity of 9. The six states with total spin 0 and 2 are designated ortho-D_2 with 66.7 percent population at high temperature. Ortho-D_2 has even J rotational states; thus, ortho-D_2 is the stable species at low temperatures and corresponds in that sense with para-H_2. Para-D_2 has three spin states, a high-temperature population of 33.3 percent and odd J rotational states.

The ortho–para conversion in D_2 is slower than in H_2; hence, there is less interference with precise measurements for the solid. It is possible by an adsorption process to enrich either H_2 or D_2 in the $J = 1$ species; this is important for D_2, which has only 33 percent of that species in the normal state. Grenier and White[10] measured the heat capacity of various compositions of D_2 from normal (33 percent para) to 87 percent over the temperature range 1.5 K to the melting point. Earlier measurements[11] extended the range down to 0.3 K for normal D_2 and included the ortho species. For percentages of 69 percent or higher of the $J = 1$ species, there is a sharp transition similar to that shown in Fig. 22-3 for H_2. For n-D_2 there is only a broad region of high heat capacity, as shown in Fig. 22-4 together with curves for pure ortho-D_2 and for 87 percent para-D_2. References 12–14 report additional heat capacity data for D_2.

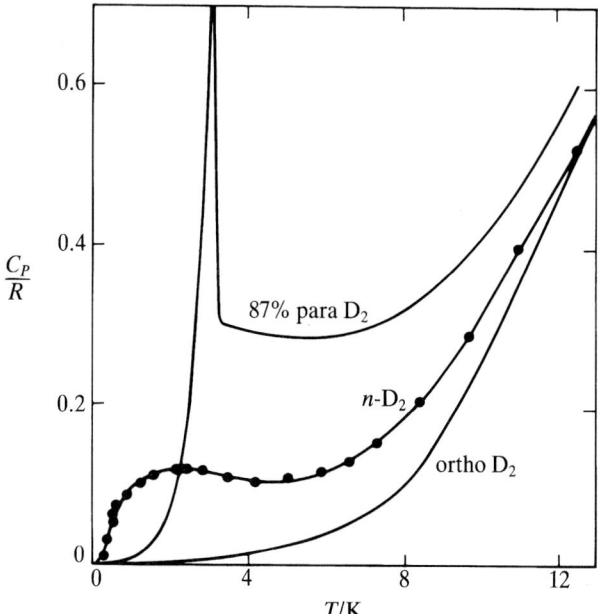

FIGURE 22-4
The heat capacity of solid D_2.

Grenier and White[10] integrated their experimental heat capacity curves for D_2 of various percentages of para and compared the results with the values calculated from spectroscopic data on the practical basis (excluding nuclear spin entropy). For n-D_2 the values agree well within the experimental uncertainty of $0.03R$. For higher percentages of para, the uncertainties are somewhat larger but the agreement is still good.

H_2, D_2, and the Third Law

The general prediction of the third law is that the entropy of a perfect crystal will approach the practical zero as $T \to 0$. This implies that all nuclear-spin entropy is still present when the extrapolation to 0 K is made from a temperature in the appropriate range, ordinarily 1–15 K. In application to H_2 or D_2, one starts with the normal ortho–para composition. If this shifts on cooling and maintains ortho–para equilibrium, the solid is pure $J = 0$ (para-H_2 or ortho-D_2) and the residual nuclear-spin entropy is zero for H_2, and is substantially reduced for D_2. But, if the spin-species composition is frozen, we saw in the preceding paragraph that the practical zero of entropy is reached for D_2 by extrapolation from 0.3 K. The present data for H_2 are less precise and there is more ortho to para conversion in any practical experiment, but there is every reason to believe that the situation is similar to that for D_2.

For all other molecules the moment of inertia is much larger, which implies that any tendency to shift from the high-temperature spin-species distribution will occur at a much lower temperature than for H_2 or D_2. Thus, the measured heat capacities, from which an extrapolation can be made, are almost always in the range where the spin-species distribution remains at the high-temperature proportion. One can, of course, make heat capacity measurements at lower temperatures where spin-species conversion may occur, and one must recognize this possibility in treating data. A subsequent section on methane presents an example of such a situation.

Before concluding the discussion of H_2 and D_2, we note again the importance of nonthermodynamic information in presenting a complete and unambiguous picture. Silvera[9] presented an excellent review covering theory, crystal structures, neutron diffraction, and spectral data.

HD

The unsymmetrical molecule HD presents some new aspects, and it is interesting to consider these briefly. In this case all J values are allowed in the rotational energy-level pattern. For "practical" calculations recognizing different isotopes, one wishes an entropy that still excludes nuclear-spin effects, and this is readily obtained from the equations of Chapter 5 by simply ignoring nuclear spin at all points. The thermal properties of HD at low temperatures were measured by Clusius, Popp, and Frank[15] and by Grenier and White.[16] An extrapolation of the heat capacity of solid HD to 0 K in the usual fashion also yields this practical entropy value.

The zero-entropy state (excluding nuclear spin) of HD at low temperatures might be thought to imply that the H and D atoms had formed an ordered array, but this seems unlikely. The properties of the various species of H_2 are best understood on the assumption that the molecules rotate freely in the solid at 3–14 K. In this case HD will be in its lowest rotational state, which is the single $J = 0$ state. Consequently, an extrapolation from this range implies free rotation even at 0 K but no rotational entropy because all the molecules are in a single quantum state.

Crystals of other diatomic molecules have properties indicating that rotation has ceased at these temperatures. Therefore, a molecule such as ^{35}Cl—^{37}Cl has a residual entropy $R \ln 2$ because of random end-for-end orientation like that found for CO and N_2O (see Chapter 21). If one is considering Cl_2 without distinguishing isotopes, then one accounts for this $R \ln 2$ as an entropy of mixing of ^{35}Cl with ^{37}Cl and there is no discrepancy for Cl_2 (but then HD is an exception).

METHANE

The low-temperature properties of methane and its deuterated derivatives have many interesting aspects. They are also important as examples showing the transition from the special effects shown by hydrogen to the normal pattern of all other nonmagnetic solids.

All of the methanes have at least two different nuclear-spin species. CH_4 and CD_4, with tetrahedral symmetry, have three species: one totally symmetric with symbol A, a triply degenerate species T with the symmetry of the three Cartesian axes, and a doubly degenerate species E. CH_3D and CHD_3 have two species, A and E (doubly degenerate), while CH_2D_2 has two nondegenerate species analogous to ortho- and para-H_2.

The abnormalities of hydrogen were associated with its very small moment of inertia and the associated large excitation energy, $\Delta\varepsilon/k = 170$ K, between the two lowest rotational states. The corresponding energy differences for CH_4 and CD_4 are 15 K and 7.5 K, respectively. These values indicate the temperatures below which appreciable shifts in spin-species composition from the normal, high-temperature values would be expected at equilibrium in the gas phase. The methanes, however, are solids at these temperatures. Also, the methanes show one or two transitions that are presumably associated with loss of molecular rotation in the solid, and the 15 K and 7.5 K temperatures lie below those transitions. Although there is still torsional vibration and rotation by quantum tunneling in the solid methanes, one expects the temperatures at which substantial shifts in spin-species composition might occur to be lower than the 15 K for CH_4 and 7.5 K for CD_4.

CD_4

The low-temperature heat capacity and related properties of CD_4 were measured by Colwell, Gill and Morrison[17] (see Fig. 22-5). The heat-capacity curve in the range 3–12 K is normal and essentially the same entropy is obtained if the extrapolation to 0 K is made from 3 K or 12 K or any intermediate temperature.

Colwell, Gill, and Morrison make their entropy comparison at 97.82 K where the vapor pressure is 208.8 mmHg. The statistical value from molecular properties is $S/R = 20.55$, while their calorimetric value is 20.57 ± 0.05, in perfect agreement.

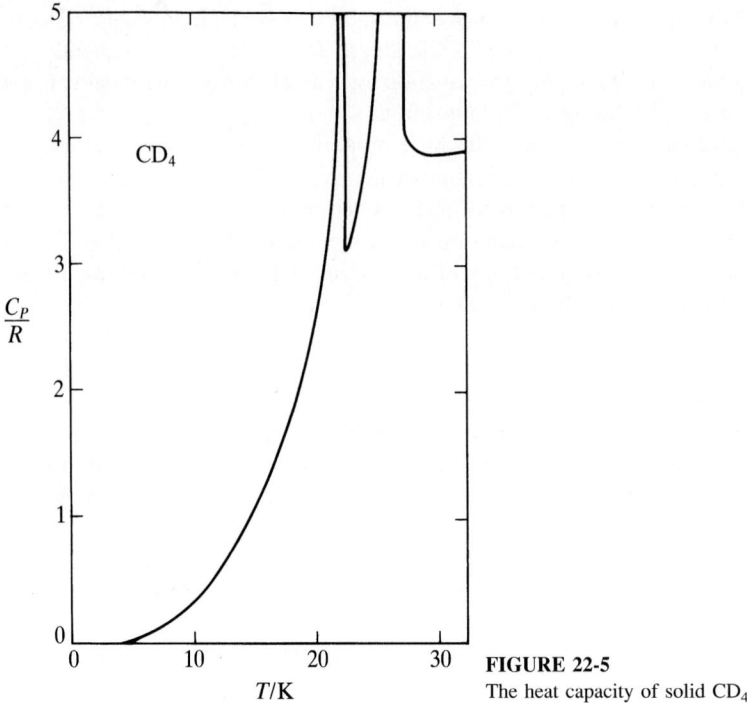

FIGURE 22-5
The heat capacity of solid CD_4.

Thus, CD_4 shows no abnormality from a third-law viewpoint on the basis of measurements down to 3 K.

Below 3 K the possibility of loss of nuclear spin entropy arises. This implies conversion of spin-species E and T to A, which is predicted to have the lowest energy. Neutron scattering and nuclear magnetic resonance can measure the A–T and A–E energy differences, and values in the 10–50 mK range have been reported. White and Morrison[18] report an increase in heat capacity of CD_4 above expected curve for the lattice in the range below 2 K. Above 0.5 K this can be explained by a 0.4 percent impurity of CHD_3, which has a region of high heat capacity with a maximum at 0.5 K. At still lower temperatures, 0.15–0.3 K, White and Morrison find high heat capacities (not explainable as CHD_3) and slow equilibrium. Thus, spin-species conversion may be taking place, although it might not have been expected to have an appreciable rate in the absence of a magnetic catalyst.

Before proceeding to other methane species, it is interesting to note the two transitions and regions of high heat capacity shown in Fig. 22-5. Similar pairs of transitions occur for CHD_3, CH_2D_2, and CH_3D at successively slightly lower temperatures. At low pressure, CH_4 shows only one transition, which corresponds to the upper one in the heavy methanes, but at high pressure CH_4 has the two transitions. The three solid phases are known as I, II, and III with decrease in temperature. The high-temperature form, I, is cubic closest-packed with rotating or randomly oriented molecules in sites of O_h symmetry. Phase II has two types of molecular sites: 3/4 have low symmetry and with definite molecular orientation, while 1/4 retain O_h symmetry with molecules either rotating or randomly distributed over several orientations. Phase III has tetragonal symmetry with nonrotating molecules,[19] but its detailed structure is still being investigated.

CH_3D, CH_2D_2, CHD_3

The properties of the three partially deuterated methanes are similar. Heat capacities and related properties were measured by Morrison and various associates.[17,19–23] Above 6 K all have properties similar to CD_4, but all three have anomalous regions of high heat capacity at lower temperatures, with maxima near 0.5 K for CH_3D and CHD_3 and near 0.25 K for CH_2D_2.

If the entropies are calculated without the low-temperature anomalies, i.e., by extrapolation from above 6 K, the statistical entropies are larger than the calorimetric entropies by $R \ln 4$ for CH_3D and CHD_3 and by $R \ln 6$ for CH_2D_2. The number 4 (or 6) is just the number of rotational reorientations of that substituted tetrahedral molecule that do not merely interchange identical atoms. Specifically, this is the ratio of the symmetry number of CX_4, 12, to the symmetry number of 3 for CX_3Y or 2 for CX_2Y_2. This analysis of the limiting entropies of the CH_nD_{4-n} species was first given by Clusius, Popp, and Frank[15] in connection with their early measurements on CH_3D in 1937.

The anomalies with maxima below 1 K arise from the splitting by rotational tunneling of groups of quantum states that would have the same energy if rotation were forbidden. The theory was developed by Nagamiya[24] and applied by Hopkins, Kasper, and Pitzer[25] to early measurements[19] on these molecules. For the partially deuterated species, these arrays of energy levels split by tunneling arise for each spin species; consequently, the resulting anomalies need not involve any conversion between spin species. The anomalies, as observed down to about 0.15 K, remove only a portion of the orientational entropies of $R \ln 4$ or $R \ln 6$. Thus, there is no need to invoke any spin-species conversion. Efforts to relate these effects to neutron scattering data and to explain the heat capacities quantitatively have been only partially successful.[19]

CH_4

Light methane at low pressure remains in phase II to 0 K and does not undergo the lower transition to phase III. The structure of phase II is known and described in detail in various papers cited here. For our purposes, we need to know only that there are two types of molecular sites: 1/4 have O_h symmetry and the molecules in these sites either rotate freely or redistribute rapidly among multiple orientations. The remaining 3/4 of the sites have D_{2d} symmetry and there the tetrahedral molecules are restricted to torsional oscillation about a single potential minimum.

Also, CH_4 differs from H_2 in that CH_4 interconverts rapidly between spin-species in the gas state.[26,27] This process involves the $J = 2$ and higher rotational states where two or more spin species have the same rotational energy. The rate would be expected to become slower with decrease in temperature but to remain fairly rapid as long as the molecules are in rotational states. Thus, one expects the CH_4 in O_h sites to show measurable interconversion in phase II, although with decreasing rate at lower T. Interconversion of molecules in D_{2d} sites is expected to be slower and possibly very slow. Early experiments[17] indicated high heat capacities for CH_4 below 10 K but with increasingly sluggish equilibrium making measurements very uncertain below 5 K.

The problem of slow equilibration was solved by doping the CH_4 with a little O_2 whose paramagnetism serves as an interconversion catalyst.[28] By this means, Vogt and Pitzer[29] were able to measure the heat capacity of CH_4 with 0.8 percent O_2 down to 0.4 K, with the results shown in Fig. 22-6. It was also possible, by carefully excluding oxygen, to make accurate measurements on a spin-species-frozen basis from 2 K to 5 K and to continue measurements above 5 K but with increasing uncertainty. The best estimate of the spin-species-frozen heat capacity is also shown on Fig. 22-6.

Entropy comparisons can be made on two bases: spin-species conversion frozen or at equilibrium. On the former basis, one compares the statistical entropy excluding nuclear spin with the calorimetric value from measurements made without a spin-conversion catalyst. Vogt and Pitzer[29] find agreement within 0.02 in S/R, but the calorimetric value is uncertain by about 0.1 because some spin conversion could not be avoided in the range 5–12 K. Alternatively, one may calculate the calorimetric entropy using experimental data above 12 K and an extrapolation from 12 K to 0 K with a

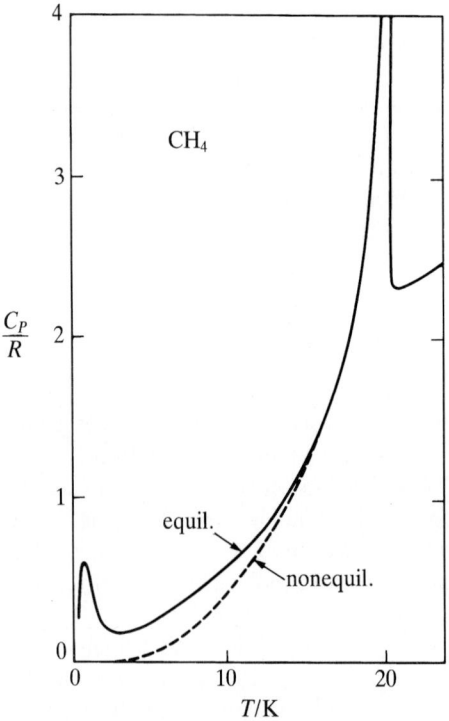

FIGURE 22-6
The heat capacity of solid CH_4 when spin conversion is catalyzed (equil.) or is not catalyzed (nonequil.).

Debye function; again, there is agreement well within an uncertainty of 0.1 in S/R on the practical basis.

If spin-species equilibrium is maintained, all of the CH_4 will convert to the A species at low temperatures, and the nuclear spin entropy will be reduced from $4R \ln 2$ at high temperature to $R \ln 5$ (since the spin of the A species is 2). The calorimetric entropy based on the equilibrium heat capacities agrees with the statistical value on this basis[29] within an uncertainty of 0.1 in S/R. Some of this uncertainty arises from the extrapolation of the heat capacity below 0.4 K. This analysis of the behavior of CH_4 at temperatures down to 0.4 K was confirmed by recent neutron scattering experiments of Heidemann et al.[30]

To recapitulate, the low-temperature thermal properties of both CH_4 and CD_4, if they are cooled rapidly and without spin-conversion catalyst to temperatures in the range 3–12 K, yield entropies in agreement with the third law on the practical basis (no loss of nuclear spin entropy). Since some spin conversion actually occurs in CH_4, however, the comparison in that case is less precise. For CH_4 one can catalyze spin-species conversion, and the entropy from calorimetric measurement then agrees with theoretical expectations.

HELIUM

Helium is a relatively unexciting substance at ordinary temperatures, but below 2.2 K its behavior provides many surprises and the two isotopes behave very differently. We shall do little more than mention some of these unique properties; other books[1,31] and reviews must be consulted for details. The abundant isotope of helium, ^{4}He, shows relatively normal behavior from its normal boiling point of 4.2 K down to 2.18 K, where it undergoes a transition to a second liquid, a superfluid, state known as He II. There is no latent heat of transition—only a region of high heat capacity. He II, although still mechanically fluid, approaches zero entropy as the temperature approaches 0 K. It remains a liquid at $T = 0$ for $P \gtrsim 25$ bar. The flow and heat-transport properties of He II are quite abnormal, but we shall not consider these.

The rare isotope ^{3}He, which became available from the radioactive decay of artificially produced ^{3}H, also remains liquid at $T = 0$ for $P \gtrsim 30$ bar. Whereas ^{4}He has no nuclear spin, ^{3}He has 1/2 unit of spin and a spin entropy of $R \ln 2$ at any ordinary temperature. This spin entropy is completely lost as $T \to 0$; this is demonstrated by heat capacity measurements[32] extending down to 1 mK. Furthermore, the limiting heat capacity depends on the first power of T for ^{3}He, in contrast to the T^3 dependence for lattice vibrations. This linear dependence on T is also found for the contribution of electrons in metals. Both electrons and ^{3}He atoms have spin $\frac{1}{2}$ and follow Fermi statistics; hence, the similarity in temperature dependence of heat capacity is reasonable.

A second liquid phase, also superfluid, appears for ^{3}He below 0.001 K at low pressure, with a transition curve increasing with pressure to 2.2 mK at 20 bar. At pressures in the 20–30 bar range, further phase changes occur between 2.0 and 2.5 mK.[32]. Both ^{3}He and ^{4}He may be solidified under pressures of approximately

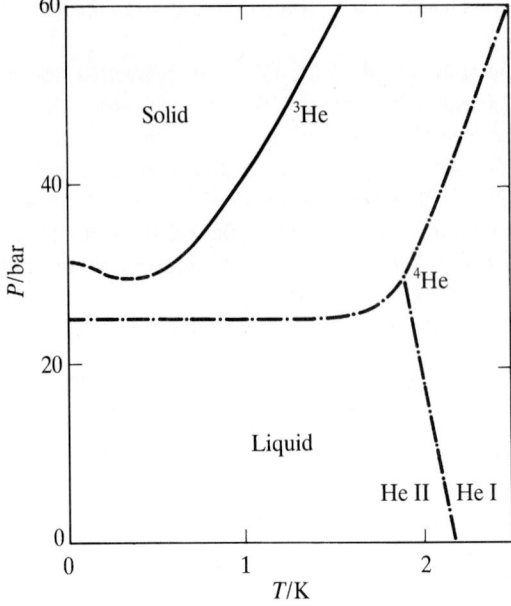

FIGURE 22-7
The solid–liquid phase diagram for ^{3}He (solid and broken lines) and for ^{4}He (dash-dot lines). The He I to II transition in liquid ^{4}He is also shown.

30 and 25 bar, respectively, near 0.3 K. The solid–liquid diagram in Fig. 22-7 shows that the equilibrium pressure of ^{4}He remains nearly constant up to 2 K and then rises sharply, but the pressure for ^{3}He shows a minimum near 0.3 K and then rises steadily above that temperature.

An excellent review by Wheatley[33] describes many of these properties as of 1975. While these phenomena are consistent with thermodynamics and are very interesting in terms of the theory of quantum fluids, they are so specialized that we will forego further discussion.

Substances other than H_2, He, and CH_4 show very interesting properties at extremely low temperatures where nuclear magnetic moments have large effects. Pobell has summarized these properties recently.[34]

REFERENCES

1. For theory specifically for hydrogen, see A. Farkas, *Orthohydrogen, Parahydrogen, and Heavy Hydrogen*, Cambridge University Press, New York, 1935; for helium see *Progress in Low Temperature Physics*, C. J. Gorter, ed., vol. I (1955); and for methane see E. B. Wilson, Jr., *J. Chem. Phys.*, **3**, 276 (1935).
2. W. F. Giauque and H. L. Johnston, *J. Am. Chem. Soc.*, **50**, 3221 (1928).
3. K. F. Bonhoeffer and P. Harteck, *Naturweiss*, **17**, 182 (1929); *Sitzber. preuss. Akad. Wiss.* **1929**, 103 (1929).
4. A. Eucken and K. Hiller, *Z. physik. Chem*, **B4**, 142 (1929).
5. K. Clusius and K. Hiller, *Z. physik. Chem.*, **B4**, 158 (1929).
6. R. W. Hill and B. W. A. Ricketson, *Phil. Mag.*, **45**(7), 277 (1954).

7. G. Ahlers, *J. Chem. Phys.*, **41**, 86 (1964).
8. G. Ahlers and W. H. Orttung, *Phys. Rev.*, **133**, A1642 (1964).
9. I. F. Silvera, *Rev. Mod. Phys.*, **52**, 393 (1980).
10. G. Grenier and D. White, *J. Chem. Phys.*, **40**, 3015 (1964).
11. O. D. Gonzalez, D. White, and H. L. Johnston, *J. Phys. Chem.*, **61**, 773 (1957).
12. D. White, *Chem. Phys.*, **14**, 301 (1976).
13. R. J. Roberts and J. G. Daunt, *J. Low Temp. Phys.*, **16**, 405 (1974).
14. R. J. Roberts, E. Rojas, and J. G. Daunt, *J. Low Temp. Phys.*, **24**, 265 (1976).
15. K. Clusius, L. Popp, and A. Frank, *Physica*, **4**, 1105 (1937).
16. G. Grenier and D. White, *J. Chem. Phys.*, **40**, 3451 (1964).
17. J. H. Colwell, E. K. Gill, and J. A. Morrison, *J. Chem. Phys.*, **39**, 635 (1963); **40**, 2042 (1964).
18. M. A. White and J. A. Morrison, *J. Chem. Phys.*, **72**, 5927 (1980).
19. K. J. Lushington, K. Maki, J. A. Morrison, A. Heidemann and W. Press, *J. Chem. Phys.*, **75**, 4010 (1981).
20. J. H. Colwell, E. K. Gill, and J. A. Morrison, *J. Chem. Phys.*, **42**, 3144 (1965).
21. K. J. Lushington and J. A. Morrison, *J. Chem. Phys.*, **69**, 4214 (1978).
22. M. A. White, K. J. Lushington, and J. A. Morrison, *J. Chem. Phys.*, **69**, 4227 (1978).
23. M. A. White and J. A. Morrison, *J. Chem. Phys.*, **70**, 5384 (1979).
24. T. Nagamiya, *Progr. Theor. Phys. (Kyoto)*, **6**, 702 (1951).
25. H. P. Hopkins, Jr., J. V. V. Kasper, and K. S. Pitzer, *J. Chem. Phys.*, **46**, 218 (1967).
26. R. F. Curl, Jr., J. V. V. Kasper, K. S. Pitzer, and K. Sathianandan, *J. Chem. Phys.*, **44**, 4636 (1966).
27. R. F. Curl, Jr., J. V. V. Kasper, and K. S. Pitzer, *J. Chem. Phys.*, **46**, 3220 (1967).
28. H. P. Hopkins, P. L. Donoho, and K. S. Pitzer, *J. Chem. Phys.*, **47**, 864 (1967).
29. G. J. Vogt and K. S. Pitzer, *J. Chem. Thermodyn.*, **8**, 1011 (1976).
30. A. Heidemann, K. J. Lushington, J. A. Morrison, K. Neumaier, and W. Press, *J. Chem. Phys.*, **81**, 5799 (1984).
31. D. Pines and P. Nozieres, *The Theory of Quantum Liquids*, W. A. Benjamin, New York, 1966.
32. D. S. Greywall, *Phys. Rev. B*, **27**, 2747 (1983); **33**, 7520 (1986).
33. J. C. Wheatley, *Rev. Mod. Phys.*, **47**, 415 (1975).
34. F. Pobell, *Matter and Methods at Low Temperatures*, Springer-Verlag, Berlin, 1992; *Physics Today* 34 (1993).

PROBLEMS

22-1. Calculate all the energy levels of H_2 within $3000\,\mathrm{cm}^{-1}$ above the lowest level. Indicate the multiplicity g_i of each level; remember that all rotational states of ortho-H_2, that is with odd J, have a triple multiplicity factor from nuclear spin. Obtain Q and Q' at 300 K by direct summation of Eqs. (5-14) and (5-20). Calculate $S°/R$ and $-(G° - H_0°)/RT$ at 300 K; remember to subtract $\ln 4$ to eliminate the nuclear-spin entropy, which is excluded on the practical scale.

22-2. By numerical summation over the rotational energy levels calculate Q_rot for ortho-H_2 and for para-H_2 at 50 K. Calculate the equilibrium constant for the reaction para-H_2 = ortho-H_2 at 50 K (remember the multiplicity factor for nuclear spin).

CHAPTER 23

SURFACE EFFECTS

Heretofore we have explicitly neglected the effects of surfaces or interfaces on the thermodynamic properties of various systems. Frequently surface effects are negligible, but in other cases they may be important and hence must be considered. The first exact treatment of surface thermodynamics, as of so many of our topics, was given by Gibbs.

The boundary between two contiguous phases, which is known as a surface or an interface, is not a two-dimensional mathematical boundary of zero volume. There is a region of small but finite thickness in which the properties differ appreciably from the properties in the interior of either phase. Unless the conditions are close to those of the critical point for the disappearance of the interface, there is every reason to believe that the thickness of the surface region is very small—a few molecular layers. Optical properties support this estimate of the thickness of the surface region. We consider initially surfaces between fluid phases and thus avoid the complication of solid surfaces, which differ for different crystal faces of the same substance. The basic equations apply, however, to a solid–fluid interface provided that the crystal face is specified.

Surface Tension, Interfacial Tension

Consider a change in shape of our system such that the interfacial area is increased but the number of moles of each constituent remains unchanged. Also the pressure on the exterior boundaries of the system is unchanged. For this change the increase in Gibbs energy of the whole system per unit increase in interfacial area, A_s, is the *surface tension*, or *interfacial tension*, γ.

$$\gamma = \left(\frac{\partial G}{\partial A_s}\right)_{T,P,n_i} \tag{23-1}$$

The general differential of Gibbs energy now contains an additional term, and instead of Eq. (10-10) we have

$$dG = -S\,dT + V\,dP + \gamma\,dA_s + \sum_i \mu_i\,dn_i \tag{23-2}$$

Surface Concentration

Before proceeding further, we must consider the concept of surface concentration and of other surface properties. Take as the system a region containing a substantial volume of phase I and phase II, together with a flat surface of area A_s between the two phases. If we plot the concentration of some component as a function of the distance perpendicular to the interface, a curve such as that in Fig. 23-1 is to be expected. Let us imagine a plane in the region of the surface and at some location such as y' in Fig. 23-1. For the present, the location is arbitrary so long as the plane is parallel to the surface; i.e., it lies at the same point y' throughout the surface region. With this plane as the boundary the volume of phase I is $V'_{\rm I}$ and of phase II is $V'_{\rm II}$.

Next we define the surface concentration of each material component as the excess of that quantity in the system over that calculated from the sum of the products of the concentration in each phase times the volume of each phase. Thus, if the molal concentration of component 1 in phase I is $c_1^{\rm I}$ and in phase II is $c_1^{\rm II}$, then the apparent surface concentration Γ'_1 is

$$\Gamma'_1 A_s = n_1 - c_1^{\rm I} V_{\rm I} - c_1^{\rm II} V_{\rm II} \tag{23-3}$$

and similar equations may be written for the other components. We shall presently define thermodynamic quantities in the analogous fashion, but let us first consider certain properties of surface concentrations.

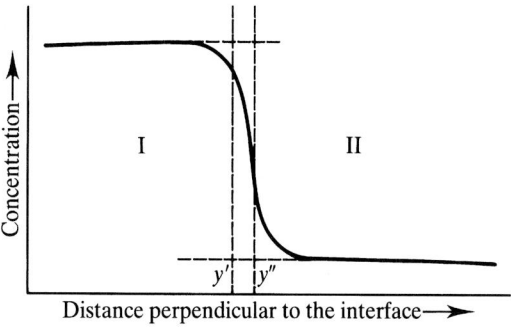

FIGURE 23-1
Actual concentration as a function of distance perpendicular to the surface is given by the solid curve. The surface concentration is the difference in area between the curve and the dashed rectangular path, which implies constant concentration on each side of a mathematical surface at y' (or y'').

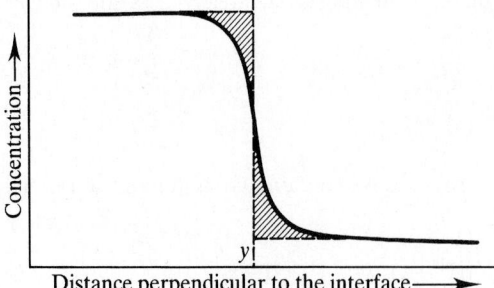

FIGURE 23-2
The surface plane at y for zero surface concentration, which requires that the indicated areas be equal.

Are the surface concentrations Γ_i intrinsically uncertain because of an arbitrary location of the dividing place at y'? Suppose we use, instead, a dividing plane at y''. Now the volume assigned to phase I is increased by $(y'' - y')A_s$, and the volume assigned to phase II is decreased by the same amount. There is no change in the real physical system, however. The change in surface concentration is

$$(\Gamma_i'' - \Gamma_i') = (c_i^{II} - c_i^{I})(y'' - y') \tag{23-4}$$

We see that Γ_i does depend on the location of the dividing plane unless $c_i^{II} - c_i^{I}$ is zero.

In a single-component system, where we have a vapor–liquid interface, $c^{liq} - c^{vap}$ differs substantially from zero, except near the critical point. But it is possible to choose that dividing plane which makes Γ zero because there is only one component present. Reference to Eq. (23-3) indicates that this location of the plane is the one which makes the two areas of Fig. 23-2 equal.

With two or more components it is not generally possible to make more than one Γ_i zero. While it is a purely arbitrary choice, it is usually preferable to make the surface concentration of the solvent zero, $\Gamma_1 = 0$. Then we expect nonzero values of $\Gamma_{i(1)}$ for all solute components, where the additional subscript indicates the component chosen to define the dividing plane. If in a particular case $c_j^{II} - c_j^{I} = 0$, then we see from Eq. (23-2) that Γ_j is unaffected by the choice of the dividing plane. Except for this special case, all individual Γ_i are subject to the one arbitrary definition and we shall usually deal with the series $\Gamma_{i(1)}$, $\Gamma_{1(1)} = 0$.

Surface Gibbs Energy, Entropy, etc.

Let us now consider thermodynamic functions for surfaces. Their definition is analogous to that for surface concentration,

$$Y_s' A_s = Y - \sum_i n_i^{I} \overline{Y}_i^{I} - \sum_i n_i^{II} \overline{Y}_i^{II} \tag{23-5}$$

where Y is any thermodynamic function, $\overline{Y}_i^{I}$ and $\overline{Y}_i^{II}$ are the usual partial molar quantities in phase I and II, and n_i^{I} and n_i^{II} are the numbers of moles in bulk phases I and II, respectively. For single-component systems, where there is no surface concentration of matter, Y_s is unambiguous and represents an additional energy, entropy, etc., associated with the presence of the surface. For multicomponent systems the arbitrary decision about the dividing surface affects surface thermodynamic properties as well as surface concentrations. It will be assumed that the choice $\Gamma_1 = 0$ is made unless otherwise specified.

We note particularly that there is no surface volume. Since all space has been assigned to bulk phases I and II, there is no space remaining to be assigned to the surface. Alternate definitions are possible which include a surface volume, but the same equations connecting observable quantities are necessarily obtained eventually. Consequently, we believe it is simpler to adopt definitions that eliminate a surface volume at the beginning.

Now we return to Eq. (23-2) and consider the increment of the Gibbs energy for the process of increasing the interfacial area by dA_s at constant T and P and total n_i. In view of Eq. (23-1), the definition of surface Gibbs energy in Eq. (23-5), and the fact that $\overline{G}_i^I = \overline{G}_i^{II} = \mu_i$,

$$G_s \, dA_s = \gamma \, dA_s - \sum_i \mu_i (dn_i^I + dn_i^{II})$$

The change $dn_i^I + dn_i^{II}$ represents the change in the number of moles of component i in the bulk phases. This is just the material added to the surface phase where there is a nonzero surface concentration. Hence

$$dn_i^I + dn_i^{II} = -\Gamma_i \, dA_s$$

and

$$G_s \, dA_s = \gamma \, dA_s + \sum_i \mu_i \Gamma_i \, dA_s$$

We may now divide by dA_s and obtain

$$\gamma = G_s - \sum_i \mu_i \Gamma_i \tag{23-6}$$

where the same definition of dividing surface must, of course, be used for G_s and the Γ_i.

The differential of Eq. (23-6) is

$$d\gamma = dG_s - \sum_i \mu_i \, d\Gamma_i - \sum_i \Gamma_i \, d\mu_i \tag{23-7}$$

but the differential of G_s for fixed unit surface area from Eq. (23-5), after simplification, is

$$dG_s = -S_s \, dT + \sum_i \mu_i \, d\Gamma_i \tag{23-8}$$

and substitution of Eq. (23-8) into (23-7) yields

$$d\gamma = -S_s \, dT - \sum_i \Gamma_i \, d\mu_i \tag{23-9}$$

This is the basic equation for the surface tension, which was first derived by Gibbs and which will be the basis of much of our further discussion. A completely general treatment beyond this point tends to be unduly complex. Hence it is more useful to consider several special cases of practical interest, which will also illustrate methods that could be extended to more complex cases as needed.

VAPOR–LIQUID INTERFACES

One-component Systems

The vapor–liquid interface is the only possible interface between two fluid phases of a one-component system.† The term *surface tension of the liquid* is used for γ in this system. We recall that in a single-component system there is no surface concentration; consequently, Eq. (23-9) simplifies to

$$d\gamma = -S_s\, dT \qquad (23\text{-}10)$$

The heat absorbed per unit of surface formed reversibly at constant amount of liquid and vapor and hence at constant volume is TS_s, where

$$S_s = -\frac{d\gamma}{dT} \qquad (23\text{-}11)$$

If surface area is lost irreversibly at constant pressure and volume, i.e., the work associated with the surface tension is not extracted, then the heat evolved is H_s, which is

$$H_s = G_s + TS_s = \gamma - T\frac{d\gamma}{dT} \qquad (23\text{-}12)$$

Since there is no volume change, $E_s = H_s$ and Eq. (23-12) gives the *total surface energy* as well as the *surface enthalpy*.

The surface tension of various pure liquids shows a similar trend with temperature. For normal fluids, the empirical equation

$$\gamma = \gamma_0(1 - T_r)^{11/9} \qquad (23\text{-}13)*$$

is quite successful.[1-3] Here T_r is the reduced temperature T/T_c, and γ_0 is an empirical constant. Appendix 3 includes a correlation of γ_0 with molar volume and T_c for normal liquids. Where Eq. (23-13)* holds, the surface entropy is

$$S_s = \frac{11}{9}\frac{\gamma_0}{T_c}(1 - T_r)^{2/9} \qquad (23\text{-}14)*$$

We note that both γ and S_s drop to zero at the critical temperature, as they must. However, while γ decreases rapidly with rising temperature over the full range below T_c, the surface entropy is more nearly constant until the temperature is close to T_c.

---

† Surface tensions of liquids are often measured in air, which may have a small but significant effect. For ethanol at 20°C the presence of air reduces γ by about 2 percent.

For other types of liquids, the behavior of the surface tension is more complex and no comparably general correlations are available. Adamson[4] gives a good general account, while Grosse[5] reports specifically on liquid metals. Extensive compilations of surface tensions for pure liquids are given by Korosi and Kovats[6] and Jasper.[7]

Two-component Systems

There are two types of interface between fluid phases in two-component systems that are of interest. First is the vapor–liquid interface analogous to that of the one-component system. Also there may be a liquid–liquid interface between immiscible components, which will be considered later. In the first case temperature and liquid composition are the natural independent variables, and there is one surface concentration. Equation (23-9) becomes

$$d\gamma = -S_s\, dT - \Gamma_2\, d\mu_2 \tag{23-15}$$

where we have chosen $\Gamma_1 = 0$ and abbreviated $\Gamma_{2(1)}$ to Γ_2. Also, the definition of surface entropy must be consistent with the choice $\Gamma_1 = 0$. It is natural to select x_2 of the liquid as the independent composition variable; hence

$$d\mu_2 = -\bar{S}_2\, dT + \bar{V}_2\, dP + \left(\frac{\partial \mu_2}{\partial x_2}\right)_{T,P} dx_2 \tag{23-16}$$

The pressure is determined by the properties of the bulk phases since it is the equilibrium vapor pressure of the system. Hence we may rewrite Eq. (23-16) as

$$d\mu_2 = -\left[\bar{S}_2 - \bar{V}_2\left(\frac{\partial P}{\partial T}\right)_{x_2,\,\text{sat}}\right] dT + \left(\frac{\partial \mu_2}{\partial x_2}\right)_{T,\text{sat}} dx_2 \tag{23-17}$$

and this may be combined with Eq. (23-15) to yield

$$d\gamma = -\left[S_s - \Gamma_2 \bar{S}_2 + \Gamma_2 \bar{V}_2\left(\frac{\partial P}{\partial T}\right)_{x_2,\,\text{sat}}\right] dT - \Gamma_2 \left(\frac{\partial \mu_2}{\partial x_2}\right)_{T,\text{sat}} dx_2 \tag{23-18}$$

The partial molar entropy and volume are for the solute in the liquid phase (since x_2 was taken for the liquid).

The quantity in brackets, which is $-(\partial \gamma/\partial T)_{x_2}$, may be shown to be the entropy increase per unit increase of surface at constant volume. The first term S_s is the entropy of the surface created, and $\Gamma_2 \bar{S}_2$ is the entropy of the Γ_2 moles of component 2 as it previously existed in the bulk liquid. The removal of the Γ_2 moles of component 2 leaves a volume $\Gamma_2 \bar{V}_2$, which must be filled with vapor. Since $\partial P/\partial T$ is just $\Delta S/\Delta V$ of vaporization, we see that the third term is the entropy of vaporization required to fill this volume $\Gamma_2 \bar{V}_2$ and thus the demonstration is complete.

This argument has shown that the heat absorbed in the reversible increase of surface by unit area at constant volume is $-T(\partial \gamma/\partial T)_{x_2}$ for the two-component system just as for the one-component system. Indeed, a relationship between macroscopic properties of this sort must be independent of the number of components present, as we shall demonstrate presently. Likewise the enthalpy change for the irreversible destruction of vapor–liquid surface area at constant volume is given by $\gamma - T(\partial \gamma/\partial T)$ for a solution as well as for a single component.

The new situation that arises for a two-component system is the possibility of determining a surface concentration from the change of γ with composition. Equation (23-18) yields

$$\Gamma_2 = -\left(\frac{\partial \mu_2}{\partial x_2}\right)^{-1}_{T,\text{sat}} \left(\frac{\partial \gamma}{\partial x_2}\right)_T \qquad (23\text{-}19)$$

The use of fugacity as the independent variable gives a simpler equation, which is also convenient because the partial pressure may be used as the fugacity at low pressures:

$$\Gamma_2 = -\frac{1}{RT}\left(\frac{\partial \gamma}{\partial \ln f_2}\right)_T \qquad (23\text{-}20)$$

In an ideal solution the fugacity is proportional to the mole fraction and

$$\Gamma_2 = -\frac{1}{RT}\left(\frac{\partial \gamma}{\partial \ln x_2}\right)_T \qquad \text{ideal solution} \qquad (23\text{-}21)*$$

TABLE 23-1
The surface tension[8] and the surface concentration for water–ethyl alcohol solutions at 25°C

x_2	γ dyne cm^{-1}	log p_2/mm	$\Gamma_2 \times 10^{-13}$ molecules cm^{-2}	$\Gamma_2 \times 10^{10}$ mol cm^{-2}
1.00	21.93	1.771		
0.90	22.59	1.722	14.0	2.32
0.80	23.26	1.679	16.5	2.74
0.70	23.93	1.639	17.5	2.91
0.60	24.67	1.600	20.4	3.39
0.50	25.43	1.565	23.0	3.82
0.40	26.43	1.529	29.5	4.90
0.30	27.60	1.492	33.5	5.56
0.25	28.49	1.467	37.8	6.28
0.20	29.97	1.428	40.3	6.69
0.15	32.20	1.372	42.3	7.02
0.12	34.42	1.316	42.2	7.01
0.10	36.72	1.256	40.7	6.76
0.064	42.13	1.097	37.2	6.18
0.04	47.86	0.908	32.2	5.35
0.02	55.57	0.602	26.8	4.45
0.00	71.97			

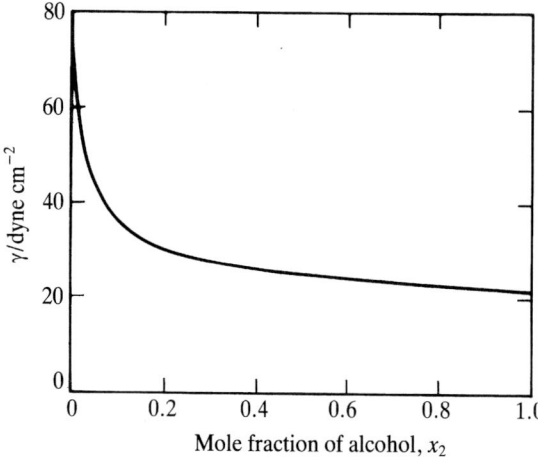

FIGURE 23-3
The surface tension of water–ethyl alcohol solutions.[8]

From any of the last three equations one sees that a solute that lowers the surface tension is concentrated in the surface, that is, $\Gamma_2 > 0$, while a solute that raises surface tension is less abundant in the surface than in the bulk phase and $\Gamma_2 < 0$. These results are readily understood in terms of molecular behavior.

Let us apply Eq. (23-20) to the data on ethyl alcohol–water solutions following Butler and Wightman.[8] In Table 23-1 the surface tension is given as a function of mole fraction of alcohol, x_2, together with the partial pressure of the alcohol. Figure 23-3 shows the general nature of the surface tension trend with composition. Note, however, that it is a graph of γ versus $\log p_2$ whose slope may be used to evaluate Γ_2 in Eq. (23-20).

From our general chemical knowledge we expect that alcohol will concentrate on the surface with the OH groups oriented into the liquid, where they may hydrogen-bond to water molecules, and with the alkyl groups on the surface. The minimum area per surface molecule is about 24 Å^2 in the 15 mole percent solution. This is only a little more than the cross-sectional area per chain in *n*-paraffin crystals; hence the surface of the water–alcohol solution may have an almost complete coverage of alcohol with most of the ethyl groups perpendicular to the surface. We emphasize, however, that this paragraph of speculation, interesting and probable as it may be, cannot be proved or disproved by thermodynamics.

Multicomponent Systems

The vapor–liquid interface of a system with more than two components has the same properties as have already been discussed. The thermal effects are similar, but the number of surface concentrations is increased. Thus for k components Eq. (23-9) yields

$$d\gamma = -\left\{ S_s - \sum_{i=2}^{k} \Gamma_{i(1)} \left[\bar{S}_i - \bar{V}_i \left(\frac{\partial P}{\partial T} \right)_{x_i,\text{sat}} \right] \right\} dT - \sum_{i=2}^{k} \Gamma_{i(1)} RT \, d\ln f_i \quad (23\text{-}22)$$

The temperature derivative of γ gives the entropy quantity in the braces, whose interpretation is analogous to that given before. In order to obtain the $k - 1$ surface concentrations, the fugacities (or activities) of the $k - 1$ solute components must be measured in k different solutions with appropriate small differences of composition but at the same temperature. The differences then yield $k - 1$ equations like Eq. (23-22), but with the first term missing since $dT = 0$. These can be solved for the surface concentrations.

AN ALTERNATE APPROACH TO SURFACE PROPERTIES

Before proceeding to consider other examples, let us take an alternate approach to the thermodynamics of flat surfaces. This method deals with observable properties of the whole system together with the surface tension and avoids introducing surface concentrations until later stages. Since we shall wish to consider processes at either constant pressure or constant volume, we need the differential expression for both Gibbs and Helmholtz energies. The former was given as Eq. (23-2), and the latter is readily obtained therefrom:

$$dG = -S\,dT + V\,dP + \gamma\,dA_s + \sum_i \mu_i\,dn_i \qquad (23\text{-}2)$$

$$dA = -S\,dT - P\,dV + \gamma\,dA_s + \sum_i \mu_i\,dn_i \qquad (23\text{-}23)$$

All quantities pertain to the entire system, including bulk phases.

Since each of these expressions is an exact differential, we may use Eq. (2-16) to obtain a series of relationships. Consider first the entropy,

$$\left(\frac{\partial S}{\partial A_s}\right)_{T,P,n_i} = -\left(\frac{\partial \gamma}{\partial T}\right)_{P,A_s,n_i} \qquad (23\text{-}24)$$

$$\left(\frac{\partial S}{\partial A_s}\right)_{T,V,n_i} = -\left(\frac{\partial \gamma}{\partial T}\right)_{V,A_s,n_i} \qquad (23\text{-}25)$$

While these two equations are very similar, the difference between constancy of pressure and volume is important. For the vapor–liquid problem, one cannot vary temperature at constant pressure and retain both phases; hence, Eq. (23-24) is not applicable. The use of the constant-volume condition [Eq. (23-25)] yields the results that we have already obtained. $T(\partial S/\partial A_s)_{T,V,n_i}$ is just the heat absorbed for the reversible increase of the surface by unit area at constant temperature, volume, and composition. For the one-component system, the result is identical with Eq. (23-11). In the case of two components, our earlier discussion was complex, and Eq. (23-25) leads much more directly to the relationship of the heat absorbed to $\partial \gamma/\partial T_{n_i}$, which was obtained eventually from Eq. (23-18).

We may also obtain the earlier result for surface concentration in the two-component vapor–liquid system as follows:

$$\left(\frac{\partial \gamma}{\partial n_2}\right)_{T,V,A_s,n_1} = -\left(\frac{\partial \mu_2}{\partial A_s}\right)_{T,V,n_1,n_2} \tag{23-26}$$

The chemical potential of component 2 is affected by the surface area under these conditions only if some material is added to or subtracted from the bulk phases by transfer from or to the surface. Consequently we consider n_2 as a function of both μ_2 and A_s, and, by Eq. (3-23),

$$\left(\frac{\partial \mu_2}{\partial A_s}\right)_{T,V,n_1,n_2} = -\frac{(\partial n_2/\partial A_s)_{\mu_2,T,V,n_1}}{(\partial n_2/\partial \mu_2)_{A_s,T,V,n_1}}$$

The increase of n_2 with surface area at constant μ_2, T, V, and n_1 is just Γ_2 as defined before. Introduction of these results into Eq. (23-26) with inversion of $\partial n_2/\partial \mu_2$ yields

$$\left(\frac{\partial \gamma}{\partial n_2}\right)_{T,V,A_s,n_1} = -\Gamma_2 \left(\frac{\partial \mu_2}{\partial n_2}\right)_{A_s,T,V,n_1} \tag{23-27}$$

which one may readily show to be equivalent to Eqs. (23-19) and (23-20).

Liquid–Liquid Interface, Two Components

The case of a liquid–liquid interface between two immiscible components such as benzene and water, although formally similar to the vapor–liquid problem, is quite different in practice. The composition of each phase is now effectively determined by the equilibrium with the other phase, and the convenient independent variables are temperature and pressure. While it is possible to apply these conditions to Eq. (23-9), the results are very complex and difficult to interpret. It seems better to use the alternate approach.

Constant pressure rather than constant volume is the appropriate condition for the temperature derivative, and Eq. (23-24) is now applicable, which gives the entropy absorbed

$$\left(\frac{\partial S}{\partial A_s}\right)_{T,P,n_i} = -\left(\frac{\partial \gamma}{\partial T}\right)_{P,A_s,n_i} \tag{23-24}$$

per unit increase of surface under constant temperature, pressure, and composition. Since the bulk phases may be made as large as desired, constant composition of the whole system may be interpreted as substantially constant mole fraction in each bulk phase. It is important to realize, however, that the increase of surface may transfer some material to or from the bulk phase and the surface region. In the absence of further information, however, it is impossible to determine how much of the entropy increase with surface, $\partial S/\partial A_s$, is associated with material transfer.

The heat effects associated with surface-area changes are related to $T(\partial S/\partial A_s)$ and $(\partial H/\partial A_s)$, as has been discussed before. For example, the heat evolved on deemulsification at constant T and P is

$$\left(\frac{\partial H}{\partial A_s}\right)_{T,P,n_i} = \gamma - T\left(\frac{\partial \gamma}{\partial T}\right)_{T,P,n_i} \tag{23-28}$$

From Eq. (23-5) we may also derive the equation

$$\left(\frac{\partial V}{\partial A_s}\right)_{T,P,n_i} = \left(\frac{\partial \gamma}{\partial P}\right)_{T,A_s,n_i} \tag{23-29}$$

This gives the volume increase per unit increase of surface area at constant temperature, pressure, and composition. While this is a sort of surface volume, it is important to realize that material is also transferred to and from bulk phases and the surface region. Thus a detailed interpretation is not possible without further information. Nevertheless, Eq. (23-29) gives a useful relationship between macroscopic quantities that are potentially observable.

Liquid–Liquid Interface, Three Components

The addition of a third, or solute, component to a system of two immiscible solvents gives a third independent variable, which we may take as n_3 or more conveniently as the concentration of component 3 in one phase or the other. If we hold temperature, pressure and the amounts of both solvent components constant, we obtain from Eq. (23-2)

$$\left(\frac{\partial \gamma}{\partial n_3}\right)_{T,P,A_s,n_1,n_2} = \left(\frac{\partial \mu_3}{\partial A_s}\right)_{T,P,n_1,n_2,n_3} \tag{23-30}$$

Now we proceed as we did from Eq. (23-26) to Eq. (23-27) for the two-component vapor–liquid case, but omitting the subscripts T, P, n_1, n_2; we obtain first

$$\left(\frac{\partial \mu_3}{\partial A_s}\right)_{n_3} = -\frac{(\partial n_3/\partial A_s)_{\mu_3}}{(\partial n_3/\partial \mu_3)_{A_s}}$$

and then on substitution into Eq. (23-30)

$$\left(\frac{\partial \gamma}{\partial n_3}\right)_{A_s} = -\left(\frac{\partial n_3}{\partial A_s}\right)_{\mu_3}\left(\frac{\partial \mu_3}{\partial n_3}\right)_{A_s} \tag{23-31}$$

The first derivative on the right side of Eq. (23-31) is related to the surface concentration of component 3. But in this system we have no simple way of obtaining $\Gamma_{2(1)}$ (or $\Gamma_{1(2)}$); hence it is difficult to relate $(\partial n_3/\partial A_s)_{\mu_3}$ to $\Gamma_{3(1)}$ or $\Gamma_{3(2)}$. However, if component 3 has a marked effect on the interfacial tension without appreciably changing the mutual solubility of the two solvents, then $(\partial n_3/\partial A_s)_{\mu_2}$ from Eq. (23-31) may be safely interpreted as a surface concentration of component 3 which will approximate $\Gamma_{3(1)}$ or $\Gamma_{3(2)}$.

CURVED INTERFACES

Let us now remove the limitation of planarity of the surface and consider the effects particularly caused by curvature. We shall assume that the surface tension is not affected by surface curvature. This is presumably justified as long as the radius of curvature is large compared with the thickness of the layer within which the properties differ from the bulk phases. We mentioned at the beginning of this chapter that the evidence indicates a thickness of only a few molecules, i.e., approximately 10^{-7} cm, under most conditions. Thus, curved interfaces with radii of curvature much greater than 10^{-7} cm may be expected to follow the equations to be derived below. If experiments are extended to extremely small droplets, however, deviations may be expected.

Interior Pressure of Bubbles and Drops

A very simple argument shows that the pressure P' within a drop is higher than that outside $P°$. Transfer of a volume dV of liquid from a bulk phase to the interior of the drop requires the work $(P' - P°)\,dV$, which must equal the work of extending the surface $\gamma\,dA_s$. Since $dV = 4\pi r^2\,dr$ and $dA_s = 8\pi r\,dr$, one obtains $dA_s = (2/r)\,dV$ and the Laplace equation†

$$P' - P° = \frac{2\gamma}{r} \qquad (23\text{-}32)$$

The excess pressure within a bubble is twice that in a drop of liquid because both the inner and outer surfaces of the thin liquid film contribute separate surface tensions. It is interesting to note that this phenomenon was treated by Kelvin[9] in 1858.

Escaping Tendency from Curved Surfaces

In 1871 Thomson[10] (Lord Kelvin) showed that the vapor pressure of a small droplet will exceed that of a plane surface of the same liquid, while the vapor pressure over a concave surface is decreased. This phenomenon is important in the behavior of finely dispersed volatile material.

We know that increase of pressure raises the fugacity or chemical potential, and we could easily combine this relationship with Eq. (23-32) for the excess pressure

---

† This equation was given in 1806 by P. S. de Laplace in his celebrated *Mécanique céleste*. The more general form for nonspherical surfaces is $P' - P° = \gamma(1/r_1 + 1/r_2)$, where r_1 and r_2 are the principal radii of curvature.

inside a drop to obtain the vapor pressure. We shall, however, follow an alternate approach, starting with the general equation for the Gibbs energy:

$$dG = -S\,dT + V\,dP + \gamma\,dA_s + \sum_i \mu_i\,dn_i \qquad (23\text{-}2)$$

Our initial applications of this equation were limited to flat surfaces, but the derivation is valid for curved surfaces provided that the surface tension is unchanged by the curvature, and provided that the interface within the system does not change the pressure on the exterior. A spherical interface completely within the system satisfies the latter condition. It is also important to note that the chemical potentials in Eq. (23-2) pertain to transfer of matter by a process that does not change A_s. Transfer across a planar interface satisfies this criterion; consequently we shall write $\mu_i^{(P)}$ to emphasize the restriction and note that

$$\mu_i^{(P)} = \left(\frac{\partial G}{\partial n_i}\right)_{T,P,A_s,n_j}$$

For a spherical drop, however, the surface area is no longer independently variable but rather is dependent on the volume and therefore the amount of material in the drop. Consequently, the term $\gamma\,dA_s$ in Eq. (23-2) must be expressed in terms of the various increments of matter, dn_i, and considered in obtaining the true escaping tendency from the drop.

The volume change on addition of increments dn_i moles of the various components is

$$dV = \sum_i \overline{V}_i\,dn_i$$

and, in view of the geometrical properties of a sphere,

$$dA_s = \frac{2\,dV}{r} = \sum_i \frac{2\overline{V}_i}{r}\,dn_i$$

Combining the last result with Eq. (23-2) yields

$$dG = -S\,dT + V\,dP + \sum_i \left(\mu_i^{(P)} + \frac{2\overline{V}_i\gamma}{r}\right)dn_i \qquad (23\text{-}33)$$

The escaping tendency of the ith component from the drop is the derivative of the Gibbs energy as given in Eq. (23-33) with respect to n_i,

$$\mu_i = \left(\frac{\partial G}{\partial n_i}\right)_{T,P,n_j} = \mu_i^{(P)} + \frac{2\overline{V}_i\gamma}{r}$$

Thus the increased chemical potential for the droplet as compared with the material with a plane surface is

$$\mu_i - \mu_i^{(P)} = \frac{2\overline{V}_i\gamma}{r} \qquad (23\text{-}34)$$

It is convenient to express this result in terms of the fugacity, which is approximated by the vapor pressure (unless the pressure is high):

$$\ln \frac{f_i}{f_i^{(P)}} = \frac{2\overline{V}_i \gamma}{rRT} \tag{23-35}$$

If the surface is concave toward the vapor, the fugacity is decreased instead of being increased. Equations (23-34) and (23-35) remain applicable except for the change in sign and the identification of r as a radius of spherical curvature. It is also possible to consider surfaces with other than spherical curvature, whereupon the factor $2/r$ is replaced by $1/r_1 + 1/r_2$, where r_1 and r_2 are the two radii of curvature. Thus, for a cylindrical surface, the cylinder radius is r_1 and $r_2 = \infty$.

These results had been accepted for many years before they were verified experimentally by Thomä[11] and La Mer and Gruen.[12] Thomä measured the change in vapor pressure for a curved meniscus of isovaleric acid in a fine capillary. In the experiments of La Mer and Gruen, uniform-sized droplets of a nonvolatile solvent were produced and then equilibrated with the vapor over a solution of the same solvent and a volatile solute. The droplets rapidly grew to a larger size, which was measured by light-scattering methods. By study of drop size as a function of the fugacity of the solute, it was possible to check Eq. (23-35). The solvent–solute systems tested were dioctyl phthalate–toluene and oleic acid–chloroform.

Measurement of Surface Tension

Most of the methods for the measurement of surface tension depend on the relationship of pressure difference to surface curvature given in Eq. (23-32). In the capillary-rise method the surface is spherical, and the theory is simple provided that the contact angle is known. Methods such as that of measuring the maximum pull to lift a plate or ring out of the surface depend primarily on the tension of flat or nearly flat surfaces, but even these methods are subject to corrections that are required by the curvature of the surface near the plate or ring. Adamson[4] gives details concerning these and other experimental methods.

FILMS OF AN INSOLUBLE COMPONENT

If one component of a system is present only on the surface or only in a vapor phase in addition to the surface, then the amount of material on the surface becomes a directly measurable quantity. There are two common examples that have received a great deal of attention. The first is the adsorption of a gas on a liquid or solid. Here the vapor pressure is a direct measure of the fugacity of the surface component, and its amount may be determined by subtracting the amount remaining as vapor from the total added. The second example is an insoluble and nonvolatile film at a liquid–gas interface. In this case the surface component is insoluble in both bulk phases, but its Gibbs energy can be determined by the film pressure, which is the difference in surface tension from that of the pure liquid. While these two examples present very different experimental problems, their interpretation is essentially similar. Let us first relate the different quantities measured in the two cases.

Let us take the fugacity as our measure of chemical potential. The surface concentration Γ will be used without subscript since it is unambiguous in these cases. Actually the surface area of the solid may be uncertain, but that is a separate problem from the one before us now. The film pressure on a liquid surface may be measured directly by a delicate film balance, which measures the force on a barrier separating a surface film on one side from a clean liquid–gas interface on the other. The force divided by the barrier length gives the film pressure π, which is the difference between the surface tension $\gamma°$ of the clean liquid and γ of the surface with the film:

$$\pi = \gamma° - \gamma \tag{23-36}$$

Let us start with the following identity,

$$\left(\frac{\partial \ln f}{\partial \Gamma}\right)_T = \left(\frac{\partial \ln f}{\partial \pi}\right)_T \left(\frac{\partial \pi}{\partial \Gamma}\right)_T \tag{23-37}$$

where f and Γ both refer to the film component, and recall the Gibbs adsorption equation, which we take as Eq. (23-15) reduced to isothermal conditions,

$$d\gamma = -\Gamma d\mu = -\Gamma RT \, d\ln f$$

or

$$\left(\frac{\partial \ln f}{\partial \pi}\right)_T = \frac{1}{\Gamma RT} \tag{23-38}$$

since $d\pi = -d\gamma$ from Eq. (23-36). The combination of Eqs. (23-37) and (23-38) yields

$$\left(\frac{\partial \pi}{\partial \Gamma}\right)_T = \Gamma RT \left(\frac{\partial \ln f}{\partial \Gamma}\right)_T \tag{23-39}$$

which gives the π–Γ behavior of an adsorbed film if the f–Γ isotherm is available, or vice versa.

Two-dimensional Gases, Liquids, and Solids

By analogy to three-dimensional phases we might expect surface films to show gaseous behavior at low surface concentrations, where the molecules are far apart, and to show liquidlike and possibly solidlike behavior as the surface molecules come close together. In some respects the film pressure is more analogous to an osmotic pressure than a gas pressure, but the same pressure is predicted on either basis.

There is every theoretical reason to expect a two-dimensional perfect gas to follow the equation

$$\pi A_s = nRT \quad \text{or} \quad \pi = \Gamma RT \tag{23-40}*$$

and to have other properties analogous to those of a three-dimensional perfect gas. While measurements of very dilute films are particularly difficult, the evidence clearly indicates that this ideal-gas law is approached at low surface concentration on both liquid and solid surfaces.

At higher concentrations one or more different surface phases are observed, which have been given various names. It is beyond the scope of our discussion to consider the particular properties of these liquidlike and solidlike phases except to note that some are still quite compressible. The work of Copeland and Harkins[13] provides excellent examples of film-pressure investigations.

Adsorption of Gases[14,15]

Our treatment is limited to reversible adsorption processes. Although the adsorption of gases on liquid–vapor interfaces has been studied,[16] much more attention has been given to adsorption on finely divided solids of large surface area. In this case the surface area is not directly measurable, but reasonably good indirect estimates are available. One does measure directly the amount of gas adsorbed as a function of the pressure at a given temperature, and the result can be expressed in a variety of ways. Adsorption isotherms are used primarily to determine thermodynamic parameters that characterize the adsorbed layer (heats of adsorption, and the entropy and heat capacity changes associated with the adsorption process), and to determine the surface area of the adsorbing solid. The latter measurement is of great technical importance because of the widespread use of porous solids of high surface area in various industrial processes. The effectiveness of participation by a porous solid in a surface reaction is often proportional to the surface area of the solid. The simplest adsorption isotherm is

$$\sigma = kP \qquad (23\text{-}41)*$$

where σ is the surface coverage (molecules cm^{-2}) and k is a constant that can be related to the heat of adsorption and other parameters. Equation (23-41)* is based on assumptions that the adsorbed gas atoms do not interact with each other and that there are an unlimited number of surface sites at which adsorption can occur. It is also assumed that the adsorption energy is the same for all of the molecules. Equation (23-41)* is unlikely to be suitable for describing the overall adsorption process. Nevertheless, it approximates the adsorption isotherms for many real systems at very low pressures, $< 10^{-8}$ bar.

Langmuir[17] derived a different adsorption isotherm by assuming that adsorption is terminated upon completion of a monomolecular adsorbed gas layer. He did this by asserting that any gas molecule that strikes an adsorbed atom must reflect from the surface. All the other assumptions (i.e., homogeneous surface and noninteracting adsorbed species) used to obtain Eq. (23-41)* were also maintained. If σ_0 is the surface coverage of a completely covered surface, the concentration of surface sites available for adsorption, after adsorbing σ molecules, is $\sigma_0 - \sigma$. Of the total flux F incident on the surface, a fraction $(\sigma/\sigma_0)F$ will strike molecules already adsorbed and therefore be reflected. Thus a fraction $[1 - (\sigma/\sigma_0)]F$ of the total incident flux will be available for adsorption. One then obtains

$$\sigma = \left(1 - \frac{\sigma}{\sigma_0}\right) F\tau \tag{23-42}*$$

where $\tau = \sigma/F$ for an empty surface. Then rearrangement yields

$$\sigma = \frac{\sigma_0 F\tau}{\sigma_0 + F\tau} = \frac{\sigma_0 bP}{\sigma_0 + bP} \tag{23-43}*$$

By writing $\theta = \sigma/\sigma_0$, where θ is often called the degree of coverage, Eq. (23-43)* can be rewritten as

$$\theta = \frac{b'P}{1 + b'P} \tag{23-44}*$$

where $b' = b/\sigma_0$. From Eq. (23-44)*, $b'P$ may be neglected at low pressures in comparison with 1 in the denominator, and Eq. (23-41)* is obtained. The adsorption isotherm of ethyl chloride on charcoal,[18] which appears to obey an equation of the form of Eq. (23-43)*, is shown in Fig. 23-4.

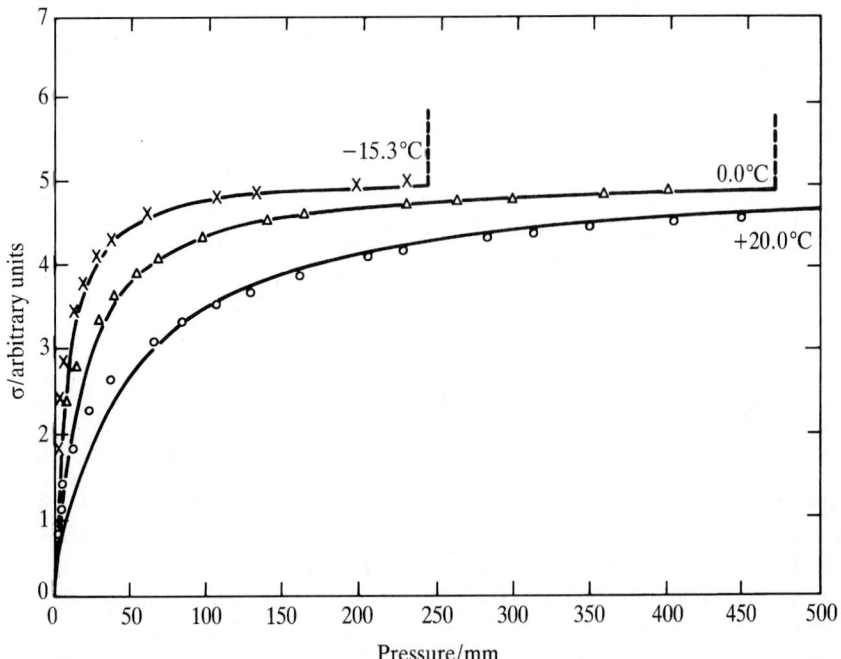

FIGURE 23-4
Adsorption isotherms of ethyl chloride on charcoal (from Somorjai[18]).

Equation (23-43)* can be rearranged to give

$$\frac{1}{\sigma} = \frac{1}{bP} + \frac{1}{\sigma_0} \qquad (23\text{-}45)^*$$

Therefore, a linear Langmuir plot is obtained by plotting $1/\sigma$ against $1/P$. Such a plot is shown for the adsorption of oxygen, carbon monoxide, and carbon dioxide on silica[18] in Fig. 23-5.

There are several other derivations of the Langmuir adsorption isotherm from statistical mechanics and thermodynamics. Although the model is physically unrealistic for describing the adsorption of gases on real surfaces, its success, just like the success of other adsorption isotherms also based on different simple adsorption models, is due to the relative insensitivity of macroscopic adsorption measurements to the atomic details of the adsorption process. Thus the adsorption isotherm provides one with useful approximate values of important adsorption parameters and permits the determination of the surface area.

Another frequently used adsorption model that allows for adsorption in multilayers where gas atoms or molecules may adsorb on top of already adsorbed

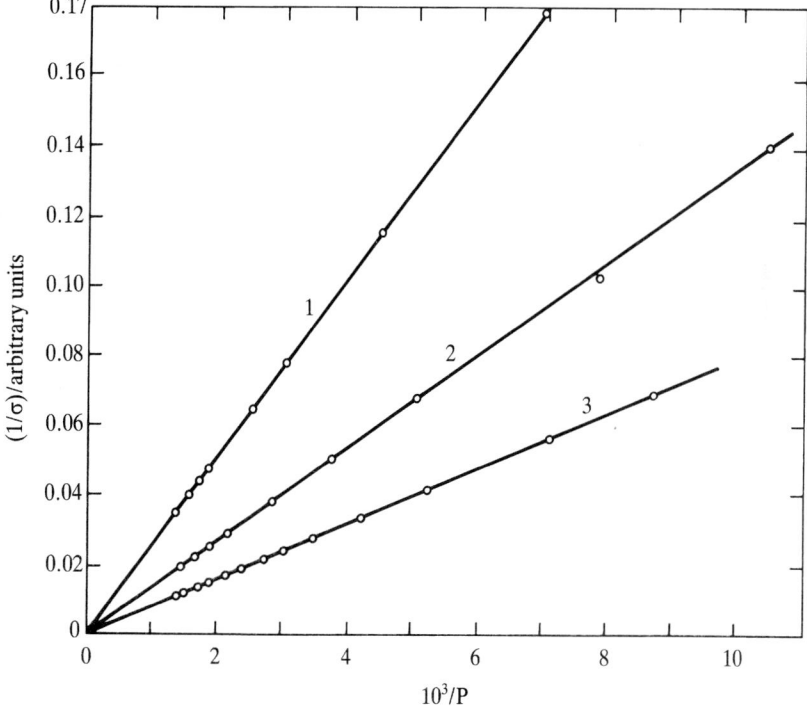

FIGURE 23-5
Adsorption isotherm of (1) oxygen, (2) carbon monoxide, and (3) carbon dioxide on silica plotted as $1/\sigma$ versus $1/P$ (from Somorjai[18]).

molecules was proposed by Brunauer, Emmett, and Teller[19] (the BET model). With the exception of the assumption that the adsorption process terminates at monolayer coverage, the BET model retains all other assumptions made in deriving the Langmuir adsorption isotherm.

Several other theoretical models have attempted to give a more realistic description than the Langmuir and BET models of the gas–surface interactions that lead to physical adsorption. The variable parameters in these models are the interaction potential, the structure of the adsorbed layer (mobile or localized monolayer or multilayer), and the structure of the surface (homogeneous or heterogeneous, number of nearest neighbors). These are described and evaluated in the monographs of Adamson[4] or Somorjai[15] on surface science.

REFERENCES

1. A. Furgeson, *Trans. Faraday Soc.*, **19**, 408 (1923).
2. E. A. Guggenheim, *J. Chem. Phys.*, **13**, 253 (1945).
3. L. Riedel, *Chem.-Ing.-Tech.*, **27**, 209 (1955).
4. A. W. Adamson, *Physical Chemistry of Surfaces*, 5th edn., Wiley, New York, 1990, Chapter 3.
5. A. V. Grosse, *J. Inorg. Nucl. Chem.*, **24**, 147 (1962).
6. G. Korosi and E. SZ. Kovats, *J. Chem. Eng. Data*, **26**, 323 (1981).
7. J. J. Jasper, *J. Phys. Chem. Ref. Data*, **1**, 841 (1972).
8. J. A. V. Butler and A. Wightman, *J. Chem. Soc.*, **1932**, 2089 (1932).
9. W. Thomson (Lord Kelvin), *Proc. Roy. Soc. (London)*, **9**, 255 (1858); *Phil. Mag.* **17**(4), 61 (1859).
10. W. Thomson, *Phil. Mag.*, **42**(4), 448 (1871).
11. M. Thomä, *Z. Physik*, **64**, 224 (1930).
12. V. K. La Mer and R. Gruen, *Trans. Faraday Soc.*, **48**, 410 (1952).
13. L. E. Copeland and W. D. Harkins, *J. Am. Chem. Soc.*, **64**, 1600 (1942); *J. Chem. Phys.*, **10**, 272 (1942).
14. The thermodynamics of gas adsorption has been treated by several authors. T. L. Hill has given especially complete and careful discussions in *J. Chem. Phys.*, **17**, 520 (1949); **18**, 246 (1950).
15. G. A. Somorjai, *Principles of Surface Chemistry*, Prentice-Hall, Englewood Cliffs, N.J., 1972.
16. For example, see C. L. Cutting and D. C. Jones, *J. Chem. Soc.*, **1955**, 4067 (1955).
17. I. Langmuir, *J. Am. Chem. Soc.*, **40**, 1361 (1918).
18. G. A. Somorjai, personal communication.
19. S. Brunauer, P. H. Emmett, and E. Teller, *J. Am. Chem. Soc.*, **60**, 309 (1938).

PROBLEMS

23-1. The following data are given for the surface tension (in dyne cm^{-1}) of dilute electrolyte solutions in water at 20°C. For pure water $\gamma = 72.75$.

Solute	Wt%	γ	Wt%	γ	Wt%	γ	Wt%	γ
HCl	1.78	72.55	3.52	72.45	6.78	72.25	12.81	71.85
KCl	0.74	72.99	3.60	73.45	6.93	74.14	13.88	75.55

(a) Is the surface concentration of each solute positive or negative? The surface concentration of water is zero by definition.

(b) Calculate the approximate value of the surface concentration of 1.0 molal HCl at 20°C. State your assumptions and approximations and also the units for your answer.

23-2. The surface concentration for the alcohol–water system was treated on the assumption that $\Gamma_{1(1)} = 0$ and $\Gamma_{2(1)}$ was calculated. Make the alternate assumption that $\Gamma_{2(2)} = 0$. Derive an equation for $\Gamma_{1(2)}$ in terms of $\Gamma_{2(1)}$ and calculate $\Gamma_{1(2)}$ at $x_2 = 0.1$ and at $x_2 = 0.5$. Discuss the interpretations on the two alternate bases.

23-3. What is the equation for the surface heat capacity C_s implied by Eq. (23-13)* for the surface tension of a normal liquid?

23-4. Calculate the vapor pressure of 0.1 μm (10^{-5} cm) diameter droplets of water at 25°C. What is the vapor-pressure ratio for these droplets to that of a plane surface of water?

23-5. What composition of aqueous KCl in droplets of 0.2 μm diameter would be in equilibrium with a plane surface of pure water? Neglect the effect of KCl on the surface tension; see Table 17-1 and equations in Chapter 17 for the properties of KCl solutions.

23-6. Derive Eq. (23-35) by considering the excess pressure within a small droplet, together with the usual equation for the change of fugacity with pressure.

CHAPTER 24

SYSTEMS INVOLVING GRAVITATIONAL, CENTRIFUGAL, ELECTRICAL, OR MAGNETIC FIELDS

In many thermodynamic calculations it can be assumed, as we have usually assumed hitherto, that the thermodynamic state of a substance is determined solely by temperature, pressure, and composition; but there are also many examples in which it is necessary to consider other independent variables. The effects of surfaces were considered in Chapter 23; in this chapter we turn our attention to problems involving centrifugal, gravitational, electrical, or magnetic fields. The situation for gravitational and centrifugal fields is relatively straightforward and is considered first because their effect on thermal or other internal properties of a substance can be neglected. In contrast, magnetic fields have very significant thermal effects; indeed, the technique of adiabatic demagnetization is very important in the attainment of very low temperatures.

GRAVITATIONAL AND CENTRIFUGAL FIELDS

For a pure substance the effect of a gravitational field contributes a term like that for pressure to the Gibbs energy,

$$dG = -S\,dT + V\,dP + mg\,dh \tag{24-1}$$

where h is the height in the field of strength g and m is the mass. Consider a vertical column of a gas or liquid at constant temperature. One increment of a given mass, m, and volume, V, will be in equilibrium with another if $dG = 0$; thus

$$V\,dP + mg\,dh = 0 \qquad (24\text{-}2)$$

The relationship of pressure to height evidently depends on the ratio of m to V, which is the density ρ. Then

$$\left(\frac{\partial P}{\partial h}\right)_T = -g\rho \qquad (24\text{-}3)$$

It is interesting to consider the variation with height of the pressure of a perfect gas of molecular mass M in a gravitational field. First we substitute MP/RT for ρ; then Eq. (24-3) may be rearranged to

$$d\ln P = -\frac{Mg}{RT}dh \qquad (24\text{-}4)^*$$

If M, g, and T are constants, integration yields

$$\ln\frac{P''}{P'} = \frac{Mg}{RT}(h' - h'') \qquad (24\text{-}5)^*$$

In the earth's atmosphere M, g, and T all vary with height; consequently Eq. (24-5)* gives only a crude approximation to atmospheric pressure over a limited altitude range.

For multicomponent systems we need a more general formulation. Since a phase is defined to be homogeneous in all respects, the regions of matter at different locations with respect to the gravitational or centrifugal field must be taken to be different phases. If the potential for that field is designated by ψ, the work to move a mass M from ψ' to ψ'' is given by

$$w = M(\psi'' - \psi') \qquad (24\text{-}6)$$

If the molar mass of species i is M_i, the work to transfer dn_i moles of species i is

$$dw = (\psi'' - \psi')M_i\,dn_i \qquad (24\text{-}7)$$

Then the general equation for the Gibbs energy for a phase becomes

$$dG' = -S'\,dT' + V'\,dP' + \sum_i(\mu_i' + M_i\psi')\,dn_i' \qquad (24\text{-}8)$$

For equilibrium of species i between two phases, as defined by their potentials ψ' and ψ'' as well as their temperatures and pressures, the general condition is then

$$\mu_i'' + M_i\psi'' = \mu_i' + M_i\psi' \qquad (24\text{-}9)$$

Also, we note that the cross-differentiation identity applied to Eq. (24-8) yields

$$\frac{\partial \mu_i}{\partial P} = \frac{\partial V}{\partial n_i} = \overline{V}_i \qquad (24\text{-}10)$$

Since $\psi = gh$, Eq. (24-3) becomes

$$\frac{\partial P}{\partial \psi} = -\rho \qquad (24\text{-}11)$$

Next, we must consider the dependencies of μ_i and $\overline{V}_i$ on composition and pressure. In general, these can be complex, but we first consider just the simple case of an ideal solution that is incompressible. Then the $\overline{V}_i$ are constants and from Eq. (11-2)

$$\left(\frac{\partial \mu_i}{\partial x_i}\right)_{T,P,\psi} = \frac{RT}{x_i} \qquad (24\text{-}12)$$

Equation (24-9) is satisfied if

$$RT \ln\left(\frac{x_i''}{x_i'}\right) = (\rho \overline{V}_i - M_i)(\psi'' - \psi') \qquad (24\text{-}13)$$

The important application of this result is for the centrifuge when it is used to determine molecular masses. For this situation

$$d\psi = -\omega^2 r\, dr \qquad (24\text{-}14a)$$

or

$$\psi'' - \psi' = -\tfrac{1}{2}\omega^2[(r'')^2 - (r')^2] \qquad (24\text{-}14b)$$

with ω the angular velocity (ordinarily in radians per second) and r the distance from the axis of rotation. Then

$$\ln\left(\frac{x_i''}{x_i'}\right) = \frac{(M_i - \rho \overline{V}_i)\omega^2}{2RT}[(r'')^2 - (r')^2] \qquad (24\text{-}15)*$$

Another simple case is that of a perfect gaseous solution. In that case it is easier to deal with the partial pressure† of each component, $p_i = x_i P$. Now, at any ψ,

$$\mu_i'' - \mu_i' = RT \ln\left(\frac{p_i''}{p_i'}\right)$$

and

$$\ln\left(\frac{p_i''}{p_i'}\right) = \frac{M_i \omega^2}{2RT}[(r'')^2 - (r')^2] \qquad (24\text{-}16)*$$

---

† We shall not use the fugacity function in gravitational or centrifugal problems, although it might be desirable to do so in the future. Presumably one would define f_i to become equal to p_i at the reference height or radius, that is, $r = 0$ or $h = 0$.

This result is equivalent to that from Eq. (24-5)* for a pure perfect gas. As we expect from the properties of perfect gases, the presence of other components has no effect on the behavior of a given component.

In order to investigate the separation of two gaseous components, we subtract Eq. (24-16)* for one component from that for the other

$$\ln\left(\frac{p_2''/p_1''}{p_2'/p_1'}\right) = \frac{(M_2 - M_1)\omega^2}{2RT}[(r'')^2 - (r')^2] \qquad (24\text{-}17)^*$$

It is interesting to note that the separation depends only on the absolute difference in molecular masses—not on the relative difference. Thus it would be as easy to separate by centrifugation heavy isotopes differing by 1 mass unit as to separate He^3 from He^4 (provided that the temperature of the centrifuge remained unchanged).

ELECTRIC AND MAGNETIC FIELDS

While electric fields have entered our discussions of galvanic cells and of electrolyte solutions, the emphasis in these cases was upon the motion and properties of charged ions. At this time we wish to turn our attention to the effect of electric fields on polarizable or dielectric material. It is convenient to consider also the similar, but not identical, effects of magnetic fields on magnetically polarizable material, which then leads to a very important topic—the production of extremely low temperatures by adiabatic demagnetization of paramagnetic materials.

In our general discussion of the thermodynamics of polarizable materials in the presence of electric and magnetic fields we shall use SI units and quantities defined in the rational system. In contrast, for electrolyte solutions in Chapter 16 the unrationalized esu system is used, because it is more convenient and almost universally used for those applications. Also, the interrelationship is explained there. Since the discussion in many texts of the work associated with electric and magnetic fields is frequently limited by simplifying assumptions, we note that Stratton[1] gives an excellent presentation for our purposes.

In our equations we shall use vector notation, although in most cases of practical interest all vectors are either parallel or antiparallel and hence their products are just the products of absolute magnitudes with the proper sign. We shall retain the boldface type for such vectors, in any case, to distinguish magnetic field **H** from enthalpy H, etc. The resulting equation for the electromagnetic work required to establish a system of fields is

$$w = \iint \mathbf{E}\cdot d\mathbf{D}\, dV + \iint \mathbf{H}\cdot d\mathbf{B}\, dV \qquad (24\text{-}18)$$

where the volume integrals are over all space and the field integrals are from zero field to the final state. **E** and **H** are the electric- and magnetic-field vectors, respectively, while **D** and **B** are known, respectively, as the electric displacement and the magnetic induction.

In free space **D** is proportional to **E** and **B** is proportional to **H** with the proportionality constants ε_0 and μ_0,† called the permittivity and permeability, respectively, of free space.

$$\mathbf{D} = \varepsilon_0 \mathbf{E} \quad \text{and} \quad \mathbf{B} = \mu_0 \mathbf{H} \tag{24-19}$$

The selection of one of these constants is arbitrary, but their product $\varepsilon_0 \mu_0$ may be shown to be the reciprocal of the square of the velocity of electromagnetic waves in free space, that is, $\varepsilon_0 \mu_0 = c^{-2}$. In SI units, $\mu_0 = 4\pi \times 10^{-7}\,\mathrm{H\,m^{-1}}$, or $\mathrm{V\,s\,A^{-1}\,m^{-1}}$, or $\mathrm{J\,s^2\,C^{-2}\,m^{-1}}$, and $\varepsilon_0 = 8.8542 \times 10^{-12}\,\mathrm{F\,m^{-1}}$, or $\mathrm{C\,V^{-1}\,m^{-1}}$, or $\mathrm{C^2\,J^{-1}\,m^{-1}}$. Here H is the unit of inductance, the henry, and F is the unit of capacitance, the farad. Other units are more familiar but are listed in Appendix 17.

In other media **D** and **B** are expressed in a formally similar manner, but ε and μ are, in general, symmetric tensors of the second rank (or dyadics). However, for isotropic media, i.e, material with uniform properties in all directions, ε and μ reduce to scalar quantities but of magnitude different from ε_0 and μ_0. Thus

$$\mathbf{D} = \varepsilon \mathbf{E} \quad \text{and} \quad \mathbf{B} = \mu \mathbf{H} \tag{24-20}$$

This same proportionality of **D** to **E** and of **B** to **H** is also obtained for crystals, provided that the field lies along an appropriate axis of the crystal. We shall deal here only with cases where ε and μ are simple scalar quantities.

The ratio $\varepsilon/\varepsilon_0$ is the *relative permittivity* ε_r, which is also called the *dielectric constant*. In this chapter we shall write equations in terms ε or $\varepsilon_r = \varepsilon/\varepsilon_0$, as is most convenient, but note, however, that in Chapters 15 and 16 on electrolytes the working equations involved only ε_r and the subscript r was omitted. The similar ratio μ/μ_0 may be described as the magnetic inductive capacity, but it is customary, instead, to use the *magnetic susceptibility* χ, which is defined as

$$\chi = \frac{\mu}{\mu_0} - 1 \tag{24-21}$$

or the molar susceptibility $\chi_M = \chi V_m$, where V_m is the molar volume.

The electric and magnetic *polarization* vectors are defined by

$$\mathbf{p} = \mathbf{D} - \varepsilon_0 \mathbf{E} \quad \text{and} \quad \mathbf{m} = \frac{\mathbf{B}}{\mu_0} - \mathbf{H} \tag{24-22}$$

which in an isotropic medium simplify to

$$\mathbf{p} = (\varepsilon - \varepsilon_0)\mathbf{E} = (\varepsilon_r - 1)\varepsilon_0 \mathbf{E} \tag{24-23}$$

$$\mathbf{m} = \left(\frac{\mu}{\mu_0} - 1\right)\mathbf{H} = \chi \mathbf{H} \tag{24-24}$$

---

† Care must be taken not to confuse μ in this usage with the chemical potential.

For thermodynamic purposes we frequently need, instead of the polarization per unit volume, the polarization of a definite amount of matter, usually 1 mole. The capital letters **P** and **M** will be used for the molar polarization and

$$\mathbf{M} = \chi_M \mathbf{M} \tag{24-25}$$

Under most conditions, ε, μ, and therefore ε_r and χ are constants for a given material at a particular temperature and pressure. Exceptions occur for ferromagnetic materials and for other materials at extremely high fields and low temperatures. Under such conditions one deals with a differential susceptibility, which we shall write explicitly as

$$\left(\frac{\partial \mathbf{m}}{\partial \mathbf{H}}\right)_{P,T} \quad \text{or per mole} \quad \left(\frac{\partial \mathbf{M}}{\partial \mathbf{H}}\right)_{P,T}$$

Additional complications arise for permanent magnetization of ferromagnetic materials, where there is a fixed magnetization $\mathbf{M}_0$ in addition to an induced magnetization **M**.

Electric Polarization

If we assume that the charges generating an electric field are fixed in position, then the change in w associated with the introduction of polarizable material may be related to the Gibbs energy of that substance. An array of fixed charges establishes the electric displacement vector **D**. If we limit ourselves to cases where either the polarizable material fills all space or its boundaries are either perpendicular or parallel to the field vector, the **D** is unchanged by the introduction of the dielectric. The change in the field **E** on introduction of dielectric from Eq. (24-25) is

$$\Delta \mathbf{E} = \frac{\mathbf{D}}{\varepsilon} - \frac{\mathbf{D}}{\varepsilon_0}$$

and from Eq. (24-18) the work of introduction of the dielectric is

$$\Delta w = -\frac{1}{\varepsilon_0} \int\!\!\int \left(1 - \frac{\varepsilon_0}{\varepsilon}\right) \mathbf{D} \cdot d\mathbf{D}\, dV \tag{24-26}$$

The integration must cover the entire volume of polarizable material, but the integrand is evidently zero elsewhere.

Let us next consider a pair of parallel, uniformly charged plates, one positive and one negative, immersed in a liquid dielectric. If we neglect edge effects, then in the space between the plates **D** is constant and has the magnitude of the surface-charge density, q/a, and the direction perpendicular to the plates. Elsewhere **D** is zero. If the volume of the space between plates is V_c, we have

$$\Delta w = -\frac{V_c}{2\varepsilon_0} \int \left(1 - \frac{\varepsilon_0}{\varepsilon}\right) d(\mathbf{D}^2)$$

Also, by regarding the magnitude of the charges as continuously variable, one may vary **D** similarly and write for an increment in Δw

$$dw = -\frac{V_c}{2\varepsilon_0}\left(1 - \frac{\varepsilon_0}{\varepsilon}\right)d(\mathbf{D}^2) \tag{24-27}$$

The differential of the Gibbs energy of all the dielectric material is obtained by adding Eq. (24-27) to Eq. (3-6):

$$dG = -S\,dT + V\,dP - \frac{V_c}{2\varepsilon_0}\left(1 - \frac{1}{\varepsilon_r}\right)d(\mathbf{D}^2) \tag{24-28}$$

The quantity ε_r, the relative permittivity (or dielectric constant), is always greater than 1 and is nearly independent of **D** at low fields. Thus we note that the Gibbs energy of polarizable material is reduced when in an electric field. Equation (24-28) is an exact differential; hence we may use Eq. (2-16) to obtain the change of entropy and of volume with the field:

$$\left(\frac{\partial S}{\partial (\mathbf{D}^2)}\right)_{P,T} = -\frac{V_c}{2\varepsilon_0}\left(\frac{\partial (1/\varepsilon_r)}{\partial T}\right)_{P,\mathbf{D}} \tag{24-29}$$

$$\left(\frac{\partial V}{\partial (\mathbf{D}^2)}\right)_{P,T} = \frac{V_c}{2\varepsilon_0}\left(\frac{\partial (1/\varepsilon_r)}{\partial P}\right)_{T,\mathbf{D}} \tag{24-30}$$

Here P is the pressure on the liquid dielectric outside the capacitor.

Since ε_r usually increases with pressure, Eq. (24-30) indicates a volume decrease in the field. This effect is called electrostriction. The dielectric constant of most substances decreases with rise in temperature; hence we see that the entropy decreases as the material is polarized. This would be expected on a statistical basis for dipolar molecules since the field will tend to reduce the randomness of molecular orientation.

The entropy effect that we just noted implies evolution of heat on polarization in the amount $-T\,dS$. The enthalpy change in the dielectric is given, not by Eq. (24-27) but rather by

$$\left(\frac{\partial H}{\partial (\mathbf{D}^2)}\right)_{P,T} = -\frac{V_c}{2\varepsilon_0}\left\{1 - \frac{1}{\varepsilon_r} + T\left(\frac{\partial (1/\varepsilon_r)}{\partial T}\right)\right\} \tag{24-31}$$

The last several equations pertain to all the dielectric fluid in between and around the capacitor plates, and, as we have seen, the charging of the plates will tend to force more material between the plates. While the changes in the molar quantities are just V_m/V_c times the results given above, V_m varies slightly with **D**, whereas V_c was a constant.

Let us also write the differential of the molar Gibbs energy of material between the plates where the field is described by the vector **D**,

$$dG_m = -S_m\,dT + V_m\,dP - \frac{V_m}{2\varepsilon_0}\left(1 - \frac{1}{\varepsilon_r}\right)d(\mathbf{D}^2) \tag{24-32}$$

where P is now the pressure on the material in the field. If the dielectric is a fluid free to flow into the space between the plates, then G must be the same in and out of the field. This equality of G requires a pressure increase dP on the material in the field, which is given by

$$dP = + \frac{1}{2\varepsilon_0}\left(1 - \frac{1}{\varepsilon_r}\right)d(\mathbf{D}^2) \tag{24-33}$$

This equation may be integrated if ε_r is known as a function of $\mathbf{D}$, or even more easily if ε_r is constant.

Solutions in Electric Fields

Another interesting possibility is a dielectric solution. Now the chemical potential of each component must remain constant throughout the solution, and the composition between the plates need not be the same as outside. Equation (24-28) may be solved for the derivative of the total Gibbs energy with respect to $\mathbf{D}^2$ as a measure of the field strength. Further differentiation with respect to the number of moles of component i between the plates yields the change of chemical potential with $\mathbf{D}^2$:

$$\frac{\partial \mu_i}{\partial (\mathbf{D}^2)} = \frac{V_c}{2\varepsilon_0}\frac{\partial (1/\varepsilon_r)}{\partial n_i} \tag{24-34}$$

In order to estimate the effect on composition, we must convert this derivative to one with respect to mole fraction. Let us assume a two-component system, whereupon

$$\frac{\partial \mu_2}{\partial (\mathbf{D}^2)} = \frac{V_m^2}{2\overline{V}_1 \varepsilon_0}\frac{\partial (1/\varepsilon_r)}{\partial x_2} \tag{24-35}$$

with V_m the molar volume of the solution and $\overline{V}_1$ the partial molal volume of the other component.

The differential of μ_2 must be zero:

$$d\mu_2 = \frac{\partial \mu_2}{\partial (\mathbf{D}^2)}d(\mathbf{D}^2) + \frac{\partial \mu_2}{\partial \ln x_2}d\ln x_2 = 0$$

Hence

$$\frac{\partial \ln x_2}{\partial (\mathbf{D}^2)} = -\left(\frac{\partial \mu_2}{\partial \ln x_2}\right)^{-1}\left(\frac{V_m^2}{2\varepsilon_0 \overline{V}_1}\right)\frac{\partial (1/\varepsilon_r)}{\partial x_2} \tag{24-36}$$

If we assume an ideal solution, this becomes

$$\frac{\partial \ln x_2}{\partial (\mathbf{D}^2)} = -\frac{V_m^2}{2\varepsilon_0 RT\overline{V}_1}\frac{\partial (1/\varepsilon_r)}{\partial x_2} \tag{24-37}*$$

Charged Sphere in a Dielectric

An ion in solution is sometimes approximated as a charged sphere in a continuous dielectric medium; hence the latter problem is of interest at this point. The field around a sphere carrying charge q is described by a displacement vector **D** of radial orientation and of magnitude

$$\mathbf{D}_r = \frac{q}{4\pi r^2} \tag{24-38}$$

The Gibbs energy of polarization of the dielectric is then the integral of Eq. (24-26) from the surface of the ion at radius r_0 to infinity and from charge zero up to ze. In general the relative permittivity ε_r is a function of the field strength, and one has for 1 mole of ions

$$\Delta G = -\frac{N_A}{\varepsilon_0} \int_{r_0}^{\infty} 4\pi r^2 \, dr \int_0^{ze/4\pi r^2} \left(1 - \frac{1}{\varepsilon_r}\right) \mathbf{D} \cdot d\mathbf{D} \tag{24-39}$$

If one assumes a constant value of ε_r, then one obtains the Born equation,

$$\Delta G = -\frac{z^2 e^2 N_A}{8\pi \varepsilon_0 r_0}\left(1 - \frac{1}{\varepsilon_r}\right) \tag{24-40}*$$

The coefficient $e^2 N_A/8\pi\varepsilon_0$ has the value 6.95×10^{-5} J m mol^{-1} or 695 kJ Å mol^{-1}. Thus the predicted ΔG of solvation of an ion of $r_0 = 3$ Å is $232z^2(1 - 1/\varepsilon_r)$ kJ mol^{-1}. While the assumption of a constant value of ε_r cannot be very accurate for the high field near an ion in solution, one notes that as long as $\varepsilon_r \gg 1$ the error will not be too serious. The choice of the appropriate value for r_0 involves the molecular nature of the solvent; it should exceed somewhat the crystallographic radius. Latimer et al.[2] found good agreement with data for aqueous ions for r_0 values greater than crystal radii by 0.1 Å for negative ions and 0.85 Å for positive ions. Various other investigations report good agreement for similar but slightly more complex radius adjustments.[3]

MAGNETIC POLARIZATION

The basic integral for the magnetic work from Eq. (24-18),

$$w = \iint \mathbf{H} \cdot d\mathbf{B} \, dV \tag{24-41}$$

is identical in form to that for the electrical work, but the generation of a magnetic field is not as simple as the use of one or more electric charges to establish an electric field. The simplest magnetic-field generator is a very long solenoid with uniform windings. If we neglect end effects, the field **H** within the magnet is given simply by the number of ampere-turns per meter of length. Suppose that we insert a long cylinder of a paramagnetic substance, that is, $\mu > \mu_0$, within the solenoid. We know that the paramagnetic material is attracted into the field and hence that its Gibbs energy should be lower in the field. But $\mathbf{B} = \mu \mathbf{H}$, and therefore **B** is increased by the paramagnetic material, yielding a greater w in Eq. (24-41). The explanation of this apparent contradiction is that additional electrical work must be supplied to keep the solenoid current and thereby **H** constant. The magnetic field thus plays a role much like the pressure in volumetric effects.

The additional work associated with 1 mole of magnetically polarizable material in a uniform field is readily obtained from Eqs. (24-22) and (24-41),

$$\Delta w = \mu_0 \int \mathbf{H} \cdot d\mathbf{M} \tag{24-42}$$

This work, expressed in differential form, is added to the usual expression for Helmholtz energy to obtain

$$dA = -S\,dT - P\,dV + \mu_0 \mathbf{H} \cdot d\mathbf{M} \tag{24-43}$$

The Gibbs energy is now defined† as

$$G = A + PV - \mu_0 \mathbf{H} \cdot \mathbf{M} \tag{24-44}$$

where the final term is added to exclude the electrical work of maintaining a constant magnetic field just as the PV term excludes the volumetric work of maintaining constant pressure. Thus

$$dG = -S\,dT + V\,dP - \mu_0 \mathbf{M} \cdot d\mathbf{H} \tag{24-45}$$

The functions U and H are related, as usual, to A and G by the addition of TS:

$$dU = T\,dS - P\,dV + \mu_0 \mathbf{H} \cdot d\mathbf{M} \tag{24-46}$$

$$dH = T\,dS + V\,dP - \mu_0 \mathbf{M} \cdot d\mathbf{H} \tag{24-47}$$

We note from the last equation that dH is the heat absorbed at constant pressure and magnetic field.

Application of Eq. (2-16) to the total differential of the Gibbs energy yields

$$\left(\frac{\partial V}{\partial \mathbf{H}}\right)_{P,T} = -\mu_0 \left(\frac{\partial \mathbf{M}}{\partial P}\right)_{T,\mathbf{H}} \tag{24-48}$$

$$\left(\frac{\partial S}{\partial \mathbf{H}}\right)_{P,T} = \mu_0 \left(\frac{\partial \mathbf{M}}{\partial T}\right)_{P,\mathbf{H}} \tag{24-49}$$

In our use of Eqs. (24-46) through (24-51)* we shall be concerned only with examples where the magnetic polarization $\mathbf{M}$ is parallel to the field $\mathbf{H}$; hence each quantity may be regarded as a scalar (rather than a vector). We shall, however, retain the boldface type for $\mathbf{H}$, etc., to distinguish these quantities from others represented by the same letters, e.g., enthalpy H. We have avoided making the further assumption that $\mathbf{M}$ is proportional to $\mathbf{H}$ because exceptions that show saturation of polarization are of importance at low temperatures.

---

† The function $A + PV$ may still be used in a magnetic field, but we shall not do so and hence prefer to retain G for the function used.

In the more restricted case where $\mathbf{M} = \chi_M \mathbf{H}$, the following approximate equations are readily derived:

$$dG = -S\,dT + V\,dP - \tfrac{1}{2}\mu_0 \chi_M d(\mathbf{H}^2) \qquad (24\text{-}45')*$$

$$\left[\frac{\partial V}{\partial (\mathbf{H}^2)}\right]_{P,T} = -\tfrac{1}{2}\mu_0 \left(\frac{\partial \chi_M}{\partial P}\right)_{T,\mathbf{H}} \qquad (24\text{-}48')*$$

$$\left[\frac{\partial S}{\partial (\mathbf{H}^2)}\right]_{P,T} = -\tfrac{1}{2}\mu_0 \left(\frac{\partial \chi_M}{\partial T}\right)_{P,\mathbf{H}} \qquad (24\text{-}49')*$$

The magnetostrictive effect given by Eq. (24-48) or (24-48')* is usually very small and of little interest, but the thermal effect given by Eq. (24-49) or (24-49')* is of major importance.

Cooling by Adiabatic Demagnetization

It was pointed out both by Giauque and by Debye in 1926 that the thermal effects associated with the magnetization of typical paramagnetic salts at approximately 1 K provide a means of producing much lower temperatures. Certain salts of paramagnetic rare-earth or transition-metal ions such as $Gd_2(SO_4)_3 \cdot 8H_2O$ are appropriate. At relatively low temperatures and moderate magnetic fields, these materials follow Curie's law, which is

$$\frac{\mathbf{M}}{\mathbf{H}} = \chi_M = \frac{C_M}{T} \qquad (24\text{-}50)*$$

where C_M is a constant. Hence $\partial \chi / \partial T$ is large and negative at low temperatures, and from Eq. (24-49')*

$$\left[\frac{\partial S}{\partial (\mathbf{H}^2)}\right]_{P,T} = -\frac{\mu_0 C_M}{2T^2} \qquad (24\text{-}51)*$$

If Curie's law holds as the field is reduced to zero, the entropy decrease is

$$\Delta S = -\frac{\mu_0 C_M \mathbf{H}^2}{2T^2} \qquad (24\text{-}52)*$$

and we see that there is a large decrease of entropy (and evolution of heat) on application of a strong magnetic field at low temperatures.

After the heat of magnetization has been conducted away, the sample is thermally isolated and the field is removed. Adiabatic demagnetization then occurs, with reduction of temperature from an initial value T_1 to a final temperature T_2. In order to treat this process, let us define the molar heat capacity at constant field (analogous to constant pressure),

$$C_H = \left(\frac{\partial H}{\partial T}\right)_{\mathbf{H}} = T\left(\frac{\partial S}{\partial T}\right)_{\mathbf{H}} \qquad (24\text{-}53)$$

Then the decrease in entropy in isothermal magnetization may be set equal to the entropy decrease on cooling from T_1 to T_2 at zero field:

$$\Delta S = \int_{T_1}^{T_2} C_{\mathbf{H}=0} d\ln T = -\frac{\mu_0 \mathbf{H}^2 C_M}{2T_1^2} \tag{24-54}*$$

Salts such as $Ce_2Mg_3(NO_3)_{12} \cdot 24H_2O$, with large magnetic moments spaced widely apart by inert material, have properties such that nearly all the entropy can be removed by attainable magnetic fields at temperatures near 1.5 K. On thermal isolation and demagnetization, a very low temperature is reached. But Curie's law can no longer hold exactly; also, the measurement of this temperature is not straightforward. These difficulties can be overcome, but the complete procedure is complex and will not be described here. Fisher et al.[4] report especially careful and accurate experiments on $Ce_2Mg_3(NO_3)_{12} \cdot 24H_2O$, in which temperatures below 0.001 K were accurately measured.

Figure 24-1 shows the heat capacity of $Ce_2Mg_3(NO_3)_{12} \cdot 24H_2O$ at zero magnetic field on the basis of 1 mole of the magnetic ion Ce^{3+}. This particular compound is notable in having the peak of the heat capacity curve as low as 1.6 mK and in following Curie's law quite accurately down to 6 mK. Other crystals with magnetic ions have similar properties but at somewhat higher temperatures.

With the availability in quantity of the helium isotope of mass 3, alternate methods became available for the attainment of temperatures near 1 mK (see Chapter 22 for the properties of ^{3}He). With material at about 1 mK the nuclear magnetic moments can be aligned by a strong magnetic field. The principles are exactly the same

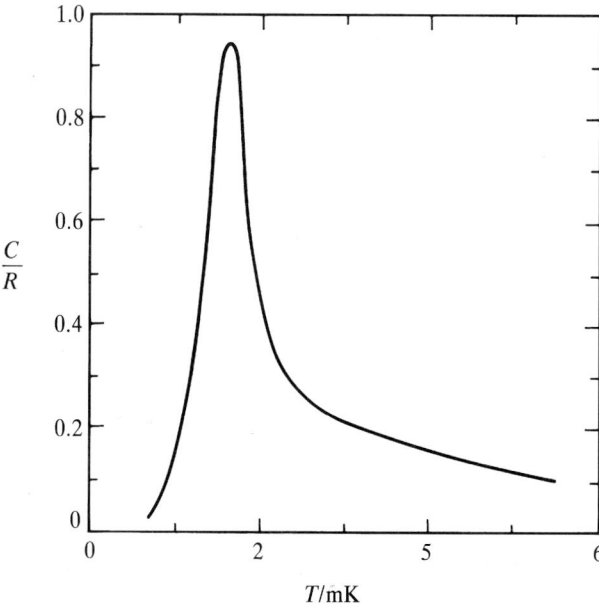

FIGURE 24-1
The heat capacity of $\frac{1}{2}$ mol of $Ce_2Mg_3(NO_3)_{12} \cdot 24H_2O$ from Fisher et al.[4] At this temperature the lattice heat capacity is negligible and the effect is entirely magnetic per mole of Ce.

as for electronic magnetic moments, but the magnitude of μ_0 is less by about three orders of magnitude. Then, adiabatic demagnetization yields still lower temperatures.

Hudson[5] and Lounasmaa[6] present comprehensive accounts of this type of research as of 1972 and 1974, respectively. Hakonen et al.[7] review recent work including record low temperatures and studies of nuclear magnetism.

SUMMARY OF THERMODYNAMICS OF SYSTEMS WITH ADDITIONAL VARIABLES

Chapters 23 and 24 have been devoted to thermodynamic relationships for systems that involve other energy terms in addition to thermal and volumetric energies. It is valuable to pause a moment and summarize these results. The change in energy content of the system may be written

$$dU = \sum_i X_i \, dx_i \qquad (24\text{-}55)$$

where each term is the product of an intensive variable or force X_i and the change of an extensive variable x_i. For the various types of energy one has the terms listed in Table 24-1.

We have intentionally omitted electric and magnetic polarization from the table because of the need to define carefully the treatment of the energy of the fields in the absence of polarizable material. These problems have been considered in this chapter, and the appropriate energy equations are given.

Since Eq. (24-55) is an exact differential, Eq. (2-16) may be applied to yield

$$\left(\frac{\partial X_i}{\partial x_j}\right)_{x_i, \ldots} = \left(\frac{\partial X_j}{\partial x_i}\right)_{x_j, \ldots} \qquad (24\text{-}56)$$

where x_i and x_j are any pair of variables. Also, by defining functions such as $U - X_i x_i$ and dealing with their differentials, one may obtain equations of the type

$$\left(\frac{\partial X_i}{\partial X_j}\right)_{x_i, \ldots} = -\left(\frac{\partial x_j}{\partial x_i}\right)_{X_j, \ldots} \qquad (24\text{-}57)$$

TABLE 24-1

Type of energy	X_i	x_i
Thermal	T	S
Volumetric	P	$-V$
Surface area	γ	A_s
Gravitational	$g(h - h_0)$[a]	M
Centrifugal	$-r^2 \omega^2/2$	M
Electric charge	$\Delta \mathcal{V}$	q
Material content of component i	μ_i	n_i

[a] More precisely $\int_{h_0}^{h} g \, dh$ if g varies significantly.

and

$$\left(\frac{\partial x_i}{\partial X_j}\right)_{X_j,\ldots} = \left(\frac{\partial x_j}{\partial X_i}\right)_{X_j,\ldots} \tag{24-58}$$

Many such equations have been given in these chapters, but others may be derived as desired for any given pair of variables.

It is possible to imagine experiments in which either the extensive or the intensive variable is held constant. Just as either P or V and either T or S may be held constant, so may either γ or A_s, either gh or M, etc. In some cases it would be extremely difficult to hold a given variable fixed, whereas the other variable is readily controlled. Thus one may readily hold a magnetic field constant but not the magnetic polarization. One may readily prevent flow of electric charge; it is also practical to hold a constant difference in electric potential between wires connected to a pair of electrodes, but there is little to be gained by doing experiments under this condition.

In the case where only thermal and volumetric energies were involved it was useful to define and use the four possible functions $U, H = U + PV, A = U - TS$, and $G = U + PV - TS$. Clearly there are many related functions when even a few of the possible additional variables are involved, that is, $U - \gamma A_s$, $U - \gamma A_s - TS$, etc. But most of these are so rarely if ever of practical use that there is no need to give them names and symbols. Indeed we have found it quite convenient to deal with most of these additional variables without modifying the definitions of U, H, A, and G, except in the single case of the magnetic field, where the term $\mu_0 \mathbf{H} \cdot \mathbf{M}$ was introduced into the definitions of H and G. For biochemical systems some authors[8,9] have defined modified functions by subtracting $\mu_i n_i$ for a species such as H^+. Strong buffers allow a system to be held to nearly constant pH and thereby constant μ_{H^+}. This has been proposed also for Mg^{2+} ion, but in that case no corresponding buffers are available. Also, we note in Chapter 27 that, although one step in a calculation for a biochemical system is made at predetermined values of the activities of H^+ and of Mg^{2+}, this is only a rough approximation, and an iterative calculation adjusting the activities of H^+ and Mg^{2+} is required to obtain acceptable accuracy. Thus, it seems to us to be best to avoid the definition of unnecessary new functions but to reserve the possibility of modification of the definition of H and G by the addition of the particular product $-X_i x_i$ whenever desired. This modification should be clearly stated, of course, whenever it is made.

REFERENCES

1. J. A. Stratton, *Electromagnetic Theory*, McGraw-Hill, New York, 1941.
2. W. M. Latimer, K. S. Pitzer, and C. M. Slansky, *J. Chem. Phys.*, **7**, 108 (1939).
3. A. A. Rashin and B. Honig, *J. Phys. Chem.*, **89**, 5588 (1985).
4. R. A. Fisher, E. W. Hornung, G. E. Brodale, and W. F. Giauque, *J. Chem. Phys.*, **58**, 5584 (1973).
5. R. P. Hudson, *Principles and Applications of Magnetic Cooling*, North-Holland, Amsterdam, 1972.
6. O. V. Lounasmaa, *Experimental Principles and Methods Below 1 K*, Academic Press, London and New York, 1974.
7. P. Hakonen, O. V. Lounasmaa, and A. Oja, *J. Magnetism and Magnetic Mater.*, **100**, 394 (1991).

8. J. Wyman and S. J. Gill, *Binding and Linkage*, University Science Books, Mill Valley, CA, 1990, chapter 8.
9. R. A. Alberty, *Biophys. Chem.*, **43**, 239 (1992).

PROBLEM

24-1. Consider the centrifugal separation of uranium isotopes as UF_6, assumed to be a perfect gas at 100°C. What angular velocity is necessary to obtain enrichment of U^{235} to 1 percent at $r = 3$ cm if its abundance is 0.7 percent at $r = 10$ cm?

CHAPTER 25

IRREVERSIBLE PROCESSES NEAR EQUILIBRIUM; NONISOTHERMAL SYSTEMS; STEADY STATES

Thermodynamics generally deals with systems that are at equilibrium with respect to some processes but not with respect to others. Ideally the forbidden processes have zero rates, but actually it suffices that their rates be very small compared with the rates of allowed processes. Then the observed properties can still be compared with the theoretical result for the ideal case.

The velocity of physical or chemical processes is not treated by thermodynamics, although the statistical theories of rate processes draw much background information from thermodynamics. Also, near equilibrium the various rate coefficients must satisfy certain restricting relationships in order that the thermodynamic equations will be fulfilled at the equilibrium state. These equations between rate coefficients then constitute a thermodynamics of near-equilibrium irreversible processes. A steady state is obtained when the properties of the system itself do not change with time but there is an irreversible flow of heat, electricity, or some substance through the system. In general, steady states are treated in terms of the theories of the rates of the various processes involved. However, when this irreversible flow causes only a small deviation from true equilibrium, thermodynamic methods are still useful.

Nonisothermal systems present many examples of steady states which are subject to thermodynamic treatment. Perfect heat insulators are not available, and hence heat flows through a system at a finite rate; but by making the temperature gradient small, one can always obtain a steady state that is only slightly shifted from the isothermal equilibrium state.

The classical theory of thermodynamics as developed by Carnot, Clausius, Kelvin, and others in the middle of the nineteenth century was clearly intended to apply to nonisothermal systems such as heat engines. The theory applies to the reversible changes that occur to the working fluid; simultaneous dissipative processes such as friction and nonisothermal heat conduction are regarded as independent and treated separately. In his classic paper "On the Dynamical Theory of Heat," Kelvin[1] assumes this separation of processes to be obvious in ordinary heat engines, where he simply specifies a "perfect engine," but he discusses the separation at length for thermoelectric phenomena as a separate postulate. Thermodynamics does not predict the absolute rate of the flow process; rather it yields relations between measurable quantities such as the thermoelectric potential and the Peltier heat effect, which occurs when electricity flows from one metal to another.

If the two or more processes can be clearly separated into noninteracting mechanisms of change and the steady irreversible flow affects only one, then a straightforward equilibrium treatment can be given for the others. For example, a slight leakage of the solute through a semipermeable membrane will not affect an osmotic-pressure experiment provided that the solution composition on each side is maintained constant by external means. The thermodynamic treatment of such isothermal systems is now so familiar that we shall devote no further attention to such examples. But in nonisothermal cases certain new features arise, and an example that illustrates these is the subject of the next section. There are, in addition, methods that avoid the identification of noninteracting processes, which will be presented later. In this chapter several examples of steady-state processes are presented, but there are others; Haase[2] provides a comprehensive treatment of this area as of 1969.

THERMOMOLECULAR PRESSURE, THERMOOSMOSIS[3]

Take a vessel divided into two compartments by a barrier resistant to heat flow but permeable to the fluid substance, gas or liquid, that is contained in the vessel. The barrier may be a capillary, a porous plate, or a material dissolving the fluid. A temperature difference is established by steadily removing heat from the cooler compartment of temperature T and adding heat at an equal rate to the warmer compartment at $T + dT$. It is assumed that the heat conductance of the barrier, including the fluid in it, is so small that each compartment remains at constant temperature and that the temperature gradient is within the barrier. The ordinary heat leakage through the barrier by the thermal conductance of the materials will have no direct interaction with the passage of fluid through the barrier. In general this system will develop a steady-state pressure difference dP that will depend not only upon the fluid but also upon the nature of the barrier. In gaseous systems this is usually termed the thermomolecular pressure difference, whereas with liquids the term thermoosmosis is more common.

Consider the transfer of a small amount of fluid from the cooler to the hotter compartment. In addition to the enthalpy of the substance itself, which increases by dH per mole, there is usually a heat effect when the fluid enters the surface of the barrier and passes into the interior. This may be thought of as primarily a heat of solution, and, of course, the reverse heat effect occurs when the fluid comes out of the barrier into the other compartment. Thus heat is transferred with the substance; and it should be emphasized that thermodynamics is not concerned with the detailed source, heat of solution, etc., but only with the total heat transferred per mole of substance. This quantity is called the *heat of transport*† and written as Q^*; this and all similar quantities are on a molar basis in this chapter and the subscript m is omitted.

The criterion for the steady-state pressure difference is just that for a reversible equilibrium: zero change in entropy for the transfer of the fluid. Per mole of fluid transferred, the change in entropy of the fluid itself may be written

$$dS_1 = \left(\frac{\partial S}{\partial T}\right)_P dT + \left(\frac{\partial S}{\partial P}\right)_T dP$$

while the heat effects defined in the previous paragraph yield the additional entropy changes

$$dS_2 = -\frac{dH}{T} - \frac{Q^*}{T} + \frac{Q^*}{T + dT} = -\left(\frac{\partial H}{\partial T}\right)_P \frac{dT}{T} - \left(\frac{\partial H}{\partial P}\right)_T \frac{dP}{T} - \frac{Q^* dT}{T^2}$$

Upon addition of $dS_1 + dS_2$, we note that the first terms cancel one another and the next two combine into a derivative of the Gibbs energy. The result is

$$dS = -\left(\frac{\partial G}{\partial P}\right)_T \frac{dP}{T} - \frac{Q^* dT}{T^2} = 0 \qquad (25\text{-}1)$$

Substitution of V for $\partial G/\partial P$ and rearrangement yields

$$\frac{dP}{dT} = -\frac{Q^*}{VT} \qquad (25\text{-}2)$$

Thermodynamics thus yields an equation relating the thermal pressure difference to the heat of transport but, of course, yields no prediction about the magnitude of the effect. While the pressure difference in thermoosmosis, etc., has been observed by various workers, there are as yet no sufficiently accurate measurements of the heat of transport to constitute any check on Eq. (25-2). The calculated heats of transport for thermoosmosis of gases through a rubber membrane do, however, approximate the heats of solution of those gases in rubber.[4]

---

† Some authors use the term *reduced heat of transport* to distinguish Q^* from the total enthalpy that moves with the substance, $Q^* + H$. We shall always write Q^* and H separately.

The kinetic theory of ideal gases predicts,[5] for capillary passages that are small compared with the mean free path of the gas, that the steady-state pressure is proportional to the square root of the temperature, or

$$\frac{d\ln P}{d\ln T} = \frac{1}{2} \qquad (25\text{-}3)$$

This requires $Q^* = -\frac{1}{2}RT$. As the size of the tube increases, this effect decreases, and for an open tube of diameter large compared with the mean free path $Q^* = 0$ and no thermal pressure differences occur.

GENERAL THEORY OF NEAR-EQUILIBRIUM PROCESSES

In the preceding section we considered an example of a near-equilibrium system in which one irreversible steady-state process of heat flow was readily distinguished from another process of matter flow at equilibrium. The latter process was affected, however, by the temperature difference between the two portions of the system. Ordinarily, steady-state systems of this sort may be treated by similar methods. In nonisothermal systems there is always some pure heat conductance in addition to whatever other processes exist that transfer some component of matter (or electricity) along with its heat of transport. Nevertheless, in some cases the selection of the independent processes may be ambiguous or inconvenient, and a more general theory is useful.

It is also necessary to consider whether or not the usual thermodynamic variables and properties are still valid when irreversible processes are proceeding at finite rates. One may picture the system as being subdivided into various portions and that suddenly all transfer of heat, matter, etc., between subdivisions is stopped. Each portion of the system will now come to equilibrium, and its temperature, pressure, entropy, etc., will be unambiguous. If it is possible to make these portions large enough to contain many molecules (and hence to have macroscopic properties) and yet small enough so that the original gradients within a given portion were small, then the finally measured properties may be assigned to the original system. There may also be regions of high gradient that contain a negligible portion of the total matter, energy, etc.

Consideration of actual methods of measurement leads to similar conditions. For example, a thermometer must have thermal contact with a macroscopic amount of matter within the system. The resulting temperature measurement will be unambiguous only if there are no significant temperature differences within the matter that is in thermal contact with the thermometer.

Prigogine[6] has considered these questions from the viewpoint of kinetic theory and statistical mechanics. He concludes that the domain of validity of thermodynamic variables when transport processes are occurring is the range of linear-rate laws. Chemical reactions must be sufficiently slow so that the Maxwellian distribution of molecular velocities is not significantly disturbed. These conditions are clearly consistent with those developed from the other points of view.

Entropy Production in Heat Flow

In earlier discussions we have considered the possibility of entropy production in irreversible processes. Let us now write down exact expressions for the entropy production in a few common situations. Take a system comprising two regions at different temperatures T_1 and T_2 but each isothermal. The entropy increase in region 1 is $dS_1 = \delta q_1/T_1$ and for region 2 is $dS_2 = \delta q_2/T_2$. If heat flows from region 1 to region 2, $\delta q_1 = -\delta q_2$ and

$$dS_{irr} = \delta q_2\left(\frac{1}{T_2} - \frac{1}{T_1}\right) = \delta q_2 \frac{T_1 - T_2}{T_1 T_2} \tag{25-4}$$

The irreversible entropy increase† dS_{irr} must, of course, be positive; hence $T_2 < T_1$ in agreement with the well-known direction of spontaneous heat flow. If the temperature difference is infinitesimal, Eq. (25-4) becomes

$$dS_{irr} = \delta q\, d\left(\frac{1}{T}\right) \tag{25-5}$$

Entropy Production in Matter Flow

Consider next the flow of δn moles of some component of matter (or of electricity) from one region of temperature T_1 where the chemical potential is μ_1, to a second region where the temperature is T_2 and the chemical potential μ_2. The partial molar entropy $\overline{S} = (\overline{H}/T) - (\mu/T)$, but we must also consider the heat of transport Q^* defined earlier in this chapter. The first law of thermodynamics requires that the sum $\overline{H} + Q^*$ be the same whether considered from region 1 or region 2. The entropy increase for transfer of δn moles is then

$$dS_{irr} = \delta n\left[\left(\frac{\mu_1}{T_1} - \frac{\mu_2}{T_2}\right) + (\overline{H} + Q^*)\left(\frac{1}{T_2} - \frac{1}{T_1}\right)\right] \tag{25-6a}$$

or, for infinitesimal differences,

$$dS_{irr} = \delta n\left[-d\left(\frac{\mu}{T}\right) + (\overline{H} + Q^*)d\left(\frac{1}{T}\right)\right] \tag{25-6b}$$

We see at once that the second term drops out if $T_2 = T_1$. This explains why the heat of transport never needs to be considered in isothermal problems.

---

† We shall sometimes omit the subscript irr when it seems obvious that only entropy increase and no entropy transfer is involved.

If there is but a single component and pressure is the only external force, the change of the ratio μ/T may be expressed as

$$d\left(\frac{\mu}{T}\right) = \frac{V}{T}dP + H\,d\left(\frac{1}{T}\right)$$

and Eq. (25-6) becomes

$$dS_{\text{irr}} = \delta n\left[-\frac{V}{T}dP + Q^*d\left(\frac{1}{T}\right)\right] \tag{25-7}$$

which is equivalent to the result derived as Eq. (25-1)†.

Entropy Production from Electric Current

An electric current may always be considered as a flow of one or more kinds of charged matter, with the term $z_i F\,d\mathcal{V}$ included in the change of chemical potential $d\mu_i$. Here z_i is the charge in protonic units, F is the Faraday constant, and $\mathcal{V}$ the electrical potential.‡ Electrons and other charged particles commonly have a heat of transport; also other factors than electrical potential may affect the chemical potential of the charged species.

In the special case of an electric current in an isothermal system, one may write

$$dS_{\text{irr}} = \frac{\delta n F}{T}d\mathcal{V} = \frac{I\,\delta t}{T}d\mathcal{V} \tag{25-8}$$

where I is the current and δt is the duration of flow.

Rate of Entropy Production

We now wish to discuss the rate of flow of heat, of some component of matter, or of electricity. It is customary to adopt the symbol J_i for ith flux, which may be calories of heat per second, moles of matter per second, etc. In each of the examples of entropy production, the resulting expression was the product of an extensive quantity δq or δn times some sort of potential difference. The latter may be regarded as the driving force for the flow in question. The symbol X_i is customary for the potential difference associated with the ith flux. The total rate of entropy production becomes

$$\frac{dS}{dt} = \dot{S} = \sum_i J_i X_i \tag{25-9}$$

with X_i defined as the coefficient of J_i in the expression for $\dot{S}$.

---

† The material must remain in the same state, gas, liquid, or solid; otherwise the finite difference $\overline{H}_2 - \overline{H}_1$ enters the result.

‡ We use the script $\mathcal{V}$ to distinguish potential from volume.

It is frequently much easier to write out Eq. (25-9) for a given system in terms of simple flow quantities, heat, components of matter, etc., than it is to analyze the actual flow processes. Thus in the case of two regions with flow of heat and of a single component of matter we may write at once

$$\dot{S} = J_h \Delta\left(\frac{1}{T}\right) + J_m\left(-\frac{V}{T}\Delta P\right) \tag{25-10}$$

where J_m is the flow of matter and J_h is the total flow of heat whether associated with the matter as a heat of transport or not. In view of our analysis above, the total heat flow J_h may be divided into that arising from heat of transport $J_m Q^*$ and a remaining pure heat flow J_h'. In these terms Eq. (25-10) may be rewritten

$$\dot{S} = J_h' \Delta\left(\frac{1}{T}\right) + J_m\left[Q^*\Delta\left(\frac{1}{T}\right) - \frac{V}{T}\Delta P\right] \tag{25-11}$$

where the first term is clearly derivable from Eq. (25-5) and the second from Eq. (25-7).

Rate Equations

Let us now postulate a reasonable form for the rate equations governing the fluxes J_i in the region near equilibrium. The rates must become zero when all potential differences X_i are zero, and the fluxes certainly reverse direction if all the potential differences change sign. Thus it is reasonable to write for the Js power-series expansions in the Xs and to retain only the first-power terms for the near-equilibrium case. If we have identified truly independent flow processes, each flux should be governed by its own potential difference and be independent of other potentials. Thus

$$J_i' = L_i' X_i' \tag{25-12}$$

where L_i' is a rate constant. If, however, we have written down fluxes and their associated potentials without regard to actual processes, as in Eq. (25-10), we may expect cross terms to arise. Since Eq. (25-9) for the entropy production and the accompanying definition of the potential differences must be valid for any set of fluxes and associated potentials, a relationship may be derived between off-diagonal rate constants. Let the fluxes and potentials J_i' and X_i' pertain to the independent processes with rate equations (25-12), while some other definition of fluxes J_i is given by the transformation

$$J_i = \sum_j \alpha_{ij} J_j' \tag{25-13}$$

Then

$$\dot{S} = \sum_i \sum_j \alpha_{ij} J_j' X_i \tag{25-14}$$

and X_j', the coefficient of J', is

$$X_j' = \sum_i \alpha_{ij} X_i \tag{25-15}$$

Substitution of these results and the rate equations for the independent processes [Eq. (25-12)] into Eq. (25-13) yields

$$J_i = \sum_j \alpha_{ij} L'_j X'_j$$

$$= \sum_j \sum_k \alpha_{ij} L'_j \alpha_{kj} X_k$$

If new rate constants are defined by

$$L_{ik} = \sum_j \alpha_{ij} L'_j \alpha_{kj} \qquad (25\text{-}16)$$

then

$$J_i = \sum_k L_{ik} X_k \qquad (25\text{-}17)$$

and it is apparent from Eq. (25-16) that

$$L_{ik} = L_{ki} \qquad (25\text{-}18)$$

This last equation is commonly called the Onsager relation.† Onsager[7] derived it from the postulate of microscopic reversibility, using statistical arguments that we shall not repeat here. Other derivations have been given.[8,9] It is a matter of personal opinion which postulates constitute the most plausible basis for Eq. (25-18). In some cases the independence of the several separate processes is clear-cut, and the derivation used here seems most straightforward. In other cases, such as simultaneous fluxes of heat and electricity in metals, the separation is less evident, and other postulates may seem preferable. In any case, the Onsager relation is believed to be generally true, and the agreement with experiment of results derived from these equations has been good.

The second law of thermodynamics requires that the entropy production $\dot{S}$ be positive (or zero). This places a limit on the magnitude of the off-diagonal rate constants $L_{ij}(i \neq j)$. For example, with two fluxes,

$$\dot{S} = L_{11} X_1^2 + (L_{12} + L_{21}) X_1 X_2 + L_{22} X_2^2 \geq 0$$

regardless of the sign and magnitude of either X_1 or X_2. This requires $L_{11} > 0$, $L_{22} > 0$, and $(L_{12} + L_{21})^2 < 4 L_{11} L_{22}$. In general the quadratic form must be positive definite. While this restriction is unquestionably valid, it does not ordinarily yield any useful information in practical cases.

Before proceeding to other systems, let us see how the Onsager equation may be used to obtain Eq. (25-2) for thermoosmosis. The problem is formulated in terms of a flux of matter J_m and a flux of heat (other than the heat content of the matter) J_h. The

---

† In the presence of a magnetic field **H** the statistical mechanical theory yields a modification of Eq. (25-18):

$$L_{ik}(\mathbf{H}) = L_{ki}(-\mathbf{H}) \qquad (25\text{-}18a)$$

entropy production is given in these terms by Eq. (25-10), which serves also to define the potentials

$$X_m = -\frac{V}{T}\Delta P \tag{25-19a}$$

$$X_h = \Delta\left(\frac{1}{T}\right) \tag{25-19b}$$

The rate equations are

$$J_m = L_{mm}X_m + L_{mh}X_h \tag{25-20a}$$

$$J_h = L_{hm}X_m + L_{hh}X_h \tag{25-20b}$$

The steady state in thermoosmosis is given by $J_m = 0$, whereby only a heat flow remains, as required to maintain the temperature difference. Therefore

$$X_m = -X_h\frac{L_{mh}}{L_{mm}} = -X_h\frac{L_{hm}}{L_{mm}} \tag{25-21}$$

where the second equality follows because $L_{mh} = L_{hm}$ by Eq. (25-18).

In order to identify the ratio L_{hm}/L_{mm} with some physical quantity, consider the processes caused by a pressure difference in an isothermal system ($X_h = 0$). The heat transferred per mole of matter flowing is the heat of transport

$$Q^* = \frac{J_h}{J_m} = \frac{L_{hm}}{L_{mm}} \tag{25-22}$$

Substitution of this result in Eq. (25-21) yields

$$X_m = -Q^*X_h$$

or

$$-\frac{V}{T}\Delta P = -Q^*\Delta\left(\frac{1}{T}\right) \tag{25-23}$$

This result is readily seen to be equivalent to Eq. (25-2) obtained before. Also, this result could have been obtained at once by setting dS_{irr} to zero in Eq. (25-7), in other words, by requiring no irreversible entropy production for infinitesimal transfer of matter.

ELECTROKINETIC EFFECTS

The general theory of near-equilibrium processes forms a convenient basis for deriving the equations relating various electrokinetic phenomena. These systems are isothermal, with two compartments separated by a porous membrane. There are two possible irreversible fluxes: the volumetric flow of liquid J and electric current I. The entropy production is readily obtained from Eq. (25-7) with $d(1/T) = 0$ and Eq. (25-8):

$$\dot{S} = J\frac{\Delta P}{T} + I\frac{\Delta \mathcal{V}}{T} \tag{25-24}$$

and the potential gradients are self-evident. The linear-rate equations are

$$J = L_{11}\frac{\Delta P}{T} + L_{12}\frac{\Delta \mathcal{V}}{T} \tag{25-25a}$$

$$I = L_{21}\frac{\Delta P}{T} + L_{22}\frac{\Delta \mathcal{V}}{T} \tag{25-25b}$$

with the Onsager relationship $L_{12} = L_{21}$.

The four electrokinetic phenomena are given in terms of these rate constants. The *streaming potential* is the potential difference per unit pressure difference with zero electrical current:

$$\left(\frac{\Delta \mathcal{V}}{\Delta P}\right)_{I=0} = -\frac{L_{21}}{L_{22}} \tag{25-26}$$

The flow of liquid at zero pressure caused by unit electrical current is called *electroosmosis* and is given by

$$\left(\frac{J}{I}\right)_{\Delta P = 0} = \frac{L_{12}}{L_{22}} \tag{25-27}$$

The *electroosmotic pressure* is the pressure difference per unit potential difference at zero flow of matter:

$$\left(\frac{\Delta P}{\Delta \mathcal{V}}\right)_{J=0} = -\frac{L_{12}}{L_{11}} \tag{25-28}$$

The *streaming current* is the electrical current per unit volumetric flow at zero electrical potential difference:

$$\left(\frac{I}{J}\right)_{\Delta \mathcal{V} = 0} = \frac{L_{21}}{L_{11}} \tag{25-29}$$

From the Onsager relationship we see that the first quantity is just the negative of the second and the third is the negative of the fourth:

$$\left(\frac{\Delta \mathcal{V}}{\Delta P}\right)_{I=0} = -\left(\frac{J}{I}\right)_{\Delta P = 0} \tag{25-30}$$

$$\left(\frac{\Delta P}{\Delta \mathcal{V}}\right)_{J=0} = -\left(\frac{I}{J}\right)_{\Delta \mathcal{V} = 0} \tag{25-31}$$

The first of these, known as Saxen's relation, was first derived from kinetic considerations.

This derivation shows the convenience and elegance of the general theory for relating rate phenomena within the linear region to the ratios of potential gradients under steady-state conditions.

A two-component system in a temperature gradient yields a situation much like that of thermoosmosis. A concentration gradient develops in the steady state. This thermal-diffusion phenomenon is also called the Soret effect when the system is a liquid system. If no pressure gradient is present, the solvent must move as required to compensate for the volume change of solute motion and one has effectively a single-component material flow. The entropy production for transfer of 1 mole of solute from Eq. (25-6b) is

$$dS_{\text{irr}} = -d\left(\frac{\mu_2}{T}\right) + (\overline{H}_2 + Q_2^*)d\left(\frac{1}{T}\right)$$

and expansion of the first term with P, T^{-1}, and composition as independent variables yields

$$dS_{\text{irr}} = -\frac{\overline{V}_2}{T}dP - R\frac{\partial \ln a_2}{\partial \ln x_2}d\ln x_2 + Q_2^*d\left(\frac{1}{T}\right) \qquad (25\text{-}32)$$

In most cases the pressure gradient is negligible and the thermal-diffusion steady state ($dS_{\text{irr}} = 0$) is given by

$$\frac{d\ln x_2}{dT} = -\frac{Q_2^*}{RT^2}\left(\frac{\partial \ln a_2}{\partial \ln x_2}\right)^{-1} \qquad (25\text{-}33)$$

where the final factor is, of course, unity for an ideal solution.

Frequently molality is used as the composition variable, and one is interested in electrolyte solutions. Also the Soret coefficient σ is defined (sometimes with opposite sign), and one finds from Eq. (25-32)

$$\sigma = \frac{d\ln m}{dT} = -\frac{Q_2^*}{RT^2}\left[(\nu_+ + \nu_-)\left(1 + \frac{\partial \ln \gamma_\pm}{\partial \ln m}\right)\right]^{-1} \qquad (25\text{-}34)$$

where ν_+ and ν_- are the numbers of positive and negative ions, respectively, per mole of electrolyte and $\gamma_\pm$ is the usual mean activity coefficient. The total heat of transport of the solute Q_2^* may be written as the sum over the component ions,

$$Q_2^* = \nu_+ Q_+^* + \nu_- Q_-^* \qquad (25\text{-}35)$$

Heat of Transport in Electrolytes

The Soret coefficient has been measured for many systems, and the corresponding heat-of-transport values have been calculated from Eqs. (25-33) or (25-34). Since electrolytes will be of interest again in connection with thermocells, we present in Table 25-1 the values of Q_2^* for several alkali halides at 25°C as summarized by Agar.[10,11]

TABLE 25-1
Heats of transport of aqueous ions at infinite dilution and 25°C[a]

Ion	Q^*/RT	Ion	Q^*/RT
Na^+	1.44	OH^-	6.80
K^+	1.11	Cl^-	(0)
Mg^{2+}	3.4	Br^-	0.12
Ba^{2+}	4.5	I^-	−0.82
Cd^{2+}	3.5	SO_4^{2-}	3.24
La^{3+}	7.3		

[a] From Agar,[11] where individual sources are cited.

The values[10,11] of the heat of transport for some other 1-1 electrolytes are considerably larger than any given in Table 25-1. For example, Q^* for both HCl and $N(C_2H_5)_4Cl$ is close to 13 kJ mol^{-1} at 25°C and 0.01 mol kg^{-1}.

The available data also show that for 1-1 electrolytes Q^* increases with temperature by about 0.17 kJ mol^{-1} K^{-1} and decreases with increasing concentration approximately linearly in $m^{1/2}$. The slope $(dQ^*/dm^{1/2})$ is approximately -8 kJ kg$^{1/2}$ mol$^{-3/2}$.

There is no fully successful interpretation as yet of the absolute magnitude of these heat-of-transport values for strong electrolytes. Indeed, the peculiar reversal of trend between Cl$^-$ and Br$^-$ in the halide series indicates that no simple interpretation will be possible.

Agar[11] described various theories in his review in 1963, while more recently Agar et al.[12] and Takeyama and Nakashima[13] discuss the status of the theory and report values on a single ion basis.

THERMOELECTRIC EFFECTS

The simplest thermoelectric device is a thermocouple comprising two different metals arranged as shown in Fig. 25-1. We shall also consider somewhat more complex systems (Fig. 25-2) where an electrolyte solution replaces one of the metals. The two electrode–electrolyte junctions now comprise half cells, which are identical except for their temperature and possibly the electrolyte concentrations. Let us first consider the thermodynamics of such units from an overall viewpoint. Later we shall trace the escaping tendency of electrons through a thermocouple and see that it yields the same result.

In each case, we shall take E (or dE), the emf of the cells, as the electrical potential of the wire from the hotter electrode or junction less that of the wire from the colder electrode or junction, that is, $E = \mathcal{V}_2 - \mathcal{V}_1$ in Figs. 25-1 and 25-2. Assume that 1 equiv of positive electricity passes through the system from terminal 1 to 2, that is, through the cold junction first and then the hot junction. Assume also that the rate of transfer, i.e., the current, is so small that resistance heating is negligible. If the system

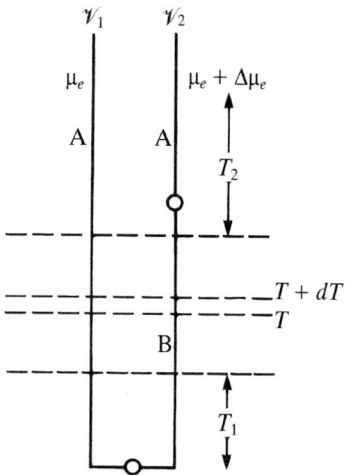

FIGURE 25-1
A thermocouple.

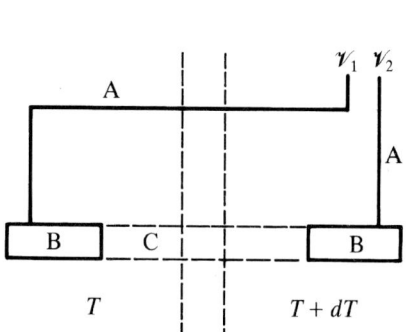

FIGURE 25-2
A thermocell. C is the electrolyte solution connecting the electrodes B, which are in turn connected to wires A.

is at equilibrium, except for spontaneous heat conduction, then no net entropy change occurs, but electrical work FE is transferred to the surroundings.

The entropy change ΔS is absorbed from the surroundings as heat at the higher temperature, $T + dT$, and an equal entropy is emitted to the surroundings at T. In the thermocell system some chemical change may occur at one electrode, but the reverse change occurs at the other electrode, and the net entropy change is an infinitesimal that may be neglected. Now from the first law we see that

$$F\,dE = (T + dT)\Delta S - T\Delta S = \Delta S\,dT$$

$$F\frac{dE}{dT} = \Delta S = \Delta S_j + \Delta S_t \tag{25-36}$$

where we have divided ΔS into two terms to emphasize that it arises from two rather distinct sources. ΔS_j is the entropy change in the junction or at the electrode itself and is an ordinary entropy of chemical reaction involving the molar or partial molar entropies of the species formed or consumed. On the other hand, ΔS_t is the entropy absorbed at the higher temperature because of the heat of transport of the species moving through the temperature gradient.

Metallic Thermocouple

In the thermocouple the heat absorbed on passage of an equivalent of positive electricity from metal B to A (see Fig. 25-1) is called the Peltier heat and is given the symbol $\Pi_{BA} = -\Pi_{AB}$. Thus, $T\Delta S = \Pi_{BA}$, and

$$\frac{dE}{dT} = \frac{\Pi_{BA}}{FT} \tag{25-37}$$

It is helpful to calculate the two terms of Eq. (25-36) separately. The electrode or junction effect is the reaction e^- (in A) = e^- (in B) and in addition

$$\Delta S_j = \bar{S}_B - \bar{S}_A \tag{25-38}$$

where $\bar{S}_A$ and $\bar{S}_B$ are the partial molar entropies of electrons in the metals indicated. The electron flow in wire B carries the heat of transport Q_B^* from the warmer to the colder region, while the electron flow in wire A carries Q_A^* in the opposite direction. Hence the net entropy absorbed at the hot junction from transfer phenomena, ΔS_t, is

$$\Delta S_t = \frac{Q_B^* - Q_A^*}{T} \tag{25-39}$$

The sum of these two effects yields ΔS, or the potential of the thermocouple,

$$F \frac{dE}{dT} = \left(\bar{S} + \frac{Q^*}{T}\right)_B - \left(\bar{S} + \frac{Q^*}{T}\right)_A \tag{25-40}$$

It is interesting to analyze the thermocouple by microscopic consideration of the chemical potential of electrons instead of the thermal effects of the whole system. The measured electrical potential difference times the Faraday constant is just the negative of the difference of chemical potential of electrons in the two wires of A at the top, that is, $FE = -\Delta \mu_e$.

Consider first a small section of one of the metals with an infinitesimal temperature difference from T to $T + dT$. For the transfer of 1 mole of electrons from T to $T + dT$ in metal B the entropy increase from Eq. (25-6) is

$$dS_B = -d\left(\frac{\mu_e}{T}\right)_B + (\bar{H}_B + Q_B^*) d\left(\frac{1}{T}\right) \tag{25-41}$$

where $\bar{H}_B + Q_B^*$ is the sum of the partial molar enthalpy of electrons and the heat of transport of electrons in metal B. We assume here, as before, that the electron flow is so slow that the resistance heating is negligible.

The steady-state situation in a thermocouple with heat flow but no electric current is entirely analogous to that of thermoosmosis. Following any of the methods already described leads to the result that $dS_B = 0$ for electron flow and

$$d\left(\frac{\mu_e}{T}\right)_B = (\bar{H}_B + Q_B^*) d\left(\frac{1}{T}\right) \tag{25-42}$$

or

$$(d\mu_e)_B = -\left(\bar{S}_B + \frac{Q_B^*}{T}\right) dT \tag{25-43}$$

At the junctions μ_e is the same in each metal; hence the thermoelectric potential difference arises purely from the difference in $d\mu_e$ between the two metals in the temperature gradient.

$$F \frac{dE}{dT} = -\frac{d\Delta \mu_e}{dT} = \Delta \bar{S} + \frac{\Delta Q^*}{T} \tag{25-44}$$

or

$$FE = -\Delta\mu_e = \int_{T_1}^{T_2}\left(\Delta\bar{S} + \frac{\Delta Q^*}{T}\right)dT \qquad (25\text{-}45)$$

where the Δ refers to the difference B minus A in each case, and we note that Eqs. (25-44) and (25-40) are equivalent.

This analysis has yielded exactly the same equation as before, but it emphasizes the conclusion that the potential difference arises from the wire in the temperature gradient. Frequently the thermocouple wire is nonuniform in composition or other properties, and it is valuable to know that the portion of the wire in the temperature gradient determines the potential and that variations in the wire in isothermal regions have no effect.

Let us now consider more carefully a small segment of wire B in the temperature gradient. If 1 mole of electrons flows in the direction from temperature T to $T + dT$, the entropy drawn in at T is $\bar{S}_B(T) + T^{-1}Q_B^*(T)$, while that passing out at $T + dT$ is $\bar{S}_B(T + dT) + (T + dT)^{-1}Q_B^*(T + dT)$. The difference is

$$\left(\frac{-\sigma_B}{T}\right)dT = \left(\frac{\bar{C}_B}{T} + \frac{1}{T}\frac{dQ_B^*}{dT} - \frac{Q_B^*}{T^2}\right)dT \qquad (25\text{-}46)$$

where $\bar{C}_B$ is the partial molar heat capacity of electrons in metal B. Since no entropy was created or destroyed in this reversible electron flow, this additional entropy $(-\sigma_B/T)\,dT$ must have flowed into the wire as heat from the surroundings. This heat absorbed, $-\sigma_B$, is the Thomson heat, which was first predicted to occur by thermodynamic argument by Thomson (Lord Kelvin) and later found experimentally. The minus sign before σ_B occurs because it was originally defined for a flow of positive electricity.

Differentiation of Eq. (25-44) with respect to temperature yields

$$F\frac{d^2E}{dT^2} = \frac{\Delta\bar{C}}{T} + \frac{1}{T}\frac{d\Delta Q^*}{dT} - \frac{\Delta Q^*}{T^2}$$

which is readily combined with Eq. (25-46) to obtain

$$\frac{d^2E}{dT^2} = \frac{-\Delta\sigma}{FT} \qquad (25\text{-}47)$$

The basic phenomenological equations of thermoelectricity are Eqs. (25-37) and (25-47), which relate the electrical potential difference to the Peltier and Thomson heats. These relationships could have been derived without separate reference to the thermal properties of the electrons in metals A and B and their heats of transport. We have presented the derivation in this more complex form to emphasize the relationships with the corresponding quantities in thermoosmosis and in electrochemical thermocells, to which we return in the next section. Before proceeding, however, we note that a thermoelectric potential difference would arise if the heats of transport were zero, provided that the partial molar heat capacities and entropies of electrons were different in the two metals. But, likewise, the effect might arise purely from differences in heat of transport. Apportionment of the experimental Peltier and Thomson heats between the differences in equilibrium thermal properties and the differences in heats of transport is possible only if one or the other can be determined separately. While the partial molar entropies, etc., of electrons are measurable in principle, the enormous magnitude of space-charge energies prevents practical measurement of such properties of charged species. Thus at present quantum-statistical theory of electron motion in metals provides the only source of information about the division between the equilibrium partial molar quantities and the heat of transport.

The quantity $\bar{S}_i + Q_i^*/T$ for electrons in a given metal is measurable, however, and we shall see that it is also measurable for individual ions in solution. Because of the importance of this quantity it is frequently given a symbol and name, and the literature contains some inconsistent or confusing selections. We shall follow Haase[2] in the symbol $^*S$ with the preceding superscript and in terminology with *transported entropy*. (Agar uses $\bar{\bar{S}}$ instead of $^*S$.) Thus for the ith species

$$^*S_i = \bar{S}_i + \frac{Q_i^*}{T}$$

It is to be noted that the *transported entropy* $^*S_i$ is the sum of the partial molar entropy $\bar{S}_i$ and the entropy associated with the heat of transport Q_i^*/T. Since there is still risk of confusion, the sum will be written explicitly in many cases.

From Eq. (25-46) one can readily deduce that for electrons in a metal

$$^*S_{T'} = \left(\bar{S} + \frac{Q^*}{T}\right)_{\text{at } T'} = \int_0^{T'} \left(-\frac{\sigma}{T}\right) dT \qquad (25\text{-}48)$$

since $\bar{S}$ and Q^*/T are zero at $0\,\text{K}$ by the third law of thermodynamics. If the total entropy of a substance is zero and negative entropies are impossible, then the partial molar entropies of components must be zero and motion of such components cannot transfer entropy. Data on metals at low temperatures are consistent with this interpretation.[14]

The individual Thomson coefficients are obtained from measurements of the change in temperature gradient in wires with the amount of current flowing. Temkin and Khoroshin[15] have integrated Eq. (28-48) for electrons in copper and platinum and find that at 298 K:

Pt $\qquad\qquad ^*S = \left(\bar{S} + \dfrac{Q^*}{T}\right) = 0.052R$

Cu $\qquad\qquad ^*S = \left(\bar{S} + \dfrac{Q^*}{T}\right) = -0.023R$

Thermocells

An electrochemical cell with a single uniform electrolyte and two identical electrodes yields zero emf, of course, if it is isothermal. But if one half-cell is at a different temperature from the other, a potential difference may develop, just as it does in a metallic thermocouple. Such systems are known as thermocells and were first studied by Gockel[16] in 1885.

The electrolyte connecting the two half-cells is subject to thermal diffusion (Soret effect), and if this is allowed to take place the two half-cells will differ in electrolyte composition as well as in temperature. It is possible, however, to make the electrical measurements before significant thermal diffusion occurs, and this yields an initial emf, E_{in}. After the thermal-diffusion steady state is established, a different emf, E_{st}, may be measured.

To consider the thermodynamics of a thermocell, we return to our first analysis of thermoelectric systems and note Eq. (25-36),

$$F \frac{dE}{dT} = \Delta S_j + \Delta S_t \qquad (25\text{-}36)$$

where ΔS_j is the entropy change of the electrode reaction of a thermocell and ΔS_t is the entropy absorbed from the hot region because of heat-of-transport effects. The former quantity is the usual ΔS of the half-cell reaction for the passage of 1 equiv in the direction of positive electricity from electrolyte to electrode.

The heat-of-transport effects become more complex in an electrolyte solution than in a metallic wire. If we assume uniform composition, i.e., neglect thermal diffusion, then various ions of charge z_i and molal heat of transport Q_i^* carry portions of the current indicated by their transference numbers t_j. The net heat of transport for 1 equiv of positive electricity is

$$Q_{in}^* = \sum_i \frac{t_i Q_i^*}{z_i} \qquad (25\text{-}49)$$

where the subscript "in" indicates that this is the initial state before thermal diffusion.

If instead we assume that thermal diffusion is rapid compared with the electrical flow in the cell, then the only ion transferred through the temperature gradient in the electrolyte is the ion (j) that enters the electrode reaction. Hence

$$Q_{st}^* = \frac{Q_j^*}{z_j} \qquad (25\text{-}50)$$

where "st" refers to the stationary state.

In either the initial state or the stationary state the total entropy absorbed at the hot electrode by transport effects includes also the electrons in the wire and is

$$\Delta S_t = -\frac{1}{T} [Q_{e^-}^* + Q_{ion}^*] \qquad (25\text{-}51)$$

where Q_{ion}^* is either Q_{in}^* or Q_{st}^* as appropriate and $Q_{e^-}^*$ is the heat of transport of the electron in the metal wire passing through the temperature gradient.

This completes our general analysis of a thermocell. Let us turn to examples with Ag–AgBr electrodes and M^+Br^- aqueous electrolytes.[17] The electrode reaction in the direction defined is

$$AgBr + e^-(\text{in wire}) = Ag + Br^-(aq)$$

$$\Delta S_j = S_{Ag} + \overline{S}_{Br^-} - S_{AgBr} - \overline{S}_{e^-} \quad (25\text{-}52)$$

The entropy from heat of transport in the initial state of uniform composition is

$$\Delta S_t = \frac{t_- Q^*_{Br^-} - t_+ Q^*_{M^+} - Q^*_{e^-}}{T} \quad (25\text{-}53)$$

and the initial cell emf is

$$F\left(\frac{dE}{dT}\right)_{in} = S_{Ag} - S_{AgBr} + \overline{S}_{Br} + \frac{t_- Q^*_{Br^-}}{T} - \frac{t_+ Q^*_{M^+}}{T} - \overline{S}_{e^-} - \frac{Q^*_{e^-}}{T} \quad (25\text{-}54)$$

In the final stationary state the result is somewhat less complex,

$$F\left(\frac{dE}{dT}\right)_{st} = S_{Ag} - S_{AgBr} + \left(\overline{S}_{Br^-} + \frac{Q^*_{Br^-}}{T}\right) - \left(\overline{S}_{e^-} + \frac{Q^*_{e^-}}{T}\right) \quad (25\text{-}55)$$

Note that the terms in parentheses in Eq. (25-55) for Br^- and e^- comprise the transported entropy $^*S_i$ in each case. Thus the stationary-state result is more nearly analogous to that for the simple thermocouple than is the case for the initial state. Also, since all other quantities are measurable independently, $^*S_{Br^-}$ may be determined from Eq. (25-55).

The difference between the initial and stationary-state emf values yields the heat of transport of the electrolyte. For the Ag–AgBr cell the result is

$$F\frac{d}{dT}(E_{st} - E_{in}) = \frac{t_+}{T}(Q^*_{M^+} + Q^*_{Br^-}) \quad (25\text{-}56)$$

One can readily see from Eqs. (25-49) and (25-50) that this is a general result.

Agar and Breck[18] measured both E_{in} and E_{st} for cells with thallium amalgam (5 percent) electrodes and various thallous salts (Tl^+X^-) as electrolytes. The equations for the emf are

$$F\left(\frac{dE}{dT}\right)_{in} = \overline{S}_{Tl} - \overline{S}_{Tl^+} - \frac{t_+ Q^*_{Tl^+}}{T} + \frac{t_- Q^*_{X^-}}{T} - {^*S}_{e^-} \quad (25\text{-}57)$$

$$F\left(\frac{dE}{dT}\right)_{st} = \overline{S}_{Tl} - \left(\overline{S}_{Tl^+} + \frac{Q^*_{Tl^+}}{T}\right) - {^*S}_{e^-} \quad (25\text{-}58)$$

The second equation shows that E_{st} should be independent of the anion X^- so long as $^*S_{Tl^+}$ for the thallous ion is unaffected. The comparison was made at $0.10\,\text{mol kg}^{-1}$, which is hardly dilute enough to make interionic effects negligible, but the values of $^*S_{Tl^+}$ in Table 25-2 show no significant differences. The partial molar entropy of

TABLE 25-2
Thermocell results[18] for thallous salts; mean temperature 25.2 ± 0.3°C and 0.100 ± 0.001 molal Tl$^+$

Salt	$^*S_{Tl^+}/R$	Q^*_{salt}/R	t_{x^-}
TlClO$_4$	17.0$_6$	0.86	0.467
TlNO$_3$	17.1$_1$	0.84	0.481
TlOAc	17.1$_1$	3.10	0.344
Tl$_2$CO$_3$	17.0$_6$	4.30	0.48
Tl$_2$SO$_4$	17.1$_1$	4.06	0.532

thallium in the amalgams is obtained by standard methods. Q^* in Table 25-2 is the total heat of transport of the entire salt, that is, $Q^*_{Tl^+} + Q^*_{X^-}$ for the 1–1 salts and $2Q^*_{Tl^+} + Q^*_{X^-}$ for the 2–1 salts, respectively. An interesting recent thermocell investigation is that of Payton et al.[19] for NaOH(aq) with hydrogen electrodes.

Thermocells may have a fused salt, or even an ion-conducting solid salt, in place of the aqueous solution as the electrolyte. In that case, there is no Soret effect; also, with no solvent, there is no transference number referenced to a solvent. One can define a transference number for a solid; indeed our interest is in solids such as AgI, which is shown to conduct by motion of the Ag$^+$ ion.

Just as for the aqueous solution in the steady-state mode, the ion moving in the thermal gradient is the ion generated at one electrode and consumed at the other electrode. We confine our attention to cells with pure metal electrodes and simple MX$_\nu$ electrolytes where X is a halogen or nitrate. Now the electrode reaction is just $\nu e^- + M^{\nu+} = M$ and the entropy change per equivalent is

$$\Delta S_j = \frac{(S_M - \overline{S}_{M^{\nu+}})}{\nu} - \overline{S}_{e^-} \tag{25-59}$$

The entropy from heat of transport is

$$\Delta S_t = -\frac{(Q^*_{e^-} + Q^*_{M^{\nu+}}/\nu)}{T} \tag{25-60}$$

Insertion of these two terms in Eq. (25-36) yields

$$F\frac{dE}{dT} = \frac{1}{\nu}(S_M - {^*S}_{M^{\nu+}}) - {^*S}_{e^-} \tag{25-61}$$

where $\overline{S}$ and Q^*/T for the ion have been combined to yield $^*S$, the transported entropy. Table 25-3 shows the results assembled from several sources for fused-salt thermocells.[20]

Using the results of Table 25-3, Pitzer[20] made a comparison with the values of the absolute entropy of an ion in a fused salt calculated from exact theory for mass effects and reasonable approximations for other aspects. It was found that the values of these ion entropies and the observed values of the transported entropy were equal

TABLE 25-3
Thermocells with fused-salt electrolytes

Salt	T/K	$\dfrac{S_M}{R}$	$-\dfrac{\nu F}{R}\dfrac{dE}{dT}$	$\dfrac{^*S_{M^{\nu+}}}{R}$
AgCl	800	8.27	4.7	13.0
AgBr	750	8.05	5.5	13.6
AgI	850	8.47	5.0	13.5
$AgNO_3$	500	6.73	3.8	10.5
$ZnCl_2$	~600	7.25	−3	4
$SnCl_2$	~600	10.36	+0.5	11

within the various uncertainties, which leads to the conclusion that the heat of transport is small in a fused salt.

Experimental values for thermocells with solid electrolytes were also assembled by Pitzer[20] and discussed in terms of theory and the probable structure of a solid showing ionic conduction.

ADDITIONAL APPLICATIONS

This chapter does not pretend to be a complete account of the applications of the Onsager equations. Haase[2] gives a comprehensive treatment as of 1969. We shall, however, describe briefly a few additional applications without derivations or equations. One interesting recent application concerns the rate of transfer of carbon dioxide between the atmosphere and the ocean when there is a temperature gradient at the surface.[21] With gradients in two properties, the off-diagonal term must be considered.

Isothermal diffusion in solutions of various types and complexities is another interesting area. Isothermal diffusion in a two-component system is described by a single diffusion constant. This is the rate constant for the single process occurring and no near-equilibrium steady state is possible. For a molecular fluid no Onsager relationship appears. For an electrolyte, however, Miller[22] has used the transference number to separate the transport of the cation from that of the anion. He then combines this information with the diffusion coefficient and the conductance to obtain Onsager transport coefficients for the individual ions. The equations relating these quantities involve the chemical potential in addition. The required osmotic or activity coefficients are usually available.

Miller[23] extended this treatment to ternary systems with two salts having a common ion and the solvent. Again there are a set of relationships between the six Onsager coefficients and six experimental quantities: three diffusion coefficients, two transference numbers, and the conductance. Again chemical potential information is required. The derivation of these equations is complex, but Miller has presented a simple summary and review.[24]

For both binaries and ternaries there is no comparison with other independently measurable quantities. But it is interesting to compare the Onsager coefficients for a given ion from different salts and for a given salt in ternaries with the value in its binary. Also, these Onsager coefficients may be more interpretable theoretically than the original experimentally measured quantities.

A recent example of the application of this method to a specific system is that for $ZnCl_2$–H_2O by Miller and Rard.[25] They compare their new results with earlier results for $MgCl_2$–H_2O. Various properties of the two systems are similar at molality below about $0.2\,\mathrm{mol\,kg^{-1}}$ but then deviate rapidly. This behavior is consistent with that observed for the activity and osmotic coefficients and suggests that $ZnCl_2$ is fully ionized in dilute solution but becomes strongly associated above 0.2 molal.

Another example is the ternary $NaCl$–$MgCl_2$–H_2O, for which a series of papers are appearing.[26]

Felmy and Weare[27] have applied the Miller treatment to complex multi-component aqueous electrolytes. Making reasonable assumptions, they develop a semi-empirical theory, evaluate parameters, and calculate diffusion effects of geological importance.

Further Theory

All of the examples above involved only linear gradients and were limited to scalar quantities or systems with axial symmetry. Relationships have been developed where more complex vectorial or tensorial properties arise; these are described by DeGroot and Mazur,[28] and others, where a theorem of Curie applies.

REFERENCES

1. W. Thomson (Lord Kelvin), *Mathematical and Physical Papers*, vol. I, art. XLVIII, University Press, Cambridge, 1882; reprints from articles in *Trans. Roy. Soc. Edinburgh*, 1851–1854.
2. R. Haase, *Thermodynamics of Irreversible Processes*, Addison-Wesley, New York, 1969.
3. E. D. Eastman, *J. Am. Chem. Soc.*, **48**, 1482 (1926); **50**, 283, 292 (1928).
4. R. J. Bearman, *J. Phys. Chem.*, **61**, 708 (1957).
5. E. H. Kennard, *Kinetic Theory of Gases*, McGraw-Hill, New York, 1938, p. 66.
6. I. Prigogine, *Physica*, **15**, 272 (1949); *J. Phys. Chem.*, **55**, 765 (1951).
7. L. Onsager, *Phys. Rev.*, **37**, 405 (1931); **38**, 2265 (1931).
8. R. B. Parlin, R. J. Marcus, and H. Eyring, *Proc. Natl. Acad. Sci. U.S.A.*, **41**, 900 (1955).
9. J. C. M. Li, *J. Chem. Phys.*, **29**, 747 (1958).
10. J. N. Agar, "Thermal Diffusion in Electrolyte Solutions," in *The Structure of Electrolyte Solutions*, W. J. Hamer, ed., Wiley, New York, 1959, p. 200.
11. J. N. Agar, "Thermogalvanic Cells," in *Advances in Electrochemistry and Electrochemical Engineering*, Interscience, New York, 1963, vol. 3, Chapter 2; "Solutions of Electrolytes," in *Annual Reviews of Physical Chemistry*, 1964, vol. 15, pp. 469–488.
12. J. N. Agar, C. Y. Mou, and J.-l. Lin, *J. Phys. Chem.*, **93**, 2079 (1989).
13. N. Takeyama and K. Nakashima, *J. Solution Chem.*, **17**, 305 (1988).
14. It is interesting to note that σ and *S are zero even at finite temperatures for electrons in superconductors. See D. Shoenberg, *Superconductivity*, Cambridge University Press, New York, 1952, p. 86.
15. M. I. Temkin and A. V. Khoroshin, *Zhur. Fiz. Khim.*, **26**, 500 (1952).

16. A. Gockel, *Wied. Ann.*, **24**, 618 (1885).
17. J. C. Goodrich et al., *J. Am. Chem. Soc.*, **72**, 4411 (1950).
18. J. N. Agar and W. G. Breck, *Trans. Faraday Soc.*, **53**, 167 (1957).
19. A. D. Payton, M. J. Brucker, and Yi Zhang, *J. Solution Chem.*, **22**, 995 (1993).
20. K. S. Pitzer, *J. Phys. Chem.*, **65**, 147 (1961).
21. L. F. Phillips, *Geophys. Res. Lett.*, **18**, 1221 (1991); *J. Chem. Soc. Faraday Trans.*, **87**, 2187 (1991).
22. D. G. Miller, *J. Phys. Chem.*, **70**, 2639 (1966).
23. D. G. Miller, *J. Phys. Chem.*, **71**, 616, 3588 (1967).
24. D. G. Miller, *Faraday Disc., Chem. Soc.*, **64**, 295 (1977).
25. D. G. Miller and J. A. Rard, *J. Mol. Liquids*, **52**, 145 (1992).
26. R. Mathew, J. G. Albright, D. G. Miller, and J. A. Rard, *J. Phys. Chem.*, **94**, 6875 (1990), and earlier papers there cited.
27. A. R. Felmy and J. H. Weare, *Geochim. Cosmochim. Acta*, **55**, 113, 133 (1991).
28. S. R. de Groot and P. Mazur, *Non-Equilibrium Thermodynamics*, North-Holland, Amsterdam, 1962.

PROBLEM

25-1. The emf in millivolts for Pt–Pt 10% Rh thermocouples with one junction at 0°C and the other at t°C is given as

t (°C)	0	10	20	30	40
E (mV)	0	0.06	0.11	0.17	0.24

Calculate the Peltier heat at 0°C; be sure to specify the units.

CHAPTER 26

MULTICOMPONENT SOLUTIONS

This chapter completes the treatment of nonelectrolyte solutions with more than two components that was initiated at the end of Chapter 12. Multicomponent electrolytes were considered in Chapter 17. In metallurgy, chemical engineering, and geology, systems of three or more components are very important, and the literature for those fields covers many examples and variations in methodology. We can include here only a summary of certain widely used working equations and a few representative examples.

With the advance of computers, numerical methods with analytical equations have largely replaced the graphical methods used previously.[1] By the use of an analytical expression for the excess Gibbs energy, the various activity coefficients are given as derivatives, and the Gibbs–Duhem equation is automatically satisfied. Equations for the excess enthalpy and entropy are also used in relation to temperature effects. A new journal *CALPHAD* (Computer Coupling of Phase Diagrams and Thermochemistry) was established for this particular field. A special review by Hillert[2] in that journal summarizes the most widely used equations with evaluations and provides additional details.

All of the equations considered below are of the Margules[3] type; some other useful equations are given in Appendix 11. For the molar Gibbs energy one has

$$\frac{g^E}{RT} = \sum_i \sum_j a_{ij} x_i x_j + \sum_i \sum_j \sum_k a_{ijk} x_i x_j x_k + \sum_i \sum_j \sum_k \sum_l a_{ijkl} x_i x_j x_k x_l + \cdots$$

(12-73)

In this chapter we use lower case g for the molar Gibbs energy instead of G_m; also, we continue the practice of Chapter 12 with dimensionless parameters and equations for

g^E/RT rather than equations and parameters with dimension energy. It is optional whether the multiple sums are defined in an unrestricted manner as above with a_{12} and a_{21} both present or with restrictions to eliminate this duplication. If unrestricted, one imposes symmetry so that $a_{12} = a_{21}$, etc., and then the restricted a_{12} is twice the unrestricted a_{12}, the restricted a_{123} is six times the unrestricted a_{123}, etc. Also, the excess Gibbs energy is zero for any pure component, and this condition is conveniently imposed by setting to zero any parameter with all subscripts the same, i.e., $a_{11} = a_{22} = a_{111} = a_{222} = \cdots = 0$.

At the two-suffix level, Eq. (12-73) reduces to a single term for a two-component system. This was given in Chapter 12 with the parameter w:

$$\frac{g^E}{RT} = 2a_{12}x_1x_2 = wx_1x_2 \tag{12-12}*$$

For this case there is no ambiguity, and the extension to three or more components is straightforward as given in Eqs. (12-71)* and (12-72)*.

At the three-suffix level, one obtains

$$\frac{g^E}{RT} = 2a_{12}x_1x_2 + 3a_{112}x_1^2x_2 + 3a_{122}x_1x_2^2 \tag{26-1}*$$

But $x_1 + x_2 = 1$ and the substitution of this equality in different ways yields the following alternate forms:

$$\frac{g^E}{RT} = x_1x_2(2a_{12} + 3a_{112}) + x_1x_2^2(3a_{122} - 3a_{112}) \tag{26-2}*$$

$$= x_1x_2(2a_{12} + 3a_{122}) + x_1^2x_2(3a_{112} - 3a_{122}) \tag{26-3}*$$

$$= x_1x_2[2a_{12} + \tfrac{3}{2}(a_{112} + a_{122})] + x_1x_2(x_1 - x_2)\tfrac{3}{2}(a_{112} - a_{122}) \tag{26-4}*$$

$$= x_1^2x_2(2a_{12} + 3a_{112}) + x_1x_2^2(2a_{12} + 3a_{122}) \tag{26-5}*$$

It is clear from any of these alternates that there is a redundancy among the three a_{12}, a_{112}, a_{122} parameters and that only two are independent. But these alternate expressions are equally valid, and one is free to choose among them for the definition of two independent parameters.

Next we derive the expressions for activity coefficients for each alternative. One rewrites the excess Gibbs energy in terms of numbers of moles, n_1, n_2, etc., for an indefinite amount of material:

$$\frac{G^E}{RT} = \frac{\sum_i\sum_j a_{ij}n_in_j}{\sum_i n_i} + \frac{\sum_i\sum_j\sum_k a_{ijk}n_in_jn_k}{(\sum_i n_i)^2} + \cdots \tag{26-6}$$

with $n_T = \sum_i n_i$. Then the usual derivative is taken,

$$\ln \gamma_m = \left[\frac{\partial(G^E/RT)}{\partial n_m}\right]_{T,P,n_{i \neq m}} \tag{12-11}$$

Since the equations for an indefinite number of components become very cumbersome, we consider next examples with only two components. At the two-suffix level, one obtains from Eq. (12-12) above the results

$$\ln \gamma_1 = w_{12} x_2^2 \qquad (12\text{-}14a)^*$$

$$\ln \gamma_2 = w_{12} x_1^2 \qquad (12\text{-}14b)^*$$

The two-suffix equation often suffices and, even when needed, the contribution at the three-suffix level is often very small. Thus, many investigators prefer to retain a term in $x_1 x_2$. Then one can simply omit either the $x_1^2 x_2$ or the $x_1 x_2^2$ term with the first or second alternative, respectively. An example is the 1961 treatment of Kleppa and Hersh[4] for the heat of mixing of alkali nitrates. This alternative was also chosen for both the solid and liquid solutions of NaCl–KCl.[5] Here the equation used was

$$\frac{g^E}{RT} = x_1 x_2 (wa_{12} + x_1 wb_{12}) \qquad (26\text{-}7)^*$$

with NaCl as 1 and KCl as 2. This follows the second alternate above, Eq. (26-3)*, with $wa_{12} = 2a_{12} + 3a_{122}$ and $wb_{12} = 3a_{112} - 3a_{122}$. The corresponding activity coefficients are

$$\ln \gamma_1 = x_2^2 wa_{12} + 2 x_1 x_2^2 wb_{12} \qquad (26\text{-}8a)^*$$

$$\ln \gamma_2 = x_1^2 wa_{12} + x_1^2 (1 - 2x_2) wb_{12} \qquad (26\text{-}8b)^*$$

While this set of equations is perfectly satisfactory and is widely used, the difference in form between $\ln \gamma_1$ and $\ln \gamma_2$ for the wb_{12} term is somewhat unattractive and one can seek a more symmetrical alternative.

The two-suffix expression is symmetric to 1–2 exchange, while the three-suffix terms allow that symmetry to be broken. This is expressed most clearly by the third alternative, Eq. (26-4)*, with the antisymmetric term in $x_1 x_2 (x_1 - x_2)$. This is just the second term, $B x_1 x_2 (x_1 - x_2)$ in the Redlich–Kister[6,7] equation for a two-component system [see Eq. (12-33)]. In the equations for short-range force effects in electrolytes in the mole-fraction system [Chapter 17, Eq. (17-49)*], this pattern was followed with the definitions

$$\frac{g^E}{RT} = x_1 x_2 [w_{12} + u_{12}(x_1 - x_2)] \qquad (17\text{-}49)^*$$

$$w_{12} = 2a_{12} + \tfrac{3}{2}(a_{112} + a_{122}) \qquad (17\text{-}50a)$$

$$u_{12} = \tfrac{3}{2}(a_{112} - a_{122}) \qquad (17\text{-}50b)$$

Hillert[2] recommends this formulation with the symbols $^0A_{12} = RT w_{12}$ and $^1A_{12} = RT u_{12}$.

For the two-component system the activity coefficients are

$$\ln \gamma_1 = x_2^2[w_{12} + u_{12}(3 - 4x_2)] \qquad (26\text{-}9)^*$$

$$\ln \gamma_2 = x_1^2[w_{12} - u_{12}(3 - 4x_1)] \qquad (26\text{-}10\text{a})^*$$

$$= x_1^2[w_{12} + u_{21}(3 - 4x_1)] \qquad (26\text{-}10\text{b})^*$$

The form is now the same for $\ln \gamma_1$ and $\ln \gamma_2$ except for the change in sign for the u_{12} term. Also, when the relationship $u_{21} = -u_{12}$ is recognized in the final line, the form is exactly the same with the 1–2 interchange.

In either of the systems described above, the w_{12} term is dominant and the additional term is very small in most cases. With the fourth alternative, Eq. (26-5)*, the two parameters are nearly equal and neither can ever be neglected. Rather the two become exactly equal at the two-suffix approximation. The appropriate definitions, as given by Wohl[8,9] are†

$$A_{12} = 2a_{12} + 3a_{112} \qquad A_{21} = 2a_{12} + 3a_{122} \qquad (26\text{-}11)$$

Then, for two components the activity coefficients are

$$\ln \gamma_1 = x_2^2[A_{12} + 2x_1(A_{21} - A_{12})] \qquad (26\text{-}12\text{a})^*$$

$$\ln \gamma_2 = x_1^2[A_{21} + 2x_2(A_{12} - A_{21})] \qquad (26\text{-}12\text{b})^*$$

In this case the form is the same with the 1-for-2 exchange converting γ_1 to γ_2 and vice versa. Also, note that at infinite dilution the second term disappears in each case and the first term gives the limiting activity coefficient. But both A_{12} and A_{21} are relatively large and become equal when the three-suffix terms become negligible. Also there is neither symmetry nor antisymmetry between A_{12} and A_{21}. It is useful to note the relationship between the w_{ij}, u_{ij} and the A_{ij}, A_{ji} quantities:

$$w_{ij} = \tfrac{1}{2}(A_{ij} + A_{ji}) \qquad (26\text{-}13\text{a})$$

$$u_{ij} = \tfrac{1}{2}(A_{ij} - A_{ji}) \qquad (26\text{-}13\text{b})$$

As one moves at the three-suffix level to three or more components, no new redundancies appear, but there is a new type of term, $a_{123}x_1x_2x_3$, in g^E/RT. All other terms in the excess Gibbs energy are of the same type that appeared for two components, but there are more significant changes for the activity coefficients. The first method, which yields different forms for the equations for $\ln \gamma_i$, can be used[5] without difficulty, as is shown in examples below. Further development on a general basis is not useful, however.

---

† Wohl's complete equations[8,9] are more complex, but they are commonly used in the Margules-related form given here.

For the second option with symmetric and antisymmetric terms the results for the general case were given in Chapter 17:

$$\frac{g^E}{RT} = \sum\sum_{j<k} x_j x_k [w_{jk} + u_{jk}(x_j - x_k)] + \sum\sum\sum_{j<k<l} x_j x_k x_l C_{jkl} \quad (17\text{-}51)*$$

with w_{jk} and u_{jk} generalized from Eqs. (17-50a) and (17-50b) to

$$w_{ij} = 2a_{ij} + \tfrac{3}{2}(a_{iij} + a_{ijj}) \quad (17\text{-}51\text{a})$$

$$u_{ij} = \tfrac{3}{2}(a_{iij} - a_{ijj}) \quad (17\text{-}51\text{b})$$

and

$$C_{jkl} = 6a_{jkl} - \tfrac{3}{2}(a_{jjk} + a_{jkk} + a_{jjl} + a_{jll} + a_{kkl} + a_{kll}) \quad (17\text{-}51\text{c})$$

Then

$$\ln \gamma_i = \sum_j{}' x_j[(1-x_i)w_{ij} + (2x_i - 2x_i^2 + 2x_i x_j - x_j)u_{ij}]$$

$$- \sum_{k>j}{}'\sum{}' x_j x_k [w_{jk} + 2(x_j - x_k)u_{jk} - (1 - 2x_i)C_{ijk}]$$

$$- \sum_{l>k>j}{}'\sum{}'\sum{}' 2 x_j x_k x_l C_{jkl}$$

The prime on the summations is a reminder that neither j nor k nor l may equal i in the multiple sums.

For three components, the Wohl formulation remains the same for the A_{ij} terms in relation to a_{ij} and a_{iij} or a_{ijj}, but the a_{123} term appears in some combination with the A_{ij} terms. The exact definition of this triple interaction term varies; hence, care is needed when comparing various investigations. Also, the A_{ij} terms are sometimes defined to be smaller by the factor 1/2.303 to yield $\log \gamma_i$ rather than $\ln \gamma_i$. A particular example is given below. We do not extend consideration of the Wohl formulation further, but note that Mukhopadhayay et al.[10] present an extensive development with various details. The (Redlich–Kister) pattern extends naturally to higher-order terms for a two-component system. With Hillert's symbols, but in dimensionless form, one has

$$\frac{g^E}{RT} = x_1 x_2 \sum_{m=0}^{n} {}^m A_{12}(x_1 - x_2)^m \quad (26\text{-}14)$$

By comparison with Eqs. (12-33)* and (17-49)*, ${}^0A_{12} = A = w_{12}$, ${}^1A_{12} = B = u_{12}$, ${}^2A_{12} = C$, etc.

While this system is generally quite satisfactory, an alternate merits consideration. It is the use of Legendre polynomials wherein a higher-order term is orthogonal to all lower-order terms. This method was discussed in detail by Bale and Pelton.[11] In terms of the parameter $z_{12} = x_1 - x_2$ the expression through fourth order is

$$\frac{g^E}{RT} = x_1 x_2 \left({}^0B_{12} + {}^1B_{12} z_{12} + {}^2B_{12} \frac{(3z_{12}^2 - 1)}{2} + {}^3B_{12} \frac{(5z_{12}^3 - 3z_{12})}{2} \right.$$

$$\left. + {}^4B_{12} \frac{(35z_{12}^4 - 30z_{12}^2 + 3)}{8} + \cdots \right) \quad (26\text{-}15)$$

Hillert[2] discusses the relative merits of Eqs. (26-14 and 26-15) and gives a table relating the $^kA_{12}$ and $^kB_{12}$ coefficients.

There is every reason to evaluate the $a_{123}x_1x_2x_3$ term or its equivalent, if there are accurate experimental data for a ternary system. But there is an extensive literature[12–17] concerning the merits of alternate methods of combining established equations for each of the three binaries to provide an estimate for a ternary without the addition of this term. The effect of these methods is to give different weights to the various binary three-suffix terms, A_{112}, etc., in the ternary. There is no difference, of course, for the two-suffix terms. Three of the methods are symmetrical with respect to the three components; the originators, and the equations for g^E are:

Muggianu[13]

$$g^E = \sum\sum_{i>j} \frac{x_i x_j}{v_{ij} v_{ji}} g^E_{ij}(v_{ij}, v_{ji}) \quad \text{where } v_{ij} = \tfrac{1}{2}(1 + x_i - x_j) \qquad (26\text{-}16)$$

Kohler[14]

$$g^E = \sum\sum_{i>j} (x_i + x_j)^2 \, g^E_{ij}\!\left(\frac{x_i}{x_i + x_j}, \frac{x_j}{x_i + x_j}\right) \qquad (26\text{-}17)$$

Colinet[15]

$$g^E = \tfrac{1}{2}\sum\sum_{i>j}\left(\frac{x_j}{1-x_i} g^E_{ij}(x_i, 1-x_i) + \frac{x_i}{1-x_j} g^E_{ij}(1-x_j, x_j)\right) \qquad (26\text{-}18)$$

A fourth method from Toop[16] is unsymmetrical and is recommended when two components, 2 and 3, are very similar and are quite different from the other component, 1. The equation is

Toop[16]

$$g^E = \sum_{j=2}^{3}\left[\frac{x_j}{1-x_1} g^E_{1j}(x_1, 1-x_1)\right] + (x_2 + x_3)^2 \, g^E_{23}\!\left(\frac{x_2}{x_2+x_3}, \frac{x_3}{x_2+x_3}\right)$$

$$(26\text{-}19)$$

Hillert[2] gives the rationale for each method and Ansara et al.[17] compare the results for typical ternary alloy systems.

METALLIC SYSTEMS

A large number of three-component metallic systems have been treated by the methods outlined above and more are reported every year. Our first example involves just the chemical potential of one component as a measured quantity. But equilibria between phases are often present and require more complex calculations.

We consider first the Bi–Cd–Pb system and its treatment for 773 K by Lee[18] using the Redlich–Kister formulation. This system was the example for the

TABLE 26-1
Coefficients for Eq. (26-20) for three binary systems at 773 K from Lee[18]

System	Coefficients			
	^0A	^1A	^2A	^3A
Cd–Bi	−0.2256	−0.1561	0.2407	0.2121
Pb–Cd	1.3123	−0.1639	0.1008	−0.0459
Bi–Pb	−0.7711	0.0094	0.0916	−0.0338

multicomponent Gibbs–Duhem treatment in the second edition of this book.[1] In the same paper, Lee[18] also treats Cd–Pb–Sn and Ag–Pb–Sn.

The coefficients $^0A_{ij}$ through $^3A_{ij}$ for each binary were evaluated from experimental data cited by Lee[18] with the results in Table 26-1. As a first approximation for the ternary, Lee used the Muggianu equation. Substitution of Eq. (26-14) into Eq. (26-16) and conversion to dimensionless form yields

$$\frac{g^E}{RT} = \sum\sum_{i>j} x_i x_j \sum_m {}^m A_{ij} (x_i - x_j)^m \qquad (26\text{-}20)$$

Then the excess chemical potential of one component can be obtained by differentiation of

$$\frac{G^E}{RT} = \sum\sum_{i>j} \frac{n_i n_j}{\sum_i n_i} \sum_m {}^m A_{ij} \left(\frac{n_i - n_j}{\sum_i n_i} \right)^m \qquad (26\text{-}21)$$

The resulting excess chemical potential can be converted to the activity of that component. Figure 26-1 shows as curves of constant activity of Cd the experimental measurements of Elliot and Chipman,[19] while open squares show the calculated values in this approximation. Lee[18] then added an $A_{123} x_1 x_2 x_3$ term to g^E/RT with the result

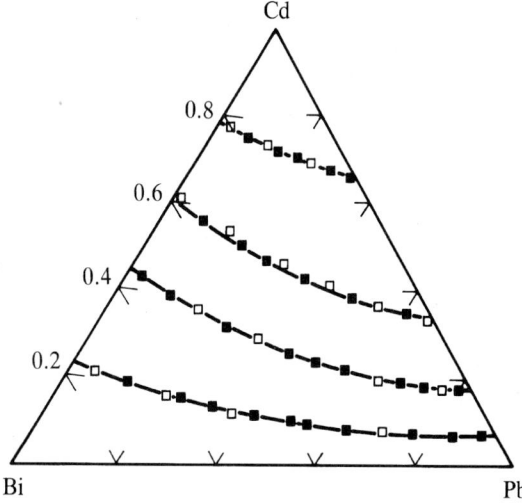

FIGURE 26-1
Curves of constant activity of Cd in the Bi–Cd–Pb system at 773 K (from Lee[18]); additional details in the text.

shown by the closed symbols in Fig. 26-1. The value of A_{123} is -0.492. Lee also reports a higher approximation with fourth-order terms such as $A_{1123}x_1^2x_2x_3$, but this was not really needed to obtain agreement within experimental uncertainty. With the parameters in g^E evaluated, the activities of Bi and Pb are readily calculated.

As a second metallic example, we consider solid–liquid phase equilibria in the system Al–In–Sb, which was carefully investigated by Ishida et al.[20] Their equations representing the data were improved by Sharma and Srivastava,[21] who also considered the similar system Al–Ga–Sb. For the binaries there are not only the phase equilibria but also enthalpy data and other information. While the solid elements In and Sb remain essentially pure and there is only very small solubility in solid Al, the pseudo binary solid (Al,In)Sb has great stability and plays a major role. Thus, the solid–liquid equilibrium usually involves the (Al,In)Sb solid solution and one of the essentially pure elements. The criteria are, of course, the equality of each of the chemical potentials as well as of T and P. Where the solid solution is present, the values of the three chemical potentials have a complex dependency for the solid as well as the liquid, and indirect, usually iterative, calculational methods must be used.

The molar Gibbs energy of the solid solution, symbol θ, is simply described as a regular solution in both investigations:[20,21]

$$\frac{g_m^\theta - y g_{AlSb}^{°s} - (1-y) g_{InSb}^{°s}}{RT} = y \ln y + (1-y)\ln(1-y) + y(1-y)\frac{433.67}{T}$$

(26-22)*

where y is the mole fraction AlSb, $g_{AlSb}^{°s}$ and $g_{InSb}^{°s}$ are the molar Gibbs energies of the pure solid compounds, and the numerical constant is that of reference 21.

For the liquid phase, four-suffix Margules equations were used. Sharma and Srivastava[21] adopt the form

$$\frac{g^E}{RT} = \sum_{i=1}^{2} \sum_{j=i+1}^{3} \left[\frac{w_{ij} + w_{ji}}{2} + \left(\frac{w_{ij} - w_{ji}}{2}\right)(x_j - x_i) - 4v_{ij}x_ix_j \right] x_ix_j + u x_1 x_2 x_3$$

(26-23)*

The different solution parameters, w_{ij}, v_{ij}, and u are, in general, functions of temperature. These parameters relate to those in Eq. (26-14) by $^0A_{ij} = (w_{ij} + w_{ji})/2 + V_{ij}$, $^1A_{ij} = (w_{ij} - w_{ji})/2$, $^2A_{ij} = v_{ij}$. The activity coefficient of component p in these terms is

$$\ln \gamma_p = \sum_{\substack{i=1 \\ i \neq p}}^{3} \left[\frac{w_{ip} + w_{pi}}{2} + (w_{ip} - w_{pi})\left(x_p - \frac{x_i}{2}\right) - 8v_{ip}x_ix_p \right] x_i$$

$$- \sum_{i=1}^{2} \sum_{j=i+1}^{3} \left(\frac{w_{ij} + w_{ji}}{2} + (w_{ij} - w_{ji})(x_j - x_i) - 12v_{ij}x_ix_j \right) x_ix_j$$

$$- u\left(2x_1x_2x_3 - \frac{x_1x_2x_3}{x_p} \right)$$

(26-24)*

TABLE 26-2
Differences between standard-state Gibbs energies (J mol^{-1})

$G_{Al}^{\circ L} - G_{Al}^{\circ S}$	$10\,792.0 - 11.56T$
$G_{In}^{\circ L} - G_{In}^{\circ S}$	$3\,263.53 - 7.5934T$
$G_{Sb}^{\circ L} - G_{Sb}^{\circ S}$	$19\,874.0 - 21.9868T$
$G_{AlSb}^{\circ S} - G_{Al}^{\circ S} - G_{Sb}^{\circ S}$	$-47\,647.06 - 36.651\,84T + 6.498T \ln T$
$G_{InSb}^{\circ S} - G_{In}^{\circ S} - G_{Sb}^{\circ S}$	$-30\,252.3 - 0.246T + 2.4284T \ln T$

TABLE 26-3
Optimized parameters[21] for the liquid phase Al(1)–In(2)–Sb(3)

Parameter	Value	Parameter	Value
w_{12}	$2755.1/T - 0.132$	w_{13}	$-1932.93/T + 0.3196$
w_{21}	$3824.2/T - 0.606$	w_{31}	$607.94/T - 0.038\,67$
v_{12}	$439.1/T - 0.086$	v_{13}	$704.31/T$
w_{23}	$-1946.6/T + 9.1611 - 1.196 \ln T$	u	$-5261.67/T$
w_{32}	$-3469.1/T + 15.431 - 1.962 \ln T$		
v_{23}	$300.3/T + 0.3230$		

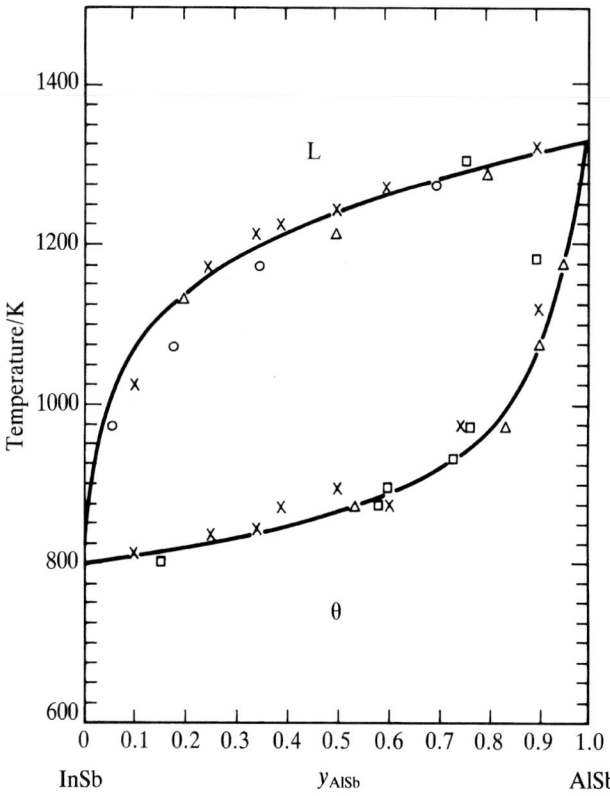

FIGURE 26-2
Liquidus and solidus compositions in the (Al,In)Sb pseudo binary. The solid curve is calculated from Sharma and Srivastava,[21] while the experimental points are from Ishida et al.[20] or earlier sources there cited (from Sharma and Srivastava[21]).

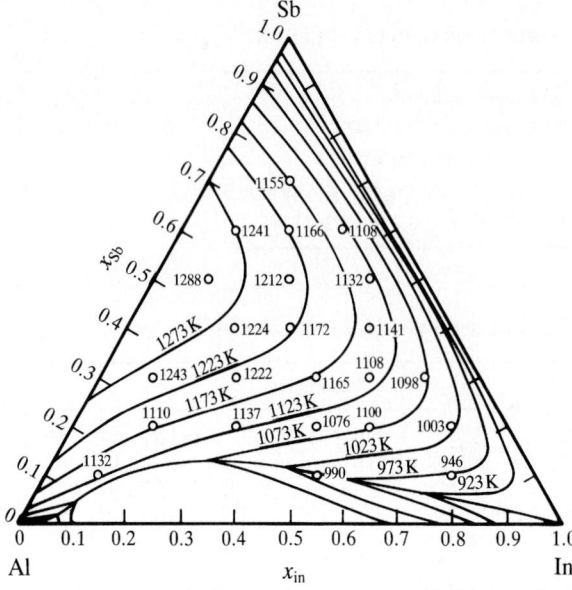

FIGURE 26-3
The liquidus for the system Al–In–Sb with calculated isothermal contours from Sharma and Srivastava[21] and experimental points[20] (from Sharma and Srivastava[21]).

The various differences between standard-state Gibbs energies are given in Table 26-2, while Table 26-3 gives the liquid solution parameters. Figure 26-2 compares the calculated curve[21] with experimental values for the pseudobinary (Al,In)Sb; here θ designates the solid solution. Figure 26-3 shows the calculated liquidus temperature contours[21] for the ternary, together with experimental values for Ishida et al.[20] The earlier calculations of Ishida et al.[20] yield similar but not quite as accurate results. Note the region of two liquid phases in and near the Al–In binary.

Before closing this section on metallic systems, we note the interesting situation if mercury is one component and the only volatile component at the given temperature. Then the activity of mercury can easily be measured from the vapor pressure. Also, the isopiestic method can be used. Wang et al.[22] report isopiestic results for the ternary system Hg–Bi–Sn and several binaries; they take the Hg–Cd system as the reference for calculating activities.

NONMETALLIC SYSTEMS

Multicomponent systems of nonmetallic nature vary widely from nearly ideal hydrocarbon mixtures to examples involving very different species where liquid-phase immiscibility may arise. As one example of systems with substantially different components, we consider solid–liquid equilibria for $LiCl$–$BaCl_2$–$SrCl_2$ in the range 750–1250 K as treated by Zhiyu et al.[23]

Sterner et al.[5] used the same type of equation for the partial treatment of H_2O–$NaCl$–KCl in the range above 650 K in which only salt-saturated aqueous solutions were included. Pressure dependency was considered for that system. At lower temperatures the system H_2O–$NaCl$–KCl is very satisfactorily treated as an electrolyte

TABLE 26-4
Parameters for Eq. (26-7)* for several binary systems

(1)	(2)	$wa_{12}(P = 1)$	$wb_{12}(P = 1)$	$wa_{12}/(P - 1)$
$BaCl_2(l)$	$LiCl(l)$	$-373/T$	0	
$SrCl_2(l)$	$LiCl(l)$	$172/T$	$24/T$	
$BaCl_2(l)$	$SrCl_2(l)$	$-219/T$	$-2/T$	
$BaCl_2(s,\alpha)$	$SrCl_2(s,\alpha)$	$162/T$	0	
$NaCl(l)$	$H_2O(l)$	$-600.1/T + 0.608$	0	$0.1292/T$
$KCl(l)$	$H_2O(l)$	$-781.4/T + 0.395$	0	$0.1545/T$
$NaCl(l)$	$KCl(l)$	$-246.6/T - 0.227$	$-32.7/T + 0.121$	$0.0310/T$
$NaCl(s)$	$KCl(s)$	$397.9/T + 9.267$ $-0.017\,728/T + 8.85 \times 10^{-6}T^2$	$573.0/T + 0.180$ $-4.20 \times 10^{-5}T$	$0.1477/T$ -1.172×10^{-4}

by the methods of Chapter 17,[24,25] but above the critical temperature of water there is little dissociation into ions, and nonelectrolyte methods are more appropriate. A complete treatment of the binary $NaCl-H_2O$ above 650 K is available,[26] but the equation is very complex. The salt-saturated $NaCl-H_2O$ and $KCl-H_2O$ systems both show simple behavior, however, which can be represented by the regular solution equation.[5]

The $BaCl_2-SrCl_2$ and $NaCl-KCl$ binaries are similar in showing solid solubility increasing rapidly with temperature to miscibility above 970 K and 790 K, respectively. In each case the liquid–solid phase diagrams have been determined as well as the heat of mixing for the liquid.[27,28]

It happens that exactly the same form of equation was used for the binaries in these two investigations, with the excess molar Gibbs energy given by Eq. (26-7)*,

$$\frac{g^E_{12}}{RT} = x_1 x_2 (wa_{12} + wb_{12} x_1) \qquad (26\text{-}7)^*$$

The expressions for the activity coefficients were given in Eqs. (26-8)* above. The values for the parameters are given in Table 26-4. For the $LiCl-BaCl_2-SrCl_2$ systems, the excess entropy parameters are all zero, since all entries have T^{-1} dependency. For the other systems the entropy terms are clearly small in all cases except solid NaCl–KCl, where the temperature dependency is complex. The changes in chemical potential on fusion were available from data on ΔH, ΔS, and ΔC_P. The results for NaCl and KCl are in different format and are as follows:

$$\Delta\mu^{NaCl}_{fus} = 11.539 - 2.6857 \times 10^{-2} T \ln T + 1.05985 \times 10^{-5} T^2$$
$$+ 1.65312 \times 10^{-1} T + 6.3608 \times 10^{-4}(P - 1)$$

$$\Delta\mu^{KCl}_{fus} = 40.438 - 9.8113 \times 10^{-2} T \ln T + 2.4261 \times 10^{-5} T^2 - 5.484 T^{1/2}$$
$$- \frac{802.7}{T} + 7.88341 \times 10^{-1} T + 6.4047 \times 10^{-4}(P - 1)$$

The fusion properties for the other three salts are given in Table 26-5.

TABLE 26-5
Temperature and Gibbs energy of phase transformation of the pure salts $\Delta G° = a + bT + cT^2 + dT \ln T$ (J mol^{-1})

Compound	Phase transformation I→II	$T°_{fus}$ or $T°_{tra}$ (K)	a	b	c × 10^3	d
LiCl	$\alpha \to \ell$	883	4 420.2	197.318	16.435	−31.966
BaCl$_2$	$\gamma \to \alpha$	1 195	−19 904.1	287.214	6.9875	−39.361
	$\alpha \to \ell$	1 235	25 261.2	−69.5957	0	6.903
SrCl$_2$	$\beta \to \alpha$	1 003	6 276.0	−6.257 23	0	0
	$\alpha \to \ell$	1 146	−10 002.1	203.288	5.1045	−28.451

Both SrCl$_2$ and BaCl$_2$ have structure changes with temperature. The high temperature, α, structure is the same for both, as is implied by the continuous range of solid solubility, but the lower temperature structures, β for SrCl$_2$ and γ for BaCl$_2$, are different. While there is no direct β-liq or γ-liq equilibrium in the two-component SrCl$_2$–BaCl$_2$ system, the addition of LiCl$_2$ to the liquid lowers the chemical potentials of SrCl$_2$ and BaCl in that phase and, thereby, the temperature of equilibrium with the solid to the range of the β or γ structures. Thus, Zhiyu et al.[23] had to evaluate the properties of these solids. Their conclusion was that solid solubility of SrCl$_2$ into γ-BaCl$_2$ or of BaCl$_2$ into β-SrCl$_2$ occurred with regular solution behavior and $g^E = 5000 x_1 x_2$ J mol^{-1} in each case. The transition temperatures and chemical-potential parameters are included in Table 26-5.

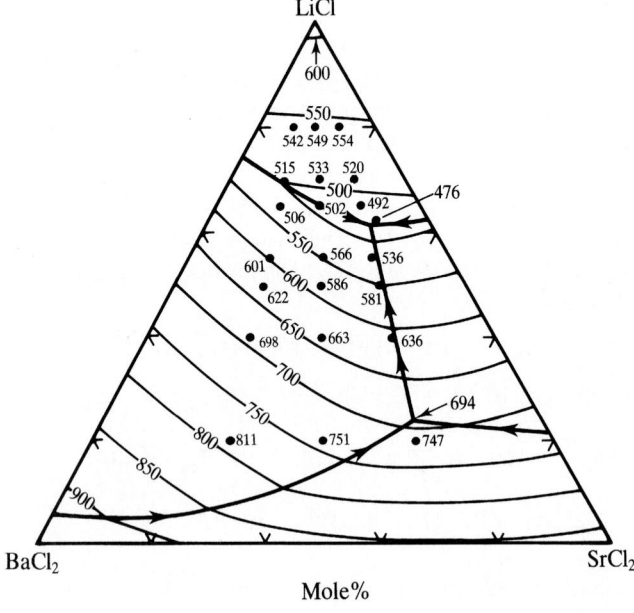

FIGURE 26-4
Calculated isotherms for the liquidus surface for LiCl–BaCl$_2$–SrCl$_2$ with experimentally measured points (from Zhiyu et al.[23]).

For the three-phase liquid, Zhiyu et al.[23] use the Toop method, Eq. (26-23), with $SrCl_2$ and $BaCl_2$ as components 2 and 3, whose properties are assumed to be closely similar. Then each of the three chemical potentials must be equal between the liquid and the solids present. Since the LiCl(s) remains pure, it determines μ_{LiCl} as a function of T. But μ_{SrCl_2} and μ_{BaCl_2} are functions of composition for both solid and liquid. A rather complex calculation is required to find mole fractions such that all three chemical potentials and the temperature are consistent. The resulting liquidus surface is shown in Fig. 26-4, while the phase diagram for 600°C is shown in Fig. 26-5. Since 600°C is only 10 K below the melting point of LiCl, the region for solid LiCl is very small; it appears in Fig. 26-4, but is omitted in Fig. 26-5. The calculated ternary eutectic is at 476°C, $x_{LiCl} = 0.61$, $x_{SrCl_2} = 0.28$, $x_{BaCl_2} = 0.11$.

For the ternary liquid H_2O–NaCl–KCl, Sterner et al.[5] retained the binary Margules terms from Eq. (26-7)* without change, and added a three-component term as follows:

$$\frac{g^E}{RT} = \sum\sum_{i>j} x_i x_j (wa_{ij} + wb_{ij}x_i) + w_{ijk}x_i x_j x_k \qquad (25\text{-}25)*$$

The value of w_{ijk} was optimized by fitting the ternary liquidus values of Chou et al.[29] with the result

$$w_{H_2O,NaCl,KCl} = \frac{392.6 + 0.1522(P-1)}{T} - 0.6153$$

In this system only the two chemical potentials of the salts are at equilibrium between solid and liquid phases at a given T and P. The chemical potential of H_2O does not enter

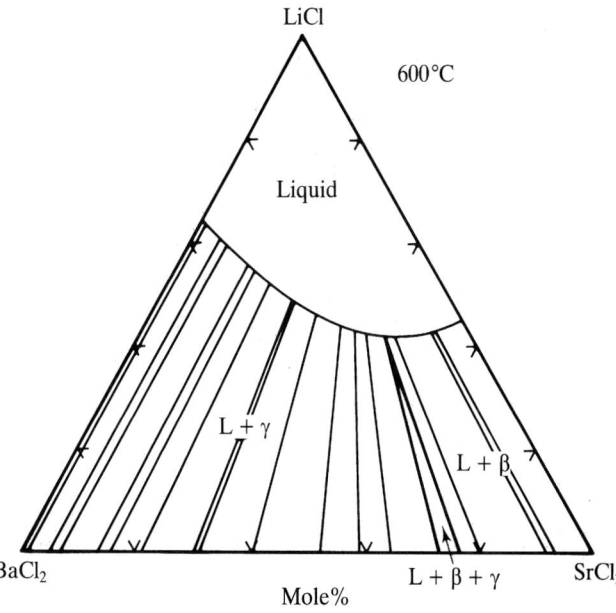

FIGURE 26-5
Calculated phase diagram for $LiCl$–$BaCl_2$–$SrCl_2$ at 600°C and low pressure (from Zhiyu et al.[23]).

the calculation of compositions, although the pressure must be high enough to prevent formation of a vapor phase.

Figure 26-6 shows the phase diagram for the NaCl–KCl binary as a function of temperature for 1 bar, while Fig. 26-7 shows the ternary liquidus for 1 kbar. The size of the symbols for experimental values indicates the deviations of the calculated surface at that point in temperature with the largest deviations 8 degrees.

There are, of course, simpler three-component systems; examples involve neutral-molecule liquids in equilibrium with a nearly ideal vapor at low pressure. Examples of this type are included in Appendix 11, where the Wilson equation is applied. Such systems can equally well be treated by a Margules equation, which can be extended to high enough order to obtain good agreement. Van Ness and associates,[30] in a series of articles, describe several treatments of this type. One example is the ternary: acetone–chloroform–methanol. A five-suffix equation was needed for the chloroform–methanol binary, while a four-suffix equation sufficed for acetone–chloroform and only a three-suffix equation was needed for acetone–methanol. Four additional terms and parameters were added for the ternary; this implies a four-suffix equation.

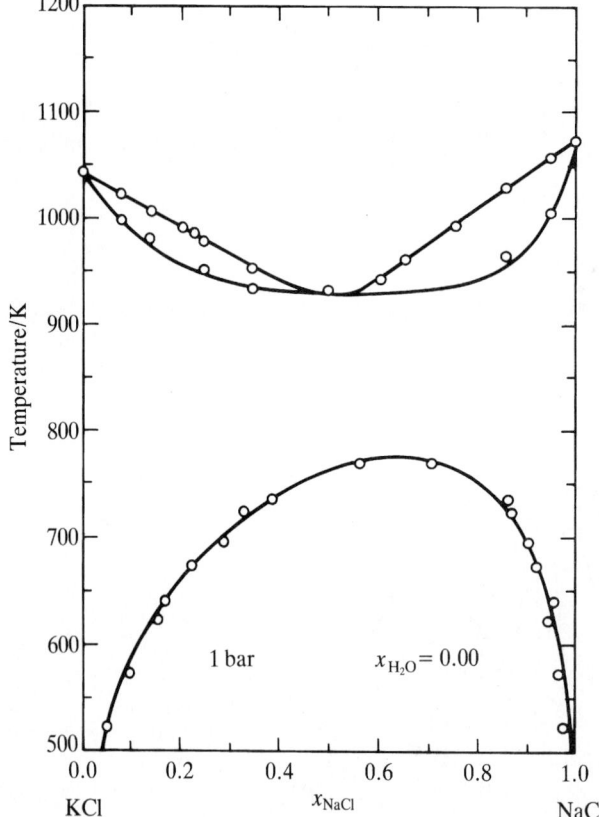

FIGURE 26-6

Calculated curves from Sterner et al.[5] and experimental points for the phase diagram for NaCl–KCl at 1 bar.

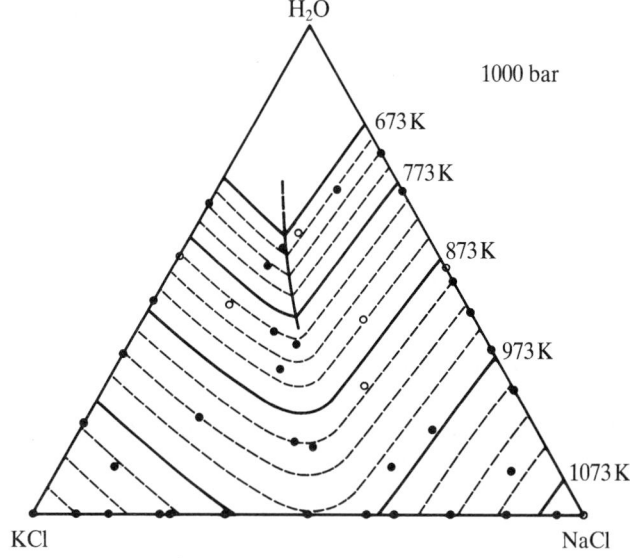

FIGURE 26-7
The liquidus surface for NaCl–KCl–H_2O at 1 kbar with calculated isotherms from Sterner et al.[5] and experimental points.[29] The diameter of each circle shows the difference of the calculated temperature from that of the experimental measurement: ○, $T_{obs} - T_{calc} = +8$ K; ○, $T_{obs} - T_{calc} = 0$ K; ●, $T_{obs} - T_{calc} = -8$ K.

SYSTEMS WITH FOUR OR MORE COMPONENTS

At the two- or three-suffix level, all Margules parameters are determined by ternary systems and the properties of more complex systems can be calculated. An example of this general type is the treatment of Talley et al.[31] for the five-component system hexane–benzene–toluene–methylcyclopentane–cyclohexane. They use the Legendre polynomial method, Eq. (26-15), for the ten binaries and the Kohler method, Eq. (26-17), for combining the results in more complex systems. They find excellent agreement, which they discuss in comparison with results of others using the Wilson equation and other equations listed in Appendix 11. Since these calculations for four or more components involve no new principles or problems different from those discussed for ternaries, no further examples or details will be given here. We note, however, that electrolytes with more than three components were considered in Chapter 17 with the results for the five-component seawater system Na–K–Mg–Cl–SO_4–H_2O shown in Table 17-6.

DETERMINATION OF OTHER CHEMICAL POTENTIALS WHEN ONLY ONE IS MEASURED

In Chapters 12, 14, and 17, we considered the problem of the determination of parameters in an expression for the excess Gibbs energy from measurements of only one chemical potential in a multicomponent system. Where the measurements cover the complete range of composition, all parameters are determined in principle, although there may be near redundancies that can be resolved only with extremely precise information.

In the molality system, a pure solute corresponds to infinite molality; hence there is only partial coverage of composition. Nevertheless, with measured activities of the solvent, all parameters can be evaluated. But in the case of two or more solutes in the molality system, it was found in Chapters 14 and 17 that certain parameters are either absent or exactly redundant in the expression for the activity coefficient of a particular solute.

For mole-fraction-based equations, it appears that the situation is more subtle and complex. At the end of Chapter 12, a simple example was given where a redundancy appeared in the first order of an expansion about zero concentration of the component whose activity was measured. But at higher order the situation became so complex that an unambiguous answer was not apparent.

One can understand the difference between the mole fraction and molality systems by considering the definitions of these quantities. An addition of any one component affects the mole fractions of all other components because the sum over all components appears in the denominator. But a molality is the ratio of the number of moles of that particular solute to the number of moles of solvent. Addition of a different solute has no effect on a molality. Thus, a derivative of a term involving one molality with respect to a different molality is zero, while the corresponding derivative with a mole fraction yields a finite result.

For cases of limited solubility of a solute component, one uses the solute standard state, which is the natural basis for solutes in the molality system. But for the mole fraction system, the ordinary expression for an activity coefficient relates to the pure liquid standard state. The expression based on the solute (infinitely dilute) standard state is readily obtained by difference, but this requirement is sometimes overlooked.

In order to illustrate these features for a moderately complex equation, we extend to the three-suffix Margules level the system considered in Chapter 12 with two solutes (2 and 3) and a solvent (component 1). The activity coefficient of component 2 on the usual, pure liquid, basis can be written

$$\ln \gamma_2 = x_1(1 - x_2)(A_{12} + 2x_2 A_{122}) + x_3(1 - x_2)(A_{23} + 2x_2 A_{223})$$
$$- x_1 x_3 [A_{13} + 2x_3 A_{133} + (2x_2 - 1)A_{123}] \qquad (26\text{-}26)*$$

At the limit $x_1 = 1$, $x_2 = x_3 = 0$, this becomes just A_{12}, which is subtracted to yield the activity coefficient on the solute state basis,

$$\ln \gamma_2^* = (x_1 - x_1 x_2 - 1)A_{12} + 2x_1 x_2(1 - x_2)A_{122} + x_3(1 - x_2)(A_{23} + 2x_2 A_{223})$$
$$- x_1 x_3 [A_{13} + 2x_3 A_{133} + (2x_1 - 1)A_{123}] \qquad (26\text{-}27)*$$

Since we are interested in the region of small x_2 and x_3, we substitute $x_1 = 1 - x_2 - x_3$ to eliminate x_1, with the result

$$\ln \gamma_2^* = (-2x_2 - x_3 + x_2 x_3 + x_2^2)A_{12} + 2x_2(1 - x_2)(1 - x_2 - x_3)A_{122}$$
$$+ x_3(1 - x_2)(A_{23} + 2x_2 A_{223}) - x_3(1 - x_2 - x_3)[A_{13} + 2x_3 A_{133}$$
$$+ (1 - 2x_2 - 2x_3)A_{123}] \qquad (26\text{-}28)*$$

To interpret this result, we collect terms having the same composition dependency:

$$\ln \gamma_2^* = -2x_2(A_{12}-A_{122}) - x_3(A_{12}-A_{23} + A_{13} + A_{123})$$
$$+ x_2^2(A_{12}-4A_{122}) + x_2 x_3(A_{12}-A_{23} + A_{13} - 2A_{122} + 2A_{133} + 3A_{123})$$
$$+ x_3^2(A_{13}-A_{133} + 3A_{123}) + 2x_2^3 A_{122} + 2x_2^2 x_3(A_{122}-A_{223}-A_{123})$$
$$+ 2x_2 x_3^2(A_{133} - 2A_{123}) + 2x_3^3(A_{133}-A_{123}) \qquad (26\text{-}29)^*$$

In Chapter 12 we noted that, with only two-suffix parameters, the terms linear in x yield a value for A_{12}, but only for the difference $A_{13} - A_{23}$. The x_2^2 term resolves this redundancy at the two-suffix level. But at the three-suffix level there are seven parameters and only five terms through the second power of x. Again, the redundancies can be resolved with x^3 terms. But then the possibility must be considered that four-suffix terms might be significant.

It is clear that, if the data for the measured $\ln \gamma_i$ are limited to a dilute range where the x-dependence is linear, then a redundancy remains and the activity coefficients of other components cannot be predicted. But it remains a complex question in the mole-fraction equations whether, with data for higher molalities, the various parameters can be determined accurately enough to make valid predictions. However, if the maximum mole fractions x_2 and x_3 remain moderate, the activity coefficient could be represented with comparable accuracy in the molality system. Then, as noted in Chapter 14, there are terms of various orders that are required for one activity coefficient that are absent from the expressions for other activity coefficients. In that case one clearly cannot calculate the activity coefficient of a different component, and one must suspect that this limitation remains when mole fractions are used.

In 1978 and 1979 there was a major controversy[32] concerning activities in multicomponent silicate systems that concerned a question similar to that discussed above. In view of the complexity and ambiguity, we find it is not surprising that the controversy remained unresolved at that time.

REFERENCES

1. For graphical methods using the Gibbs–Duhem equation for three or more components, see Chapter 34 in the second edition of this book: K. S. Pitzer and L. Brewer, *Thermodynamics*, revision of 1st edn by G. N. Lewis and M. Randall, McGraw-Hill, New York, 1961.
2. M. Hillert, *CALPHAD*, **4**, 1 (1980).
3. M. Margules, *Sitzber. Akad. Wiss. Wien., Math-naturw. Kl., Abt. IIa*, **104**, 1243 (1895).
4. O. J. Kleppa and L. S. Hersh, *J. Chem. Phys.*, **34**, 351 (1961).
5. S. M. Sterner, I.-M. Chou, R. T. Downs, and K. S. Pitzer, *Geochim. Cosmochim. Acta*, **56**, 2295 (1992).
6. O. Redlich and A. T. Kister, *Ind. Eng. Chem.*, **40**, 345 (1948).
7. O. Redlich, A. T. Kister, and C. E. Turnquist, *Chem. Eng. Prog. Series*, **48**(2), 49 (1952).
8. K. Wohl, *Trans. Am. Inst. Chem. Eng.*, **42**, 215 (1946).
9. K. Wohl, *Chem. Eng. Prog.*, **46**, 218 (1953).
10. B. Mukhopadhayay, S. Basu, and M. J. Holdaway, *Geochim. Cosmochim. Acta*, **57**, 277 (1993).
11. C. W. Bale and A. D. Pelton, *Metal. Trans.*, **5**, 2323 (1974).
12. S. D. Choi, *CALPHAD*, **12**, 25 (1988).

13. Y. M. Muggianu, M. Gambino, and J. P. Bros, *J. Chim. Phys.*, **72**, 83 (1975).
14. F. Kohler, *Monatsh. Chem.*, **91**, 738 (1960).
15. C. Colinet, D.E.S. Fac. des Sci., University Grenoble, France (1967).
16. G. W. Toop, *Trans. AIME*, **233**, 850 (1965).
17. I. Ansara, C. Bernard, L. Kaufman, and P. Spencer, *CALPHAD*, **2**, 1 (1978).
18. H. M. Lee, *CALPHAD*, **16**, 47 (1992).
19. J. F. Elliot and J. Chipman, *J. Am. Chem. Soc.*, **73**, 2682 (1951).
20. K. Ishida, T. Shumiya, H. Ohtani, M. Hasebe, and T. Nishizawa, *J. Less Common Metals*, **143**, 279 (1988).
21. R. C. Sharma and M. Srivastava, *CALPHAD*, **16**, 387, 409 (1992).
22. Z-C. Wang, X-H. Zhang, T-Z. He, and Y-H. Bao, *J. Chem. Thermodyn.*, **21**, 653 (1989).
23. Q. Zhiyu, J. Sangster, and A. D. Pelton, *CALPHAD*, **11**, 277 (1987).
24. R. T. Pabalan and K. S. Pitzer, *Geochim. Cosmochim. Acta*, **51**, 2429 (1987).
25. R. T. Pabalan and K. S. Pitzer, in *Chemical Modeling of Aqueous Systems II*, D. C. Melchior and R. L. Bassett, eds., A.C.S. Symposium Series 416, American Chemical Society, Washington, 1990, Chapter 4.
26. A. Anderko and K. S. Pitzer, *Geochim. Cosmochim. Acta*, **57**, 1657 (1993).
27. G. A. Papatheodorou and O. J. Kleppa, *J. Chem. Phys.*, **47**, 2014 (1967).
28. L. S. Hersh and O. J. Kleppa, *J. Chem. Phys.*, **42**, 1309 (1965).
29. I.-M. Chou, S. M. Sterner, and K. S. Pitzer, *Geochim. Cosmochim. Acta*, **56**, 2281 (1992).
30. M. M. Abbott, J. K. Floess, G. E. Walsh, Jr., and H. C. Van Ness, *Am. Inst. Chem. Eng. J.*, **21**, 72 (1975) and preceding papers in a series.
31. P. K. Talley, J. Sangster, A. D. Pelton, and C. W. Bale, *CALPHAD*, **16**, 93 (1992).
32. A. C. Lasaga and C. W. Burnham, *Geochim. Cosmochim. Acta*, **43**, 643 (1979) and earlier papers there cited.

PROBLEMS

26-1. The three binaries carbon tetrachloride–benzene, carbon tetrachloride–cyclohexane, and benzene–cyclohexane follow the simple two-suffix Margules or "regular" solution behavior of Eq. (12-12)*. Table 12-1 gives parameters values for each binary that can be interpolated to 313 K. For the carbon tetrachloride–benzene–cyclohexane ternary, give the numerical expressions as functions of composition for (a) g^E/RT, (b) $\ln \gamma_{CCl_4}$, and (c) the activity a_{CCl_4}. Then make a graph like Fig. 26-1 with curves of a_{CCl_4} = 0.2, 0.4, 0.6, and 0.8.

26-2. Differentiate Eq. (26-21) to obtain the chemical potential μ_i and the activity a_i.

26-3. Calculate the chemical potential and the activity of bismuth for x_{Bi} = 0.30 and the series of values x_{Pb} = 0.10, 0.35, 0.60. Use the parameters in Table 26-1, the equation obtained in Problem 26-2, and omit the term in A_{123}.

26-4. For the system pyridine(1)–acetone(2)–dichloromethane(3) at 303 K, calculate g^E/RT and each $\ln \gamma_i$ for the equimolal composition $x_1 = x_2 = x_3 = 1/3$. Use the values in Table 12-2 and Eq. (26-20), which implies the Muggianu method. Note that $A = {}^0A_{12}$, $B = {}^1A_{12}$, $C = {}^2A_{12}$, etc.

26-5. Substitute Eq. (26-14) into Eqs. (26-17), (26-18), and (26-19); then compare the resulting expressions with Eq. (26-20) for the terms with $m = 0$, $m = 1$, and $m = 2$, both in algebraic form and for the composition $x_1 = x_2 = x_3 = 1/3$.

26-6. Zhiyu et al.[22] used Eq. (26-19), the Toop[16] method, to combine the binary expressions Eq. (26-7)* for the ternary LiCl(1)–BaCl$_2$(2)–SrCl$_2$(3). How would the result differ if Eq. (26-16), the Muggianu[13] method, had been used instead? What would this difference be numerically at 600°C?

CHAPTER 27

BIOCHEMICAL SYSTEMS; SPECIAL ASPECTS

Many systems important in biochemistry do not differ in the basic methods of treatment from examples described in other chapters. There is no need to repeat such treatments here with the mere substitution of biologically active substances. There are, however, other systems where special methods have been developed to handle the complexities present. We present here two examples that illustrate the methods that have been adopted. Some of these methods may also be useful for nonbiological systems.

Reactions at rapid equilibrium are commonly called "binding." We first consider multiple binding sites on a macromolecule. Secondly, an example with large numbers of species and many simultaneous equilibria is described.

OXYGEN BINDING TO MACROMOLECULES

Binding to biological macromolecules has been treated thoroughly and comprehensively by Wyman and Gill[1] as of 1990. We present here only two examples of oxygen binding. Although a full description of the properties of a macromolecule would be very complex, for our problem these molecules remain in dilute aqueous solution and are not significantly changed in their interaction with the water.

The first case is that of myoglobin, My, which is an oxygen storage protein found in some living species. It has only a single oxygen binding site per molecule of molecular mass 16 000. The reaction, written as a simple association, is

$$\text{My} + \text{O}_2 = \text{MyO}_2 \tag{A}$$

with the equilibrium constant

$$K = \frac{m(\text{MyO}_2)}{m(\text{My})\,P_{\text{O}_2}} \tag{27-1}$$

If the fraction θ has O_2 bound, then

$$\frac{m(\text{myO}_2)}{m(\text{My})} = \frac{\theta}{1-\theta} = P_{\text{O}_2} K \tag{27-2a}$$

$$\log\left(\frac{\theta}{1-\theta}\right) = \log P_{\text{O}_2} + \log K \tag{27-2b}$$

Thus, a graph on the log–log basis is a straight line, as is shown in Fig. 27-1. This is known as a Hill plot, after a 1910 paper of A. V. Hill.[2] This simple case was presented only as a reference for the more complex case to follow.

The oxygen storage and transport protein in humans and some other species is hemoglobin, Hb; its binding curve, also shown in Fig. 27-1, is more complex. The hemoglobin molecule, of molecular mass 64 000, is known to have four binding sites. Now there are four reactions of the type

$$\text{Hb} + i\text{X} = \text{HbX}_i \tag{B}$$

$$\beta_i = \frac{m(\text{HbX}_i)}{m(\text{Hb})x^i} \tag{27-3}$$

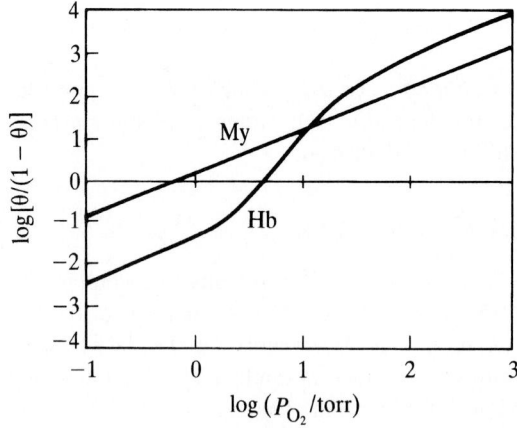

FIGURE 27-1
Log–log (Hill) plot of fractional binding θ as $[\theta/(1-\theta)]$ vs. activity of ligand for myoglobin, My, and hemoglobin, Hb.

TABLE 27-1
Binding constants for oxygen to hemoglobin at 25°C and heme concentration of 0.004 mol dm^{-3}

pH	β_1/torr	β_2/torr^2	β_3/torr^3	β_4/torr^4
9.1	0.61 ± 0.2	0.23 ± 0.05	0(0.03)[a]	0.091 ± 0.01
6.9	0.054 ± 0.01	0.0031 ± 0.001	0(7 × 10^{-5})[a]	9.4 × 10^{-5} ± 5 × 10^{-6}

[a] Maximum uncertainty for β_3 above zero.

where X represents O_2 or another ligand and x is its activity, or, for O_2 at low pressure, its partial pressure. It is also interesting to consider stepwise binding or association with

$$HbX_{i-1} + X = HbX_i \tag{C}$$

$$K_i = \frac{\beta_i}{\beta_{i-1}} \tag{27-4}$$

A simple model assumes that the intrinsic binding tendency at each site is the same, whereupon the K_i values differ only by the appropriate statistical factors. This model yields the same straight line of unit slope on a Hill plot, where θ is now the fraction of binding sites occupied by O_2. The observed curve for hemoglobin in Fig. 27-1 is much steeper in the middle range. This implies cooperativity, i.e., binding is stronger if nearby binding sites are occupied than if they are vacant.

Until recently it was difficult to obtain accurate values for all four binding constants for hemoglobin from the information on the total binding as a function of P_{O_2}. Recently, Gill et al.[3] reported very accurate optical density measurements from which all four β_i values were obtained. Various conditions of pH and heme (binding site) concentration in range of human physiology were used, and the full set of β values was reported. The calculations are quite complex and the original paper[3] should be consulted for the details. Two sets of β's are given in Table 27-1 that show the strong pH dependence. But most striking is the zero value of β_3, which implies that the concentration of Hb with 3 O_2 is negligible. This was unexpected and remains surprising.

There is an extensive literature on binding to hemoglobin by carbon monoxide and by CO–O_2 mixtures[1,4] as well as by O_2. As the molecular structure becomes better known, the theory of the cooperativity and other properties advances.

SMALL-MOLECULE SYSTEMS

Systems of biochemical interest often include several components and, with ionization, hydrolysis, and metal-ion binding reactions, an even larger number of species. All of the appropriate equilibrium expressions apply and must be satisfied. It is useful, however, to arrange this multitude of equations in a particular way that relates to the available experimental information. Thus, if the biologically important component is a weak acid or base, the sum of all species such as H_2R^+, HR, R^-, etc., is one primary variable with the relative amounts of each species determined by separate, subsidiary equations. Similarly, if phosphate is involved in the primary reaction, one considers the sum of H_3PO_4, $H_2PO_4^-$, HPO_4^{2-}, and PO_4^{3-} as one coordinate with separate, subsidiary equations for the concentrations of each species. Such a component may also bind metal atoms, and the metal-containing species can be added to the total.

In other respects, the treatment of such biochemical systems can be simplified by approximations. Except for the ionic-strength (Debye–Hückel) effect on charged species, activity coefficients are usually omitted and equilibrium constant expressions written in molalities or molar concentrations. Also the ionic strength is relatively small, usually 0.1 to 0.25 mol kg^{-1}, and often can be taken to be constant. Then an effective equilibrium constant is used that is dependent on ionic strength in general but is constant for the problem at hand. The temperature range of primary interest is limited to 25–38°C. ΔC_P values are often taken to be constant over this range, whereupon the temperature dependency is simple for a ΔG or a $\ln K$. Indeed, it is shown below for a typical example that even a rather large value of ΔC_P can be neglected; then ΔH is constant and the temperature dependency of $\ln K$ is even simpler.

As our second example, we illustrate these methods for a typical, widely studied system. The entire area of biochemical thermodynamics is much too extensive to be treated here. References are given to sources that describe other systems. The methods used in many of these cases are similar to those in our example.

Adenosine Phosphate Disproportionation

The adenosine phosphates are important in various biological processes that often involve enzymes or other complex organic species. We consider just the disproportionation of adenosine diphosphate to the mono- and triphosphates. Goldberg and Tewari[5] reported recently a very thorough and complete study.

The phosphate portions of adenosine phosphate structures are

AMP (adenosine monophosphate)

$$HO-\underset{\underset{O}{\|}}{\overset{\overset{OH}{|}}{P}}-O-R$$

ADP (adenosine diphosphate)

$$HO-\underset{\underset{O}{\|}}{\overset{\overset{OH}{|}}{P}}-O-\underset{\underset{O}{\|}}{\overset{\overset{OH}{|}}{P}}-O-R$$

ATP (adenosine triphosphate)

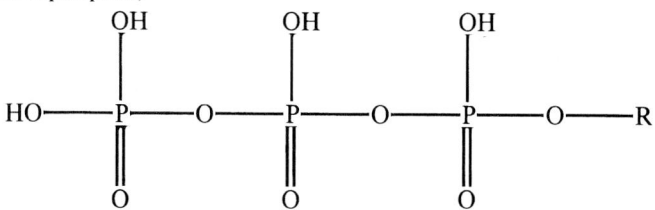

where the adenosine radical R comprises an adenine unit bonded to a D-ribose ring to which the phosphate is also bonded.

The basic reaction is

$$2\sum \text{ADP(aq)} = \sum \text{ATP(aq)} + \sum \text{AMP(aq)} \tag{D}$$

where the $\sum$ denotes the sum of species involving proton or metal ion binding. The total amount of such proton or metal-ion (magnesium in this case) binding will change with this reaction. Thus, a more complete equation is

$$2\sum \text{ADP(aq)} = \sum \text{ATP(aq)} + \sum \text{AMP(aq)} + \nu_H \text{H}^+(\text{aq}) + \nu_{Mg}\text{Mg}^{2+}(\text{aq}) \tag{E}$$

The quantity ν_H is the number of protons produced per unit of $\sum$ATP or $\sum$AMP; ν_{Mg} is the corresponding quantity for Mg^{2+} ions. Both ν_H and ν_{Mg} are, in general, nonintegral and may be either positive or negative. Although electrical charges are not shown explicitly in reaction (E), they are considered implicitly to be included and balanced.

When the individual species are considered, they will be designated ADP^{3-}, HADP^{2-}, H_2ADP^-, H_3ADP, MgADP^-, MgHADP, and similarly for AMP and ATP. For partially protonated ADP or ATP, we do not distinguish particular locations of the protons on different phosphates and consider only the total molality.

Equations Representing Equilibrium

Goldberg and Tewari[5] begin their detailed analysis by selecting one equation involving individual species as a reference reaction, specifically

$$2\text{ADP}^{3-}(\text{aq}) = \text{AMP}^{2-}(\text{aq}) + \text{ATP}^{4-}(\text{aq}) \tag{F}$$

The choice of the reference reaction is arbitrary and a different choice could have been made. This is an ordinary chemical reaction with the corresponding equilibrium constant

$$K^\circ_{\text{ref}} = \frac{[a(\text{AMP}^{2-})][a(\text{ATP}^{4-})]}{[a(\text{ADP}^{3-})]^2} \tag{27-5}$$

Here, the a's are the usual activities in the molality system based on the hypothetical ideal solution at $1\,\text{mol}\,\text{kg}^{-1}$. The pressure is assumed to be the standard pressure of 1 bar (0.1 MPa).

Next, we need the equilibrium constants for the various reactions of proton or magnesium-ion binding to the adenosine phosphate species in the reference equation (F). Goldberg and Tewari[5] evaluated the extensive array of published information and adopted the values of $K°$ (or $pK° = -\log K°$) given in Table 27-2. They also selected values of $\Delta H°$ and $\Delta S°$ for each reaction which allow the conversion of the $K°$ values to different temperatures.

Activity coefficients of the various species are needed both for the evaluation of the values in Table 27-2 and for the calculations that are to follow. Goldberg and Tewari[5] considered the full use of an equation of the Guggenheim[6] type [see Chapter 17, Eq. (17-1)*]. For a single ion of charge z_i this can be written

$$\ln \gamma_i = \frac{-z_i^2 A_\gamma I^{1/2}}{1 + BI^{1/2}} + 2\sum_j \lambda_{ij} m_j \tag{27-6}$$

Here, A_γ is the Debye–Hückel parameter, with $A_\gamma = 3A_\phi = 1.1745 \,\text{kg}^{1/2}\,\text{mol}^{-1/2}$ at 298.15 K. In his original work, Guggenheim[6] chose a value 1.0 for the general parameter B and individual values for all of the λ_{ij} interactions. At present, there is not enough information to evaluate the multitude of λ_{ij} quantities for the present system or for many biochemical systems. But as more accurate and more detailed information becomes available, the methods of Chapter 17, including the more advanced ion-interaction or Pitzer equations as well as the Guggenheim equations, are being used for biochemical systems.[7]

For the adenosine phosphate system, Goldberg and Tewari[5] omitted the second term on the right in Eq. (27-6) and chose the value $B = 1.5 \,\text{kg}^{1/2}\,\text{mol}^{-1/2}$ as optimum

TABLE 27-2
Thermodynamic parameters at 298.15 K relevant to the disproportionation of ADP to AMP and ATP in aqueous solution from Goldberg and Tewari[5]

Reaction	Equilibrium constant or pK	$(\Delta H°/R)/K$	$-\Delta S°/R$
$2\,ADP^{3-} = ATP^{4-} + AMP^{2-}$	$K°_{\text{ref}} = 0.225 \pm 0.010$	-180 ± 180	2.0 ± 0.6
$HAMP^- = H^+ + AMP^{2-}$	$pK°_{1AMP} = 6.74 \pm 0.07$	-650 ± 200	17.7 ± 0.7
$H_2AMP = H^+ + HAMP^-$	$pK°_{2AMP} = 3.99 \pm 0.05$	2160 ± 220	1.9 ± 0.7
$MgAMP = Mg^{2+} + AMP^{2-}$	$pK°_{MgAMP} = 2.81 \pm 0.08$	-1370 ± 460	11.1 ± 1.6
$HADP^{2-} = H^+ + ADP^{3-}$	$pK°_{1ADP} = 7.20 \pm 0.10$	-670 ± 190	18.9 ± 0.7
$H_2ADP^- = H^+ + HADP^{2-}$	$pK°_{2ADP} = 4.37 \pm 0.06$	2120 ± 170	3.0 ± 0.6
$MgADP^- = Mg^{2+} + ADP^{3-}$	$pK°_{MgADP} = 4.68 \pm 0.14$	-2290 ± 190	18.4 ± 0.7
$MgHADP = Mg^{2+} + HADP^{2-}$	$pK°_{MgHADP} = 2.52 \pm 0.09$	-1500 ± 1100	10.8 ± 3.6
$HATP^{3-} = H^+ + ATP^{4-}$	$pK°_{1ATP} = 7.62 \pm 0.09$	-760 ± 140	20.1 ± 0.5
$H_2ATP^{2-} = H^+ + HATP^{3-}$	$pK°_{2ATP} = 4.70 \pm 0.05$	1790 ± 140	4.8 ± 0.5
$MgATP^{2-} = Mg^{2+} + ATP^{4-}$	$pK°_{MgATP} = 6.22 \pm 0.16$	-2770 ± 220	23.6 ± 0.8
$MgHATP^- = Mg^{2+} + HATP^{3-}$	$pK°_{MgHATP} = 3.65 \pm 0.17$	-2030 ± 1000	15.3 ± 3.4
$Mg_2ATP = Mg^{2+} + MgATP^{2-}$	$pK°_{Mg_2ATP} = 2.72 \pm 0.22$	-1310 ± 180	10.7 ± 0.8

on that basis. Then the activity coefficient for any species is predicted from its charge and the ionic strength.

For further calculations it is convenient to define molality equilibrium quotients K by combining the activity coefficients with the standard-state equilibrium constants $K°$. Thus, for reaction (F) one has

$$K_{\text{ref}} = \frac{[m(\text{AMP}^{2-})][m(\text{ATP}^{4-})]}{[m(\text{ADP}^{3-})]^2} \tag{27-7}$$

and

$$\ln K_{\text{ref}} = \ln K_{\text{ref}}° + 2\ln \gamma(z=-3) - \ln \gamma(z=-2) - \ln \gamma(z=-4)$$

$$= \ln K_{\text{ref}}° - \frac{2A_\gamma I^{1/2}}{1 + BI^{1/2}} \tag{27-8}$$

Similar relationships apply to all of the other reactions and K values in Table 27-2. The conversion from $K°$ to K in every case is a simple function of ionic strength, and the resulting Ks are, of course, functions of ionic strength as well as temperature. The usual matter of dimensionality arises for the equilibrium expressions on the molality basis (see Chapter 14). The required power of $m_0 = 1.0\,\text{mol kg}^{-1}$ is implied wherever needed to make the full expression dimensionless.

Introduction of equilibrium quotients for the appropriate proton and magnesium binding reactions (see Table 27-2) leads to the following expression for the total molality of ADP:

$$m(\Sigma\text{ADP}) = m(\text{ADP}^{3-})\left(1 + \frac{m(\text{H}^+)}{[K_{1\text{ADP}}]} + \frac{[m(\text{H}^+)]^2}{[K_{1\text{ADP}}K_{2\text{ADP}}]}\right.$$

$$\left. + \frac{m(\text{Mg}^{2+})}{[K_{\text{MgADP}}]} + \frac{m(\text{Mg}^{2+})m(\text{H}^+)}{[K_{\text{MgHADP}}K_{1\text{ADP}}]}\right) \tag{27-9}$$

Also, the fraction in the species ADP^{3-} is

$$f(\text{ADP}^{3-}) = \frac{m(\text{ADP}^{3-})}{m(\Sigma\text{ADP})}$$

$$= \left(1 + \frac{m(\text{H}^+)}{[K_{1\text{ADP}}]} + \frac{[m(\text{H}^+)]^2}{[K_{1\text{ADP}}K_{2\text{ADP}}]} + \frac{m(\text{Mg}^{2+})}{[K_{\text{MgADP}}]} + \frac{m(\text{Mg}^{2+})m(\text{H}^+)}{[K_{\text{MgHADP}}K_{1\text{ADP}}]}\right)^{-1}$$

$$\tag{27-10}$$

In Eqs. (27-9) and (27-10) the term for H_3ADP has been omitted since it is not significant for the pH range of interest; the corresponding line is absent in Table 27-2.

Expressions for the fractions of ATP and AMP in the forms ATP^{4-} and AMP^{2-}, respectively, can also be derived. The results are

$$f(ATP^{4-}) = \frac{m(ATP^{4-})}{m(\sum ATP)}$$

$$= \left(1 + \frac{m(H^+)}{K_{1ATP}} + \frac{[m(H^+)]^2}{[K_{1ATP}K_{2ATP}]} + \frac{m(Mg^{2+})}{[K_{MgATP}]} + \frac{[m(Mg^{2+})]^2}{[K_{MgATP}K_{Mg2ATP}]} \right.$$

$$\left. + \frac{m(Mg^{2+})m(H^+)}{[K_{MgHATP}K_{1ATP}]}\right)^{-1} \tag{27-11}$$

$$f(AMP^{2-}) = \frac{m(AMP^{2-})}{m(\sum AMP)}$$

$$= \left(1 + \frac{m(H^+)}{K_{1AMP}} + \frac{[m(H^+)]^2}{[K_{1AMP}K_{2AMP}]} + \frac{m(Mg^{2+})}{[K_{MgAMP}]}\right)^{-1} \tag{27-12}$$

In a similar manner, one can calculate the fractions of the other species present in solution, i.e., $f(HATP^{3-})$, $f(MgATP^{2-})$, $f(MgHATP^-)$, etc. In his excellent treatment of this type of system, Alberty[8] uses the symbol Z here rather than f; he also refers to this quantity as a partition function in analogy to molecular partition functions. Since this quantity is a mole fraction, the symbol x could also be used.

For calculations at other temperatures, one needs the value of each $\ln K°$ at that temperature, which we can write as $T + t$ where T is the reference temperature. Then, if $\Delta C_P°$ is constant,

$$\Delta H°_{T+t} = \Delta H°_T + t\Delta C_P° \tag{27-13a}$$

$$\Delta S°_{T+t} = \Delta S°_T + \Delta C_P° \ln\left(1 + \frac{t}{T}\right) \tag{27-13b}$$

$$\ln K°_{T+t} = \frac{\Delta S°_T}{R} - \frac{\Delta H°_T}{R(T+t)} - \frac{\Delta C_P°}{R}\left[\frac{t}{T+t} - \ln\left(1 + \frac{t}{T}\right)\right] \tag{27-13c}$$

It is interesting to expand the $\Delta C_P°$ term in powers of t/T and to rearrange the expression to yield the change in $\ln K°$. Then one obtains

$$\ln K°_{T+t} - \ln K°_T = \left(\frac{\Delta H°_T}{R}\right)\left[\frac{t}{T(T+t)}\right] + \frac{\Delta C_P°}{2R}\left(\frac{t}{T}\right)^2\left(1 - \frac{4t}{3T} + \cdots\right) \tag{27-14}$$

In the present case the largest t/T of interest is $13/298 = 0.044$. Then we see that the correction from the ΔC_P term is $0.0010\,\Delta C_P°/R$ for $\ln K°$. The smallest uncertainties in pK in Table 27-2 are 0.05 or 0.11 in $\ln K$. Thus, the effect of ΔC_P is negligible unless $\Delta C_P^\circ/R$ exceeds 100. There are estimates[5] for $\Delta C_P^\circ/R$ for the reactions in Table 27-2, but the largest is about 30, which indicates that their effect on $\ln K$ would be negligible over the 13 degree range of present interest. Then one needs only the first term on the right side of Eq. (27-14).

Although the correction of each reaction is simple, the full treatment of temperature dependency for this system is complex because the molalities of H^+ and Mg^{2+} are not fixed but rather vary according to the various equilibrium relationships. Thus, an iterative calculation scheme is required and this must be designed to converge rather than diverge. Goldberg and Tewari[5] describe their flow chart and provide many other details.

In contrast with the H^+ and Mg^{2+} binding reactions, which are rapid, the disproportionation reaction (D) is very slow in the absence of a catalyst. Tewari et al.[9] measured the equilibrium (D) with adenylate kinase as the catalyst. They determine the total amounts of AMP, ADP, and ATP chromatographically. They report results for temperatures from 286 K to 311 K, at ionic strengths from 0.06 to 0.33 mol kg^{-1}, over

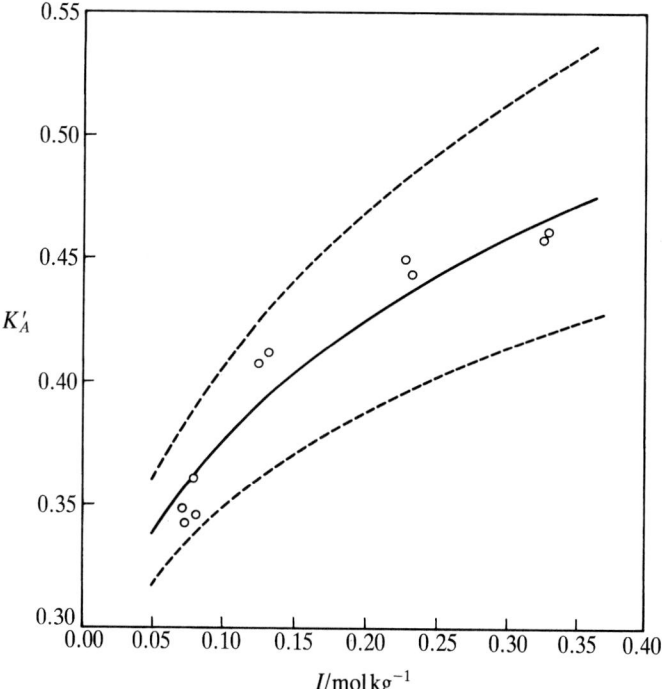

FIGURE 27-2

The equilibrium quotient K'_A [see Eq. (27-15)] as a function of ionic strength at 298.15 K. The measurements shown were obtained for the pH range 8.4–8.9 and pMg range 6.4–7.1. The solid line is the calculated curve, while dashed lines indicate the range allowed by the uncertainties in various parameters.

the pH range 6.04 to 8.87, and over the pMg range 2.22 to 7.16. Their results are first presented as the equilibrium quotient K'_A for reaction (D)

$$K'_A = \frac{m(\sum \text{AMP})m(\sum \text{ATP})}{[m(\sum \text{ADP})]^2} \tag{27-15}$$

Figure 27-2 shows the ionic-strength dependence of K'_A at 298.15 K. The agreement within the estimated range of uncertainty indicates that the treatment of activity coefficients was adequate for this problem. They also present their final results in the form of $K°$, $\Delta H°$, and $\Delta S°$ for the reference reaction (F), given in the first line of Table 27-2.

Tewari et al.[9] present additional figures showing the dependency of K'_A on pH and pMg; again, there is agreement within the estimated error ranges. They were also able to measure the heat of reaction for this disproportionation. The calculation of the predicted ΔH for reaction (D) from the model is complex in detail but straightforward in principle; it is fully described in the original papers.[5,9]

REFERENCES

1. J. Wyman and S. J. Gill, *Binding and Linkage*, University Science Books, Mill Valley, CA, 1990.
2. A. V. Hill, *J. Physiol. (London)*, **40**, iv–vii (1910).
3. S. J. Gill, E. DiCera, M. L. Doyle, G. A. Bishop, and C. H. Robert, *Biochemistry*, **26**, 3995 (1987).
4. E. DiCera, M. L. Doyle, P. R. Connelly, and S. J. Gill, *Biochemistry*, **26**, 6494 (1987).
5. R. N. Goldberg and Y. B. Tewari, *Biophys. Chem.*, **40**, 241 (1991).
6. E. A. Guggenheim, *Phil. Mag.*, **19**(7), 558 (1935).
7. C. M. Baumgarten, *Comput. Biol Med.*, **11**, 189 (1981).
8. R. A. Alberty, *Physical Chemistry*, 7th edn., Wiley, New York, 1987.
9. Y. B. Tewari, R. N. Goldberg, and J. V. Advani, *Biophys. Chem.*, **40**, 263 (1991).

CHAPTER 28

MULTICOMPONENT SOLID–VAPOR SYSTEMS

There are interesting and important systems involving equilibria of vapor and solid phases where the complexity lies in the various molecular species that might be formed in the vapor by various reactions. The condensed phase can also be a liquid, but we do not consider here any cases where the liquid-phase properties introduce complexity. In these systems, which usually involve high temperatures, the pressure is low and the gas can be taken to be ideal. However, there are several chemical reactions and associations or dissociations that might occur; thus a number of different molecular species might be present in the vapor phase. Two examples are given. First is a method whereby the molecular species present were determined initially by macroscopic thermodynamic measurements and only later verified by mass spectrometry and other molecular-level methods.

VOLATILIZATION BY REACTION WITH A GAS

Brewer and Lofgren[1] investigated copper chloride vapor by a very interesting method. In order to determine the particular species present, they studied the vapor prepared by equilibrating a mixture of H_2 and HCl with copper. They showed that all copper chloride vapor species had formulas Cu_xCl_x with equal numbers of copper and chlorine atoms. Hence one has the set of equations

$$x\text{Cu(s)} + x\text{HCl(g)} = \text{Cu}_x\text{Cl}_x(\text{g}) + \frac{x}{2}\text{H}_2(\text{g})$$

with the equilibrium constants

$$K_x = \frac{[Cu_xCl_x][H_2]^{x/2}}{[HCl]^x}$$

in which the quantities in brackets are the fugacities (or partial pressures) of the gaseous species. They measured the amount of copper volatilized, which may be related to

$$\sum x[Cu_xCl_x] = K_1\frac{[HCl]}{[H_2]^{1/2}} + 2K_2\frac{[HCl]^2}{[H_2]} + 3K_3\frac{[HCl]^3}{[H_2]^{3/2}} + \cdots \quad (28\text{-}1)$$

Measurements were made at 1309 K with H_2–HCl gas mixtures such that $[HCl]/[H_2]^{1/2}$ varied from 0.0722 to 0.432 bar$^{1/2}$. Each measurement yields an equation of the type (28-1), and simultaneous solution of the various equations for K_1, K_2, K_3, etc., gave the result $K_1 = 6.8 \times 10^{-4}$, $3K_3 = 43 \times 10^{-4}$, and all other Ks zero within experimental error.

Less extensive series of measurements were made at other temperatures from 988 to 1340 K, and all results were consistent with the result at 1309 K, indicating just the two species CuCl and Cu_3Cl_3. From these data the heats and entropies of formation as well as the Gibbs energies of CuCl and Cu_3Cl_3 could be calculated.

At 1119 K they found the vapor pressure to be 0.067 ± 0.008 bar and that 0.212 g was evaporated to saturate 0.0105 mole of argon carrier gas at 1 bar. From these data one calculates that the 0.212 g constituted 7.04×10^{-4} mole of cuprous chloride vapor, or that the mean molecular mass was 300 ± 36. This result is consistent with an average composition very close to Cu_3Cl_3 (mol. wt. = 297). The ratio of CuCl to Cu_3Cl_3 decreases with decrease in temperature and would be very small at 1119 K.

Subsequent to this determination of Brewer and Lofgren[1] in 1950, electron diffraction,[2] infrared spectra,[3] and other molecular-level experiments confirmed the species assignment and provided the parameters for a third-law treatment. The JANAF tables[4] give the Gibbs-energy function, as reproduced in Appendix 6, Table A6-4; also, they report for the enthalpy of sublimation of CuCl(s) to Cu_3Cl_3 at 298 K the values $\Delta_{sub}H°_{298} = 18\,800R$ (second law) and $18\,740R$ (third law). The agreement of these two values offers confirmation for the molecular parameters adopted. For the enthalpy of formation from the elements, the recommended value is

$$\frac{\Delta_f H}{R} = -31\,100 + 300\,K$$

TUNGSTEN OXIDES

In three different investigations[5–7] ranging from 1300 K to 2500 K, the dominant vapor species of tungsten oxide varied from W_3O_9 to WO_2 with substantial WO under some conditions. Ackermann and Rauh[5] examined the vapor in equilibrium with various solid tungsten oxides in the 1300–1500 K range and found W_3O_9 to be the dominant species, with smaller amounts of W_2O_6 and other species. Measurements were by mass spectrometry of the vapor effusing from a cell containing, in most cases, two solids. For the simplest case the solids were W and WO_2; in this case the cell itself could be tungsten metal with solid WO_2 inside. In other cases a platinum cell was used with either W + WO_2, WO_2 + $W_{18}O_{49}$, or even higher oxides. The results were treated by the second-law method, a plot of $\log P_i$ vs. $1/T$, with the partial pressure determined by the mass spectrometric intensity, the ionization cross section, and appropriate calibration factors.

For the W–WO_2 solids the reaction to form W_3O_9 is

$$\tfrac{9}{2} WO_2(s) = W_3O_9(g) + \tfrac{3}{2} W(s)$$

for which they report

$$\frac{\Delta H°}{R} = 72\,020 \pm 1500 \,\text{K} \qquad \frac{\Delta S°}{R} = 35.1$$

The uncertainty in $\Delta H°$ is just that of the slope on the $\log P_i$ vs. $1/T$ plot and can be judged from these measurements, but uncertainties in cross sections and other temperature-independent factors affect the $\Delta S°$.

The standard enthalpy and entropy of formation of $WO_2(s)$ in the 1300–1500 K range are

$$\frac{\Delta_f H°}{R} = -69\,020 \,\text{K} \qquad \frac{\Delta_f S°}{R} = -20.1$$

One can then calculate for the formation of $W_3O_9(g)$ from W(s) and $O_2(g)$ the values

$$\frac{\Delta_f H°}{R} = -238\,570 \,\text{K} \qquad \frac{\Delta_f S°}{R} = -55.4$$

Ackermann and Rauh[5] also investigated the solid oxide $WO_{2.96}$, which evaporates congruently, with measurements of total weight loss as well as by mass spectrometry. In this case the vapor-phase information was used to calculate the formation properties of the solid. For both solid sources, $WO_{2.96}$ and W + WO_2, vapor properties of W_2O_6 were determined in addition to those for W_3O_9. For W_2O_6 a subsequent study[6] is probably more reliable, however, because it is less subject to error from a contribution to the mass spectrometric $W_2O_6^+$ signal due to fragmentation of $W_3O_9^+$ in addition to that from the simple ionization of W_2O_6.

In order to obtain better measurements on oxides simpler than W_3O_9, Norman and Staley[6] reduced the activity of WO_3 in the condensed phase by dissolution in a melt of CaO–Al_2O_3–SiO_2. Only the relative properties of different vapor species were measured; thus, the exact characteristics of the liquid phase were not needed. This WO_3-containing sample was placed in an iridium Knudsen cell, which is a cell with a very small hole such that equilibrium is attained for the vapor that effuses into the surrounding vacuum where it is analyzed. These authors used mass spectrometry with a low ionizing voltage, usually 20 eV, to minimize ion-fragmentation effects.

Measurements were made for WO_3, W_2O_6, and W_3O_9 with two different source samples. Sample A was measured over the range 1700–1900 K, while sample B, richer in WO_3, was measured from 1920 K to 2060 K. The resulting enthalpies and entropies of reaction from the second-law analysis are given in Table 28-1. The authors report that a separate analysis was made for the disproportionation reaction $2\,W_2O_6 = WO_3 + W_3O_9$, shown on the last line.

For these molecules WO_3, W_2O_6, W_3O_9, there is not enough molecular information to yield really independent statistical calculations of the entropies and Gibbs energy functions. The JANAF Project[4] does report such calculations, but on an empirical basis. For W_3O_9 a plausible structure was assumed in which each W is surrounded tetrahedrally by four O; then the WO_4 units are linked by sharing oxygens in a six-membered ring. The distances can be estimated, but the choices of the large number of vibration frequencies were arbitrarily made to yield the observed entropy of 110.3R at 1450 K. Thus, this is essentially a method of estimating the heat capacity of W_3O_9 in order to obtain its properties at 298 K.

The situation for W_2O_6 is entirely parallel to that for W_3O_9. In these cases one does not have a real third-law treatment, but rather an interesting estimation of properties consistent with, but not determined by, the second-law treatment. For WO_3 the situation is different; it is discussed below.

De Maria et al.[7] report interesting results concerning the vapor in equilibrium with solid tungsten and aluminum oxide at temperatures in the range 2100–2500 K. They measured by mass spectrometry the concentrations of various species effusing from a tungsten Knudsen cell containing some Al_2O_3. At low pressure and this very high temperature, the oxygen is largely monatomic, and the tungsten species are WO, WO_2, and WO_3. Aluminum oxide species were also observed, but will not be considered here.

Both De Maria et al.[7] and JANAF[4] present third-law treatments for this system; we summarize briefly the latter. For WO(g), Weltner and McLeod[8] measured the vibration frequency, 1065 cm^{-1}, and the ground-state electronic configuration, $^3\Sigma^-$, which implies a multiplicity $g_0 = 3$. The bond distance and the small anharmonicity effects were estimated. Then the methods of Chapters 5 and 20 yield the Gibbs energy function. The calculations for $WO_2(g)$ and $WO_3(g)$ are similar; the ground electronic state of WO_3 is $^1\Sigma$ with $g_0 = 1$, but that for WO_2, $g_0 = 3$, was an estimate based on

TABLE 28-1
Enthalpies and entropies of reaction determined for 1700–1900 K (A) and 1920–2060 K (B)

Reaction	$[\Delta H°/R]/10^3$ K		$\Delta°S/R$	
	(A)	(B)	(A)	(B)
$2\,WO_3 = W_2O_6$	−63.9 ± 0.5	−63.4	−20.6	−19.9
$3\,WO_3 = W_3O_9$	−121.8 ± 1	−119.8	−40.3	−36.7
$2\,W_2O_6 = WO_3 + W_3O_9$	7.0 ± 2		3.0 ± 1.5	

W^{4+} ion. The experimental vibration frequency data are incomplete for WO_2 and WO_3, but there is a reasonable basis for the estimates used.

For the reaction forming WO_n from $O(g)$ and $W(s)$,

$$W(s) + n\,O(g) = WO_n(g)$$

the Gibbs-energy relationships are

$$-\frac{\Delta_r G}{RT} = \ln P_{WO_n} - n \ln P_O$$

$$= -\Delta_r\left(\frac{G^\circ_T - H^\circ_{298}}{RT}\right) - \frac{\Delta_r H^\circ_{298}}{RT}$$

The results for $\Delta_r H^\circ_{298}$ can be converted to values for formation from $O_2(g)$. Unfortunately, the resulting values of $\Delta_f H^\circ_{298}$ are not constant but show significant trends with temperature for each of the three oxides. The average values and the uncertainties (trends) are shown in Table 28-2. Other reaction systems[9,10] yield values for WO_2 and WO_3, which are also shown in Table 28-2; the uncertainties in these values are similarly large.

Surprisingly, De Maria et al.[7] did not present a second-law analysis of their data. The large trends in ΔH°_{298} from the third-law treatment indicate that there were either large errors in the Gibbs-energy functions or large and systematic errors in the equilibrium compositions. For WO, at least, the molecular properties appear to be quite certain; thus, systematic errors appear to be indicated. Nevertheless, a second-law presentation, discussed in the next section, shows the expected pattern. Also second- and third-law treatments of the data of Ackermann and Thorn[10] are in reasonable agreement for the reaction

$$3\,MgO(cr) + W(cr) = 3\,Mg(g) + WO_3(g)$$

This favors their value of -36 for $\Delta_f H^\circ_{298}$. Thus, we believe that the results shown in Table 28-2 are probably reliable within the large uncertainties indicated, but new measurements are needed to reduce those uncertainties.

TABLE 28-2
Enthalpies of formation at 298.15 K by the third-law method[4]

	($\Delta_f H^\circ_{298}/R/10^3$ K)	
Source	De Maria et al.[7]	Other
WO	51 ± 10	
WO_2	9 ± 8	7[9]
WO_3	−33 ± 11	−37,[9] −36[10]

ENTROPIES AND ENTHALPIES IN VAPORIZATION REACTIONS

In the vaporization of a pure liquid or solid the saturation pressure increases so rapidly with temperature that the vapor, ideal at low T and P, becomes very nonideal as the critical point is approached. If one considers this imperfection as the association of single molecules to dimers, trimers, etc., the association increases with temperature. We now consider, instead, the reaction of a gas with a solid at varying temperature but constant pressure.

Consider the processes

$$A(s) + \tfrac{1}{2}B_2(g) = AB(g)$$

$$A(s) + B_2(g) = AB_2(g)$$

$$A(s) + \tfrac{3}{2}B_2(g) = AB_3(g)$$

If we examine these reactions, we note that the first one corresponds to the production of $\tfrac{1}{2}$ mole of gas. The second corresponds to no net change of moles of gas, and the third corresponds to loss of $\tfrac{1}{2}$ mole of gas. Since the translational contribution to the entropy of the gas is the largest of all the factors contributing to the entropy, we predict that the entropy change for the first reaction would be positive; the entropy change for the second reaction would be close to zero and the entropy change for the third reaction negative. Typical averages[11] for a series of halide reactions are $\Delta S°/R = 12$ for the first reaction, 2 for the second, and -5 for the third. The second reaction usually has a small positive entropy increase rather than a zero entropy change because vibrational and rotational terms are larger for the AB_2 species. This information allows one to estimate the pattern of behavior for the various reactions in terms of their enthalpies.

We first recall that

$$R \ln K = -\frac{\Delta G°}{T} = \Delta S° - \frac{\Delta H°}{T}$$

Then, if $\Delta H°$ is positive (endothermic), $\Delta G°$ will be positive and K very small unless $\Delta S°$ is significantly positive. Thus, only AB is expected to be formed with significantly positive $\Delta H°$ and AB_2 marginally with $\Delta H°$ near zero. If $\Delta H°$ is negative (exothermic), however, any of the AB_x species can be formed. But AB_3 will be formed only with $-\Delta H°$ large, and then the equilibrium constant will decrease rapidly with increase in temperature. The K for AB, however, may change very little with temperature.

Next, consider the case where the reacting gas is monatomic, as is oxygen at very high temperature and low pressure. Then, using tungsten as the example, the reactions are

$$W(s) + O(g) = WO(g)$$

$$W(s) + 2O(g) = WO_2(g)$$

$$W(s) + 3O(g) = WO_3(g)$$

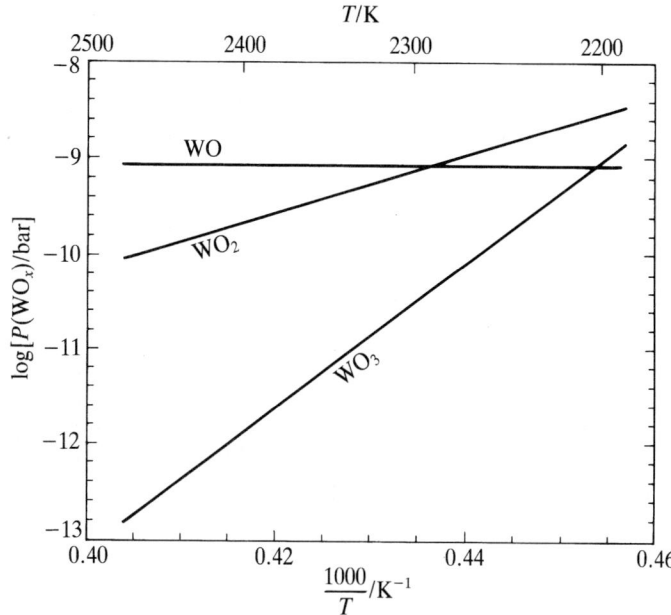

FIGURE 28-1
Log pressures in bars of tungsten oxides in equilibrium with solid tungsten and monatomic oxygen at 10^{-7} bar. Values calculated from measurements of De Maria et al.[7] by conversion of $P(O)$ to 10^{-7} bar.

$\Delta S°$ for the formation of WO will be small but positive and one expects little change in P_{WO} with temperature at constant partial pressure of oxygen. For the formation of WO_2, $\Delta S°$ will be negative, and even a larger negative quantity for WO_3. Then a significant formation of WO_2 is expected only if $-\Delta H°$ is substantial, whereupon P_{WO_2} will decrease with rise in temperature. For P_{WO_3} one expects a more rapid decrease. The experimental results of De Maria et al.[7] for the tungsten–oxygen system, shown in Fig. 28-1, illustrate this pattern clearly.

REFERENCES

1. L. Brewer and N. L. Lofgren, *J. Am. Chem. Soc.*, **72**, 3038 (1950).
2. C. H. Wong and V. Schomaker, *J. Phys. Chem.*, **61**, 358 (1957).
3. W. Klemperer, S. A. Rice, and R. S. Berry, *J. Am. Chem. Soc.*, **79**, 1810 (1957).
4. M. W. Chase, Jr., et al., "JANAF Thermochemical Tables," 3d edn., *J. Phys. Chem. Ref. Data*, **14** (supplement No. 1) (1985).
5. R. J. Ackermann and E. G. Rauh, *J. Phys. Chem.*, **67**, 2596 (1963).
6. J. H. Norman and H. G. Staley, *J. Chem. Phys.*, **43**, 3804 (1965).
7. G. De Maria, R. P. Burns, J. Drowart, and M. G. Inghram, *J. Chem. Phys.*, **32**, 1373 (1960).
8. W. Weltner, Jr. and D. McLeod, Jr., *J. Mol. Spectrosc.*, **17**, 276 (1965).
9. W. A. Chuptka, J. Berkowitz, and C. F. Giese, *J. Chem. Phys.*, **30**, 827 (1959).

10. R. J. Ackermann and R. J. Thorn, *J. Phys. Chem.*, **64**, 350 (1960), and references there cited.
11. L. Brewer, L. A. Bromley, P. W. Gilles, and N. L. Lofgren, in *The Chemistry and Metallurgy of Miscellaneous Materials: Thermodynamics*, L. L. Quill, ed., McGraw-Hill, New York, 1950, p. 184.

PROBLEMS

28-1. Calculate $\Delta S°_{298}/R$ for each of the following reactions from data in Appendix 6. Compare the values for the more complex cases with that for KCl.

$$KCl(cr) = KCl(g)$$

$$\tfrac{1}{2} Hg_2Cl_2(cr) = \tfrac{1}{2} Hg(g) + \tfrac{1}{2} HgCl_2(g)$$

$$2 FeCl_3(cr) = 2 FeCl_2(cr) + Cl_2(g)$$

$$4 CuO(cr) = 2 Cu_2O(cr) + O_2(g)$$

28-2. Compare the equilibrium partial pressures calculated for Al(g), O_2(g), O(g), WO_2(g), and WO_3(g) when Al_2O_3(cr) is heated in a tungsten Knudsen cell at 2000 K. Data are available in Appendix 6. This system is discussed briefly in the text.

APPENDIX 1

BOLTZMANN DISTRIBUTION LAW AND THE TRANSLATIONAL ENTROPY OF AN IDEAL GAS

We first consider the Boltzmann distribution law that governs the distribution of particles among quantum states of differing energy. The reader may be familiar with this expression and may choose to proceed directly to Eq. (A1-5). However, we shall give a simple derivation of the Boltzmann distribution law at this point.

We assume that there are N particles and an array of one-particle quantum states of varying energy. The particles are assumed not to interact with one another; hence the energy for an N-particle system is just the sum of the energies of the N one-particle states. Each different set of one-particle quantum numbers yields a different N-particle quantum state. In order to make the problem tractable, we segregate the N particles into groups of N_i particles, each of which has the same energy ε_i.

Now we must obtain the number of individual quantum states that a set of N_i particles of a given energy may have if there are g_i single-particle states of that energy. Let us first consider a very simple example of two particles and four single-particle states in the accompanying diagram. The two-particle states fall into two groups: those with only one particle in any one-particle state and those with both particles in the same one-particle state. In our example there are 12 states of the first type and 4 of the second type, or 16 in all. In a more general case with g_i single-particle states but still $N_i = 2$,

Quantum states for system of two particles and four one-particle states

One-particle state			
1	2	3	4
Type 1			
A	B		
B	A		
A		B	
B		A	
A			B
B			A
	A	B	
	B	A	
	A		B
	B		A
		A	B
		B	A
Type 2			
AB			
	AB		
		AB	
			AB

the result would be a total of g_i^2 states, of which $g_i^2 - g_i$ would be of the first type and g_i would be of the second type.

Atoms and molecules of the same sort, however, are not labeled A, B, etc., but rather are completely indistinguishable from one another. There is only one two-particle quantum state with one particle in state 1 and one in state 2. Thus the first two states in the diagram are really the same state; likewise the third and fourth are a single state, and so on. Thus the actual number of states of the first type is $\frac{1}{2}(g_i^2 - g_i)$. For reasons we shall not discuss, the second type of state exists for some atoms or molecules and not for others. Thus, the total number is either $\frac{1}{2}(g_i^2 + g_i)$ or $\frac{1}{2}(g_i^2 - g_i)$. We readily see, however, that, if g_i is a large number, the second term is negligible and the result is $g_i^2/2$.

Now we must generalize this result for N_i particles. The total number of states of both types, if the particles were distinguishable, would clearly be $g_i^{N_i}$. The number of rearrangements of N_i particles is $N_i!$; consequently, the desired result is

$$W_i = \frac{g_i^{N_i}}{N_i!} \qquad (\text{A1-1})$$

This result is valid only in the approximation $g_i \gg N_i$, where states of the second type have negligible effect.

Actually we shall use the logarithm of the probability, and it may be simplified by the use of Stirling's approximation† for a factorial as follows:

$$\ln W_i = N_i \ln g_i - \ln N_i!$$
$$= N_i \ln g_i - N_i \ln N_i + N_i$$
$$= N_i\left(1 + \ln \frac{g_i}{N_i}\right) \tag{A1-2}$$

The probability of the complete array of N particles is

$$\ln W = \sum_i \ln W_i$$
$$= \sum_i N_i\left(1 + \ln \frac{g_i}{N_i}\right) \tag{A1-3}$$

where this sum, and those to follow, cover all values of i.

The most probable distribution of N_i values will be that which maximizes $\ln W$ subject to the conditions of a fixed total number of particles and fixed total energy. This maximum is found by the use of Lagrange's method of undetermined multipliers. We maximize the function

$$\ln W - \alpha \sum_i N_i - \beta \sum_i \varepsilon_i N_i$$

where α and β are the undetermined multipliers. The partial derivative with respect to each N_i vanishes,

$$\frac{\partial}{\partial N_i}\left[\sum_i N_i\left(1 - \alpha - \beta\varepsilon_i + \ln \frac{g_i}{N_i}\right)\right] = 0$$

$$-\alpha - \beta\varepsilon_i + \ln \frac{g_i}{N_i} = 0$$

and there results

$$N_i = g_i e^{-\alpha} e^{-\beta\varepsilon_i} \tag{A1-4}$$

Since there are N particles in all and the energy is fixed, α and β can be evaluated from the two equations

$$N = \sum_i N_i = e^{-\alpha} \sum_i g_i e^{-\beta\varepsilon_i}$$
$$U = \sum_i \varepsilon_i N_i = e^{-\alpha} \sum_i g_i \varepsilon_i e^{-\beta\varepsilon_i}$$

---

† In case the reader is unfamiliar with Stirling's formula, we note that

$$\ln N! \cong \int_1^N \ln y \, dy = [y \ln y - y]_1^N = N \ln N - N + 1$$

If N is large, the 1 may be neglected.

Since β controls the total energy by the relative distribution over high and low energy states, it is clearly a sort of temperature. We shall presently evaluate the energy of translation for an ideal gas, which will serve to determine β. In the meantime we shall avoid rewriting formulas by stating that the relationship is $\beta = (kT)^{-1}$. Equation (A1-4) is then the desired result.

The Boltzmann distribution law may be written in the form

$$N_i = e^{-\alpha} g_i e^{-\varepsilon_i/kT} \tag{A1-5}$$

The factor $e^{-\alpha}$ may be evaluated from the condition that the sum of N_i must be the total number of particles, as we show below and in Chapter 5.

We turn now to the calculation of the translational entropy of an ideal gas. The basic relation between entropy and probability was given in Eq. (5-2)

$$S = k \ln W \tag{5-2}$$

Next we need the basic quantum-mechanical equation for a particle in a box.[1] Before proceeding, however, we remind the reader of Eq. (5-3) and the associated discussion, which indicates that only a very approximate calculation of W suffices to yield a very accurate entropy.

The quantum mechanics of a particle moving freely within a cubical box[1] yields the following equation for the energies of the quantum states,

$$\varepsilon = \frac{h^2(n_1^2 + n_2^2 + n_3^2)}{8ml^2} \tag{A1-6}$$

where m is the mass of the particle, l is the edge length of the box, h is Planck's constant, and n_1, n_2, n_3 are three quantum numbers, each of which can have any positive integral values: 1, 2, 3, etc. We shall be interested in the statistical distribution of these states with respect to energy. For this purpose it is convenient to consider a three-dimensional cartesian-coordinate space with integral spacing of points in each direction. Each point in the octant of all positive coordinates then represents a quantum state, and the energy of the state is proportional to the square of the distance from the origin. Thus

$$r^2 = n_1^2 + n_2^2 + n_3^2 \tag{A1-7}$$

$$\varepsilon = r^2 \frac{h^2}{8ml^2} \tag{A1-8}$$

We shall need to have the number of quantum states g_i that have, within narrow limits, the same energy. This is readily seen to be the volume of the octant of a spherical shell of radius r and thickness δr:

$$g_i = \frac{\pi}{2} r^2 \delta r \tag{A1-9}$$

We now substitute the expressions for ε and for g into the Boltzmann distribution law, Eq. (A1-5), and proceed to evaluate the quantity α. We set the sum over all particles to the Avogadro number N_A:

$$\sum_i N_i = N_A = e^{-\alpha} \sum_i g_i e^{-\varepsilon_i/kT}$$

$$e^\alpha = \frac{\pi}{2N_A} \int_0^\infty r^2 \, dr \exp\left(\frac{-r^2 h^2}{8ml^2 kT}\right)$$

$$= \frac{l^3}{N_A} \left(\frac{2\pi mkT}{h^2}\right)^{3/2}$$

Here the sum is replaced by an integral, and we also note that l^3 is the volume of the box V. Thus

$$\alpha = \ln\frac{V}{N_A} + \frac{3}{2} \ln\frac{2\pi mkT}{h^2} \tag{A1-10}$$

Also the energy may be calculated as follows:

$$U = \sum_i N_i \varepsilon_i$$

$$= e^{-\alpha} \frac{\pi h^2}{16ml^2} \int_0^\infty r^4 \, dr \exp\left(\frac{-r^2 h^2}{8ml^2 kT}\right)$$

$$= \tfrac{3}{2} N_A kT = \tfrac{3}{2} RT \tag{A1-11}$$

This is the familiar result, which confirms the relationship of β to T assumed above.

We may now calculate the logarithm of the total probability from Eq. (A1-3) and the Boltzmann law:

$$\ln\frac{g_i}{N_i} = \alpha + \frac{\varepsilon_i}{kT}$$

$$\ln W = \sum_i N_i \left(1 + \alpha + \frac{\varepsilon_i}{kT}\right)$$

But from Eq. (A1-1) this becomes

$$\ln W = N_A(\alpha + \tfrac{5}{2}) \tag{A1-12}$$

and we may now calculate the entropy for 1 mole of gas from Eq. (5-2) with the substitution of α from Eq. (A1-10):

$$S = R(\alpha + \tfrac{5}{2})$$

$$= R\left(\frac{5}{2} + \ln\frac{V}{N_A} + \frac{3}{2}\ln\frac{2\pi mkT}{h^2}\right) \tag{A1-13}$$

To convert this to the usual form of the Sackur–Tetrode equation, one substitutes the atomic or molecular mass $M = N_A m$, whence

$$S = R(\ln V + \tfrac{3}{2}\ln T + \tfrac{3}{2}\ln M) + C \tag{A1-14}$$

$$C = R\left(\frac{5}{2} + \frac{3}{2}\ln\frac{2\pi k}{h^2} - \frac{5}{2}\ln N_A\right) \tag{A1-15}$$

This result was given without proof in Chapter 5. Note that this expression for the entropy is not valid at $T = 0$, but it is accurate at any temperature where a real gas can be considered to be ideal.

As a final comment, we note that we calculated only the number of quantum states represented by the most probable distribution of particles N_i over the possible energies ε_i. All the other states of energy U constitute a very large number, but the addition of this number to the value of W obtained above has a negligible effect on the calculated entropy. Similar refinements, such as allowing fluctuations in U about its most probable value, also have no significant effect on the entropy of a macroscopic amount of gas.

REFERENCE

1. This result is given in various elementary texts on quantum mechanics, for example, L. Pauling and E. B. Wilson, Jr., *Introduction to Quantum Mechanics*, McGraw-Hill, New York, 1935.

APPENDIX 2

DEBYE FUNCTIONS FOR THE THERMODYNAMIC PROPERTIES OF SOLIDS

The chief features of the derivation of the Debye heat capacity theory are outlined very briefly, after which the final equations and tables of values are given.

The vibrations of continuous solid media are distributed with a density of modes of oscillation proportional to the square of the frequency. Thus the number of modes of frequency between ν and $\nu + d\nu$ may be written

$$\rho(\nu)\,d\nu = c\nu^2\,d\nu \tag{A2-1}$$

where c is a constant. We know, however, that in a solid composed of N atoms there are $3N$ total modes of vibration. This follows because the interaction of the atoms can only rearrange the modes and frequencies of motion but cannot change the total number. Debye assumes that Eq. (A2-1) holds up to a certain maximum frequency of vibration ν_D that may be evaluated by calculating the total number of modes:

$$3N = \int_0^{\nu_D} c\nu^2\,d\nu = \frac{c\nu_D^3}{3} \tag{A2-2}$$

where

$$c = \frac{9N}{\nu_D^3}$$

TABLE A2-1
Debye heat-capacity function, $C_V/3R$, as a function of θ_D/T

When $\dfrac{\theta_D}{T} \geq 16$, $\dfrac{C_V}{3R} = 77.927\left(\dfrac{T}{\theta_D}\right)^3$

θ_D/T	0.0	0.1	0.2	0.3	0.4	0.5	0.6	0.7	0.8	0.9	1.0
0.0	1.000 0	0.999 5	0.998 0	0.995 5	0.992 0	0.987 6	0.982 2	0.975 9	0.968 7	0.960 6	0.951 7
1.0	0.951 7	0.942 0	0.931 5	0.920 3	0.908 5	0.896 0	0.882 8	0.869 2	0.855 0	0.840 4	0.825 4
2.0	0.825 4	0.810 0	0.794 3	0.778 4	0.762 2	0.745 9	0.729 4	0.712 8	0.696 1	0.679 4	0.662 8
3.0	0.662 8	0.646 1	0.629 6	0.613 2	0.596 8	0.580 7	0.564 7	0.549 0	0.533 4	0.518 1	0.503 1
4.0	0.503 1	0.488 3	0.473 8	0.459 5	0.445 6	0.432 0	0.418 7	0.405 7	0.393 0	0.380 7	0.368 6
5.0	0.368 6	0.356 9	0.345 5	0.334 5	0.323 7	0.313 3	0.303 1	0.293 3	0.283 8	0.274 5	0.265 6
6.0	0.265 6	0.256 9	0.248 6	0.240 5	0.232 6	0.225 1	0.217 7	0.210 7	0.203 8	0.197 2	0.190 9
7.0	0.190 9	0.184 7	0.178 8	0.173 0	0.167 5	0.162 2	0.157 0	0.152 1	0.147 3	0.142 6	0.138 2
8.0	0.138 2	0.133 9	0.129 7	0.125 7	0.121 9	0.118 2	0.114 6	0.111 1	0.107 8	0.104 6	0.101 5
9.0	0.101 5	0.098 47	0.095 58	0.092 80	0.090 11	0.087 51	0.085 00	0.082 59	0.080 25	0.078 00	0.075 82
10.0	0.075 82	0.073 72	0.071 69	0.069 73	0.067 83	0.066 00	0.064 24	0.062 53	0.060 87	0.059 28	0.057 73
11.0	0.057 73	0.056 24	0.054 79	0.053 39	0.052 04	0.050 73	0.049 46	0.048 23	0.047 05	0.045 90	0.044 78
12.0	0.044 78	0.043 70	0.042 65	0.041 64	0.040 66	0.039 70	0.038 78	0.037 88	0.037 01	0.036 17	0.035 35
13.0	0.035 35	0.034 55	0.033 78	0.033 03	0.032 30	0.031 60	0.030 91	0.030 24	0.029 59	0.028 96	0.028 35
14.0	0.028 35	0.027 76	0.027 18	0.026 61	0.026 07	0.025 53	0.025 01	0.024 51	0.024 02	0.023 54	0.023 07
15.0	0.023 07	0.022 62	0.022 18	0.021 74	0.021 32	0.020 92	0.020 52	0.020 13	0.019 75	0.019 38	0.019 02

TABLE A2-2
Debye function for energy content, $(U - U_0)/3RT$, as a function of θ_D/T

When $\dfrac{\theta_D}{T} \geq 16$, $\dfrac{U - U_0}{3RT} = 19.482 \left(\dfrac{T}{\theta_D}\right)^3$

θ_D/T	0.0	0.1	0.2	0.3	0.4	0.5	0.6	0.7	0.8	0.9	1.0
0.0	1.000 0	0.963 0	0.927 0	0.892 0	0.858 0	0.825 0	0.792 9	0.761 9	0.731 8	0.702 6	0.674 4
1.0	0.674 4	0.647 1	0.620 8	0.595 4	0.570 8	0.547 1	0.524 3	0.502 3	0.481 1	0.460 7	0.441 1
2.0	0.441 1	0.422 3	0.404 2	0.386 8	0.370 1	0.354 1	0.338 8	0.324 1	0.310 0	0.296 5	0.283 6
3.0	0.283 6	0.271 2	0.259 4	0.248 1	0.237 3	0.226 9	0.217 0	0.207 6	0.198 6	0.190 0	0.181 7
4.0	0.181 7	0.173 9	0.166 4	0.159 2	0.152 4	0.145 9	0.139 7	0.133 8	0.128 1	0.122 7	0.117 6
5.0	0.117 6	0.112 7	0.108 0	0.103 6	0.099 30	0.095 24	0.091 37	0.087 68	0.084 15	0.080 79	0.077 58
6.0	0.077 58	0.074 52	0.071 60	0.068 81	0.066 15	0.063 60	0.061 18	0.058 86	0.056 64	0.054 53	0.052 51
7.0	0.052 51	0.050 57	0.048 73	0.046 96	0.045 27	0.043 66	0.042 11	0.040 63	0.039 21	0.037 86	0.036 56
8.0	0.036 56	0.035 32	0.034 13	0.032 98	0.031 89	0.030 84	0.029 83	0.028 87	0.027 94	0.027 05	0.026 20
9.0	0.026 20	0.025 38	0.024 59	0.023 84	0.023 11	0.022 41	0.021 74	0.021 09	0.020 47	0.019 87	0.019 30
10.0	0.019 30	0.018 74	0.018 21	0.017 69	0.017 20	0.016 72	0.016 26	0.015 81	0.015 38	0.014 97	0.014 57
11.0	0.014 57	0.014 18	0.013 81	0.013 45	0.013 11	0.012 77	0.012 45	0.012 13	0.011 83	0.011 53	0.011 25
12.0	0.011 25	0.010 98	0.010 71	0.010 45	0.010 20	0.009 96	0.009 73	0.009 50	0.009 28	0.009 07	0.008 86
13.0	0.008 86	0.008 66	0.008 46	0.008 27	0.008 09	0.007 91	0.007 74	0.007 57	0.007 41	0.007 25	0.007 10
14.0	0.007 10	0.006 95	0.006 80	0.006 66	0.006 52	0.006 39	0.006 26	0.006 13	0.006 01	0.005 89	0.005 77
15.0	0.005 77	0.005 66	0.005 55	0.005 44	0.005 33	0.005 23	0.005 13	0.005 03	0.004 94	0.004 85	0.004 76

TABLE A2-3
Debye function for Helmholtz energy, $(A - U_0)/3RT$, as a function of θ_D/T

When $\dfrac{\theta_D}{T} \geq 16$, $-\dfrac{A - U_0}{3RT} = 6.494 \left(\dfrac{T}{\theta_D}\right)^3$

θ_D/T	0.0	0.1	0.2	0.3	0.4	0.5	0.6	0.7	0.8	0.9	1.0
0.0	∞	2.673 2	2.016 8	1.647 6	1.395 6	1.207 7	1.060 2	0.940 3	0.840 5	0.756 0	0.683 5
1.0	0.683 5	0.620 5	0.565 3	0.516 6	0.473 4	0.434 9	0.400 3	0.369 2	0.341 0	0.315 6	0.292 5
2.0	0.292 5	0.271 4	0.252 2	0.234 6	0.218 5	0.203 7	0.190 1	0.177 6	0.166 1	0.155 4	0.145 6
3.0	0.145 6	0.136 5	0.128 1	0.120 3	0.113 0	0.106 3	0.100 0	0.094 23	0.088 82	0.083 77	0.079 06
4.0	0.079 06	0.074 67	0.070 57	0.066 74	0.063 16	0.059 81	0.056 67	0.053 73	0.050 97	0.048 39	0.045 96
5.0	0.045 96	0.043 68	0.041 54	0.039 52	0.037 63	0.035 84	0.034 16	0.032 58	0.031 08	0.029 67	0.028 34
6.0	0.028 34	0.027 09	0.025 90	0.024 77	0.023 71	0.022 71	0.021 75	0.020 85	0.020 00	0.019 18	0.018 41
7.0	0.018 41	0.017 68	0.016 99	0.016 33	0.015 70	0.015 10	0.014 54	0.014 00	0.013 48	0.012 99	0.012 52
8.0	0.012 52	0.012 08	0.011 65	0.011 24	0.010 85	0.010 48	0.010 13	0.009 79	0.009 46	0.009 15	0.008 86
9.0	0.008 86	0.008 57	0.008 30	0.008 04	0.007 79	0.007 55	0.007 31	0.007 09	0.006 88	0.006 67	0.006 48
10.0	0.006 48	0.006 29	0.006 11	0.005 93	0.005 76	0.005 60	0.005 44	0.005 29	0.005 15	0.005 01	0.004 87
11.0	0.004 87	0.004 74	0.004 62	0.004 50	0.004 38	0.004 27	0.004 16	0.004 05	0.003 95	0.003 85	0.003 76
12.0	0.003 76	0.003 66	0.003 57	0.003 49	0.003 40	0.003 32	0.003 25	0.003 17	0.003 10	0.003 02	0.002 96
13.0	0.002 96	0.002 89	0.002 82	0.002 76	0.002 70	0.002 64	0.002 58	0.002 53	0.002 47	0.002 42	0.002 37
14.0	0.002 37	0.002 32	0.002 27	0.002 22	0.002 17	0.002 13	0.002 09	0.002 04	0.002 00	0.001 96	0.001 92
15.0	0.001 92	0.001 89	0.001 85	0.001 81	0.001 78	0.001 74	0.001 71	0.001 68	0.001 65	0.001 62	0.001 59

TABLE A2-4
Debye entropy function $S/3R$ as a function of θ_D/T

When $\dfrac{\theta_D}{T} \geq 16$, $\dfrac{S}{3R} = 25.976 \left(\dfrac{T}{\theta_D}\right)^3$

θ_D/T	0.0	0.1	0.2	0.3	0.4	0.5	0.6	0.7	0.8	0.9	1.0
0.0	∞	3.636 2	2.943 8	2.539 6	2.253 6	2.032 7	1.853 1	1.702 2	1.572 3	1.458 7	1.357 9
1.0	1.357 9	1.267 6	1.186 1	1.112 0	1.044 2	0.982 0	0.924 6	0.871 4	0.822 2	0.776 3	0.733 6
2.0	0.733 6	0.693 7	0.656 4	0.621 4	0.588 6	0.557 8	0.528 9	0.501 7	0.476 1	0.451 9	0.429 2
3.0	0.429 2	0.407 7	0.387 5	0.368 3	0.350 3	0.333 2	0.317 1	0.301 8	0.287 4	0.273 7	0.260 8
4.0	0.260 8	0.248 6	0.237 0	0.226 0	0.215 6	0.205 7	0.196 4	0.187 5	0.179 1	0.171 1	0.163 6
5.0	0.163 6	0.156 4	0.149 6	0.143 1	0.136 9	0.131 1	0.125 5	0.120 3	0.115 2	0.110 5	0.105 9
6.0	0.105 9	0.101 6	0.097 50	0.093 58	0.089 86	0.086 31	0.082 93	0.079 71	0.076 64	0.073 71	0.070 92
7.0	0.070 92	0.068 26	0.065 72	0.063 29	0.060 97	0.058 76	0.056 65	0.054 63	0.052 70	0.050 85	0.049 08
8.0	0.049 08	0.047 39	0.045 78	0.044 23	0.042 74	0.041 32	0.039 96	0.038 66	0.037 41	0.036 21	0.035 06
9.0	0.035 06	0.033 95	0.032 89	0.031 87	0.030 90	0.029 96	0.029 05	0.028 18	0.027 35	0.026 55	0.025 77
10.0	0.025 77	0.025 03	0.024 31	0.023 62	0.022 96	0.022 32	0.021 70	0.021 11	0.020 53	0.019 98	0.019 44
11.0	0.019 44	0.018 93	0.018 43	0.017 95	0.017 49	0.017 04	0.016 60	0.016 18	0.015 78	0.015 39	0.015 01
12.0	0.015 01	0.014 64	0.014 28	0.013 94	0.013 61	0.013 28	0.012 97	0.012 67	0.012 37	0.012 09	0.011 81
13.0	0.011 81	0.011 55	0.011 29	0.011 03	0.010 79	0.010 55	0.010 32	0.010 10	0.009 88	0.009 67	0.009 46
14.0	0.009 46	0.009 26	0.009 07	0.008 88	0.008 70	0.008 52	0.008 34	0.008 18	0.008 01	0.007 85	0.007 70
15.0	0.007 70	0.007 54	0.007 40	0.007 25	0.007 11	0.006 97	0.006 84	0.006 71	0.006 59	0.006 46	0.006 34

Since the contribution of each single vibrational mode of frequency ν is well known [Eqs. (5-41) to (5-44)], one need only integrate over the Debye frequency distribution. The Debye characteristic temperature is defined by $\theta_D = h\nu_D/k$, where h and k are the Planck and Boltzmann constants, respectively. Also let $u = h\nu/kT$. Then the frequency distribution becomes

$$\rho(\nu)\,d\nu = 9N\left(\frac{T}{\theta_D}\right)^3 u^2\,du \qquad (A2\text{-}3)$$

and the thermodynamic functions are

$$\frac{C_V}{3R} = 3\left(\frac{T}{\theta_D}\right)^3 \int_0^{\theta_D/T} \frac{u^4 e^u\,du}{(e^u - 1)^2} \qquad (A2\text{-}4)$$

$$\frac{U - U_0}{3RT} = 3\left(\frac{T}{\theta_D}\right)^3 \int_0^{\theta_D/T} \frac{u^3\,du}{e^u - 1} \qquad (A2\text{-}5)$$

$$-\frac{A - U_0}{3RT} = 3\left(\frac{T}{\theta_D}\right)^3 \int_0^{\theta_D/T} \ln(1 - e^{-u})u^2\,du \qquad (A2\text{-}6)$$

At low pressure $H \approx U$, and $G \approx A$; hence the equations may be used also in good approximation for those more commonly used functions. The entropy is given by $(U - A)/T$. The numerical values for the functions as calculated by Beattie[1] and Shomate[2] are tabulated.

At low temperatures the upper limit of the integral in Eqs. (A2-4) to (A2-6) may be approximated by infinity. This yields the equations under the table titles, which are valid at large θ_D/T.

REFERENCES

1. J. A. Beattie, *J. Math. Phys.*, **6**, 1 (1926).
2. C. H. Shomate, private communication.

APPENDIX 3

ESTIMATION OF PROPERTIES OF NORMAL FLUIDS

In Chapter 9, thermodynamic principles are applied to fluids deviating from ideal-gas behavior. Also, the continuity of gas and liquid states is noted. While the properties of many substances have been measured over broad ranges of pressure and temperature, far more substances have not been studied as completely and a system of prediction is to be desired. The brief discussion of predictive methods in Chapter 9 is extended here. Molecular theory is used to guide forms of equations in some cases, but the final results are based upon the observed properties of many substances. It is impractical to repeat here the numerous comparisons with observed data—for that the reader must refer to the original papers cited below.

The prediction of the volumetric and thermodynamic properties of pure fluids has been the subject of many studies since 1860. Although the underlying principles in terms of intermolecular forces are now well understood, the calculation or even the quantitative empirical representation of the resulting macroscopic properties has proved to be unusually difficult. While with modern computers complex equations of state now provide a satisfactory method for dealing with fluid properties, there are many occasions when an approximate value of a given quantity will suffice. In that case, the programming of an equation with a large number of terms is unnecessarily burdensome. Thus, it still seems worthwhile to present a simple method for the estimation of the properties of normal fluids. The general treatments, where these estimates are useful, are usually on a P–T basis rather than a V–T or ρ–T basis; hence, the present discussion will be limited to the case with pressure and temperature as independent variables. This correlation was developed simultaneously by two separate groups[1-4] in forms that were superficially different but fundamentally the same.

Statistical theory[5] shows that a group of substances will conform to the principle of corresponding states only if their intermolecular potentials are identical except for distance and energy-scaling factors characteristic of each substance. Also, their intermolecular motion must be classical, i.e., quantum effects must be negligible. The only group of substances that may be expected to conform to these criteria are the heavier rare gases Ar, Kr, and Xe, which do conform quite accurately to corresponding-states behavior. These are called *simple fluids*. Various types of molecular shapes and molecular dipole moments might be expected to cause different deviations from the macroscopic properties of simple fluids. It is found, however, that the reduced theoretical second virial coefficients for a wide variety of these molecular types fall into a single family of curves that may be characterized by a single parameter.[2] The primary exception noted is that of molecules with large dipole moments, although there are other special types of abnormality. The molecules falling into this single family are just those commonly called normal liquids or fluids. Normal fluids are defined more precisely below.

ACENTRIC FACTOR

The theory thus suggests an extension of the corresponding-states correlation involving a third parameter.† The slope of the reduced vapor-pressure curve is the most sensitive property upon which to base the third parameter, and it has the additional advantage that vapor pressures are readily measured with high accuracy. An arbitrary but convenient definition[3]‡ is based upon the reduced vapor pressure at a point well removed from the critical point and takes the form

$$\omega = -\log \frac{P_s}{P_c} - 1.000 \tag{A3-1}$$

where

$$P_s = \text{vapor pressure at } T_r = 0.700$$

---

† A third parameter was first suggested by Nernst in 1907 and by various authors since. See papers by Riedel[1] and by Pitzer et al.[2,3] for references and for the advantages of the present system over others.

‡ The parameter α_k chosen by Riedel[1] is equivalent but is different in detail of definition. $\alpha_k = 5.808 + 4.93\omega$.

The form is chosen to make $\omega = 0$ for the simple fluids Ar, Kr, and Xe, with simple spherical molecules. Other normal fluids have small positive values of ω. The name *acentric factor* was adopted to indicate that the factor measures the deviation of the intermolecular potential function from that of the simple spherical molecules.

Any property of the fluid, in reduced or dimensionless form, is assumed to be given by a function of the three variables reduced pressure, reduced temperature, and acentric factor. For example, the compression factor z may be written

$$z = z(P_r, T_r, \omega) \tag{A3-2}$$

It is found that a linear equation in ω is usually adequate:

$$z = z^{(0)}(P_r, T_r) + \omega z^{(1)}(P_r, T_r) \tag{A3-3}$$

Analytical representation of the functions $z^{(0)}$ and $z^{(1)}$ is not simple, and they are tabulated in most cases.

Vapor Pressure, Heat and Entropy of Vaporization

Where two phases exist in equilibrium, the vapor pressure and other properties are given as functions of two variables, T_r and ω:

$$\log P_r = (\log P_r)^{(0)} + \omega \left(\frac{\partial \log P_r}{\partial \omega}\right)_T \tag{A3-4}$$

These two functions are tabulated in Table A3-1. From the definition of the acentric factor, $(\log P_r)^{(0)} = -1.000$ and $(\partial \log P_r / \partial \omega)_T = -1.000$ at $T_r = 0.7$.

Also included in Table A3-1 are values of the compression-factor functions for the vapor. These are to be used in Eq. (A3-3). A method of estimation of the liquid volume is presented below. By combination of the compression-factor data with the temperature derivative of the vapor pressure, the enthalpy and entropy of vaporization may be calculated. The Clapeyron equation rearranges into

$$\frac{\Delta S}{R} = -\frac{\Delta z}{T_r} \frac{\partial \ln P}{\partial (1/T_r)} \tag{A3-5}$$

where Δz is the change in compression factor on vaporization. The values of $\Delta S^{(0)}$ and $\Delta S^{(1)}$ in Table A3-1 come from this source. The enthalpy of vaporization is, of course, $T \Delta S$.

The full vapor-pressure functions of Table A3-1 make it possible to evaluate the acentric factor from any vapor-pressure value at a temperature well below the critical point.

TABLE A3-1
Data for vapor pressures and vaporization

T_r	Vapor pressure[a]		Vaporization[b]		Vapor	
	$-(\log P_r)^{(0)}$	$-\left(\dfrac{\partial \log P_r}{\partial \omega}\right)_T$	$\dfrac{\Delta S^{(0)}}{R}$	$\dfrac{\Delta S^{(1)}}{R}$	$z^{(0)}$	$z^{(1)}$
1.00	0.000 0	0.000 0	0.00	0.00	0.291	−0.080
0.99	0.025 1	0.021 4	1.29	1.42	0.43	−0.030
0.98	0.050 6	0.043 2	1.70	1.97	0.47	0.000
0.97	0.076 4	0.065 4	2.01	2.38	0.51	0.020
0.96	0.102 7	0.088 2	2.27	2.71	0.54	0.035
0.95	0.129 3	0.111 5	2.52	3.00	0.565	0.045
0.94	0.156 4	0.135 3	2.74	3.28	0.59	0.055
0.92	0.211 8	0.184 8	3.13	3.79	0.63	0.075
0.90	0.269 2	0.237 0	3.50	4.29	0.67	0.095
0.88	0.328 7	0.292 1	3.81	4.73	0.70	0.110
0.86	0.390 5	0.350 6	4.12	5.2	0.73	0.125
0.84	0.454 8	0.412 9	4.42	5.6	0.756	0.135
0.82	0.521 9	0.479 4	4.72	6.1	0.781	0.140
0.80	0.592 0	0.550 6	5.02	6.5	0.804	0.144
0.78	0.665 4	0.627 0	5.32	7.0	0.826	0.144
0.76	0.742 6	0.709 5	5.64	7.5	0.846	0.142
0.74	0.823 7	0.798 5	5.96	8.1	0.864	0.137
0.72	0.909 4	0.895 1	6.29	8.6	0.881	0.131
0.70	1.000 0	1.000 0	6.64	9.1	0.897	0.122
0.68	1.096 1	1.114 4	6.99	9.7	0.911	0.113
0.66	1.198 3	1.239 4	7.36	10.3	0.922	0.104
0.64	1.307 3	1.376 4	7.73	11.0	0.932	0.097
0.62	1.423 9	1.526 9	8.11	11.7	0.940	0.090
0.60	1.549 0	1.692 9	8.51	12.4	0.947	0.083
0.58	1.683 6	1.876 2	8.93	13.2	0.953	0.077
0.56	1.828 9	2.079 5	9.38	14.0	0.959	0.070

[a] These values for vapor pressure are based on the recent equation of Schreiber and Pitzer[6] and differ slightly from those of reference 3.

[b] Strictly there is a small $\Delta S^{(2)}$, which is almost always negligible. See reference 3.

Second Virial Coefficient

Equations for the second virial coefficient in the acentric factor system have been proposed by several authors in the past and most recently by Schreiber and Pitzer.[6] Their equation, which is quite simple, is given in Chapter 9 as Eq. (9-66)*. Additional details are given in Appendix 4. Figure 9-6 shows a comparison with experimental data.

Volumetric Data at High Pressure

The volumetric data for a number of substances were interpolated to a series of even reduced pressures and temperatures and plotted against the acentric factor. Figure A3-1 shows a sample of these plots at $P_r = 3.0$ and a series of reduced temperatures. It is seen that the linear dependence on acentric factor that was assumed in Eq. (A3-3) is a good approximation. The intercepts on the $\omega = 0$ axis constitute the function $z^{(0)}$ and the slopes the function $z^{(1)}$. The values from the plots were checked for smoothness of variation with P_r and T_r and adjusted within experimental error if necessary. The resulting functions are given in Tables A3-2 and A3-3. Additional values near the two-phase region and the critical point are given in reference 3.

Subsequently, Lee and Kessler[7] presented similar tables based on an eBWR equation fitted to data for several fluids. For the paraffin hydrocarbons, their tables are generally more accurate than those given here, but for a broader range of normal fluids the advantage is less clear. Since both sets of tables are now superseded where maximum accuracy is desired, more detailed consideration of the relative merits of these tables is not important and either can be used to obtain simple estimates.

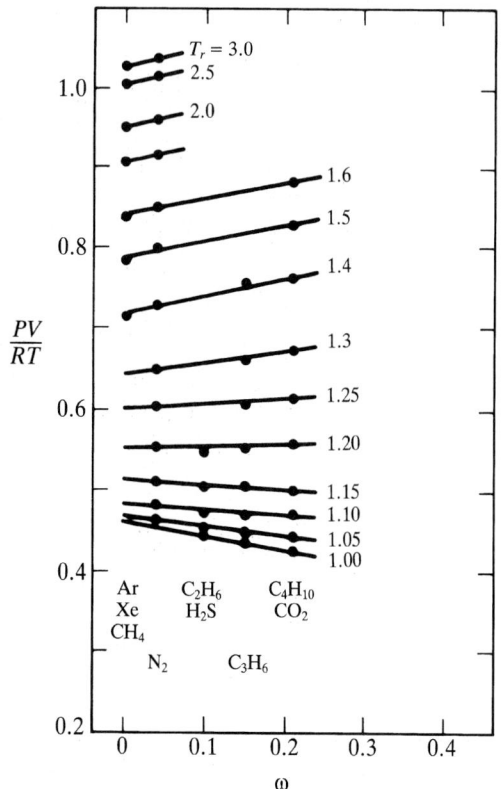

FIGURE A3-1
The compressibility factor as a function of the acentric factor for $P_r = 3.0$ and the indicated values of T_r. The substances are indicated below the columns of points. [K. S. Pitzer, D. Z. Lippmann, R. F. Curl, Jr., C. M. Huggins, and D. E. Petersen, *J. Am. Chem. Soc.*, **77**, 3433 (1955), fig. 2.]

TABLE A3-2
Values of $z^{(0)}$ for compressibility-factor calculation[a]

T_r	\	P_r																								
	0.2	0.4	0.6	0.8	1.0	1.2	1.4	1.6	1.8	2.0	2.2	2.4	2.6	2.8	3.0	3.2	3.4	3.6	3.8	4.0	4.5	5.0	6.0	7.0	8.0	9.0
0.80	0.851	0.066	0.100	0.133	0.164	0.192	0.225	0.258	0.287	0.318	0.347	0.376	0.405	0.433	0.461	0.490	0.519	0.547	0.576	0.605	0.675	0.746	0.883	1.017	1.15	1.28
0.85	0.882	0.067	0.101	0.134	0.165	0.194	0.226	0.258	0.287	0.316	0.345	0.374	0.403	0.431	0.459	0.487	0.515	0.542	0.569	0.597	0.663	0.730	0.861	0.990	1.115	1.24
0.90	0.904	0.778	0.102	0.135	0.167	0.198	0.229	0.258	0.288	0.316	0.345	0.373	0.402	0.430	0.458	0.485	0.512	0.538	0.565	0.591	0.655	0.718	0.842	0.966	1.089	1.21
0.95	0.920	0.819	0.697	0.145	0.176	0.205	0.235	0.262	0.292	0.321	0.347	0.375	0.403	0.430	0.457	0.484	0.510	0.536	0.561	0.587	0.647	0.709	0.828	0.947	1.066	1.185
1.00	0.932	0.849	0.756	0.638	0.291	0.231	0.250	0.278	0.304	0.329	0.356	0.381	0.407	0.433	0.458	0.484	0.509	0.534	0.557	0.582	0.642	0.702	0.819	0.932	1.048	1.166
1.05	0.942	0.874	0.800	0.714	0.609	0.470	0.341	0.320	0.332	0.350	0.372	0.393	0.417	0.441	0.466	0.489	0.512	0.535	0.557	0.580	0.639	0.700	0.814	0.923	1.032	1.147
1.10	0.950	0.893	0.833	0.767	0.691	0.607	0.512	0.442	0.408	0.402	0.405	0.420	0.440	0.462	0.484	0.504	0.525	0.547	0.567	0.589	0.643	0.699	0.810	0.916	1.019	1.129
1.15	0.958	0.908	0.858	0.805	0.746	0.684	0.620	0.562	0.514	0.484	0.477	0.478	0.485	0.498	0.513	0.529	0.546	0.563	0.581	0.600	0.651	0.705	0.809	0.911	1.008	1.113
1.20	0.963	0.921	0.879	0.835	0.788	0.737	0.690	0.640	0.598	0.568	0.553	0.545	0.544	0.548	0.554	0.563	0.574	0.587	0.601	0.618	0.664	0.714	0.810	0.907	1.000	1.100
1.25	0.968	0.930	0.896	0.858	0.820	0.778	0.740	0.702	0.664	0.636	0.618	0.606	0.599	0.597	0.598	0.602	0.609	0.618	0.629	0.643	0.682	0.726	0.816	0.907	0.994	1.088
1.30	0.971	0.940	0.909	0.878	0.846	0.811	0.780	0.749	0.718	0.691	0.671	0.657	0.649	0.644	0.642	0.642	0.645	0.651	0.659	0.668	0.701	0.740	0.824	0.910	0.992	1.078
1.4	0.977	0.952	0.929	0.908	0.883	0.859	0.838	0.817	0.795	0.777	0.759	0.745	0.734	0.725	0.720	0.718	0.718	0.722	0.727	0.734	0.754	0.781	0.844	0.921	0.994	1.071
1.5	0.982	0.963	0.945	0.927	0.909	0.892	0.875	0.859	0.844	0.831	0.819	0.808	0.800	0.794	0.790	0.785	0.784	0.784	0.786	0.790	0.805	0.826	0.877	0.934	1.000	1.070
1.6	0.985	0.971	0.957	0.944	0.930	0.917	0.904	0.893	0.882	0.872	0.863	0.855	0.848	0.843	0.840	0.836	0.834	0.833	0.834	0.835	0.844	0.860	0.904	0.953	1.010	1.075
1.7	0.988	0.977	0.966	0.956	0.946	0.936	0.926	0.919	0.911	0.903	0.896	0.889	0.883	0.879	0.875	0.873	0.872	0.872	0.873	0.874	0.882	0.895	0.930	0.972	1.023	1.082
1.8	0.991	0.982	0.974	0.966	0.958	0.950	0.944	0.937	0.931	0.926	0.921	0.916	0.913	0.910	0.908	0.907	0.906	0.906	0.907	0.908	0.914	0.925	0.955	0.993	1.039	1.091
1.9	0.993	0.986	0.980	0.974	0.968	0.962	0.958	0.952	0.948	0.944	0.940	0.936	0.933	0.931	0.930	0.929	0.929	0.930	0.932	0.934	0.941	0.950	0.976	1.010	1.051	1.097
2.0	0.995	0.989	0.984	0.979	0.975	0.971	0.968	0.964	0.961	0.959	0.956	0.954	0.953	0.953	0.952	0.952	0.953	0.954	0.954	0.956	0.962	0.972	0.996	1.027	1.064	1.106
2.5	1.000	0.999	0.998	0.998	0.998	0.998	0.998	0.997	0.999	1.000	1.001	1.001	1.002	1.004	1.006	1.008	1.009	1.012	1.014	1.018	1.026	1.035	1.055	1.079	1.105	1.136
3.0	1.001	1.002	1.003	1.004	1.005	1.007	1.008	1.010	1.012	1.014	1.016	1.019	1.022	1.025	1.028	1.030	1.033	1.036	1.038	1.041	1.049	1.058	1.077	1.10	1.124	1.150
3.5	1.002	1.004	1.006	1.008	1.011	1.013	1.015	1.018	1.020	1.022	1.024	1.027	1.030	1.033	1.036	1.039	1.042	1.045	1.048	1.051	1.058	1.067	1.086	1.105	1.226	1.148
4.0	1.003	1.005	1.008	1.010	1.013	1.015	1.017	1.020	1.022	1.024	1.026	1.029	1.032	1.035	1.038	1.041	1.044	1.047	1.050	1.053	1.060	1.068	1.086	1.104	1.124	1.143

[a] K. S. Pitzer, D. Z. Lippmann, R. F. Curl, Jr., C. M. Huggins, and D. E. Petersen, *J. Am. Chem. Soc.*, **77**, 3427 (1955), table II. See reference 3 for additional data in the region enclosed by dashed lines.

TABLE A3-3
Values of $z^{(1)}$ for compressibility-factor calculation[a]

T_r	P_r 0.2	0.4	0.6	0.8	1.0	1.2	1.4	1.6	1.8	2.0	2.2	2.4	2.6	2.8	3.0	4.0	5.0	6.0	7.0	8.0	9.0
0.80	−0.095	−0.028	−0.044	−0.058	−0.07	−0.08	−0.10	−0.11	−0.12	−0.13	−0.14	−0.15	−0.16	−0.17	−0.18	−0.23	−0.26	−0.29	−0.32	−0.35	−0.37
0.85	−0.067	−0.031	−0.049	−0.064	−0.08	−0.09	−0.11	−0.12	−0.13	−0.14	−0.15	−0.16	−0.17	−0.18	−0.18	−0.22	−0.25	−0.28	−0.31	−0.34	−0.36
0.90	−0.042	−0.09	−0.053	−0.068	−0.085	−0.10	−0.11	−0.12	−0.13	−0.14	−0.15	−0.16	−0.17	−0.17	−0.18	−0.21	−0.24	−0.27	−0.30	−0.32	−0.35
0.95	−0.025	−0.050	−0.10	−0.072	−0.091	−0.10	−0.11	−0.12	−0.12	−0.13	−0.14	−0.15	−0.15	−0.16	−0.17	−0.20	−0.22	−0.25	−0.28	−0.31	−0.34
1.00	−0.012	−0.016	−0.020	−0.05	−0.080	−0.090	−0.099	−0.108	−0.115	−0.123	−0.13	−0.13	−0.14	−0.14	−0.15	−0.17	−0.20	−0.23	−0.26	−0.30	−0.33
1.05	0.000	+0.001	+0.005	+0.015	+0.02	+0.01	−0.01	−0.04	−0.06	−0.07	−0.08	−0.09	−0.10	−0.10	−0.11	−0.14	−0.17	−0.20	−0.24	−0.28	−0.31
1.10	+0.002	0.008	0.016	0.030	0.055	0.082	+0.11	+0.082	+0.035	0.000	−0.02	−0.03	−0.05	−0.06	−0.07	−0.10	−0.13	−0.16	−0.21	−0.25	−0.28
1.15	0.004	0.012	0.012	0.040	0.064	0.093	0.12	0.140	0.136	+0.100	+0.07	+0.04	+0.02	0.00	−0.01	−0.04	−0.08	−0.12	−0.16	−0.20	−0.24
1.20	0.009	0.018	0.028	0.044	0.069	0.10	0.13	0.16	0.17	0.17	0.16	0.14	0.12	+0.09	+0.07	0.00	−0.04	−0.08	−0.12	−0.16	−0.19
1.25	0.011	0.023	0.036	0.050	0.069	0.10	0.13	0.16	0.18	0.19	0.19	0.18	0.16	0.14	0.12	+0.05	0.00	−0.03	−0.07	−0.11	−0.13
1.30	0.013	0.027	0.041	0.055	0.072	0.10	0.13	0.16	0.18	0.20	0.20	0.20	0.20	0.19	0.18	0.10	+0.04	0.00	−0.04	−0.07	−0.09
1.4	0.016	0.032	0.049	0.065	0.082	0.10	0.13	0.16	0.18	0.19	0.20	0.21	0.21	0.21	0.20	0.15	0.11	+0.07	+0.04	+0.01	−0.01
1.5	0.017	0.035	0.052	0.070	0.088	0.10	0.13	0.15	0.17	0.18	0.20	0.20	0.21	0.21	0.21	0.20	0.17	0.14	0.11	0.09	+0.07
1.6	0.018	0.036	0.054	0.07	0.08	0.10	0.12	0.14	0.16	0.17	0.18	0.19	0.20	0.20	0.21	0.22	0.21	0.19	0.17	0.15	0.14
1.7	0.018	0.036	0.054	0.07	0.09	0.10	0.11	0.13	0.15	0.16	0.17	0.18	0.19	0.20	0.21	0.24	0.25	0.26	0.25	0.24	0.22
1.8	0.018	0.036	0.054	0.07	0.09	0.10	0.11	0.13	0.15	0.16	0.17	0.18	0.19	0.20	0.21	0.26	0.29	0.31	0.32	0.32	0.30
1.9	0.018	0.035	0.05	0.07	0.09	0.10	0.11	0.13	0.15	0.16	0.17	0.18	0.19	0.20	0.21	0.26	0.30	0.35	0.38	0.40	0.40
2.0	0.016	0.031	0.05	0.07	0.08	0.10	0.11	0.13	0.14	0.15	0.16	0.17	0.19	0.20	0.21	0.26	0.30	0.35	0.40	0.43	0.45
2.5	0.01	0.02	0.04	0.05	0.07	0.08	0.10	0.11	0.12	0.13	0.15	0.16	0.18	0.19	0.20	0.25	0.30	0.35	0.40	0.45	0.50
3.0	0.01	0.02	0.03	0.05	0.06	0.07	0.08	0.09	0.10	0.11	0.13	0.14	0.15	0.16	0.17	0.23	0.28	0.34	0.38	0.45	0.50
3.5	0.01	0.02	0.03	0.04	0.05	0.06	0.07	0.08	0.08	0.09	0.10	0.11	0.12	0.13	0.14	0.19	0.24	0.28	0.33	0.38	0.42
4.0	0.01	0.02	0.02	0.03	0.04	0.05	0.06	0.06	0.07	0.08	0.09	0.10	0.10	0.11	0.12	0.16	0.20	0.23	0.27	0.31	0.35

[a] K. S. Pitzer, D. Z. Lippmann, R. F. Curl, Jr., C. M. Huggins, and D. E. Petersen, *J. Am. Chem. Soc.*, **77**, 3427 (1955), table IV. See reference 3 for additional data in the region enclosed by dashed lines.

Fugacity, Enthalpy, and Entropy

The fugacity and other thermodynamic functions may be calculated from the equation of state. For the low-pressure region, the second virial coefficient is adequate and the equations are given in Chapter 9.

At higher pressures the second virial coefficient does not suffice, and we must use the general equation of state defined in Eq. (A3-3) and the tables. The thermodynamic properties are obtained from the usual equations. In terms of the compression factor, they are

$$\ln \frac{f}{P} = \int_0^{P_r} \frac{z-1}{P'_r} dP'_r \tag{A3-6}$$

$$\frac{H° - H}{RT_c} = T_r^2 \int_0^{P_r} \left(\frac{1}{P'_r}\right)\left(\frac{\partial z}{\partial T_r}\right) dP'_r \tag{A3-7}$$

$$\frac{S° - S}{R} = \frac{1}{T_r}\frac{H° - H}{RT_c} + \ln\frac{f}{P} + \ln P \tag{A3-8}$$

In Eqs. (A3-7) and (A3-8) the resulting fugacity and enthalpy pertain to the reduced pressure P_r, which is the upper limit of integration. The superscript °, as usual, denotes the ideal-gas standard state. The integration and differentiation was carried out by graphical and numerical methods. A problem arises when the path of integration crosses the condensation curve because, as ω increases, the condensation occurs at a lower pressure for the same temperature. The appropriate methods are given in reference 4.

The resulting functions are given in Tables A3-4 to A3-9. Again, Lee and Kessler[7] give similar tables. Because of the lower vapor pressure for liquids of high acentric factor at a given temperature, it is sometimes necessary to extrapolate the functions on the liquid side. This extrapolation should be taken with respect to pressure at constant temperature. The complete equations, which have the form of Eq. (A3-3), are as follows:

$$\log\frac{f}{P} = \left(\log\frac{f}{P}\right)^{(0)} + \omega\left(\log\frac{f}{P}\right)^{(1)} \tag{A3-9}$$

$$\frac{H° - H}{RT_c} = \left(\frac{H° - H}{RT_c}\right)^{(0)} + \omega\left(\frac{H° - H}{RT_c}\right)^{(1)} \tag{A3-10}$$

$$\frac{S° - S}{R} = \left(\frac{S° - S}{R}\right)^{(0)} + \omega\left(\frac{S° - S}{R}\right)^{(1)} + \ln P \tag{A3-11}$$

TABLE A3-4
Values of $[\log(f/P)]^{(0)}$

T_r	P_r												
	0.2	0.4	0.6	0.8	1.0	1.2	1.4	1.6	1.8	2.0	2.2	2.4	2.6
0.80	−0.060	−0.262	−0.425	−0.535	−0.618	−0.683	−0.736	−0.780	−0.817	−0.849	−0.877	−0.901	−0.922
0.85	−0.046	−0.120	−0.281	−0.392	−0.474	−0.539	−0.592	−0.636	−0.673	−0.705	−0.733	−0.757	−0.779
0.90	−0.042₅	−0.087₅	−0.163	−0.273	−0.356	−0.421	−0.474	−0.517	−0.554	−0.587	−0.614	−0.639	−0.680
0.95	−0.033	−0.070	−0.112	−0.173	−0.255	−0.319	−0.372	−0.415	−0.452	−0.483	−0.511	−0.535	−0.557
1.00	−0.028	−0.059	−0.094	−0.131	−0.175	−0.237	−0.287	−0.330	−0.367	−0.398	−0.425	−0.449	−0.470
1.05	−0.024	−0.051	−0.079	−0.109	−0.142	−0.178	−0.218	−0.257	−0.292	−0.322	−0.349	−0.372	−0.393
1.10	−0.021	−0.044	−0.067	−0.093	−0.120	−0.147	−0.177	−0.207	−0.237	−0.264	−0.289	−0.311	−0.331
1.15	−0.018	−0.037	−0.058	−0.079	−0.101	−0.123	−0.146	−0.170	−0.194	−0.217	−0.238	−0.258	−0.276
1.20	−0.016	−0.032	−0.050	−0.067	−0.086	−0.104	−0.124	−0.143	−0.163	−0.182	−0.200	−0.217	−0.233
1.25	−0.014	−0.029	−0.044	−0.059	−0.075	−0.091	−0.107	−0.123	−0.139	−0.155	−0.171	−0.186	−0.199
1.3	−0.012	−0.025	−0.038	−0.051	−0.065	−0.078	−0.092	−0.106	−0.119	−0.133	−0.146	−0.159	−0.171
1.4	−0.010	−0.021	−0.031	−0.041	−0.052	−0.062	−0.072	−0.082	−0.092	−0.102	−0.111	−0.120	−0.130
1.5	−0.008	−0.016	−0.024	−0.032	−0.040	−0.047	−0.055	−0.063	−0.070	−0.078	−0.085	−0.092	−0.099
1.6	−0.007	−0.013	−0.019	−0.026	−0.032	−0.038	−0.044	−0.050	−0.056	−0.062	−0.067	−0.072	−0.077
1.7	−0.005	−0.010	−0.015	−0.020	−0.025	−0.030	−0.034	−0.039	−0.043	−0.047	−0.051	−0.056	−0.059
1.8	−0.004	−0.008	−0.012	−0.015	−0.019	−0.022	−0.026	−0.030	−0.033	−0.036	−0.039	−0.042	−0.045
1.9	−0.003	−0.006	−0.009	−0.012	−0.015	−0.018	−0.020	−0.023	−0.025	−0.028	−0.030	−0.033	−0.035
2.0	−0.002	−0.004	−0.007	−0.009	−0.011	−0.013	−0.015	−0.017	−0.019	−0.021	−0.023	−0.025	−0.026
2.5	0.000	0.000	0.000	0.000	−0.001	−0.001	−0.001	−0.001	−0.001	−0.001	−0.001	−0.001	−0.001
3.0	0.000	+0.001	+0.001	+0.002	+0.002	+0.003	+0.003	+0.004	+0.004	+0.005	+0.005	+0.006	+0.007
3.5	+0.001	0.002	0.003	0.003	0.004	0.005	0.006	0.007	0.008	0.009	0.010	0.011	0.012
4.0	0.001	0.002	0.003	0.005	0.006	0.007	0.008	0.009	0.010	0.011	0.012	0.013	0.014

TABLE A3-4 Continued

T_r	\multicolumn{14}{c}{P_r}													
	2.8	3.0	3.2	3.4	3.6	3.8	4.0	4.5	5.0	6.0	7.0	8.0	9.0
0.80	−0.941	−0.957	−0.972	−0.985	−0.997	−1.007	−1.016	−1.035	−1.048	−1.064	−1.067	−1.063	−1.052
0.85	−0.797	−0.814	−0.829	−0.842	−0.854	−0.864	−0.874	−0.893	−0.907	−0.924	−0.929	−0.926	−0.917
0.90	−0.679	−0.696	−0.710	−0.724	−0.736	−0.746	−0.756	−0.775	−0.789	−0.807	−0.814	−0.813	−0.805
0.95	−0.575	−0.592	−0.607	−0.621	−0.632	−0.643	−0.652	−0.672	−0.687	−0.706	−0.713	−0.713	−0.707
1.00	−0.489	−0.505	−0.520	−0.534	−0.545	−0.556	−0.566	−0.586	−0.601	−0.620	−0.629	−0.630	−0.624
1.05	−0.411	−0.428	−0.442	−0.455	−0.467	−0.478	−0.488	−0.508	−0.523	−0.543	−0.552	−0.553	−0.549
1.10	−0.348	−0.364	−0.378	−0.391	−0.403	−0.413	−0.422	−0.442	−0.457	−0.477	−0.487	−0.489	−0.486
1.15	−0.293	−0.307	−0.321	−0.333	−0.344	−0.354	−0.363	−0.383	−0.397	−0.417	−0.427	−0.429	−0.426
1.20	−0.247	−0.261	−0.273	−0.285	−0.295	−0.305	−0.314	−0.332	−0.346	−0.366	−0.375	−0.378	−0.376
1.25	−0.212	−0.224	−0.236	−0.246	−0.256	−0.264	−0.273	−0.290	−0.304	−0.322	−0.331	−0.334	−0.332
1.3	−0.182	−0.193	−0.203	−0.212	−0.221	−0.229	−0.237	−0.253	−0.266	−0.283	−0.292	−0.295	−0.294
1.4	−0.138	−0.146	−0.154	−0.162	−0.169	−0.175	−0.181	−0.194	−0.205	−0.220	−0.228	−0.231	−0.229
1.5	−0.104	−0.112	−0.117	−0.124	−0.129	−0.134	−0.139	−0.149	−0.158	−0.170	−0.176	−0.178	−0.176
1.6	−0.082	−0.087	−0.092	−0.096	−0.100	−0.104	−0.108	−0.116	−0.123	−0.132	−0.137	−0.138	−0.136
1.7	−0.063	−0.067	−0.071	−0.074	−0.077	−0.080	−0.083	−0.089	−0.094	−0.101	−0.105	−0.105	−0.102
1.8	−0.048	−0.051	−0.053	−0.056	−0.058	−0.060	−0.063	−0.067	−0.071	−0.076	−0.078	−0.077	−0.074
1.9	−0.037	−0.039	−0.041	−0.043	−0.045	−0.046	−0.048	−0.051	−0.054	−0.057	−0.057	−0.055	−0.051
2.0	−0.028	−0.029	−0.031	−0.032	−0.033	−0.034	−0.035	−0.037	−0.039	−0.040	−0.039	−0.038	−0.034
2.5	−0.001	−0.001	−0.001	−0.001	0.000	0.000	0.000	+0.001	+0.003	+0.006	+0.011	+0.016	+0.022
3.0	+0.007	+0.008	+0.009	+0.010	+0.011	+0.012	+0.012	0.015	0.017	0.023	0.028	0.035	0.042
3.5	0.013	0.014	0.015	0.016	0.017	0.018	0.020	0.022	0.025	0.031	0.038	0.044	0.051
4.0	0.015	0.016	0.017	0.019	0.020	0.021	0.022	0.025	0.028	0.034	0.040	0.047	0.054

TABLE A3-5
Values of $[\log(f/P)]^{(1)}$

T_r	\multicolumn{19}{c}{P_r}
	0.2	0.4	0.6	0.8	1.0	1.2	1.4	1.6	1.8	2.0	2.2	2.4	2.6	2.8	3.0	4.0	5.0	6.0	7.0	8.0	9.0
0.80	−0.04	−0.47	−0.48	−0.48	−0.48	−0.49	−0.50	−0.50	−0.51	−0.51	−0.52	−0.52	−0.53	−0.53	−0.54	−0.56	−0.59	−0.61	−0.63	−0.65	−0.67
0.85	−0.03	−0.31	−0.31	−0.32	−0.33	−0.33	−0.34	−0.35	−0.35	−0.36	−0.37	−0.37	−0.38	−0.38	−0.39	−0.41	−0.44	−0.46	−0.48	−0.50	−0.51
0.90	−0.02	−0.04	−0.18	−0.20	−0.20	−0.21	−0.21	−0.22	−0.23	−0.23	−0.24	−0.24	−0.25	−0.26	−0.26	−0.29	−0.31	−0.33	−0.35	−0.36	−0.38
0.95	−0.01	−0.02	−0.03	−0.09	−0.10	−0.11	−0.12	−0.12	−0.13	−0.13	−0.14	−0.15	−0.15	−0.16	−0.16	−0.18	−0.20	−0.22	−0.24	−0.26	−0.27
1.00	−0.01	−0.01	−0.01	−0.02	−0.03	−0.03	−0.04	−0.05	−0.05	−0.06	−0.06	−0.07	−0.07	−0.08	−0.08	−0.10	−0.12	−0.13	−0.15	−0.17	−0.18
1.05	0.00	0.00	0.00	0.00	+0.01	+0.01	+0.01	+0.01	0.00	0.00	0.00	0.00	−0.01	−0.01	−0.01	−0.03	−0.05	−0.06	−0.07	−0.09	−0.11
1.10	0.00	0.00	0.00	+0.01	0.01	0.02	0.02	0.03	+0.03	+0.03	+0.03	+0.03	+0.03	+0.03	+0.03	+0.02	0.00	−0.01	−0.02	−0.03	−0.05
1.15	0.00	0.00	0.00	0.01	0.02	0.02	0.03	0.04	0.04	0.05	0.05	0.05	0.06	0.06	0.06	0.05	+0.05	+0.04	+0.02	+0.01	0.00
1.20	0.00	+0.01	+0.01	0.01	0.02	0.03	0.04	0.05	0.05	0.06	0.07	0.07	0.08	0.08	0.08	0.09	0.09	0.08	0.07	0.07	+0.06
1.25	0.00	0.01	0.01	0.02	0.03	0.03	0.04	0.05	0.06	0.07	0.07	0.08	0.09	0.09	0.10	0.11	0.11	0.11	0.10	0.10	0.09
1.3	+0.01	0.01	0.02	0.02	0.03	0.04	0.04	0.05	0.06	0.07	0.08	0.08	0.09	0.10	0.10	0.12	0.13	0.13	0.13	0.12	0.12
1.4	0.01	0.01	0.02	0.03	0.04	0.04	0.05	0.06	0.07	0.08	0.08	0.09	0.10	0.11	0.11	0.13	0.15	0.15	0.16	0.16	0.16
1.5	0.01	0.02	0.02	0.03	0.04	0.05	0.05	0.06	0.06	0.07	0.08	0.08	0.09	0.10	0.11	0.13	0.15	0.16	0.17	0.17	0.18
1.6	0.01	0.02	0.02	0.03	0.04	0.05	0.05	0.06	0.06	0.07	0.08	0.08	0.09	0.10	0.11	0.14	0.16	0.18	0.19	0.20	0.21
1.7	0.01	0.02	0.02	0.03	0.04	0.05	0.05	0.06	0.06	0.07	0.08	0.08	0.09	0.10	0.11	0.14	0.16	0.18	0.20	0.21	0.23
1.8	0.01	0.02	0.02	0.03	0.04	0.05	0.05	0.06	0.06	0.07	0.08	0.08	0.09	0.10	0.11	0.14	0.16	0.19	0.21	0.23	0.24
1.9	0.01	0.02	0.02	0.03	0.04	0.05	0.05	0.06	0.06	0.07	0.08	0.08	0.09	0.10	0.11	0.14	0.16	0.19	0.21	0.23	0.25
2.0	0.01	0.01	0.02	0.03	0.04	0.05	0.05	0.06	0.07	0.07	0.08	0.08	0.09	0.09	0.10	0.13	0.16	0.19	0.21	0.23	0.26
2.5	0.01	0.01	0.02	0.02	0.03	0.04	0.04	0.05	0.06	0.06	0.07	0.07	0.08	0.08	0.09	0.12	0.14	0.17	0.19	0.22	0.24
3.0	0.00	0.01	0.01	0.02	0.02	0.03	0.04	0.04	0.05	0.05	0.05	0.06	0.06	0.07	0.07	0.10	0.12	0.15	0.17	0.20	0.22
3.5	0.00	0.01	0.01	0.02	0.02	0.02	0.03	0.03	0.04	0.04	0.04	0.05	0.05	0.06	0.06	0.08	0.10	0.13	0.15	0.17	0.19
4.0	0.00	0.01	0.01	0.02	0.02	0.02	0.02	0.03	0.03	0.03	0.04	0.04	0.04	0.05	0.05	0.07	0.09	0.10	0.12	0.14	0.15

TABLE A3-6
Values of $\left(\dfrac{H^\circ - H}{RT_c}\right)^{(0)}$

T_r	0.2	0.4	0.6	0.8	1.0	1.2	1.4	1.6	1.8	2.0	2.2	2.4	2.6	2.8	3.0	3.2	3.4	3.6	3.8	4.0	4.5	5.0	6.0	7.0	8.0	9.0
0.80	0.37	4.52	4.52	4.52	4.52	4.52	4.52	4.52	4.52	4.52	4.52	4.52	4.52	4.52	4.52	4.51	4.51	4.50	4.50	4.49	4.46	4.43	4.38	4.34	4.29	4.24
0.85	0.32	4.35	4.35	4.35	4.34	4.34	4.33	4.32	4.32	4.32	4.32	4.32	4.32	4.32	4.32	4.32	4.31	4.31	4.31	4.30	4.28	4.24	4.20	4.16	4.10	4.04
0.90	0.27	0.60	4.06	4.10	4.14	4.14	4.15	4.15	4.14	4.13	4.13	4.12	4.12	4.12	4.11	4.10	4.10	4.10	4.10	4.10	4.09	4.05	4.02	3.99	3.92	3.85
0.95	0.23	0.52	0.86	3.69	3.80	3.85	3.87	3.88	3.89	3.90	3.90	3.90	3.90	3.90	3.90	3.90	3.90	3.89	3.89	3.89	3.85	3.85	3.84	3.81	3.76	3.71
1.00	0.21	0.45	0.76	1.15	2.3$_5$	3.09	3.32	3.44	3.52	3.57	3.60	3.63	3.65	3.68	3.70	3.71	3.71	3.70	3.70	3.70	3.69	3.68	3.67	3.64	3.60	3.57
1.05	0.19	0.40	0.64	0.95	1.35	1.94	2.54	2.86	3.07	3.21	3.30	3.36	3.39	3.42	3.44	3.46	3.47	3.49	3.50	3.51	3.51	3.50	3.49	3.48	3.46	3.45
1.10	0.17	0.36	0.57	0.82	1.10	1.44	1.83	2.25	2.55	2.75	2.89	3.00	3.08	3.15	3.20	3.24	3.27	3.29	3.30	3.32	3.34	3.34	3.34	3.33	3.32	3.32
1.15	0.14	0.30	0.49	0.70	0.93	1.19	1.48	1.78	2.07	2.33	2.52	2.67	2.78	2.86	2.93	2.98	3.02	3.05	3.09	3.12	3.17	3.18	3.18	3.19	3.19	3.20
1.20	0.13	0.27	0.44	0.63	0.83	1.03	1.25	1.49	1.73	1.95	2.13	2.30	2.44	2.56	2.66	2.73	2.78	2.82	2.87	2.91	2.99	3.02	3.05	3.07	3.07	3.08
1.25	0.12	0.25	0.39	0.56	0.73	0.91	1.09	1.29	1.50	1.70	1.87	2.03	2.17	2.29	2.39	2.48	2.55	2.61	2.67	2.72	2.82	2.87	2.92	2.93	2.95	2.98
1.3	0.11	0.23	0.36	0.50	0.66	0.81	0.97	1.14	1.32	1.49	1.64	1.79	1.93	2.05	2.16	2.24	2.32	2.39	2.45	2.52	2.63	2.72	2.79	2.81	2.84	2.88
1.4	0.09	0.19	0.31	0.42	0.54	0.67	0.80	0.94	1.08	1.23	1.36	1.47	1.59	1.70	1.79	1.88	1.96	2.04	2.11	2.18	2.32	2.43	2.53	2.58	2.62	2.65
1.5	0.09	0.18	0.29	0.39	0.49	0.59	0.70	0.80	0.93	1.04	1.15	1.26	1.36	1.45	1.53	1.61	1.68	1.75	1.82	1.88	2.01	2.12	2.25	2.33	2.39	2.41
1.6	0.09	0.18	0.27	0.36	0.45	0.54	0.62	0.71	0.81	0.91	1.00	1.09	1.18	1.26	1.33	1.39	1.44	1.51	1.57	1.64	1.78	1.87	2.01	2.12	2.19	2.21
1.7	0.08	0.16	0.25	0.33	0.41	0.48	0.56	0.64	0.71	0.80	0.87	0.95	1.02	1.08	1.15	1.21	1.27	1.33	1.39	1.45	1.57	1.66	1.79	1.91	2.00	2.03
1.8	0.07	0.15	0.23	0.30	0.37	0.44	0.51	0.58	0.64	0.71	0.78	0.84	0.90	0.95	1.01	1.06	1.12	1.18	1.24	1.31	1.42	1.50	1.62	1.74	1.83	1.86
1.9	0.06	0.13	0.19	0.26	0.33	0.40	0.46	0.51	0.57	0.63	0.68	0.73	0.78	0.82	0.87	0.92	0.97	1.02	1.08	1.13	1.23	1.32	1.44	1.54	1.63	1.67
2.0	0.06	0.12	0.18	0.24	0.30	0.36	0.42	0.46	0.51	0.55	0.59	0.64	0.69	0.74	0.78	0.82	0.85	0.89	0.93	0.97	1.07	1.14	1.25	1.33	1.43	1.46
2.5	0.04	0.08	0.12	0.16	0.19	0.22	0.25	0.28	0.31	0.34	0.37	0.40	0.43	0.45	0.47	0.50	0.52	0.54	0.56	0.58	0.63	0.67	0.75	0.81	0.87	0.91
3.0	0.03	0.05	0.07	0.09	0.11	0.14	0.16	0.18	0.20	0.22	0.24	0.26	0.28	0.30	0.31	0.33	0.34	0.36	0.37	0.38	0.41	0.44	0.50	0.53	0.56	0.58
3.5	0.02	0.04	0.05	0.06	0.07	0.09	0.10	0.11	0.12	0.13	0.15	0.16	0.17	0.18	0.19	0.20	0.21	0.22	0.23	0.23	0.25	0.26	0.28	0.30	0.30	0.29
4.0	0.01	0.02	0.03	0.04	0.04	0.05	0.06	0.07	0.08	0.08	0.09	0.10	0.10	0.11	0.11	0.11	0.12	0.12	0.12	0.13	0.13	0.14	0.14	0.14	0.13	0.12

TABLE A3-7

Values of $\left(\dfrac{H° - H}{RT_c}\right)^{(1)}$

T_r	0.2	0.4	0.6	0.8	1.0	1.2	1.4	1.6	1.8	2.0	2.2	2.4	2.6	2.8	3.0	4.0	5.0	6.0	7.0	8.0	9.0
0.80	0.44	5.05	5.04	5.02	4.98	4.97	4.94	4.93	4.90	4.88	4.87	4.86	4.84	4.83	4.82	4.83	4.85	4.86	4.88	4.90	4.91
0.85	0.37	4.74	4.71	4.70	4.67	4.65	4.63	4.63	4.63	4.60	4.57	4.57	4.56	4.56	4.56	4.57	4.61	4.63	4.65	4.68	4.71
0.90	0.31	0.71	4.33	4.32	4.31	4.31	4.30	4.30	4.30	4.30	4.29	4.28	4.29	4.30	4.30	4.36	4.39	4.43	4.43	4.50	4.53
0.95	0.25	0.55	1.01	3.83	3.80	3.81	3.84	3.84	3.87	3.87	3.88	3.90	3.90	3.93	3.93	4.04	4.12	4.19	4.23	4.28	4.32
1.00	0.20	0.41	0.68	0.95	2.66	3.17	3.27	3.33	3.38	3.41	3.45	3.49	3.53	3.58	3.61	3.76	3.88	3.97	4.02	4.07	4.10
1.05	0.14	0.27	0.40	0.54	0.68	1.22	1.77	2.19	2.45	2.59	2.71	2.79	2.88	2.94	2.99	3.26	3.48	3.57	3.66	3.75	3.81
1.10	0.12	0.22	0.29	0.36	0.42	0.52	0.69	0.92	1.32	1.71	1.97	2.11	2.24	2.30	2.39	2.74	3.01	3.17	3.30	3.42	3.53
1.15	0.09	0.17	0.23	0.28	0.32	0.36	0.39	0.45	0.58	0.81	1.08	1.32	1.52	1.67	1.78	2.24	2.52	2.73	2.93	3.09	3.23
1.20	0.08	0.14	0.20	0.24	0.28	0.29	0.29	0.32	0.37	0.47	0.59	0.74	0.91	1.08	1.21	1.70	1.98	2.21	2.41	2.59	2.79
1.25	0.06	0.13	0.18	0.21	0.25	0.25	0.26	0.28	0.28	0.30	0.36	0.43	0.51	0.63	0.76	1.26	1.56	1.79	1.98	2.15	2.41
1.3	0.03	0.06	0.10	0.13	0.17	0.19	0.24	0.23	0.24	0.25	0.27	0.32	0.39	0.46	0.53	0.85	1.11	1.43	1.60	1.83	2.10
1.4	0.02	0.04	0.06	0.08	0.10	0.11	0.11	0.12	0.12	0.12	0.14	0.16	0.19	0.22	0.25	0.45	0.73	0.98	1.15	1.40	1.66
1.5	0.01	0.01	0.02	0.03	0.04	0.04	0.04	0.04	0.04	0.04	0.05	0.06	0.07	0.08	0.09	0.20	0.47	0.68	0.86	1.10	1.36
1.6	0.00	0.00	0.00	0.00	0.00	0.00	0.00	0.00	−0.01	−0.01	−0.01	−0.01	−0.02	−0.02	−0.02	0.05	0.28	0.45	0.64	0.88	1.10
1.7	−0.01	−0.01	−0.02	−0.03	−0.04	−0.04	−0.05	−0.05	−0.06	−0.06	−0.07	−0.08	−0.09	−0.10	−0.10	−0.07	0.11	0.25	0.43	0.63	0.85
1.8	−0.01	−0.02	−0.03	−0.04	−0.06	−0.07	−0.08	−0.09	−0.10	−0.10	−0.11	−0.12	−0.13	−0.14	−0.16	−0.16	−0.03	0.08	0.23	0.42	0.61
1.9	−0.02	−0.03	−0.05	−0.07	−0.09	−0.11	−0.13	−0.15	−0.17	−0.19	−0.20	−0.20	−0.21	−0.21	−0.21	−0.21	−0.15	−0.07	0.06	0.22	0.39
2.0	−0.02	−0.04	−0.06	−0.08	−0.10	−0.13	−0.16	−0.19	−0.22	−0.25	−0.25	−0.25	−0.26	−0.26	−0.26	−0.28	−0.26	−0.19	−0.10	0.03	0.18
2.5	−0.03	−0.06	−0.10	−0.13	−0.17	−0.20	−0.23	−0.26	−0.29	−0.32	−0.35	−0.39	−0.43	−0.45	−0.48	−0.53	−0.55	−0.57	−0.59	−0.63	−0.61
0	−0.04	−0.08	−0.13	−0.17	−0.21	−0.25	−0.29	−0.33	−0.37	−0.41	−0.45	−0.49	−0.53	−0.57	−0.60	−0.74	−0.85	−0.95	−1.05	−1.14	−1.21
3.5	−0.04	−0.08	−0.13	−0.17	−0.21	−0.25	−0.29	−0.34	−0.37	−0.42	−0.47	−0.51	−0.55	−0.59	−0.63	−0.83	−1.02	−1.19	−1.37	−1.50	−1.67
4.0	−0.04	−0.08	−0.13	−0.17	−0.21	−0.25	−0.29	−0.34	−0.37	−0.42	−0.46	−0.50	−0.54	−0.58	−0.62	−0.84	−1.03	−1.25	−1.47	−1.66	−1.86

TABLE A3-8
Values of $\left(\dfrac{S^\circ - S}{R}\right)^{(0)}$

T_r	0.2	0.4	0.6	0.8	1.0	1.2	1.4	1.6	1.8	2.0	2.2	2.4	2.6	2.8	3.0	3.2	3.4	3.6	3.8	4.0	4.5	5.0	6.0	7.0	8.0	9.0
0.80	0.33	5.04	4.66	4.41	4.23	4.08	3.94	3.85	3.76	3.69	3.63	3.56	3.53	3.48	3.44	3.39	3.36	3.33	3.30	3.26	3.19	3.12	3.01	2.96	2.91	
0.85	0.27	4.85	4.47	4.21	4.01	3.86	3.73	3.62	3.53	3.46	3.39	3.34	3.29	3.25	3.21	3.18	3.13	3.11	3.08	3.05	2.98	2.90	2.82	2.76	2.70	
0.90	0.20	0.47	4.13	3.92	3.78	3.63	3.52	3.42	3.32	3.23	3.18	3.11	3.06	3.01	2.97	2.92	2.89	2.86	2.83	2.81	2.76	2.68	2.61	2.56	2.48	2.42
0.95	0.17	0.39	0.64	3.48	3.41	3.32	3.22	3.13	3.05	2.99	2.93	2.87	2.82	2.78	2.74	2.70	2.67	2.64	2.61	2.58	2.57	2.48	2.43	2.37	2.31	2.29
1.00	0.14	0.31	0.54	0.85	1.95	2.55	2.66	2.68	2.68	2.65	2.62	2.60	2.57	2.55	2.54	2.51	2.48	2.44	2.42	2.40	2.34	2.30	2.24	2.19	2.14	2.13
1.05	0.12	0.27	0.43	0.66	0.96	1.44	1.91	2.13	2.25	2.31	2.34	2.34	2.32	2.31	2.30	2.28	2.26	2.25	2.23	2.22	2.17	2.13	2.08	2.05	2.02	2.02
1.10	0.11	0.23	0.36	0.53	0.73	0.97	1.25	1.57	1.77	1.89	1.96	2.01	2.04	2.06	2.07	2.07	2.07	2.06	2.05	2.05	2.02	1.98	1.94	1.91	1.89	1.90
1.15	0.08	0.17	0.30	0.43	0.57	0.75	0.95	1.16	1.36	1.53	1.64	1.73	1.18	1.82	1.84	1.85	1.86	1.86	1.87	1.88	1.88	1.85	1.81	1.79	1.78	1.80
1.20	0.07	0.15	0.25	0.37	0.49	0.62	0.76	0.91	1.07	1.21	1.32	1.42	1.50	1.57	1.62	1.64	1.66	1.67	1.69	1.70	1.72	1.72	1.70	1.69	1.68	1.70
1.25	0.07	0.14	0.21	0.31	0.41	0.52	0.62	0.75	0.88	1.00	1.10	1.20	1.28	1.34	1.39	1.44	1.47	1.50	1.53	1.55	1.59	1.60	1.59	1.58	1.59	1.62
1.30	0.06	0.12	0.19	0.27	0.36	0.44	0.53	0.63	0.74	0.84	0.92	1.01	1.09	1.15	1.22	1.25	1.29	1.33	1.35	1.39	1.44	1.48	1.49	1.48	1.51	1.54
1.4	0.04	0.09	0.15	0.21	0.26	0.34	0.41	0.49	0.56	0.64	0.71	0.77	0.84	0.89	0.94	0.99	1.03	1.07	1.11	1.14	1.21	1.26	1.30	1.32	1.34	1.37
1.5	0.04	0.09	0.14	0.19	0.23	0.28	0.34	0.39	0.45	0.51	0.57	0.63	0.68	0.72	0.76	0.81	0.83	0.87	0.91	0.93	0.99	1.05	1.11	1.15	1.18	1.20
1.6	0.04	0.08	0.12	0.17	0.21	0.25	0.29	0.33	0.38	0.43	0.47	0.51	0.56	0.60	0.63	0.66	0.68	0.71	0.74	0.78	0.84	0.89	0.95	1.01	1.05	1.07
1.7	0.03	0.07	0.11	0.15	0.18	0.21	0.25	0.29	0.32	0.36	0.39	0.43	0.46	0.49	0.52	0.55	0.58	0.61	0.64	0.66	0.72	0.76	0.82	0.88	0.94	0.96
1.8	0.03	0.06	0.09	0.13	0.16	0.19	0.22	0.26	0.28	0.31	0.34	0.37	0.39	0.42	0.44	0.47	0.49	0.52	0.55	0.58	0.63	0.67	0.73	0.79	0.84	0.86
1.9	0.03	0.05	0.08	0.11	0.14	0.17	0.19	0.22	0.24	0.27	0.28	0.31	0.33	0.35	0.37	0.39	0.41	0.43	0.46	0.48	0.53	0.57	0.63	0.68	0.73	0.76
2.0	0.03	0.05	0.08	0.10	0.13	0.15	0.18	0.19	0.21	0.23	0.24	0.27	0.29	0.31	0.33	0.34	0.35	0.37	0.39	0.41	0.45	0.48	0.54	0.58	0.63	0.65
2.5	0.02	0.03	0.05	0.06	0.07	0.09	0.10	0.12	0.13	0.14	0.15	0.16	0.17	0.18	0.19	0.20	0.21	0.22	0.23	0.23	0.25	0.27	0.31	0.35	0.37	0.41
3.0	0.01	0.02	0.03	0.04	0.04	0.05	0.06	0.07	0.08	0.08	0.09	0.10	0.11	0.12	0.12	0.13	0.14	0.15	0.16	0.16	0.18	0.19	0.22	0.24	0.27	0.29
3.5	0.01	0.02	0.02	0.03	0.03	0.04	0.04	0.05	0.05	0.06	0.07	0.08	0.08	0.09	0.09	0.10	0.10	0.11	0.11	0.11	0.12	0.13	0.15	0.17	0.19	0.20
4.0	0.005	0.01	0.01	0.02	0.02	0.03	0.03	0.04	0.04	0.05	0.05	0.05	0.06	0.06	0.06	0.07	0.07	0.07	0.08	0.08	0.09	0.10	0.11	0.13	0.14	0.15

TABLE A3-9

Values of $\left(\dfrac{S^\circ - S}{R}\right)^{(1)}$

T_r	\multicolumn{21}{c}{P_r}
	0.2	0.4	0.6	0.8	1.0	1.2	1.4	1.6	1.8	2.0	2.2	2.4	2.6	2.8	3.0	4.0	5.0	6.0	7.0	8.0	9.0
0.80	0.46	5.23	5.20	5.16	5.12	5.08	5.03	5.00	4.96	4.92	4.89	4.86	4.83	4.81	4.79	4.74	4.71	4.68	4.65	4.62	4.60
0.85	0.37	4.86	4.82	4.79	4.74	4.70	4.66	4.63	4.61	4.58	4.54	4.52	4.50	4.48	4.46	4.43	4.41	4.39	4.37	4.36	4.35
0.90	0.30	0.70	4.38	4.36	4.33	4.31	4.29	4.27	4.26	4.24	4.22	4.20	4.19	4.19	4.18	4.18	4.17	4.17	4.16	4.16	4.15
0.95	0.24	0.53	0.98	3.81	3.77	3.76	3.76	3.76	3.76	3.76	3.76	3.76	3.76	3.77	3.77	3.83	3.87	3.90	3.90	3.91	3.92
1.00	0.19	0.39	0.65	0.90	2.60	3.09	3.18	3.23	3.26	3.28	3.31	3.34	3.37	3.41	3.42	3.53	3.61	3.66	3.67	3.68	3.68
1.05	0.13	0.26	0.38	0.51	0.67	1.18	1.71	2.11	2.34	2.47	2.58	2.65	2.72	2.77	2.82	3.03	3.20	3.26	3.32	3.36	3.39
1.10	0.11	0.20	0.27	0.35	0.41	0.51	0.68	0.91	1.27	1.63	1.86	1.99	2.11	2.16	2.23	2.53	2.75	2.86	2.96	3.04	3.10
1.15	0.09	0.16	0.22	0.28	0.32	0.36	0.41	0.48	0.60	0.81	1.06	1.28	1.45	1.58	1.68	2.07	2.30	2.46	2.60	2.71	2.81
1.20	0.08	0.14	0.20	0.24	0.28	0.31	0.33	0.38	0.44	0.53	0.65	0.79	0.94	1.09	1.21	1.62	1.85	2.03	2.18	2.31	2.38
1.25	0.06	0.12	0.17	0.22	0.26	0.28	0.30	0.33	0.35	0.39	0.46	0.53	0.61	0.71	0.83	1.25	1.50	1.68	1.82	1.95	2.02
1.3	0.04	0.07	0.10	0.13	0.16	0.18	0.20	0.22	0.24	0.26	0.31	0.36	0.41	0.46	0.51	0.77	0.98	1.23	1.36	1.53	1.74
1.4	0.03	0.05	0.07	0.09	0.11	0.13	0.14	0.15	0.16	0.17	0.20	0.22	0.24	0.26	0.29	0.45	0.67	0.85	0.98	1.16	1.3
1.5	0.02	0.03	0.05	0.06	0.07	0.07	0.08	0.09	0.09	0.10	0.11	0.12	0.13	0.15	0.17	0.26	0.46	0.61	0.74	0.90	1.09
1.6	0.01	0.00	0.03	0.04	0.04	0.04	0.05	0.05	0.06	0.06	0.07	0.08	0.08	0.09	0.10	0.17	0.33	0.46	0.59	0.74	0.90
1.7	0.00	0.01	0.01	0.02	0.02	0.02	0.02	0.03	0.03	0.03	0.04	0.04	0.04	0.05	0.05	0.10	0.22	0.33	0.45	0.59	0.73
1.8	0.00	0.00	0.01	0.01	0.01	0.01	0.01	0.01	0.01	0.01	0.01	0.01	0.01	0.02	0.02	0.05	0.14	0.23	0.34	0.46	0.58
1.9	0.00	0.00	−0.01	−0.01	−0.01	−0.01	−0.02	−0.02	−0.03	−0.03	−0.02	−0.02	−0.01	−0.01	0	0.03	0.08	0.15	0.24	0.35	0.46
2.0	0.00	0.00	−0.01	−0.01	−0.01	−0.02	−0.03	−0.04	−0.05	−0.06	−0.05	−0.05	−0.04	−0.04	−0.03	−0.01	0.03	0.09	0.16	0.25	0.35
2.5	−0.01	−0.02	−0.02	−0.03	−0.04	−0.04	−0.05	−0.06	−0.06	−0.07	−0.08	−0.09	−0.09	−0.10	−0.10	−0.09	−0.08	−0.06	−0.05	−0.03	0
3.0	−0.01	−0.02	−0.03	−0.04	−0.05	−0.06	−0.07	−0.08	−0.09	−0.09	−0.10	−0.11	−0.11	−0.12	−0.13	−0.15	−0.16	−0.17	−0.18	−0.18	−0.18
3.5	−0.01	−0.02	−0.02	−0.03	−0.04	−0.05	−0.06	−0.07	−0.08	−0.08	−0.09	−0.09	−0.10	−0.11	−0.12	−0.16	−0.19	−0.21	−0.24	−0.26	−0.29
4.0	−0.01	−0.01	−0.02	−0.02	−0.03	−0.04	−0.05	−0.06	−0.07	−0.08	−0.08	−0.09	−0.10	−0.10	−0.11	−0.14	−0.17	−0.21	−0.25	−0.28	−0.32

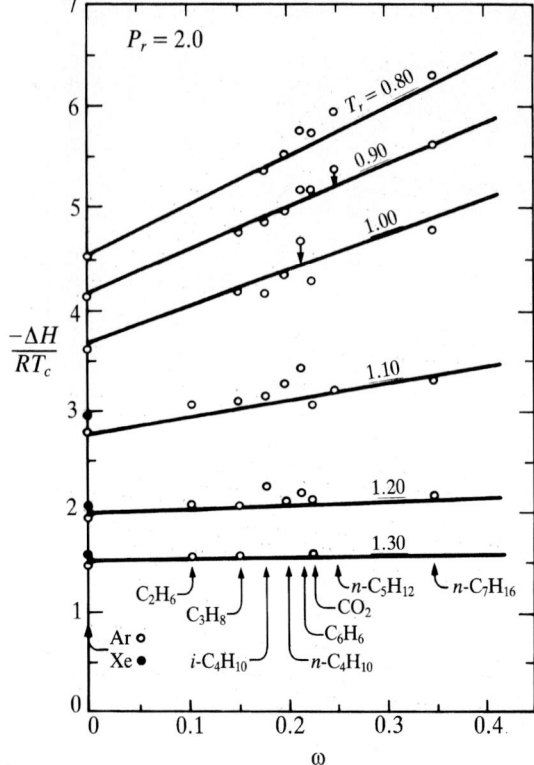

FIGURE A3-2
A comparison of calculated and observed values of the enthalpy difference from the ideal gas at the same pressure and temperature. All values are for reduced pressure of 2.0 and at the reduced temperatures indicated. The acentric factor is a constant for a particular substance; consequently the points lie in vertical columns as labeled. [R. F. Curl, Jr., and K. S. Pitzer, *Ind. Eng. Chem.*, **50**, 265 (1958), fig. 3.]

Figure A3-2 shows a comparison of the calculated enthalpy difference from the ideal gas with experimental values from Joule–Thomson expansion and other methods. The fugacity functions are interpolated easily in the critical region. Although reference 4 includes more detailed tables for the enthalpy in the critical region, other methods should be used for that region unless very rough estimates suffice.

Internal-energy and Helmholtz-energy values can be obtained by appropriate combination of the functions tabulated. Also, temperature derivatives of $(H - H°)$ will yield rough values of $C_P - C_P°$.

Liquid Density

The properties of the liquid at low pressure are especially accessible to measurement and have particular interest. Strictly, we consider the liquid under its own vapor pressure, but usually the properties under atmospheric pressure are not significantly different. The following results are based upon the correlations of Riedel.[1] For the ratio of the density of the liquid to that of the critical point, the equation is

$$\frac{d}{d_c} = 1 + 0.85(1 - T_r) + (1 - T_r)^{1/3}(1.89 + 0.91\omega) \qquad \text{(A3-12)}$$

TABLE A3-10
Values of d/d_0

T_r	0	0.01	0.02	0.03	0.04	0.05	0.06	0.07	0.08	0.09	0.10
0.3	0.875	0.870	0.866	0.862	0.858	0.853	0.849	0.844	0.840	0.835	0.830
0.4	0.830	0.826	0.821	0.816	0.812	0.807	0.802	0.797	0.792	0.787	0.782
0.5	0.782	0.777	0.772	0.767	0.762	0.757	0.752	0.747	0.742	0.737	0.731
0.6	0.731	0.726	0.720	0.715	0.709	0.703	0.698	0.692	0.686	0.680	0.674
0.7	0.674	0.668	0.662	0.656	0.649	0.642	0.636	0.629	0.622	0.613	0.608

If we substitute $T_r = 0$, we find a hypothetical density d_0 for the liquid at 0 K. It turns out that the ratio of the actual density to d_0 is effectively independent of the acentric factor at temperatures lower than $T_r = 0.8$. The resulting values are listed in Table A3-10. This table, together with Eq. (A3-12), summarizes the correlation of liquid density. They allow prediction of the density variation over the full range of temperature from a single measured value. Thus, the coefficient of expansion is determined and may be obtained by differentiating Eq. (A3-12).

Surface Tension, a Criterion of a Normal Fluid

Another important property of the liquid is the surface tension. Again following Riedel,[1] its temperature dependence is well represented by the equation

$$\frac{\gamma}{\gamma_o} = (1 - T_r)^{11/9} \qquad (A3\text{-}13)$$

where γ_0 is a hypothetical surface tension at 0 K. The constant γ_0 may be predicted from the equation

$$\frac{\gamma_0 V_0^{2/3}}{T_c} = 1.86 + 1.18\omega \qquad (\text{dyne cm}^{-1})(\text{cm}^3\,\text{mol}^{-1})^{2/3}\,\text{K}^{-1} \qquad (A3\text{-}14)$$

The quantity V_0 is the hypothetical molal volume at 0 K as calculated from d_0. Given a density at a known reduced temperature, one may obtain d/d_0 from Table A3-10; then from the definitions one has

$$V_0 = \frac{M}{d}\frac{d}{d_0} \qquad (A3\text{-}15)$$

A test of Eq. (A3-14) is shown in Fig. A3-3. We note that typical normal liquids agree with the predicted line quite well, whereas the values of $\gamma_0 V_0^{2/3}/T_c$ for hydrogen-bonding liquids are as low as 50 percent of that predicted. Thus, the surface tension represents a sensitive test for a normal liquid. A deviation of over 5 percent from Eq. (A3-14) appears to indicate significant abnormality. This relationship for γ_0 is recommended as the operational test of a normal liquid.

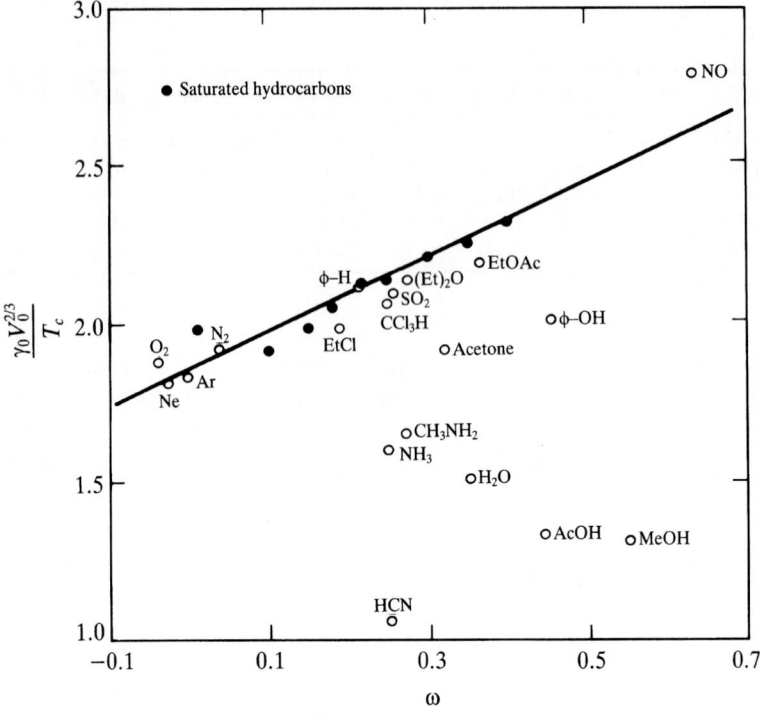

FIGURE A3-3
A criterion for a normal fluid. The reduced surface-tension parameter $\gamma_0 V_0^{2/3}/T_c$ as a function of the acentric factor ω. Points for normal fluids fall within 5 percent of the line of Eq. (A3-14). [R. F. Curl, Jr., and K. S. Pitzer, *Ind. Eng. Chem.*, **50**, 265 (1958), fig. 4.]

ESTIMATION OF CRITICAL PROPERTIES

From this general knowledge of the properties of normal fluids, it is possible to estimate critical properties from measured properties at lower pressures and temperatures. Such a procedure is described in reference 4.

Methods for the estimation of critical properties have also been proposed that are based on the number of atoms or atomic groups in the molecule and their structural arrangement. Somayajulu[8] presented a system of this type and gives references to other estimation methods. A somewhat different method has been proposed by Bae et al.[9] and applied to over 200 fluids. A method of Tsonopoulos and Tan[10] was designed to be valid for the full range of normal alkanes and to extrapolate to an even larger carbon number.

TABLES OF CRITICAL CONSTANTS AND ACENTRIC FACTORS

Table A3-11 presents a selection of critical constants and acentric factors. Many of the values are from a very extensive table of Reid et al.[11]

TABLE A3-11
Critical constants and acentric factors for selected fluids

	T_c/K	P_c/bar	$V_c/\text{cm}^3\,\text{mol}^{-1}$	ω
Elements				
Argon[a]	150.86	48.979	74.57	−0.004
Bromine	584	103.4	127	0.132
Chlorine	417	77.0	124	0.073
Fluorine	144.3	52.2	66.2	0.048
Hydrogen	33.2	13.0	65.0	−0.22
Krypton[a]	209.39	54.96	92.0	−0.002
Neon	44.4	27.6	41.7	0
Nitrogen[a]	126.20	34.00	89.2	0.037
Oxygen	154.6	50.5	73.4	0.021
Xenon[a]	289.74	58.40	119.5	0.002
Hydrocarbons				
Acetylene	308.3	61.4	113	0.184
Benzene	562.1	48.9	259	0.212
n-Butane[a]	425.16	37.96	255.1	0.200
1-Butene	419.6	40.2	240	0.187
Cyclobutane	459.9	49.8	210	0.209
Cyclohexane	553.4	40.7	308	0.213
Cyclopropane	397.8	54.9	170	0.264
Ethane[a]	305.34	48.714	145.5	0.100
Ethylene[a]	282.345	50.403	131.0	0.087
n-Heptane[a]	540.15	27.36	426.4	0.350
n-Hexane[a]	507.85	30.58	368.9	0.303
Isobutane	408.1	36.5	263	0.176
Isobutylene	417.9	40.0	239	0.190
Isopentane	460.4	33.8	306	0.227
Methane[a]	190.53	45.980	98.4	0.011
Naphthalene	748.4	40.5	410	0.302
Neopentane[a]	433.75	31.963	311.1	0.197
n-Octane[a]	568.76	24.87	492.4	0.398
n-Pentane[a]	469.69	33.64	311.0	0.252
Propadiene	393	54.7	162	0.313
Propane[a]	369.85	42.477	200.0	0.153
Propylene[a]	365.57	46.646	188.4	0.141
Toluene	591.7	41.1	316	0.257
m-Xylene	617.0	35.5	376	0.331
Miscellaneous inorganic compounds				
Ammonia	405.6	112.8	72.5	0.250
Carbon dioxide[a]	304.21	73.825	94.43	0.223
Carbon disulfide	552	79.0	170	0.115
Carbon monoxide	132.9	35.0	93.1	0.049
Carbon tetrachloride	556.4	45.6	276	0.194
Carbon tetrafluoride	227.6	37.4	140	0.191
Chloroform	536.4	54.7	239	0.216
Hydrogen chloride	324.6	83.1	81.0	0.12
Hydrogen sulfide	373.2	89.4	98.5	0.100
Nitrous oxide	309.6	72.4	97.4	0.160
Sulfur dioxide	430.8	78.8	122	0.251
Sulfur hexafluoride[a]	318.70	37.590	197.4	0.208

TABLE A3-11 Continued

	T_c/K	P_c/bar	$V_c/\text{cm}^3\,\text{mol}^{-1}$	ω
	Miscellaneous organic compounds			
Acetaldehyde	461	56	154	0.303
Acetone	508.1	47.0	209	0.309
Acetonitrile	548	48.3	173	0.321
Chlorobenzene	632.4	45.2	308	0.249
Dichlorodifluoromethane	385.0	41.2	217	0.176
Diethyl ether	466.7	36.9	280	0.281
Dimethyl ether	400.0	53.7	178	0.192
Ethylene oxide	469	72.8	140	0.200
Isopropyl alcohol	508.3	47.6	220	
Methyl chloride	416.3	67.7	139	0.156
Methyl ethyl ketone	535.6	41.5	267	0.329
Pyridine	620.0	56.3	254	0.24
Tetrafluoromethane[a]	227.527	37.45	140.65	0.177
Trichlorofluoroethane	487.2	34.1	304	0.252
Trichlorofluoromethane	471.2	44.1	248	0.188
Trimethylamine	433.2	40.7	254	0.195

[a] Values taken from Schreiber and Pitzer[6]; other values taken from a more complete list given by Reid et al.[11]

REFERENCES

1. L. Riedel, *Chem. -Ing. Tech.*, **26**, 83, 259, 679 (1954); **27**, 209, 475 (1955); **28**, 557 (1956).
2. K. S. Pitzer, *J. Am. Chem. Soc.*, **77**, 3427 (1955).
3. K. S. Pitzer, D. Z. Lippmann, R. F. Curl, Jr., C. M. Huggins, and D. E. Petersen, *J. Am. Chem. Soc.*, **77**, 3433 (1955).
4. R. F. Curl, Jr. and K. S. Pitzer, *Ind. Eng. Chem.*, **50**, 265 (1958).
5. K. S. Pitzer, *J. Chem. Phys.*, **7**, 583 (1939).
6. D. R. Schreiber and K. S. Pitzer, *Fluid Phase Equil.*, **46**, 113 (1989).
7. B. I. Lee and M. G. Kessler, *Am. Inst. Chem. Eng. J.*, **21**, 510 (1975).
8. G. R. Somayajulu, *J. Chem. Eng. Data*, **34**, 106 (1989).
9. H.-K. Bae, S.-Y. Lee, and A. S. Teja, *Fluid Phase Equil.*, **66**, 225 (1991).
10. C. Tsonopoulos and Z. Tan, *Fluid Phase Equil.*, **83**, 127 (1993).
11. R. C. Reid, J. M. Prausnitz, and T. K. Sherwood, *The Properties of Gases and Liquids*, 3d edn., McGraw-Hill, New York, 1977.

APPENDIX 4

EQUATIONS OF STATE FOR NORMAL FLUIDS

The prediction of the volumetric and thermodynamic properties of fluids has been the subject of many studies for over a century. As described in Chapter 9, the principle of corresponding states was an important advance but was not adequate. Specifying T_c and P_c and the use of reduced properties is an essential first step, but at least one more parameter is required to select groups of fluids that follow corresponding states accurately within each group.

The acentric factor has proven to be very useful in the representation or estimation of fluid properties. In comparison with earlier three-parameter systems extending corresponding states, there were two important advances. The first was the use of theory to indicate the characteristics of molecules that could be expected to follow accurately a pattern of behavior defined by T_c, P_c, and a third parameter.[1] The second was the choice of an easily and accurately measurable quantity, the slope of the vapor pressure curve, as the basis for the third parameter.[2] Initially, the $P-V-T$ and other properties were given in the form of tables because in 1955 the best analytical equation of state was inconvenient to use and still failed to represent the $P-V-T$ data within experimental uncertainty. These tables are presented in Appendix 3. More complex equations are now reasonably convenient with modern computers and are much more accurate. Several equations of state with the acentric factor as a third parameter have been developed. In this appendix, we first discuss certain features that are important for any equation of state in a three-parameter system. Next, we describe some improved equations of the Redlich–Kwong type. Then we describe a few equations of higher accuracy and discuss the deviations from conformance to the acentric factor system that limit the accuracy attainable by any equation of this type. And finally, the details are given for a particular equation that is especially useful for mixed fluids.

Definition

The acentric factor definition is

$$\omega = -\log P_r(\text{at } T_r = 0.7) - 1.00 \tag{A4-1}$$

where P_r is the reduced vapor pressure at $T_r = 0.7$, but ω can be evaluated from any vapor pressure datum well removed from the critical point by use of the vapor-pressure tables or equation (Appendix 3). Since vapor pressures are easily measured to high precision, the uncertainty in ω is small provided the critical pressure and temperature are accurately known.

Choice of form for reduced density

While the usual definition of reduced density is the simple ratio to the experimental critical density $\rho_r = \rho/\rho_c$, the latter quantity is not directly measured and is subject to considerably greater uncertainty than either T_c or P_c. Thus, there is significant advantage in the use of a reduced density based on T_c and P_c instead of ρ_c. The original presentation of the acentric factor system was in terms of the compressibility factor as a function of T_r and P_r, which avoided the use of ρ_r. But we now wish to use ρ_r, along with T_r, as an independent variable in our equation. Schreiber and Pitzer[3] adopted a procedure to accomplish this purpose. They first selected a "best" expression for the compressibility factor as a function of the acentric factor. A linear expression sufficed:

$$z_c^* = 0.2905 - 0.0787\omega \tag{A4-2}$$

Then the adopted critical density is

$$\rho_c^* = \frac{P_c}{RT_c z_c^*} \tag{A4-3}$$

Yamada[4] adopted a different procedure that accomplished the same purpose. While it is probably best to use the method of obtaining ρ_c^* that was used in the original generation of a particular equation of state, the differences are small, and the most convenient method can be chosen. Indeed, the reported ρ_c for a particular fluid can also be used to calculate $\rho_r = \rho/\rho_c$ without serious error, but the use of ρ_c^* is preferable.

IMPROVED EQUATIONS OF THE REDLICH–KWONG TYPE

While it is easy, with modern computers, to use complex and accurate equations for pure fluids, there are substantial advantages in the use of a simpler basic equation for the calculation of vapor–liquid equilibria or similar properties of multicomponent systems. Good results have been reported, especially when the simple equation of state was fitted to the particular properties of interest in a given calculation. In this section we list a few of the equations developed as improvements on the Redlich–Kwong (RK) equation. In all cases, the acentric factor is used as a third parameter.

We write the general equation of the RK type in the form

$$P = \frac{RT\rho}{1 - b\rho} - \frac{a\alpha(T)\rho^2}{f(\rho)} \tag{A4-4}$$

For the RK equation[5] itself,

$$\alpha(T) = T^{-1/2} \tag{A4-5a}$$

$$f(\rho) = 1 + b\rho \tag{A4-5b}$$

Soave[6] retained the $f(\rho)$ of RK but showed that a more complex temperature dependency was required to fit the experimental vapor pressure for several fluids. It is

$$\alpha(T) = [1 + m(1 - T_r^{1/2})]^2 \tag{A4-6a}$$

$$m = 0.480 + 1.574\omega - 0.176\omega^2 \tag{A4-6b}$$

where ω is the acentric factor. The Redlich–Kwong–Soave (RKS) equation has been widely and successfully used for high-pressure vapor–liquid equilibrium calculations. In 1993 Soave[7] presented an interesting account of the history, successful applications, and later refinements of the Soave–Redlich–Kwong equation.

The other change that Redlich and Kwong made in the van der Waals equation was the factor $f(\rho) = 1 + b\rho$ in the denominator of the second term. This is adequate for fluids with a small acentric factor. But for fluids with larger acentric factors, a more substantial modification gives better results. A popular equation of this type is that of Peng and Robinson.[8] Their equation is of the form of Eq. (A4-4) with

$$f(\rho) = 1 + 2b\rho - b^2\rho^2 \tag{A4-7a}$$

$$\alpha(T) = [1 + m(1 - T_r^{1/2})]^2 \tag{A4-7b}$$

$$m = 0.374\,64 + 1.542\,26\omega - 0.269\,92\omega^2 \tag{A4-7c}$$

Their temperature dependent function has the same form as Soave's and was similarly designed to fit measured vapor pressures.

If Soave's equation is best for small ω and the Peng–Robinson equation for large ω, then the next step is the selection of an acentric-factor-dependent expression for $f(\rho)$ in Eq. (A4-4). Several equations of this type were presented nearly simultaneously[9-12] and others will doubtless follow. The equations have similar properties in that their properties vary continuously between those of or near the RKS equation for small ω and those of or near the PR equation for ω near $\frac{1}{3}$. They differ in detail and their expressions for $\alpha(T)$ are rather complex. They retain the advantage of cubic form in density, however, which is useful in many calculations.

MORE ACCURATE EQUATIONS

There are several examples where the Benedict–Webb–Rubin (BWR) equation[13] [Eq. (9-62*)] was used in an acentric-factor system. In the form adopted by Lee and Kessler,[14] the results were presented as tables to be used with P_r instead of ρ_r as the variable in addition to T_r and ω. For most purposes, Yamada's system[4] is preferable, since it uses ρ_r as the independent variable with an excellent method of selection of that variable from P_c, T_c, and ω. Yamada presented quadratic expressions in ω for each of the eight BWR coefficients; thus there are 24 parameters in all. He found that this system gave good results up to a reduced density of about 1.8. For higher densities and above critical temperatures, he recommends an extended BWR equation with 44 parameters in all. Schreiber and Pitzer[3] selected a somewhat similar extended BWR equation with 22 terms whose ω-dependencies are given by 57 parameters.

The Schreiber–Pitzer (SP) equation is accurate enough to show clearly the departures of individual fluids from the acentric factor system in the range 1.02 to 2.0 in T_r. In particular, Ar and Xe, both with $\omega \cong 0$, follow corresponding states accurately up to $\rho_r \cong 1.4$, but at higher densities Ar deviates in one direction from the SP equation, while Xe deviates in the opposite direction. Similar results were found for fluids with other values of ω.

There are dangers in the use of many-term equations when fitted to a limited database. Extrapolation beyond the range of the database can yield grossly erroneous values. This has been observed for the SP equation for the liquid at large ω, low T, and low P. Thus, there is no advantage in using any more complex equation within a normal-fluid, acentric-factor correlation, and there are many reasons for using a somewhat simpler equation. The equation of Anderko and Pitzer[15] was designed on this basis and with a particular objective of convenient use for mixed fluids.

Anderko-Pitzer Equation[15]

This equation, which was presented briefly in Chapter 9 as Eq. (9-61)*, includes accurate expressions for the second, third, and fourth virial coefficients, together with a two-coefficient repulsive term. In reduced form, it is

$$z = \frac{1 + c\rho_r}{1 - b\rho_r} + \alpha\rho_r + \beta\rho_r^2 + \gamma\rho_r^3 \tag{A4-8}$$

For normal fluids, the acentric-factor-dependent expressions for each coefficient are

$b = 0.251\,896 + 0.048\,788\omega$

$c = -0.080\,968 - 0.879\,256\omega + (1.802\,846 + 0.671\,581\omega)/T_r$
$\qquad - (0.387\,457 + 0.226\,976\omega)/T_r^2$

$$\alpha = 0.271\,331 + 1.556\,118\omega - (2.783\,816 + 0.452\,867\omega)/T_r - (0.223\,685 + 1.022\,78\omega)/T_r^2 - (0.005\,1562 + 0.189\,187\omega)/T_r^6$$

$$\beta = 0.088\,425 - 0.516\,256\omega - (0.619\,446 + 0.319\,606\omega)/T_r + (0.694\,687 + 1.270\,432\omega)/T_r^2 + (0.006\,874 + 0.273\,364\omega)/T_r^6$$

$$\gamma = 0.099\,362 + 0.816\,731\omega - (0.259\,215 + 0.742\,537\omega)/T_r - (0.099\,234 + 0.295\,499\omega)/T_r^2 - (0.001\,311 + 0.071\,180\omega)/T_r^6$$

For optimum accuracy in use of this equation, the critical volume of the fluid should be obtained from Eq. (A4-3) above. With no density dependency higher than ρ_r^3, it is practical to use rigorous composition-dependency relationships for mixed fluids. These are considered briefly in Chapter 13 and are developed and applied in reference 15.

REFERENCES

1. K. S. Pitzer, *J. Am. Chem. Soc.*, **77**, 3427 (1955).
2. K. S. Pitzer, D. Z. Lippmann, R. F. Curl, Jr., C. M. Huggins, and D. E. Petersen, *J. Am. Chem. Soc.*, **77**, 3433 (1955).
3. D. R. Schreiber and K. S. Pitzer, *Fluid Phase Equil.*, **46**, 113 (1989).
4. T. Yamada, *Am. Inst. Chem. Eng. J.*, **19**, 286 (1973).
5. O. Redlich and J. N. S. Kwong, *Chem. Rev.*, **44**, 233 (1949).
6. G. Soave, *Chem. Eng. Sci.*, **27**, 1197 (1972).
7. G. Soave, *Fluid Phase Equil.*, **82**, 345 (1993).
8. D. Y. Peng and D. B. Robinson, *Ind. Eng. Chem. Fundam.*, **15**, 59 (1976).
9. G. Schmidt and H. Wenzel, *Chem. Eng. Sci.*, **35**, 1503 (1980).
10. A. Harmens and H. Knapp, *Ind. Eng. Chem. Fundam.*, **19**, 291 (1980).
11. G. Heyen, "A Cubic Equation of State with Extended Range of Application," 2d World Congress Chem. Eng., Montreal (1981); *Chemical Engineering Thermodynamics*, S. A. Newman, ed., Ann Arbor Sci., Ann Arbor, Mich., 1983, pp. 175–185.
12. N. C. Patel and A. S. Teja, *Chem. Eng. Sci.*, **37**, 463 (1982).
13. M. Benedict, G. B. Webb, and L. C. Rubin, *J. Chem. Phys.*, **8**, 334 (1940).
14. B. I. Lee and M. G. Kessler, *Am. Inst. Chem. Eng. J.*, **21**, 510 (1975).
15. A. Anderko and K. S. Pitzer, *Am. Inst. Chem. Eng. J.*, **37**, 1379 (1991).

APPENDIX 5

FLUID PROPERTIES VERY NEAR THE CRITICAL POINT

For the van der Waals or any similar equation of state, the vapor–liquid coexistence curve has a parabolic shape near the critical point, i.e.,

$$(\rho_l - \rho_v)^2 = (\text{const})(T_c - T) \tag{A5-1}$$

where ρ_l and ρ_v are the densities of the liquid and vapor, respectively. For real, normal fluids, however, the actual curve is broader and close to cubic in shape. This was first noted by Verschaffelt[1] in 1896. He defined what we now call the effective exponent β_e as

$$\beta_e = \frac{\Delta \ln(\rho_l - \rho_v)}{\Delta \ln(T_c - T)} \tag{A5-2}$$

and showed that, for the best data then available, its value was close to 1/3 instead of 1/2.

Although this result was published in Holland, a center of research on fluids, it was ignored. Many years later, this discrepancy was rediscovered and a very interesting and elaborate theory has been developed concerning the properties of fluids or disordered crystals near a critical point.[2] Several exponents are defined as follows for the limiting behavior near T_c. With $t = (T - T_c)/T_c = T_r - 1$, the heat capacity at constant volume diverges as

$$C_V = At^{-\alpha} \quad \text{at } \rho = \rho_c \tag{A5-3}$$

Other properties for fluids have limiting behavior as follows:

$$\frac{\rho_l - \rho_v}{2\rho_c} = B(-t)^\beta \tag{A5-4}$$

$$\left(\frac{\partial P}{\partial \rho}\right)_T = \Gamma t^\gamma \quad \text{at } \rho = \rho_c \tag{A5-5}$$

$$|P - P_c| = D(\rho - \rho_c)^\delta \quad \text{at } T = T_c \tag{A5-6}$$

The amplitudes A, B, Γ and D are specific to each system, although there are relations between them. Some authors add refinements or use alternate definitions that need not concern us for our present purposes.

For a critical point where the separation is into two liquids, a reduced composition difference such as $(x' - x'')/2x_c$ or $(\phi' - \phi'')/2\phi_c$, with x or ϕ mole or volume fractions, respectively, replaces the density expression in Eq. (A5-4) with the exponent β.

The theory for near-critical phenomena concerns fluctuations in density, composition, or disorder, and in particular, the range in distance of these fluctuations that approaches infinity as $t \to 0$ ($T \to T_c$). At temperatures away from T_c, the characteristics of these fluctuations depend on the details of each particular system, but as $t \to 0$, the range of the fluctuations exceeds the range of intermolecular forces in most cases. Then the details of the intermolecular potentials become irrelevant and universal exponents apply.[2] There are different "universality classes"; two-dimensional systems, for example, have different exponents than those for three dimensions.

In general, all of the exponents α, β, γ, δ may be irrational numbers. There is no violation of thermodynamics in this singular behavior near a critical point. An expression can be written for the Helmholtz energy that yields each of Eqs. (A5-3) to (A5-6) upon appropriate differentiation and manipulation. This expression is, in general, not a mathematically analytic function at the critical point. But for certain universality classes, the Helmholtz energy may be an analytic function that yields simple rational exponents. In particular, an example of primary interest is that of sufficiently long-range potentials where the exponents (for three dimensions) become $\alpha = 0$, $\beta = 1/2$, $\gamma = 1$, and $\delta = 3$. These are the values one obtains from any mathematically analytic equation of state (with a most improbable exception that need not concern us). For inverse power potentials R^{-n}, the common case of $n = 6$ is short range, as is also true for any larger n, while the long-range exponents apply for n less than 4.5 (the dimensionality times 3/2).[3,4] Experimental data[5–7] for ionic systems are consistent with the long-range exponents, $\beta = 1/2$, etc., but more research is needed for both theory and experiment for the ionic fluids.

The theoretical values for the "short-range" exponents require elaborate calculations involving series evaluations. Recent advances yield the relatively precise values[8] $\alpha = 0.110 \pm 0.003$, $\beta = 0.326 \pm 0.002$, $\gamma = 1.239 \pm 0.002$, and $\delta = 4.80 \pm 0.02$. These theoretical values lie well outside the limits of uncertainty for the experimental values of β and δ determined[9] for pure fluids in the range $|t| \gtrsim 0.005$;

those values are $\beta = 0.355 \pm 0.007$ and $\delta = 4.35 \pm 0.010$. Recent experiments with special optical methods now obtain measurements down to $|t| \cong 10^{-5}$ that yield values of β in agreement with the recent theoretical value of 0.326. Thus, the effective β must decrease very suddenly from 0.355 to 0.326 in the range exceedingly close to the critical point.[10] The complete near-critical theory includes expansions in series with leading terms based on Eqs. (A5-3) through (A5-6). Thus, a rapid change of β_e with t falls within the theory but implies large values of the coefficients of higher-order terms. The range of convergence of these series expansions is problematical; thus, these expressions are useful only quite close to the critical point if the higher-order terms are large.

The theory as summarized above ignores any effect of gravity. But, if the two phases have different densities, gravity tends to suppress fluctuations in the same manner as long-range forces and could yield the exponents $\alpha = 0$, $\beta = 1/2$, $\gamma = 1$, $\delta = 3$ for three dimensions. Indeed, Wagner, Kurzeja, and Pieperbeck[11] have reported these values for β, γ, and δ for SF_6 in the limit at the critical point. With a molecular mass of 146 and a T_c of only 318.7 K, the gravitational effect would be large in this case. In view of the possible effect of gravity and the interesting results of Wagner et al.,[11] the theory of limiting exponents merits further study.

For systems with separation into two liquid phases,[10] the effective β approaches the value 0.326 very smoothly from the side of lower (rather than higher) values, and the higher-order terms are quite small. Reasons for this difference between liquid–liquid and vapor–liquid systems and other aspects of near-critical behavior were discussed by Singh and Pitzer.[10]

Various proposals[12–14] have been made for equations of state involving a cross-over from an analytic equation for the wide range of T and ρ to a nonanalytic exponential expression consistent with Eqs. (A5-3) to (A5-6) near the critical point. When it was thought that the limiting exponents were reached at $|t| \cong 0.005$, this was not too difficult, but even then there was little practical advantage in the cross-over. With many terms required in any case for the wide-range equation, it is simpler to add a few additional terms to attain a near-critical accuracy sufficient for any practical purpose. The introduction of the cross-over was more a matter of aesthetics. But with the more accurate limiting exponents that differ substantially from the effective exponents at $|t| \cong 0.005$, the cross-over problem becomes much more complex. Also with the possible effect of gravity on the limiting behavior, there is further reason to avoid complex cross-over functions that may assume incorrect limiting exponents. For the representation of measured properties within experimental uncertainty, it is more convenient to use enough well-selected terms in an analytical equation to attain the desired accuracy in the near-critical region. For example, the expressions of Wagner and associates[15,16] yield good agreement with the precise measurements for O_2 and H_2O to very small values of $|t|$.

The theory for the behavior of fluids and disordered crystals very near a critical point is an interesting and continuing scientific development. But from the point of view of chemical thermodynamics broadly, this is a rather isolated topic. Except for a very limited and special type of calculation involving properties very close to the critical point, there is no need to consider these details. Nevertheless, it is important to be sure that an equation of state fits the real fluid properties to sufficiently high accuracy at all temperatures and pressures of interest in a particular application.

REFERENCES

1. J. E. Verschaffelt, *Comm. Phys. Lab., Leiden*, **28**, 1 (1896).
2. See, for example, Chapter 3 of J. S. Rowlinson and F. L. Swinton, *Liquids and Liquid Mixtures*, 3d edn., Butterworths, London, 1982; or for greater detail the chapters in *Phase Transitions and Critical Phenomena*, C. Domb and M. S. Green, eds., Academic Press, New York, 1976.
3. M. E. Fisher, S.-k. Ma, and B. G. Nickel, *Phys. Rev. Lett.*, **29**, 1917 (1972).
4. M. Suzuki, *Prog. Theor. Phys.*, **49**, 424 (1973).
5. R. R. Singh and K. S. Pitzer, *J. Am. Chem. Soc.*, **110**, 8723 (1988); *J. Chem. Phys.*, **90**, 5742 (1989).
6. K. S. Pitzer, *Acc. Chem. Res.*, **23**, 333 (1990).
7. K. C. Zhang, M. E. Briggs, R. W. Gammon, and J. M. H. Levelt Sengers, *J. Chem. Phys.*, **97**, 8692 (1992).
8. J. V. Sengers and J. M. H. Levelt Sengers. *Annu. Rev. Phys. Chem.*, **37**, 189 (1986).
9. J. M. H. Levelt Sengers and J. V. Sengers, *Phys. Rev. A*, **12**, 2622 (1975).
10. R. R. Singh and K. S. Pitzer, *J. Chem. Phys.*, **90**, 5742 (1989).
11. W. Wagner, N. Kurzeja and B. Pieperbeck, *Fluid Phase Equil.*, **79**, 151 (1992).
12. G. A. Chapela and J. S. Rowlinson, *J. Chem. Soc., Faraday Trans.*, **70**(1), 584 (1974).
13. P. G. Hill, *Proc. 10th Intl. Conf. Prop. of Steam*, **1**, 117 (1986); *J. Phys. Chem. Ref. Data*, **19**, 1233 (1990).
14. P. C. Albright, Z. Y. Chen, and J. V. Sengers, *Phys. Rev. B*, **36**, 877 (1987).
15. R. Schmidt and W. Wagner, *Fluid Phase Equil.*, **19**, 175 (1985).
16. A. Saul and W. Wagner, *J. Phys. Chem. Ref. Data*, **18**, 1537 (1989).

APPENDIX 6

TABLES OF THERMODYNAMIC PROPERTIES

At the end of Chapter 8, a list was given of compilations from which a wide variety of data can be obtained for various thermodynamic calculations. The abbreviated tables in this appendix are presented for convenience in making approximate calculations without having to consult other sources. Where the highest accuracy is desired, however, the best source should be sought for each property required.

It is customary in thermodynamic tabulations to give more significant figures than are warranted by the absolute accuracy when differences between thermodynamic quantities are known very accurately. Thus, if high-temperature C_p° values are known accurately, $(G_T^\circ - H_{298}^\circ)/RT$ values above 298.15 K are given to 0.001, even when the value at 298.15 K can be given only to 0.1, to retain full accuracy for the increments above 298.15 K. Where the abbreviated 298 appears, 298.15 is implied.

The quantities presented in the following tables consist of three types. The major part of the tables is devoted to Gibbs-energy functions based either on H_0° or on H_{298}°. The use of these Gibbs-energy functions has been discussed in Chapter 8. For those substances where the Gibbs-energy functions are based on 298 K, values of the enthalpy of formation ΔH_{298}° are also tabulated. This is the enthalpy change on formation of the indicated substance from the elements in their standard reference states. The standard reference state for an element is the stable form of the element at 1 bar pressure and the temperature of interest, for example, 25°C. Thus, with a few exceptions such as hydrogen, oxygen, nitrogen, fluorine, and chlorine, which are gaseous diatomic molecules, and mercury and bromine, which are liquids, all the other elements that are included in the following tables are solids in their standard reference states at 25°C. Where two possible forms exist, such as for sulfur, the more stable form,

TABLE A6-1
Solid and liquid elements

	\-$(G_T°-H_{298}°)/RT$					$(H_{298}°-H_0°)/R$
	298.15 K	500 K	1000 K	1500 K	2000 K	K
Ag	5.133	5.482	6.742 m	7.979	9.066	
Al	3.401	3.744 m	5.123	6.633	7.713	545.9
As	4.294	4.637	5.874			
Au	5.713	6.061	7.310 m	8.448	9.524	
B	0.702	0.892	1.760	2.561	3.242	146.0
Ba	7.514	7.938 tr	9.756 m	11.538	12.832	831.3
Be	1.135	1.389	2.420	3.326 tr	4.345	232.4
Bi	6.824	7.180 m	9.640	11.117		
Br$_2$(l)	18.306					2948.
C	0.690	0.834	1.523	2.191	2.767	126.4
Ca	5.002	5.363 tr	6.788 m	8.277		
Cd	6.230	6.594 m	8.445			
Co	3.616	3.969 tr	5.340 tr	6.578 m	7.756	573.8
Cr	2.841	3.175	4.438	5.524	6.483	487.9
Cs	10.241 m	11.006	(13.00)			928.1
Cu	3.989	4.329	5.564 m	6.687	7.787	602.2
Fe	3.286	3.647	5.093 tr	6.479 m	7.587	541.8
Hf	5.239	5.598	6.920	8.023	8.937	702.6
Hg(l)	9.144	9.521				1123.7
I$_2$	13.969 m	(16.84)				1587.
K	7.778 m	8.477	10.216 m			851.8
La	6.844	7.216 tr	8.579	9.924	11.033	
Li	3.498 m	3.931	5.646	6.824		555.9
Mg	3.929	4.279 m	5.666			601.1
Mn	3.850	4.228 tr	5.685 tr	7.052 m	8.476	600.6
Mo	3.440	3.776	5.003	6.004	6.82	551.5
Na	6.189 m	6.816	8.561			775.4
Ni	3.593	3.967 tr	5.411	6.573 m	7.685	575.6
P(ß)	4.940 m	5.384	6.71			644.7
Pb	7.792	8.165 m	9.908	11.175	12.099	827.2
Pt	5.007	5.364	6.651	7.696	8.547	
Rb	9.234 m	9.985	(11.74)			900.8
S	3.855 m	4.378				530.6
Sb	5.475	5.825 m	7.352	9.210		
Se	5.083 m	5.469	7.672			
Si	2.264	2.554	3.655	4.568 m	5.875	387.0
Sn	6.157	6.540 m	8.739	10.075	11.034	
Ta	4.988	5.337	6.591	7.593	8.394	683.3
Te	5.953	6.322 m	8.588			
Th	6.421	6.804	8.238	9.465 tr	10.547	
Ti	3.699	4.057	5.376 tr	6.559 tr	7.545	580.9
Tl	7.719	8.087 m	9.874	11.139	11.196	
U	6.049	6.443 tr	8.033 m	9.733		
V	3.480	3.831	5.110	6.163	7.043	558.1
W	3.928	4.264	5.473	6.449	7.242	598.1
Zn	5.017	5.371 m	7.083			681.8
Zr	4.675	5.025	6.297 tr	7.447	8.359	661.1

TABLE A6-2
Gaseous elements

	$-(G_T^\circ - H_{298}^\circ)/RT$					$(H_{298}^\circ - H_0^\circ)/R$	$\Delta_f H_{298}^\circ/R$
	298.15 K	500 K	1000 K	1500 K	2000 K	K	10^3 K
Ag	20.806	21.090	22.077	22.842	23.437		34.1$_7$
Al	19.791	20.080	21.080	21.850	22.448	832.2	39.7 ± 0.5
Au	21.710	21.993	22.980	23.746	24.345		44.2$_8$
B	18.454	18.737	19.725	20.490	21.086	759.6	67.4 ± 1.4
Ba	20.476	20.759	21.747	22.514	23.126	745.4	21.5 ± 0.6
Be	16.390	16.673	17.661	18.426	19.021	745.4	39.0 ± 0.6
Bi	22.492	22.775	23.763	24.528	25.123		25.2$_1$
Bi$_2$	32.924	33.429	35.197	36.570	37.638		26.4$_7$
Br	21.050	21.333	22.323	23.100	23.715	745.4	13.45 ± 0.01
Br$_2$	29.514	30.011	31.769	33.147	34.226	1169.3	3.72 ± 0.02
C	19.015	19.299	20.287	21.053	21.649	786.1	86.20 ± 0.06
C$_2$	23.980	24.541	26.346	27.704	28.763	1272.3	100.8 ± 0.5
C$_3$	28.534	29.004	30.900	32.442	33.708	1414.4	98.6 ± 2.0
Ca	18.629	18.912	19.899	20.665	21.260	745.4	21.38 ± 1.0
Cd	20.175	20.458	21.446	22.211			13.45
Cl	19.868	20.171	21.242	22.067	22.701	754.4	14.59 ± 0.01
Cl$_2$	26.830	27.304	29.010	30.362	31.425	1104.2	0.0
Co	21.591	21.917	23.112	24.061	24.808	764.9	51.1 ± 0.5
Cr	20.965	21.248	22.236	23.003	23.610	745.4	47.8 ± 0.5
Cs	21.120	21.403	22.391	23.156	23.752	745.4	9.2 ± 0.1
Cu	20.013	20.296	21.284	22.049	22.645	745.4	40.6 ± 0.2
F	19.093	19.401	20.451	21.249	21.863	783.9	9.55 ± 0.05
F$_2$	24.390	24.831	26.452	27.764	28.809	1061.4	0.0
Fe	21.708	22.057	23.232	24.102	24.765	823.9	50.0 ± 0.2
H	13.797	14.081	15.068	15.833	16.428	745.4	26.219 ± 0.001
H$_2$	15.717	16.113	17.504	18.601	19.477	1018.4	0.0
Hf	22.479	22.764	23.798	24.676	25.413	745.5	74.4 ± 0.8
Hg	21.044	21.327	22.315	23.080	23.675	745.4	7.382 ± 0.005
I	21.744	22.027	23.015	23.780	24.377	745.4	12.841 ± 0.005
I$_2$	31.353	31.859	33.638	35.030	36.130	1216.7	7.508 ± 0.009
K	19.285	19.568	20.555	21.321	21.916	745.4	10.70 ± 0.05
Li	16.692	16.975	17.962	18.728	19.323	745.4	19.16 ± 0.12
Mg	17.878	18.162	19.149	19.914	20.509	745.4	17.69 ± 0.10
Mn	20.893	21.177	22.164	22.929	23.524	745.4	34.1 ± 0.5
Mo	21.884	22.167	23.155	23.920	24.516	745.4	79.2 ± 0.5
N	18.438	18.721	19.709	20.474	21.069	745.4	56.85 ± 0.01
N$_2$	23.045	23.433	24.861	26.012	26.942	1042.8	0.0
Na	18.482	18.765	19.753	20.518	21.113	745.4	12.91 ± 0.08
Ni	21.913	22.235	23.389	24.294	24.992	820.9	51.7 ± 1.0
O	19.371	19.666	20.679	21.456	22.057	808.8	29.97 ± 0.01
O$_2$	24.674	25.080	26.565	27.783	28.765	1044.3	0.0
Pb	21.093	21.376	22.364	23.131	23.737	745.4	23.48 ± 0.10
Pt	23.141	23.503	24.753	15.671	26.360		67.9$_4$
Rb	20.458	20.741	21.728	22.494	23.089	745.4	9.73 ± 0.05
S	20.185	20.504	21.585	22.399	23.022	745.4	33.31 ± 0.03
S$_2$	27.442	27.898	29.558	30.890	31.951	1097.4	15.47 ± 0.04
Sb	21.680	21.964	22.952	23.718	24.316		31.8$_2$

TABLE A6-2
(Continued)

	$-(G_T°-H_{298}°)/RT$					$(H_{298}°-H_0°)/R$	$\Delta_f H_{298}°/R$
	298.15 K	500 K	1000 K	1500 K	2000 K	K	10^3 K
Sb_2	30.659	31.159	32.914	34.281	35.347		27.8_1
Se	21.254	21.544	22.580	23.404	24.057		28.3_1
Se_2	29.301	29.868	31.865	33.407	34.592		16.62
Si	20.204	20.501	21.518	22.298	22.906	908.1	54.1 ± 1.0
Sn	20.264	20.567	21.791	22.871	23.732	747.5	36.2_3
Ta	22.277	22.564	23.634	24.558	25.339	745.7	94.1 ± 1.0
Te	21.974	22.256	23.240	24.015	24.628		25.4_6
Te_2	31.148	31.649	33.453	34.913	36.086		19.2_9
Th	22.872	23.155	24.179	25.069	25.848		69.1_9
Ti	21.685	22.007	23.077	23.885	24.516	906.7	56.9 ± 0.5
Tl	21.765	22.048	23.036	23.802			21.7_6
U	24.029	24.355	25.479	26.365	27.100		$62._9$
V	21.926	22.269	23.439	24.354	25.066	951.0	62.0 ± 1.0
W	20.922	21.226	22.432	23.465	24.298	747.6	102. ± 1
Zn	19.363	19.646	20.633	21.399	21.994	745.4	15.69 ± 0.03
Zr	22.013	22.356	23.536	24.447	25.192	1612. ?	73.4 ± 1.0

rhombic sulfur, is the standard reference state except for the use of white (β) phosphorus. Except for the elements that are gaseous at 25°C and are retained in the gaseous state for expressing $\Delta H_0°$ for the formation of a compound, $\Delta H_0°$ is based on the solid state of the elements at 0 K. In the tabulations of Gibbs-energy functions for condensed substances, the state of the condensed substances is either solid or liquid depending upon whether the indicated temperatures are above or below the melting point. The symbol "m" is inserted to indicate that melting took place between the tabulated temperatures. Similarly, the symbol "tr" indicates a transition to a different solid structure. When gaseous substances are specifically indicated, all the data presented at all temperatures are for the gaseous state of the substance, whether or not the gaseous state is the stable form at 1 bar.

The data are tabulated at rather wide temperature intervals, and interpolations are required for values at intermediate temperatures. It is very commonly found that the function $(\Delta G_T° - \Delta H_{298}°)/RT$ or the corresponding function based on 0 K varies more slowly with temperature than the Gibbs-energy functions of the individual substances. Thus it is usually expedient to calculate the value of $(\Delta G_T° - \Delta H_{298}°)/RT$ at the even temperatures given in the tables and then to interpolate to obtain the value of the $(\Delta G_T° - \Delta H_{298}°)/RT$ function at the desired temperature. If one of the elements undergoes a transition, such as the melting of sulfur or the vaporization of bromine, the rate of change with temperature of $(\Delta G_T° - \Delta H_{298}°)/RT$ for the formation of a compound will show a discontinuity at the transition point, which makes interpolation difficult. Under those circumstances it is often expedient to use the gaseous standard state even at room temperature for elements such as iodine, bromine, and sulfur, so that the Gibbs-energy functions for the formation of compounds will be expressed in terms of the same

TABLE A6-3
Solid and liquid halides

	\multicolumn{5}{c}{$-(G_T^\circ - H_{298}^\circ)/RT$}	$(H_{298}^\circ - H_0^\circ)/R$	$\Delta_f H_{298}^\circ /R$
	298.15 K	500 K	1000 K	1500 K	2000 K	K	10^3 K
AgCl	11.570	12.314 m	15.665	18.353			-15.2_8
AgBr	12.882	13.648 m	17.152	19.830			-12.1_0
AgI	13.891 tr	14.971 m	18.701				-7.4_4
AlF$_3$	7.996	9.114 tr	13.487	17.076	19.949	1398.2	-181.7 ± 0.2
BaF$_2$	11.594	12.609	16.341 tr	16.659 m	22.849	1738.	-145.4 ± 0.5
BaCl$_2$	14.874	15.916	19.682 m	23.461	(26.80)	2009.	$-103. \pm 2$
BiCl$_3$	21.286	22.704					-45.5_9
CaF$_2$	8.247	9.229	12.867 tr	15.969 m	19.06	1402.	-147.4 ± 1.0
CaCl$_2$	12.581	13.597	17.262 m	21.346	24.51	1851.	-95.7 ± 0.2
CdCl$_2$	13.864	14.906 m	19.498				-47.0_9
CoCl$_2$	13.142	14.238	18.192 m	(23.16)		1900.	-37.6 ± 0.2
CrF$_3$	11.291	12.443	16.798				-141.1
CrCl$_3$	14.795	16.100	20.919				-68.1_3
FeCl$_2$	14.186	15.255 m	19.402	24.386	(27.92)	1957.	-41.11 ± 0.05
FeCl$_3$	17.119	18.519 m	27.916	(34.31)		2370.	-48.04 ± 0.1
HfCl$_4$	22.932	24.615					-119.1_1
Hg$_2$Cl$_2$	23.157	24.580	(29.78)			2824.	-31.86 ± 0.03
Hg$_2$Br$_2$	26.310	27.776					-24.56
KF	8.004	8.689	11.205 m	14.05	(16.4)	1203.	-68.39 ± 0.05
KCl	9.929	10.643	13.263 m	16.437	(18.8)	1367.	-52.52 ± 0.03
KBr	11.539	12.264	14.905 m	18.150	(20.5)	1469.	-47.36 ± 0.05
LiF	4.289	4.900	7.242 m	9.988	(12.2)	778.5	-74.20 ± 0.1
LiCl	7.132	7.812 m	10.679	13.61		119.	-49.10 ± 0.2
MgF$_2$	6.886	7.782	11.195	14.005 m	17.37	1192.	-135.2 ± 0.2
MgCl$_2$	10.780	11.789 m	15.556	20.34	(23.7)	1655.	-77.17 ± 0.05
MnCl$_2$	14.221	15.249 m	19.407	24.07			-57.8_9
NaF	6.159	6.821	9.262 m	11.803	(14.2)	1021.	-69.20 ± 0.1
NaCl	8.674	9.377	11.959 m	15.041	(17.4)	1276.	-49.45 ± 0.04
NaBr	10.442	11.159	13.770 m	16.889	(19.1)	1394.	-43.47 ± 0.05
NaI	11.826	12.551 m	15.391	18.50			-34.62
NiF$_2$	8.852	9.771	13.153	15.93			-79.1_1
NiCl$_2$	11.806	12.821	16.523 m	(20.47)		1735.	-36.67 ± 0.03
PbCl$_2$	16.355	17.430 m	22.236				-43.2_5
PbBr$_2$	19.379	20.479 m	25.83	(30.23)		2313.	-33.36 ± 0.3
SbCl$_3$	22.041 m						-45.8_4
TiCl$_4$ (l)	26.692	28.678	35.68				-96.7 ± 0.5
TlCl	13.386	14.096 m	17.60				-24.5_6
UF$_4$	18.242	19.871	25.773				-231.0
VCl$_3$	15.647	16.966					-69.8_9
ZnCl$_2$	13.406	14.407 m	18.95				-49.9_2
ZnBr$_2$	16.355	17.286					-39.6_5
ZrCl$_4$	21.820	23.50				2998.	-117.9 ± 0.2

elemental form at all temperatures. This is easily done with the following tables since Gibbs-energy functions are given not only for the condensed states of these elements but also for the gaseous states. The heats of formation of compounds at 298.15 K are given with respect to the solid states of the element for sulfur and iodine and the liquid state for bromine, but the heats of sublimation and vaporization at 298 K are given in Table A6-2 for these elements so that one can convert the heats of formation with respect to the condensed states of the elements to heats of formation with respect to the gaseous states. While careful interpolation by these methods will give satisfactory results, it is often more convenient to use the tables cited in Chapter 8, which give values at close intervals in temperature. Then interpolation is very simple.

TABLE A6-4
Gaseous halides

	$-(G_T^\circ - H_{298}^\circ)/RT$					$(H_{298}^\circ - H_0^\circ)/R$	$\Delta_f H_{298}^\circ / R$
	298.15 K	500 K	1000 K	1500 K	2000 K	K	10^3 K
AlCl	27.417	27.899	29.624	30.985	32.053	1121.3	-6.2 ± 0.8
AlCl$_3$	38.825	38.84	42.53	45.47	47.79	1993.5	-70.3 ± 0.4
AsCl$_3$	39.365	40.420	44.206	47.19			-31.4_5
CF$_4$	31.442	32.368	36.161	39.459	42.163	1531.2	-112.2 ± 0.2
CF$_2$Cl$_2$	36.190	37.260	41.442	44.949	47.775	1788.0	-59.1 ± 1.0
CCl$_4$	37.262	38.468	43.009	46.705	49.64	2073.3	-11.54 ± 0.3
CsCl	30.798	31.305	33.088	34.481	35.57	1218.4	-28.88 ± 0.5
CuCl	28.530	29.018	30.758	32.126	33.20	1139.1	10.96 ± 0.2
Cu$_3$Cl$_3$	51.660	53.389	59.533	64.35	68.11	3455.	-31.10 ± 0.3
FeCl$_2$	35.996	36.796	39.656	41.915	43.70	1717.	-17.0 ± 0.3
FeCl$_3$	41.399	42.477	46.314	49.323	51.67	2191.	-30.4 ± 0.6
Fe$_2$Cl$_4$	55.867	57.608	63.780	68.620	72.42	3590.	-51.9 ± 0.5
Fe$_2$Cl$_6$	64.579	66.979	75.479	82.126	87.31	4865.	-78.7 ± 1.0
HCl	22.479	22.877	24.278	25.403	26.312	1039.2	-11.10 ± 0.03
HgCl	31.272	31.773	33.545	34.934	36.02	1184.2	9.4 ± 1.2
HgCl$_2$	35.456	36.262	39.134	41.388	43.15	1749.1	-17.6 ± 0.8
KCl	28.756	29.258	31.031	32.419	33.506	1188.9	-25.82 ± 0.05
K$_2$Cl$_2$	42.440	43.551	47.455	50.495	52.86	2421.	-74.3 ± 0.5
MgCl$_2$	33.319	34.114	36.957	39.197	40.95	1718.	-47.2 ± 0.3
MoF$_6$	42.181	43.928	50.531	55.917	60.20	2886.	-187.34 ± 0.1
NaCl	27.638	28.133	29.888	31.267	32.35	1156.5	-21.8 ± 0.3
Na$_2$Cl$_2$	39.141	40.230	44.092	47.112	49.67	2219.	-68.1 ± 1.0
NiCl$_2$	35.388	36.203	39.157	41.510	43.36	1708.6	-8.89 ± 0.03
PbCl$_2$	38.151	38.914	41.616	43.730	45.38	1688.4	-20.93 ± 0.2
SF$_6$	35.064	36.548	42.537	47.622	51.73	2034.4	-146.79 ± 0.1
SiCl$_4$	39.804	41.089	45.817	49.605	52.59	2334.	-79.71 ± 0.2
TiCl$_4$	42.684	44.028	48.890	52.741	55.76	2599.	-91.8 ± 0.5
UF$_6$	45.451	47.314	54.179	59.682	64.03		-258.3
WF$_6$	41.028	42.764	49.344	54.721	59.00	2735.	-207.1 ± 0.2
ZnCl$_2$	33.276	34.072	36.926	39.171	40.928		-31.9_5
ZrCl$_4$	44.226	45.599	50.523	54.405	57.44	2713.	-104.6 ± 0.3

TABLE A6-5
Solid, liquid and gaseous oxides

	$-(G_T^\circ - H_{298}^\circ)/RT$					$(H_{298}^\circ - H_0^\circ)/R$	$\Delta_f H_{298}^\circ / R$
	298.15 K	500 K	1000 K	1500 K	2000 K	K	10^3 K
Al_2O_3	6.128	7.348	12.297	16.53	19.99	1205.	-201.5 ± 0.2
B_2O_3	6.489	7.470 m	12.898	17.936	21.78	1117.7	-153.0 ± 0.3
BaO	8.668	9.334	11.789	13.796	15.419	1200.7	-65.92 ± 0.3
BeO	1.656	2.073	3.882	5.497	6.846	341.0	-73.2 ± 0.4
CO(g)	23.772	24.171	25.600	26.764	27.704	1042.9	-13.29 ± 0.02
CO_2(g)	25.714	26.254	28.373	30.196	31.70	1126.2	-47.33 ± 0.01
CaO	4.596	5.208	7.513	9.399	10.915	811.7	-76.38 ± 0.1
CoO	6.374	7.102	9.664	11.693	13.34	1135.	-28.59 ± 0.05
Cr_2O_3	9.761	11.285	16.849	21.351	24.95	1838.	-136.5 ± 1.0
Cu_2O	11.108	12.004	15.367	18.184 m	21.80	1515.	-20.5 ± 0.3
CuO	5.123	5.737	8.060	9.982		853.	-18.8 ± 0.3
$Fe_{0.947}O^a$	6.926	7.601	10.062	12.066	13.69	1138.	-32.03 ± 0.1
Fe_3O_4	17.472	19.676 tr	28.924	36.583	42.47	2955.	-134.81 ± 0.1
Fe_2O_3	10.512	12.061 tr	18.325 tr	23.599	27.70	1871.	-99.3 ± 0.2
H_2O (l)	8.413	(9.5)					-34.378 ± 0.005
HgO	8.452	9.087	(11.53)			1095.0	-10.92 ± 0.02
Li_2O	4.557	5.378	8.660	11.53		871.7	-72.0 ± 0.3
MgO	3.238	3.790	5.925	7.700	9.134	620.5	-72.31 ± 0.1
MoO_2	5.588	6.415	9.624	12.379	14.72	1001.	-70.7 ± 0.4
Na_2O	9.026	10.023	13.874 m	17.64		1491.	-50.3 ± 0.5
Nb_2O_5	16.513	18.423	25.703	31.763 m	37.44	2681.	-228.5 ± 0.5
PbO	7.976	8.638 tr	11.148 m	13.82		1096.	-26.39 ± 0.1
Pb_3O_4	25.493	27.755	(36.4)			3631.	-86.4 ± 0.8
PbO_2	8.635	9.523	(12.9)			1319.	-33.0 ± 0.5
SO_2 (g)	29.853	30.427	32.635	34.496	36.01	1269.1	-35.70 ± 0.02
SO_3 (g)	30.882	31.631	34.611	37.169	39.26	1406.8	-47.60 ± 0.1
SiO_2	4.987	5.668 tr	8.499	10.929 m	12.99	831.8	-109.55 ± 0.2
SrO	6.678	7.322	9.715	11.68		1043.	-71.2 ± 0.4
Ta_2O_5	17.213	19.159	26.598	32.846	37.93	2771.	-246.1 ± 0.5
TiO_2	6.049	6.865	10.013	12.617	14.71	1039.	-113.6 ± 0.2
V_2O_5	15.70	17.65 m	25.7	34.7		2587.	-186.5 ± 0.8
WO_2	6.077	6.901	10.103	12.789	15.00	1047.	-70.92 ± 0.1
WO_3	9.130	10.202	14.370 tr	17.893 m	21.40	1485.	-101.38 ± 0.1
ZnO	5.19	5.79 tr	8.04			832.	-42.15 ± 0.05
ZrO_2	6.057	6.887	10.067 tr	17.708	14.93	1052.	-132.0 ± 0.2

[a] See Chapter 21 concerning possible entropy of disorder.

For various reasons it seemed best to present most of the tables on the H_{298}° basis. One reason is that most $\Delta_f H^\circ$ information is based on measurements at or near 298 K. For various gases and for hydrocarbons of moderate molecular mass, however, the published literature is almost entirely on the H_0° basis and Table A6-7 is presented on that basis. Conversion from one basis to the other is simple, however, provided accurate values of $(H_{298}^\circ - H_0^\circ)$ are available. Such values are always available

TABLE A6-6
Miscellaneous solids

	$-(G_T°-H_{298}°)/RT$					$(H_{298}°-H_0°)/R$	$\Delta_f H_{298}°/R$
	298.15 K	500 K	1000 K	1500 K	2000 K	K	10^3 K
Ca(OH)$_2$	10.029	11.315	(16.21)			1703.	-118.60 ± 0.2
CaCO$_3$[a]	11.030	12.403	17.21				-145.21 ± 0.2
CaSO$_4$	12.832	14.278	20.08				-172.5 ± 0.5
CaSiO$_3$	9.853	11.145					-196.67 ± 0.2
Ca$_2$SiO$_4$	14.49	16.35	23.52				-278.6 ± 0.5
Mg(OH)$_2$	7.606	8.779	(13.43)			1579.	-111.2 ± 0.2
MgCO$_3$	7.829	8.988	13.72			1399.	-133.90 ± 0.2
MgSO$_4$	10.992	12.417	18.092 m	23.179	27.62	1851.	-151.8 ± 3
MgSiO$_3$	8.151	9.371 tr	14.175 tr	18.297 m	22.00	1457.	-186.3 ± 0.5
Mg$_2$SiO$_4$	11.443	13.220	20.209	26.150	31.03	2078.	-261.8 ± 0.5
NaOH	7.75	8.61 m	13.26			1261.5	-51.21 ± 0.01
Na$_2$CO$_3$	16.694	18.321 tr	25.134 m	31.890	37.40	2503.	-136.00 ± 0.2
Na$_2$SO$_4$	17.992 tr	19.885	28.786 m	36.411	42.44	2792.	-169.92 ± 0.1
Na$_2$SiO$_3$	13.693	15.351	21.911 m	28.006	33.70	2275.	-187.8 ± 0.5

[a] Calcite.

whenever the $(G_T° - H_0°)/RT$ values were statistically calculated; hence, one can always convert such tables to the $H_{298}°$ basis:

$$-\frac{G_T° - H_{298}°}{RT} = -\frac{G_T° - H_0°}{RT} + \frac{H_{298}° - H_0°}{RT}$$

But when $(H_{298}° - H_0°)$ values are not available, as is the case for some solids, the reverse conversion is not possible. This is the most important reason for using the $H_{298}°$ basis for most of the tables. Where $(H_{298}° - H_0°)$ is not given in the tables, it had not been included in the source from which the Gibbs-energy function was taken. Provided the heat capacity has been measured at low temperatures, $(H_{298}° - H_0°)$ can always be calculated, but many of the older compilations do not include it.

Values of $(G_T° - H_{298}°)$ or $(G_T° - H_0°)$ have been divided by RT to convert from $J\,mol^{-1}$ to dimensionless; the reasons for this change are discussed at the end of Chapter 3. All of the Gibbs-energy-function values for ideal gases were originally calculated as dimensionless quantities, and it is unfortunate that they were not so reported. For the reconversion, the original value of R was used, if available (rather than the current best value). The differences in values of R are unimportant, however, for the purpose of these tables.

Most of the entries in the tables referenced to $H_{298}°$ were taken from the final summary report of the JANAF Project: "JANAF Thermochemical Tables," 3d edn., M. W. Chase, Jr., C. A. Davies, J. R. Downey, Jr., D. J. Frurip, R. A. McDonald, and A. N. Syverud, *J. Phys. Chem. Ref. Data*, **14** (supplement No. 1) (1985). The procedures used are clearly stated and a brief summary is given of the particular input data for each

TABLE A6-7
Gaseous elements and compounds with values referenced to H_0°

	$-(G_T^\circ - H_0^\circ)/RT$					$(H_{298}^\circ - H_0^\circ)/R$	$\Delta_f H_0^\circ / R$
	298.15 K	500 K	1000 K	1500 K	2000 K	K	10^3 K
H_2	12.301	14.076	16.485	17.921	18.968	1018.5	–
O_2	21.173	22.992	25.521	27.088	28.243	1044.0	–
S_2	23.758	25.702	28.462	30.162	31.409	1098.3	–
CO	20.275	22.086	24.558	26.069	27.183	1042.9	-13.69 ± 0.02
CO_2	21.934	24.001	27.246	29.445	31.138	1126.4	-47.29 ± 0.01
H_2O	18.716	20.802	23.674	25.493	26.881	1191.3	-28.736 ± 0.005
H_2S	20.736	22.836	25.827	27.768	29.263	1197.6	-2.13 ± 0.06
CH_4	18.376	20.531	24.00	26.63	28.82	1204.7	-7.999
C_2H_2	20.127	22.409	26.20	28.86	30.96	1204.	27.51_3
C_2H_4	22.127	24.507	28.76	32.08	34.85	1265.	7.34_0
C_2H_6	22.780	25.538	30.71	34.92		1428.	-8.20_3

substance. Other entries referenced to H_{298}° were taken from the more comprehensive compilation, *Thermochemical Data of Pure Substances*, I. Barin et al., VHS Publishers, New York, 1989. This is a compilation of other compilations and gives only references to the source compilations, which must be consulted for further details. Values for some minerals were taken from "Thermodynamic Properties of Minerals," R. A. Robie, B. S. Hemingway, and J. R. Fisher, *Geol. Survey Bull. 1492*, U. S. Govt. Printing Office, Washington D.C., 1979.

The values in Table A6-7 are referenced to H_0°; some were taken from *CODATA Key Values for Thermodynamics*, J. D. Cox, D. D. Wagman, and V. A. Medvedev, Hemisphere, New York, 1989. Other values referenced to H_0° for hydrocarbons were taken from *TRC Thermodynamic Tables: Hydrocarbons*, Thermodynamics Research Center, K. N. Marsh, Director, Texas A&M University, College Station, Texas.

For convenience, values for H_2, O_2, S_2, CO, and CO_2 are included on both reference bases. They were taken from different sources, as indicated above, and are not always exactly consistent, although there are no significant differences.

APPENDIX 7

VALUES OF THE DEBYE–HÜCKEL PARAMETER FOR VARIOUS TEMPERATURES

In addition to the Debye–Hückel parameter for the Gibbs energy as well as activity and osmotic coefficients A_ϕ, the parameters for volume A_V, enthalpy A_L, and heat capacity A_J are given in Table A7-1. The defining equations were given in Chapter 15 and subsequent chapters, but are reproduced here for convenient reference.

$$A_\phi = \frac{1}{3}\left(\frac{2\pi N_A d_w}{1000}\right)^{1/2}\left(\frac{e^2}{4\pi\varepsilon_0 \varepsilon kT}\right)^{3/2} \tag{15-35}$$

$$A_V = 2A_\phi RT\left[3\left(\frac{\partial \ln \varepsilon}{\partial P}\right)_T + \left(\frac{\partial \ln V_w}{\partial P}\right)_T\right] \tag{18-11}$$

$$\frac{A_L}{RT} = -6A_\phi\left[1 + T\left(\frac{\partial \ln \varepsilon}{\partial T}\right)_P + \frac{T}{3}\left(\frac{\partial \ln V_w}{\partial T}\right)_P\right] \tag{18-15}$$

$$\frac{A_J}{R} = \left[\frac{\partial (A_L/R)}{\partial T}\right]_P \tag{18-25}$$

Here d_w and V_w are the density and volume of pure water, respectively, and ε is its relative permittivity (dielectric constant). The density is so accurately known that no comment is needed. The table below is taken from a more extensive table of Bradley and Pitzer,[1] who developed an equation for ε as a function of T and P for the range 0–350°C, 1–1000 bar. The present table is for 1 bar below 100°C and saturation pressure above. Bradley and Pitzer give tables also for higher pressures.

TABLE A7-1
Debye–Huckel parameters for the osmotic coefficient, volume, enthalpy, and heat capacity

t (°C)	A_ϕ $kg^{1/2} mol^{-1/2}$	A_V $cm^3 kg^{1/2} mol^{-3/2}$	A_L/RT $kg^{1/2} mol^{-1/2}$	A_J/R $kg^{1/2} mol^{-1/2}$
0.0	0.3767	1.504	0.556	2.95
10.0	0.3821	1.643	0.649	3.39
20.0	0.3882	1.793	0.749	3.76
25.0	0.3915	1.875	0.801	3.94
30.0	0.3949	1.962	0.854	4.13
40.0	0.4023	2.153	0.965	4.51
50.0	0.4103	2.372	1.081	4.92
60.0	0.4190	2.622	1.203	5.37
70.0	0.4283	2.909	1.331	5.86
80.0	0.4384	3.238	1.467	6.40
90.0	0.4491	3.615	1.611	7.00
100.0	0.4606	4.050	1.764	7.66
110.0	0.4727	4.550	1.927	8.40
120.0	0.4857	5.127	2.102	9.24
130.0	0.4994	5.795	2.290	10.17
140.0	0.5140	6.572	2.492	11.23
150.0	0.5295	7.477	2.712	12.45
160.0	0.5460	8.536	2.951	13.84
170.0	0.5634	9.779	3.213	15.47
180.0	0.5820	11.25	3.500	17.38
190.0	0.6017	12.99	3.819	19.65
200.0	0.6228	15.07	4.175	22.38
210.0	0.6453	17.6	4.576	25.72
220.0	0.6694	20.6	5.032	29.85
230.0	0.6953	24.3	5.556	35.05
240.0	0.7232	28.8	6.165	41.73
250.0	0.7535	34.4	6.885	50.46
260.0	0.7865	41.5	7.749	62.15
270.0	0.823	50.5	8.806	78.18
280.0	0.863	62.3	10.13	100.8
290.0	0.908	77.8	11.82	133.7
300.0	0.960	98.7	14.05	183.4
310.0	1.02	127.	17.1	261.
320.0	1.09	169.	21.4	391.
330.0	1.18	231.	28.0	622.
340.0	1.29	330.	38.6	1060.
350.0	1.44	493.	57.3	1920.

Recently Archer and Wang[2] presented a more complex equation for the dielectric constant. The difference between the two treatments for the parent function A_ϕ is very small, but significant differences appear for the functions involving derivatives of ε and especially for A_J, which involves the second temperature derivative. Since all of the other parameters for electrolytes given in this book were derived using the Bradley and Pitzer values of the Debye–Hückel parameters, it seemed best to show those values here.

REFERENCES

1. D. J. Bradley and K. S. Pitzer, *J. Phys. Chem.*, **83**, 1599 (1979); **87**, 3798 (1983).
2. D. G. Archer and P. Wang, *J. Phys. Chem. Ref. Data*, **19**, 371 (1990); D. G. Archer, *J. Chem. Eng. Data*, **35**, 340 (1990).

APPENDIX 8

PARAMETERS FOR AQUEOUS ELECTROLYTE PROPERTIES AT 25°C

Presented here is an expansion of Table 17-1, which lists ion-interaction (Pitzer) equation parameters for several very common aqueous electrolytes. Most of the parameters given below are from Pitzer and Mayorga.† For other entries a reference number is given in parentheses and the citation appears in the references. An even longer list of parameters is given in a recent review‡ where various related aspects are discussed in detail.

In most cases the experimental basis was the osmotic coefficient and both the maximum molality and the standard deviation of fit are given. In other cases the data analysis was more complex and the source paper must be consulted for further information on range and accuracy. All of these parameters are of comparable accuracy, however. For the lanthanides, values are available for each member, but the parameters vary smoothly from La to Lu; hence, only the end members are given here.

---

† K. S. Pitzer and G. Mayorga, *J. Phys. Chem.*, **77**, 2300 (1973); **78**, 2698 (1974); *J. Solution Chem.*, **3**, 539 (1974).

‡ K. S. Pitzer, "Ion-interaction Approach: Theory and Data Correlation," in *Activity Coefficients in Electrolyte Solutions*, 2d edn., K. S. Pitzer, ed., CRC Press, Boca Raton, Fla, 1991, chapter 3.

TABLE A8-1
Acids, bases, and salts of 1–1 type

	$\beta^{(0)}$	$\beta^{(1)}$	$C^\phi = 2C$	Max m	σ
HCl	0.177 5	0.294 5	0.000 80	6	
HBr (1)	0.208 5	0.347 7	0.001 52	6.2	0.003
HI (2)	0.221 1	0.490 7	0.004 82	6	
HClO$_4$	0.174 7	0.293 1	0.008 19	5.5	0.002
HNO$_3$ (2)	0.116 8	0.345 6	−0.005 39	6	
H(HSO$_4$) (3)	0.210 3	0.471 1			
H(HSO$_4$) (4)	0.206 5	0.555 6			
LiCl	0.149 4	0.307 4	0.003 59	6	0.001
LiBr	0.174 8	0.254 7	0.005 3	2.5	0.002
LiI	0.210 4	0.373		1.4	0.006
LiOH	0.015	0.14		4	
LiClO$_3$ (5)	0.170 5	0.229 4	−0.005 24	4.2	0.002
LiClO$_4$	0.197 3	0.399 6	0.000 8	3.5	0.002
LiBrO$_3$ (5)	0.089 3	0.215 7	0.000 0	5	0.001
LiNO$_2$	0.133 6	0.325	−0.005 3	6	0.003
LiNO$_3$	0.142 0	0.278 0	−0.005 51	6	0.001
NaF	0.021 5	0.210 7		1	0.001
NaCl	0.076 5	0.266 4	0.001 27	6	0.001
NaBr	0.097 3	0.279 1	0.001 16	4	0.001
NaI	0.119 5	0.343 9	0.001 8	3.5	0.001
NaOH	0.086 4	0.253	0.004 4	6	b
NaClO$_3$	0.024 9	0.245 5	0.004	3.5	0.001
NaClO$_4$	0.055 4	0.275 5	−0.001 18	6	0.001
NaBrO$_3$	−0.020 5	0.191 0	0.005 9	2.5	0.001
NaCNS	0.100 5	0.358 2	−0.003 03	4	0.001
NaNO$_2$	0.064 1	0.101 5	−0.004 9	5	0.005
NaNO$_3$	0.006 8	0.178 3	−0.000 72	6	0.001
NaHCO$_3$ (6)	0.028	0.044			
NaHSO$_4$ (4)	0.045 4	0.398			
NaH$_2$PO$_4$	−0.053 3	0.039 6	0.007 95	6	0.003
NaH$_2$AsO$_4$	−0.044 2	0.289 5		1.2	0.001
NaB(OH)$_4$	−0.052 6	0.110 4	0.015 4	4.5	0.004
NaBF$_4$	−0.025 2	0.182 4	0.002 1	6	0.006
Na formate	0.082 0	0.287 2	−0.005 23	3.5	0.001
Na acetate	0.142 6	0.323 7	−0.006 29	3.5	0.001
Na propionate	0.187 5	0.278 9	−0.012 77	3	0.001
NaH malonate	0.022 9	0.160 0	−0.001 06	5	0.002
NaH succinate	0.035 4	0.160 6	0.000 40	5	0.001
NaH adipate	0.047 2	0.316 8		0.7	0.001
KF	0.080 89	0.202 1	0.000 93	2	0.001
KCl	0.048 35	0.212 2	−0.000 84	4.8	0.000 5
KBr	0.056 9	0.221 2	−0.001 80	5.5	0.001
KI	0.074 6	0.251 7	−0.004 14	4.5	0.001
KOH	0.129 8	0.320	0.004 1	5.5	b
KClO$_3$	−0.096 0	0.248 1		0.7	0.001
KBrO$_3$	−0.129 0	0.256 5		0.5	0.001
KCNS	0.041 6	0.230 2	−0.002 52	5	0.001
KNO$_2$	0.015 1	0.015	0.000 7	5	0.003

TABLE A8-1 *Continued*

	$\beta^{(0)}$	$\beta^{(1)}$	$C^{\phi} = 2C$	Max m	σ
KNO_3	−0.081 6	0.049 4	0.006 60	3.8	0.001
$KHCO_3$ (7)	−0.010 7	0.047 8			
$KHSO_4$ (4)	−0.000 3	0.173 5			
KH_2PO_4	−0.067 8	−0.104 2		1.8	0.003
KH_2AsO_4	−0.058 4	0.062 6		1.2	0.003
KSCN (5)	0.038 9	0.253 6	−0.001 92	5	0.001
KPF_6	−0.163	−0.282		0.5	0.001
K acetate	0.158 7	0.325 1	−0.006 60	3.5	0.001
KH malonate	−0.009 5	0.142 3	0.001 67	5	0.004
KH succinate	0.011 1	0.156 4	0.002 74	4.5	0.002
KH adipate	0.041 9	0.252 3		1	0.001
RbF	0.114 1	0.284 2	−0.010 5	3.5	0.002
RbCl (8)	0.043 1_9	0.153 9_8	−0.001 09_8	7.8	0.003
RbBr	0.039 6	0.153 0	−0.001 44	5	0.001
RbI	0.039 7	0.133 0	−0.001 08	5	0.001
$RbNO_2$	0.026 9	−0.155 3	−0.003 66	5	0.002
$RbNO_3$	−0.078 9	−0.017 2	0.005 29	4.5	0.001
CsF	0.130 6	0.257 0	−0.004 3	3.2	0.002
CsCl (9)	0.034 7_8	0.039 7_4	−0.000 49_6	7.4	0.002
CsBr	0.027 9	0.013 9	0.000 04	5	0.002
CsI	0.024 4	0.026 2	−0.003 65	3	0.001
CsOH	0.150	0.30			
$CsNO_2$	0.042 7	0.060	−0.005 1	6	0.004
$CsNO_3$	−0.075 8	−0.066 9		1.4	0.002
$AgNO_3$	−0.085 6	0.002 5	0.005 91	6	0.001
$TlClO_4$	−0.087	−0.023		0.5	0.001
$TlNO_3$	−0.105	−0.378		0.4	0.001
NH_4Cl	0.052 2	0.191 8	−0.003 01	6	0.001
Me_4NCl	0.043 0	−0.029	0.007 8	3.4	0.005
Et_4NCl	0.061 7	−0.099	0.010 5	3	0.002
Pr_4NCl	0.134 6	−0.300	0.011 9	2.5	0.002
Bu_4NCl	0.233 9	−0.410	−0.056 7	2.5	0.001
NH_4Br	0.062 4	0.194 7	−0.004 36	2.5	0.001
NH_4I (5)	0.057 0	0.315 7	−0.003 08	7.5	0.002
NH_4HCO_3 (10)	−0.038	0.070		0.7	
$NH_4H_2PO_4$ (11)	−0.070 4	−0.415 6	0.006 69	3.5	0.003
NH_4ClO_4	−0.010 3	−0.019 4		2	0.004
NH_4NO_3	−0.015 4	0.112 0	−0.000 03	6	0.001
(MgOH)Cl (4)	−0.10	1.658			

TABLE A8-2
2–1 Electrolytes

	$\frac{4}{3}\beta^{(0)}$	$\frac{4}{3}\beta^{(1)}$	$\frac{16}{3}C = \frac{2^{5/2}}{3}C^{\phi}$	Max m	σ
$MgCl_2$	0.469 8	2.242	0.009 79	4.5	0.003
$MgCl_2$ (12)	0.469 9$_1$	2.201	0.012 27	4	0.003
$MgBr_2$	0.576 9	2.337	0.005 89	5	0.004
MgI_2	0.653 6	2.405 5	0.014 96	5	0.003
$Mg(ClO_4)_2$	0.661 5	2.678	0.018 06	2	0.002
$Mg(NO_3)_2$	0.489 5	2.113	−0.038 89	2	0.003
$Mg(HCO_3)_2$ (13)	0.044	1.133			
$Mg(HSO_4)_2$ (4)	0.632 8	2.305			
$CaCl_2$	0.421 2	2.152	−0.000 64	2.5	0.003
$CaCl_2$ (13)	0.407 1	2.278	0.004 06	4.3	0.003
$CaBr_2$	0.508 8	2.151	−0.004 85	2	0.002
CaI_2	0.583 9	2.409	−0.001 58	2	0.001
$Ca(ClO_4)_2$	0.601 5	2.342	−0.009 43	2	0.005
$Ca(NO_3)_2$	0.281 1	1.879	−0.037 98	2	0.002
$Ca(HCO_3)_2$ (4)	0.533	3.97			
$Ca(HSO_4)_2$ (4)	0.286	3.37			
$SrCl_2$	0.381 0	2.223	−0.002 46	4	0.003
$SrCl_2$ (9)	0.377 9$_2$	2.167$_5$	−0.001 68	3.8	0.002
$SrBr_2$	0.441 5	2.282	0.002 31	2	0.001
SrI_2	0.535 0	2.480	0.005 01	2	0.001
$Sr(ClO_4)_2$	0.569 2	2.089	−0.024 72	2.5	0.003
$Sr(NO_3)_2$	0.179 5	1.840	−0.037 57	2	0.002
$BaCl_2$	0.350 4	1.995	−0.036 54	1.8	0.001
$BaBr_2$	0.419 4	2.093	−0.030 09	2	0.001
BaI_2	0.562 5	2.249	−0.032 86	1.8	0.003
$Ba(OH)_2$	0.229	1.60		0.1	
$Ba(ClO_4)_2$	0.481 9	2.101	−0.058 94	2	0.003
$Ba(NO_3)_2$	−0.043	1.07		0.4	0.001
$MnCl_2$ (8)	0.442 9$_7$	2.019$_5$	−0.042 78	4	0.003
$FeCl_2$	0.447 9	2.043	−0.016 23	2	0.002
$Fe(HSO_4)_2$ (14)	0.569 7	4.64			
$CoCl_2$	0.485 7	1.967	−0.028 69	3	0.004
$CoBr_2$	0.569 3	2.213	−0.001 27	2	0.002
CoI_2	0.695	2.23	−0.008 8	2	0.01
$Co(NO_3)_2$	0.415 9	2.254	−0.014 36	5.5	0.003
$NiCl_2$ (15)	0.466 5$_5$	2.040	−0.008 88$_1$	2.5	0.002
$CuCl_2$	0.395 5	1.855	−0.067 92	2	0.002
$Cu(NO_3)_2$ (16)	0.374 3	2.310	−0.015 80	8	0.003
$Cu(NO_3)_2$	0.422 4	1.907	−0.041 36	2	0.002
$ZnCl_2$ (17)	0.034 3$_3$	2.308$_5$	−0.123 5$_6$	1.5	0.007
$ZnBr_2$	0.621 3	2.179	−0.203 5	1.6	0.007
ZnI_2	0.642 8	2.594	−0.026 9	0.8	0.002
$Zn(ClO_4)_2$	0.674 7	2.396	0.021 34	2	0.003
$Zn(NO_3)_2$	0.464 1	2.255	−0.029 55	2	0.001
$Cd(NO_3)_2$	0.382 0	2.224	−0.048 36	2.5	0.002
$Pb(ClO_4)_2$	0.444 3	2.296	−0.016 67	6	0.004
$Pb(NO_3)_2$	−0.048 2	0.380	0.010 05	2	0.002
UO_2Cl_2	0.569 8	2.192	−0.069 51	2	0.001

TABLE A8-2 *Continued*

	$\frac{4}{3}\beta^{(0)}$	$\frac{4}{3}\beta^{(1)}$	$\frac{16}{3}C = \frac{2^{5/2}}{3}C^\phi$	Max m	σ
$UO_2(ClO_4)_2$	0.815 1	2.859	0.040 89	2.5	0.003
$UO_2(NO_3)_2$	0.614 3	2.151	−0.059 48	2	0.002
Li_2SO_4	0.181 7	1.694	−0.007 53	3	0.002
Na_2SO_4	0.026 1	1.484	0.009 38	4	0.003
Na_2SO_4 (18)	0.024 9_2	1.466	0.010 463	4	0.003
$Na_2S_2O_3$	0.088 2	1.701	0.007 05	3.5	0.002
Na_2CrO_4	0.125 0	1.826	−0.004 07	2	0.002
Na_2CO_3 (6)	0.048 3	2.013	0.009 8		
Na_2HPO_4	−0.077 7	1.954	0.055 4	1	0.002
Na_2HAsO_4	0.040 7	2.173	0.003 4	1	0.001
Na_2 fumarate	0.308 2	1.203	−0.037 8	2	0.003
Na_2 maleate	0.186 0	0.575	−0.017 0	3	0.004
K_2SO_4	0.066 6	1.039		0.7	0.002
K_2CrO_4	0.101 1	1.652	−0.001 47	3.5	0.003
K_2CO_3 (7)	0.171 7	1.911	0.000 94		
$K_2Pt(CN)_4$	0.088 1	3.164	0.024 7	1	0.005
K_2HPO_4	0.033 0	1.699	0.030 9	1	0.002
K_2HAsO_4	0.172 8	2.198	−0.033 6	1	0.001
Na_2 succinate (19)	0.417 5	2.391 5	−0.092 4	1.4	0.004
K_2 succinate (20)	0.167 3	2.185 1		1.5	0.003
Rb_2SO_4	0.077 2	1.481	−0.000 19	1.8	0.001
Cs_2SO_4 (21)	0.095 2	1.601	0.005 49	5	
Cs_2SO_4	0.118 4	1.481	−0.011 31	1.8	0.001
$(NH_4)_2SO_4$ (11)	0.052 1	0.885 1	−0.001 56	5.8	0.002
$(NH_4)_2SO_4$	0.054 5	0.878	−0.002 19	5.5	0.004

TABLE A8-3
3–1 Electrolytes

	$\frac{3}{2}\beta^{(0)}$	$\frac{3}{2}\beta^{(1)}$	$\frac{3^{3/2}}{2}C^{\phi} = 9C$	Max m	σ
$AlCl_3$	1.049 0	8.767	0.007 1	1.6	0.005
$ScCl_3$	1.050 0	7.978	−0.084 0	1.8	0.005
YCl_3 (22)	0.939_6	8.40	-0.040_6	4.1	0.006
$LaCl_3$	0.883_4	8.40	-0.061_9	3.9	0.006
$LuCl_3$	0.922_8	8.40	-0.033_2	4.1	0.005
$CrCl_3$	1.104 6	7.883	−0.117 2	1.2	0.005
$Cr(NO_3)_3$	1.056 0	7.777	−0.153 3	1.4	0.004
$Ga(ClO_4)_3$	1.238 1	9.794	0.090 4	2	0.008
$InCl_3$	−1.68	−3.85		0.01	
$Y(NO_3)_3$ (22)	0.915_8	7.70	-0.189_8	2.0	0.008
$La(NO_3)_3$ (23)	0.737_4	7.70	-0.198_9	1.5	0.007
$Pr(NO_3)_3$ (15)	0.724_5	7.70	-0.173_4	1.5	0.005
$Nd(NO_3)_3$ (24)	0.702_3	7.70	-0.142_7	2.0	0.008
$Sm(NO_3)_3$	0.701	7.70	−0.131	1.5	0.007
$Eu(NO_3)_3$ (22)	0.713_3	7.70	-0.125_7	2.0	0.007
$Gd(NO_3)_3$	0.776	7.70	−0.170	1.4	0.005
$Tb(NO_3)_3$	0.838	7.70	−0.202	1.4	0.005
$Dy(NO_3)_3$ (25)	0.848_4	7.70	−0.180 9	2.0	0.008
$Ho(NO_3)_3$ (25)	0.876_9	7.70	−0.185 2	2.0	0.009
$Er(NO_3)_3$	0.938	7.70	−0.226	1.5	0.006
$Tm(NO_3)_3$	0.952	7.70	−0.222	1.5	0.006
$Yb(NO_3)_3$	0.948	7.70	−0.208	1.5	0.006
$Lu(NO_3)_3$ (25)	0.926_4	7.70	-0.174_9	2.0	0.008
$La(ClO_4)_3$	1.15_8	9.80	0.001_6	2.0	0.009
$Lu(ClO_4)_3$	1.18_6	9.80	0.029_0	2.0	0.006
Na_3PO_4	0.267 2	5.777	−0.133 9	0.7	0.003
Na_3AsO_4	0.358 2	5.895	−0.124 0	0.7	0.001
K_3PO_4	0.559 4	5.958	−0.225 5	0.7	0.001
$K_3P_3O_9$	0.486 7	8.349	−0.088 6	0.8	0.004
K_3AsO_4	0.749 1	6.511	−0.337 6	0.7	0.001
$K_3Fe(CN)_6$	0.503 5	7.121	−0.117 6	1.4	0.003
$K_3Co(CN)_6$	0.560 3	5.815	−0.160 3	1.4	0.008

TABLE A8-4
4–1 and 5–1 Electrolytes

4–1 solute	$\frac{8}{5}\beta^{(0)}$	$\frac{8}{5}\beta^{(1)}$	$\frac{16}{5}C^\phi = \frac{64}{5}C$	Max m	σ
$Na_4P_2O_7$	0.699	17.16		0.2	0.01
$K_4P_2O_7$	0.977	17.88	$-0.241\ 8$	0.5	0.01
$K_4Fe(CN)_6$	1.021	16.23	$-0.557\ 9$	0.9	0.008
$K_4Mo(CN)_8$	0.854	18.53	$-0.349\ 9$	0.8	0.01
$K_4W(CN)_8$	1.032	18.49	$-0.493\ 7$	1	0.005
$(Me_4N)_4Mo(CN)_8$	0.938	15.91	$-0.333\ 0$	1.4	0.01

5–1 solute	$\frac{5}{3}\beta^{(0)}$	$\frac{5}{3}\beta^{(1)}$	$\frac{5^{3/2}}{3}C^\phi = \frac{50}{3}C$		
$Na_5P_3O_{10}$	1.869	36.10	-1.630	0.4	0.01
$K_5P_3O_{10}$	1.939	39.64	-1.055	0.5	0.015

TABLE A8-5
2–2 Electrolytes ($\alpha_1 = 1.4\,kg^{1/2}\,mol^{-1/2}$, $\alpha_2 = 12.0\,kg^{1/2}\,mol^{-1/2}$ throughout)

Electrolyte	$\beta^{(0)}$	$\beta^{(1)}$	$\beta^{(2)}$	$C^\phi = 4C$	Range	σ	Ref.
$MgSO_4$	0.221 0	3.343	-37.23	0.025 0	0.006–3.0	0.004	
	0.215 0	3.365	-32.74	0.028 0	0.006–3.6	0.004	19
$NiSO_4$	0.170 2	2.907	-40.06	0.036 6	0.005–2.5	0.005	
$MnSO_4$	0.213	2.938	-41.91	0.015 5$_1$	0.1–5.0	0.005	8
$FeSO_4$	0.256 8	3.063	(-42)	0.020 9	0.1–2.0		14
$CoSO_4$	0.163 1	3.346	-30.7	0.037 04	0.2–2.4	0.001	26
$CuSO_4$	0.234 0	2.527	-48.33	0.004 4	0.005–1.4	0.003	27
$ZnSO_4$	0.194 9	2.883	-32.81	0.029 0	0.005–3.5	0.004	
$CdSO_4$	0.205 3	2.617	-48.07	0.011 4	0.005–3.5	0.002	
$CaSO_4$	0.20	3.197$_3$	-54.24		0.004–0.011	0.003	4
$SrSO_4$	0.220	2.88	-41.8	0.019			28
$BeSO_4$	0.317	2.914	?	0.006 2	0.1–4.0	0.004	
UO_2SO_4	0.322	1.827	?	$-0.017\ 6$	0.1–5.0	0.003	

REFERENCES

1. J. B. Macaskill and R. G. Bates, *J. Solution Chem.*, **12**, 607 (1983).
2. S. L. Clegg (Calculated from data of W. J. Hamer and Y.-C Wu), *J. Phys. Chem. Ref. Data*, **1**, 1047 (1972).
3. K. S. Pitzer, R. N. Roy, and L. F. Silvester, *J. Am. Chem. Soc.*, **99**, 4930 (1977).
4. C. E. Harvie, N. Moller, and J. H. Weare, *Geochim. Cosmochim. Acta*, **48**, 723 (1984).
5. H.-T. Kim and W. J. Frederick, Jr., *J. Chem. Eng. Data*, **33**, 177 (1988).
6. J. C. Peiper and K. S. Pitzer, *J. Chem. Thermodyn.*, **14**, 613 (1982).
7. R. N. Roy, J. J. Gibbons, R. Williams, L. G. Godwin, G. Baker, J. M. Simonson, and K. S. Pitzer, *J. Chem. Thermodyn.*, **16**, 303 (1984).
8. J. A. Rard, *J. Chem. Eng. Data*, **29**, 443 (1984).

9. J. A. Rard and D. G. Miller, *J. Chem. Eng. Data*, **27**, 169 (1982).
10. R. N. Roy, K. Hufford, P. J. Lord, D. R. Mrad, L. N. Roy, and D. A. Johnson, *J. Chem. Thermodyn.*, **20**, 63 (1988).
11. V. K. Filippov, M. V. Charykova, and Yu. M. Trofimov, *J. Appl. Chem. U.S.S.R.*, **58**, 1807 (1985).
12. J. A. Rard and D. G. Miller, *J. Chem. Eng. Data*, **26**, 38 (1981).
13. R. C. Phutela and K. S. Pitzer, *J. Solution Chem.*, **12**, 201 (1983).
14. E. J. Reardon and R. D. Beckie, *Geochim. Cosmochim. Acta*, **15**, 2355 (1987).
15. J. A. Rard, *J. Chem. Eng. Data*, **32**, 334 (1987).
16. V. K. Filippov, D. S. Barkov, and Ju. A. Federov, *Z. Phys. Chem. Leipzig*, **266**, 129 (1985).
17. J. A. Rard and D. G. Miller, *J. Chem. Thermodyn.*, **21**, 463 (1989).
18. J. A. Rard and D. G. Miller, *J. Chem. Eng. Data*, **26**, 33 (1981).
19. M. A. Esteso, L. Fernandez-Merida, and F. F. Hernandez-Luis, *J. Electroanal. Chem.*, **230**, 69 (1987).
20. M. A. Esteso, L. Fernandez-Merida, D. Gonzalez-Diaz, and F. F. Hernandez-Luis, *J. Chem. Soc. Faraday Trans.*, **1**, 85, 2575 (1989).
21. V. K. Filippov, A. M. Kalinkin, and S. K. Vasin, *J. Chem. Thermodyn.*, **19**, 185 (1987).
22. J. A. Rard and F. H. Spedding, *J. Chem. Eng. Data*, **27**, 454 (1982).
23. J. A. Rard, *J. Chem. Eng. Data*, **32**, 92 (1987).
24. J. A. Rard, D. G. Miller, and F. H. Spedding, *J. Chem. Eng. Data*, **24**, 348 (1979).
25. J. A. Rard and F. H. Spedding, *J. Chem. Eng. Data*, **26**, 391 (1981).
26. V. K. Filippov, G. V. Dmitriev, and S. I. Yakovleva, *Dokl. Akad. Nauk SSSR Fiz. Khim.*, **252**, 156 (1980); English transl., **252**, 359 (1980).
27. C. J. Downes and K. S. Pitzer, *J. Solution Chem.*, **5**, 389 (1976).
28. E. J. Reardon and D. K. Armstrong, *Geochim. Cosmochim. Acta*, **51**, 63 (1987).

APPENDIX 9

THEORY FOR UNSYMMETRICAL MIXING OF IONS OF THE SAME SIGN

Here we give the theory and a practical method of calculation for the higher-order electrostatic term for unsymmetrical mixing.[1,2] First we take from Friedman[3] the statistical equation at the microscopic level that includes the Debye–Hückel and virial coefficient terms.

$$\frac{A^{ex}}{VkT} = -\frac{\kappa^3}{12\pi} + \sum_i \sum_j c_i c_j B_{ij}(\kappa) + \sum_i \sum_j \sum_k c_i c_j c_k C_{ijk}(\kappa) \cdots \quad \text{(A9-1)}$$

Here the first term on the right is the Debye–Hückel limiting law and the B_{ij}, C_{ijk}, ... are the second, third, ... virial coefficients. After conversion to a macroscopic basis and from the Helmholtz to the Gibbs energy, this provided a basis for the postulation of Eq. (17-5). One notes that B_{ij} was shown to be a function of the Debye kappa (κ), which corresponds to the ionic strength dependency of λ_{ij} in Eq. (17-5).

The second virial coefficient $B_{ij}(\kappa)$ of Eq. (A9-1) is shown by Friedman[3] to be given by

$$B_{ij}(\kappa) = \left(\frac{2\pi z_i z_j l}{\kappa^2}\right) J_{ij}(\kappa, z_i, z_j \ldots) \quad \text{(A9-2)}$$

with the electrostatic length

$$l = \frac{e^2}{\varepsilon kT} \tag{A9-3}$$

We note that the interionic potential of mean forces can be written as

$$v_{ij} = u_{ij} + \frac{kT z_i z_j l}{r} \tag{A9-4}$$

where the second term is the electrostatic interaction and u_{ij}, a function of the interionic distance r, is the short-range potential. Then the function J_{ij} of Eq. (A9-2) is

$$J_{ij} = -\left(\frac{\kappa^2}{z_i z_j l}\right) \int_0^\infty \left[\exp\left(q_{ij} - \frac{u_{ij}}{kT}\right) - 1 - q_{ij} - \frac{q_{ij}^2}{2}\right] r^2 \, dr \tag{A9-5}$$

$$q_{ij} = -\left(\frac{z_i z_j l}{r}\right) \exp(-\kappa r) \tag{A9-6}$$

The integral in Eq. (A9-5) cannot be evaluated, in general, without knowledge of the short-range potential u_{ij}. Since the quantity is not known accurately, the entire second virial coefficient is treated as an empirical quantity. However, for the particular case of ions of the same sign, an approximation yields useful results.

Ions of the same sign repel one another strongly enough that they seldom approach one another closely; hence, the short-range potential should have little or no effect. This can be seen mathematically in Eq. (A9-5). If q_{ij} is large and negative for the range of r for which u_{ij} differs from zero, then the value of $\exp(q_{ij})$ is extremely small throughout this range. Thus, provided u_{ij} is positive (or if negative, is small), the effect of u_{ij} will be negligible.

In view of this situation, one can evaluate the effect of electrostatic forces on the difference terms Φ_{ij} without making any detailed assumption about short-range forces. We write

$$\Phi_{ij} = \theta_{ij} + {}^E\theta_{ij}(I) \tag{17-34}$$

where the first term on the right arises from the combined effects of short-range forces acting directly or through the solvent, of the use of molalities instead of concentration, and of the difference in the Debye–Hückel term in Eq. (A9-1) from that in (17-8). The second term ${}^E\theta_{MN}$ will be calculated from the corresponding terms of the cluster-integral theory with the omission of short-range forces. From the definition of Φ_{MN} we have

$${}^E\theta_{MN} = {}^E\lambda_{MN} - \left(\frac{z_N}{2z_M}\right){}^E\lambda_{MM} - \left(\frac{z_M}{2z_N}\right){}^E\lambda_{NN} \tag{A9-7}$$

$${}^E\lambda_{ij} = \left(\frac{z_i z_j}{4I}\right) J_{ij} \quad \text{with } u_{ij} = 0 \tag{A9-8}$$

$$J_{ij} = \frac{\kappa^2}{z_i z_j I} \int_0^\infty (1 + q_{ij} + \tfrac{1}{2} q_{ij}^2 - e^{q_{ij}}) r^2 \, dr \tag{A9-9}$$

With the substitutions

$$y = \kappa r \tag{A9-10}$$

$$x = z_i z_j I \kappa \tag{A9-11}$$

$$q = -(x/y) e^{-y} \tag{A9-12}$$

$$J(x) = x^{-1} \int_0^\infty (1 + q + \tfrac{1}{2} q^2 - e^q) y^2 \, dy \tag{A9-13}$$

In our working units,

$$x_{ij} = 6 z_i z_j A_\phi I^{1/2} \tag{A9-14}$$

where for ions of the same sign x_{ij} is always positive. Also,

$$^E\theta_{MN} = \left(\frac{z_M z_N}{4I}\right) [J(x_{MN}) - \tfrac{1}{2} J(x_{MM}) - \tfrac{1}{2} J(x_{NN})] \tag{A9-15}$$

We also need the ionic strength derivative of $^E\theta$ and therefore of J. If $J' = \partial J / \partial x$, we find for $^E\theta'$ the expression

$$^E\theta'_{MN} = -\frac{^E\theta_{MN}}{I} + \left(\frac{z_M z_N}{8I^2}\right) [x_{MN} J'(x_{MN}) - \tfrac{1}{2} x_{MM} J'(x_{MM}) - \tfrac{1}{2} x_{NN} J'(x_{NN})] \tag{A9-16}$$

For J the integrals of the second and third terms in the parentheses in Eq. (A9-13) are straightforward, with the results

$$J = \tfrac{1}{4} x - 1 + J_2 \tag{A9-17}$$

$$J' = \tfrac{1}{4} - \left(\frac{J_2}{x}\right) + J_3 \tag{A9-18}$$

$$J_2 = x^{-1} \int_0^\infty (1 - e^q) y^2 \, dy \tag{A9-19}$$

$$J_3 = x^{-1} \int_0^\infty \exp(q - y) y \, dy \tag{A9-20}$$

There are no simple integrals for J_2 and J_3 but they are readily evaluated numerically with modern computers. The resulting functions $^E\theta$ and $^E\theta'$ for 2–1, 3–1, and 4–1 mixing are shown in Fig. A9-1. Reference 1 includes a table of values of J and J'. A convenient although approximate equation is[1]

$$J = x[4 + 4.581(x^{-0.7237}) \exp(-0.0120 x^{0.528})]^{-1} \tag{A9-21}*$$

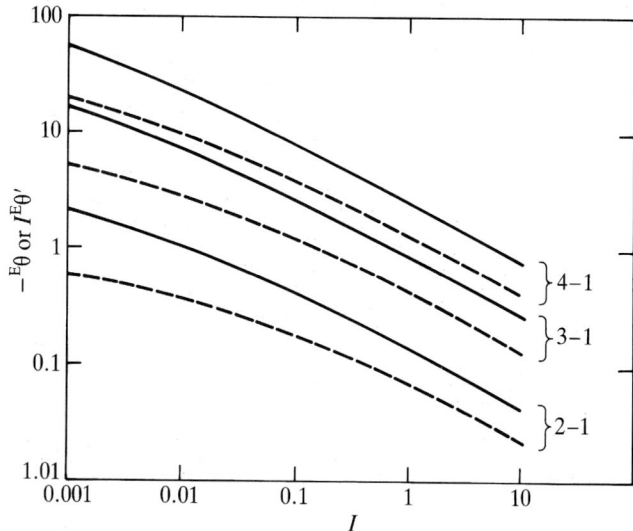

FIGURE A9-1
The functions $-^E\theta$ (solid curves) and $I^E\theta'$ (dashed curves) for mixing ions of charge types 2–1, 3–1, and 4–1.

This equation is correct in the limit of large x, is accurate to 2 per cent or better for x greater than 0.03, while its maximum deviation for smaller x is 6×10^{-6}. Equation (A9-21)* can be differentiated to obtain J'.

REFERENCES

1. K. S. Pitzer, *J. Solution Chem.*, **4**, 249 (1975).
2. K. S. Pitzer, *J. Phys. Chem.*, **87**, 2360 (1983).
3. H. L. Friedman, *Ionic Solution Theory*, Wiley-Interscience, New York, 1962.

APPENDIX 10

PARAMETERS FOR AQUEOUS ELECTROLYTES AT HIGH TEMPERATURES

Given in this appendix are the numerical parameters for the temperature dependency for the ion-interaction quantities $\beta_{MX}^{(0)}$, $\beta_{MX}^{(1)}$, and C_{MX}^{ϕ} (also $\beta_{MX}^{(2)}$ if included) and for the standard state heat capacity $C_{P,MX}^{\circ}$. Table A10-1 pertains to LiCl, KCl, and CsCl and is taken from Holmes and Mesmer.[1]. Table A10-2 is for Li_2SO_4, Na_2SO_4, K_2SO_4, and Cs_2SO_4 and is taken from Holmes and Mesmer,[2] who have also reported some high-temperature values for $SrCl_2$,[3] $BaCl_2$,[3] $NiCl_2$,[4] and $CoCl_2$.[4] Table A10-3 contains an equation for NaCl based on the comprehensive treatment of Pitzer et al.[5] but simplified for saturation pressure only; an equation for $MgCl_2$ from de Lima and Pitzer[6] based primarily on isopiestic data of Holmes et al.;[7] an equation of Møller[8] for $CaCl_2$; and an equation of Phutela and Pitzer[9] for $MgSO_4$. Equations for $\overline{C}_{P,2}^{\circ}$ for $MgCl_2$ and $CaCl_2$ are from Phutela et al.[10]

More comprehensive equations are listed in Table 18-3 for several systems, including HCl, NaBr, and NaOH, in addition to NaCl, KCl, Na_2SO_4, and $MgSO_4$ for which simplified equations are given here. For NH_4Cl,[11] H_2SO_4,[12] and $NaHSO_4$[13] somewhat similar treatments are available, but they are more complex, because of ionization or hydrolysis equilibria, and are not given here.

Mixing parameters, $\Phi_{cc'}$ and $\psi_{cc'a}$, for temperatures up to about 523 K, are also available from isopiestic measurements for the following pairs of cations, all in Cl^- solution: (Li^+-Na^+),[14] (Na^+-K^+),[15] (Na^+-Cs^+),[16] (Na^+-Ca^{2+}),[17] (Cs^+-Ba^{2+}).[18] Mixing parameters for high temperatures are also available from treatments of mineral solubility data.[19]

TABLE A10-1
Parameters for LiCl(aq), KCl(aq), and CaCl(aq)[1]
Equation for $\beta^{(0)}$, $\beta^{(1)}$, and $C^\phi = 2C$:

$$f(T) = p_1 + p_2\left(\frac{1}{T} - \frac{1}{T_r}\right) + p_3 \ln\left(\frac{T}{T_r}\right) + p_4(T - T_r) + p_5(T^2 - T_r^2)$$
$$+ p_6 \ln(T - 260) \text{ with } T_r = 298.15 \text{ K}.$$

Equation for standard state heat capacity:

$$\overline{C}_{P,2}^\circ = q_1 + q_2 T + q_3 T^2 + \frac{q_4}{T - 270}$$

Quantity	Parameter	LiCl(aq)	KCl(aq)	CsCl(aq)
$\beta^{(0)}$	p_1	0.148 47	0.048 08	0.033 52
$\beta^{(0)}$	p_2	0	−758.48	−1290.0
$\beta^{(0)}$	p_3	0	−4.706 2	−8.427 9
$\beta^{(0)}$	p_4	-1.546×10^{-4}	0.010 072	0.018 502
$\beta^{(0)}$	p_5	0	$-3.759\,9 \times 10^{-6}$	$-6.794\,2 \times 10^{-6}$
$\beta^{(0)}$	p_6	0	0	0
$\beta^{(1)}$	p_1	0.307	0.047 6	0.042 9
$\beta^{(1)}$	p_2	0	303.9	−38.0
$\beta^{(1)}$	p_3	0	1.066	0
$\beta^{(1)}$	p_4	6.36×10^{-4}	0	0.001 306
$\beta^{(1)}$	p_5	0	0	0
$\beta^{(1)}$	p_6	0	0.047 0	0
C^ϕ	p_1	0.003 710	-7.88×10^{-4}	-2.62×10^{-4}
C^ϕ	p_2	4.115	91.270	157.13
C^ϕ	p_3	0	0.586 43	1.086 0
C^ϕ	p_4	0	−0.001 298 0	−0.002 524 2
C^ϕ	p_5	-3.71×10^{-9}	$4.956\,7 \times 10^{-7}$	9.840×10^{-7}
C^ϕ	p_6	0	0	0
$\overline{C}_{P,2}^\circ$	q_1	−47.96	−991.51	−1474.13
$\overline{C}_{P,2}^\circ$	q_2	4.823 3	5.564 5	7.684 7
$\overline{C}_{P,2}^\circ$	q_3	−0.007 389 6	−0.008 530 0	−0.010 823 1
$\overline{C}_{P,2}^\circ$	q_4	0	−686	0

TABLE A10-2
Parameters for Li$_2$SO$_4$(aq), Na$_2$SO$_4$(aq), K$_2$SO$_4$(aq), and Cs$_2$SO$_4$(aq)^2

Equation for $\beta^{(0)}$, $\beta^{(1)}$, and $C^\phi = 2^{3/2}C$; note that $\alpha = 1.4\,\text{kg}^{1/2}\,\text{mol}^{-1/2}$ instead of the normal value of $2.0\,\text{kg}^{1/2}\,\text{mol}^{-1/2}$.

$$f(T) = p_1 + p_2\left(T_r - \frac{T_r^2}{T}\right) + p_3\left(T^2 + \frac{2T_r^3}{T} - 3T_r^2\right) + p_4\left(T + \frac{T_r^2}{T} - 2T_r\right)$$

$$+ p_5\left[\ln\left(\frac{T}{T_r}\right) + \left(\frac{T}{T_r}\right) - 1\right] + p_6\left(\frac{1}{T - 263} + \frac{263T - T_r^2}{T(T_r - 263)^2}\right)$$

$$+ p_7\left(\frac{1}{680 - T} + \frac{T_r^2 - 680T}{T(680 - T_r)^2}\right)$$

with $T_r = 298.15\,\text{K}$.

Equation for standard state heat capacity for Na$_2$SO$_4$. For K$_2$SO$_4$ only q_1 is listed, which gives the value at 25°C. No heat capacities were given for Li$_2$SO$_4$ or Cs$_2$SO$_4$.

$$\bar{C}_{P,2}^\circ = q_1 + q_2T + q_3T^2 + \frac{q_4}{T - 263}$$

Parameter	Constant	Li$_2$SO$_4$(aq)	Na$_2$SO$_4$(aq)	K$_2$SO$_4$(aq)	Cs$_2$SO$_4$(aq)
$\beta^{(0)}$	p_1	9.152×10^{-2}	-1.727×10^{-2}	0	6.358×10^{-2}
	p_2	2.797×10^{-4}	1.7828×10^{-3}	7.476×10^{-4}	-4.0908×10^{-3}
	p_3	-8.27×10^{-7}	9.133×10^{-6}	0	-1.11177×10^{-4}
	p_4	0	0	4.265×10^{-3}	0.310 174
	p_5	0	-6.552	-3.088	-144.047
	p_6	0	0	0	20.864 6
	p_7	0	-96.90	0	0
$\beta^{(1)}$	p_1	0.836 6	0.753 4	0.617 9	0.696 3
	p_2	1.88×10^{-3}	5.61×10^{-3}	6.85×10^{-3}	1.734×10^{-2}
	p_3	0	-5.7513×10^{-4}	5.576×10^{-5}	-5.29905×10^{-3}
	p_4	0	1.110 68	-5.841×10^{-2}	10.918 0
	p_5	-11.17	-378.82	0	-3862.82
	p_6	0	0	-0.90	0
	p_7	899.7	1 861.3	0	9 721.8
C^ϕ	p_1	4.689×10^{-3}	1.1745×10^{-2}	9.15467×10^{-3}	-2.471×10^{-3}
	p_2	-1.7240×10^{-4}	-3.3038×10^{-4}	0	-1.488×10^{-5}
	p_3	2.284×10^{-7}	1.85794×10^{-5}	0	0
	p_4	0	-3.9200×10^{-2}	-1.81×10^{-4}	0
	p_5	0	14.213 0	0	0
	p_6	-7.01×10^{-2}	0	0	0
	p_7	0	-24.950	0	0
$\bar{C}_{P,2}^\circ$	q_1	—	-1206.2	-257.7	—
	q_2	—	7.6405	0	—
	q_3	—	-1.23672×10^{-2}	0	—
	q_4	—	$-6\,045$	0	—

TABLE A10-3
Parameters for NaCl(aq), MgCl$_2$(aq), CaCl$_2$(aq), and MgSO$_4$(aq)

NaCl(aq)[5]

Equation for $\beta^{(0)}$, $\beta^{(1)}$, and $C^\phi = 2C$; note that this is taken from an equation valid to 1000 bar pressure with the omission of terms that are negligible for pressures less than 40 bar (approximately the saturation pressure at 250°C). P refers to pressure in bars.

$$f(T) = \frac{Q1}{T} + Q2 + Q3P + Q4\ln(T) + (Q5 + Q6P)T$$

$$+ (Q7 + Q8P)T^2 + \frac{Q9 + Q10P}{T - 227} + \frac{Q11 + Q12P}{680 - T}$$

	$\beta^{(0)}$	$\beta^{(1)}$	$C^\phi = 2C$
Q1	−656.815 18	119.319 66	−6.108 458 9
Q2	24.869 129 50	−0.483 093 27	4.021 779 3 × 10^{-1}
Q3	5.381 275 267 × 10^{-5}	0	2.290 283 7 × 10^{-5}
Q4	−4.464 095 2	0	−0.075 354 649
Q5	0.011 109 913 83	1.406 809 5 × 10^{-3}	1.531 767 295 × 10^{-4}
Q6	−2.657 339 906 × 10^{-7}	0	−9.055 090 1 × 10^{-8}
Q7	−5.307 012 889 × 10^{-6}	0	−1.538 600 820 × 10^{-8}
Q8	8.634 023 325 × 10^{-10}	0	8.692 660 0 × 10^{-11}
Q9	−1.579 365 943	−4.234 581 4	0.353 104 136 0
Q10	2.202 282 079 × 10^{-3}	0	−4.331 425 2 × 10^{-4}
Q11	9.706 578 079	0	−0.091 871 455 29
Q12	−0.026 860 396 22	0	5.190 477 7 × 10^{-4}

Standard state heat capacity: the following equation was fitted to values from 273 to 573 K and at 1 bar or saturation pressure tabulated by Pitzer et al.[5]

$$\overline{C}^\circ_{P,2} = -1.848\ 175 \times 10^6 + \frac{4.411\ 878 \times 10^7}{T} + 3.390\ 654 \times 10^5 \ln(T)$$

$$- 8.893\ 249 \times 10^2\ T + 4.005\ 770 \times 10^{-1}\ T^2 - \frac{7.244\ 279 \times 10^4}{T - 227}$$

$$- \frac{4.098\ 218 \times 10^5}{647 - T}$$

MgCl$_2$(aq)

Ion-interaction parameters: de Lima and Pitzer,[6] with equation for C^ϕ_{MX} modified to fit the solubility data, 298–473 K:

$$f(T) = Q1T^2 + Q2T + Q3$$

	$\beta^{(0)}$	$\beta^{(1)}$	$C^\phi = 2^{3/2} C$
Q1	5.939 15 × 10^{-7}	2.601 69 × 10^{-5}	2.418 31 × 10^{-7}
Q2	−9.316 54 × 10^{-4}	−1.094 38 × 10^{-2}	−2.499 49 × 10^{-4}
Q3	0.576 066	2.601 35	5.953 20 × 10^{-2}

Standard state heat capacity: Phutela et al.,[10] 298–453 K:

$$\overline{C}^\circ_{P,2} = \frac{-7.398\ 72 \times 10^6}{T} + 7.964\ 87 \times 10^4 - 3.258\ 68 \times 10^2 T$$

$$+ 5.987\ 22 \times 10^{-1} T^2 - 4.211\ 87 \times 10^{-4} T^3$$

TABLE A10-3 Continued
CaCl$_2$(aq)
Ion-interaction parameters: Møller,[8] 298–523 K, 0–4 mol kg^{-1}:

$$f(T) = Q1 + Q2T + \frac{Q3}{T} + Q4 \ln T + \frac{Q5}{T - 263} + Q6T^2 + \frac{Q7}{680 - T}$$

	$\beta^{(0)}$	$\beta^{(1)}$	$C^\phi = 2^{3/2} C$
Q1	$-9.418\,958\,32 \times 10^1$	$3.478\,7$	$-3.035\,787\,31 \times 10^1$
Q2	$-4.047\,500\,2 \times 10^{-2}$	$-1.541\,7 \times 10^{-2}$	$-1.362\,647\,28 \times 10^{-2}$
Q3	$2.345\,503\,68 \times 10^3$	0	$7.645\,822\,38 \times 10^2$
Q4	$1.709\,123\,00 \times 10^1$	0	$5.504\,580\,61$
Q5	$-9.228\,858\,41 \times 10^{-1}$	0	$-3.273\,777\,82 \times 10^{-1}$
Q6	$1.514\,881\,22 \times 10^{-5}$	$3.179\,1 \times 10^{-5}$	$5.694\,058\,69 \times 10^{-6}$
Q7	$-1.390\,820\,00$	0	$-5.362\,311\,06 \times 10^{-1}$

Standard state heat capacity: Phutela et al.,[10] 298–373 K:

$$\overline{C}^\circ_{P,2} = \frac{-1.267\,21 \times 10^6}{T} + 7.410\,13 \times 10^3 - 11.522\,2T$$

MgSO$_4$(aq)[9]
Ion-interaction parameters: 298–473 K. Note that the final column gives the temperature coefficients for C. For MgSO$_4$, this is related to C^ϕ by: $C^\phi = 4C$.

$$f(T) = Q1\left(\frac{T}{2} + \frac{298^2}{2T} - 298\right) + Q2\left(\frac{T^2}{6} + \frac{298^3}{3T} - \frac{298^2}{2}\right)$$
$$+ Q3\left(\frac{T^3}{12} + \frac{298^4}{4T} - \frac{298^3}{3}\right) + Q4\left(\frac{T^4}{20} + \frac{298^5}{5T} - \frac{298^4}{4}\right)$$
$$+ Q5\frac{298 - 298^2}{T} + Q6$$

	$\beta^{(0)}$	$\beta^{(1)}$	$\beta^{(2)}$	$C = C^\phi/4$
Q1	$-1.028\,2$	$-2.959\,6 \times 10^{-1}$	$-1.376\,4 \times 10^{-1}$	$1.054\,1 \times 10^{-1}$
Q2	$8.479\,0 \times 10^{-3}$	$9.456\,4 \times 10^{-4}$	$1.212\,1 \times 10^{-1}$	$-8.931\,6 \times 10^{-4}$
Q3	$-2.336\,67 \times 10^{-5}$	0	$-2.764\,2 \times 10^{-4}$	2.51×10^{-6}
Q4	$2.157\,5 \times 10^{-8}$	0	0	$-2.343\,6 \times 10^{-9}$
Q5	$6.840\,2 \times 10^{-4}$	1.028×10^{-2}	$-2.151\,5 \times 10^{-1}$	$-8.789\,9 \times 10^{-5}$
Q6	$0.214\,99$	$3.364\,6$	-32.743	$0.006\,993$

Standard state heat capacity:

$$\overline{C}^\circ_{P,2} = \frac{-6.254\,3 \times 10^6}{T} + 6.527\,7 \times 10^4 - 2.604\,4 \times 10^2 T + 4.693\,0 \times 10^{-1}T^2 - 3.265\,6 \times 10^{-4}T^3$$

REFERENCES

1. H. F. Holmes and R. E. Mesmer, *J. Phys. Chem.*, **87**, 1242 (1983).
2. H. F. Holmes and R. E. Mesmer, *J. Solution Chem.*, **15**, 495 (1986).
3. H. F. Holmes and R. E. Mesmer, *J. Chem. Thermodyn.*, **13**, 1025 (1981).
4. H. F. Holmes and R. E. Mesmer, *J. Chem. Thermodyn.*, **13**, 131 (1981).
5. K. S. Pitzer, J. C. Peiper, and R. H. Busey, *J. Phys. Chem. Ref. Data*, **13**, 1 (1984).
6. M. C. P. de Lima and K. S. Pitzer, *J. Solution Chem.*, **12**, 187 (1983).
7. H. F. Holmes, C. F. Baes, Jr., and R. E. Mesmer, *J. Chem. Thermodyn.*, **10**, 983 (1978).
8. N. Møller, *Geochim. Cosmochim. Acta*, **52**, 821 (1988).
9. R. C. Phutela and K. S. Pitzer, *J. Phys. Chem.*, **90**, 895 (1986).
10. R. C. Phutela, K. S. Pitzer, and P. P. S. Saluja, *J. Chem. Eng. Data*, **32**, 76 (1987).
11. W. E. Thiessen and J. M. Simonson, *J. Phys. Chem.*, **94**, 7794 (1990).
12. H. F. Holmes and R. E. Mesmer, *J. Chem. Thermodyn.*, **24**, 317 (1992).
13. H. F. Holmes and R. E. Mesmer, *J. Chem. Thermodyn.*, **25**, 99 (1993).
14. H. F. Holmes and R. E. Mesmer, *J. Chem. Thermodyn.*, **20**, 1049 (1988).
15. H. F. Holmes and R. E. Mesmer, *J. Chem. Thermodyn.*, **11**, 1035 (1979).
16. H. F. Holmes and R. E. Mesmer, *J. Phys. Chem.*, **94**, 7800 (1990).
17. H. F. Holmes, C. F. Baes, Jr., and R. E. Mesmer, *J. Chem. Thermodyn.*, **13**, 101 (1981).
18. H. F. Holmes and R. E. Mesmer, *J. Chem. Thermodyn.*, **24**, 829 (1992).
19. R. T. Pabalan and K. S. Pitzer, *Geochim. Cosmochim. Acta*, **51**, 2429 (1987); also "Mineral Solubilities in Electrolyte Solutions," in *Activity Coefficients in Electrolyte Solutions*, 2d edn., K. S. Pitzer, ed., CRC Press, Boca Raton, Fla, 1991, chapter 7.

APPENDIX 11

EQUATIONS FOR MIXED LIQUIDS

This appendix considers equations for mixed, neutral-molecule liquids in addition to those described in Chapter 12. For such systems the Redlich–Kister edition of the Margules equation, Eq. (12-33), is the most generally useful. But for mixtures with quite different components, three parameters are required to obtain a satisfactory fit—see Table 12-2. The two-parameter van Laar equation is also inadequate for most of these cases. While with modern computers no problem arises with three or even four parameters, it is an advantage to have a two-parameter equation provided it can represent the data.

WILSON EQUATION

A particularly widely used two-parameter equation is that originally proposed by Wilson[1] and reformulated and applied by Orye and Prausnitz.[2] The two- and multicomponent forms of the excess Gibbs energy were given in Chapter 12 as

$$\frac{G_m^E}{RT} = -x_1 \ln(x_1 + \Lambda_{12}x_2) - x_2 \ln(x_2 + \Lambda_{21}x_1) \qquad (12\text{-}43)^*$$

and

$$\frac{G_m^E}{RT} = -\sum_{i=1}^{m} x_i \ln\left(\sum_{j=1}^{m} \Lambda_{ij}x_j\right) \qquad (12\text{-}43\text{a})^*$$

where $\Lambda_{ii} = 1.0$ and Λ_{ij} with $i \neq j$ are empirical parameters. If all $\Lambda_{ij} = 1.0$, $G_m^E = 0$ and the solution is ideal.

The corresponding activity coefficients for the binary are

$$\ln \gamma_1 = -\ln(x_1 + \Lambda_{12}x_2) + x_2\left(\frac{\Lambda_{12}}{x_1 + \Lambda_{12}x_2} - \frac{\Lambda_{21}}{\Lambda_{21}x_1 + x_2}\right) \quad \text{(A11-1a)*}$$

$$\ln \gamma_2 = -\ln(x_2 + \Lambda_{21}x_1) - x_1\left(\frac{\Lambda_{12}}{x_1 + \Lambda_{12}x_2} - \frac{\Lambda_{21}}{\Lambda_{21}x_1 + x_2}\right) \quad \text{(A11-1b)*}$$

For more than two components one obtains

$$\ln \gamma_k = -\ln\left(\sum_{j=1}^{m} \Lambda_{kj}x_j\right) + 1 - \sum_{i=1}^{m} \frac{x_i \Lambda_{ik}}{\sum_{j=1}^{m} \Lambda_{ij}x_j} \quad \text{(A11-2)*}$$

One notes that all of the parameters in the multicomponent Eqs. (12-43a)* and (A11-2)* are the binary parameters Λ_{ij}. Thus, if all of these are known from the binary systems, the properties of ternaries or more complex systems are predicted.

Orye and Prausnitz[2] offered a rationale for the Wilson equation in terms of the local composition of the solution near a given type of molecule, and on this basis gave temperature-dependent expressions for the Λ_{ij} parameters:

$$\Lambda_{ij} = \frac{v_j}{v_i} \exp\left(\frac{-(\lambda_{ij} - \lambda_{ii})}{RT}\right) \quad \text{(A11-3a)}$$

$$\Lambda_{ji} = \frac{v_i}{v_j} \exp\left(\frac{-(\lambda_{ji} - \lambda_{jj})}{RT}\right) \quad \text{(A11-3b)}$$

Here v_i is the molar liquid volume of pure component i that is independently known; also $\lambda_{ij} = \lambda_{ji}$. Thus the two parameters can be taken as $(\lambda_{ij} - \lambda_{ii})$ and $(\lambda_{ij} - \lambda_{jj})$ or as Λ_{ij} and Λ_{ji}, optionally. Sometimes the temperature dependency of the Λ_{ij} is ignored for calculations over a small range of T.

For three systems with quite different components, ethanol–hexane, acetone–water, and CH_3NO_2–CCl_4, Orye and Prausnitz[2] showed that the Wilson equation fitted the data better than the van Laar equation. Table A11-1 compares, for the CH_3NO_3–CCl_4 system at 45°C, the calculated vapor mole fractions of CH_3NO_2, y_1, from the Wilson and van Laar equations with the experimental values; the liquid mole fractions x_1 are also given. While neither equation gives a perfect fit, the standard deviation for the Wilson equation is considerably smaller. Others, including Holmes and Van Winkle,[3] used the Wilson equation for very many such systems with similar results, which are better than those for the van Laar equation or the Redlich–Kister (Margules) equation at the two-parameter level.

Ohe[4] has calculated or collected Wilson parameters for a very large number of systems and demonstrated the quality of agreement with experiment. Table A11-2 shows Ohe's parameters by different optimization procedures for two systems 2-methyl-2-butene–acetonitrile and chloroform–benzene. The former is a system of

TABLE A11-1
Calculated and experimental vapor compositions for $CH_3NO_2(1)$–$CCl_4(2)$ at 45°C[a]

x_1 (liq) exp.	y_1 (vapor)		
	Exp.	Wilson	van Laar
0.045 9	0.130	0.147	0.117
0.091 8	0.178	0.191	0.183
0.195 4	0.222	0.225	0.247
0.282 9	0.237	0.236	0.262
0.365 6	0.246	0.243	0.264
0.465 9	0.253	0.251	0.261
0.536 6	0.260	0.258	0.259
0.606 5	0.266	0.266	0.259
0.683 5	0.277	0.279	0.266
0.804 3	0.314	0.318	0.304
0.903 9	0.408	0.410	0.411
0.948 8	0.528	0.524	0.540
Error		±0.004	±0.011

[a] From Orye and Prausnitz;[2] also in Prausnitz et al.,[11] table 6-8.

TABLE A11-2
Wilson parameters for two systems optimized by two different procedures[a]

Optimization technique	2-Methyl-2-butene–acetonitrile		Chloroform–benzene	
	Λ_{12}	Λ_{21}	Λ_{12}	Λ_{21}
A	0.253 8	0.084 5	1.027 7	1.128 5
B	0.253 6	0.084 6	2.593 0	0.203 0

[a] From Ohe,[4] table 1; technique A was "pattern search," technique B was "nonlinear least-squares."

quite different components for which the Wilson equation has advantages and the parameters differ very substantially from the value 1.0 for ideality. Also, the two sets are almost exactly the same. Chloroform–benzene is a nearly ideal system, however, and the first set of parameters have each value a little larger than unity. For this system the second set of parameters differ grossly from the first set but yield an equally good fit to the data.

This very large difference between sets of parameters, both giving good fits, indicates a near redundancy of the two parameters for that data set. This is not surprising; if the system is nearly ideal, one parameter should suffice to describe the small nonideality.

In one respect, it is an advantage that no new parameters appear when the Wilson equation is extended to multicomponent systems. If a component binary shows the redundancy illustrated by chloroform–benzene, however, it is not clear whether the two sets will yield equal results when used for a ternary. It seems safest in such a case to choose a set with each parameter near 1.0. Interactions among three different molecules do occur, however, and it can be an advantage if the equation provides for such a term as the Margules equation does.

Holmes and Van Winkle[3] report calculations for 19 ternary systems, comparing results for the Wilson equation with those for the two-parameter Margules (Redlich–Kister) equation and with two multicomponent extensions of the van Laar equation. No adjustment was made for a possible difference of the 1–2–3 interaction from the appropriate average of 1–1–2, 1–2–2, etc., interactions in the Margules system. For the three systems reported in detail, the Wilson equation is slightly superior to the two-parameter Margules equation, and both are greatly superior to either edition of the van Laar equation.

There are important limitations to the Wilson equation. It is unable to describe liquid–liquid phase separation. With $\Lambda_{12} = \Lambda_{21} = 0.0$ the equation for G_m^E yields just the negative of the ideal Gibbs energy of mixing. Then the $\Delta_{\mathrm{mix}}G$ is zero and no phase separation is predicted, and no other values of Λ_{ij} give immiscibility. A second but less important limitation is the inability to represent a maximum or minimum in an activity coefficient as a function of mole fraction.

A few treatments have been reported[5,6] for systems with four or five components, including the quinary benzene–chloroform–methanol–methyl acetate–acetone.[5] In spite of the limitation of the Wilson equation for the benzene–methyl acetate binary, which shows an activity coefficient minimum, the overall agreement for this five-component system is good. The quinary, hexane–methylcyclopentane–cyclohexane–benzene–toluene, was measured by Weatherford and Van Winkle,[6] who obtained good agreement with the Wilson equation. As noted in Chapter 26, this system was later treated with the Margules equation at the 3-parameter level with even more perfect agreement.

Shallcross et al.[7] describe an interesting treatment of multicomponent ion exchange equilibria in which the Wilson equation is used for the nonideality of the resin phase. In this calculation the activity coefficients for the aqueous phase, containing Mg^{2+}, Ca^{2+}, and Na^+ with Cl^-, were obtained from the electrolyte equations represented in Chapter 17. Then equilibrium constants for interphase transfer were evaluated with good agreement over a wide range in ratios of ion molalities and of total molalities.

NRTL AND UNIQUAC EQUATIONS

Following the extensive application of the Wilson equation to various systems, Prausnitz and his associates[8,9] proposed two new equations that have advantageous features for certain types of applications. First was the NRTL (nonrandom, two-liquid) equation of Renon and Prausnitz,[8] which is an alternate development of the local composition concept; it is presented briefly below. An advantage of the UNIQUAC equation is its applicability to polymer solutions as well as ordinary nonelectrolytes. Also a "group-contribution" method, called UNIFAC, has been developed in which parameters are estimated on the basis of various portions of the molecules involved. Detailed information is given by Reid et al.[10] Prausnitz et al.[11] give the full background and various examples for the UNIQUAC as well as other equations.

An important advantage of the NRTL equation, in comparison with the Wilson equation, is the representation of partially miscible as well as fully miscible systems. For two components the excess molar Gibbs energy is

$$\frac{G_m^E}{RT} = x_1 x_2 \left[\frac{\tau_{12} G_{12}}{x_2 + x_1 G_{12}} + \frac{\tau_{21} G_{21}}{x_1 + x_2 G_{21}} \right] \qquad (A11\text{-}4)^*$$

$$G_{12} = \exp(-\alpha_{12} \tau_{12}) \qquad G_{21} = \exp(-\alpha_{12} \tau_{21}) \qquad (A11\text{-}4a)$$

Now there are three parameters, τ_{12}, τ_{21}, α_{12}, instead of two for the Wilson equation. If α_{12} is zero, $G_{12} = G_{21} = 1.0$ and one obtains the single-parameter Margules equation with $w = \tau_{12} + \tau_{21}$. If τ_{12} and τ_{21} are zero, the solution is ideal.

The NRTL activity coefficients are

$$\ln \gamma_1 = x_2^2 \left[\tau_{21} \left(\frac{G_{21}}{x_1 + x_2 G_{21}} \right)^2 + \frac{\tau_{12} G_{12}}{(x_2 + x_1 G_{12})^2} \right] \qquad (A11\text{-}4b)^*$$

$$\ln \gamma_2 = x_1^2 \left[\tau_{12} \left(\frac{G_{12}}{x_2 + x_1 G_{12}} \right)^2 + \frac{\tau_{21} G_{21}}{(x_1 + x_2 G_{21})^2} \right] \qquad (A11\text{-}4c)^*$$

The τ_{ij} quantities can be related to interaction energies

$$\tau_{12} = \frac{g_{12} - g_{22}}{RT} \qquad \tau_{21} = \frac{g_{21} - g_{11}}{RT} \qquad (A11\text{-}5)$$

Here the role of the g_{ij} is similar to that of the Wilson parameters λ_{ij} in Eq. (A11.3).

For multicomponent systems, the NRTL equation for the molar excess Gibbs energy is

$$\frac{G_m^E}{RT} = \sum_{i=1}^{m} x_i \frac{\sum_{j=1}^{m} \tau_{ji} G_{ji} x_j}{\sum_{l=1}^{m} G_{li} x_l} \qquad (A11\text{-}6)^*$$

The activity coefficient for a component i is given by

$$\ln \gamma_i = \frac{\sum_{j=1}^{m} \tau_{ji} G_{ji} x_j}{\sum_{l=1}^{m} G_{li} x_l} + \sum_{j=1}^{m} \frac{x_j G_{ij}}{\sum_{l=1}^{m} G_{lj} x_l} \left(\tau_{ij} - \frac{\sum_{r=1}^{m} x_r \tau_{rj} G_{rj}}{\sum_{l=1}^{m} G_{lj} x_l} \right) \qquad (A11\text{-}6a)^*$$

The generalization of τ_{12} to τ_{ij} and G_{12} to G_{ij} is straightforward. There are three parameters for each pair of components, τ_{ij}, τ_{ji}, and $\alpha_{ij} = \alpha_{ji}$, and no new parameters for interactions of three different species.

In practical applications, it is often found that the third parameter can be held constant for many or even for all pairs of components. A typical choice is $\alpha_{12} = 0.3$. Prausnitz et al.[11] give examples of multicomponent applications and discuss the types of systems where the NRTL equation has shown advantages over other equations.

REFERENCES

1. G. M. Wilson, *J. Am. Chem. Soc.*, **86**, 127 (1964).
2. R. V. Orye and J. M. Prausnitz, *Ind. Eng. Chem.*, **57**(5), 19 (1965).
3. M. J. Holmes and M. Van Winkle, *Ind. Eng. Chem.*, **62**(1), 21 (1970).
4. S. Ohe, *Vapor–Liquid Equilibrium Data*, Elsevier, Amsterdam, 1989.
5. J. W. Hudson and M. Van Winkle, *J. Chem. Eng. Data*, **14**, 310 (1969).
6. R. M. Weatherford and M. Van Winkle, *J. Chem. Eng. Data*, **15**, 386 (1970).
7. D. C. Shallcross, C. C. Herman, and B. J. McCoy, *Chem. Eng. Sci.*, **43**, 279 (1988).
8. H. Renon and J. M. Prausnitz, *Am. Inst. Chem. Eng. J.*, **14**, 135 (1968).
9. D. Adams and J. M. Prausnitz, *Am. Inst. Chem. Eng. J.*, **21**, 116 (1975).
10. R. C. Reid, J. M. Prausnitz, and E. B. Poling, *The Properties of Gases and Liquids*, 4th edn., McGraw-Hill, New York, 1987, Chapter 8.
11. J. M. Prausnitz, R. N. Lichtenthaler, and E. G. de Azevedo, *Molecular Thermodynamics of Fluid Phase Equilibria*, 2d edn., Prentice-Hall, Englewood Cliffs, N. J., 1986.

APPENDIX 12

FUNCTIONS FOR RESTRICTED INTERNAL ROTATION

In Chapter 20 the contribution of internal rotation to various thermodynamic functions was considered for simple cases. This appendix extends the treatment to more complex cases and provides numerical tables for the resulting functions. The basis for the appendix is given in a series of four papers,[1-4] of which the first (paper I) gives results adequate for many cases, while the other three consider exceptional situations.

We consider initially a more general definition of the reduced moment of inertia. If there is one symmetrical top attached to a less symmetrical structure, as in the case of H_3C—CH=CH_2, one may write

$$I_r = A\left(1 - \sum_{i=1}^{3} \frac{\alpha_i^2 A}{I_i}\right) \qquad \text{(A12-1)}$$

where A is the moment of inertia of the symmetrical top and the sum covers the three principal axes for overall rotation. Then α_i is the direction cosine between the axis of internal rotation and the ith principal axis of the entire molecule, and I_i is the moment of inertia for overall rotation about the ith axis.

If neither of the two parts of the molecule has symmetry, then there is less clear separation of overall rotation from the internal rotation. Paper II[2] gives methods of treating such cases. Paper III[3] treats cases of compound rotation, i.e., internal rotations within larger internal rotating groups. Paper IV[4] concerns cases of a very small reduced moment of inertia at very low temperatures; also, it includes further information

TABLE A12-1
Heat capacity C/cal K^{-1}

V/RT	0.0	0.05	0.10	0.15	0.20	0.25	0.30	0.35	0.40	0.45	0.50	0.55	0.60	0.65	0.70	0.75	0.80	0.85	0.90	0.95
											$1/Q_f$									
0.0	0.994	0.994	0.994	0.994	0.994	0.994	0.994	0.994	0.994	0.994	0.994	0.994	0.994	0.994	0.994	0.994	0.994	0.994	0.994	0.994
0.2	1.003 5	1.003	1.003	1.002	1.001	1.000	0.999	0.998	0.998	0.998	1.000	1.000	1.000	1.000	1.000	1.000	1.000	0.999	0.999	0.999
0.4	1.032 8	1.033	1.032	1.030	1.028	1.025	1.024	1.021	1.019	1.017	1.018	1.017	1.015	1.013	1.012	1.010	1.008	1.007	1.005	1.004
0.6	1.080 1	1.080	1.079	1.076	1.073	1.068	1.065	1.060	1.056	1.051	1.049	1.046	1.041	1.036	1.031	1.026	1.021	1.017	1.014	1.011
0.8	1.143 5	1.143	1.141	1.138	1.133	1.128	1.121	1.114	1.106	1.099	1.092	1.084	1.075	1.067	1.058	1.049	1.040	1.031	1.025	1.020
1.0	1.220 3	1.219	1.217	1.212	1.206	1.199	1.190	1.180	1.169	1.157	1.144	1.131	1.118	1.105	1.091	1.078	1.065	1.052	1.040	1.031
1.5	1.450 8	1.449	1.444	1.435	1.423	1.408	1.391	1.370	1.348	1.324	1.299	1.273	1.247	1.218	1.192	1.165	1.141	1.115	1.090	1.070
2.0	1.677 8	1.695	1.687	1.673	1.655	1.632	1.606	1.574	1.541	1.505	1.465	1.424	1.382	1.341	1.300	1.258	1.218	1.180	1.146	1.113
2.5	1.921 3	1.917	1.908	1.888	1.866	1.840	1.801	1.756	1.717	1.670	1.619	1.562	1.504	1.448	1.393	1.341	1.289	1.238	1.190	1.146
3.0	2.098 9	2.095	2.082	2.062	2.033	1.996	1.952	1.900	1.846	1.794	1.732	1.663	1.597	1.532	1.466	1.401	1.337	1.276	1.217	1.164
3.5	2.222 6	2.218	2.204	2.180	2.146	2.106	2.054	1.995	1.934	1.869	1.803	1.727	1.654	1.580	1.506	1.432	1.361	1.293	1.226	1.165
4.0	2.298 9	2.294	2.276	2.249	2.213	2.168	2.110	2.048	1.980	1.907	1.834	1.754	1.674	1.593	1.513	1.435	1.359	1.286	1.215	1.148
4.5	2.335 8	2.330	2.312	2.280	2.238	2.190	2.129	2.062	1.990	1.911	1.832	1.749	1.664	1.578	1.496	1.413	1.333	1.259	1.185	1.115
5.0	2.344 7	2.338	2.318	2.285	2.241	2.186	2.120	2.056	1.972	1.890	1.808	1.718	1.631	1.543	1.457	1.373	1.292	1.214	1.140	1.068
6.0	2.315 8	2.307	2.283	2.245	2.192	2.130	2.059	1.979	1.893	1.803	1.711	1.614	1.520	1.429	1.342	1.255	1.173	1.096	1.022	0.954
7.0	2.265 0	2.256	2.228	2.185	2.126	2.055	1.973	1.883	1.787	1.688	1.588	1.487	1.390	1.296	1.207	1.120	1.040	0.962	0.890	0.826
8.0	2.216 0	2.205	2.174	2.125	2.058	1.979	1.888	1.788	1.684	1.576	1.468	1.366	1.262	1.164	1.074	0.988	0.908	0.834	0.765	0.704
9.0	2.176 2	2.164	2.130	2.074	1.999	1.909	1.808	1.699	1.587	1.474	1.362	1.250	1.144	1.048	0.956	0.869	0.789	0.717	0.652	0.593
10.0	2.145 7	2.133	2.094	2.033	1.951	1.854	1.745	1.630	1.507	1.382	1.262	1.151	1.045	0.943	0.850	0.765	0.688	0.618	0.556	0.499
12.0	2.105 3	2.089	2.043	1.972	1.877	1.763	1.636	1.502	1.365	1.233	1.107	0.989	0.877	0.774	0.682	0.600	0.528	0.463	0.407	0.358
14.0	2.081 3	2.063	2.009	1.923	1.814	1.686	1.546	1.400	1.254	1.112	0.978	0.855	0.744	0.644	0.554	0.479	0.411	0.352	0.303	0.262
16.0	2.065 7	2.044	1.983	1.887	1.764	1.622	1.468	1.311	1.156	1.009	0.873	0.749	0.639	0.542	0.457	0.387	0.324	0.272	0.229	0.194
18.0	2.054 7	2.031	1.961	1.853	1.717	1.562	1.397	1.232	1.070	0.919	0.780	0.657	0.549	0.456	0.378	0.312	0.259	0.215	0.175	0.144
20.0	2.046 5	2.020	1.944	1.827	1.678	1.510	1.333	1.158	0.991	0.837	0.701	0.580	0.477	0.389	0.316	0.256	0.208	0.168	0.135	0.109

TABLE A12-2
Enthalpy $[(H_T-H_O)/T]/$cal K^{-1}

V/RT	\multicolumn{20}{c}{$1/Q_f$}
	0.0	0.05	0.10	0.15	0.20	0.25	0.30	0.35	0.40	0.45	0.50	0.55	0.60	0.65	0.70	0.75	0.80	0.85	0.90	0.95
0.0	0.994	0.994	0.994	0.994	0.994	0.994	0.994	0.994	0.994	0.994	0.994	0.994	0.994	0.994	0.994	0.994	0.994	0.994	0.994	0.994
0.2	1.182 4	1.142	1.106	1.074	1.050	1.032	1.022	1.015	1.008	1.004	1.000	0.996	0.994	0.994	0.994	0.992	0.992	0.991	0.990	0.989
0.4	1.351 5	1.300	1.249	1.200	1.151	1.106	1.073	1.051	1.036	1.025	1.015	1.006	0.999	0.994	0.992	0.990	0.988	0.988	0.986	0.985
0.6	1.501 3	1.437	1.374	1.311	1.251	1.190	1.138	1.099	1.072	1.049	1.030	1.014	1.004	0.995	0.990	0.987	0.984	0.982	0.980	0.979
0.8	1.632 6	1.556	1.482	1.411	1.340	1.272	1.211	1.157	1.114	1.077	1.048	1.026	1.009	0.996	0.984	0.980	0.976	0.974	0.972	0.971
1.0	1.746 3	1.660	1.576	1.495	1.418	1.344	1.275	1.211	1.155	1.106	1.065	1.038	1.014	0.996	0.982	0.972	0.965	0.962	0.960	0.959
1.5	1.961 0	1.856	1.753	1.654	1.561	1.472	1.385	1.306	1.230	1.164	1.103	1.059	1.019	0.987	0.962	0.945	0.932	0.922	0.916	0.915
2.0	2.093 7	1.971	1.854	1.742	1.636	1.536	1.440	1.350	1.265	1.190	1.120	1.057	1.005	0.962	0.928	0.904	0.886	0.873	0.864	0.860
2.5	2.166 0	2.031	1.900	1.779	1.662	1.550	1.448	1.351	1.260	1.179	1.104	1.032	0.972	0.922	0.882	0.850	0.827	0.811	0.801	0.796
3.0	2.197 4	2.049	1.909	1.777	1.651	1.535	1.426	1.321	1.224	1.140	1.060	0.988	0.924	0.870	0.828	0.791	0.763	0.744	0.732	0.728
3.5	2.203 3	2.043	1.893	1.753	1.621	1.497	1.382	1.275	1.176	1.088	1.006	0.933	0.868	0.811	0.765	0.727	0.697	0.676	0.663	0.659
4.0	2.194 7	2.024	1.864	1.715	1.577	1.448	1.329	1.221	1.121	1.030	0.947	0.872	0.806	0.749	0.701	0.661	0.630	0.609	0.595	0.590
4.5	2.179 1	1.998	1.829	1.673	1.529	1.394	1.273	1.162	1.061	0.968	0.884	0.810	0.744	0.687	0.638	0.599	0.567	0.545	0.531	0.526
5.0	2.161 0	1.971	1.794	1.631	1.481	1.344	1.218	1.104	1.002	0.909	0.824	0.750	0.685	0.628	0.580	0.540	0.508	0.485	0.470	0.465
6.0	2.126 4	1.918	1.727	1.552	1.392	1.247	1.115	0.999	0.893	0.799	0.714	0.644	0.580	0.523	0.476	0.437	0.406	0.383	0.368	0.361
7.0	2.098 7	1.875	1.670	1.484	1.315	1.164	1.029	0.908	0.802	0.708	0.624	0.554	0.491	0.437	0.392	0.354	0.324	0.302	0.286	0.279
8.0	2.078 4	1.840	1.623	1.427	1.251	1.095	0.955	0.833	0.725	0.631	0.549	0.480	0.420	0.368	0.326	0.290	0.261	0.239	0.223	0.215
9.0	2.063 7	1.811	1.583	1.379	1.196	1.035	0.892	0.768	0.661	0.569	0.488	0.421	0.363	0.312	0.273	0.240	0.211	0.191	0.176	0.168
10.0	2.052 9	1.787	1.548	1.335	1.147	0.982	0.838	0.715	0.608	0.515	0.437	0.370	0.314	0.269	0.231	0.200	0.174	0.154	0.140	0.132
12.0	2.038 5	1.749	1.492	1.264	1.067	0.896	0.745	0.624	0.519	0.431	0.356	0.296	0.244	0.202	0.170	0.143	0.121	0.104	0.091	0.084
14.0	2.029 5	1.717	1.441	1.202	0.997	0.823	0.672	0.551	0.450	0.365	0.297	0.240	0.195	0.158	0.127	0.103	0.084	0.072	0.062	0.056
16.0	2.023 2	1.690	1.401	1.150	0.937	0.760	0.613	0.493	0.394	0.314	0.249	0.198	0.157	0.127	0.098	0.076	0.061	0.051	0.044	0.038
18.0	2.018 5	1.666	1.363	1.102	0.886	0.707	0.561	0.443	0.347	0.271	0.211	0.164	0.128	0.099	0.077	0.060	0.047	0.036	0.029	0.026
20.0	2.015 0	1.646	1.329	1.061	0.841	0.660	0.515	0.399	0.307	0.236	0.181	0.138	0.105	0.080	0.061	0.047	0.036	0.028	0.022	0.018

TABLE A12-3
Gibbs energy $(-G/T)/\text{cal K}^{-1}$

V/RT	0.25	0.30	0.35	0.40	0.45	0.50	0.55	$1/Q_f$ 0.60	0.65	0.70	0.75	0.80	0.85	0.90	0.95
0.0	2.754	2.392	2.086	1.821	1.587	1.377	1.190	1.014	0.856	0.710	0.575	0.443	0.323	0.208	0.102
0.2	2.710	2.359	2.061	1.803	1.574	1.368	1.182	1.009	0.852	0.707	0.570	0.441	0.321	0.207	0.101
0.4	2.623	2.296	2.014	1.765	1.543	1.342	1.164	0.997	0.842	0.699	0.565	0.438	0.318	0.206	0.099
0.6	2.518	2.208	1.944	1.708	1.498	1.309	1.136	0.974	0.826	0.687	0.555	0.431	0.315	0.204	0.097
0.8	2.406	2.106	1.856	1.636	1.442	1.266	1.099	0.947	0.804	0.670	0.543	0.424	0.310	0.200	0.096
1.0	2.296	2.004	1.764	1.559	1.379	1.214	1.056	0.912	0.777	0.647	0.526	0.411	0.302	0.195	0.094
1.5	2.040	1.770	1.548	1.370	1.210	1.069	0.937	0.815	0.700	0.588	0.481	0.379	0.277	0.178	0.084
2.0	1.819	1.563	1.360	1.193	1.052	0.927	0.817	0.713	0.615	0.521	0.428	0.338	0.249	0.160	0.074
2.5	1.630	1.389	1.197	1.043	0.912	0.802	0.705	0.616	0.534	0.454	0.375	0.298	0.219	0.141	0.063
3.0	1.473	1.240	1.059	0.914	0.793	0.695	0.608	0.530	0.458	0.390	0.324	0.258	0.191	0.122	0.053
3.5	1.340	1.117	0.943	0.802	0.694	0.603	0.525	0.457	0.395	0.336	0.278	0.222	0.165	0.105	0.042
4.0	1.225	1.013	0.847	0.713	0.613	0.527	0.455	0.393	0.339	0.288	0.239	0.190	0.140	0.088	0.034
4.5	1.133	0.925	0.764	0.637	0.543	0.463	0.398	0.340	0.290	0.247	0.205	0.162	0.117	0.074	0.027
5.0	1.053	0.849	0.696	0.577	0.483	0.408	0.347	0.297	0.253	0.214	0.177	0.139	0.102	0.063	0.020
6.0	0.919	0.728	0.586	0.477	0.393	0.325	0.273	0.230	0.193	0.161	0.131	0.103	0.074	0.045	0.012
7.0	0.819	0.636	0.503	0.402	0.325	0.267	0.218	0.181	0.149	0.123	0.100	0.078	0.056	0.032	0.008
8.0	0.735	0.564	0.440	0.346	0.275	0.221	0.179	0.145	0.118	0.096	0.078	0.060	0.042	0.024	0.005
9.0	0.667	0.504	0.388	0.300	0.235	0.186	0.149	0.120	0.095	0.078	0.062	0.047	0.032	0.019	0.004
10.0	0.610	0.456	0.345	0.264	0.203	0.159	0.124	0.100	0.079	0.063	0.049	0.037	0.026	0.015	0.002
12.0	0.521	0.380	0.280	0.209	0.157	0.120	0.092	0.071	0.054	0.042	0.033	0.025	0.018	0.010	0.001
14.0	0.452	0.321	0.232	0.169	0.124	0.092	0.069	0.052	0.038	0.030	0.023	0.016	0.012	0.007	0.000
16.0	0.396	0.276	0.195	0.139	0.100	0.072	0.053	0.039	0.028	0.021	0.016	0.012	0.008	0.004	0.000
18.0	0.351	0.240	0.166	0.117	0.082	0.058	0.042	0.030	0.022	0.016	0.012	0.008	0.006	0.003	0.000
20.0	0.315	0.211	0.144	0.098	0.068	0.047	0.033	0.023	0.017	0.012	0.009	0.006	0.004	0.002	0.000

TABLE A12-4
Gibbs energy increase from free rotation $[(G-G_f)/T]/\text{cal K}^{-1}$

$G_f T = -R \ln Q_f$

V/RT	0	0.05	0.10	0.15	0.20	0.25	1/Q_f 0.30	0.35	0.40	0.45	0.50	0.55
0.0	0.000	0.000	0.000	0.000	0.000	0.000	0.000	0.000	0.000	0.000	0.000	0.000
0.2	0.193 7	0.154	0.117	0.085	0.061	0.044	0.033	0.025	0.018	0.013	0.009	0.005
0.4	0.377 6	0.326	0.274	0.225	0.176	0.131	0.096	0.072	0.056	0.044	0.035	0.026
0.6	0.551 6	0.489	0.424	0.361	0.298	0.236	0.184	0.142	0.113	0.089	0.068	0.054
0.8	0.716 1	0.640	0.566	0.493	0.420	0.348	0.286	0.230	0.185	0.145	0.111	0.088
1.0	0.871 1	0.784	0.699	0.617	0.537	0.461	0.389	0.322	0.262	0.208	0.163	0.129
1.5	1.220 0	1.114	1.010	0.909	0.809	0.714	0.622	0.538	0.451	0.375	0.308	0.250
2.0	1.518 2	1.395	1.276	1.159	1.045	0.935	0.829	0.726	0.628	0.535	0.450	0.371
2.5	1.772 4	1.635	1.501	1.371	1.246	1.124	1.004	0.889	0.778	0.675	0.575	0.479
3.0	1.989 3	1.839	1.693	1.552	1.415	1.282	1.152	1.027	0.907	0.794	0.682	0.576
3.5	2.175 6	2.013	1.856	1.704	1.557	1.414	1.275	1.143	1.019	0.893	0.774	0.660
4.0	2.336 6	2.163	1.996	1.833	1.676	1.525	1.379	1.239	1.108	0.974	0.850	0.732
4.5	2.477 2	2.293	2.117	1.945	1.780	1.621	1.467	1.322	1.184	1.044	0.914	0.791
5.0	2.601 2	2.408	2.221	2.042	1.868	1.703	1.543	1.392	1.244	1.104	0.969	0.841
6.0	2.810 8	2.599	2.396	2.202	2.015	1.836	1.664	1.500	1.344	1.194	1.052	0.916
7.0	2.983 3	2.755	2.537	2.328	2.129	1.936	1.757	1.583	1.418	1.262	1.111	0.971
8.0	3.129 4	2.886	2.653	2.432	2.220	2.020	1.828	1.646	1.474	1.312	1.157	1.011
9.0	3.256 3	2.998	2.753	2.520	2.298	2.087	1.888	1.698	1.520	1.351	1.192	1.039
10.0	3.368 6	3.097	2.839	2.594	2.362	2.144	1.936	1.741	1.557	1.383	1.219	1.063
12.0	3.560 2	3.263	2.982	2.718	2.468	2.233	2.013	1.806	1.612	1.429	1.258	1.096
14.0	3.720 5	3.400	3.099	2.816	2.551	2.303	2.071	1.854	1.651	1.462	1.285	1.119
16.0	3.858 4	3.517	3.197	2.897	2.618	2.358	2.116	1.891	1.682	1.486	1.305	1.135
18.0	3.979 3	3.618	3.280	2.965	2.674	2.403	2.152	1.920	1.704	1.505	1.319	1.146
20.0	4.087 2	3.707	3.353	3.024	2.720	2.440	2.181	1.942	1.722	1.519	1.331	1.155

TABLE A12-5
Entropy S/cal K^{-1}

										$1/Q_f$						
V/RT	0.25	0.30	0.35	0.40	0.45	0.50	0.55	0.60	0.65	0.70	0.75	0.80	0.85	0.90	0.95
0.0	3.748	3.386	3.079	2.814	2.580	2.371	2.182	2.009	1.850	1.703	1.567	1.438	1.316	1.203	1.097
0.2	3.743	3.382	3.076	2.811	2.578	2.369	2.180	2.003	1.848	1.701	1.563	1.433	1.312	1.196	1.091
0.4	3.730	3.370	3.065	2.801	2.568	2.359	2.170	1.996	1.837	1.691	1.555	1.428	1.307	1.193	1.085
0.6	3.709	3.347	3.043	2.780	2.547	2.340	2.151	1.980	1.823	1.677	1.541	1.415	1.295	1.184	1.076
0.8	3.679	3.318	3.013	2.750	2.519	2.315	2.125	1.957	1.800	1.654	1.523	1.399	1.284	1.171	1.068
1.0	3.638	3.279	2.974	2.714	2.485	2.279	2.094	1.928	1.774	1.629	1.499	1.377	1.262	1.153	1.052
1.5	3.512	3.156	2.854	2.600	2.376	2.173	1.997	1.833	1.685	1.552	1.428	1.310	1.201	1.094	1.000
2.0	3.355	3.004	2.709	2.458	2.241	2.048	1.874	1.718	1.578	1.450	1.332	1.224	1.122	1.024	0.936
2.5	3.180	2.836	2.548	2.303	2.091	1.907	1.739	1.589	1.456	1.335	1.224	1.126	1.031	0.942	0.860
3.0	3.008	2.667	2.380	2.138	1.933	1.756	1.576	1.456	1.330	1.217	1.114	1.021	0.936	0.855	0.779
3.5	2.838	2.500	2.218	1.978	1.782	1.610	1.458	1.323	1.206	1.100	1.004	0.919	0.841	0.769	0.703
4.0	2.678	2.343	2.069	1.834	1.643	1.475	1.328	1.199	1.087	0.988	0.901	0.821	0.748	0.683	0.623
4.5	2.528	2.199	1.926	1.698	1.511	1.348	1.209	1.086	0.978	0.884	0.804	0.730	0.662	0.607	0.551
5.0	2.396	2.068	1.798	1.579	1.392	1.233	1.097	0.982	0.881	0.794	0.716	0.648	0.588	0.535	0.486
6.0	2.166	1.844	1.585	1.370	1.192	1.040	0.915	0.808	0.715	0.637	0.568	0.509	0.457	0.412	0.372
7.0	1.983	1.665	1.411	1.204	1.033	0.891	0.774	0.672	0.588	0.516	0.453	0.401	0.357	0.319	0.285
8.0	1.830	1.519	1.272	1.071	0.906	0.770	0.660	0.566	0.486	0.422	0.366	0.320	0.281	0.248	0.220
9.0	1.703	1.397	1.156	0.962	0.804	0.674	0.570	0.483	0.407	0.350	0.300	0.258	0.223	0.195	0.171
10.0	1.593	1.295	1.060	0.872	0.719	0.596	0.496	0.414	0.348	0.293	0.248	0.211	0.180	0.154	0.134
12.0	1.417	1.125	0.904	0.728	0.588	0.476	0.388	0.315	0.255	0.213	0.176	0.146	0.122	0.101	0.084
14.0	1.275	0.994	0.783	0.620	0.492	0.388	0.309	0.247	0.196	0.157	0.126	0.100	0.084	0.069	0.056
16.0	1.157	0.890	0.688	0.533	0.414	0.322	0.251	0.196	0.155	0.119	0.092	0.075	0.059	0.048	0.038
18.0	1.058	0.801	0.609	0.464	0.353	0.270	0.205	0.158	0.121	0.093	0.072	0.056	0.042	0.034	0.026
20.0	0.975	0.727	0.542	0.405	0.303	0.228	0.170	0.129	0.097	0.073	0.056	0.042	0.032	0.024	0.018

TABLE A12-6
Entropy decrease from free rotation $(S_f - S)/\text{cal K}^{-1}$
$S_f = R(\frac{1}{2} + \ln Q_f)$

V/RT	0.0	0.05	0.10	0.15	0.20	0.25	0.30	0.35	0.40	0.45	0.50	0.55
							$1/Q_f$					
0.0	0.000 0	0.000	0.000	0.000	0.000	0.000	0.000	0.000	0.000	0.000	0.000	0.000
0.2	0.004 9	0.005	0.004	0.004	0.004	0.004	0.004	0.003	0.003	0.002	0.002	0.002
0.4	0.019 8	0.020	0.018	0.018	0.018	0.018	0.016	0.014	0.013	0.012	0.012	0.010
0.6	0.044 0	0.044	0.043	0.043	0.040	0.039	0.039	0.036	0.034	0.033	0.031	0.028
0.8	0.077 1	0.077	0.077	0.075	0.072	0.069	0.068	0.066	0.064	0.061	0.056	0.053
1.0	0.118 5	0.118	0.117	0.115	0.112	0.110	0.107	0.105	0.100	0.095	0.092	0.086
1.5	0.252 7	0.252	0.250	0.248	0.242	0.236	0.230	0.225	0.214	0.204	0.198	0.189
2.0	0.418 2	0.417	0.415	0.410	0.402	0.393	0.382	0.370	0.356	0.339	0.323	0.308
2.5	0.600 1	0.599	0.594	0.585	0.577	0.568	0.550	0.531	0.511	0.489	0.464	0.440
3.0	0.785 6	0.783	0.777	0.768	0.757	0.740	0.719	0.699	0.676	0.647	0.615	0.581
3.5	0.966 0	0.964	0.957	0.944	0.929	0.910	0.886	0.861	0.836	0.798	0.761	0.722
4.0	1.135 6	1.133	1.126	1.111	1.094	1.070	1.043	1.011	0.980	0.937	0.896	0.855
4.5	1.291 8	1.289	1.280	1.265	1.244	1.220	1.187	1.153	1.116	1.069	1.023	0.977
5.0	1.433 9	1.431	1.421	1.404	1.380	1.352	1.318	1.281	1.235	1.188	1.138	1.086
6.0	1.678 1	1.674	1.662	1.643	1.616	1.582	1.542	1.494	1.444	1.388	1.331	1.268
7.0	1.878 3	1.874	1.860	1.837	1.807	1.765	1.721	1.668	1.610	1.547	1.480	1.411
8.0	2.044 7	2.040	2.024	1.998	1.962	1.918	1.867	1.807	1.743	1.674	1.601	1.525
9.0	2.186 4	2.180	2.163	2.134	2.095	2.045	1.989	1.923	1.852	1.776	1.697	1.612
10.0	2.309 5	2.303	2.284	2.252	2.208	2.155	2.091	2.019	1.942	1.861	1.775	1.686
12.0	2.515 5	2.508	2.485	2.447	2.394	2.331	2.261	2.175	2.086	1.992	1.895	1.793
14.0	2.684 7	2.676	2.650	2.607	2.547	2.473	2.392	2.296	2.194	2.088	1.983	1.872
16.0	2.828 9	2.819	2.788	2.740	2.674	2.591	2.496	2.391	2.281	2.166	2.049	1.930
18.0	2.954 5	2.943	2.910	2.855	2.781	2.690	2.585	2.470	2.350	2.227	2.101	1.976
20.0	3.065 9	3.054	3.017	2.956	2.872	2.773	2.659	2.537	2.409	2.277	2.143	2.011

concerning the exact range of validity of the tables of paper I. Only the tables from paper I are presented below; the other papers should be consulted for the exceptional situations there considered.

The thermodynamic functions are given in Tables A12-1 to A12-6. It is more convenient to give a portion of the entropy and Gibbs energy tables in terms of the difference between the function for restricted rotation and for free rotation. The formulas for the free-rotation values are given in the table headings. From the present viewpoint, it is unfortunate that the tables were presented in terms of calories instead of the dimensionless ratio to R. At this time it does not seem worthwhile to convert, with slight loss in accuracy, to either joules or dimensionless ratios C/R, etc. The value of R used in paper I was $1.9869 \, \text{cal} \, \text{K}^{-1} \, \text{mol}^{-1}$. Thus the values in the tables can be divided by 1.9869 for dimensionless C/R, etc., or multiplied by $(8.3145/1.9869) = 4.1847$ for $J \, K^{-1} \, \text{mol}^{-1}$.

REFERENCES

1. K. S. Pitzer and W. D. Gwinn, *J. Chem. Phys.*, **10**, 428 (1942).
2. K. S. Pitzer, *J. Chem. Phys.*, **14**, 239 (1946).
3. J. E. Kilpatrick and K. S. Pitzer, *J. Chem. Phys.*, **17**, 1064 (1949).
4. J. C. M. Li and K. S. Pitzer, *J. Phys. Chem.*, **60**, 466 (1956).

# APPENDIX 13

ESTIMATION OF ENTROPIES

We have seen in various typical thermodynamic calculations that enthalpy, entropy, and heat-capacity data are required but that the sensitivity of the final answer to errors in these data varies widely. Also we have remarked that the heat of formation of a vast number of chemical substances is known, at least to the accuracy attained in the work of Thomsen, Berthelot, and others in the latter part of the nineteenth century. While our knowledge of entropy values has increased rapidly since the third law became established and generally applied, nevertheless there are many substances for which the heat of formation is known but the entropy has not been measured. Thus there is particular need to estimate entropy values to an accuracy corresponding to the enthalpies measured by Thomsen. Since an uncertainty of $1-2\,\text{kJ}\,\text{mol}^{-1}$ is typical of the older heat-of-reaction data, an uncertainty of $0.5R$ to R in the entropy has the equivalent effect at room temperature on an equilibrium constant or the Gibbs energy of reaction in view of the equation

$$\Delta G = \Delta H - T\Delta S \tag{8-3}$$

In order to calculate the Gibbs energy at other temperatures, the heat capacities of reactants and products are required, and it is frequently both necessary and feasible to estimate the heat capacity near or above room temperature if it has not been measured. The heat capacity of pure substances was considered in Chapter 5 and the theory of the heat capacity of ideal gases was developed further in Chapter 20. The heat capacities of various types of solutions were considered in Chapters 12 and 18.

Many of the ideas given in the discussions of heat capacity are also applicable to calculations of entropies. Indeed the equation

$$S_T = \int_0^T C_p \, d\ln T' \qquad (6\text{-}4)$$

shows the close relationship between the two quantities. But we recall from the rule of Dulong and Petit that the heat capacity of most solid elements is the same at and above room temperature, whereas we know that at lower temperatures the heat capacity of each element falls to low values along a different curve. Hence the entropies of solid elements are not equal, and we see that estimation of entropy involves factors in addition to those affecting the heat capacity at room temperature and above. If these factors are all known accurately, then the entropy may be calculated by precise statistical methods such as those given in Chapters 5 and 20. But in many cases only the major factors affecting the entropy are identified, or the system may be so complex that precise statistical treatment is impractical. In such cases it is still possible to make a useful estimate of the entropy by a combination of theory and empiricism. As many important factors as possible are considered theoretically, and then the remaining contributions to the entropy are evaluated by comparison with experimental values for similar substances.

A different but related situation arises when the heat capacity has been measured but only from a temperature in the range 50–70 K, attainable with liquid air or nitrogen. Then the extrapolation to 0 K is subject to considerable uncertainty. If closely similar substances have recently been measured to 15 K or lower, it may be possible to improve the extrapolation originally reported.

ENTROPIES OF SOLID ELEMENTS

In Chapter 5 we noted the characteristic behavior of the heat-capacity curve for monatomic solids and that the Debye theory gives a useful approximation. This theory gives the entropy as a function of T/θ_D; the function is tabulated in Appendix 2. At high temperatures, where $C_V \cong 3R$, the entropy is[†]

$$\frac{S}{R} = 3(\ln T - \ln \theta_D + 1.333) \qquad (A13\text{-}1)*$$

Thus the principal factors affecting the entropy are the temperature and those factors that determine θ_D. We noted in Chapter 5 that θ_D is proportional to a frequency of atomic vibration and therefore is proportional to the square root of an interatomic force constant and inversely proportional to the square root of the atomic mass. Thus the

---

[†] Since all quantities are on a molar basis, unless stated otherwise, the subscript m is omitted.

difference in entropy between two related solid elements at the same high temperature, where both have attained a heat capacity $\cong 3R$, is approximated by

$$S_B - S_A = \tfrac{3}{2} R \left(\ln \frac{M_B}{M_A} - \ln \frac{k_B}{k_A} \right) \qquad (A13\text{-}2)^*$$

The atomic masses yield the ratio M_B/M_A at once. By choosing a comparison element A of known entropy whose hardness or compressibility is about the same as that of the element B, one may assume the force constants to be equal, that is, $k_B \cong k_A$, and the second term drops out, yielding

$$S_B = S_A + \tfrac{3}{2} R \ln \frac{M_B}{M_A} \qquad (A13\text{-}3)^*$$

If the element forms molecules that are only weakly bound into a solid lattice, as is the case for I_2 or S_8, then the problem is more complex. The mass is still the predominant factor, but Eq. (A13-3)* is accurately applicable only if both the intramolecular and intermolecular forces are equal. If there are electronic or magnetic contributions to the entropy, these must be added; Chapter 6 includes a very brief discussion of this aspect. Actually the entropies of most elements have been measured, and this discussion is primarily an introduction to more complex solids.

ENTROPIES OF SOLID COMPOUNDS

Kopp proposed the rule that the heat capacity of a solid compound was equal to the sum of the heat capacities of the component elements. If this were true at all temperatures, the corresponding summation rule would apply for the entropy. Actually Kopp's rule is only a rough approximation, but, if used with judgment, it is a very useful method of estimation.

One should keep in mind the factors found to be important in the preceding section—the atomic mass and the force constant restraining atomic motion. It is not usually feasible to obtain numerical values for the force constants; rather one selects reference substances for which the force constant should be about the same. The atomic-mass effect can then be taken into account by use of Eq. (A13-3)*, which may be generalized for a pair of compounds X and Y differing by the replacement of n atoms of A in X by n atoms of B in Y:

$$S_Y = S_X + \tfrac{3}{2} Rn \ln \frac{M_B}{M_A} \qquad (A13\text{-}4)^*$$

If Eq. (A13-3)* holds for the elements A and B, it is clear that the entropy additivity rule will yield Eq. (A13-4)* for the compounds. However, Eq. (A13-4)* may be applied whenever the force constants for the A and B atoms are about the same in the compound even if Eq. (A13-3)* would not hold for the elements. For example, chlorine is gaseous and bromine is liquid at room temperature; hence no simple relationship is expected between their entropies. But for the entropy of metal chlorides and bromides (Eq. A13-4)* predicts that the bromide entropy will be larger by $1.2R$ per

TABLE A13-1
Entropies of metal bromides and chlorides[a]

$\dfrac{S°}{R}$ for $\dfrac{1}{n}$ MX$_n$(s) at 298 K

	X = Cl	X = Br	Difference
AgX	11.6	12.9	1.3
KX	9.9	11.5	1.6
NaX	8.7	10.4	1.7
PbX$_2$	8.2	9.7	1.5
Hg$_2$X$_2$	11.6	13.2	1.6

[a] Values from Appendix 6.

halogen in the formula. The values in Table A13-1 show reasonable agreement with this prediction. One may also use the data of Table A13-1 to check Eq. (A13-4)* with respect to metal-ion mass by comparing NaCl with AgCl, etc.

An equation of the type of (A13-4)* was first given by Latimer[1] in 1921, when he proposed for compounds with the Dulong and Petit heat capacity of $3R$ per atom at 298 K the formula

$$\dfrac{S°_{298}}{R} = \sum_i (\tfrac{3}{2} n_i \ln M_i - q)$$

where the sum covers all atoms in the compound. The numerical constant q can be selected for a group of similar compounds. Subsequently Latimer[2] presented extensive tables checking rules of this general type for the entropies of inorganic solids. These tables provide a convenient basis for further estimates.

IMPROVED EXTRAPOLATIONS FOR ENTROPY BELOW 50 K

For a considerable number of solids there are measured heat capacities down to 50 K, which is accessible with liquid air or nitrogen as the refrigerant. Most of these measurements were made in a program of the U.S. Bureau of Mines directed by K. K. Kelley. For compounds containing only very light atoms, the uncertainty in the extrapolation below 50 K is less than the experimental uncertainty at higher temperatures, but for CaCl$_2$ and similar salts a reexamination of the original extrapolation is worthwhile. More recent data at lower temperatures for similar substances allow comparisons and improved extrapolations. In many cases, no revision is indicated. For CaSO$_4$(cr), new measurements of Robie et al.[3] down to 8 K gave for the entropy at 50 K the value $0.753R$. Anderson[4] gives $0.760R$ for his extrapolation for CaSO$_4$, which agrees well within the uncertainty of measurements at higher temperatures.

For CaCl$_2$(cr), however, a comparison of the extrapolated entropy at 50 K of 2.30R with the values for MgCl$_2$,[5] SrCl$_2$,[6] and BaCl$_2$[7] is not satisfactory, as Fig. A13-1 indicates. For this low temperature range, it is the total formula mass of the salt that is most relevant; consequently, that mass is taken as the variable for comparison. In the very low temperature range of the T^3 law, there would be a theoretical dependence on $M^{3/2}$ for a graph like Fig. A13-1, but at this intermediate temperature there is no simple theoretical guidance. There is no conflict, however, with the slope and linear dependence shown in Fig. A13-1.

A reasonable curve through the values for MgCl$_2$, SrCl$_2$, and BaCl$_2$ on Fig. A13-1 indicated a value of $(1.05 \pm 0.2)R$ for S_{50} for CaCl$_2$. Then adding the experimental value of 11.36R for $S_{298} - S_{50}$ from Kelley and Moore,[5] one obtains $(12.41 \pm 0.4)R$ for S at 298.15 for CaCl$_2$(s). From a treatment of the properties of aqueous CaCl$_2$ and its equilibrium with the solid at 523 K, Pitzer and Oakes[8] obtained a value for the entropy of CaCl$_2$(s) that, when converted to lower temperatures, becomes $S°/R = 12.6_8 \pm 0.3$ at 298.15 K or 1.32 ± 0.4 for $S°/R$ at 50 K. The estimates above agree with these indirectly calculated experimental values within the various uncertainties.

This need for revision of the extrapolation for CaCl$_2$ was noted by Stull and Prophet,[9] who chose 1.2R for S_{50}; this agrees well within the uncertainties with our values. Garvin et al.[10] also revised the extrapolation for CaCl$_2$(cr) but chose $S_{50} = 1.7R$, which appears to be still too large a value. The data for this series of compounds can be examined with a plot of C_P/T vs. T^2, as described in Chapter 6 and illustrated

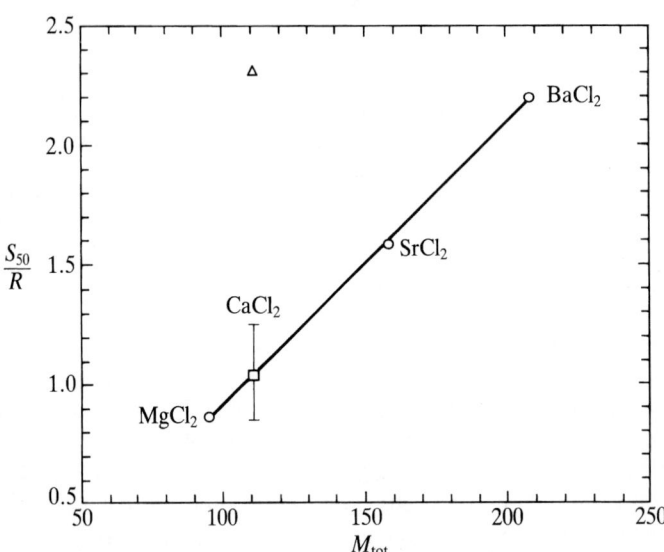

FIGURE A13-1
The entropy of solid MgCl$_2$, CaCl$_2$, SrCl$_2$, and BaCl$_2$ at 50 K from measured or extrapolated heat capacities. The values for all except CaCl$_2$ are reasonable on the basis of various criteria (see text), but that shown by the triangle for CaCl$_2$ is not. A probable value for CaCl$_2$ is shown by the square with probable error limits.

in Fig. 6-3. This type of presentation for $CaCl_2$ clearly indicates that the original extrapolation was too large and that the apparent slope $\partial C_P/\partial T$ in the range 50–60 K is too small. But that method does not yield a new value of S_{50} as clearly as Fig. A13-1 does.

ENTROPY OF FUSION

The entropy of fusion of a monatomic solid is usually in the range $1.0R$ to $1.7R$. The rare gases have values near $1.7R$, while most metals have values nearer $1.3R$. Metallic elements of complex structure such as bismuth have larger entropies of fusion, for example, $\Delta S/R = 2.4$.

Substances with infinite lattices composed of single ions also behave as monatomic solids with molal entropies of fusion in the range of $1.2R$ to $1.7R$ per ion. However, when there are drastic changes in coordination number in going from the solid to the liquid, the entropy of fusion may be much larger. The occurrence of these abnormally high entropies of fusion can be predicted from the sizes of the anions and cations involved. As an example, the entropy of fusion of $AlCl_3$ is $9.2R$ instead of the predicted value of $5R$ to $7R$. The size of the aluminum cation is such that it can accommodate a coordination number of 6 in the solid, but the increased thermal agitation of the liquid causes reduction of the coordination number and formation of Al_2Cl_6 molecules. On the other hand CCl_4, which has a molecular lattice, has an entropy of fusion of $1.2R$. Thus each CCl_4 molecule is acting as a single unit.

More complex solids have larger entropies of fusion, and it is difficult to predict the value for a particular solid unless data are available for other solids of similar structure. Even then, one must consider whether there are solid–solid transitions below the melting point. It is found that molecular substances of similar structure have approximately equal values of the sum of the entropy of fusion and of any entropies of transition. Thus there are two solid–solid phase transitions in oxygen (see Table 6-1) at which the entropy changes are $0.477R$ and $2.042R$. The entropy of fusion is $0.983R$, and the total is $3.50R$. The entropies of fusion of the halogens, which are also diatomic molecules but have no solid–solid transitions, are Cl_2 $4.47R$, Br_2 $4.77R$, and I_2 $4.91R$.

ENTROPIES OF VAPORIZATION

The Trouton[11] rule for entropies of vaporization is probably the oldest and best known of all rules for estimating entropies. It states that the entropy of vaporization at the boiling point (at 1 bar) is approximately constant for various substances. The value about $10.5R$ is commonly selected for the Trouton constant. Like the other rules we have been considering, the Trouton rule is valid only if certain other factors are very nearly constant. Thus deviations occur for substances that have hydrogen bonding, for substances boiling at very low or very high temperatures, etc. Hildebrand[12] showed that the entropies of vaporization for a variety of normal liquids were more nearly equal if compared at equal molal volume of vapor instead of equal vapor pressure. But it is found[13] that the entropies of vaporization of the extremely simple and similar substances Ar, Kr, Xe are almost exactly equal when compared at equal reduced temperatures (or equal reduced pressures). In this case the principle of corresponding states holds very precisely, whereas the range of entropies of vaporization at the boiling point is 0.2R and the range at constant vapor volume is 0.5R.

So long as one compares similar substances, it makes little difference whether the basis is at constant vapor pressure, constant reduced temperature, or constant vapor volume. One may then assume that, for such similar substances, the entropies of vaporization will be approximately equal.

ENTROPIES OF GASES

The entropies of gases with simple molecules can be calculated accurately by methods given in Chapters 5 and 20. For more complex molecules, however, the necessary spectroscopic or molecular-structure data may not be available, and estimates are then useful. Sometimes it is best to estimate the missing vibration frequencies or other molecular parameters; indeed high-frequency vibrations (such as C—H stretching) contribute little to the entropy at moderate temperatures, and their terms can even be omitted without serious error.

There are many regularities in the structure of organic compounds that aid in the estimation of entropies. For example, it is found that the entropy of paraffin isomers is lowered 1.8R, on the average, for each branch in the carbon skeleton. Also it is easy to calculate certain symmetry factors that may be introduced to improve the accuracy of estimate. On this basis, Pitzer and Scott[14] proposed the formula

$$\frac{S°}{R} = \frac{S°_n}{R} + \ln 2 + \ln \frac{I}{\sigma_e \sigma_i} - 1.8B \qquad (A13\text{-}5)*$$

where $S°_n$ is the molar entropy of the isomeric n-paraffin at the same temperature and pressure, I is the number of isomers included (two for a racemic mixture, otherwise one), σ_e and σ_i are the symmetry numbers for external rotation and for internal rotation of the carbon skeleton, respectively, and B is the number of chain branchings. The $\ln 2$ corrects for the symmetry number of the n-paraffin. The molar entropy of the n-paraffins above butane at 298.15 K in the ideal-gas standard state is well represented by the formula

$$\frac{S°_n}{R} = 4.68n + 18.65 \qquad (A13\text{-}6)*$$

where n is the number of carbon atoms per molecule.

The tables of properties,[15] including entropies, of gaseous hydrocarbons and similar compounds are now so extensive that there is little need to discuss more elaborate methods here. Many tabulated values were obtained by such methods, however.

ENERGIES AND ENTHALPIES

There is, of course, interest in estimating enthalpies as well as entropies. The energies of substances are, however, much more dependent than entropies on the detailed electronic structure of the atoms, the type of bonding, etc. Within specific groups of elements or compounds there are useful generalizations, of which a few were included in the second edition of this book.[16] With the increase in recent years in experimental knowledge, however, there is less need for estimates. And there are important advances in theory that are too complex to include here but are available in the literature.

REFERENCES

1. W. M. Latimer, *J. Am. Chem. Soc.*, **43**, 818 (1921).
2. W. M. Latimer, *Oxidation Potentials*, 2d edn., Prentice-Hall, Englewood Cliffs, N.J., 1952, app. III.
3. R. A. Robie, S. Russell-Robinson, and B. S. Hemingway, *Thermochim. Acta*, **139**, 67 (1989).
4. K. K. Kelley, J. C. Southard, and C. T. Anderson, "Thermodynamic Properties of Gypsum and its Dehydration Products," U.S. Bureau of Mines Technical Paper 625, Washington, 1941, pp. 17–19.
5. K. K. Kelley and G. E. Moore, *J. Am. Chem. Soc.*, **65**, 1264 (1943).
6. D. F. Smith, T. E. Gardner, B. B. Letson, and A. R. Taylor, Jr., Report of Investigations 6316, U.S. Bureau of Mines, 1963.
7. R. M. Goodman and E. F. Westrum, Jr., *J. Chem. Eng. Data*, **11**, 294 (1966).
8. K. S. Pitzer and C. S. Oakes, *J. Chem. Eng. Data*, **39**, 553 (1994).
9. D. R. Stull and H. Prophet, "JANAF Thermochemical Tables," 2d edn., NSRDS-NBS-37, 1971.
10. D. Garvin, V. B. Parker, and H. J. White, Jr., *CODATA Thermodynamic Tables*, Hemisphere, New York, 1987.
11. F. T. Trouton, *Phil. Mag.* **18**(5), 54 (1884).
12. J. H. Hildebrand, *J. Am. Chem. Soc.*, **37**, 970 (1915); **40**, 45 (1918).
13. K. S. Pitzer, *J. Chem. Phys.*, **7**, 583 (1939).
14. K. S. Pitzer and D. W. Scott, *J. Am. Chem. Soc.*, **63**, 2419 (1941).
15. K. N. Marsh, ed., *TRC Thermodynamic Tables*, Texas A&M University, College Station, Texas; looseleaf tables revised frequently; see item 4 at the end of Chapter 8.
16. K. S. Pitzer and L. Brewer, *Thermodynamics*, revision of 1st edition by G. N. Lewis and M. Randall, McGraw-Hill, New York, 1961, chapter 32.

APPENDIX 14

TABULAR SUMMARY OF THERMODYNAMIC FORMULAS

In this appendix we give first a list of the most widely used formulas, together with the equation numbers that will locate their derivation in the text. The text should be consulted for the conditions for validity of each equation. In a second portion we give a shorthand system by which additional formulas may be obtained for any desired partial derivative. Except for the sections on the chemical potential and on partial molar quantities, a constant amount and composition of substance is assumed.

Differential energy expressions

$$dU = T\,dS - P\,dV \tag{2-26}$$

$$dH = T\,dS + V\,dP \tag{3-4}$$

$$dA = -S\,dT - P\,dV \tag{3-5}$$

$$dG = -S\,dT + V\,dP \tag{3-6}$$

Maxwell relations

$$\left(\frac{\partial T}{\partial V}\right)_S = -\left(\frac{\partial P}{\partial S}\right)_V \qquad (3\text{-}14)$$

$$\left(\frac{\partial T}{\partial P}\right)_S = \left(\frac{\partial V}{\partial S}\right)_P \qquad (3\text{-}13)$$

$$\left(\frac{\partial S}{\partial V}\right)_T = \left(\frac{\partial P}{\partial T}\right)_V \qquad (3\text{-}11)$$

$$\left(\frac{\partial S}{\partial P}\right)_T = -\left(\frac{\partial V}{\partial T}\right)_P \qquad (3\text{-}12)$$

Energy-function derivatives

$$\left(\frac{\partial U}{\partial S}\right)_V = \left(\frac{\partial H}{\partial S}\right)_P = T \qquad (3\text{-}10)$$

$$\left(\frac{\partial U}{\partial V}\right)_S = \left(\frac{\partial A}{\partial V}\right)_T = -P \qquad (3\text{-}7)$$

$$\left(\frac{\partial H}{\partial P}\right)_S = \left(\frac{\partial G}{\partial P}\right)_T = V \qquad (3\text{-}8)$$

$$\left(\frac{\partial G}{\partial T}\right)_P = \left(\frac{\partial A}{\partial T}\right)_V = -S \qquad (3\text{-}9)$$

Heat-capacity relations

$$C_V = \left(\frac{\partial U}{\partial T}\right)_V = T\left(\frac{\partial S}{\partial T}\right)_V \qquad (2\text{-}7)\ (3\text{-}26)$$

$$C_P = \left(\frac{\partial H}{\partial T}\right)_P = T\left(\frac{\partial S}{\partial T}\right)_P \qquad (3\text{-}27)\ (3\text{-}28)$$

$$C_P - C_V = -T\left(\frac{\partial V}{\partial T}\right)_P^2 \left(\frac{\partial P}{\partial V}\right)_T \qquad (3\text{-}30)$$

$$= -T\left(\frac{\partial P}{\partial T}\right)_V^2 \left(\frac{\partial V}{\partial P}\right)_T \qquad (3\text{-}34)$$

$$\left(\frac{\partial C_P}{\partial P}\right)_T = -T\left(\frac{\partial^2 V}{\partial T^2}\right)_P \qquad (3\text{-}35)$$

Effect of P on H or V on U at constant T

$$\left(\frac{\partial H}{\partial P}\right)_T = V - T\left(\frac{\partial V}{\partial T}\right)_P \tag{3-15}$$

$$\left(\frac{\partial U}{\partial V}\right)_T = T\left(\frac{\partial P}{\partial T}\right)_V - P \tag{3-7) (3-11}$$

Chemical potential

$$\mu_i = \left(\frac{\partial G}{\partial n_i}\right)_{T,P,n_j} = \left(\frac{\partial A}{\partial n_i}\right)_{T,V,n_j} \tag{10-11) (10-14}$$

$$dA = -S\,dT - P\,dV + \sum_i \mu_i\,dn_i \tag{10-13}$$

$$dG = -S\,dT + V\,dP + \sum_i \mu_i\,dn_i \tag{10-10}$$

Partial molar quantities, Y any extensive quantity

$$\overline{Y}_1 = \left(\frac{\partial Y}{\partial n_1}\right)_{P,T,n_2,n_3,\ldots} \tag{10-19}$$

$$Y = n_1 \overline{Y}_1 + n_2 \overline{Y}_2 + \cdots \tag{10-27}$$

$$x_1 \left(\frac{\partial \overline{Y}_1}{\partial x_1}\right) + x_2 \left(\frac{\partial \overline{Y}_2}{\partial x_1}\right) + \cdots = 0 \tag{10-33}$$

$$\left(\frac{\partial \overline{Y}_i}{\partial n_j}\right) = \frac{\partial^2 Y}{\partial n_i \partial n_j} = \left(\frac{\partial \overline{Y}_j}{\partial n_i}\right) \tag{10-34}$$

Temperature effect on $\Delta G°/T = -R \ln K$:

$$\left[\frac{\partial (\Delta G°/T)}{\partial T}\right]_P = -R\frac{\partial \ln K}{\partial T} = -\frac{\Delta H°}{T^2} \tag{3-20) (7-29}$$

Several schemes of shorthand notation for a wide variety of thermodynamic formulas have been proposed; we follow the method of Shaw.† Use is made of mathematical functions known as Jacobians, which are defined as follows:

---

† A. N. Shaw, *Phil. Trans. Roy. Soc. London*, **A234**, 299 (1935).

$$\frac{\partial(x,y)}{\partial(\alpha,\beta)} = \begin{vmatrix} \left(\frac{\partial x}{\partial \alpha}\right)_\beta & \left(\frac{\partial y}{\partial \alpha}\right)_\beta \\ \left(\frac{\partial x}{\partial \beta}\right)_\alpha & \left(\frac{\partial y}{\partial \beta}\right)_\alpha \end{vmatrix} = \left(\frac{\partial x}{\partial \alpha}\right)_\beta \left(\frac{\partial y}{\partial \beta}\right)_\alpha - \left(\frac{\partial x}{\partial \beta}\right)_\alpha \left(\frac{\partial y}{\partial \alpha}\right)_\beta$$
(A14-1)

If α and β remain unchanged throughout, as will be the case here, a shorter notation may be used:

$$J(x,y) = \frac{\partial(x,y)}{\partial(\alpha,\beta)} \qquad (A14\text{-}2)$$

The following properties can be easily established by reference to the basic definition of Jacobians:

$$J(y,x) = -J(x,y) \qquad (A14\text{-}3)$$

$$\left(\frac{\partial y}{\partial x}\right)_z = \frac{J(y,z)}{J(x,z)} \qquad (A14\text{-}4)$$

$$J(\alpha,\beta) = 1 \qquad (A14\text{-}5)$$

We present two sets of equations selected alternatively for use with T and P or T and V as independent variables. In each case a constant amount and composition of substance is assumed. For the first and more generally useful case, we express the results in terms of T, P, and S and the three derivatives that are most readily capable of experimental measurement, namely, $(\partial V/\partial T)_P$, $(\partial V/\partial P)_T$, and $C_P = (\partial H/\partial T)_P$. Since all results will arise from the ratio of two Jacobians to yield a partial derivative, following Eq. (14-4), we shall use a notation immediately suggestive of this property, $(\partial y)_z = J(y,z)$. Then Eq. (14-4) becomes

$$\left(\frac{\partial y}{\partial x}\right)_z = \frac{(\partial y)_z}{(\partial x)_z} \qquad (A14\text{-}6)$$

Formula summary for T and P as independent variables

$$(\partial T)_P = -(\partial P)_T = 1$$

$$(\partial V)_P = -(\partial P)_V = \left(\frac{\partial V}{\partial T}\right)_P$$

$$(\partial S)_P = -(\partial P)_S = \frac{C_P}{T}$$

$$(\partial U)_P = -(\partial P)_U = C_P - P\left(\frac{\partial V}{\partial T}\right)_P$$

$$(\partial H)_P = -(\partial P)_H = C_P$$

$$(\partial G)_P = -(\partial P)_G = -S$$

$$(\partial A)_P = -(\partial P)_A = -\left[S + P\left(\frac{\partial V}{\partial T}\right)_P\right]$$

$$(\partial V)_T = -(\partial T)_V = -\left(\frac{\partial V}{\partial P}\right)_T$$

$$(\partial S)_T = -(\partial T)_S = \left(\frac{\partial V}{\partial T}\right)_P$$

$$(\partial U)_T = -(\partial T)_U = T\left(\frac{\partial V}{\partial T}\right)_P + P\left(\frac{\partial V}{\partial P}\right)_T$$

$$(\partial H)_T = -(\partial T)_H = -V + T\left(\frac{\partial V}{\partial T}\right)_P$$

$$(\partial G)_T = -(\partial T)_G = -V$$

$$(\partial A)_T = -(\partial T)_A = P\left(\frac{\partial V}{\partial P}\right)_T$$

$$(\partial S)_V = -(\partial V)_S = \frac{1}{T}\left[C_P\left(\frac{\partial V}{\partial P}\right)_T + T\left(\frac{\partial V}{\partial T}\right)_P^2\right]$$

$$(\partial U)_V = -(\partial V)_U = C_P\left(\frac{\partial V}{\partial P}\right)_T + T\left(\frac{\partial V}{\partial T}\right)_P^2$$

$$(\partial H)_V = -(\partial V)_H = C_P\left(\frac{\partial V}{\partial P}\right)_T + T\left(\frac{\partial V}{\partial T}\right)_P^2 - V\left(\frac{\partial V}{\partial T}\right)_P$$

$$(\partial G)_V = -(\partial V)_G = -\left[V\left(\frac{\partial V}{\partial T}\right)_P + S\left(\frac{\partial V}{\partial P}\right)_T\right]$$

$$(\partial A)_V = -(\partial V)_A = -S\left(\frac{\partial V}{\partial P}\right)_T$$

$$(\partial U)_S = -(\partial S)_U = \frac{P}{T}\left[C_P\left(\frac{\partial V}{\partial P}\right)_T + T\left(\frac{\partial V}{\partial T}\right)_P^2\right]$$

$$(\partial H)_S = -(\partial S)_H = -\frac{VC_P}{T}$$

$$(\partial G)_S = -(\partial S)_G = -\frac{1}{T}\left[VC_P - ST\left(\frac{\partial V}{\partial T}\right)_P\right]$$

$$(\partial A)_S = -(\partial S)_A = \frac{1}{T}\left\{P\left[C_P\left(\frac{\partial V}{\partial P}\right)_T + T\left(\frac{\partial V}{\partial T}\right)_P^2\right] + ST\left(\frac{\partial V}{\partial T}\right)_P\right\}$$

$$(\partial H)_U = -(\partial U)_H = -V\left[C_P - P\left(\frac{\partial V}{\partial T}\right)_P\right] - P\left[C_P\left(\frac{\partial V}{\partial P}\right)_T + T\left(\frac{\partial V}{\partial T}\right)_P^2\right]$$

$$(\partial G)_U = -(\partial U)_G = -V\left[C_P - P\left(\frac{\partial V}{\partial T}\right)_P\right] + S\left[T\left(\frac{\partial V}{\partial T}\right)_P + P\left(\frac{\partial V}{\partial P}\right)_T\right]$$

$$(\partial A)_U = -(\partial U)_A = P\left[(C_P + S)\left(\frac{\partial V}{\partial P}\right)_T + T\left(\frac{\partial V}{\partial T}\right)_P^2\right] + ST\left(\frac{\partial V}{\partial T}\right)_P$$

$$(\partial G)_H = -(\partial H)_G = -V(C_P + S) + TS\left(\frac{\partial V}{\partial T}\right)_P$$

$$(\partial A)_H = -(\partial H)_A = -\left[S + P\left(\frac{\partial V}{\partial T}\right)_P\right]\left[V - T\left(\frac{\partial V}{\partial T}\right)_P\right] + PC_P\left(\frac{\partial V}{\partial P}\right)_T$$

$$(\partial A)_G = -(\partial G)_A = -S\left[V + P\left(\frac{\partial V}{\partial P}\right)_T\right] - PV\left(\frac{\partial V}{\partial T}\right)_P$$

For work with highly compressible fluids and equations of state expressed with T and V or T and ρ as independent variables, an alternate set of equations is more useful. These are expressed in terms of T, V, and S and the three derivatives $(\partial P/\partial V)_T$, $(\partial P/\partial T)_V$, and $C_V = (\partial U/\partial T)_V$. These equations are easily converted for a change $V = n/\rho$. Since n is constant, the condition constant V is the same as constant ρ. Also,

$$dV = -\frac{n}{\rho^2}d\rho$$

$$\frac{\partial}{\partial V} = -\frac{n}{V^2}\frac{\partial}{\partial \rho} = -\frac{\rho^2}{n}\frac{\partial}{\partial \rho}$$

Formula summary for T and V as independent variables

$$(\partial V)_T = -(\partial T)_V = 1$$

$$(\partial P)_T = -(\partial T)_P = \left(\frac{\partial P}{\partial V}\right)_T$$

$$(\partial S)_T = -(\partial T)_S = \left(\frac{\partial P}{\partial T}\right)_V$$

$$(\partial U)_T = -(\partial T)_U = T\left(\frac{\partial P}{\partial T}\right)_V - P$$

$$(\partial H)_T = -(\partial T)_H = T\left(\frac{\partial P}{\partial T}\right)_V + V\left(\frac{\partial P}{\partial V}\right)_T$$

$$(\partial A)_T = -(\partial T)_A = -P$$

$$(\partial G)_T = -(\partial T)_G = V\left(\frac{\partial P}{\partial T}\right)_V$$

$$(\partial P)_V = -(\partial V)_P = -\left(\frac{\partial P}{\partial T}\right)_V$$

$$(\partial S)_V = -(\partial V)_S = -\frac{C_V}{T}$$

$$(\partial U)_V = -(\partial V)_U = -C_V$$

$$(\partial H)_V = -(\partial V)_H = -C_V - V\left(\frac{\partial P}{\partial T}\right)_V$$

$$(\partial A)_V = -(\partial V)_A = S$$

$$(\partial G)_V = -(\partial V)_G = S - V\left(\frac{\partial P}{\partial T}\right)_V$$

$$(\partial S)_P = -(\partial P)_S = \left(\frac{\partial P}{\partial T}\right)_V^2 - \frac{C_V(\partial P/\partial V)_T}{T}$$

$$(\partial U)_P = -(\partial P)_U = T\left(\frac{\partial P}{\partial T}\right)_V^2 - C_V\left(\frac{\partial P}{\partial V}\right)_T - P\left(\frac{\partial P}{\partial T}\right)_V$$

$$(\partial H)_P = -(\partial P)_H = T\left(\frac{\partial P}{\partial T}\right)_V^2 - C_V\left(\frac{\partial P}{\partial V}\right)_T$$

$$(\partial A)_P = -(\partial P)_A = S\left(\frac{\partial P}{\partial V}\right)_T - P\left(\frac{\partial P}{\partial T}\right)_V$$

$$(\partial G)_P = -(\partial P)_G = S\left(\frac{\partial P}{\partial V}\right)_T$$

$$(\partial U)_S = -(\partial S)_U = \frac{PC_V}{T}$$

$$(\partial H)_S = -(\partial S)_H = V\left[\frac{C_V(\partial P/\partial V)_T}{T} - \left(\frac{\partial P}{\partial T}\right)_V^2\right]$$

$$(\partial A)_S = -(\partial S)_A = S\left(\frac{\partial P}{\partial T}\right)_V + \frac{PC_V}{T}$$

$$(\partial G)_S = -(\partial S)_G = S\left(\frac{\partial P}{\partial T}\right)_V + V\left[C_V\left(\frac{\partial P}{\partial V}\right)_T - \left(\frac{\partial P}{\partial T}\right)_V^2\right]$$

$$(\partial H)_U = -(\partial U)_H = TV\left[\frac{C_V(\partial P/\partial V)_T}{T} - \left(\frac{\partial P}{\partial T}\right)_V^2\right] + P\left[V\left(\frac{\partial P}{\partial T}\right)_V + C_V\right]$$

$$(\partial A)_U = -(\partial U)_A = TS\left(\frac{\partial P}{\partial T}\right)_V + P(C_V - S)$$

$$(\partial G)_U = -(\partial U)_G = TS\left(\frac{\partial P}{\partial T}\right)_V + V\left[\left(\frac{\partial P}{\partial T}\right)_V^2 - C_V\left(\frac{\partial P}{\partial V}\right)_T\right] - P\left[S - V\left(\frac{\partial P}{\partial T}\right)_V\right]$$

$$(\partial A)_H = -(\partial H)_A = PC_V + (TS - PV)\left(\frac{\partial P}{\partial T}\right)_V + SV\left(\frac{\partial P}{\partial V}\right)_T$$

$$(\partial G)_H = -(\partial H)_G = TS\left(\frac{\partial P}{\partial T}\right)_V + V\left[T\left(\frac{\partial P}{\partial T}\right)_V^2 + (S - C_V)\left(\frac{\partial P}{\partial V}\right)_T\right]$$

$$(\partial G)_A = -(\partial A)_G = P\left[V\left(\frac{\partial P}{\partial T}\right)_V - S\right] - SV\left(\frac{\partial P}{\partial V}\right)_T$$

APPENDIX 15

SYMBOLS†

Italic capital letters

1. Extensive thermodynamic quantities
   - A Helmholtz energy
   - C Heat capacity
   - G Gibbs energy
   - H Enthalpy (heat content)
   - J Relative heat capacity
   - L Relative enthalpy
   - S Entropy
   - U Internal energy (energy content)
   - V Volume
   - Y Any thermodynamic function
2. Molar thermodynamic quantities are designated by the subscript m, as A_m, H_m, etc. The subscript is omitted when the molar basis has been stated clearly or is obvious by context. Heat capacities are almost always molar and the subscript is seldom used.
3. Partial molar properties are designated by a bar and a subscript indicating the species, as $\bar{H}_1$, $\bar{V}_2$, etc.

---

† Symbols and usages confined to a single section of text are omitted here.

4. Other quantities

A	Area; moment of inertia
$A_\phi, A_\gamma, \ldots$	Debye–Hückel parameter
A_{ij}	Nonideal-solution parameter
B	Spectroscopic rotational constant; aqueous ion-interaction parameter
B, C, D	Virial coefficients for gases
C_M	Curie constant (magnetic)
D	Determinant of inertia; dielectric constant; spectroscopic rotational stretching constant
E	Electrode potential
F	Faraday constant
H_{ij}	Henry's-law constant
I	Moment of inertia; constant of integration; ionic strength
J	Rotational quantum number; flux
K	Equilibrium constant; quantum number
L	Rate constant
M	Molecular mass
M_J	Quantum number
N	Number of molecules
N_A	Avogadro's number
P	Pressure
Q	Partition function
Q^*	Heat of transport
R	Gas constant
T	Temperature
U_{ij}	Constant in equation
V	Electrical potential, potential energy in a molecule
W	Multiplicity (statistical)
W_{ij}	Constant in equation
Z	Coordination number

Script capital letter

$\mathscr{P}$	Relative probability (statistical)
$\mathscr{V}$	Electrical potential (to distinguish from volume)

Boldface capital letters (vectors)

A	Electromagnetic potential
B	Magnetic induction
D	Electric displacement
E	Electric field
H	Magnetic field
M	Molal magnetic polarization
P	Molal electric polarization

Lower-case italic

a	Activity
a, b, c	Constants in equations
b	Volume parameter, van der Waals equation, etc.
c	Concentration; velocity of light
d	Density
e	Charge on the electron; eccentricity of spheroid
f	Fugacity
g	Acceleration of gravity; magnetic-moment factor; number of quantum states of equal energy
h	Planck constant; height
k	Boltzmann constant
k_{ij}	Henry's-law constant
l	Length
m	Molality; mass of single atom
n	Number of moles; quantum number
p_i	Partial pressure
q	Quantity of heat absorbed
r	Radius; mole ratio
t	Time; transference number; temperature, °C
u	Statistical quantity for molecular vibration
v	Vibrational quantum number
w	Quantity of work absorbed; nonideal-solution parameter
x	Mole fraction; spectroscopic anharmonicity constant
y	Statistical quantity for molecular rotation
z	Compression (compressibility) factor
z_+, z_-	Charge on a positive, negative ion
x, y, z	Linear coordinates

Boldface lower case (vectors)

m	Magnetic polarization per unit volume
p	Electric polarization per unit volume

Greek letters

α	Coefficient of thermal expansion; degree of ionization; spectroscopic rotation–vibration interaction constant
β	Bohr magneton
β_{ij}	Aqueous ion-interaction parameter
γ	Activity coefficient; surface tension
Γ	Surface concentration; activity of a reference state at pressure other than 1 bar
ε	Molecular energy level; electric permittivity
ε_0	Electric permittivity of free space

Θ	Freezing-point depression; characteristic temperature for molecular rotation or for Debye crystal
κ	Debye–Hückel ion atmosphere parameter; compressibility
λ	Molal freezing-point lowering
Λ	Equivalent conductivity
μ	Chemical potential; Joule–Thomson coefficient; magnetic permeability
μ_0	Magnetic permeability of free space
ν	Frequency; number of ions per molecule of electrolyte
Π	Osmotic pressure
π	Film pressure; 3.14159
ρ	Density; electric charge density
σ	Symmetry number
ϕ	Osmotic coefficient; volume fraction
χ	Magnetic susceptibility
χ_m	Molar magnetic susceptibility
ψ	Electric potential
ω	Vibration frequency in reciprocal centimeters; acentric factor; angular velocity

Superscripts, other than mathematical exponents and prime marks

$\circ$	Standard state
$*$	Ideal-gas state
E	Excess over value for ideal solution
Em	Excess quantity for solution in molality
id	Ideal solution, ideal gas
$(0), (1)$	Functions in acentric-factor theory for fluids, parameters for electrolytes
ϕ	Apparent molal quantity, $^{\phi}V$, $^{\phi}H$, etc.

Subscripts

The absolute temperature, that is, 0 for 0 K, 298 for 298.15 K. Property, P, T, etc., held constant.

1, 2	To represent components in a solution (if a solvent is defined, it is component 1)
ads	Adsorption
c	Critical
D	Debye, as in Θ_D
dil	Dilution
f	Reaction of formation from the elements
fus	Fusion, melting
ij	Parameter for interaction between i and j species; also ijk, etc.
irr	Contribution from irreversibility, as S_{irr}
m	Molar or molality basis
mix	Mixing

r	Reduced, i.e., ratio to value for critical state
rev	Quantity for reversible process
s	Surface quantity
sol	Solution of a solid (or gas)
sub	Sublimation
t.p.	Triple-point property
trs	Transition (between two crystal phases)
vap	Vaporization
w	To represent the component water, as in a_w
x	Mole-fraction basis
±	Mean quantity for ions of an electrolyte
23	Coefficient for component 2 as a result of interaction by component 3, as in k_{23}

APPENDIX 16

EQUATIONS INVOLVING MORE THAN TWO VARIABLES

While all of the mathematics used in this book will be familiar to many readers, and each particular equation is derived and discussed in the text when it is first needed, nevertheless, it seems desirable to summarize certain of these equations here for convenient reference. We consider first the case of V as a function of the two independent variables T and P. Then a change in V is given by

$$dV = \left(\frac{\partial V}{\partial T}\right)_P dT + \left(\frac{\partial V}{\partial P}\right)_T dP \tag{A16-1}$$

This states that the change in V is equal to the rate of change of V with T alone, multiplied by the change in T, plus the rate of change with P alone, multiplied by the change in P.

There are special forms of the general equation (A16-1) that are frequently used. Thus, if we impose the condition that V is constant, i.e., if we move along a contour line of the surface, $V(T,P)$,

$$\left(\frac{\partial V}{\partial T}\right)_P dT + \left(\frac{\partial V}{\partial P}\right)_T dP = 0 \tag{3-22}$$

and expressing the constancy of V in the equation itself,

$$\left(\frac{\partial P}{\partial T}\right)_V = -\frac{(\partial V/\partial T)_P}{(\partial V/\partial P)_T} \tag{3-23}$$

599

Again, if we have some other dependent variable, i.e., some other quantity that, like the volume, depends only upon the temperature and pressure, let us say the internal energy U, we may impose the condition that U is constant and obtain

$$\left(\frac{\partial V}{\partial T}\right)_U = \left(\frac{\partial V}{\partial T}\right)_P + \left(\frac{\partial V}{\partial P}\right)_T \left(\frac{\partial P}{\partial T}\right)_U \qquad (3\text{-}21)$$

This equation states that, when we proceed upon the surface along a line of constant internal energy, the change in V corresponding to a given infinitesimal change in T is the sum of two terms, namely, the change in V that would be caused by this same change in T alone, and the change in V, caused by such a change in P as is necessary to keep the energy constant. It is also often convenient to introduce derivatives involving dependent variables by means of expressions such as

$$\left(\frac{\partial V}{\partial P}\right)_T = \left(\frac{\partial V}{\partial U}\right)_T \left(\frac{\partial U}{\partial P}\right)_T$$

Three of these equations or their equivalents were derived in a different sequence in Chapter 3 and given the numbers shown above.

Also, there is a familiar equation involving second derivatives that we have frequently employed, namely,

$$\frac{\partial}{\partial P}\left(\frac{\partial V}{\partial T}\right)_P = \frac{\partial^2 V}{\partial P \, \partial T} = \frac{\partial}{\partial T}\left(\frac{\partial V}{\partial P}\right)_T$$

According to this equation, the rate of change with P of the V–T coefficient is equal to the rate of change with T of the V–P coefficient. The intermediate member of the equation is merely a shorthand method of expressing either of the others.

When a property depends upon three or more independent variables, the equations assume a similar form. Thus, if the volume of a given quantity of material depends not only upon temperature and pressure but also upon one or more other independent variables, such as the intensity of an electric field, we write

$$dV = \frac{\partial V}{\partial T} dT + \frac{\partial V}{\partial P} dP + \frac{\partial V}{\partial X} dX + \cdots$$

Here also subscripts are employed to show the independent variables that remain constant during the differentiation. Thus for $\partial V / \partial T$ we mean $(\partial V / \partial T)_{P,X}, \ldots$.

PERFECT DIFFERENTIALS

We had occasion in Chapter 2 to consider expressions of the type

$$\delta Z = L(x,y) \, dx + M(x,y) \, dy \qquad (2\text{-}15)$$

where δZ represents an infinitesimal quantity and $L(x,y)$ and $M(x,y)$ are functions of both independent variables, as indicated. This type of expression may or may not be the total differential of another function of x and y. If there is such a function $Z(x,y)$, then

$$L(x,y) = \left(\frac{\partial Z}{\partial x}\right)_y \quad \text{and} \quad M(x,y) = \left(\frac{\partial Z}{\partial y}\right)_x$$

and

$$\frac{\partial L}{\partial y} = \frac{\partial^2 Z}{\partial x \partial y} = \frac{\partial M}{\partial x} \tag{2-16}$$

The equality $\partial L/\partial y = \partial M/\partial x$ is a necessary and sufficient condition that an expression of the type of Eq. (2-15) is a perfect differential of a function Z. Suppose we wish to integrate the total change ΔZ from the state x_1, y_1 to the state x_2, y_2. If Eq. (2-16) holds, the differential may be integrated and $\Delta Z = Z(x_2,y_2) - Z(x_1,y_1)$. However, if Eq. (2-16) does not hold, then it is impossible to integrate Eq. (2-15) unless the path from x_1, y_1 to x_2, y_2 is specified. Also the result will depend on the path chosen, and it is not possible to regard this quantity as a property of the system.

In order to help keep this distinction in mind, we have used the symbol d (as in dZ) for a perfect differential of a function and the lower-case delta δ (as in δQ) for other infinitesimal quantities that are not perfect differentials.

APPENDIX 17

PHYSICAL CONSTANTS AND NUMERICAL FACTORS

The following values were recommended by the CODATA Task Group on Fundamental Constants.† For each constant the standard deviation uncertainty in the least significant digits is given in parentheses.

Quantity	Symbol	Value
Avogadro constant	N_A	$6.022\,137(4) \times 10^{23}$ mol^{-1}
Boltzmann constant	k	$1.380\,66(1) \times 10^{-23}$ J K^{-1}
Gas constant	R	$8.314\,51(7)$ J K^{-1} mol^{-1}
		$1.987\,22(2)$ cal K^{-1} mol^{-1}
		$83.145\,1$ bar cm^3 K^{-1} mol^{-1}
Defined calorie		4.184 J exactly
Kelvin temperature of the triple point of H_2O		273.16 exactly
Zero of the Celsius scale		273.15 K exactly
Molar volume, ideal gas, $P = 1$ bar, $T = 273.15$ K		$22.711\,1(2)$ L mol^{-1}
Faraday constant	F	$9.648\,531(3) \times 10^4$ C mol^{-1}
Elementary charge	e	$1.602\,177\,3(5) \times 10^{-19}$ C
Planck constant	h	$6.626\,075\,5(40) \times 10^{-34}$ J s
	$\hbar = h/2\pi$	$1.054\,572\,7(6) \times 10^{-34}$ J s
Speed of light in vacuum	c_0	$299\,792\,458$ m s^{-1} exactly
Permeability of vacuum	μ_0	$4\pi \times 10^{-7}$ H m^{-1} exactly
Permittivity of vacuum	$\epsilon_0 = 1/\mu_0 c_0^2$	$8.854\,187\,82 \times 10^{-12}$ F m^{-1}
Standard atmosphere	atm	$101\,325$ Pa exactly
Standard gravity	g_n	$9.806\,65$ m s^{-2} exactly
Bohr magneton	$\mu_B = e\hbar/2m_e$	$9.274\,015(3) \times 10^{-24}$ J T^{-1}
Nuclear magneton	$\mu_N = (m_e/m_p)\,\mu_B$	$5.050\,787(2) \times 10^{-27}$ J T^{-1}
Bohr radius	$a_0 = 4\pi\epsilon_0\hbar^2/m_e e^2$	$5.291\,772\,5(2) \times 10^{-11}$ m

---

† E. R. Cohen and B. N. Taylor, *CODATA Bull.*, **63**, 1 (1986); *Physics Today*, August 1992, BG9.

Basic and derived SI units

Quantity	SI unit	Symbol
Length	meter	m
Mass	kilogram	kg
Time	second	s
Thermodynamic temperature	kelvin	K
Amount of substance	mole	mol
Force	newton (kg m s^{-2})	N
Energy	joule (N m)	J
Pressure	pascal (N m^{-2})	Pa
Electric charge	coulomb	C
Electric current	ampere	A
Electric potential	volt	V
Power	watt (J s^{-1})	W
Volume	cubic meter	m^3
Density	kilogram/cubic meter	kg m^{-3}

In SI, mass is expressed in terms of kilograms (not grams), force in newtons, and pressure in newtons per square meter (pascals).

SI prefixes

Submultiple	Prefix	Symbol	Multiple	Prefix	Symbol
10^{-1}	deci	d	10	deca	da
10^{-2}	centi	c	10^2	hecto	h
10^{-3}	milli	m	10^3	kilo	k
10^{-6}	micro	μ	10^6	mega	M
10^{-9}	nano	n	10^9	giga	G
10^{-12}	pico	p	10^{12}	tera	T
10^{-15}	femto	f	10^{15}	peta	P

Conversion factors to SI for selected quantities

To convert from	To	Multiply by[a]
Angstrom	meter	1.000 000 0* $\times$ 10^{-10}
atmosphere (standard)	pascal	1.013 250 0* $\times$ 10^5
bar	pascal	1.000 000 0* $\times$ 10^5
calorie (thermochemical)	joule	4.184 000 0*
centimeter of mercury (0°C)	pascal	1.333 223 7 $\times$ 10^3
dyne	newton	1.000 000 0* $\times$ 10^{-5}
erg	joule	1.000 000 0* $\times$ 10^{-7}
electron volt (eV)	joule	1.602 09 $\times$ 10^{-19}
kilogram-force	newton	9.806 650 0*
millimeter of mercury (0°C)	pascal	1.333 223 7 $\times$ 10^2
torr (mmHg, 0°C)	pascal	1.333 223 7 $\times$ 10^2

[a] An asterisk after the seventh decimal place indicates that the conversion factor is exact and all subsequent digits are zero.

Energy conversion factors

$$E = h\nu = hc\tilde{\nu} = kT; \quad E_m = N_A E$$

	Wavenumber $\tilde{\nu}$	Frequency ν	Energy E	Molar energy E_m		Temperature T
	cm^{-1}	MHz	eV	kJ mol^{-1}	kcal mol^{-1}	K
$\tilde{\nu}$: cm^{-1}	1	$2.997\,925 \times 10^4$	$1.239\,842 \times 10^{-4}$	$11.962\,66 \times 10^{-3}$	$2.859\,14 \times 10^{-3}$	1.438 769
ν: MHz	$3.335\,64 \times 10^{-5}$	1	$4.135\,669 \times 10^{-9}$	$3.990\,313 \times 10^{-7}$	$9.537\,08 \times 10^{-8}$	$4.799\,22 \times 10^{-5}$
E: eV	8065.54	$2.417\,988 \times 10^8$	1	96.485 3	23.060 5	$1.160\,45 \times 10^4$
E_m: kJ mol^{-1}	83.593 5	$2.506\,069 \times 10^6$	$1.036\,427 \times 10^{-2}$	1	0.239 006	120.272
kcal mol^{-1}	349.755	$1.048\,539 \times 10^7$	$4.336\,411 \times 10^{-2}$	4.184	1	503.217
T: K	0.695 039	$2.083\,67 \times 10^4$	$8.617\,38 \times 10^{-5}$	$8.314\,51 \times 10^{-3}$	$1.987\,22 \times 10^{-3}$	1

Example of the use of this table: 1 eV corresponds to 96.4853 kJ mol^{-1}.

ANSWERS TO SELECTED PROBLEMS

2-1. $0.00432R$
2-2. $0.440R$
2-3. $\Sigma \Delta S/R = 0.086$ for combined system
3-1. $-BRT/(V-B)^2$
3-4. (a) $(\partial V/\partial T)_P = R/P + dB/dT$; $(\partial V/\partial P)_T = -RT/P^2$
3-6. (a) $w = 3.06$, $\Delta U = 37.65$, $\Delta H = 40.71$ kJ; (b) $w = 0$
3-11. -5.6 percent
4-3. 0.547
4-4. 225 K
4-5. 9.4 kJ
4-6. $5 \times 10^3 \text{ m s}^{-1}$
5-2. $5.45R$
6-2. 20.0 kJ
7-3. 3.1×10^{-3} bar
8-2. 757, 1.5 percent
8-5. 0.162 bar, 0.057 bar
9-5. 0.56
10-1. $x_2 = 0.0177$
10-2. 6.258, 0.1013
11-3. 2.92
11-4. $4.88°C$
11-6. $RT \ln 2$, $2RT \ln 2$
12-7. 1.075

12-8. 0.92
12-9. $\gamma = 83$ for analine, 3.7 for water
12-10. $P = 0.562 P°_{C_6H_6} + 0.445 P°_{C_2H_4Cl_2}$
12-11. $x(SnI_4) = 0.0026$
12-16. $x_{Ag} = 0.09$ in the U-rich phase
12-17. (c) $f_B/x_B = 247$ torr
15-3. For $Al_2(SO_4)_3$ at molality m, $a_2 = 108\, m^5$
15-4. $E = 0.114$ V
19-1. 10^3, 10^{-41}
20-1. $S°/R = 29.6 \pm 0.3$
20-3. 23% Na_2
23-4. Pressure ratio 1.021
23-5. $0.31\,\text{mol}\,\text{kg}^{-1}$
24-1. $\omega = 0.9 \times 10^4\,\text{radians}\,\text{s}^{-1}$

NAME INDEX

Abbot, M. M., 153, 221, 474
Ackermann, R. J., 486, 489, 491, 492
Adams, D., 569
Adams, L. H., 260, 261, 264, 273
Adamson, A. W., 405, 413, 418
Advani, J. V., 484
Agar, J. N., 445, 446, 452, 455, 456
Ahlberg, J. E., 88
Ahlers, G., 389, 399
Alberty, R. A., 434, 482, 484
Al-Bizreh, N., 62, 64
Albright, J. G., 456
Albright, P. C., 533
Alcock, C. B., 119
Amagat, E. H., 239, 242
Ananthaswamy, A., 309, 320
Anderko, A., 138, 148, 152, 243, 334, 338, 339, 474, 528, 529
Anderson, C. T., 120, 581, 585
Anderson, P. W., 383, 385
Andrews, T., 7, 14
Ansara, I., 462, 474
Anstiss, R. G., 309, 320
Archer, D. G., 274, 297, 320, 329, 338, 545
Arculus, R. J., 385
Armstrong, D. K., 553
Arnett, R. L., 118
Atkinson, G., 309, 320
Aziz, R. A., 152

Bae, H.-K., 522, 524
Baes, C. F., Jr., 563
Bailey, P. L., 362
Baker, G., 552
Bale, C. W., 461, 473, 474
Bao, Y-H., 474
Bard, A. J., 349, 361, 362, 363
Barin, I., 117, 542
Barkema, G. T., 383, 385
Barker, J. A., 147, 152, 196, 221, 289
Barkov, D. S., 553
Bassett, R. L., 474
Basu, S., 473
Bates, R. G., 274, 317, 320, 341, 348, 354, 357, 362, 552
Bates, S. J., 270, 274
Baumgarten, C. M., 484
Bawn, C. E. H., 217
Bayes, C. F., Jr., 273, 349, 362
Bearman, R. J., 455
Beattie, J. A., 504
Beckie, R. D., 553
Becktold, M. F., 264, 273
Beebe, C. W., 200
Benedict, M., 138, 152, 243, 529
Berkowitz, J., 492
Berman, R. G., 117, 120
Bernard, C., 474
Berry, R. S., 491

Bichowsky, F. R., 111, 120
Bird, R. B., 63, 64, 152
Bischoff, J. L., 242
Bishop, G. A., 484
Bitter, F., 97
Bjerrum, N., 276, 287, 289
Black, J., 3, 16
Blander, M., 310, 320
Bodnar, R. J., 339
Boltzmann, L., 69, 77, 81
Bonhoeffer, K. F., 387, 398
Bonnickson, K. R., 384, 385
Born, M., 58, 64, 428
Bousfield, W. R., 273
Bower, V. E., 274, 348, 362
Bradley, D. J., 544, 545
Braitsch, O., 307, 320
Braun, R. M., 118
Breck, W. G., 452, 456
Brewer, L., 49, 51, 72, 81, 120, 214, 215, 221, 253, 338, 362, 473, 485, 491, 492, 585
Briggs, M. E., 533
Brimblecombe, P., 320
Brodale, G. E., 97, 384, 433
Bromley, L. A., 120, 492
Bronsted, J. N., 3, 273, 277, 289, 290, 320
Bros, J. P., 474
Brown, A. S., 350, 362
Brown, I., 200
Brown, O. L. I., 264, 273
Brown, P. G. M., 274
Brucker, M. J., 456
Brunauer, S., 418
Buckdahl, H. A., 64
Bukshtein, V. M., 335, 339
Burnham, C. W., 474
Burns, R. P., 491
Busey, R. H., 62, 64, 93, 94, 97, 274, 338, 339, 362, 563
Butler, J. A. V., 407, 418
Butler, J. N., 270, 274, 357, 362
Byers, S. M., 221

Card, D. N., 285, 289
Carnot, S., 3, 23, 54, 58, 64
Carter, R. W., 62, 64
Chandrasekhar, S., 64
Chao, K. C., 152, 243
Chapela, G. A., 533
Charykov, N. A., 320
Charykova, M. V., 553
Chase M. W., Jr., 116, 118, 120, 367, 378, 491, 541
Cheesman, C. H., 186

Chen, Z. Y., 533
Childs, C. W., 354, 362
Chipman, J., 463, 474
Chirico, R. D., 120, 121
Choi, S. D., 473
Chou, I.-M., 469, 473, 474
Christensen, J. H., 362
Chuptka, W. A., 492
Clapeyron, E., 46, 51
Clark, C. W., 87, 88
Clausius, R. J. E., 3, 23, 27, 33, 58, 64
Clayton, J. O., 381, 384
Clegg, S. L., 309, 320, 552
Clusius, K., 388, 392, 398, 399
Cohen, E. R., 602
Cohen, J. B., 385
Colinet, C., 462, 474
Colwell, J. H., 393, 399
Connelly, P. R., 484
Conway, B. E., 362
Cooke, D. L., 120
Coon, J. E., 152
Coops, J., 120
Copeland, L. E., 415, 418
Corak, W. S., 88
Corti, H. R., 319, 320
Coughlin, J. P., 119
Coulter, L. V., 87, 384
Covington, A. K., 318, 320
Cox, J. D., 118, 361, 363, 542
Cundall, J. T., 206, 221
Cunningham, J. R., 152
Curl, R. F., Jr., 152, 381, 384, 399, 522, 524, 529
Curtiss, C. F., 63, 64, 152
Cutting, C. L., 418

Danon, F., 145, 152
Darken, L. S., 7, 14
Das, A., 121
Daubert, T. E., 64
Daunt, J. G., 399
Davidson, N., 289, 378
Davies, C. A., 120, 378, 541
Davis, W. B., 320
Davison, S. G., 289
de Azevedo, E. G., 152, 221, 242, 243, 569
de Boer, J., 152, 383, 385
Debye, P., 77, 81, 254, 262, 273, 275, 277, 289, 430, 579
DeGraff, W., 151
DeGroot, S. R., 455, 456
Delaney, C. M., 264, 273
de Laplace, P. S., 411
de Lima, M. C. P., 558, 563

De Maria, G., 488, 491
de Pablo, J. J., 320
Desnoyers, J. E., 61, 64
DeSorbo, W., 97
Dewar, J., 77, 81
Diaz-Laviada, A. M., 152
DiCera, E., 484
Dmitriev, G. V., 553
Domalski, E. S., 120, 121
Domb, C., 533
Donoho, P. L., 399
Douslin, D. R., 127, 151
Downes, C. J., 354, 362, 553
Downey, J. R., Jr., 120, 378, 541
Downs, R. T., 473
Doyle, M. L., 484
Drowart, J., 491
du'Gay, A. P., 152
Duhem, P., 165, 171
Dulmage, W. J., 384
Dulong, P. L., 77, 81
Dunlap, R. D., 186
Dymond, J. C., 127, 141, 151, 228, 230, 242

Eastman, E. D., 97, 455
Ehlers, R. W., 274, 350, 362
Einstein, A., 17, 36, 76, 77, 81
Eisenschitz, R., 64
Elkaim, G., 186
Elliot, J. F., 463, 474
Elmore, K. L., 357, 362
Ely, J. F., 138, 152
Emmett, P. H., 418
Esteso, M. A., 553
Eucken, A., 387, 398
Ewald, A. H., 200
Eyring, H., 289, 455

Faber, J., Jr., 385
Faulkner, L. R., 349, 362
Federov, Ju. A., 553
Felmy, A. R., 320, 455, 456
Fender, B. E. F., 127, 151
Fernandez-Merida, L., 553
Ferra, M. I., 318, 320
Filippov, V. K., 309, 320, 553
Finke, H. L., 384
Fisher, J. R., 118, 542
Fisher, M. E., 533
Fisher, R. A., 97, 152, 431, 433
Floess, J. K., 474
Flory, P. J., 216, 217, 222
Forsythe, W. R., 97
Foster, A. H., 200

Frank, A., 392, 399
Frederick, W. J., Jr., 552
Freeman, R. F. J., 217
Frenkel, M., 121
Friedman, H. L., 286, 289, 319, 320, 554, 557
Frurip, D. J., 120, 378, 541
Fuger, J., 119
Furgeson, A., 418
Furukawa, G. T., 19, 36

Gadalla, N. A. M., 121
Gallagher, J. S., 242
Galobardes, J. F., 242
Gambino, M., 474
Gammon, R. W., 533
Gardner, T. E., 585
Garvin, D., 582, 585
Gee, G., 215, 217, 222
Gerke, R. H., 90, 97
Giauque, W. F., 4, 19, 36, 80, 89, 90, 91, 93, 94,
  95, 97, 381, 382, 384, 398, 430, 433
Gibbons, J. J., 552
Gibbs, J. W., 3, 14, 23, 39, 51, 107, 110, 158, 163,
  165, 171, 344, 362, 403
Gibbs, R. E., 221
Gibson, R. E., 264, 273
Giese, C. F., 492
Gill, E. K., 381, 384, 393, 399
Gill, S. J., 434, 475, 477, 484
Gilles, P. W., 120, 492
Gingerich, K. A., 120
Gockel, A., 451, 456
Godwin, L. G., 552
Goldberg, R. N., 478, 480, 484
Gonzalez, O. D., 399
Gonzalez-Diaz, D., 553
Good, C. E., 320
Goodman, R. M., 585
Goodrich, J. C., 456
Gordon, A. R., 264, 273, 350, 362
Gorter, C. J., 398
Gotoh, S., 152
Greeley, R. S., 269, 274
Green, M. S., 533
Greenberg, J. P., 334, 339
Grenier, G., 391, 392, 399
Greywall, D. S., 399
Gronvold, F., 384, 385
Gross, M. E., 384
Grosse, A. V., 405, 418
Gruen, R., 413, 418
Gucker, F. T., 348, 362
Guggenheim, E. A., 195, 221, 290, 291, 317, 320,
  342, 354, 362, 418, 480, 484

Guldberg, C. M., 203, 221
Güntleberg, E., 280, 289
Gupta, S. R., 268, 274
Gurry, R. W., 7, 14
Gurvich, L. V., 119
Gwinn, W. D., 378, 577

Haar, L., 238, 242
Haase, R., 436, 454, 455
Haber, F., 3, 14
Hakonen, P., 432, 433
Hall, D. L., 335, 339
Halsey, G. D., Jr., 127, 151
Hamer, W. J., 351, 362, 455, 552
Han, S. J., 243
Harkins, W. D., 415, 418
Harmens, A., 529
Harned, H. S., 274, 350, 354, 362
Hart, K. R., 186
Harteck, P., 387, 398
Harvie, C. E., 300, 304, 305, 307, 320, 552
Hasebe, M., 474
Haugen, G. R., 319, 320
He, T-Z., 474
Hearing, E. D., 121
Heidemann, A., 397, 399
Heidemann, R. A., 148, 152
Heidman, J. L., 153, 230, 242
Heitler, W., 221
Helmholtz, H. von, 3, 39, 51
Hemingway, B. S., 118, 120, 542, 585
Henderson, D., 147, 152, 289
Herman, C. C., 569
Hermsen, R. W., 221
Hernandez-Luis, F. F., 553
Herschbach, D. R., 378
Hersh, L. S., 316, 320, 459, 473, 474
Herzberg, G., 378
Herzfeld, K. F., 221
Heyen, G., 529
Hildebrand, J. H., 198, 200, 212, 214, 221, 583, 585
Hill, A. V., 476, 484
Hill, P. G., 138, 152, 533
Hill, R. W., 389, 398
Hill, T. L., 289, 377, 378, 418
Hiller, K., 387, 388, 398
Hillert, M., 457, 459, 462, 473
Hills, G. J., 268, 274
Hirschfelder, J. O., 63, 64, 147, 152, 378
Hirshberg, J., 200
Holdaway, M. J., 473
Holland, T. J. B., 117, 120
Hollenberg, J. L., 378

Holmes, H. F., 62, 64, 273, 274, 329, 331, 338, 339, 362, 558, 563
Holmes, M. J., 565, 567, 569
Holmes, R. D., 383, 385
Honig, B., 433
Hopkins, H. P., Jr., 384, 395, 399
Hornung, E. W., 97, 433
Huber, K. P., 378
Hückel, E., 254, 262, 273, 275, 277, 289
Hudson, J. W., 569
Hudson, R. P., 432, 433
Huffman, H. M., 152
Hufford, K., 553
Huggins, C. M., 152, 524, 529
Huggins, M. L., 216, 222
Hultgren, R., 119

Inghram, M. G., 491
Ishida, K., 464, 465, 474
Ives, D. J. G., 268, 274, 320, 362

Jackson, K. J., 320
Janz, G. J., 320, 350, 362
Jasper, J. J., 405, 418
Johnson, D. A., 320, 553
Johnston, H. L., 89, 97, 384, 398, 399
Jones, D. C., 418
Jordan, J., 363
Joule, J. P., 3, 16, 59, 64
Jura, G., 81

Kaarsemaker, S., 120
Kalinkin, A. M., 553
Kamaliddin, A. R., 217
Kasper, J. V. V., 395, 399
Kaufman, A. R., 97
Kaufman, L., 474
Kayser, R., Jr., 378
Keenan, J. H., 64
Keesom, W. H., 87, 88
Kell, G. S., 242
Kelley, K. K., 116, 119, 120, 581, 585
Kelvin, Lord (W. Thomson), 3, 18, 23, 27, 59, 64, 411, 418, 436, 449, 455
Kemp, J. D., 382, 384
Kennard, E. H., 455
Kessler, M. G., 509, 524, 528, 529
Khoroshin, A. V., 450, 455
Kihara, T., 145, 152
Kilpatrick, J. E., 378, 577
Kim, H.-T., 552
Kim, J. J., 300, 320
King, A. D., Jr., 242
King, E. G., 119

Kirschman, H. D., 270, 274
Kister, A. T., 195, 221, 459, 473
Kistiakowsky, G. B., 49, 51
Kittel, C., 378
Kiukkola, K., 362
Klemperer, W., 491
Kleppa, O. J., 316, 320, 459, 473, 474
Knapp, H., 529
Knauss, K. G., 318, 320
Kobe, K. A., 371, 378
Kohler, F., 462, 474
Kok, J. A., 88
Korosi, G., 405, 418
Kovats, E. SZ., 405, 418
Kroemer, H., 378
Ksiazczak, A., 152
Kubaschewski, O., 119, 120, 384, 385
Kudchadker, S., 121
Kurzeja, N., 532, 533
Kwong, J. N. S., 137, 151, 243, 527, 529

Lacey, W. N., 242
Ladner, W. R., 186
Laity, R. W., 310, 320
La Mer, V. K., 413, 418
Lamoreaux, R. H., 214, 221
Langmuir, I., 415, 418
Lasaga, A. C., 474
Latimer, W. M., 113, 120, 346, 361, 362, 428, 433, 585
Le Chatelier, H., 2, 14
Leduc, P.-A., 61, 64
Lee, B. I., 509, 524, 528, 529
Lee, H. M., 462, 474
Lee, S.-Y., 524
Le Neindre, B., 151, 152
Lennard-Jones, J. E., 144
Letson, B. B., 585
Levedahl, B. H., 373
Levelt, J. M., 151
Levelt Sengers, J. M. H., 533
Levitin, N. E., 88
Lewis, G. N., 3, 15, 101, 107, 221, 239, 242, 254, 262, 273, 277, 281, 289, 338, 362, 473
Li, J. C. M., 373, 378, 455, 577
Li, Y.-g., 320
Lichtenthaler, R. N., 152, 221, 242, 243, 569
Lide, D. R., 362
Lietzke, M. H., 274
Lin, H. M., 243
Lin, J.-l., 455
Lindsay, W. T., Jr., 264, 265, 273
Lindsley, C. H., 252, 253
Linke, W. F., 331, 339

Lippmann, D. Z., 152, 524, 529
Lipscomb, W. N., 384
Liu, C.-t., 264, 265, 273
Lofgren, N. L., 120, 485, 491, 492
Long, E. A., 382, 384
Long, F. A., 318, 319, 320
Lord, P. J., 553
Lorentz, H. A., 171
Lounasmaa, O. V., 432, 433
Lummer, O., 49, 51
Lushington, K. J., 399
Lutz, F., 385

Ma, S.-k., 533
Macaskill, J. B., 552
McCoy, B. J., 569
McCullough, J. P., 97, 381, 384
McDevit, W. F., 318, 319, 320
McDonald, R. A., 120, 378, 541
MacInnes, D. A., 350, 362
McLeod, D., Jr., 488, 491
McMillan, W. G., 249, 253
McQuarrie, D. A., 289
Maki, K., 399
Malanowski, S., 243
Mangum, B. W., 19, 36
Mannweiler, G. E., 362
Marcus, R. J., 455
Margules, M., 165, 171, 184, 221, 457, 473
Marsh, K. N., 118, 120, 121, 542, 585
Marshall, W. L., 274, 362
Martin, J. J., 153
Martynova, O. J., 242
Mason, C. M., 362
Massieu, M. F., 39, 51
Massucci, M., 152
Mathew, R., 456
Matsuo, T., 383, 385
Mayer, J. E., 81, 249, 253, 378
Mayer, J. R., 3, 16
Mayer, M. G., 81, 378
Mayorga, G., 296, 320, 546
Mazur, P., 455, 456
Meads, P. F., 97
Medvedev, V. A., 118, 363, 542
Melchior, D. C., 474
Melhuish, M. W., 381, 384
Mesmer, R. E., 62, 64, 273, 274, 318, 320, 329, 331, 338, 339, 349, 351, 362, 558, 563
Messerly, J. F., 384
Michels, A., 127, 134, 151, 152
Millen, D. J., 231, 242
Miller, D. G., 273, 274, 296, 454, 455, 456, 553
Miller, F., 120

Mills, I., 51, 362
Milner, R. T., 97
Milner, S. R., 275, 276, 289
Mines, G. W., 231, 242
Mochel, J. M., 186, 242
Møller, N., 320, 334, 339, 552, 558, 563
Monk, C. B., 362
Monnin, C., 326, 327, 338
Montgomery, R. L., 91, 97
Moore, C. E., 72, 81
Moore, G. E., 582, 585
Morrison, J. A., 381, 384, 393, 394, 399
Mou, C. Y., 455
Mrad, D. R., 553
Muggianu, Y. M., 462, 474
Mukhopadhayay, B., 461, 473
Myers, E. A., 384

Nagamiya, T., 395, 399
Najour, G. C., 242
Nakashima, K., 446, 455
Nedostup, V. I., 97
Nelson, R. A., 87
Nernst, W., 3, 152, 506
Neumaier, K., 399
Newing, M. J., 217
Newman, S. A., 243, 529
Newton, R. F., 264, 273
Nickel, B. G., 533
Nishizawa, T., 474
Norman, J. H., 487, 491
Nozieres, P., 399

Oakes, C. S., 339, 582, 585
Oetting, F. L., 119
Ohe, S., 565, 569
Ohtani, H., 474
Oja, A., 433
Olds, R. H., 242
Oliver, G. D., 88
Olovsson, I., 384
O'Neill, H. St. C., 383, 385
Onsager, L., 442, 455
Orr, W. J. C., 217, 222
Orttung, W. H., 389, 399
Orye, R. V., 200, 221, 564, 569
Osterberg, H., 60, 64
Otto, J., 134, 151

Pabalan, R. T., 242, 304, 331, 334, 338, 339, 474, 563
Palmer, D. A., 274, 362

Papatheodorou, G. A., 474
Parkas, A., 398
Parker, V. B., 585
Parlin, R. B., 455
Parsons, R., 363
Patel, N. C., 529
Patterson, C. S., 362
Paulaitis, M. E., 243
Pauling, L., 81, 382, 384, 498
Payton, A. D., 453, 456
Peiper, J. C., 338, 552, 563
Pel'sh, A. D., 339
Pelton, A. D., 461, 473, 474
Peng, D. Y., 243, 527, 529
Pennington, R. E., 371, 378
Petersen, D. E., 152, 524, 529
Petit, A. T., 77, 81
Pfeffer, W., 201, 221
Philip, P. R., 61, 64
Phillips, L. F., 456
Phutela, R. C., 296, 339, 553, 558, 563
Picker, P., 61, 64
Pieperbeck, B., 532, 533
Pimentel, G. C., 118
Pines, D., 399
Pitzer, K. S., 49, 51, 62, 64, 81, 87, 118, 120, 138, 141, 145, 149, 151, 179, 181, 221, 242, 253, 273, 274, 289, 291, 296, 300, 304, 309, 320, 326–328, 331, 334, 338, 354, 357, 362, 378, 382, 384, 395, 396, 399, 433, 453, 456, 473, 506, 508, 522, 524, 526, 528, 532, 533, 544–546, 552, 557, 563, 577, 582, 584
Planck, M., 58, 64, 77, 81
Platford, R. F., 273, 354, 357, 362
Pobell, F., 398, 399
Poincaré, H., 64
Poling, E. B., 569
Polissar, J., 382, 385
Popp, L., 392, 399
Posnjak, E., 116, 120
Powell, R., 117, 120
Powell, T. M., 90, 97, 117
Prausnitz, J. M., 146, 148, 152, 196, 198, 200, 221, 226, 230, 242, 243, 320, 524, 564, 567, 569
Prentiss, S. S., 261, 273
Press, W., 399
Preuner, G., 111, 120
Prigogine, I., 438, 455
Pringsheim, E., 49, 51
Prophet, H., 242, 384, 385, 582, 585
Prue, J. E., 274

Quill, L. L., 120, 492

NAME INDEX

Rabinovich, V. A., 97
Radler, M. J., 385
Randall, M., 15, 111, 120, 221, 239, 242, 262, 273, 277, 281, 289, 338, 360, 362, 473
Raoult, F. M., 173, 181
Rard, J. A., 273, 274, 296, 455, 456, 552, 553
Rasaiah, J. C., 286, 289
Rashin, A. A., 433
Rathjens, G. W., 120
Rauh, E. G., 486, 491
Raymond, C. L., 196, 221
Reamer, H. H., 242
Reardon, E. J., 553
Redlich, O., 137, 151, 195, 221, 243, 459, 473, 527, 529
Reid, R. C., 522, 524, 569
Renon, H., 567, 569
Rice, S. A., 491
Rice, W. W., 49, 51
Ricketson, B. W. A., 389, 398
Riedel, L., 140, 152, 418, 506, 520, 521, 524
Robert, C. H., 484
Roberts, R. J., 399
Robie, R. A., 116, 118, 120, 542, 581, 585
Robinson, D. B., 527, 529
Robinson, R. A., 264, 267, 273, 362
Robinson, R. L., Jr., 152, 243
Rodgers, A. S., 121
Roebuck, J. R., 60, 64
Rogers, L. B., 242
Rogers, P. S. Z., 62, 64, 328, 338
Rojas, E., 399
Rollet, A. P., 186
Rosenbauer, R. J., 242
Rosenstein, R. D., 384
Rossini, F. D., 118, 120, 221, 362
Rowlinson, J. S., 234, 242, 533
Roy, L. N., 320, 553
Roy, R. N., 270, 274, 320, 357, 362, 552, 553
Ruben, H. W., 384
Rubin, L. C., 138, 152, 243, 529
Rubin, T. R., 373
Rumford, Count, 3, 16
Rumyantsev, V., 320
Russell-Robinson, S., 120, 585
Ruzicka, V., Jr., 121

Sackur, O., 68, 80
Sage, B. H., 242
Saluja, P. P. S., 563
Sangster, J., 474
Sathianandan, K., 399
Saul, A., 138, 152, 533

Scatchard, G., 186, 196, 198, 213, 214, 221, 229, 242, 261, 273
Schmidt, G., 529
Schmidt, R., 139, 152, 533
Schminke, K. H., 348, 362
Schomaker, V., 491
Schreiber, D. R., 141, 151, 508, 524, 526, 529
Schroder, I., 178, 179, 181
Schupp, W., 111, 120
Scott, D. W., 88, 97, 179, 181, 584, 585
Scott, R. L., 234, 242, 381, 384
Seidell, A., 331, 339
Seki, S., 385
Senez, M., 186
Sengers, J. V., 533
Shallcross, D. C., 567, 569
Shankman, S., 264, 273
Sharma, R. C., 464, 465, 474
Shaw, A. N., 588
Sherwood, T. K., 524
Shi, Y., 339
Shimada, T., 385
Shoenberg, D., 455
Shomate, C. H., 504
Shumiya, T., 474
Silvera, I. F., 392, 399
Silvester, L. F., 326, 327, 338, 354, 357, 362, 552
Simons, J. H., 186
Simonson, J. M., 274, 320, 338, 339, 362, 552, 563
Singh, R. R., 532, 533
Sinha, S. P., 221
Slaman, M. J., 152
Slansky, C. M., 433
Smith, D. F., 585
Smith, E. B., 127, 141, 151, 228, 230, 242
Smith, J. M., 64
Smith, W. T., Jr., 274
Smith-Magowan, D., 62, 64, 338
Soave, G., 153, 527, 529
Somayajulu, G. R., 522, 524
Somorjai, G. A., 416, 418
Sørensen, T. S., 284, 285, 289
Southard, J. C., 87, 120, 585
Spedding, F. H., 553
Speedy, R. J., 335, 339
Spencer, P., 384, 385, 474
Spencer, R. J., 335, 339
Squires, T. G., 243
Srivastava, M., 464, 465, 474
Staley, H. G., 487, 491
Starr, C., 97
Stavely, L. A. K., 186
Steele, W. V., 120, 121

Stephenson, C. C., 91, 97
Sterner, S. M., 339, 466, 469, 473, 474
Stewart, W. E., 152
Stokes, R. H., 267, 273, 362
Stolen, S., 385
Stortenbeker, W., 171
Stoughton, R. W., 274
Stout, W., 382, 384
Stratton, J. A., 423, 433
Strehlow, H., 362
Stull, D. R., 242, 384, 385, 582, 585
Suga, H., 383, 385
Suzuki, M., 533
Swinton, F. L., 234, 242, 533
Syverud, A. N., 120, 378, 541

Tajima, Y., 385
Takeyama, N., 446, 455
Talley, P. K., 471, 474
Tan, Z., 522, 524
Tanger, J. C., IV, 338
Taylor, A. R., Jr., 585
Taylor, B. N., 602
Taylor, M. D., 151, 152
Tee, L. S., 146, 152
Teja, A. S., 524, 529
Teller, E., 418
Temkin, M. I., 450, 455
Templeton, D. H., 384
Tetrode, H., 68, 80
Tewari, Y. B., 478, 480, 483, 484
Thiessen, W. E., 563
Thomä, M., 413, 418
Thomsen, J., 110
Thomson, W. (Lord Kelvin), 59, 64, 411, 418, 449, 455
Thorn, R. J., 489, 492
Todd, S. S., 152, 384, 385
Toleedano, P., 186
Tolmach, P., 385
Toop, G. W., 462, 474
Treloar, L. R. G., 217, 222
Trofimov, Yu. M., 553
Trouton, F. T., 84, 97, 583, 585
Tsonopoulos, C., 152, 153, 230, 242, 522, 524
Tupman, W. I., 186
Turgeon, J. C., 291, 320, 354, 362
Turnquist, C. E., 221, 473
Twu, C. H., 151, 152
Tyler, W. W., 97

Valleau, J. P., 285, 289
Valyashko, M. G., 339

van der Waals, J. D., 126, 135, 136
Van Hare, D. R., 242
van Konynenburg, P. H., 234, 242
van Laar, J. J., 197, 213, 221
Van Ness, H. C., 64, 221, 470, 474
van't Hoff, J. H., 3, 203, 221
Van Winkle, M., 565, 567, 569
Vanysek, P., 361, 362
Vasin, S. K., 553
Vasserman, A. A., 97
Vekslu, L. S., 97
Verduch, A. G., 362
Verschaffelt, J. E., 530, 533
Veyts, I. V., 119
Vodar, B., 151, 152
Vogel, K. M., 320
Vogt, G. J., 396, 399

Waage, P., 203, 221
Waddington, G., 142, 152, 384
Wagman, D. D., 118, 119, 221, 304, 363, 542
Wagner, C., 362
Wagner, W., 138, 139, 152, 532, 533
Walsh, G. E., Jr., 474
Wang, P., 545
Wang, Z.-C., 466, 474
Watts, R. O., 152
Weare, J. H., 300, 304, 305, 307, 320, 339, 455, 456, 552
Weatherford, R. M., 567, 569
Webb, G. B., 138, 152, 243, 529
Webb, T. J., 252, 253
Weissman, S., 198, 221
Weltner, W., Jr., 149, 151, 152, 488, 491
Wenzel, H., 529
Westrum, E. F., Jr., 385, 585
Wheatley, J. C., 398, 399
White, D., 391, 392, 399
White, H. J., Jr., 585
White, M. A., 394, 399
Wightman, A., 407, 418
Wilhoit, R. C., 121
Williams, R., 552
Wilson, E. B., Jr., 81, 398, 498
Wilson, G. M., 200, 221, 564, 569
Wohl, K., 460, 473
Wolery, T. J., 320
Wong, C. H., 491
Wood, R. H., 62, 64, 274, 338
Wood, S. E., 186, 198, 214, 221, 242
Woolley, H. W., 371, 378
Wormald, C. J., 62, 64, 152
Wouters, H., 134, 151

Wu, Y.-C., 552
Wyman, J., 434, 475, 484
Wynne-Jones, W. F. K., 362

Yakovleva, S. I., 553
Yamada, T., 526, 528, 529
Yasamuro, O., 385
Yost, D. M., 373

Young, L. E., 360, 362
Younglove, B. A., 138, 152

Zawidzki, J. von, 165, 171
Zhang, K. C., 533
Zhang, X-H., 474
Zhang, Y., 456
Zhiyu, Q., 466, 468, 474

SUBJECT INDEX

Absolute entropies, 66, 82, 89
Absolute temperature, 18
Acentric factor, 140, 505, 526
   mixed fluids, 229
Acetic acid, 255
   dissociation, 350–352
Acetone solution, 165, 166
Acids:
   carboxylic, vapor, 150
   weak, dissociation, 255, 350
Activity and activity coefficients, 183, 187, 191, 194
   calculation from activity of other components, 194, 219, 250, 300, 471
   electrochemical cells and, 267, 348
   electrolytes, 256, 260, 295
      empirical equations, 272
      Guggenheim equations, 290
      hydrochloric acid, aqueous, 269
      ion-interaction equations, 292
      ions, individual, 316
      limiting law, 262, 280
      marginally weak, 271
      measurements, 260
      mixed, 297
      Pitzer equations, 292
      semi-empirical equations, 290
      stoichiometric, 259
      strong, 256
      vapor pressure, 264, 270
      weak, 259
   freezing point and, 250, 260
   gas, 187
   isopiestic method, 266
   molality basis, 245
   neutral salt-electrolyte, 318
   osmotic coefficient and, 250, 260
   pressure effect, 192
   quotient, 204
   solute, 190, 245, 256
   solvent, 189, 246, 258
   standard states, 187, 191
Adenosine phosphate disproportionation, 478
Adiabatic demagnetization, cooling effect, 430
Adiabatic expansion, 57
Adiabatic process, 21
Adiabatic reversible processes, 47
Adsorption of gases, 415
Alcohol, vapor, 148
Aluminum, 78, 79, 96, 361, 535
Aluminum chloride, entropy of fusion, 583
Amagat law of additive volumes, 239
Anderko–Pitzer equation, 528
Anharmonic oscillator equation, 367
Anode, 346
Apparent molar properties, 323

617

Argon, 92, 127, 129, 135, 139, 145, 146, 523
Association, effect on standard state, 205
　gas properties, 147, 237
Atmosphere, 17

B:
　ion–interaction quantity, 293
　second virial coefficient, 124, 225
　spectroscopist's quantity, 73
Bar, 17
Benedict–Webb–Rubin equation, 528
Benzene, 88
　critical constants, 523
　in gas mixture, 229
　solutions with, 185, 186, 196, 199, 214, 215
Bernoulli equation, 63
Binary solution equations, 162
Biochemical systems, 475
Blackbody radiation, 377
Boltzmann constant, 35
Boltzmann distribution law, 69, 493
Born equation, 428
Boyle point, 128
Bubbles, interior pressure, 411
Butane–carbon dioxide, 240
Butanol, freezing point, 252

Calcium chloride:
　solid, 538, 582
　solution, 170, 296, 304, 327, 329, 562
Calorie, 50
Calorimeter, 20, 61, 151
Capacity, electrical, 425
Carbon, 78, 86, 96, 113, 535
Carbon dioxide:
　adsorption on silica, 417
　critical constants, 523
　fluid (liquid and vapor), 126
　heat capacity, 49
　ideal gas, standard state, 96, 540, 542
　mixed with n-butane, 240
Carbon disulfide solution, 165, 200
Carbon tetrachloride, 523, 566
　entropy of fusion, 583
Carnot cycle, 57
Catalyst, 11
Cathode, 346
Cell (galvanic), 108, 340
　concentration, 350
　half, 345
　pressure effect, 347
　temperature effect, 347

Celsius (centigrade) temperature, 18
Centrifugal field, 420
Centrifugal stretching of a molecule, 367
Chemical equations, 13
Chemical equilibria, 104, 202
Chemical potential, 98, 157
　composition dependency, 163
　electrochemical cell, 341
　electrolytes, 258
　mixed fluid, 226
　pressure dependency, 159
　solutions, 156
　temperature dependency, 159
　(*see also* Activity *and* Fugacity)
Chemical symbols, 13
Chlorine, 90, 96, 346, 361
Chloroform solution, 166
Clapeyron equation, 46, 100
Components, 6, 154
Composition:
　mass fraction, 155
　molality, 155, 244
　molarity, 155
　mole fraction, 155, 309
Compression coefficient, 129
Compression (compressibility) factor, 123
Concentration cell, 350
Concentration, surface, 401
Conductance, 255
Conservation of energy, 15
Conversion factors, 604
Cooling by adiabatic demagnetization, 430
Copper, 79, 96, 361, 450, 485, 535
Copper chloride vapor, 485
Copper sulfate, activity, 271
Corresponding states, 139
Critical mixing point, 166, 211
Critical point of pure fluid:
　criteria, 124
　table of values, 523
Crystals, 7
Curved interfaces, 411
Cycles, 54

Dalton's law of additive pressures, 239
Debye functions, 499
Debye–Hückel parameters, 543
Debye–Hückel theory, 276, 277
Debye theory for solids, 77, 499
Demagnetization, adiabatic, 430
Density, liquid, 520
Deuterium, 387, 391
Diamond, 78, 113

Diatomic molecules, entropy, etc., 73, 366
Dibromobenzene, p-, solubility, 180
Dielectric constant, 424
  of water, 544
Dissociable solute, solubility, 170
Dissociation:
  constant for weak acid, 350
  degree of, 255, 350
  effect on standard state, 205
Distribution law, Boltzmann, 69, 493
Drops, interior pressure, 411
Dulong and Petit law, 77

Efficiency, thermal, 55
Einstein equation for heat capacity, 77
Electric current, entropy production, 440
Electric field, 423
  solution in, 427
Electrochemical cells, 340
Electrochemical potential, 342
Electrode potentials, 340, 345, 346
  sign of, 344
Electrokinetic effects, 443
Electrolytes, 254
  equations of state, comprehensive, 328
  heat of transport, 445
  ion-interaction equations, 292
  limiting law, 262, 280
  mixed, 290, 297
  mole fraction basis, 309
  neutral salt activity, 318
  pressure dependency, 322
  semi-empirical equations, 290
  solid, 357
  solid solubility in, 304, 330
  strong, 255
  temperature dependency, 322, 329
  theory, 275
  thermal properties, 322
  unsymmetrical, 257
  volumetric properties, 322
  weak, 255, 259
  (*see also* Activity and activity coefficients)
Electromotive force, 108, 340–350
Electroosmosis, 444
Electroosmotic pressure, 444
Elements, properties, 535–537
Energy, 15
  conservation of, 15
  electrical, 16
  internal, 16, 158
  kinetic, 15
  and mass, 17

  potential, 15
  thermal, 16
  units of, 50
Energy content, 16, 21
Enthalpy, 39, 158
  apparent molar, 323
  of formation, tables, 304, 536–542
    list of compilations, 117
  fusion, 84, 177
  measurements, 83
  mixing, 175, 183
  normal fluid, 516
  reaction, 111
  surface, 402
  units of, 50
  (*see also* Entropy *and* Activity)
Entropy, 15, 23, 27, 33, 66, 82
  apparent molar, 323
  bromide–chloride comparison, 581
  calculations of, 83
  electric current, 440
  electrolytes, 322
  estimation of, 578
  extrapolation to zero K, 86, 581
  fluid, 518
  fusion, 83, 583
  heat capacity and, 42, 85
  heat flow, 439
  ideal gas:
    internal energy levels, 69, 364
    internal rotation, 371, 570
    rotation, overall, 73, 365
    translation, 68, 493
    vibration, 74
  magnetic effects, 93
  matter flow, 439
  partition function, 364
  and probability, 34
  residual at zero Kelvin, 380
  surface, 402
  table of values, 96, 535–542
  temperature effect, 42, 85
  thermocells, 450
  third law of thermodynamics, 82, 92
  transported, 450
  units of, 49
  vaporization, 84, 507, 583
    reactions, 490
Equation of state, 33, 122, 224, 525
  alcohols, 148
  Anderko–Pitzer, 138, 528
  Benedict–Webb–Rubin, 138
  carboxylic acids, 150
  normal fluids, 525

Equation of state (*cont.*)
  Redlich–Kwong, 137, 527
  reduced, 139
  van der Waals, 135
Equations for electrolytes, 290
Equations, mixed liquids, 184, 195, 200, 212, 457, 564
Equations with many variables, 599
Equilibrium, 9
  cell potentials, 348
  chemical, 104
  concept of, 4
  conditions for, 98
  partial, 11
  between phases, 168
  pressure effect, 99
  temperature effect, 99
Equilibrium constant, 105, 202
  electrode potentials, 348
  temperature dependency, 106, 111
Equilibrium quotient, 204, 352
Equivalent, 12
Equivalent conductance, 255
Escaping tendency, 98
  in solutions, 156
  (*see also* Chemical potential *and* Fugacity)
Estimation of entropies, 578
Eutectic temperature, pressure effect, 168
Excess functions, 182, 194
Excess Gibbs energy, 182, 194, 247, 292
  (*see also* Activity and activity coefficients)
Expansion coefficient, 129
Experimental methods, 49, 61, 83, 151, 264, 349
Extensive properties, 8, 13, 17, 155, 160

Fermi–Dirac gas, 79
First law of thermodynamics, 15
Flory–Huggins theory, 216
Flow calorimeters, 61
Flow processes, 58
  general equation for, 62
Fluids, 122
  criteria for critical point, 124
  mixed, 224
  near critical properties, 530
  normal, 140, 505
Formula mass, 12
Free expansion, 59
Freezing-point lowering, 177, 250, 260, 335
Fugacity, 98, 101
  composition dependency, 164
  mixed fluid, 226
  normal fluid, 512
  pressure dependency, 102

  standard states and, 103
  temperature dependency, 102
  (*see also* Activity, Chemical potential, *and* Gibbs energy)
Fusion, entropy of, 83, 583

Galvanic cell (*see* Cell)
Gas, 7
  ideal (perfect), 22, 46, 61, 67
  at low pressure, 134
  mixed, 224–242
  real (imperfect), 122–151, 505–520, 525
Gas constant, 23
Gibbs–Duhem equation, 165, 194, 249
Gibbs energy, 39, 158
  electrolytes, 258
  excess, 182, 194
  fluid, 130
  galvanic cell, 108
  heat capacity equations, 116
  heat of reaction, 110
  mixed gas, 225
  mixing, ideal, 174
  molality basis, 247
  solutions, 156
  tables of, 534
  third-law entropy and, 112
Gibbs energy function, 114
  compilations of, 117
  table of, 534–542
Gibbs–Helmholtz equation, 41, 109
Graphite, 86, 96, 113, 535
Gravitational field, 420

Half cells, 345
Halogens and halides, properties, 535, 539
Harmonic vibration, 75, 367
Heat, 20
  expansion, isothermal, 22
  fusion, 84, 177
  latent, 16
  mechanical equivalent, 16
  mixing, 175, 183
  Peltier, 57, 447
  reaction, 111
  Thomson, 449
  vaporization, 84
  (*see also* Enthalpy)
Heat capacity, 22, 42
  apparent molar, 323
  Debye theory for solids, 78
  electrolytes, 322
  equation, empirical, 116, 117
  extrapolation to zero K, 86, 581

of gases, 48, 135, 149
  measurements, 83
  (*see also* Enthalpy *and* Entropy)
Heat content, 42
Heat engines, 54
Heat pumps, 56
Heat of reaction, 110
Heat of transport, 437
  electrolytes, 445
Helium, 386, 397
Helmholtz energy, 39, 44, 158
  fluid, 130
  mixed fluid, 226
Henry's law, 187
Hildebrand equation, 212
Hydrates, 205
Hydrochloric acid:
  activity and activity coefficient, 268–270, 296, 547
  chemical potential, etc., of formation, 304
  comprehensive equation of state, 329
  in electrochemical cell, 267, 342, 361
  enthalpy, 327
  in mixed electrolytes, 301
  volume, 327
Hydrogen:
  electrode, 342, 345, 361
  gas, 96, 111, 368, 523, 536
  ortho and para, 386
  solid, 389
Hydrogen chloride gas, 368, 539

Ideal gas, 22, 46, 61, 67
Ideal solutions, 173
  solubility in, 177
Intensive properties, 8, 13, 17, 155, 160
Interface, vapor–liquid, 404
Interfacial tension, 400
  liquid–liquid, 409
Interionic-attraction theory, 276
Internal energy, 16, 21
Internal rotation, 371
  tables of functions, 570
Inversion temperature, 59
Ion in a dielectric, 427
Ion-interaction equations, 292
  parameters:
    for 25°C, 296, 301, 305, 327, 546–552
    for other temperatures, 330, 336, 558–562
Ionic strength, 262, 281
Ionization, degree of, 255, 350
Irreversible processes, 435
  general equations, 440
Isentropic processes, 47

Isooctane, solution, 198, 199
Isopiestic measurements, 266
Isothermal expansion, 57
Isothermal Joule–Thomson coefficient, 62

Jet engine, 63
Joule, 50
Joule–Thomson coefficient, 59
  isothermal, 62
Joule–Thomson expansion, 59, 520

Kopp's rule, 580

LaPlace equation, 411
Latent energy (or heat), 16
Lead, 78, 361, 535
  dissociation equilibrium, 115
Limiting properties at great dilution, 186
Liquid, 7
  normal, 140, 505
Liquid junction in a cell, 349

Macroscopic state, 7, 12
Magnetic field, 423
Magnetic polarization, 428
Magnetic susceptibility, 424
Mannitol, freezing point, 252
Margules equation, 184, 194, 219, 457
Mass action law, 203
Mass and energy, 17
Mass fraction, 155
Maxwell relations, 40
Mechanical energy, 15
Metallic systems, 462
Metals, electrode potentials, 359
Methane, 124, 139, 146, 230, 523
  ideal gas, standard state, 96, 542
  solid, 393
Microscopic state, 7
Mixed electrolytes, 290, 297
  solid solubilities in, 304
  (*see also* Electrolytes)
Mixtures, 154
  (*see also* Multicomponent systems *and* Solutions)
Molal, 13
Molality, 155, 244
Molar properties, 13
Molarity, 155
Mole, 12
Mole fraction, 155
  in electrolyte, 309
Molecular mass, 12

Molecular models:
  alcohols, 148
  association reaction, 147
  carboxylic acids, 150
  fluid properties, 142
  Kihara, 145
  Lennard-Jones, 144
  square well, 143
  (see also Statistical calculations)
Multicomponent systems, 154, 219, 297, 457, 485

Newton (unit of force), 17
Nitrogen, 123, 126, 134, 146
  critical constants, 523
  ideal gas, standard state, 96, 368, 371, 536
Nonisothermal systems, 435
Normal fluids, 140, 505–524
  surface tension, 404
NRTL equation, 567
Nuclear spin, 80, 431
Numerical factors, 602

Onsager relation, 442
Osmotic coefficient, 246
  electrolytes, 258
  freezing point basis, 250
  (see also Activity coefficient)
Osmotic pressure, 201
Oxidation potential, 346
Oxides, properties, 540
Oxygen, 89, 361, 417, 523
  ideal gas, standard state, 96, 368, 536, 542
Oxygen binding to macromolecules, 475

Paraffin hydrocarbons, properties, 542, 584
Partial molar equations, 160, 161
Partial molar quantities, 159
Partition function, 70, 364
Pascal, 17
Peltier heat, 57, 447
Peng–Robinson equation, 527
Perfect crystal, 9, 92, 380
Perfect differential, 31, 600
Perfect gas (see Ideal gas)
Permittivity, 424
pH, 316
Phase, 5
Phase equilibrium, 168
Phase rule, 106
Phonons, 78
Physical constants, 602
Pitzer equations, 292
  for parameters, see Ion-interaction equations
Plasmas, 378

Polarization:
  electric, 424
  magnetic, 424, 428
Polymer solutions, 215
Potassium chloride:
  solid, 87, 538, 581
  solution, mixed, 301, 332–338
  solution, pure, at 25°C, 261, 296, 304, 327, 361
    at other temperatures, 329, 330, 336–338, 446, 467, 559
Pressure, 17
  critical, criteria, 124
    table of values, 523
  effect of (see under property affected)
  interior, of bubbles and drops, 411
  reduced, 126, 140
  standard reference, 17
  thermomolecular, 436
  units of, 17, 51
  vapor, normal fluid, 507
  (see also Fugacity and Osmotic pressure)
Probability and entropy, 34, 66
Process, 10
  adiabatic reversible, 47
  cyclic, 57
  flow, 58
  irreversible, 24, 26, 435
  isentropic, 47
  reversible, 21, 25
Processes at constant pressure, 42
Properties, 7
  compilations of, 117 (see also Tables)
  equations for electrolytes (see Electrolytes)
  equations for mixed liquids, 184, 194–201, 212–220, 249, 457–462, 564
  molar, 13
  normal fluids, 505, 525
  simple systems, 108
  state functions, 16, 38
Pure substances, properties of:
  compilations of, 117
  tables of, 534
Pyridine solution, 166

Quantum statistics, 66

Radiation, 376
Radical distribution function, 283
Raoult's law, 173
Reaction, 10
Redlich–Kister equation, 195, 459
Redlich–Kwong equation, 137, 240, 527
Reduced variables and equation of state, 126
Reference state, 193

Regular solutions, 184, 212
Relative molar properties, 322
Residual entropies, 380
Reversible process, 21, 25, 98
Rocket engine, 63
Rotational energy, entropy, etc., 73, 365, 371

Sackur–Tetrode equation, 68, 498
Saxen's relation, 444
Scatchard equations, 213
Second law of thermodynamics, 15, 23
Second virial coefficient, 49, 128, 141, 508
Seebeck effect, 57
Silicon, 72, 96, 535
Silver, 78, 88, 361, 535
Silver chloride, 90, 454, 538, 581
Silver–silver chloride electrode, 342–347, 361
Soave–Redlich–Kwong equation, 527
Sodium chloride:
   solid, pure, 96, 538, 581
   solution, liquid, mixed, 301, 334–338
      solid in steam, 237
   solution, liquid, pure, at 25°C, 296, 304, 327, 361
      at other temperatures, 265, 329, 446, 467, 561
Sodium sulfate, 87, 306, 331
Solid, 7
Solid electrolyte, 357
Solubility, solids in gases, 235
Solute, neutral, 190, 245
Solution, 6
Solutions (nonelectrolyte):
   components, 6, 154
   dilute, 186
   in electric fields, 427
   ideal, 173
   molality basis, 244
   multicomponent, 239, 457
   nonideal, 182, 224, 244, 457
   polymer, 215
   Redlich–Kister parameters, 196
   simple, 183–186
   theories of nonideality, 210–215
Solvates, 205
Solvent, 189, 246
Soret effect, 445
Spin, nuclear, 80, 431
Standard states, 187, 191, 193
   effect of association or dissociation, 205
   and fugacity, 103
   gas, 188
   solute, 190

solvent, 189
State functions, 38 (*see also* Properties)
States, 7
Statistical calculations, 66, 364
   Bose–Einstein, 376
   conduction electrons, 79
   diatomic molecules, higher approximations, 366
   electrolytes, 275, 282, 286
   electronic states, 69, 72, 374
   Fermi–Dirac, 376
   internal rotation, 371
   ionized gases, 378
   plasmas, 378
   polyatomic molecules, 371
   radiation, 376
   rotation, linear molecules, 73
      nonlinear molecules, 365
   solids, 77
   translation, 68
   vibration, 74
Streaming current, 444
Streaming potential, 444
Substances, classification of, 5
Sulfur, 96, 111, 358, 535, 542
Surface:
   concentration, 401
   energy or enthalpy, 404
   entropy, 402
   Gibbs energy, 402
   tension, 400
   vapor–liquid, 404
Surface effects, 400
   adsorption of gases, 415
   curved interfaces, 411
   films of an insoluble component, 413
   liquid–liquid, 409
   multicomponent systems, 407
   normal fluids, 404
   two-component systems, 405
Surface tension, 400, 521
   measurement, 413
System:
   adenosine phosphate, 478
   Al–In–Sb, 465
   $BaCl_2$–$SrCl_2$–LiCl, 466
   biochemical, 475
   $CaSO_4$–$MgSO_4$–H_2O, 306
   Cd–Bi–Pb, 463
   CH_3NO_2–CCl_4, 566
   closed, 5
   hemoglobin–O_2, 477
   heterogeneous, 5
   homogeneous, 5
   KCl–K_2SO_4–H_2O, 333

System (cont.)
  KCl–MgCl$_2$–H$_2$O, 306, 332
  KCl–MgCl$_2$–NaCl–H$_2$O, 334
  KCl–NaCl–H$_2$O, 306, 338, 466
  KNO$_3$–LiNO$_3$–H$_2$O, 316
  metallic, 462
  MgSO$_4$–H$_2$O, 332
  multicomponent, 154, 219, 462, 485
  NaCl–Na$_2$SO$_4$–H$_2$O, 306
  Na$_2$SO$_4$–H$_2$O, 331
  nonisothermal, 435
  nonmetallic, multicomponent, 466
  open, 5
  solid–vapor, 485
  stable, 10
  steady state, 435
  weight-reservoir, 27
  xylenes, 179

Tables:
  acentric factors, 523
  acetic acid ionization, 352
  additional variables, functions with, 432
  adenosine phosphate, 480
  aqueous electrolyte parameters:
    for 25°C, 296, 301, 305, 327, 546–552
    for other temperatures, 330, 336, 558–562
  argon, 92
  cadmium dimethyl, 373
  chlorine, 90
  compression factor, 510
  critical constants, 523
  Debye temperature, elements, 78
  Debye–Hückel parameters, 543, 544
  diatomic molecule constants, 368
  electrode potentials, 361
  electrolyte-neutral salt, 319
  electronic heat capacity, 79
  elements, properties of, 535
  energy content, Debye functions, 501
  enthalpy:
    elements, 535–537, 542
    formation, ions and minerals, 304
    halides, 538, 539
    hydrocarbons, 542
    internal rotation contribution, 572
    normal fluid, 516
    oxides, 540, 542
  entropy:
    bromide–chloride comparison, 581
    cadmium dimethyl, 373
    chlorides, 538
    chlorine, 90, 96
    compounds, 96
    Debye functions, 503
    elements, 96, 535–537, 542
    halides, 538, 539
    hydrocarbons, 542
    internal rotation contribution, 575, 576
    ions, 304
    metal bromides and chlorides, 581
    methyl chloroform, 373
    minerals, 304
    normal fluid, 518
    oxides, 96, 540, 542
    oxygen, 89
    phosphine, 91
    silver chloride, 90
    solids, miscellaneous, 541
  equilibrium between H$_2$, S$_2$, and H$_2$S, 111
  fluid properties, 505
  formulas, thermodynamics functions, 586
  fugacity, 513
  Gibbs energy:
    elements, 535–537, 542
    formation, ions and minerals, 304
    halides, 538, 539
    hydrocarbons, 542
    internal rotation contribution, 573, 574
    oxides, 540, 542
    solids, miscellaneous, 541
  half-cell potentials, 361
  halides, properties of, 538
  heat capacity:
    carbon dioxide, 49
    Debye functions, 500
    electronic, 79
    internal rotation contribution, 571
    pressure variation, n-heptane, 142
  heats of transport of ions, 446
  Helmholtz energy, Debye functions, 502
  heptane, n-, gas, heat capacity, 142
  hydrochloric acid activity, 269, 270
  internal rotation, restricted, 570
  ion-interaction equation parameters:
    for 25°C, 296, 301, 305, 327, 546–552
    for other temperatures, 330, 336, 558–562
  ion-neutral interactions, 319
  ions, 304
  Kihara core potentials, 146
  krypton, 92
  lead, gas, 115
  liquid density, 521
  mineral–seawater solubility, 308
  minerals, 304
  molecular constants, diatomic, 368
  N$_2$O$_4$ dissociation, 207

neon, 92
neutral salt–electrolyte activity, 319
nitromethane–carbon tetrachloride, 566
nonideality of simple solutions, 186, 196, 200, 214
normal fluid:
    acentric factors, 523
    compressibility-factor functions, 510, 511
    critical constants, 523
    enthalpy functions, 516, 517
    entropy functions, 508, 518, 519
    fugacity functions, 513–515
    vapor pressure, 508
numerical factors, 602
    oxides, 540, 542
oxides, properties of, 540
oxygen, 89
phosphine, 91
phosphoric acid solutions, 356
physical constants, 602
Pitzer equation parameters:
    for 25°C, 296, 301, 305, 327, 546–552
    for other temperatures, 330, 336, 558–562
potentials of half cells, 361
Redlich–Kister parameters, 196
seawater–mineral solubility, 308
silicon gas, 72
silver chloride, 90
solids, miscellaneous, 541
solutions, simple, 186
surface tension and surface concentration, water–ethyl alcohol solutions, 406
symbols, summary, 594
systems for which comprehensive EOS are available, 329
thermocell, 453, 454
thermodynamic properties, 534
tungsten oxides, 488, 489
vapor pressure, 508
xenon, 92
xylenes, 179
Temperature, 17
    absolute, 18
    Celsius, 18
    Kelvin, 18
    thermodynamic, 31
Theories:
    electrolytes, 275
    near-equilibrium processes, 438
    solution nonideality, 210
    unsymmetrical mixing of ions, 554
    (see also Equations and Statistical Calculations)
Thermal efficiency, 55

Thermocells, 450
Thermochemical calorie, 51
Thermocouple, 447
Thermodynamic formulas, summary, 586
Thermodynamic functions, definitions, 33, 38, 157
Thermodynamic temperature, 31
Thermoelectric effects, 446
Thermomolecular pressure, 436
Thermoosmosis, 436
Third law of thermodynamics, 82, 92, 380
Thomson heat, 449
Transference:
    in a liquid junction, 349
    in a thermocell, 451
Translational entropy, 68, 493
    heat capacity, etc., 69
Transported entropy, 450
Triple point, 19, 106
Tritium, 387
Trouton's rule, 84, 583
Tungsten oxides, 486

UNIQUAC equation, 567
Units:
    energy, 50
        conversion factors, 604
    entropy, 49
    pressure, 17, 51
Unsymmetrical mixing of ions, 554

van der Waals equation, 135, 232
van Laar equation, 197, 213
Vapor pressure:
    Clapeyron equation, 46, 100
    normal fluid, table of values, 507
Vaporization, entropy of, 84, 507, 583
Variables, additional, functions with, 432
Virial coefficients, 124
    equation, 129, 141
    evaluation, 127
    mixed gas, 225–230
Virial equations of state, 124

Water:
    dissociation to ions, 353
    fluid (liquid and vapor), 126
    ideal gas, standard state, 540, 542
    liquid, 96, 540
    solid (ice), residual entropy, 382
    surface tension, solution with alcohol, 406
Weak acids, 350–357
Weak electrolytes, 255
Wilson equation, 200, 564

Work, 16, 20, 44
  adiabatic process, 21
  Carnot cycle, 57
  electrical, 45, 109, 345
  electromagnetic, 423–430
  expansion, 20
  Gibbs energy, 45, 109
  heat engine, 54

Helmholtz energy, 39, 44
  maximum, 44
  sign convention, 20
Work content, 39, 44

Xylenes, 179

Zero-point energy, 19